T0310053

Biomedical Mass Transport and Chemical Reaction

Biomedical Mass Transport and Chemical Reaction

Physicochemical Principles and Mathematical Modeling

James S. Ultman
Harihara Baskaran
Gerald M. Saidel

Published by John Wiley & Sons, Inc., Hoboken, New Jersey
Published simultaneously in Canada

Library of Congress Cataloging-in-Publication Data:

Names: Ultman, James S., author. | Baskaran, Harihara, author. | Saidel, Gerald M., author.
Title: Biomedical mass transport and chemical reaction : physicochemical principles and mathematical modeling / James S. Ultman, Harihara Baskaran and Gerald M. Saidel.
Description: Hoboken, New Jersey : John Wiley & Sons, 2016. | Includes index.
Identifiers: LCCN 2015048303| ISBN 9780471656326 (cloth) | ISBN 9781119184652 (epub)
Subjects: LCSH: Biological transport. | Biomedical engineering.
Classification: LCC QH509 .U48 2016 | DDC 610.28–dc23
LC record available at http://lccn.loc.gov/2015048303

Printed in the United States of America

10 9 8 7 6 5 4 3 2 1

We dedicate this book to our wives—Deena Ultman, Lakshmi Balasubramanyan, and Mina Saidel—whose love, support, and continual encouragement sustained us during this endeavor that has lasted a decade.

Contents

Preface

The impact of engineering on medicine and biology continues to grow significantly. Not only has this resulted in an impressive worldwide increase in educational biomedical engineering programs, but many traditional chemical and agricultural engineering departments have changed their names to include "bio-." Recognizing the importance of biomedical engineering research and development to human welfare and the global economy, we have written this book to enhance the education of those students who will establish the biomedical technologies of the future.

Engineers who work in "bio" areas use analytical methods and quantitative modeling of physical, chemical, and mathematical sciences that distinguish them from those who are trained primarily in biological and medical sciences. This textbook is designed for students whose educational emphasis involves physicochemical aspects of biomedical systems. This requires instruction in principles of thermodynamics, mass transfer, chemical reaction kinetics, and fluid mechanics.

A major objective of this textbook is to integrate engineering principles with relevant biomedical applications at the cellular, tissue, organ, and whole-body levels. These applications incorporate basic as well as more sophisticated and complex concepts, which are appropriate for graduate as well as advanced undergraduate engineering students. Another major goal of this book is to teach students how to develop mathematical models and analyses associated with medical diagnostics and therapeutics.

In order to accomplish this, the book is divided into seven parts. The chapters in Part I present basic biological and mathematical modeling concepts. Part II provides an overview of the thermodynamics that relate to interfacial, membrane, and chemical reaction equilibria. In Part III, rate equations are developed to analyze the mass diffusion and chemical reaction that take place in homogeneous and heterogeneous media. The application of convection-diffusion and reaction equations to membrane transport and chemical separation devices are discussed in Part IV. In Part V, multidimensional transport of molecules and cell population dynamics are presented in the context of complex biomedical problems. Part VI develops general compartment models and analyses to represent dynamic and nonlinear responses of biomedical systems. More detailed mathematical models related to treatment of tissue and organ dysfunction, distribution and delivery of drugs, and interpretation of biomedical measurements are developed in Part VII. Key mathematical aspects related to model development and analyses are presented in appendices.

Guidance to Instructors

This textbook is especially intended for students in chemical and in biomedical engineering. Parts I–IV are presented mostly at an undergraduate level assuming knowledge of basic physics, chemistry, and mathematics (including calculus, differential equations, and elements of linear algebra). Parts V–VII include more advanced physical, chemical, and mathematical concepts (e.g., vector–tensor representations).

With its diversity of material, this book can serve as a basis for various university courses: (i) a single course for students with different backgrounds, (ii) distinct courses for undergraduate and graduates students, or (iii) a sequence of lower- and higher-level courses. In designing a particular course, instructors can choose from the wide variety of topics in different chapters to best serve specific student groups.

Chapter 1 provides students who have limited biological and physiological knowledge with a context for the applications found in later chapters. The basics of mass transport analysis in Chapter 2 with simple biomedical applications are worthwhile for all students, even as a partial review. Those who have studied chemical thermodynamics may skip Chapter 3, but the material on electrochemical potential and equilibrium should be reviewed. Most of the development of interfacial and membrane equilibrium in Chapter 4 and ligand–receptor binding and blood–gas relationships in Chapter 5 provide a basis of topics in later chapters for all students. Concepts of nonequilibrium thermodynamics in Chapter 6 may be of more interest to advanced students, but their application to membrane transport should interest all. With the exception of diffusion through multiphase materials, the theory of diffusion mechanisms in Chapter 7 is primarily aimed at advanced students. Students at every level would benefit from the sections in Chapter 8 on chemical reaction rates with biomedical applications, but their theoretical basis would mainly interest some graduate students.

The general presentation of one-dimensional transport in Chapter 9, which is not commonly found elsewhere, is intended for all students. Also recommended for all students are the early sections of Chapter 10 on membrane transport and of Chapter 11 on facilitated and secondary active transport. More complex aspects of membrane processes in the later sections of these two chapters are intended for graduate students. Mass transfer coefficients and their application to blood oxygenators and dialyzers, which are covered in Chapter 12, are particularly valuable to students interested in device development.

The topics on multidimensional transport of molecules in Chapters 13–16 require a higher level of sophistication expected of graduate students. The early sections on cell population dynamics in Chapter 17 and on compartment models in Chapter 18 are appropriate for all students. The more complex models in the later sections of these two chapters and the more comprehensive compartmental modeling in Chapter 19 would be appreciated by graduate students, especially in relation to the complex biomedical applications in Chapters 20–22. These final chapters address three distinct application areas—medical treatment, drug delivery, and diagnosis—that provide a variety of choices for the instructor.

For convenient reference, symbols and notations in this book are defined where introduced, and common symbols are also defined in a final nomenclature section. Symbols in italicized fonts represent dimensionless quantities, whereas bolded symbols refer to vectors and tensors. Standard international units (specified in Appendix A) are used in computations.

Homework problems related to each chapter are available from a supplementary website (http://engineering.case.edu/BMTR). These problems provide practice in basic computations, model development, and simulations using analytical and numerical methods.

Methods for Solving Model Equations

Whenever possible, the model equations developed in this book are solved using standard analytical techniques. Models of the more realistic systems presented in Parts V, VI, and VII require numerical methods for computer simulation and quantitative characterization. These models are composed of linear and nonlinear ordinary differential equations (ODE), partial differential equations (PDE), and mixed algebraic-differential equations. Given this diversity of model structure, we used several commercial software packages that are common for engineering education and research. Since instructors may prefer other packages, this book does not provide any instruction related to software applications.

To solve initial-value problems, we applied MathCAD (viz., rkfixed) or Matlab (viz., ode15s for stiff ODE). For interacting dynamic compartmental models, we used Matlab and Simulink. Models of dynamic, one-dimensional boundary-value problems represented by PDEs were solved with Matlab (viz., pdepe). For models involving a one-dimensional PDE–ODE combination, we discretized the spatial derivatives to transform the model into an initial-value problem (i.e., method of lines). For numerical solution of models involving a hyperbolic PDE with a first-order spatial derivative, we avoided instability by adding a second-order spatial derivative multiplied by an arbitrarily small coefficient. COMSOL Multiphysics was especially useful for model simulations involving a combination of fluid flow and mass transport in complex geometries. Mathematica was used to solve complex algebraic functions. Parameter estimation for analysis of data and simple algebraic functions was accomplished with the Solver function in Microsoft Excel.

When performing simulations to assess system dynamics or demonstrate device design, numerical values for model parameters were frequently obtained from the literature. When this was not possible, a range of dimensionless parameter values was used to explore model behavior under a variety of circumstances.

Acknowledgments

Many aspects of this book have benefited from our interactions with numerous colleagues and students. JSU is especially indebted to his undergraduate mentor, Herbert Weinstein, who first turned him on to biomedical engineering research; to his doctoral mentor, Morton Denn, who further encouraged him to apply his chemical engineering training to biomedical problems; and to his postdoctoral mentor, Kenneth Keller, who inspired him to create bioengineering courses. HB is grateful to his high school mentor Swarupanthan for introduction to mathematical methods and to his graduate school mentor JSU for research training in biotransport applications. GMS is particularly thankful to his postdoctoral mentor, the late Professor Stanley Katz of the City College of New York, whose guidance in mathematical modeling and analysis was very influential for career development. To our special families and friends, who shared our enthusiasm for this project, we also express our sincere appreciation.

James S. Ultman
University Park, PA
Harihara Baskaran
Cleveland, OH
Gerald M. Saidel
Cleveland, OH

About the Companion Website

This book is accompanied by a companion website:

http://engineering.case.edu/BMTR

The website includes:

- Homework problems
- Solution to homework problems for instructors
- Tables of data and equations
- Powerpoint slides of figures and tables for instructors

Part I

Introduction

Chapter 1

Biological Structure and Function

Living organisms operate much like a complicated chemical factory in which raw materials from the surrounding environment are distributed to chemical reactors that produce desired products along with waste that must be discarded back to the environment. And just like a chemical factory, a living organism must be capable of purifying raw materials and separating desired products from wastes. In humans, nutrients are separated from food in the upper gastrointestinal tract, oxygen is separated from air in the lungs, and many of the critical reactions that utilize these raw materials occur in the liver. Waste products are eliminated through the lower gastrointestinal tract and the kidneys as well as the lungs. It is the collection of all these chemical processes that enable an organism to maintain itself, perform work, grow, and reproduce.

The human body consists of about 100 trillion cells, each bathed in its own fluid microenvironment. In an adult, there is about 40 L of fluid, a third of which are extracellular (located outside of cells) and two-thirds of which are intracellular (located within cells). Various chemical species are nonuniformly distributed between the extracellular and intracellular fluids (Fig. 1.0-1), and there is a constant movement of ions, nutrients, waste products, and other substances between these fluid compartments. A function required of all cells is the regulation of these dynamics such that the chemical and energy needs of an organism are met.

Nature has provided for this by enclosing cells in a specialized membrane that supports a variety of transport processes, some of which are passive and others that are active in nature. During passive diffusion, the movement of a substance across a membrane occurs spontaneously in the direction of decreasing chemical potential. The uptake of O_2 from a relatively high concentration in extracellular fluid to a lower concentration in intracellular fluid is an example of passive diffusion. During active transport, energy from an independent chemical source is harnessed, allowing a substance to cross a membrane in the direction of increasing chemical potential. The maintenance of a low intracellular sodium level, for example,

Biomedical Mass Transport and Chemical Reaction: Physicochemical Principles and Mathematical Modeling,
First Edition. James S. Ultman, Harihara Baskaran, and Gerald M. Saidel.

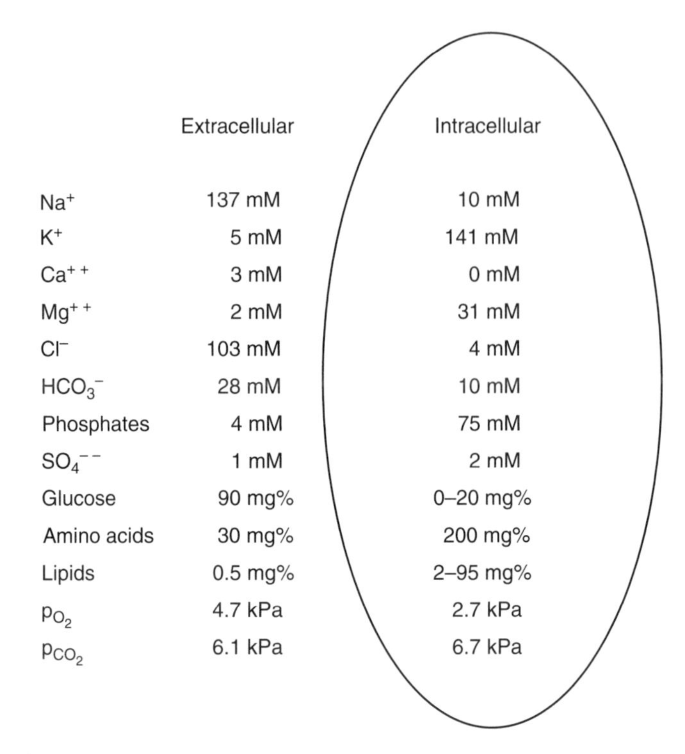

Figure 1.0-1 Homeostatic concentration conditions.

requires active transport of sodium out of cells to compensate for the passive leakage of sodium into cells.

Several energy-requiring processes in addition to active transport are necessary if an organism is to function properly and maintain its structural integrity. Energy is consumed by many of the metabolic reactions that synthesize essential molecules within cells. Energy is needed for muscular contraction in the heart, lungs, and limbs. Also, energy dissipation as heat is necessary to maintain a normal body temperature of about 37°C. The ultimate source of energy for all these tasks is the controlled oxidation of nutrients. In the remainder of this chapter, we will discuss how metabolic reactions and chemical transport are coordinated at different levels of structural organization, beginning at the whole-body level, progressing to the organ level, and ending at the cellular level.

1.1 Cell Energy Related to Whole-Body Function

Food entering the mouth is degraded to simpler molecules by hydrolytic reactions that occur in the oral cavity and the stomach. Additional chemical reactions occur downstream in the small intestine where the digested nutrients are absorbed into the bloodstream, ultimately reaching the intracellular space where they are oxidized to liberate energy. Humans eat foods containing a variety of different carbohydrates, fats, and proteins. However, the chemical selectivity of transport and reaction processes in the gastrointestinal tract produces a limited number of digestive products, such as glucose, fructose, and galactose, resulting from carbohydrate breakdown; triglycerides from fat breakdown; and amino acids from protein breakdown. Most of the galactose and fructose absorbed by the intestines are rapidly converted to glucose in the liver. Thus, glucose, triglycerides, and amino acids are the principal substrates for energy metabolism and chemical synthesis in the body.

During resting conditions, the minimal power requirement of an adult is about 70 W. This rate of energy production is provided primarily by the complete oxidation of glucose and triglycerides into CO_2 and water. These combustion reactions require about 250 ml/min of O_2 uptake through the respiratory tract. Even when a person's energy requirement is greater than 70 W, a suitable O_2 supply is usually available to sustain this aerobic metabolism. When a person is involved in very strenuous exercise, however, the energy demand can be so great that glucose is incompletely oxidized, forming a waste product, lactic acid, by anaerobic metabolism. Whether energy metabolism is aerobic or anaerobic, amino acids can never be completely oxidized. Rather, they are partially oxidized and form nitrogenous waste products, urea and creatinine, that are excreted by the kidneys.

1.1.1 Energy Generation

Suppose we burn a nutrient with O_2 in a closed vessel at an initial temperature of 37°C and we measure the heat that must be removed to reach a final temperature of 37°C. According to thermodynamics, the heat extracted from this calorimeter is identical to the total energy extracted as heat and work when the same reaction is carried out in a person at a constant body temperature of 37°C. Thus, the thermal information obtained from an inanimate calorimeter experiment is directly applicable to energy metabolism in a living organism, even though the mechanisms of nutrient oxidation are quite different.

The complete combustion of glucose with a stoichiometric amount of O_2 carried out in a calorimeter at atmospheric pressure and body temperature conditions yields the following information about carbohydrate metabolism:

$$C_6H_{12}O_6 + 6O_2 \rightarrow 6CO_2 + 6H_2O \quad \begin{cases} RQ = 6/6 = 1.00 \\ \Delta H_r(310°K;\ 101\ kPa) = -15.6\ kJ/g\ glucose \\ CE = 21.0\ kJ/L\ O_2 \end{cases} \qquad (1.1\text{-}1)$$

The respiratory quotient (RQ), defined as the molar output of CO_2 relative to the molar input of O_2 (equivalent to the CO_2 production volume relative to the O_2 consumption volume), is a direct result of the reaction stoichiometry. The heat of combustion ΔH_r is defined as heat that must be added per gram of glucose that is consumed. Because heat must actually be removed from a calorimeter to maintain a fixed pressure and temperature, ΔH_r is a negative quantity. The calorific equivalent (CE) represents the value of $-\Delta H_r$ relative to the volume of O_2 consumed. Therefore, the combustion of 1 g of glucose produces 15.6 kJ of energy, burns 15.6/21.0 = 0.743 L of O_2, and produces 0.743(1.00) = 0.743 L of CO_2.

Although many triglycerides participate in energy metabolism, we can model these reactions by focusing on a single triglyceride with a relative number of carbon–hydrogen–oxygen atoms similar to most other triglycerides. Calorimetric measurements of the complete combustion of triolein ($C_{57}H_{104}O_6$), one such model of triglyceride, result in the following data:

$$C_{57}H_{104}O_6 + 80O_2 \rightarrow 57CO_2 + 52H_2O \quad \begin{cases} RQ = 57/80 = 0.713 \\ \Delta H_r(310°K, 101\text{ kPa}) = -16.0\text{ kJ/g triolein} \\ CE = 18.5\text{ kJ/L}\,O_2 \end{cases} \tag{1.1-2}$$

We see from these values that 1 g of triolein produces 16.0 kJ of energy, burns 16.0/18.5 = 0.865 L of O_2, and produces 0.865(0.713) = 0.617 L of CO_2. Thus, on a per gram basis, fat metabolism produces slightly more energy, requires substantially more O_2, and produces significantly less CO_2 than carbohydrate metabolism.

Because of diversity in the structure of amino acids, we cannot model their energy metabolism with a specific compound. However, calorimetric measurements of a typical mix of foodstuffs have established representative parameter values for the oxidation of ingested proteins: RQ = 0.81, $\Delta H_r = -18.4$kJ/g protein, and CE = 19.2 kJ/L O_2. Although these values indicate that proteins are a favorable energy source, the amino acids created during protein digestion are normally more important for synthesizing new proteins than for producing energy.

Indirect calorimetry is a convenient procedure for evaluating energy metabolism from the rates at which a person excretes CO_2 to and extracts O_2 from the surroundings. The RQ associated with this CO_2–O_2 exchange can identify the types of nutrients being metabolized. Values of CE can then be used to predict energy production, and nutrient consumption rate can be estimated from ΔH_r. Consider the example of a person who consumes 300 ml/min of O_2 and excretes CO_2 at 300 ml/min as determined from respired gas measurements. With RQ = 300/300 = 1, it is likely that carbohydrates are the primary substrates being consumed. It follows that energy is produced at (21,000 J/L)(0.300 L/min) = 6300 J/min and the person metabolizes carbohydrates at a rate of (6300 J/min)(1440 min/day)/(15,600 J/g) = 582 g/day ≈ 1 lb/day.

1.1.2 Energy Transfer

By virtue of their high-energy phosphate groups, several nucleotides act as intermediates between the chemical reactions that generate energy and those that utilize energy. The most abundant of these nucleotides is adenosine triphosphate (ATP). At the fairly neutral pH conditions in physiological systems, ATP has a valence of −4 and a structure that is essentially $C_{10}H_{12}N_5O_4 - PO_3^- \sim PO_3^- \sim PO_3^{-2}$. Two of the three terminal phosphate groups are linked to the molecule by high-energy bonds (~) that can be broken in sequence to form adenosine diphosphate (ADP) ($C_{10}H_{16}N_5O_4 - PO_3^- \sim PO_3^{-2}$) and adenosine monophosphate (AMP) ($C_{10}H_{16}N_5O_4 - PO_3^{-2}$) according to the following reversible reactions:

$$\begin{aligned} ATP^{-4} + H_2O &\rightleftharpoons ADP^{-3} + HPO_4^{-2} + H^+ \\ ADP^{-3} + H_2O &\rightleftharpoons AMP^{-2} + HPO_4^{-2} + H^+ \end{aligned} \tag{1.1-3a,b}$$

where HPO_4^{-2} is the hydrogen phosphate ion. When these reactions proceed in the forward direction, the hydrolysis of either ATP or ADP cleaves one high-energy phosphate bond, liberating about 50 kJ of energy per mole. Conversely, when these reactions occur in the reverse direction, the phosphorylation of ADP or AMP requires an energy donation of 50 kJ/mol. The exact amount of energy depends on the specific concentrations of reactants and products.

Metabolic oxidation is frequently coupled with the phosphorylation of ADP. For example, the intracellular metabolism of glucose is initiated by a multistep process called glycolysis, resulting in a splitting of glucose into two pyruvic acid ($C_3H_4O_3$) molecules:

$$C_6H_{12}O_6 + 2ADP^{-3} + 2HPO_4^{-2} + 2H^+ \rightarrow 2C_3H_4O_3 + 2H_2O + 2ATP^{-4} + 4H \qquad (1.1\text{-}4)$$

During this partial oxidation of glucose, the energy released from glucose is captured in the terminal phosphate bond of the ATP product. Conversely, biological reactions that require energy are often coupled with the hydrolysis of ATP. Consider the extraction of energy from ATP molecules that is required to transform multiple glucose molecules into glycogen, a polysaccharide used for the short-term storage of carbohydrate. If n represents the number of monomeric repeating units $(-C_6H_{10}O_5-)$ initially present in a particular glycogen molecule $H(C_6H_{10}O_5)_nOH$, then the overall reaction for the incorporation of one more glucose molecule is

$$C_6H_{10}O_6 + H(C_5H_{10}O_5)_nOH + ATP^{-4} \rightarrow H(C_6H_{10}O_5)_{n+1}OH + ADP^{-3} + HPO_4^{-2} + H^+ \qquad (1.1\text{-}5)$$

In other words, the addition of each monomeric unit into a glycogen molecule requires the removal of an —OH and a —H group from one glucose molecule. The energy for this dehydration is supplied by the hydrolysis of ATP to ADP.

To summarize, a person viewed from a whole-body perspective (Fig. 1.1-1) digests food by specialized enzymatic reactions, ultimately generating energy by the oxidation of glucose, triglycerides, and amino acids. By cycling between the capture of this energy by ADP and its release from ATP, the body efficiently transfers the energy necessary to perform chemical

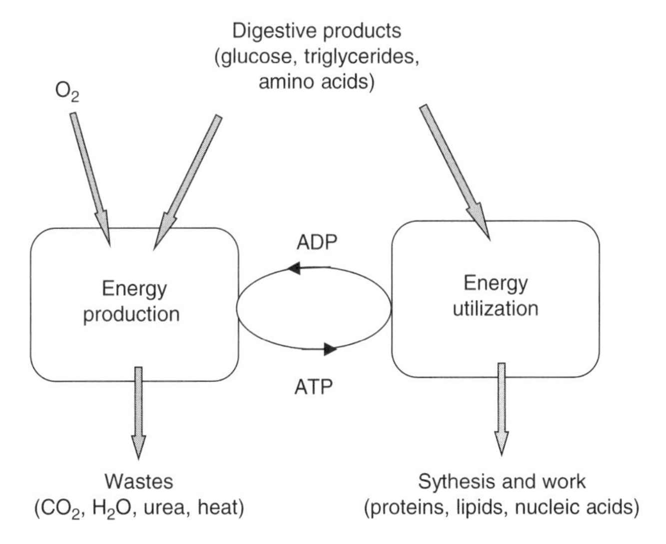

Figure 1.1-1 Energy production and utilization.

and mechanical work and synthesize new molecules while wasting a minimum amount of energy as heat.

1.2 Tissue and Organ Systems

Metabolism is intimately associated with mass transport and chemical reactions that are regulated both within and between a number of internal or visceral organs located in the thoracic, abdominal, and pelvic cavities of the body (Fig. 1.2-1). The organs primarily associated with mass transport and chemical reaction processes are the lungs (A), which exchange oxygen from the surroundings with carbon dioxide formed by nutrient oxidation; the stomach (B) and intestines (C), which digest and absorb nutrients and eliminate a semisolid waste; the liver (D), which stores glucose while synthesizing and degrading myriad substances; and the kidneys (E), which remove toxic metabolic products and excess water from the blood and expel them as a liquid waste. These organs are interconnected by the cardiovascular system and the lymphatic system that together provide continuous circulation of extracellular fluid.

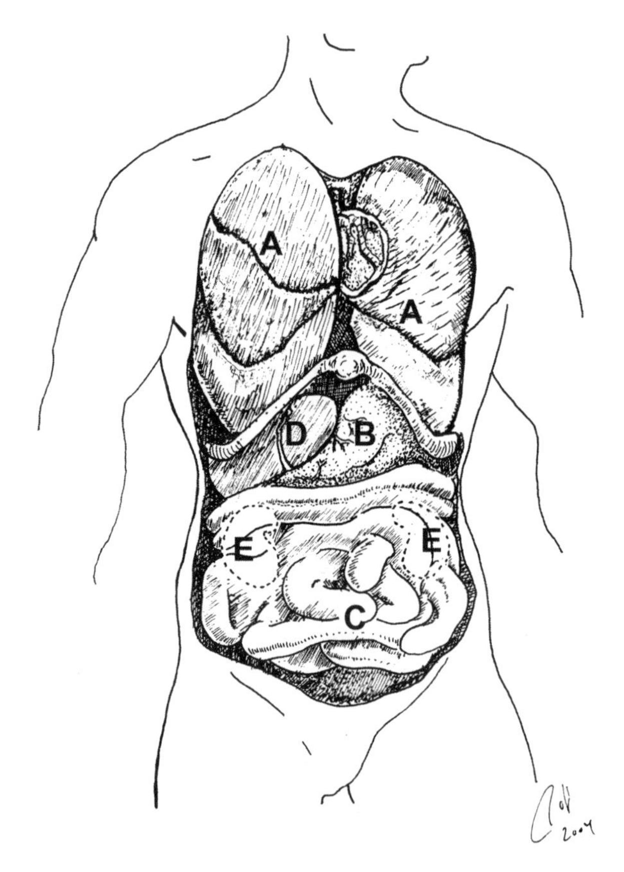

Figure 1.2-1 The visceral organs. A. lungs, B. stomach, C. intestines, D. liver, and E. kidneys (Courtesy of Joshua Stulman).

1.2.1 Circulation of Extracellular Fluid

Biological tissues consist of specialized groups of cells surrounded by interstitial fluid, a gel-like substance containing some rivulets of liquid flow (Fig. 1.2-2). Also embedded in the interstitial fluid is a network of narrow capillaries through which blood flows. About 45% of the blood volume is composed of suspended cells, and the remaining 55% is blood plasma. Most blood cells are erythrocytes or red cells whose main function is to transport O_2 from the lungs to all other organs and tissues. Blood plasma is a fluid similar in composition to interstitial fluid except that plasma is rich in proteins, especially albumin (45 g/L) and globulins (25 g/L). Of the 40 L of fluid in the body, about 2 L is in blood cells, 3 L is plasma, 20 L is in tissue cells, and the remaining 15 L is interstitial fluid. Small solutes and water are transported between blood and interstitial fluid and between interstitial fluid and tissue cells. Plasma proteins and erythrocytes are normally retained within the capillary blood. An exception occurs in the sinusoids of the liver where a large fraction of plasma proteins leak through the capillary walls into the interstitial fluid.

As illustrated in Figure 1.2-3, capillary beds of all organs are interconnected by large blood vessels. This allows the blood to flow in continuous loops through the cardiovascular system. What is not shown in this figure is the simultaneous flow of fluid through the lymphatic system. Lymphatic vessels collect fluid from interstitial spaces in tissues throughout the body and conduct this fluid back to the cardiovascular system into the veins of the neck. Although both the cardiovascular and lymphatic systems circulate extracellular fluid, they play different roles in maintaining stable (homeostatic) conditions in the body.

Total blood flow is about 5 L/min and blood volume is 5 L for a typical adult human. Thus, the cardiovascular system is capable of distributing substances to all tissues within a minute. By comparison, lymph flow is only 2 L/day, and it can take more than a day for the interstitial fluid to transit the lymphatic vessels. Although the lymphatic system is not effective in distributing most substances, it is the route by which absorbed fats are transported from the small intestine to the blood. Via the lymphatic system, waste products, dead blood cells, pathogens, toxins, and cancer cells are removed and destroyed. The lymphatics are also responsible for returning proteins that leak into interstitial fluid back to blood plasma.

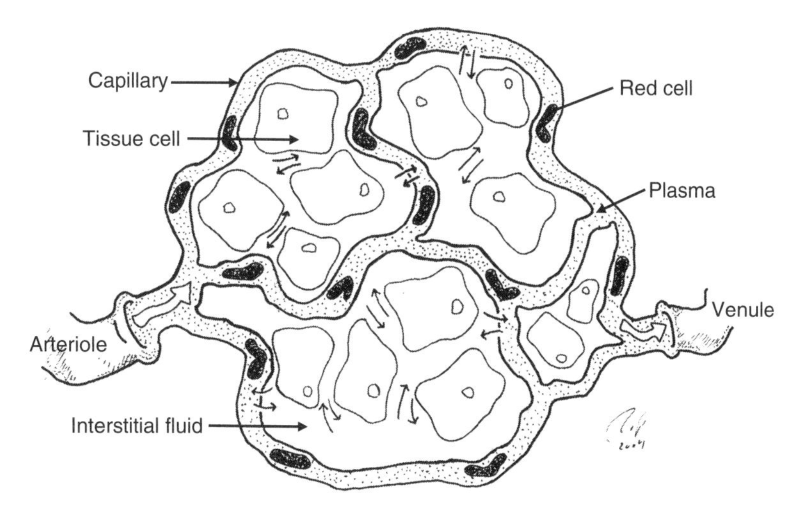

Figure 1.2-2 Compartmentalization of fluids and cells in tissue (Courtesy of Joshua Stulman).

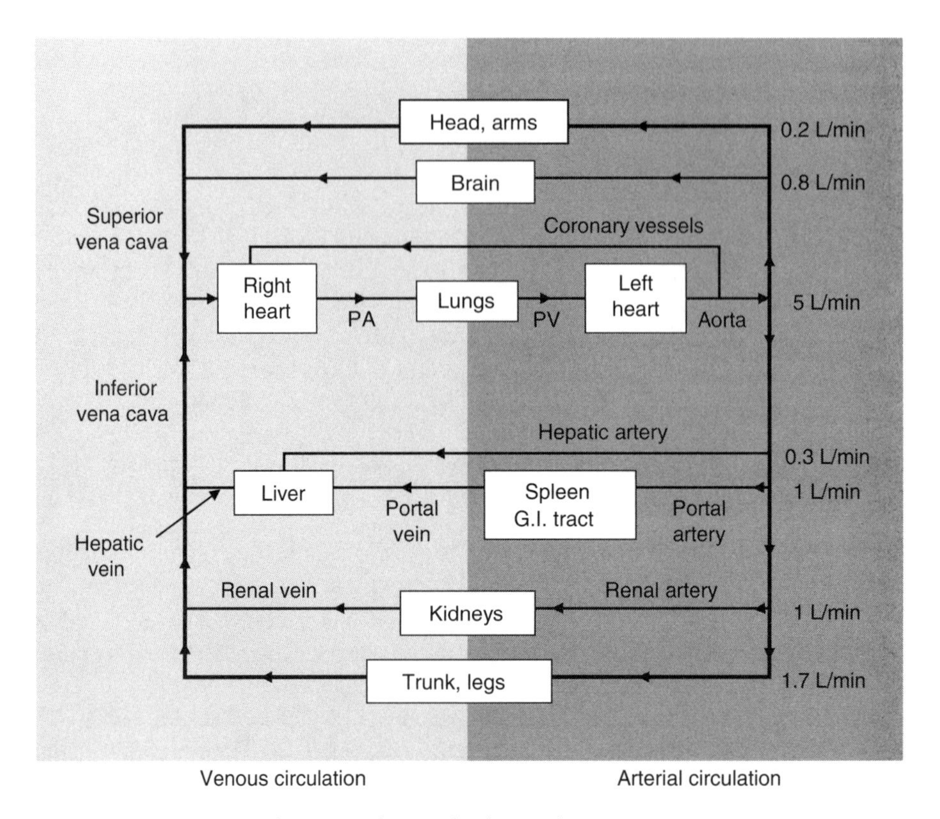

Figure 1.2-3 Interconnection of organs and tissues by the circulatory system.

Referring again to Figure 1.2-3, the heart is a double pump whose right side generates the pulmonary circulation of blood through the lungs and whose left side produces the systemic circulation through the remaining tissues of the body. The right heart pumps about 5 L of venous (deoxygenated) blood per minute through the pulmonary artery (PA) to the lungs. The arterial (oxygenated) blood leaving the lungs is returned to the left heart through the pulmonary vein (PV). The left heart pumps arterial blood via the aorta through a number of parallel paths that feed the visceral organs as well as the brain. The venous bloodstreams that exit these organs return to the right heart via the inferior vena cava and superior vena cava. The liver is a somewhat unusual organ in that its tissues are fed with a combination of oxygenated blood from the hepatic artery and partially deoxygenated blood from the portal vein.

In later chapters, we will show how flow sheets such as Figure 1.2-3 are useful when analyzing the distribution and metabolism of nutrients, pharmaceuticals, and other substances throughout the entire body. For now, let us look more closely at the structure of a few visceral organs and the role of structure in their desired function.

1.2.2 Lungs

The adult respiratory system consists of a pair of lungs, weighing about 1 kg, that is suspended in the thoracic cavity by the trachea. At maximal inhalation for a typical healthy subject, the total capacity of the lungs for air is 5 L, but the air volume at the end of a normal inhalation

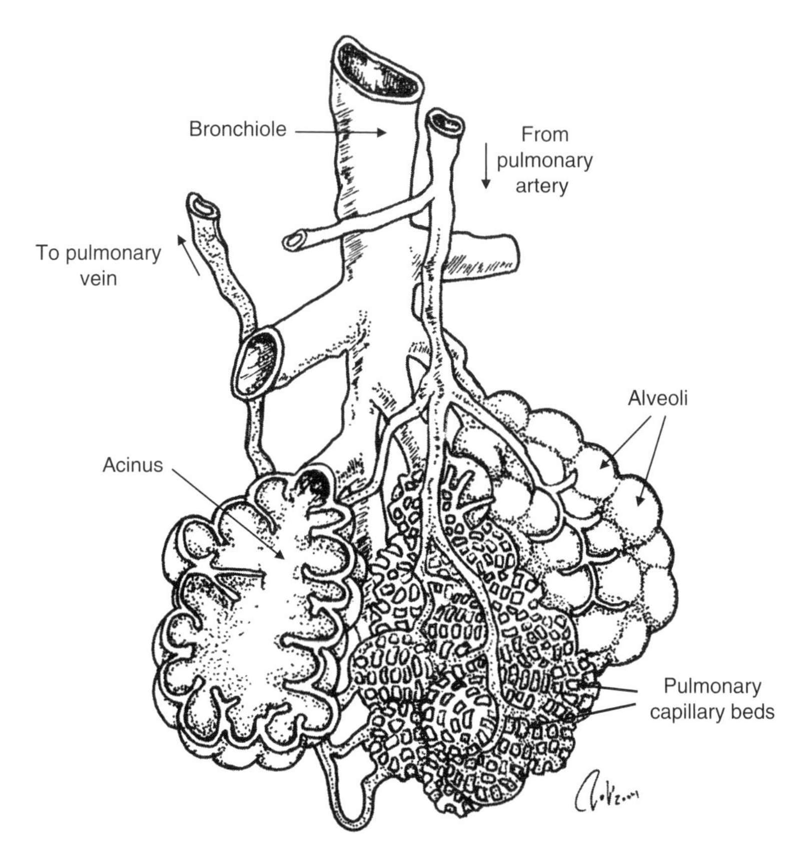

Figure 1.2-4 Terminal respiratory units in the lungs (Courtesy of Joshua Stulman).

is about half that value. Because more than 80% of this volume is occupied by millions of very small airspaces, the inside of the lungs has a spongy appearance. The lungs have several functions including exchange of CO_2 with O_2 from the environment, short-term regulation of acid–base balance by modulation of CO_2 excretion, and storage of blood in the pulmonary capillaries. At a resting metabolic rate, the lungs of an adult absorb about 250 ml/min of O_2 from inhaled air and excrete close to the same volume of CO_2 into exhaled air. To quantify this gas exchange process, the underlying mass transport and reaction processes must be considered.

Breathing occurs in a tidal (or cyclic) process driven by an alternating contraction and relaxation of the diaphragm and intercostal muscles. During quiet breathing, air is inhaled and exhaled at a volume of about 500 ml per breath at a typical breathing frequency of 15 breaths per minute. After entering the nose or mouth and reaching the trachea, inhaled air splits into a multitude of small flow streams as it passes through about 16 successive generations of bifurcating airways. These conducting airways have relatively rigid walls that are perfused by a small blood flow, about 2% of the cardiac output. The total gas-filled volume of the conducting airway region is only 150 ml.

Each of the $2^{16} \approx 70{,}000$ terminal conducting airways connects to an acinus containing about seven generations of airspaces composed mainly of elastic alveolar sacs and connective tissue (Fig. 1.2-4). While an individual gas-filled alveolus has a diameter of only 0.3 mm, the total surface area of the 300 million alveoli in the lungs is an incredible 50–100 m^2. The external surfaces of adjoining alveoli share a 0.02 mm thick sheath of capillaries that are collectively fed with virtually all of the cardiac output through the PA.

An alveolus with its surrounding capillary bed is the functional unit of the lung. Counter-diffusion of O_2 and CO_2 between alveolar air and the pulmonary capillary blood occurs across a sandwich of epithelial cells and endothelial cells separated by a layer of interstitial fluid. This alveolar–capillary membrane has a total thickness on the order of 1 μm. Because the collective surface of the alveoli is so large and the alveolar–capillary membrane is so thin, the lungs are highly efficient gas exchangers. For healthy lungs under resting conditions, for example, O_2 and CO_2 concentrations in alveolar air and capillary blood become equilibrated within the first third of the capillary length.

The blood oxygenator has been one of most successful biomedical mass transport devices. Sometimes called the artificial lung, this type of device was first used in the 1960s to replace pulmonary gas exchange for a few hours during open-heart surgery. Continuing improvements in device design and transport properties of synthetic materials have resulted in blood oxygenators that can now be used for days to weeks as a means of life support during lung failure.

1.2.3 Kidneys

The adult human kidneys consist of a pair of crescent-shaped lobes that together weigh about 0.25 kg. The macrostructure of the kidney is divided into the cortex (outer region) and the medulla (inner region). The kidneys excrete daily about 20 g of urea and 2 g of creatinine originating from the metabolic breakdown of proteins. They also maintain desired body fluid volumes and solute concentrations by typically excreting more than 1000 g of water and 0.4 equivalents of ions. In response to a change in pH outside of the normal range, the kidneys excrete acids or bases for compensation. When carbohydrate availability is limited, the kidneys can convert amino acids and triglycerides into carbohydrates via gluconeogenesis.

Each kidney lobe is supplied with a blood flow of about 500 ml through a renal artery. The renal artery branches into smaller arteries, eventually forming about one million afferent arterioles. Blood from each of these arterioles enters the glomerular capillary bed of a nephron, the functional unit of a kidney (Fig. 1.2-5). About 20% of the blood plasma entering the Bowman's capsule of the nephron is filtered through the capillary walls into the proximal tubule. The remaining blood leaves the glomerulus via an efferent arteriole. Filtered fluid flows from a proximal tubule through the loop of Henle, a distal tubule, and a collecting duct from which the urine exits to the bladder. The transport characteristics along the walls of these tubule segments change such that different combinations of solutes and water are "reabsorbed" from the tubular fluid into the capillary blood or "secreted" from the capillary blood into the tubular fluid.

The two major types of nephrons, cortical and juxtamedullary, differ with respect to their loops of Henle and their capillary beds. Cortical nephrons, those with glomeruli located in the peripheral cortex, have short loops of Henle that do not penetrate very far into the medulla. Juxtamedullary nephrons, whose glomeruli reside at the border of the cortex and medulla, have longer loops of Henle that can reach deep within the medulla. The tubules of the cortical nephrons are surrounded by a dense capillary network or "plexus" originating directly from the efferent arterioles. The efferent arterioles of the juxtamedullary nephrons divide into long straight peritubular capillaries or "vasa recta" that extend into the medulla along with the

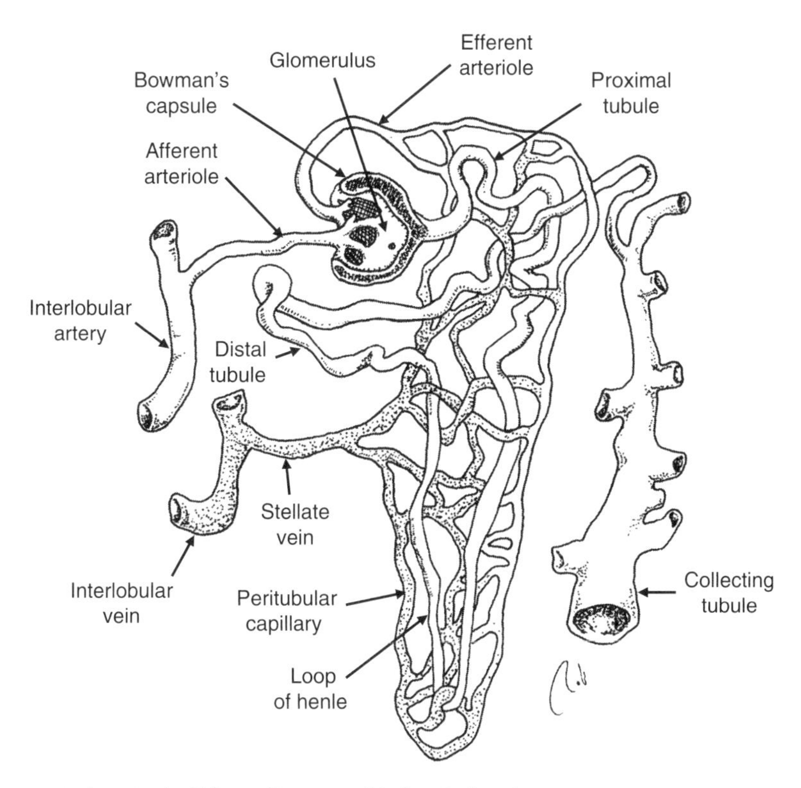

Figure 1.2-5 Nephron in the kidneys (Courtesy of Joshua Stulman).

loops of Henle. Blood from a descending limb of the vasa recta flows through a capillary plexus (less dense than in the cortex) and returns to the cortex through an ascending limb of the vasa recta.

The hydrostatic pressure in glomerular capillaries is higher than that in the proximal tubule. This is the primary driving force for filtration. In a healthy kidney, blood cells and most proteins cannot be transported through the glomerular capillary membrane. Consequently, an osmotic pressure difference produced between capillary blood and glomerular filtrate provides some opposition to the hydrostatic pressure difference. As filtrate flows along a renal tubule, salt is actively reabsorbed into the peritubular capillaries. This produces an added osmotic pressure difference that causes water to be passively transported out of the tubule. In addition to reducing the flow of renal fluid, a loss of water concentrates the urea and creatinine reaching the collection tubules. Loss of important nutrients such as amino acids is avoided by active reabsorption processes that also occur primarily in the proximal portion of the renal tubules.

When the kidneys are unable to remove metabolic waste products at a sufficient rate, an extracorporeal device, a hemodialyzer, can be used to supplement or replace this aspect of renal function. As blood from a subject continuously flows through a hemodialyzer, urea and creatinine are transported from the blood into a saline solution that also flows through the dialyzer. Although this device cannot fully substitute for natural kidneys, it is a lifesaving treatment for patients with inadequate renal function.

1.2.4 Small Intestine

Ingested food moves along the gastrointestinal tract from the mouth through the esophagus, the stomach, the small intestine, and finally the colon. The gastrointestinal tract functions to mechanically and chemically extract nutrients from the ingested food by digestion, which is followed by absorption of the nutrients into the circulatory and lymphatic system by absorption. Most of the mechanical processing is performed in the mouth and stomach, producing a watery solid suspension called chyme. The majority of the chemical processing and absorption of nutrients occurs in the small intestine, while the large intestine serves to absorb water and salts.

The small intestine is a seven meter-long muscular tube with an inner mucosal lining. Coordinated periodic contractions of longitudinal and circumferential muscles propel the chyme by peristalsis through three consecutive sections of the intestines: first the duodenum, then the jejunum, and finally the ileum. An epithelial cell layer at the luminal boundary of the mucosa secretes digestive enzymes into the chyme and also controls the transport of nutrients into the blood. Additional digestive enzymes in pancreatic juices and in liver bile flow into the intestinal lumen through small ducts located near the proximal end of the duodenum.

The anatomy of the intestinal mucosa is designed to maximize the absorption rate of nutrients. Numerous folds of Kerckring, up to 8 mm in depth, triple the mucosal area relative to that in a smooth tube. Millions of 1 mm-long villi protruding from the mucosal surface extend its area by another factor of 10 (Fig. 1.2-6). A brush border on the surface of each epithelial cell is composed of 1000 microvilli, 0.001 mm in length, that extend the surface area by yet another factor of 20. All together, the folds of Kerckring, the villi, and the microvilli provide a potential surface of more than 250 m^2 for the absorption of nutrients. This is three times more than is available for gas exchange in the lungs.

Digestion of all three types of nutrients—carbohydrates, fats, and proteins—occurs in the intestinal lumen by enzymatic reactions. Carbohydrates are hydrolyzed 80% into glucose and 20% into a mixture of fructose and galactose by amylase from the pancreas and other enzymes from the mucosal epithelium. Proteins are hydrolyzed into dipeptides and tripeptides by proteolytic enzymes supplied by the pancreas. Ingested fats are hydrolyzed primarily into fatty acids and monoglycerides by pancreatic lipase. Since lipase is a water-soluble enzyme and fats tend to collect in relatively large water-insoluble globules, the globules must be emulsified into small droplets before fat can be digested. Emulsification occurs by the addition of surface-active liver bile salts and lecithin in concert with mechanical mixing caused by peristalsis.

Depending on the diet, daily uptake into the large and small intestines consists of about 8 L of water, 100 g of ions, several hundred grams of carbohydrates, more than 100 g of fat, and 100 g of amino acids. Water and fats are absorbed from the chyme into the mucosal epithelium by diffusion. Most other substances are absorbed by specialized active transport processes. During their movement through the epithelium, several digestive products are further transformed. Tripeptides and dipeptides are converted to amino acids, which are then transported to a blood capillary. Fatty acids are resynthesized into emulsified triglycerides before passing into a lymphatic capillary, the "central lacteal."

Blood flow to the gut by the mesenteric circulation is directly related to local digestive and absorptive activity. When there is active absorption of nutrients, blood flow in a villum can be increased by as much as a factor of eight. Arteries ascend toward the tip of a villum, while

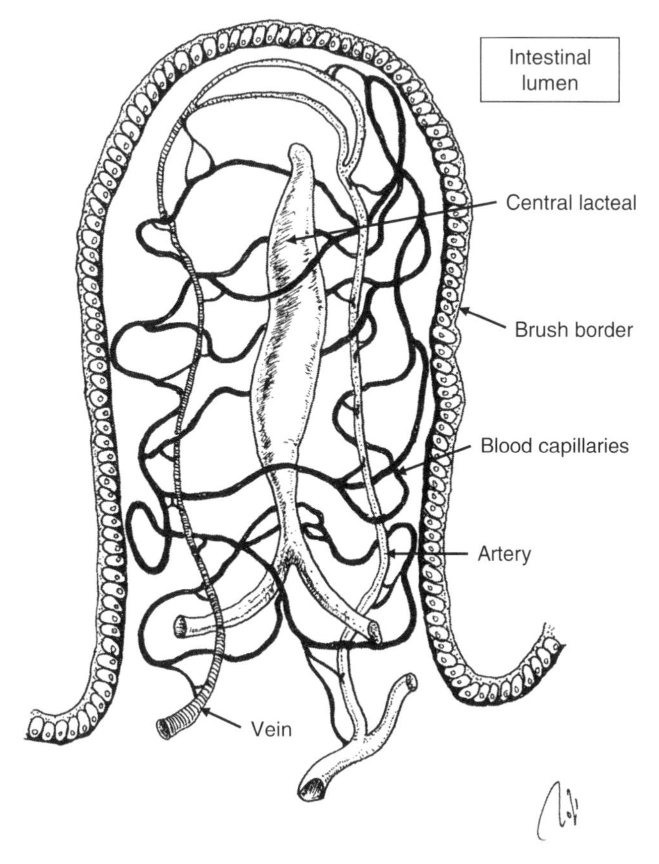

Figure 1.2-6 Villus in the small intestine (Courtesy of Joshua Stulman).

veins descend toward its base. The resulting countercurrent blood flow promotes efficient transport of absorbed substances.

1.2.5 Liver

The liver is part of the gastrointestinal tract. At a weight of about 1.6 kg, the liver is the heaviest visceral organ in the body. The functions of the liver include filtration and storage of blood, metabolism of nutrients and foreign chemicals, formation of bile, storage of vitamins and iron, and formation of blood coagulation factors. We will focus primarily on the metabolic functions of the liver.

The functional unit of the liver is the liver lobule, a cylindrical structure that is several millimeters long and about 1 mm in diameter (Fig. 1.2-7). The 100,000 lobules in the human liver each contain many hepatic cell plates that radiate from a central vein like the spokes of a wheel. The cell plates are separated by blood-filled sinusoids that are lined with special endothelium called Kupffer cells. The liver receives 1000 ml/min of blood flow from the portal vein and 300 ml/min of additional blood flow from the hepatic artery. After reaching the periphery

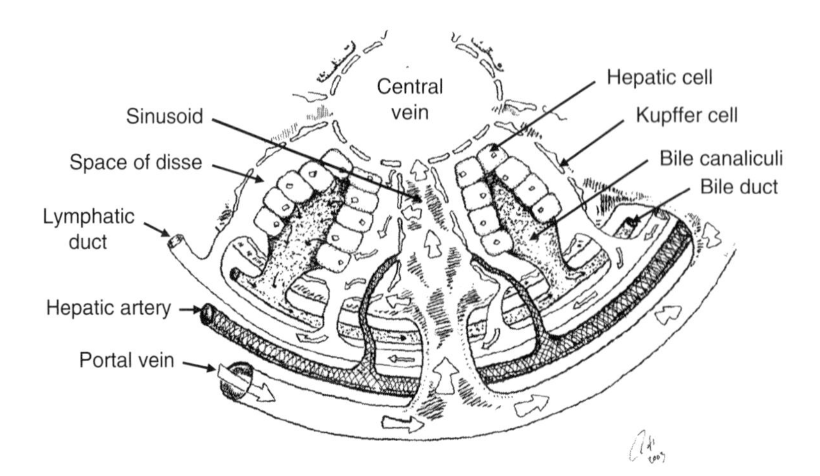

Figure 1.2-7 Liver lobule. Only two of the many hepatic cell plates are shown (Courtesy of Joshua Stulman).

of a liver lobule, blood from both of these sources flows through the sinusoids and collects in a central vein. Hepatic blood also perfuses the tissues between liver lobules.

Unlike other capillaries in the body, the Kupffer cell sheet forming the walls of the sinusoids are so permeable that protein concentrations in blood filtrate reaching the adjacent space of Disse are similar to their concentrations in blood plasma. This interstitial fluid is collected in lymphatic ducts and provides about half of the lymph flow to the entire body. Intestinal bacteria are often present in portal venous blood entering the liver, and it is important to remove these microorganisms before they can reach other organs. This function is performed by the Kupffer cells that phagocytize foreign bodies.

Hepatic cells have such a high metabolic rate that the liver dominates all other organs and tissues in the chemical processing of nutrients, drugs, and toxic materials. The liver provides most of the body's stores of carbohydrates in the form of glycogen. Along with the kidneys, the liver is also responsible for most of gluconeogenesis. The liver plays a major role in metabolizing fat to produce ATP. Protein metabolism occurs in the liver by the deamination of amino acids with subsequent conversion of ammonia to urea. The liver is crucial for the synthesis of plasma proteins and also synthesizes large quantities of cholesterol, phospholipids, and lipoproteins.

The efficiency of the liver as a chemical reactor is largely due to the inherent biochemistry of the hepatic cells. In addition, the thinness of and high fluid velocity in the space of Disse minimize limitations in mass transport to and from the hepatic cell surface.

As mentioned earlier, synthetic blood oxygenators and hemodialyzers have been successful in substituting for organ functions involving the transport of substances. An artificial liver must also act as a sophisticated biochemical reactor. The development of a bioartificial liver containing both synthetic materials and living hepatic cells is an active area of biomedical research.

1.3 Cell Structure and Energy Metabolism

We have come full circle from the beginning of this chapter, where the human body was described as a collection of a vast number of cells, to the question: just what is the structure of individual cells and how do they function? There are many different cell types such as muscle, nerve, blood, and fat cells. Rather than describing the specialized structure and function that distinguish these cell types, this section will focus on those features possessed by almost every cell.

1.3.1 Cell Composition

Animal cells, with a 5–10 μm characteristic length, contain a nucleus suspended in a gelatinous material called cytoplasm (Fig. 1.3-1). The contents of the nucleus are enclosed by the nuclear membrane, and the entire cell is surrounded by the plasma membrane. Also suspended in the cytoplasm are subcellular organelles whose functions will be discussed in the next section. The most abundant substances found in a cell are water, electrolytes, proteins, lipids, and carbohydrates. With the exception of fat cells, water is the solvent that supports the mass transport and chemical reactions that occur within cells. The most important ions in a cell are potassium, magnesium, phosphate, sulfate, bicarbonate, and the smaller amounts of sodium, chloride, and calcium. Regulating the concentration of these ions is especially important for the transmission of neural impulses.

Structural and globular proteins constitute up to 20% of the cell mass. Structural proteins are usually assembled in the form of long, insoluble filaments and microtubules that constitute the cytoskeleton of the cell. Globular proteins are more rounded, soluble molecules. Most globular proteins are enzymes that are either dissolved in the cytoplasm or adherent to membranes. The most important lipids in the cell, phospholipids and cholesterol, are used to form the membranes of both cells and organelles. Although a variety of fats are present in

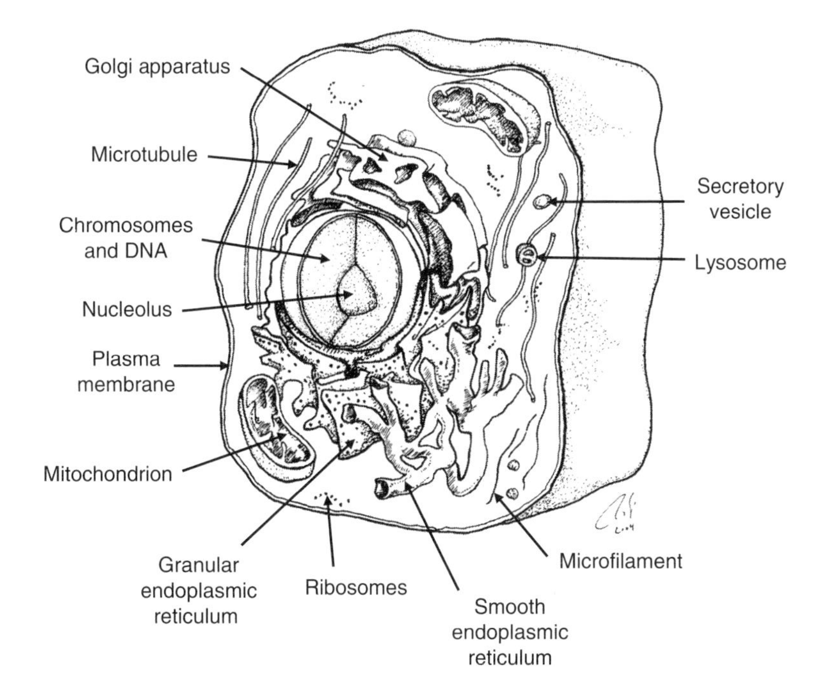

Figure 1.3-1 Eukaryotic cell (Courtesy of Joshua Stulman).

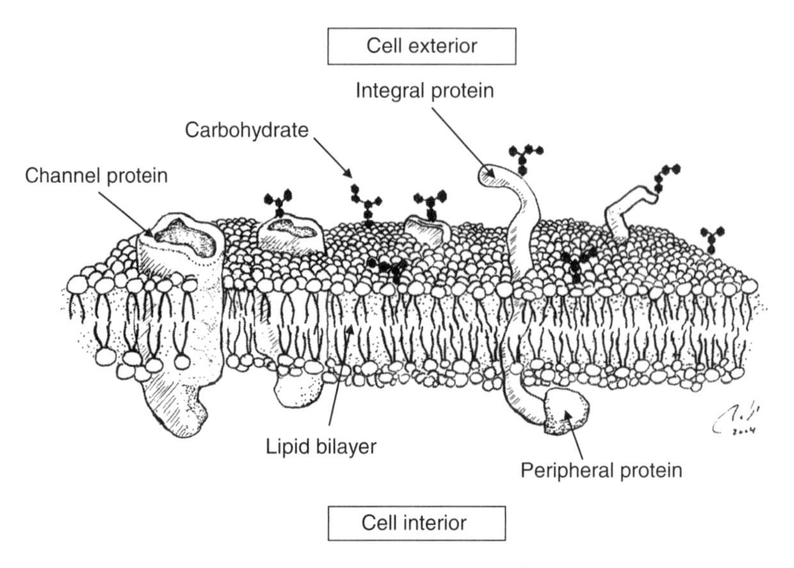

Figure 1.3-2 Fluid mosaic model of the cell membrane (Courtesy of Joshua Stulman).

all cell types, 95% of the mass of fat cells consists of triglycerides. Excess glucose in all cells is stored as insoluble glycogen particles, particularly in hepatic cells that contain up to 8% of their mass in glycogen and in muscle cells that contain up to 3% of this polysaccharide.

The plasma membrane surrounding the cell is a 7–10 nm-thick sheet composed of about 55% proteins, 25% phospholipids, 13% cholesterol, 4% other lipids, and 3% carbohydrates (Fig. 1.3-2). The foundation of the membrane is the phospholipid molecule that contains a hydrophilic head group and a hydrophobic tail group. Phospholipids are aligned in a bilayer with their tails pointed toward the center of the membrane to form a hydrophobic core while their heads face the membrane surfaces in contact with aqueous fluid. The hydrophobic core provides a barrier to the movement of water-soluble substances through the membrane, but it is relatively permeable to fat-soluble substances. Because the lipid bilayer is semifluid, portions of the membrane can flow from one point to another. The presence of cholesterol in the phospholipid bilayer modulates both its permeability and its fluidity.

Glycoproteins, which are proteins containing carbohydrate groups, are dispersed throughout the membrane. They are called integral proteins if they penetrate through the membrane or peripheral proteins if they are merely attached to one surface of the membrane. Some integral proteins form channels or pores through which ions and other small water-soluble substances are transported in a selective manner. Other integral proteins act as carrier molecules that participate in the transport of specific substances. Still other integral proteins serve as chemical transducers that transmit information about extracellular conditions to the cell interior. The peripheral proteins, generally found on the inside of the cell, are often part of this cellular signaling process.

In addition to glycoproteins, the plasma membrane contains small amounts of glycolipids and proteoglycans. The carbohydrate portions of these molecules usually protrude on the outside surface of the membrane, forming a negatively charged structure known as the glycocalyx. The glycocalyx is thought to play a role in the selective transport of substances through the membrane.

1.3.2 Cellular Organelles

Suspended in cell cytoplasm are various fat globules, glycogen granules, ribosomes, secretory vesicles, and five important organelles: the endoplasmic reticulum (ER), Golgi apparatus, mitochondria, lysosomes, and peroxisomes. The ER is a network of interconnecting tubules and vesicles that are also connected to the nuclear membrane. The large surface area of the ER is covered with enzymes that provide much of the biosynthesis in the cell. The granular ER is dotted with ribosomes that are responsible for protein synthesis, whereas the agranular ER is the site of lipid synthesis.

The Golgi apparatus, composed of several stacked layers of flat vesicles, functions in a complementary fashion with the ER. Small vesicles that continually pinch off from the ER are transported to the Golgi apparatus where they are processed to form lysosomes, peroxisomes, and secretory vesicles. Lysosomes are vesicles that contain hydrolases for digesting organic species by hydrolysis. Peroxisomes are vesicles containing catalase and oxidases that are capable of oxidizing toxic substances. The peroxisomes in the liver, for example, detoxify ingested alcohol. The secretory vesicles contain granules of high molecular weight substances that must be transported out of the cell. For instance, mucin granules secreted by goblet cells are necessary components of the protective mucous layer in the respiratory tract.

The nucleus regulates cell function by controlling the amounts and types of enzymes that are produced. The nucleus contains tens of thousands of protein-encoding genes, each of which contain unique sequences of deoxyribonucleic acid (DNA) bases. Every three DNA bases along the sequence is the code for one of the amino acids in a particular protein. The first step in producing a protein is transcription, the synthesis of a messenger ribonucleic acid (mRNA) containing a sequence of RNA bases that complement a DNA base sequence in the original gene. Therefore, each of the three RNA bases along an mRNA represents an amino acid and is called a codon.

Protein is assembled after mRNA is transported from the nucleus to the ribosomes of the granular ER. Protein assembly also requires a multitude of transfer RNA (tRNA) molecules, each consisting of an anticodon that recognizes one codon on an mRNA molecule as well as the amino acid corresponding to that codon. During protein assembly, amino acids from the tRNA are strung together, one at a time, by matching their anticodons with codons on the mRNA. Ribosomes direct this process with their own ribosomal RNA.

The mitochondria are the principal sites of energy metabolism. Each mitochondrion is enclosed in a double lipid bilayer. The many infoldings of the inner bilayer form shelves on which nutrients are enzymatically oxidized with dissolved O_2. The matrix between these shelves contains additional enzymes that preprocess nutrients before they reach the inner membrane. A unique feature of the mitochondria is self-replication. As is the case for a cell nucleus, the mitochondria contain the DNA necessary for controlling their own reproduction.

1.3.3 Mechanism of Cellular Energy Metabolism

Earlier in this chapter, the oxidation of nutrients and the simultaneous storage of energy in ATP molecules were discussed in broad terms. We conclude this chapter with a brief description of how this comes about within an individual cell, focusing on glucose, the major source of energy metabolism.

When glucose is transported across the cell membrane into the cell, it is metabolized aerobically in four stages as represented by the following overall reactions:

$$\text{Glucose} + 2\text{ADP}^{-3} + 2\text{HPO}_4^{-2} + 2\text{H}^+ \rightarrow 2\text{Pyruvic acid} + 2\text{H}_2\text{O} + 2\text{ATP}^{-4} + 4\text{H} \quad \text{Stage 1}$$
$$2\text{Pyruvic acid} + 2\text{CoA} \rightarrow 2\text{Acetyl-CoA} + 2\text{CO}_2 + 4\text{H} \quad \text{Stage 2}$$
$$2\text{Acetyl-CoA} + 4\text{H}_2\text{O} + 2\text{ADP} + 2\text{HPO}_4^{-2} + 2\text{H}^+ \rightarrow 4\text{CO}_2 + 2\text{CoA} + 2\text{ATP} + 16\text{H} \quad \text{Stage 3}$$
$$24\text{H} + 6\text{O}_2 + 34\text{ADP}^{-3} + 34\text{HPO}_4^{-2} + 34\text{H}^+ \rightarrow 34\text{ATP}^{-4} + 46\text{H}_2\text{O} \quad \text{Stage 4}$$
$$(1.3\text{-}1\text{a-d})$$

Stage 1, the initial breakdown of glucose, occurs in cell cytoplasm. In muscles cells, this occurs almost exclusively by glycolysis, the splitting of a glucose molecule into two pyruvic acid molecules according to the overall reaction scheme in Equation 1.3-1a (see also Eq. 1.1-4). Glycolysis actually consists of a sequence of 10 enzymatically regulated reaction steps in which 3% of the bond energy in a glucose molecule is transferred to 2 ATP molecules and 4 high-energy hydrogen atoms (H) are formed. In some cells, glycolysis is supplemented by other mechanisms of glucose breakdown such as the pentose phosphate pathway in hepatic cells

Stage 2, the conversion of pyruvic acid and coenzyme A (CoA) into acetyl coenzyme A (acetyl-CoA), occurs within the mitochondrial matrix by a two-step reaction. While no energy is transferred to ATP, 4 H atoms are produced. Stage 3, also occurring in the mitochondrial matrix, is referred to as the citric acid cycle or the tricarboxylic acid cycle or the Krebs cycle. The citric acid cycle consists of the oxidation of acetyl-CoA in a sequence of nine reaction steps in which 2 ATP molecules and 16 H atoms are formed.

Overall, the first three stages of energy metabolism produce 4 ATP molecules and 24 H atoms by the partial oxidation of a glucose molecule. During stage 4, known as oxidative phosphorylation, oxidation is completed by electron transfer reactions occurring along the inner membrane of the mitochondria. This "chemiosmotic process" consists of the following steps: each H atom is split into H^+ and an electron; the electron combines with dissolved O_2 and H_2O to form OH^- ions; and the energy from this oxidation is transferred to 34 ATP molecules.

The overall reaction corresponding to Equations 1.3-1a-d is given by

$$\text{Glucose} + 6\text{O}_2 + 38\text{ADP} + 38\text{HPO}_4^{-2} + 38\text{H}^+ \rightarrow 38\text{ATP} + 6\text{CO}_2 + 44\text{H}_2\text{O} \quad (1.3\text{-}2)$$

Accordingly, the four stages of energy metabolism convert 38 molecules of ADP into 38 molecules of ATP using the energy derived from the complete oxidation of one glucose molecule. The chemical efficiency of this process is very high because it consists of numerous coupled reaction steps, each of which has small energy transfer. In particular, the energy transferred to 38 moles of ATP is about (38)(50) = 1900 kJ, whereas the direct combustion of a mole of glucose releases 2810 kJ of heat. Thus, the efficiency of energy transfer from glucose to ATP approaches 70%. The remaining energy is dissipated as heat.

Oxidative phosphorylation forms about 90% of the ATP derived from glucose during aerobic metabolism. This is the only stage of metabolism that utilizes molecular O_2. When cells are deprived of O_2, oxidative phosphorylation diminishes, and an organism's energy demands must rely, at least in part, on the limited ATP produced during glycolysis and the citric acid cycle. The rate of this anaerobic metabolism would quickly diminish if the intermediate metabolites—pyruvic acid, acetyl-CoA, and H—were allowed to build up. This is avoided by the presence of lactate dehydrogenase, an enzyme that promotes the conversion of pyruvic acid and H to lactic acid.

Chapter 2

Modeling Concepts for Biological Mass Transport

This chapter gives an overview of mass transport in biological systems. We characterize the nature of biological materials and quantify their chemical composition. Mass transport mechanisms are described with the application to cellular transport processes. We end the chapter with a discussion of mathematical models that can be used to simulate biomedical systems.

2.1 Representation of Biological Media

2.1.1 Continuum Point of View

Matter is composed of atoms and molecules that are separated by relatively large voids. Over a spatial region with a very large number of these individual particles, matter can be considered continuously distributed. We visualize such a continuum as a contiguous assembly of many material points, each having locally averaged molecular properties (e.g., density, concentration, and velocity). Because these properties are spatially continuous and differentiable, mathematical models of transport processes can be expressed by differential equations. Regarding matter as a continuum is also consistent with the limited spatial resolution afforded by most experimental measurements that can only sample a large group of atoms and molecules rather than the individual entities. Analysis of mass transport processes in a solid, liquid, or gaseous continuum depends on the concentration distribution of chemical species.

Biomedical Mass Transport and Chemical Reaction: Physicochemical Principles and Mathematical Modeling, First Edition. James S. Ultman, Harihara Baskaran, and Gerald M. Saidel.

The appropriateness of a continuum representation of transport processes is gauged by the ratio of the intermolecular distance to the length scale of the physical region of interest, a dimensionless parameter called the Knudsen number. When the Knudsen number is much less than one, the region contains a sufficient number of molecules and atoms to define local material properties that vary continuously in space.

For example, consider the diffusion of potassium chloride (KCl) within an animal cell. With an intracellular KCl concentration of about 0.14 mol/L or 8×10^7 KCl ion pairs per μm^3, their intermolecular distance is about $\sqrt[3]{1/8 \times 10^7} \approx 2 \times 10^{-3}\,\mu m$. Relative to a cell with a diameter of 10 μm, the Knudsen number is about 2×10^{-4} for which a continuum model is appropriate. However, if we consider KCl diffusion through a cell membrane of thickness 0.01 μm, the Knudsen number is roughly 0.2, indicating that the continuum assumption can lead to errors.

2.1.2 Homogeneous and Heterogeneous Materials

A homogeneous phase is a region containing the same type of material, either a gas, liquid, or solid. Transport processes in biological systems most often occur in the heterogeneous mixtures of immiscible phases, each with its own physical and chemical properties. For example, intracellular fluid is a suspension of organelles, vesicles, and structural proteins in cell cytoplasm. Blood is a suspension of cells, platelets, and lipoprotein particles in plasma. Even respired gas can be a multiphase mixture of inhaled solid particles and liquid droplets suspended in air. The analysis of transport processes in these materials is complicated by the abrupt changes in microscopic properties that occur at internal boundaries between the phases. Nevertheless, as shown in Chapter 7, a heterogeneous medium can be modeled as a continuum of elemental volumes, each containing a small portion of the interspersed phases and each with macroscopic properties obtained by spatial averaging of microscopic properties among the phases.

2.1.3 Composition Variables

Molar density c, the total number of moles per unit volume, and mass density ρ, the mass per unit volume, are important material properties used in the analysis of mass transport processes. Both c and ρ can depend on temperature and pressure in addition to chemical composition. For a homogeneous phase containing i = 1,2,...I different substances, we can determine density from their molecular weights M_i and molar concentrations C_i:

$$c = \sum_{i=1}^{I} C_i, \quad \rho = \sum_{i=1}^{I} M_i C_i \tag{2.1-1a,b}$$

The equations governing mass transport are simplified in materials of constant c or ρ. However, because of spatial changes in C_i that are usually necessary to promote the transport of individual substances, some variation in c and ρ is inevitable.

The mean molecular weight M of a solution is the total mass of all species relative to their total number of moles. This is equivalent to the ratio of a solution's mass density ρ to its molar density c. The relationship between these variables is given by

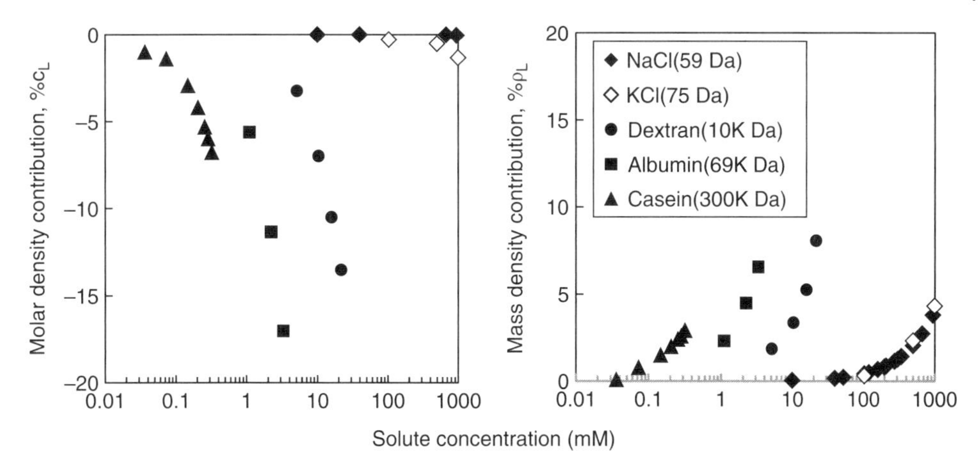

Figure 2.1-1 Densities of binary aqueous solutions relative to pure water (Data from Akashi *et al.* (2000), Chick and Martin (1913), and Millero (1970)).

$$M = \frac{\rho}{c} = \frac{1}{c}\sum_{i=1}^{I} M_i C_i = \sum_{i=1}^{I} M_i x_i \tag{2.1-2}$$

Liquid Phase

The densities of liquid solutions generally decrease with increasing temperature but are independent of pressure for almost all practical biomedical applications. During transport and reaction processes, density also varies because of spatial- or time-dependent composition changes. Most biological liquids are aqueous solutions that contain many solutes with widely different molecular weights. Because these multicomponent solutions are dilute, we can estimate the density contribution of an individual solute from its behavior with water alone. Figure 2.1-1 shows the percent contributions to molar density $\%c_L$ and mass density $\%\rho_L$ of some biologically relevant solutes in binary aqueous solutions at temperatures between 15 and 25°C (Akashi *et al.*, 2000; Chick and Martin, 1913; Millero, 1970):

$$\%c_L \equiv 100(c_L - c_w)/c_w, \quad \%\rho_L \equiv 100(\rho_L - \rho_w)/\rho_w \tag{2.1-3a, b}$$

In these equations, c_L and ρ_L are the molar and mass densities of the binary solution, while c_w and ρ_w are the corresponding densities of pure water.

Sodium and potassium chloride are the principal species in extracellular and intracellular fluids, respectively, where their concentration is about 140 mM. In binary aqueous solutions of either NaCl or KCl, molar solution density is essentially unchanged over a sevenfold increase in their concentration, from 140 to 1000 mM. While the corresponding mass solution densities do change, they increase by only a few percent. Data from three macromolecules are also shown in the figure: albumin, a plasma protein critical for osmotic balance; dextran, a glucopolysaccharide used medicinally as an antithrombotic agent; and casein (sodium caseinogenate), a phosphoprotein present in milk. Although these high molecular weight substances are present in biological fluids at much lower concentrations than NaCl and KCl, they can have a larger impact on c_L and ρ_L. For example, albumin is present in blood plasma at a concentration of about 1 mM. In a binary aqueous solution, a threefold increase in

Table 2.1-1 Universal Constants

Ideal gas constant ($\mathcal{R}$)	8.314 J/(mol-°K) or kPa-L/(mol-°K)
Avogadro's number ($\mathcal{A}$)	6.023×10^{23} molecules/mol
Boltzmann's constant ($\mathcal{K} = \mathcal{R}/\mathcal{A}$)	1.380×10^{-23} J/(molecule-°K)
Gravitational acceleration ($\mathcal{G}$)[a]	9.80665 m/s^2
Planck constant ($\mathcal{H}$)	6.626×10^{-34} kg-m^2/s
Faraday constant ($\mathcal{F}$)	96,480 A-s/eq

[a] Standard magnitude for an object in a vacuum near the Earth's surface.

this normal albumin concentration decreases molar density by about 10% and increases mass density by about 5%.

Generally, the binary solution data indicate that (i) macromolecules make a larger contribution to c_L and ρ_L than lower molecular weight solutes, (ii) a change in solute concentration causes an inverse change in c_L but a direct change in ρ_L, and (iii) for departures from typical solute concentrations, changes in ρ_L are less than the changes in c_L. Consequently, in the analysis of transport processes, it is usually more reasonable to assume a constant mass density than to assume a constant molar concentration.

Gas Phase

Transport of O_2 and CO_2 through the airspaces of the respiratory system is an essential life process. At pressures P and temperatures T under which physiological processes normally occur, interactions between the molecules in such a gas mixture are sufficiently weak that the ideal gas law is followed by each gas component i:

$$C_i \equiv \frac{P y_i}{\mathcal{R}T} \tag{2.1-4}$$

where y_i is mole fraction and $\mathcal{R}$ is the universal gas constant (Table 2.1-1 contains the value of this constant along with several other universal constants). According to the definitions in Equations 2.1-1a,b, the molar and mass densities for a mixture of I in such ideal gas components are

$$c_G = \frac{P}{\mathcal{R}T}\sum_{i=1}^{I} y_i = \frac{P}{\mathcal{R}T}, \quad \rho_G = \frac{P}{\mathcal{R}T}\sum_{i=1}^{I} M_i y_i = c_G \sum_{i=1}^{I} M_i y_i \tag{2.1-5a,b}$$

Equation 2.1-5a indicates that c_G is proportional to P/T but is independent of gas composition. Thus, when spatial variations in P/T are small, c_G is approximately constant. In many problems, it is common to express gas molar density at a standard temperature and pressure (STP) of $T^\circ = 273°K$ and $P^\circ = 101.3$ kPa. At this condition, the standard molar density of an ideal gas is $c_G^o = P^o/\mathcal{R}T^o = 0.0446\,\text{mol/L(STP)}$.

Example 2.1-1 Density of Respired Gas

As air is inhaled, it is progressively heated and humidified until it reaches the alveoli where gas exchange of O_2 and CO_2 occurs. Compare the mass and molar densities of inhaled air and alveolar gas. Typical conditions of inlet air are T = 295°K, P = 101 kPa, $y_{N_2} = 0.78$, $y_{O_2} = 0.20$, and $y_{H_2O} = 0.02$; and typical conditions of alveolar gas are T = 310°K, P = 96 kPa, $y_{N_2} = 0.76$, $y_{O_2} = 0.12$, $y_{CO_2} = 0.06$, and $y_{H_2O} = 0.06$.

Solution

The molar density under these conditions is

$$(c_G)_{inlet} = \frac{101}{8.314(295)} = 0.0412\ \text{mol/m}^3$$
$$(c_G)_{lung} = \frac{96}{8.314(310)} = 0.0372\ \text{mol/m}^3 \qquad (2.1\text{-}6a,b)$$

The corresponding mass densities are

$$(\rho_G)_{inlet} = \frac{101}{8.314(295)}[28(0.78) + 32(0.20) + 18(0.02)] = 1.18\ \text{kg/m}^3$$
$$(\rho_G)_{lung} = \frac{96}{8.314(310)}[28(0.76) + 32(0.12) + 44(0.06) + 18(0.06)] = 1.07\ \text{kg/m}^3 \qquad (2.1\text{-}7a,b)$$

Thus, the spatial variation in gas density over the respiratory system is about 10% whether density is expressed in a molar or mass form.

Chemical composition of a material can be expressed in different ways (Appendix A2). Composition of gas mixtures as well as liquid solutions may be specified in terms of mole fraction, molar concentration, or partial pressure. For a liquid, the choice of a concentration measure also depends on whether electrical or osmotic effects are considered. When gases are dissolved in liquids, it is sometimes convenient to express concentration in terms of the volumetric gas content, $\hat{C}_i \equiv C_i/c_G^o$, which is the gas volume dissolved per unit volume of liquid. For example, the molar concentration in a typical arterial blood sample is $C_{O_2} = 1.3 \times 10^{-4}$ moles O_2 per liter of blood; the corresponding value of $\hat{C}_i$ is $1.3 \times 10^{-4}/0.0446 = 0.00292$ liters (STP) O_2 gas per liter.

2.2 Mechanisms of Mass Transport

The distribution of dissolved substances in any living organism results from a coordination of passive and active processes. Passive transport may occur by convection, ordinary diffusion, diffusion facilitated by chemical carrier proteins, and migration of charged species in an electric field. Active transport utilizes specialized carrier proteins in addition to a chemical source of energy. This process occurs in cell membranes and organelles and along microtubules within cells.

2.2.1 Convection and Diffusion

Convective transport of a chemical species occurs by solution movement commonly produced by hydrostatic and osmotic pressure differences. The rate of convective transport is proportional to the concentration of the species in the solution and the flow rate of the solution. Examples of convective transport include the movement of gases between the

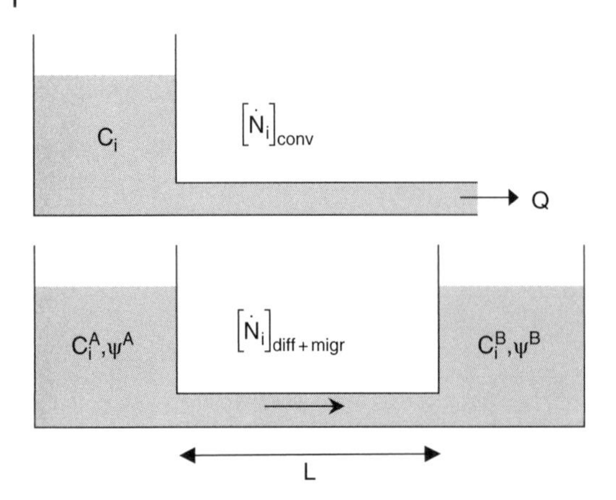

Figure 2.2-1 Transport by convection (top) and by diffusion and electrical migration (bottom).

environment and the body by respiratory airflow and the distribution of substances among the visceral organs by blood flow. On a smaller scale, examples of convective processes are the filtration of substances through the capillary walls of a kidney glomerulus and through the Kupffer cell layer of a liver sinusoid.

A simple example of convective transport is illustrated in the top of Figure 2.2-1 where a fluid is being drained from a reservoir through a tube at a volumetric flow rate $Q[m^3/s]$. If the reservoir contains an aqueous solution of substance i at a uniform molar concentration $C_i[mol/m^3]$, then the transport of i along the tube occurs by convection alone. At any position along the tube, the axial transport rate of substance i in [mol/s] is

$$\left[\dot{N}_i\right]_{conv} = QC_i \tag{2.2-1}$$

Ordinary diffusion is the movement of a chemical species by a chemical potential difference often expressed as a concentration difference. Electrical migration refers to the movement of a charged substance by an electrical potential difference. To visualize these processes, consider the system shown in the bottom of Figure 2.2-1 in which liquid solutions in reservoirs A and B are connected through a straight tube. A solute i with charge number z_i is present in the two reservoirs at different concentrations, C_i^A and C_i^B. The reservoirs are also at different electrical potentials, ψ^A and ψ^B. Because the reservoirs are open to the atmosphere and filled with solution to the same level, the pressures at the ends of the connecting tube are equal.

In the absence of a pressure difference, there is no convective transport through the tube. Rather, transport of species i occurs simultaneously by ordinary diffusion in proportion to $\left(C_i^A - C_i^B\right)$ and by electrical migration in proportional to $\left(\psi^A - \psi^B\right)$. The transport rates for both diffusion and migration are directly proportional to the cross-sectional area A of the tube and inversely proportional to the tube length L. The combined rate of solute diffusion and migration from reservoir A to reservoir B is as follows:

$$\left[\dot{N}_i\right]_{diff+migr} = \frac{\mathcal{D}_i A}{L}\left[\left(C_i^A - C_i^B\right) + \left(\frac{z_i \mathcal{F}\bar{C}_i}{\mathcal{R}T}\right)\left(\psi^A - \psi^B\right)\right] \tag{2.2-2}$$

Here, $\bar{C}_i$ is an intermediate concentration between C_i^A and C_i^B, and $\mathcal{D}_i$ is the diffusion coefficient, a parameter that depends on the molecular properties of substance i and its interaction with other substances in the solution. According to this equation, diffusion occurs in the direction of decreasing concentration, whereas electrical migration occurs in the direction of decreasing voltage for a solute with a positive charge number and in the opposite direction when the charge number is negative.

Example 2.2-1 Transmembrane Ion Transport

Sodium and potassium ions are the major contributors to electrical effects across cell membranes. What are the directions of diffusion and electrical migration of these two ions across the membrane of a resting human cell? According to Figure 1.0-1, their intracellular concentrations are $C_{Na}^{A} = 10$ mM and $C_{K}^{A} = 141$ mM, and their extracellular concentrations are $C_{Na}^{B} = 137$ mM and $C_{K}^{B} = 5$ mM. The transmembrane (inside to outside) resting potential is about $\psi^A - \psi^B = -10$ mV.

Solution

Since these ions have positive charge numbers and the cell potential is negative, $z_i(\psi^A - \psi^B) < 0$. Thus, the electrical migration of Na^+ and K^+ is both directed from the outside B to the inside A of the cell. The concentration driving force of Na^+, $(C_{Na}^{A} - C_{Na}^{B}) < 0$, indicates that diffusion of this ion is also toward the inside of the cell. For K^+, diffusion is outward, opposite to electrical migration.

Comparing Equations 2.2-1 and 2.2-2 reveals that $\left[\dot{N}_i\right]_{conv}$ is independent of L but is directly proportional to Q, whereas $\left[\dot{N}_i\right]_{diff+migr}$ is independent of Q but inversely proportion to L. Over a long path length, $L \gg 0$, diffusion and migration rates are small, and substances are more efficiently transported by convection. Over short path lengths, however, diffusion and migration are sufficient to provide the required transport. For example, oxygen transport along the long arterial path that carries blood from the lungs to the visceral organs occurs primarily by convection, but O_2 uptake through the thin alveolar–capillary membranes within the lungs is dominated by diffusion.

2.2.2 Transport through Cell Membranes

The plasma membranes surrounding cells provide the limiting barrier to the transport of solutes between extracellular and intracellular fluids. Similarly, the membranes that enclose cell organelles control the transport of substances between cell cytoplasm and the interior of the organelle. To support normal cell function, these membranes must be selective, allowing some substances to pass through easily, some to be transported more slowly, and others to not penetrate at all. The degree of selectivity depends on the properties of the transported substance in relation to the structure and properties of the membrane.

Consider plasma membranes that consist largely of a bilayer of phospholipids, sparsely interspersed with integral proteins. As depicted in Figure 2.2-2, any molecule can, in theory, be transported in the direction of a decreasing electrochemical (EC) driving force by diffusion and migration. In practice, the transport rate varies widely among different molecules depending on their diffusion coefficient and their ability to dissolve in the hydrophobic core of the bilayer. Small nonpolar molecules such as oxygen (18 Da) or carbon dioxide (44 Da) readily dissolve in and rapidly diffuse through the bilayer. Even some small uncharged polar

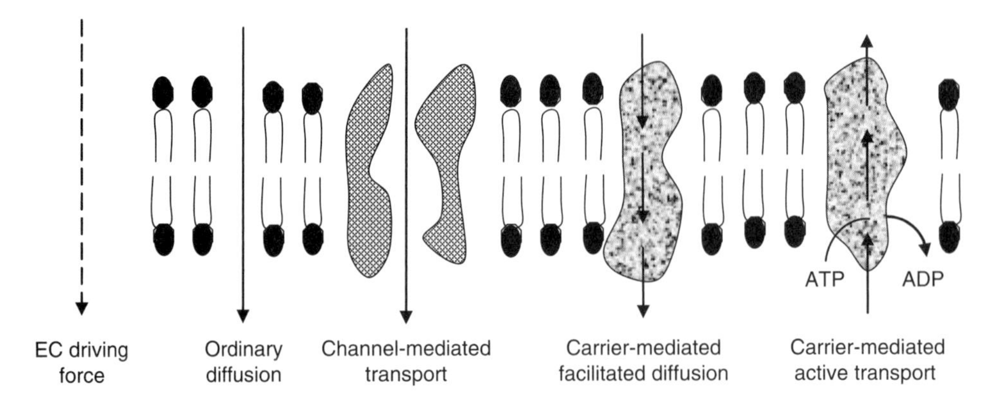

Figure 2.2-2 Mechanisms of molecular transport through cell membranes.

molecules such as water (18 Da), ethanol (46 Da), and urea (60 Da) can diffuse across the bilayer by squeezing between the hydrophobic tails of the phospholipids.

Ions such as H^+, Na^+, K^+, Ca^{2+}, and Cl^- are incapable of penetrating the lipid bilayer without special help. Instead, transmembrane channel proteins assist these solutes in crossing a cell membrane by forming continuous aqueous-filled passages less than 1 nm in diameter. During this channel-mediated transport, spontaneous but selective diffusion and migration occur at rates that depend on the geometry and internal charge distribution in the passages.

Large molecules such as glucose (180 Da) and amino acids are also unable to cross the lipid bilayer. Rather, they undergo carrier-mediated transport by binding to a specific surface site on a transmembrane protein. Some ions are also transported by means of carrier-mediated processes. Once solute is bound, the carrier protein changes its conformation in such a manner that the solute is translocated to the opposite side of the membrane. Carrier proteins that transport a single solute across a cell membrane are called uniporters. Some carrier proteins simultaneously translocate two solutes across the membrane. When the solutes are translocated in the same direction, the carrier protein is known as a symporter. When the solutes are translocated in opposite directions, the carrier protein is an antiporter.

Carrier-mediated processes can be passive or active. Passive carrier-mediated transport occurs in the direction of decreasing solute concentration as occurs in ordinary diffusion. The uptake of glucose into the muscle cells by a uniporter is an example of facilitated transport. Since glucose is metabolized within the cells, its extracellular concentration is higher than its intracellular concentration, and uniporter-bound glucose spontaneously moves into cells.

Active transport processes utilize a separate source of energy to translocate a solute-carrier protein complex along the direction of increasing (rather than decreasing) EC driving force. In primary active transport, a carrier protein relies directly on the hydrolysis of a high-energy phosphate, typically ATP, as its energy source. For example, active transport by the "Na^+/K^+ pump" simultaneously transports two K^+ into a cell and three Na^+ out of a cell. This process counteracts the passive diffusion of Na^+ into and K^+ out of a cell, thereby maintaining homeostatic conditions in the cell cytoplasm. The antiporter associated with the Na^+/K^+ pump has three receptor sites for Na^+ on the intracellular membrane surface, two receptor sites for K^+ on the extracellular surface, and another site near the intracellular surface that catalyzes ATP hydrolysis.

In secondary active transport, the EC driving force of one solute created by primary active transport is used as a source of chemical energy to overcome the adverse EC gradient of another solute. For intestinal epithelial cells to accumulate glucose from extracellular fluid, active transport is essential. In this case, extracellular glucose is cotransported with Na^+ into intestinal epithelial cells, and the Na^+ is recycled back into the extracellular space by the Na^+/K^+ pump. The protein symporter in this process has binding sites for both Na^+ and glucose on the extracellular membrane surface and does not change its conformation until both sites are occupied. This prevents unnecessary translocation of Na^+ into the cell when extracellular glucose is unavailable.

Carrier-mediated processes possess several properties that distinguish them from ordinary diffusion. Carrier-mediated processes are more selective than diffusion because binding sites on the protein carrier are tailored to a particular solute. Carrier-mediated transport rates reach a limiting value when solute concentration is so large that the binding sites on the carrier proteins are all occupied. Transport of one solute on a carrier protein can be inhibited by another solute that either competes for the same binding site or distorts the structure of the binding site. Exclusive to active transport, metabolic inhibitors that interfere with the utilization of high-energy phosphate compounds attenuate transport rates.

Very large molecules and particles that are incapable of diffusion or carrier-mediated transport can still be carried into a cell by endocytosis and out of a cell by exocytosis. Two mechanisms for these processes are as follows: phagocytosis, the direct ingestion of large particles such as bacteria or tissue debris; and pinocytosis, the ingestion of large molecules within small fluid-filled vesicles. Some proteins such as the LDL form of cholesterol undergo selective endocytosis by a receptor-mediated version of pinocytosis. During its circulation in the blood, LDL binds to membrane receptors on the intravascular surface of vascular endothelium. The resulting LDL-receptor complexes migrate to surface pits coated with contractile filaments of actin and myosin. After several complexes have aggregated in a pit, the contractile filaments shorten to form a closed pinocytotic vesicle containing the LDL with a small volume of blood plasma. Because actin–myosin contraction requires chemical energy, receptor-mediated pinocytosis is a type of active transport.

2.2.3 Transport across Cell Sheets

The transport of substances between compartments of the body is often controlled by cells assembled into sheetlike structures rather than by individual cells. Examples of these structures are the endothelial cell layer forming the inner walls of blood vessels, the epithelial cell layer forming the inner wall of the intestines, and the adjacent layers of endothelial and epithelial cells forming the alveolar membranes in the lungs.

Generally, cell sheets consist of a monolayer of cells supported on a basement membrane (Fig. 2.2-3). The perimeter of each cell is in close proximity of its neighboring cells, and strands of junctional proteins are wrapped around adjacent cells to maintain tight junctions between the cells. The tight junctions impede leakage of water-soluble molecules through the sheet. They also prevent channel and carrier proteins from migrating along a cell membrane between the apical and basal surfaces of the sheet. This segregation of apical and basal receptors allows transport mechanisms across the two surfaces to be different.

Consider, for example, the absorption of glucose and amino acids through the epithelial cell sheet from the inside (i.e., apical side) of the intestines. Tight junctions minimize paracellular transport of these molecules between cells, while transcellular transport occurs through cell

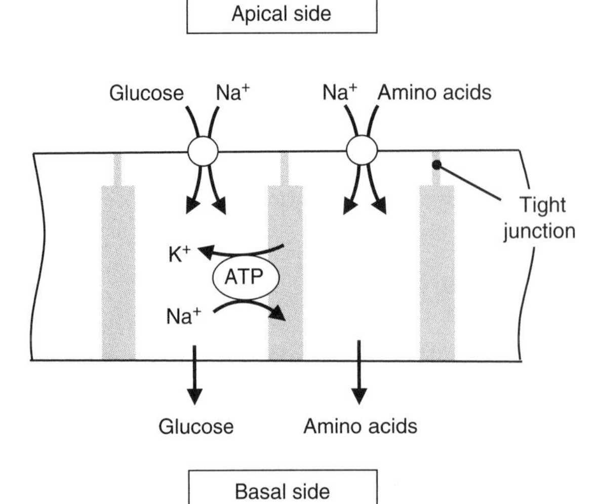

Figure 2.2-3 Transcellular uptake through the intestinal cell sheet.

membranes by a combination of active and passive processes. The apical cell surfaces contain symport proteins that facilitate the cotransport of glucose and amino acids with sodium ions into the epithelial cells from the intestinal lumen. These secondary active transport processes utilize an extracellular-to-intracellular gradient of Na^+ that is sustained by a Na^+/K^+ pump on the basolateral surfaces of the cells. The membrane on the basal surface of a cell contains carrier proteins that separately transport glucose and amino acids from the inside of the cell to the capillary blood. This uniport of the solutes is passive since glucose and amino acids are able to accumulate within cells at higher concentrations than in the basal extracellular space.

2.3 Formulation of Material Balances

The basic principle applied for quantitative analysis of mass transport is that mass can neither be created nor be destroyed in the absence of nuclear reactions. For any control volume containing a mixture of substances, this leads to a dynamic material balance that must be satisfied by each substance as well as by the mixture as a whole. When expressed in terms of molar quantities, the material balance can be stated as

$$\left\{\begin{matrix}\text{Molar rate of}\\ \text{accumulation}\end{matrix}\right\} = \left\{\begin{matrix}\text{Molar}\\ \text{input rate}\end{matrix}\right\} - \left\{\begin{matrix}\text{Molar}\\ \text{output rate}\end{matrix}\right\} + \left\{\begin{matrix}\text{Molar rate}\\ \text{of formation}\end{matrix}\right\} \tag{2.3-1}$$

To develop a mathematical formulation of this equation, we must specify a control volume, identify input and output streams through the boundaries of the control volume, and incorporate chemical reactions within the control volume.

Consider the control volume V shown in Figure 2.3-1 that contains a homogeneous mixture in which a chemical species i is present at a molar concentration $C_i[\text{mol/m}^3]$. The number of moles of i inside the control volume is C_iV. The inputs and outputs are represented by a molar transport rate $\dot{N}_{i,j}[\text{mol/s}]$ of species i in the jth stream. The rate of formation of species i

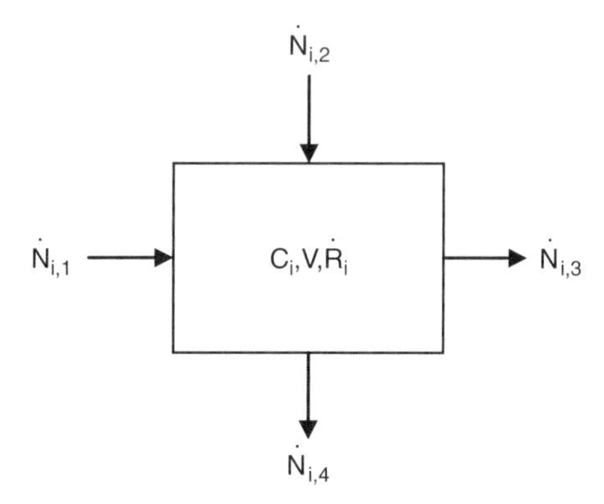

Figure 2.3-1 Material balance control volume.

by chemical reaction is represented by $\dot{R}_i[\text{mol/s}]$, which has a negative value if the species is consumed. In symbolic form, the terms in Equation 2.3-1 for a control volume with J input and output streams are as follows:

$$\left\{\begin{matrix}\text{Molar rate of}\\ \text{accumulation}\end{matrix}\right\} = \frac{\partial(C_i V)}{\partial t}$$

$$\left\{\begin{matrix}\text{Molar}\\ \text{input rate}\end{matrix}\right\} - \left\{\begin{matrix}\text{Molar}\\ \text{output rate}\end{matrix}\right\} = \sum_{j=1}^{J} \varsigma_j \dot{N}_{i,j} \tag{2.3-2a-c}$$

$$\left\{\begin{matrix}\text{Molar rate}\\ \text{of formation}\end{matrix}\right\} = \dot{R}_i$$

where ς_j is an indicator variable:

$$\varsigma_j \equiv \begin{cases} +1 & \text{when } j = \text{input stream} \\ -1 & \text{when } j = \text{output stream} \end{cases} \tag{2.3-3}$$

Combining Equation 2.3-2a-c according to Equation 2.3-1, we obtain a molar balance on species i:

$$\frac{\partial(C_i V)}{\partial t} = \sum_{j=1}^{J} \varsigma_j \dot{N}_{i,j} + \dot{R}_i \tag{2.3-4}$$

For the control volume in Figure 2.3-1, $J = 4$, $\varsigma_1 = \varsigma_2 = +1$ and $\varsigma_3 = \varsigma_4 = -1$. An overall molar balance is obtained by summing Equation 2.3-4 over all I chemical species in the mixture:

$$\sum_{i=1}^{I} \frac{\partial(C_i V)}{\partial t} = \sum_{j=1}^{J} \varsigma_j \sum_{i=1}^{I} \dot{N}_{i,j} + \sum_{i=1}^{I} \dot{R}_i \tag{2.3-5}$$

Since $\sum_i C_i$ is equivalent to the molar density c within the control volume (Eq. 2.1-1a), the overall molar balance becomes

$$\frac{\partial(cV)}{\partial t} = \sum_{j=1}^{J} \varsigma_j \sum_{i=1}^{I} \dot{N}_{i,j} + \sum_{i=1}^{I} \dot{R}_i \tag{2.3-6}$$

In general, the number of molecules produced and consumed by a reaction is not equal so that the reaction term $\sum_i \dot{R}_i$ does not vanish. For instance, the complete oxidation of glucose produces 12 moles of product for 7 moles of material that react (Eq. 1.1-1).

We can formulate a species mass balance by multiplying Equation 2.3-4 by the molecular weight of solute i:

$$\frac{\partial(\rho_i V)}{\partial t} = \sum_{j=1}^{J} \varsigma_j \dot{n}_{i,j} + M_i \dot{R}_i \qquad (2.3\text{-}7)$$

Here $\rho_i[kg/m^3] \equiv M_i C_i$ is the mass concentration of species i in the control volume and $\dot{n}_{i,j}[kg/s] \equiv M_i \dot{N}_{i,j}$ is its mass transport rate. By summing this mass balance over all I species and employing Equation 2.1-1b to specify the density ρ in the control volume, we obtain an overall mass balance equation:

$$\frac{\partial(\rho V)}{\partial t} = \sum_{j=1}^{J} \sum_{i=1}^{I} \varsigma_j \dot{n}_{i,j} \qquad (2.3\text{-}8)$$

In this case, the reaction term $\sum_i M_i \dot{R}_i$ vanishes because, excluding nuclear reactions, chemical reactions cannot change the mass of a system. Thus, when analyzing chemical reaction processes, it is usually more convenient to use an overall mass balance than an overall molar balance equation.

2.4 Spatially Lumped and Distributed Models

Mathematical models of mass transport and reaction in biomedical systems are useful for understanding the mechanisms underlying physiological function, predicting biological responses to medical interventions, and interpreting data from diagnostic tests and clinical experiments. Such models are based on mass conservation as well as on material-specific relationships describing chemical equilibrium, transport rates, and reaction rates. Because biomedical processes are so complex, the associated models must incorporate simplifying assumptions regarding both the structure of a system and its physicochemical behavior. Indeed, a mathematical model represents only some aspects of the real system, and thus a solution of model equations can only simulate a system to a limited extent.

Assumptions regarding the spatial distribution of substances within a biological system are of particular importance in developing mathematical models. In a spatially distributed, dynamic model, substance concentrations depend on time and at least one continuous spatial variable. The material balances are then expressed as partial differential equations. In some systems, spatial variations are either not significant, or they can be represented by a set of discrete concentration changes. In spatially lumped models of such systems, concentrations depend only on time, and material balances are expressed as ordinary differential or differential-difference equations. Spatially distributed models correspond more closely to physiological systems than do spatially lumped models. However, because of insufficient experimental data or a complexity of spatial distributions, models are often assumed to be spatially lumped.

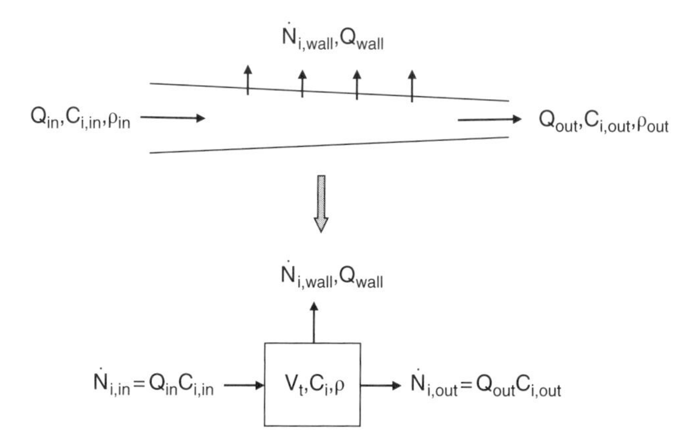

Figure 2.4-1 Spatially lumped model of transport in a rigid tube with a permeable wall.

2.4.1 Spatially Lumped Models

To illustrate a spatially lumped model, consider a solution flowing through a tube (or more generally a container of unspecified dimensions) that is permeable to solutes and solvent (Fig. 2.4-1, top). Mass transport at the entrance and exit of the tube is dominated by convection. Within the tube, mixing of the fluid by convection and diffusion is so intense that the composition of the solution is independent of spatial position. Consequently, for such a well-mixed system, the solution exiting the tube is identical in composition to the solution within the tube.

We select the entire tube as a control volume V_t (Fig. 2.4-1, bottom). Chemical species i flows into and out of the tube at molar rates $\dot{N}_{i,in}$ and $\dot{N}_{i,out}$, respectively, and is transported through the tube wall at a molar rate $\dot{N}_{i,wall}$. Species i can also be formed by chemical reaction within the tube at a molar rate $\dot{R}_i$. Substituting these rates into Equation 2.3-4, we obtain a species molar balance:

$$\frac{d(C_i V_t)}{dt} = \left(\dot{N}_{i,in} - \dot{N}_{i,out} - \dot{N}_{i,wall}\right) + \dot{R}_i \tag{2.4-1}$$

Here we have written the transient term as an ordinary derivative since the internal species concentration C_i can only depend time. With transport in the flowing streams dominated by convection, $\dot{N}_{i,in} = Q_{in}C_{i,in}$ and $\dot{N}_{i,out} = Q_{out}C_{i,out}$. Because the internal fluid composition is the same as its outflow composition, $C_i = C_{i,out}$ and the balance on species i become

$$\frac{d(C_{i,out} V_t)}{dt} = Q_{in}C_{i,in} - Q_{out}C_{i,out} - \dot{N}_{i,wall} + \dot{R}_i \tag{2.4-2}$$

Multiplying this equation by the molecular weight M_i of species i and summing over all species leads to a solution mass balance:

$$\frac{d}{dt}\left(V_t \sum_{i=1}^{I} M_i C_{i,out}\right) = \sum_{i=1}^{I} (M_i Q_{in} C_{i,in} - M_i Q_{out} C_{i,out}) - \rho Q_{wall} \tag{2.4-3}$$

Here we have defined the volumetric flow of solution through the tube wall as

$$Q_{wall} = \frac{1}{\rho}\sum_{i=1}^{I} M_i \dot{N}_{i,wall} \tag{2.4-4}$$

where ρ is the mass density of the internal fluid. Since the mass density of the inlet fluid is $\rho_{in} = \Sigma_i M_i C_{i,in}$ and of the outlet fluid is $\rho_{out} = \Sigma_i M_i C_{i,out}$, we can also write Equation 2.4-3 as

$$\frac{d(\rho V_t)}{dt} = Q_{in}\rho_{in} - Q_{out}\rho_{out} - Q_{wall}\rho_{out} \tag{2.4-5}$$

For a fluid of constant and uniform mass density in a constant volume system, $\rho = \rho_{in} = \rho_{out}$ and V_t are constants so that the balance equations reduce to

$$V_t\frac{dC_i}{dt} = Q_{in}C_{i,in} - Q_{out}C_{i,out} - \dot{N}_{i,wall} + \dot{R}_i \tag{2.4-6}$$

$$Q_{out} = Q_{in} - Q_{wall} \tag{2.4-7}$$

Example 2.4-1 Spatially Lumped Model of Kidney Tubule Function

Consider a representative renal tubule with a constant volume V_t and surface area S_t. The volumetric flow rate and urea concentration in glomerular filtrate entering the tubule are Q_{in} and $C_{u,in}$. The corresponding quantities in the urine exiting the tubule are Q_{out} and $C_{u,out}$. Urea and water are reabsorbed from the tubule into the surrounding capillaries at molar rates $\dot{N}_{u,wall}$ and $\dot{N}_{w,wall}$. The objectives of this problem are to formulate the governing equations for a spatially lumped model of urea transport and to find the steady-state relationship. Assume that glomerular filtrate has a constant mass density.

Solution

Since urea does not react within a tubule, a molar balance on urea according to Equation 2.4-6 is

$$V_t\frac{dC_{u,out}}{dt} = Q_{in}C_{u,in} - Q_{out}C_{u,out} - \dot{N}_{u,wall} \tag{2.4-8}$$

We now combine Equation 2.4-7 with 2.4-8 and get

$$V_t\frac{dC_{u,out}}{dt} + (Q_{in} - Q_{wall})C_{u,out} = Q_{in}C_{u,in} - \dot{N}_{u,wall} \tag{2.4-9}$$

To complete this dynamic model, we need to specify $\dot{N}_{u,wall}$ and Q_{wall}. Typically, this requires knowledge of how these quantities are related to pressure and concentration differences across the tubule wall. For simplicity, we assume that Q_{wall} is constant. We also assume that transport across the tube wall is proportional to the internal urea concentration C_u:

$$\dot{N}_{u,wall} = (P_u S_t)C_u = (P_u S_t)C_{u,out} \tag{2.4-10}$$

where the urea permeability coefficient P_u and surface area S_t of the tube wall are constants. Equation 2.4-10 then becomes

$$\frac{V_t}{Q_{in}}\frac{dC_{u,out}}{dt}+\left(1-\frac{Q_{wall}}{Q_{in}}+\frac{P_uS_t}{Q_{in}}\right)C_{u,out}=C_{u,in} \tag{2.4-11}$$

This linear, first-order ordinary equation with constant coefficients can be integrated with an initial condition, $C_{u,out}(0)$. To find the steady-state solution, we set the time derivative equal to zero with the result that

$$\frac{C_{u,out}}{C_{u,in}}=\left[1+\frac{1}{Q_{in}}(P_uS_t-Q_{wall})\right]^{-1} \tag{2.4-12}$$

By defining the dimensionless quantities,

$$C\equiv\frac{C_{u,out}}{C_{u,in}},\quad Q\equiv\frac{Q_{wall}}{Q_{in}},\quad P\equiv\frac{P_uS_t}{Q_{wall}} \tag{2.4-13a-c}$$

the model equation can be expressed in a more compact form:

$$C=\frac{1}{1-Q+P} \tag{2.4-14}$$

Here, C specifies the degree to which urea is concentrated during its convection through the tubule; Q represents the volumetric rate of water reabsorption through the tubule wall relative to its entering flow; and P is the characteristic rate of urea reabsorption relative to water reabsorption. As one would expect, the reabsorption of water and of urea has opposite effects on urea concentration. Increasing Q results in a more concentrated urine to be produced, while increasing P causes a dilution of the tubular fluid.

2.4.2 One-Dimensional Spatially Distributed Model

To illustrate a spatially distributed model, we consider mass transport of species i in a solution of constant mass density flowing in the axial direction z through a rigid tube whose cross section does not vary with time but can change with z (Fig. 2.4-2, top). In addition to the time dependence of the internal solute concentration C_i, incomplete mixing of solution leads to spatial variations in the z direction. In the lateral directions, however, we assume that C_i is uniform. Transport in the z direction is primarily by convection.

We select a rigid control volume consisting of a short tube section between a flow inlet at z and a flow outlet at z + Δz (Fig. 2.4-2 bottom). Molar input and output rates of species i through the lumen of the tube section are $[\dot{N}_i]_z$ and $[\dot{N}_i]_{z+\Delta z}$, respectively. Outward transport of species i through the tube wall is $\Delta\dot{N}_{i,wall}$, and its internal rate of formation by chemical reaction is $\Delta\dot{R}_i$. For any tube section, the volume ΔV does not change with time, and a molar balance on species i analogous to Equation 2.4-1 is

$$\frac{\partial(C_i\Delta V)}{\partial t}=\Delta V\frac{\partial C_i}{\partial t}=[\dot{N}_i]_z-[\dot{N}_i]_{z+\Delta z}-\Delta\dot{N}_{i,wall}+\Delta\dot{R}_i \tag{2.4-15}$$

Because C_i depends on z and t, we have expressed its time rate of change as a partial derivative. With longitudinal transport dominated by convection, $[\dot{N}_i]_z=[QC_i]_z$ and $[\dot{N}_i]_{z+\Delta z}=[QC_i]_{z+\Delta z}$ so that

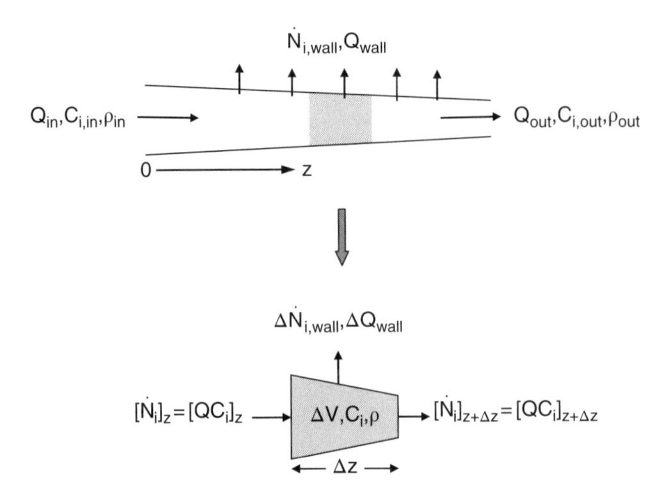

Figure 2.4-2 One-dimensional model of transport in a rigid tube with a permeable wall.

$$\frac{\partial C_i}{\partial t} + \left\{ \frac{[QC]_{z+\Delta z} - [QC_i]_z}{\Delta V} \right\} = -\left(\frac{\Delta \dot{N}_{i,wall}}{\Delta V} \right) + \left(\frac{\Delta \dot{R}_i}{\Delta V} \right) \tag{2.4-16}$$

Mass input and output rates of solution through the lumen of the tube section are $[Q\rho]_z$ and $[Q\rho]_{z+\Delta z}$, respectively. The outward rate of solution flow through the tube wall is $\rho\Delta Q_{wall}$. In terms of these quantities, an overall mass balance analogous to Equation 2.4-5 is

$$\frac{\partial(\rho\Delta V)}{\partial t} = \Delta V \frac{\partial \rho}{\partial t} = [Q\rho]_z - [Q\rho]_{z+\Delta z} - \rho\Delta Q_{wall} \tag{2.4-17}$$

For a fluid of constant density,

$$\left\{ \frac{[Q]_{z+\Delta z} - [Q]_z}{\Delta V} \right\} = -\frac{\Delta Q_{wall}}{\Delta V} \tag{2.4-18}$$

Taking the limits of these equations as $\Delta V \rightarrow 0$, the difference ratios become partial derivatives such that

$$\frac{\partial C_i}{\partial t} + \frac{\partial (QC_i)}{\partial V} = -\frac{\partial \dot{N}_{i,wall}}{\partial V} + \frac{\partial \dot{R}_i}{\partial V} \tag{2.4-19}$$

$$\frac{\partial Q}{\partial V} = -\frac{\partial Q_{wall}}{\partial V} \tag{2.4-20}$$

Using the chain rule, we rewrite the volume derivatives in terms of z derivatives and surface derivatives:

$$\begin{aligned} \frac{\partial Q}{\partial V} &= \frac{\partial Q}{\partial z}\frac{dz}{dV}, & \frac{\partial (QC_i)}{\partial V} &= \frac{\partial (QC_i)}{\partial z}\frac{dz}{dV} \\ \frac{\partial \dot{N}_{i,wall}}{\partial V} &= \frac{\partial \dot{N}_{i,wall}}{\partial S}\frac{dS}{dV}, & \frac{\partial Q_{wall}}{\partial V} &= \frac{\partial Q_{wall}}{\partial S}\frac{dS}{dV} \end{aligned} \tag{2.4-21a-d}$$

Here, we have used ordinary derivatives for the geometric variables since volume V and surface area S of a rigid tube only depend on z. The balance equations now become

$$\frac{\partial C_i}{\partial t} + \left(\frac{dz}{dV}\right)\frac{\partial (QC_i)}{\partial z} = -\left(\frac{dS}{dV}\right)\frac{\partial \dot{N}_{i,wall}}{\partial S} + \frac{\partial \dot{R}_i}{\partial V} \tag{2.4-22}$$

$$\frac{1}{A_t}\left(\frac{dz}{dV}\right)\frac{\partial Q}{\partial z} = -\left(\frac{dS}{dV}\right)\frac{\partial Q_{wall}}{\partial S} \tag{2.4-23}$$

By defining the following derivatives

$$\frac{\partial \dot{N}_{i,wall}}{\partial S} \equiv N_{i,wall}, \quad \frac{\partial Q_{wall}}{\partial S} \equiv u_{wall}, \quad \frac{\partial \dot{R}_i}{\partial V} \equiv R_i, \quad \frac{dV}{dz} \equiv A_t, \quad \frac{dS}{dV} \equiv \phi_t \tag{2.4-24a-e}$$

the balance equations become

$$\frac{\partial C_i}{\partial t} + \frac{1}{A_t}\frac{\partial (QC_i)}{\partial z} = -\phi_t N_{i,wall} + R_i \tag{2.4-25}$$

$$\frac{1}{A_t}\frac{\partial Q}{\partial z} = -\phi_t u_{wall} \tag{2.4-26}$$

In Equations 2.4-24a-c, we introduced the new dependent variables: $N_{i,wall}(z,t)$, the local transport rate of i per unit surface area across the tube wall (or molar flux); $u_{wall}(z,t)$, the local mass average velocity of fluid across the wall; and $R_i(z, t)$, the local formation rate of species i per unit volume (intensive reaction rate). The intensive variables $N_{i,wall}(z,t)$ and $R_i(z, t)$ are independent of system size. Therefore, they are more fundamental quantities than the absolute rates $\dot{N}_{i,wall}$ and $\dot{R}_i$.

In Equations 2.4-24d-e, we defined two geometric quantities: $A_t(z)$ is the local cross-sectional area of the tube, and $\phi_t(z)$ is its local surface-to-volume ratio, or equivalently, the local perimeter/cross-sectional area ratio. Although both of these quantities generally depend on z, they are constant in a tube of uniform cross section. For example, for a cylindrical conduit of length L and constant radius a, $A_t = \pi a^2$ and $\phi_t = 2\pi aL/\pi a^2 L = 2/a$.

Example 2.4-2 Spatially Distributed Model of Kidney Tubule Function

We now develop a steady-state model related to kidney function in which the volumetric flow of tubular fluid Q and the concentration of urea C_u vary with axial position z along a tubule. So that we can compare this spatially distributed model to the lumped parameter model of Example 2.4-1, we assume that the parameters A_t, ϕ_t, u_{wall}, and P_u are all constant.

Solution

We integrate the flow equation (Eq. 2.4-26) with the boundary condition $Q = Q_{in}$ at $z = 0$ to obtain

$$Q = Q_{in} - \phi_t A_t u_{wall} z \tag{2.4-27}$$

This indicates that the reabsorption of water at a uniform velocity across the tube wall results in a linear decrease in liquid flow along the length of the tubule. Eliminating the reaction rate and time derivative term from Equation 2.4-25 and introducing the wall flux from Equation 2.4-10, we obtain a steady-state molar balance in which urea concentration only depends on z:

$$\frac{d(QC_u)}{dz} = -(A_t\phi_t)N_{u,wall} = -(A_t\phi_t)P_u C_u \tag{2.4-28}$$

Expanding the z derivative in this equation and combining the result with Equation 2.4-27, we obtain the relation

$$[Q_{in}-(\phi_t A_t)u_{wall}z]\frac{dC_u}{dz}+\phi_t A_t(P_u-u_{wall})C_u=0 \tag{2.4-29}$$

For a tubule of length L_t and surface area S_t, the total volumetric flow through the tubule wall is $Q_{wall}=u_{wall}S_t$ and the surface-to-volume ratio is $\phi_t=S_t/L_tA_t$. The species concentration distribution then changes according to

$$\left(\frac{Q_{in}}{Q_{wall}}-\frac{z}{L_t}\right)\frac{dC_u}{dz}+\left(\frac{P_uS_t}{Q_{wall}}-1\right)\frac{C_u}{L_t}=0 \tag{2.4-30}$$

We solve this ordinary differential equation by separating variables and using the boundary condition $C_u=C_{u,in}$ at $z=0$ to obtain

$$\frac{C_u}{C_{u,in}}=\left[1-\left(\frac{Q_{wall}}{Q_{in}}\right)\frac{z}{L_t}\right]^{\left(\frac{P_uS_t}{Q_{wall}}-1\right)} \tag{2.4-31}$$

At the tube outlet where $z=L_t$ and $C_u=C_{u,out}$, this equation becomes

$$\frac{C_{u,out}}{C_{u,in}}=\left(1-\frac{Q_{wall}}{Q_{in}}\right)^{\left(\frac{P_uS_t}{Q_{wall}}-1\right)} \tag{2.4-32}$$

In terms of the dimensionless quantities defined in Equation 2.4-13, this equation is given by

$$C=\frac{1}{(1-Q)^{1-P}} \tag{2.4-33}$$

In the adult kidneys, about 99.2% of the water in the tubular fluid is reabsorbed into the surrounding capillaries so that $Q\approx 0.992$. Figure 2.4-3 compares the steady-state behaviors of C predicted by this model and the corresponding spatially lumped model (Eq. 2.4-14) when

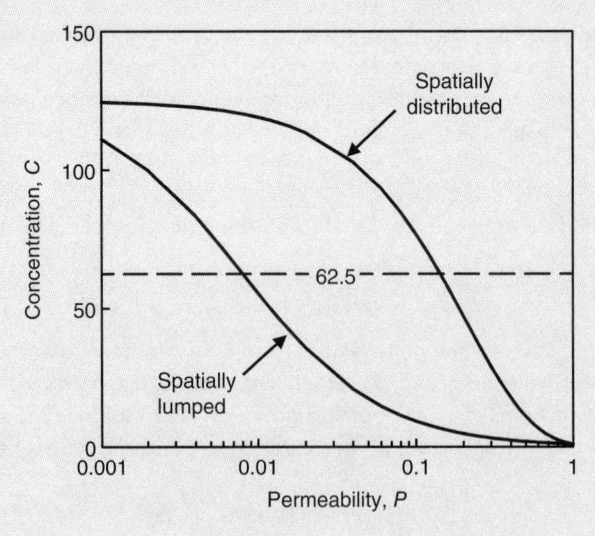

Figure 2.4-3 Dimensionless output concentration with 99.2% water reabsorption.

$Q = 0.992$. Clearly, the two models behave quite differently, even though they both depend on the same parameters. At any given P, the spatially lumped model always predicts a smaller C than the spatially distributed model. Therefore, at fixed values of P and Q, urea reabsorption is greater for the lumped model than for the distributed model. This occurs because urea reabsorption is proportional to internal urea concentration. In the lumped model, the internal concentration of urea is equal to its outflow concentration, whereas in the distributed model, internal urea concentration is everywhere less than its outflow concentration.

The adult kidneys reabsorb about 50% of the urea from the glomerular filtrate. Applying flow and species balance equations over an entire tubule allows us to compute the corresponding value of C:

$$\left.\begin{array}{r} Q_{out} = Q_{in} - 0.992 Q_{in} \\ Q_{out} C_{u,out} = Q_{in} C_{u,in} - 0.5 Q_{in} C_{u,in} \end{array}\right\} \Rightarrow C = \frac{C_{u,out}}{C_{u,in}} = 62.5 \qquad (2.4\text{-}34a\text{-}c)$$

To achieve this C value, the spatially lumped model requires a $P = 0.008$, whereas the spatially distributed model requires a $P = 0.14$ (Fig. 2.4-3). Thus, to reach the C and Q performance goals, the lumped parameter model underpredicts the required leakiness of a tubule with respect to urea diffusion.

References

Akashi N, Kushibiki J-I, Dunn F. Measurements of acoustic properties of aqueous dextran solutions in the VHF/UHF range. Ultrasonics. 2000; 38:915–919.

Chick H, Martin CJ. The density and solution volume of some proteins. Biochemistry. 1913; 7(part 1):92–96.

Millero FJ. The apparent and partial molal volume of aqueous sodium chloride solutions at various temperatures. J Phys Chem. 1970; 74:356–362.

Part II

Thermodynamics of Biomedical Processes

Chapter 3

Basics of Equilibrium Thermodynamics

Thermodynamics is the study of equilibrium and the transfer of energy that accompanies changes from one equilibrium state to another. Although a living organism is never actually at equilibrium, thermodynamics is useful for prescribing the limits on the performance of an organism. Thermodynamics can also predict whether a particular process will occur spontaneously. Thermodynamics cannot, however, specify how rapidly such a process will occur. Models that quantify rates of chemical transport and reaction processes are presented in later chapters.

Traditional thermodynamics associated with mechanical and chemical systems is largely based on two postulates that are supported by many years of experimental evidence: (i) energy accumulates in a body when heat or work are transferred from the surroundings; and (ii) any natural process cannot be reversed in its entirety without changing some other part of the universe. In this chapter, the mathematical embodiment of these postulates expressed by the first and second laws of thermodynamics is used to develop a general criterion for electrochemical equilibrium. In the following two chapters, we show how this criterion is applied to substances present in immiscible phases or phases separated by a membrane and to substances participating in a chemical reaction.

3.1 Thermodynamic Systems and States

A system is defined in a region of space enclosed by a set of real or imaginary boundaries. Everything in the universe outside the system is known as the surroundings or environment.

Biomedical Mass Transport and Chemical Reaction: Physicochemical Principles and Mathematical Modeling,
First Edition. James S. Ultman, Harihara Baskaran, and Gerald M. Saidel.

An isolated system exchanges neither mass nor energy (typically as heat) with its surroundings. A closed system exchanges energy but not mass with its environment. Since it is always composed of the same collection of matter, a closed system is often referred to as a body. An open system is capable of exchanging both mass and energy with its environment.

Early theoretical development of thermodynamics was limited to closed, uniform systems under static conditions. Realistic problems most often deal with open systems governed by spatially distributed variables that have time-dependent behavior. When changes in time and space are not too sharp, however, equilibrium can be assumed at local points within such systems (Slattery, 1972).

The state of a system is a unique condition defined by a set of measurable properties known as state variables. State variables may be either extensive or intensive. Extensive variables, such as volume and mass, are directly related to the size of a system and therefore can only be specified for the system as a whole. Intensive variables, such as temperature, pressure, and density, are independent of size and can thus be specified at each point of a system. State functions are quantities that depend on the state variables.

A phase is a system or part of a system that has spatially uniform properties. For a nonreacting system, the number of independent intensive variables F necessary to define the equilibrium state of a system depends on the number of distinct chemical components I and the number of phases φ according to the Gibbs phase rule:

$$F = I - \varphi + 2 \tag{3.1-1}$$

When extensive variables are used to define the state of a system, an additional system variable must be specified.

Let us apply the Gibbs phase rule to a few different systems. For a single-phase system of a pure substance like pure liquid water, $I = 1$ and $\varphi = 1$ so that $F = 2$. The state can thus be defined by the temperature and pressure of this system. For a system containing two phases and one component, such as water in equilibrium with its vapor, $I = 1$ and $\varphi = 2$ so that $F = 1$. The state can now be defined solely in terms of its temperature, while the pressure corresponds to the equilibrium vapor pressure of water at that temperature. For a single-phase system containing two components, such as glucose in water, $I = 2$ and $\varphi = 1$ so that $F = 3$, and the mole fraction of glucose in the aqueous solution must be specified in addition to temperature and pressure.

According to the second law, spontaneous processes, such as the equilibration of temperature between a hot and cold body or the mixing of two pure gases to form a binary mixture, cannot be entirely returned to their original state without altering their surroundings. These changes of state are considered to be irreversible. In principle, any change of state could occur so slowly that its path consists of an infinite number of intermediate equilibrium states. Although such a process could be reversed without changing some other part of the universe, it would require an infinite time to occur. Therefore, a truly reversible change of state cannot exist.

3.2 Heat, Work, and the First Law

The first law of thermodynamics, a statement of energy conservation, stipulates that the net exchange of heat and work between a system and its surroundings leads to a change in a state function called the internal energy. Heat represents energy transport from a hotter body to a cooler body. Work is the product of a force with the displacement induced by that force. For example, the mechanical work performed by a system when it expands by a differential volume dV against the pressure P of its surroundings is PdV.

For a differential change between two equilibrium states of a closed system, the first law is expressed as

$$dE = dq - dw \tag{3.2-1}$$

where dE is the differential increase in internal energy in a system caused by a differential amount of heat dq transferred from the surroundings to the system and a differential amount of work dw performed by the system on its surroundings.

For a finite change of state, Equation 3.2-1 must be integrated between the initial state 1 and the final state 2 as follows:

$$\Delta E \equiv E_2 - E_1 = \int_1^2 dq - \int_1^2 dw \equiv q - w \tag{3.2-2}$$

Because it is a state variable, E has a unique value once the state of a system is specified by the appropriate number of state variables. By extension, ΔE has a unique value once the variables that define the initial state and the final states are specified. Unlike ΔE, the values of q and w depend on which of a multitude of reversible or irreversible paths are taken between initial and final states. Thus, q and w are not state properties, but are defined as path functions. According to the first law, q and w can take on any combination of values as long as their difference is equal to ΔE.

3.3 Enthalpy and Heat Effects

Enthalpy, closely related to E, is another state function that expresses the energy within a system:

$$H \equiv E + PV \tag{3.3-1}$$

It follows from this definition that the change in enthalpy associated with a differential change of state is given by

$$dH = dE + PdV + VdP \tag{3.3-2}$$

For a change of state occurring in a closed system at a constant pressure with work accomplished only because of volume changes,

$$dH = dE + PdV + \cancel{VdP} = dE + dw = dq \tag{3.3-3}$$

Integrating this expression between an initial state 1 and a final state 2, we obtain

$$\Delta H \equiv H_2 - H_1 = q \tag{3.3-4}$$

Thus, the enthalpy change for this type of process is purely a heat effect that is classified as sensible heat when it causes a change in temperature, latent heat when it results in a change of phase, and compositional heat when separate substances are either mixed together or chemically react.

Our principal interest is the heat effect associated with chemical reactions. For example, Equation 1.1-1 lists −15.6 kJ as the heat of reaction for the complete combustion of 1 g of solid glucose with a stoichiometric amount of O_2 gas to form gaseous CO_2 and liquid water at P = 101.3 kPa and T = 37°C. With a molecular weight of 180 Da for glucose, its heat of reaction on a molar basis is $\Delta H_r = -(180)15.6 = -2810$ kJ. Since $\Delta H_r < 0$, glucose combustion results in heat transfer to the surroundings and is therefore an exothermic reaction. When a heat of reaction is positive such that the reaction requires positive heat transfer, the reaction is endothermic.

3.4 Entropy and the Second Law

The second law of thermodynamics is embodied in a state function called the entropy S. The differential change in entropy of a closed system during a change in state with heat transfer dq is given by

$$dS \geq \frac{dq}{T} \tag{3.4-1}$$

where the equality sign applies to a reversible change of state and the inequality is necessary for an irreversible process. Since S is a state function, the dS associated with a particular change of state is the same whether it is carried out in a reversible or an irreversible manner. Therefore, when an entropy change dS occurs at temperature T, heat transfer must be smaller along an irreversible path for which $dq < TdS$ than along a reversible path for which $dq = TdS$.

For an isolated system, $dq = 0$ such that Equation 3.4-1 becomes

$$dS_{isolated} \geq 0 \tag{3.4-2}$$

This indicates that the entropy of an isolated system cannot decrease. Rather, it increases in an irreversible process and remains constant in a reversible process.

On a microscopic level, entropy is directly related to the number of distinguishable molecular states of a system or, equivalently, to the disorder of the system. Therefore, another interpretation of Equation 3.4-2 is that isolated systems naturally tend to a state of less order. How is it then that biological organisms can maintain their high degree of organization and even replicate themselves? Does life exist in defiance of the second law of thermodynamics? To answer these questions, we must recognize that biological organisms are not isolated. Instead, they are open systems that exchange raw materials and waste products with their surroundings. Thus, a biological organism decreases its entropy (increases its order) at the expense of an even greater increase in entropy (decrease in order) of the rest of the universe.

3.5 Gibbs Free Energy and Equilibrium

3.5.1 Gibbs Free Energy Changes in Closed Systems

When specifying equilibrium conditions and predicting the spontaneity of various processes, it is helpful to define energy functions that depend on S. In many biomedical applications,

T and P are reasonably constant, and it is convenient to represent an energy state in terms of the Gibbs free energy:

$$G \equiv H - TS = E + PV - TS \tag{3.5-1}$$

For a differential change of state in any system, the change in Gibbs free energy is

$$dG = dE + (PdV + VdP) - (TdS + SdT) \tag{3.5-2}$$

This equation is independent of the nature of the system and the path taken to achieve the state change. For a closed system, we can introduce Equation 3.2-1 so that Equation 3.5-2 becomes

$$dG = VdP - SdT + (dq - TdS) - (dw - PdV) \tag{3.5-3}$$

We define the work performed by the system in addition to that required to change its volume as nonpressure–volume work or simply as added work:

$$dw' \equiv dw - PdV \tag{3.5-4}$$

Recognizing that heat transfer from a closed system must follow the second law, $(dq - TdS) \leq 0$, we rewrite Equation 3.5-3 as

$$dG \leq VdP - SdT - dw' \tag{3.5-5}$$

For any state change in a closed system, dG has a specific value, no matter what path is followed. However, dw' is dependent on path, giving rise to the inequality in this expression. In the special case of a reversible path, the added work has a specific value dw'_{rev}, and the equation becomes an equality:

$$dG = VdP - SdT - dw'_{rev} \tag{3.5-6}$$

Consider a change of state at constant T and P during which a Gibbs free energy change $(dG)_{T,P} < 0$ is found to occur. If the state change is reversible, then Equation 3.5-6 indicates that $dw'_{rev} = -(dG)_{T,P} > 0$ so that added work is performed by the system on its surroundings. If the state change is irreversible, then Equation 3.5-5 indicates that $dw' < -(dG)_{T,P}$, which is less added work than that occurring during a reversible change. Thus, provided that T and P are constant, dw'_{rev} is the maximum nonpressure–volume work that can be extracted from a closed system by decreasing its Gibbs free energy.

3.5.2 Chemical Potential Changes in Open Systems

State functions in multicomponent systems depend on the number of moles m_i of each substance i as well as on other variables such as T, P, and V. For a single-phase system containing I chemical substances, the phase rule indicates that $(I + 1)$ intensive variables are necessary to specify a value for an intensive state function. For an extensive state function such as G, an additional variable is necessary to account for the size of the system. By selecting T, P, and m_i $(i = 1,2,\ldots I)$ as the $(I + 2)$ variables of interest, we expand the differential of G as

$$dG(P,T,m_i) = \left(\frac{\partial G}{\partial P}\right)_{T,m_i} dP + \left(\frac{\partial G}{\partial T}\right)_{P,m_i} dT + \sum_{i=1}^{I} \left(\frac{\partial G}{\partial m_i}\right)_{T,P,m_j} dm_i \tag{3.5-7}$$

where it is understood that the subscript m_i on the first two derivatives refers to all species present in the solution, while the subscript j on the third derivative refers to all species except species i.

Of the three terms that comprise dG, only $(\partial G/\partial m_i)$ changes with the composition of the system. In addition to molecular transformations during chemical reactions, this term can account for mass transfer between an open system and its environment. Because of its importance in prescribing equilibrium, we define $(\partial G/\partial m_i)$ as the electrochemical potential:

$$\mu_i \equiv \left(\frac{\partial G_i}{\partial m_i}\right)_{T,P,m_j} \tag{3.5-8}$$

This intensive state variable depends on the temperature, pressure, and chemical composition of a solution as well as the interaction of charged species with electric fields.

Because the other two derivatives in Equation 3.5-7 are specified at constant moles m_i of all substances, they can be evaluated by comparing the dP and dT terms in Equation 3.5-7 to those appearing in Equation 3.5-6, which applies to a closed system:

$$\left(\frac{\partial G}{\partial P}\right)_{T,m_i} = V, \quad \left(\frac{\partial G}{\partial T}\right)_{P,m_i} = -S \tag{3.5-9a,b}$$

Substituting Equations 3.5-8 and 3.5-9a,b into Equation 3.5-7, we arrive at the Gibbs equation that applies to either an open or a closed system:

$$dG = VdP - SdT + \sum_{i=1}^{I} \mu_i dm_i \tag{3.5-10}$$

Combining this equation with Equation 3.5-2, we obtain another form of the Gibbs equation:

$$dE = TdS - PdV + \sum_{i=1}^{I} \mu_i dm_i \tag{3.5-11}$$

Comparing this Equation 3.5-10 to Equation 3.5-6, we find that

$$dw'_{rev} = -\sum_{i=1}^{I} \mu_i dm_i \tag{3.5-12}$$

when a state change affects the composition or total mass of a system. If T and P are constant, dw'_{rev} only depends on m_i and can be expanded as

$$dw'_{rev} = \sum_{i=1}^{I} \left(\frac{\partial w'_{rev}}{\partial m_i}\right)_{T,P,m_j} dm_i \tag{3.5-13}$$

Comparing this relation to Equation 3.5-12 indicates that electrochemical potential can be directly related to the reversible added work:

$$\mu_i = -\left(\frac{\partial w'_{rev}}{\partial m_i}\right)_{T,P,m_j} \tag{3.5-14}$$

Thus, μ_i represents the maximum nonpressure–volume work that a system is able to perform on its surroundings per mole decrease of substance i when T, P, and moles m_j of other substances are constant.

To determine electrical work, consider a species i whose molecules with a charge number z_i are initially separated by such a large distance that there is negligible repulsive force between them. When m_i moles of this substance are transported into a system with finite boundaries, the development of an electrical potential ψ requires that an amount of reversible work $z_i\mathcal{F}\psi m_i$ be added to the system ($\mathcal{F}$ is Faraday's constant). Since work is positive when it is performed by the system rather than on the system, $w'_{rev} = -z_i\mathcal{F}\psi m_i$ for this process. Taking the derivative of this expression according to Equation 3.5-14 indicates that electrical work makes a contribution $z_i\mathcal{F}\psi$ to the electrochemical potential. With the remaining portion of μ_i defined as the chemical potential $\breve{\mu}_i$,

$$\mu_i = \breve{\mu}_i + z_i\mathcal{F}\psi \tag{3.5-15}$$

3.5.3 Gibbs–Duhem Equation

We now consider a process occurring in which the moles of each substance in a system, and thus the total amount of the system, undergo an increase by the same factor f. In that case, the amount of material in the system increases, but its composition does not change. Therefore, intensive variables such as T, P, and μ_i will not change from their initial values, while extensive variables such as G and m_i will increase in proportion to f. Integrating the Gibbs equation between an initial state (T,P,G,m_i) to a final state (T,P,fG,fm_i), we obtain

$$\int_{G}^{fG} dG = \int_{P}^{P} V dP - \int_{T}^{T} S dT + \int_{m_i}^{fm_i} \sum_{i=1}^{I} \mu_i dm_i = \sum_{i=1}^{I} \mu_i \int_{m_i}^{fm_i} dm_i \tag{3.5-16}$$

Completing the integration, we get

$$fG - G = \sum_{i=1}^{I} \mu_i (fm_i - m_i) \Rightarrow G = \sum_{i=1}^{I} \mu_i m_i \tag{3.5-17a,b}$$

Since G is a state function, this is a general result whose differential is

$$dG = \sum_{i=1}^{I} \mu_i dm_i + \sum_{i=1}^{I} m_i d\mu_i \tag{3.5-18}$$

Substituting this expression into Equation 3.5-10, we obtain the Gibbs–Duhem equation for an open system:

$$SdT - VdP + \sum_{i=1}^{I} m_i d\mu_i = 0 \tag{3.5-19}$$

3.5.4 Spontaneous Processes and Electrochemical Equilibrium

We now consider changes in G that occur in a closed system when T and P are constant, and only PV work is performed on the environment. In that case, $dw' = 0$ and Equation 3.5-5 indicates that

$$(dG)_{T,P} \le 0 \tag{3.5-20}$$

depending on whether the process is irreversible (inequality) or reversible (equality). During a spontaneous process, which is irreversible by definition, $(dG)_{T,P} < 0$ so that G will decrease. However, if a process is sufficiently slow over the long term, then all changes become reversible so that $(dG)_{T,P} = 0$ and G will be constant. Interpreted in another way, $(dG)_{T,P} = 0$ indicates that a small perturbation of a system from an equilibrium state, when carried out at constant T and P, does not change its Gibbs free energy. This criterion for electrochemical equilibrium can be expressed in terms of electrochemical potentials by making use of the Gibbs equation at constant T and P:

$$(dG)_{T,P} = \sum_{i=1}^{I} \mu_i dm_i = 0 \tag{3.5-21}$$

Note that this equation can also be obtained by specifying that nonpressure–volume work cannot be performed by a system during perturbations from a reversible equilibrium state (i.e., $dw'_{rev} = 0$ in Eq. 3.5-12). By employing Equation 3.5-1, we can write Equation 3.5-20 in an alternative form as

$$(dH - TdS)_{T,P} \leq 0 \tag{3.5-22}$$

The inequality in this equation indicates that a change in state tends to be spontaneous when it is exothermic ($dH < 0$) and when it causes a disordering of the system ($dS > 0$). The equality specifies that a process reaches equilibrium when the heat effect dH matches the entropic change TdS.

As an example of chemical equilibrium, consider the reaction between a heme group Hb of a hemoglobin molecule with a molecule of oxygen to form an oxyheme group HbO_2:

$$Hb + O_2 \rightleftharpoons HbO_2 \tag{3.5-23}$$

The double arrow in this stoichiometric equation indicates that this is reversible reaction. When the reaction is displaced from equilibrium, it can occur in either the forward or reverse direction depending on temperature, pressure, and the mole fractions of reactants, x_{Hb} and x_{O_2}, relative to the mole fraction of product, x_{HbO_2} (Fig. 3.5-1). At relatively large initial x_{Hb} and x_{O_2} values, the free energy of the mixture is so great that the reaction spontaneously occurs in the forward direction. This results in simultaneous increases in x_{HbO_2} and decreases in x_{Hb} and x_{O_2},

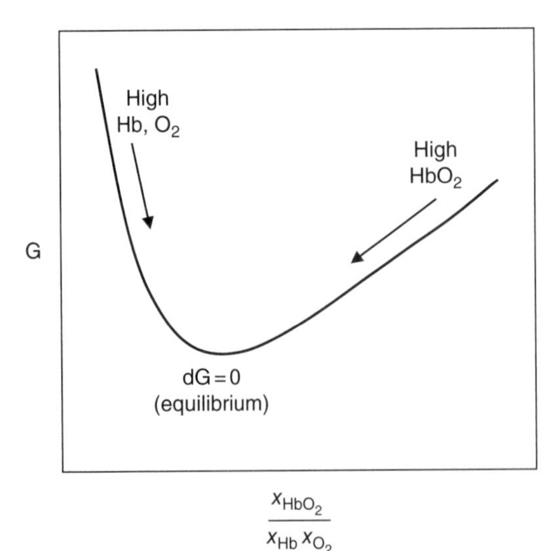

Figure 3.5-1 Gibbs free energy changes for the oxyhemoglobin reaction.

which reduce G. When x_{HbO_2} is initially large, G will also be large and the reverse reaction will be promoted. This causes a decrease in x_{HbO_2} and increases in x_{Hb} and x_{O_2} that also reduce G. Whether the process begins with an excess of reactants or an excess of product, the composition of the mixture will eventually reach an equilibrium value of $x_{HbO_2}/x_{Hb}x_{O_2}$. At that point, dG = 0 (or equivalently, G is minimized), and no further reaction occurs.

By integrating the inequality in Equation 3.5-20, we obtain a criterion for the spontaneous, finite, change of state carried out at constant T and P:

$$\Delta G \equiv G_2 - G_1 = \int_1^2 dG < 0 \tag{3.5-24}$$

For a biological process consisting of K coupled steps, the overall free energy change ΔG can be expressed in terms of the free energy changes of the individual steps ΔG_k by generalizing Equation 3.5-24:

$$\Delta G = \int_1^2 dG + \int_2^3 dG + \cdots + \int_{K-1}^{K} dG = \sum_{k=1}^{K} \Delta G_k < 0 \tag{3.5-25}$$

This equation indicates that ΔG can be negative even though some of the ΔG_k are positive. In other words, a multistep process can occur spontaneously even when one or more of its component steps are not spontaneous.

For example, during active transport of sodium ions out of a cell, there is a positive free energy change associated with the movement of sodium ions opposite to their natural diffusion direction. This is more than compensated by a negative free energy change associated with a simultaneous hydrolysis of ATP to ADP. Another example is the incorporation of a glucose molecule into a glycogen containing n repeating units (Eq. 1.1-5), which can be represented by two reaction steps:

$$\begin{array}{lr}
C_6H_{12}O_6 + ATP^{-4} \rightarrow C_6H_{11}O_6PO_3^{-2} + ADP^{-3} + H^{+} & \Delta G_{r1}(37^{\circ}C) \approx -50\ kJ \\
C_6H_{11}O_6PO_4^{-2} + H(C_5H_{10}O_5)_nOH & \\
\qquad \rightarrow H(C_6H_{10}O_5)_{n+1}OH + HPO_4^{-2} & \Delta G_{r2}(37^{\circ}C) \approx +30\ kJ \\
\hline
C_6H_{12}O_6 + H(C_6H_{10}O_5)_nOH + ATP^{-4} \rightarrow & \\
\qquad H(C_6H_{10}O_5)_{n+1}OH + ADP^{-3} + HPO_4^{-2} + H^{+} & \Delta G_{r1}(37^{\circ}C) \approx -20\ kJ
\end{array} \tag{3.5-26}$$

The high-energy glucose-6-phosphate $C_6H_{11}O_6PO_3^{-2}$ produced in the first reaction supplies the energy and the $-C_6H_{10}O_5-$ repeating unit necessary to produce the (n + 1) polysaccharide from the n polysaccharide. Even though $\Delta G_{r2} > 0$, the overall reaction is spontaneous since $\Delta G_{r1} + \Delta G_{r2} = -20$ kJ.

3.6 Properties of the Chemical Potential

3.6.1 Constitutive Equations

For the equilibrium criterion of Equation 3.5-21 to be useful, we need a constitutive equation that relates chemical potential to measurable variables such as pressure, temperature, and concentration. For a gas species i that follows the ideal gas law (Denbigh, 1964; Eqs. 3.19 and 3.20),

$$\breve{\mu}_i(T,p_i) = \mu^o(T) + \mathcal{R}T\ln p_i; \quad p_i \equiv y_i P \tag{3.6-1a,b}$$

where y_i is mole fraction in a gas phase, p_i is the partial pressure of component i, and μ^o is a function that depends on T alone. By analogy to the constitutive equation for species i in an ideal gas, $\breve{\mu}_i$ in a nonideal liquid solution is related to its mole fraction (Denbigh, 1964; Eq. 9.1) as

$$\breve{\mu}_i(T,P,x_i) = \mu_i^*(T,P) + \mathcal{R}T\ln(\gamma_i x_i) \tag{3.6-2}$$

where x_i is mole fraction, μ_i^* is the standard state chemical potential, and γ_i is an activity coefficient that corrects for the nonideality of species i in a liquid solution. Both γ_i and μ_i^* depend on P and T; γ_i also depends on x_i. The value of γ_i approaches one when substance i behaves in an ideal fashion. We will associate ideal behavior with an infinitely dilute solution of solute s in nearly pure solvent w. This means that $\gamma_s \to 1$ as $x_s \to 0$ for the solute, while $\gamma_w \to 1$ as $x_w \to 1$ for the solvent.

3.6.2 Temperature and Pressure Dependence

To find the sensitivity of chemical potential to pressure in a gas mixture or liquid solution, we first take the derivative of Equation 3.5-9a with respect to the moles m_i of species i:

$$\left[\frac{\partial}{\partial m_i}\left(\frac{\partial G}{\partial P}\right)_{T,m_i}\right]_{T,P,m_j} = \left(\frac{\partial V}{\partial m_i}\right)_{T,P,m_j} \equiv \hat{V}_i \tag{3.6-3}$$

where $\hat{V}_i$ is the partial molar volume of the solution with respect to component i. Interchanging the order of differentiation with respect to m_i and P also indicates that

$$\left[\frac{\partial}{\partial m_i}\left(\frac{\partial G}{\partial P}\right)_{T,m_i}\right]_{T,P,m_j} = \left[\frac{\partial}{\partial P}\left(\frac{\partial G}{\partial m_i}\right)_{T,P,m_j}\right]_{T,m_i} = \left(\frac{\partial \mu_i}{\partial P}\right)_{T,m_i} \tag{3.6-4}$$

Since the electrical contribution to μ_i does not depend on P, we conclude that

$$\left(\frac{\partial \mu_i}{\partial P}\right)_{T,m_i} = \left(\frac{\partial \breve{\mu}_i}{\partial P}\right)_{T,m_i} \equiv \hat{V}_i \tag{3.6-5}$$

Similarly, we can show with Equation 3.5-9b that the sensitivity of chemical potential to T in a gas mixture or liquid solution is given by

$$\left(\frac{\partial \mu_i}{\partial T}\right)_{P,m_i} = \left(\frac{\partial \breve{\mu}_i}{\partial T}\right)_{P,m_i} \equiv -\hat{S}_i \tag{3.6-6}$$

For an ideal species i in a liquid solution, $\gamma_i = 1$ so that the P and T dependence of $\breve{\mu}_i$ is associated with μ_i^*. Equations 3.6-5 and 3.6-6 then reduce to

$$\left(\frac{\partial \mu_i^*}{\partial P}\right)_T = \hat{V}_i, \quad \left(\frac{\partial \mu_i^*}{\partial T}\right)_P \equiv -\hat{S}_i \tag{3.6-7a,b}$$

Because ideal solutions are dilute, the molar density of the solvent in the solution, $1/\hat{V}_w$, is essentially the same as the molar density of the pure solvent, c_w, and Equation 3.6-7a can be approximated for solvent as

$$\left(\frac{\partial \mu_w^*}{\partial P}\right)_T = \frac{1}{c_w} \tag{3.6-8}$$

3.6.3 Composition Dependence

To explore the sensitivity of chemical potential to composition, we apply the Gibbs–Duhem equation (Eq. 3.5-19) to a binary gas mixture or to a binary liquid solution consisting of m_1 moles of species 1 and m_2 moles of species 2:

$$SdT - VdP + m_1 d\mu_1 + m_2 d\mu_2 = 0 \tag{3.6-9}$$

According to the phase rule, the chemical potentials of the two substances in a single-phase binary solution depend on three intensive variables. Selecting T, P, and x_1 as these variables, we expand the total differentials $d\mu_1$ and $d\mu_2$ to obtain

$$d\mu_i = \left(\frac{\partial \mu_i}{\partial T}\right)_{P,x_1} dT + \left(\frac{\partial \mu_i}{\partial P}\right)_{T,x_1} dP + \left(\frac{\partial \mu_i}{\partial x_1}\right)_{T,P} dx_1 \quad (i = 1,2) \tag{3.6-10}$$

In a closed system, holding x_1 constant also means that m_1 is constant. Thus, we can substitute Equations 3.6-5 and 3.6-6 for the first two derivatives in this equation:

$$d\mu_i = -\hat{S}_i dT + \hat{V}_i dP + \left(\frac{\partial \mu_i}{\partial x_1}\right)_{T,P} dx_1 \quad (i = 1,2) \tag{3.6-11}$$

By the same reasoning used to derive Equation 3.5-17b, it can be shown that $m_1\hat{S}_1 + m_2\hat{S}_2 = S$ and $m_1\hat{V}_1 + m_2\hat{V}_2 = V$. Combining Equation 3.6-11 with Equation 3.6-9 according to these relations for S and V, we get

$$m_1\left(\frac{\partial \mu_1}{\partial x_1}\right)_{T,P} + m_2\left(\frac{\partial \mu_2}{\partial x_1}\right)_{T,P} = 0 \tag{3.6-12}$$

From Equation 3.5-15, we see that electrical contributions to μ_1 and μ_2 do not depend on x_1 and x_2. Also realizing that $m_1 = mx_1$, $m_2 = mx_2$, and $dx_1 = -dx_2$, we can relate changes in the chemical potentials $\breve{\mu}_1$ and $\breve{\mu}_2$ to changes in the mole fractions:

$$x_1\left(\frac{\partial \breve{\mu}_1}{\partial x_1}\right)_{T,P} - x_2\left(\frac{\partial \breve{\mu}_2}{\partial x_2}\right)_{T,P} = 0 \tag{3.6-13}$$

Employing Equation 3.6-2, we can also relate changes in activity coefficients to changes in mole fractions:

$$\left[\frac{\partial(\ln\gamma_1)}{\partial(\ln x_1)}\right]_{T,P} - \left[\frac{\partial(\ln\gamma_2)}{\partial(\ln x_2)}\right]_{T,P} = 0 \tag{3.6-14}$$

References

Denbigh K. The Principles of Chemical Equilibrium. Cambridge: Cambridge University Press; 1964.

Slattery JC. Momentum, Energy, and Mass Transfer in Continua. New York: McGraw-Hill, Inc; 1972, p 280.

Chapter 4

Interfacial and Membrane Equilibria

In this chapter, we analyze the equilibrium distribution of chemical species between immiscible phases such as a gas mixture in contact with an aqueous solution or an organic solution in contact with an aqueous solution. Also in this chapter, we discuss electrochemical equilibrium between homogeneous phases that are separated by a membrane. An example of this in biological systems is the separation of intracellular and extracellular fluid by individual cell membranes. On a larger scale, respiratory gas in mammalian lungs is separated from pulmonary blood by a pair of cell layers that form the alveolar–capillary membrane. Interfacial and membrane equilibria are also important considerations in the design of many medical devices.

At the end of this chapter, we explore the electrostatic equilibrium between charges fixed on a surface and mobile ions that are spatially distributed in an adjacent solution. Unlike our analyses of phase and membrane equilibria in which the phases are well mixed, we model the ion distribution by enforcing electrochemical equilibrium locally, at each point in the solution.

4.1 Equilibrium Criterion

Before examining interfacial or membrane equilibria, we need to establish a specific electrochemical equilibrium criterion. Consider a closed system consisting of immiscible phases A and B maintained at a constant temperature and pressure. An abrupt transition between the phases typically occurs in a thin interfacial region that we model as a dividing surface (Fig. 4.1-1). A general equilibrium constraint for the I different chemical species i capable of crossing this surface is given by Equation 3.5-21 as

Biomedical Mass Transport and Chemical Reaction: Physicochemical Principles and Mathematical Modeling,
First Edition. James S. Ultman, Harihara Baskaran, and Gerald M. Saidel.

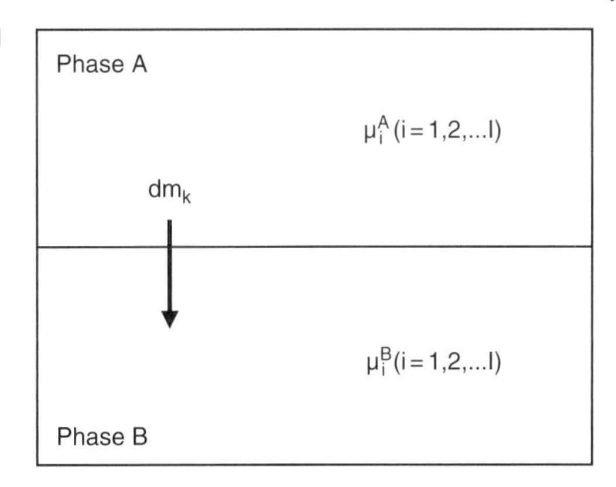

Figure 4.1-1 Deviation from interfacial equilibrium by small mass transfer.

$$(dG)_{T,P} = \sum_{i=1}^{I} \left(\mu_i^A dm_i^A + \mu_i^B dm_i^B\right) = 0 \tag{4.1-1}$$

where μ_i^A and μ_i^B represent the electrochemical potentials of species i in the two phases and dm_i^A and dm_i^B are deviations in the moles of species i from equilibrium.

Now consider the particular change in state occurring when dm_k moles of a single component k transfers from phase A to phase B without modifying the amounts of the other substances in the two phases. The resulting increase in the number of moles of species i within the two phases is then given by

$$dm_i^A = \begin{cases} -dm_k & \text{when } i = k \\ 0 & \text{when } i \neq k \end{cases} \quad \text{and} \quad dm_i^B = \begin{cases} +dm_k & \text{when } i = k \\ 0 & \text{when } i \neq k \end{cases} \tag{4.1-2a, b}$$

Since dm_k is arbitrarily small, the composition of the two phases will deviate very little from their original equilibrium values. Consequently, μ_k^A and μ_k^B will be virtually unchanged and Equation 4.1-1 reduces to

$$\mu_k^A(-dm_k) + \mu_k^B(+dm_k) = 0 \Rightarrow \left(\mu_k^A - \mu_k^B\right) dm_k = 0 \tag{4.1-3}$$

Given that dm_k is nonzero, this equation can only be satisfied if

$$\mu_k^A = \mu_k^B \tag{4.1-4}$$

This is the electrochemical equilibrium criterion for component k. Since we can repeat this analysis for each of the I components, Equation 4.1-4 implies that

$$\mu_i^A = \mu_i^B \quad (i = 1,\ 2,\ \ldots I) \tag{4.1-5}$$

Thus, in a closed two-phase system maintained at constant T and P, all chemical species that are in equilibrium must have electrochemical potentials that are equal between the phases. For uncharged substances or in the absence of electrical potential differences, this equation reduces to an equilibrium criterion between the chemical potentials:

$$\breve{\mu}_i^A = \breve{\mu}_i^B \tag{4.1-6}$$

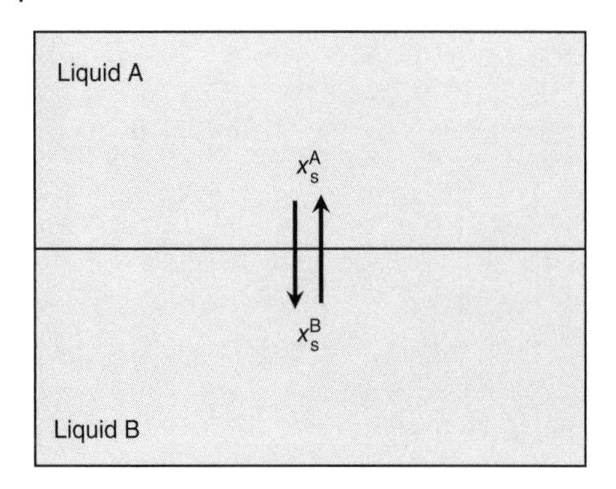

Figure 4.2-1 Equilibrium of a solute between two immiscible liquids.

4.2 Interfacial Equilibria

4.2.1 Immiscible Liquid Phases

We first apply the interfacial equilibrium criterion to an uncharged solute s distributed between liquid phases A and B at mole fractions x_s^A and x_s^B (Fig. 4.2-1). Because we consider the solvents in these phases to be immiscible, they are not subject to interfacial equilibrium. By utilizing Equation 3.6-2, the chemical potentials of a solute s in the two phases are

$$\breve{\mu}_s^A = \mu_s^{*A}(T,P) + \mathcal{R}T\ln(\gamma_s^A x_s^A) \quad (4.2\text{-}1)$$

$$\breve{\mu}_s^B = \mu_s^{*B}(T,P) + \mathcal{R}T\ln(\gamma_s^B x_s^B) \quad (4.2\text{-}2)$$

Since the solute is uncharged, the chemical potentials must be equal at equilibrium:

$$\mu_s^{*A}(T,P) + \mathcal{R}T\ln(\gamma_s^A x_s^A) = \mu_s^{*B}(T,P) + \mathcal{R}T\ln(\gamma_s^B x_s^B) \quad (4.2\text{-}3)$$

We define the Nernst partition coefficient $\eta_s^{A,B}$ as the ratio of the mole fractions of solute s in the two phases:

$$x_s^A = \eta_s^{A,B} x_s^B; \quad \eta_s^{A,B} = \frac{\gamma_s^B}{\gamma_s^A}\exp\left(\frac{\mu_s^{*B} - \mu_s^{*A}}{\mathcal{R}T}\right) \quad (4.2\text{-}4a,b)$$

In theory, the μ_s^* and γ_s depend on T and P as does $\eta_s^{A,B}$. However, liquids are essentially incompressible, so that the effect of P is very small. Because the γ_s also depends on composition, $\eta_s^{A,B}$ can vary with x_s^A and x_s^B. If $\gamma_s^A \approx \gamma_s^B$, then $\eta_s^{A,B}$ is influenced primarily by T, and x_s^A will be proportional to x_s^B. This simplification is true of biological solutions that are so dilute that they approach ideal behavior in which case $\gamma_s^A \approx \gamma_s^B \approx 1$.

As an alternative to $\eta_s^{A,B}$, it is often convenient to use a concentration partition coefficient $\lambda_s^{A,B}$ defined as the ratio of the molar concentrations in the two liquid phases:

$$C_s^A = \lambda_s^{A,B} C_s^B; \quad \lambda_s^{A,B}(T) = \left(\frac{c_L^A}{c_L^B}\right)\eta_s^{A,B} \quad (4.2\text{-}5a,b)$$

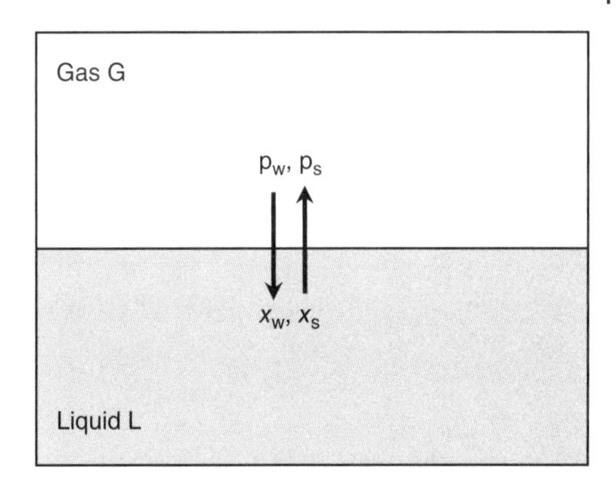

Figure 4.2-2 Equilibrium of a gas mixture with a liquid solution.

Here c_L^A and c_L^B are the molar densities of liquid phases A and B at temperature T. Based on these definitions, we see that $\eta_s^{B,A} = 1/\eta_s^{A,B}$ and $\lambda_s^{B,A} = 1/\lambda_s^{A,B}$.

4.2.2 Gas–Liquid Interfaces

We now consider the equilibrium of components in a gas mixture that are soluble in an adjoining liquid solution. Figure 4.2-2 illustrates a system consisting of a liquid phase L containing a solvent w and dissolved gas s that can both be transported into an adjoining gas phase G. Usually, phase G behaves as an ideal gas mixture, and the gas species are uncharged in both G and L phases.

Solute Equilibrium

Solute s is in equilibrium when its chemical potential in phase G, as given by Equation 3.6-1a, is equal to its chemical potential in phase L, as given by Equation 3.6-2:

$$\mu^o(T) + \mathcal{R}T\ln p_s = \mu_s^*(T,P) + \mathcal{R}T\ln(\gamma_s x_s) \tag{4.2-6}$$

where p_s is the partial pressure of component s in phase G and x_s is its mole fraction in phase L. Solving Equation 4.2-6 for p_s gives rise to Henry's law:

$$p_s = h_s x_s; \quad h_s = \gamma_s \exp\left(\frac{\mu_s^* - \mu^o}{\mathcal{R}T}\right) \tag{4.2-7a,b}$$

Because gases are so poorly soluble in most liquids, γ_s is close to one, and as we discussed for the Nernst partition coefficient, this means that h_s depends primarily on T. Notice that h_s is inversely related to the gas solubility in a liquid. That is, for a fixed value of p_s, larger values of h_s correspond to smaller values of x_s.

Other parameters related to h_s are often used as measures of solubility. As in liquid–liquid systems, the concentration partition coefficient $\lambda_s^{L,G}$ is defined as the ratio of liquid-phase molar concentration C_s^L to the gas-phase molar concentration C_s^G. Interfacial equilibrium is then expressed as

$$C_s^L = \lambda_s^{L,G} C_s^G \tag{4.2-8}$$

Gas content $\widehat{C}_s$ is defined as the STP volume of solute s in gaseous form that is dissolved in a unit volume of a liquid. This concentration measure is directly related to molar concentration: $\widehat{C}_s = C_s/c_G^o$ where c_G^o is the molar density of a gas phase at STP conditions. The Bunsen solubility coefficient α_s is defined as the liquid-phase gas content $\widehat{C}_s$ divided by the gas-phase partial pressure p_s so that interfacial equilibrium is given by

$$\widehat{C}_s = \alpha_s p_s \tag{4.2-9}$$

These three gas solubility coefficients depend on temperature and are related as follows:

$$\lambda_s^{L,G}(T) = \frac{\mathcal{R}Tc_L(T)}{h_s(T)}, \quad \alpha_s(T) = \frac{\lambda_s^{L,G}(T)}{\mathcal{R}Tc_G^o} \tag{4.2-10a,b}$$

A useful special case is the equilibrium distribution of a component between two gas phases separated by a physical partition such as a permeable membrane. In that case, $\lambda^{G,G} = 1$ and Equation 4.2-10b indicates that $\alpha_s = 1/\mathcal{R}Tc_G^o$.

Example 4.2-1 Pulmonary Oxygen Uptake

Venous blood entering the lungs through the pulmonary artery has an O_2 partial pressure of $p_{O_2,v} = 5.3$ kPa. Arterial blood leaving the lungs through the pulmonary vein has an O_2 partial pressure of $p_{O_2,a} = 12.7$ kPa. If O_2 were present in blood only in dissolved form, what would be the O_2 uptake in the lungs at a typical cardiac output of 5 L/min?

Solution

We select pulmonary capillary blood as a control volume (Fig. 4.2-3). This control volume has a molar O_2 input rate $\dot{N}_{O_2,v}$ by convection in venous blood and a molar O_2 output rate $\dot{N}_{O_2,a}$ by convection in arterial blood. There is also a molar O_2 input rate $\dot{N}_{O_2,A}$ by diffusion across the alveolar–capillary membrane. According to Equation 2.4-1, the O_2 molar balance equation in a pulmonary capillary volume V_{cap} is

$$\frac{d(C_{O_2}V_{cap})}{dt} = \dot{N}_{O_2,v} + \dot{N}_{O_2,A} - \dot{N}_{O_2,a} + \dot{R}_{O_2} \tag{4.2-11}$$

We assume that this system is in steady state such that the time derivative is zero. We also assume that there is no chemical reaction between O_2 and other substances in blood such that $\dot{R}_{O_2} = 0$. The O_2 molar balance then reduces to

$$\dot{N}_{O_2,v} + \dot{N}_{O_2,A} - \dot{N}_{O_2,a} = 0 \tag{4.2-12}$$

For a constant volumetric blood flow Q, we can specify the convective transport rates as $\dot{N}_{O_2,v} = QC_{O_2,v}$ and $\dot{N}_{O_2,a} = QC_{O_2,a}$:

$$QC_{O_2,v} + \dot{N}_{O_2,A} - QC_{O_2,a} = 0 \tag{4.2-13}$$

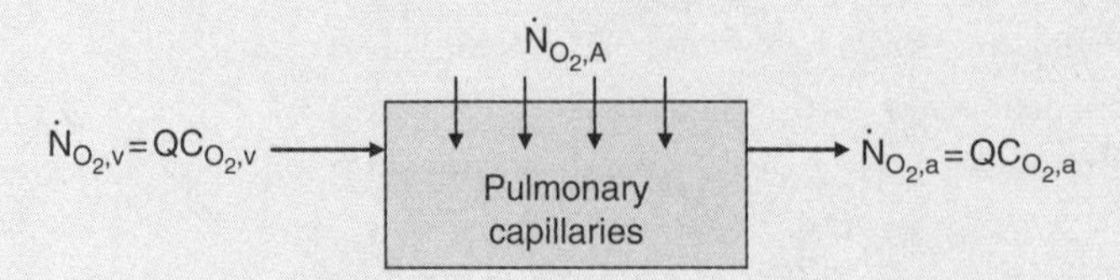

Figure 4.2-3 Control volume for oxygen absorption model.

In respiratory system analysis, it is traditional to express material balance equations in terms of standard gas volumes rather than in moles of gas. This conversion is achieved by dividing each term in Equation 4.2-13 by the standard value of the gas molar density c_G^o:

$$Q\left(\frac{c_{O_2,v}}{c_G^o}-\frac{c_{O_2,a}}{c_G^o}\right)+\frac{\dot{N}_{O_2,A}}{c_G^o}=0 \tag{4.2-14}$$

With O_2 content $\hat{C}_{O_2}$ and volumetric O_2 uptake rate $\dot{V}_{O_2}$ given as

$$\hat{C}_{O_2}\equiv\frac{c_{O_2}}{c_G^o},\quad \dot{V}_{O_2}\equiv\frac{\dot{N}_{O_2,A}}{c_G^o} \tag{4.2-15a,b}$$

Equation 4.2-14 becomes

$$\dot{V}_{O_2}=Q\left(\hat{C}_{O_2,a}-\hat{C}_{O_2,v}\right) \tag{4.2-16}$$

The problem statement provides values for the p_{O_2} of venous and arterial blood. This variable represents the partial pressure that would be required for an O_2 gas phase to be in equilibrium with a blood sample. Using Equation 4.2-9, to convert from $\hat{C}_{O_2}$ to p_{O_2}, we write Equation 4.2-16 in an alternative form:

$$\dot{V}_{O_2}=Q\alpha_{O_2}\left(p_{O_2,a}-p_{O_2,v}\right) \tag{4.2-17}$$

where we have assumed that α_{O_2} is the same in venous and arterial blood. Based on a value of $\alpha_{O_2}=0.0219$ ml(STP)/(dL-kPa) for whole blood at 37°C (Table A3-3), we compute the volumetric O_2 uptake rate as

$$\dot{V}_{O_2}=50(0.0219)(12.7-5.3)=8\,\text{ml(STP)}/\text{min} \tag{4.2-18}$$

The metabolic demand for O_2 in a resting adult, about 250 ml/min, is much greater than this. This discrepancy occurs because we have not accounted for the binding of oxygen to hemoglobin, a reversible chemical reaction that dramatically increases the content of O_2 in blood. We will analyze this process in the next chapter where reaction equilibrium is discussed.

Gas–Solvent Equilibrium

Because the solvent in a liquid solution is volatile to some degree, its vapor will appear in an adjoining gas space. The criterion for equilibrium between liquid solvent w and its vapor is

$$\mu^o(T)+\mathcal{R}T\ln p_w=\mu_w^*(T,P)+\mathcal{R}T\ln(\gamma_w x_w) \tag{4.2-19}$$

We define the pure component vapor pressure p_w^* as the value of p_w in equilibrium with pure solvent. In that case, $x_w=\gamma_w=1$, and it follows from Equation 4.2-19 that

$$p_w^*(T)\equiv\exp\left(\frac{\mu_w^*-\mu^o}{\mathcal{R}T}\right) \tag{4.2-20}$$

When solutes are present, these two equations indicate that

$$p_w=\gamma_w p_w^* x_w \tag{4.2-21}$$

Realizing that the sum of x_w and all the solute mole fractions x_s(s = 1,2...S) must equal to one, we write Equation 4.2-21 as

$$p_w = \gamma_w p_w^* \left(1 - \sum_{s=1}^{S} x_s\right) \approx p_w^* \left(1 - \sum_{s=1}^{S} x_s\right) \tag{4.2-22}$$

where the approximation applies to dilute solutions for which $\gamma_w \rightarrow 1$. This equation indicates that the addition of a solute to a solvent causes a reduction in the equilibrium vapor pressure that depends on the mole fraction of the solute but not on its chemical properties.

The temperature dependence of the pure component vapor pressure of water can be computed from the semiempirical Antoine relationship (Lange, 1967):

$$\log\left(p_w^*[\text{kPa}]\right) = 7.23245 - \frac{1750.286}{T[°\text{K}] - 38} \tag{4.2-23}$$

where the constants apply over a temperature range $333 > T > 273°K$. Percent relative humidity expresses the partial pressure of water vapor in air relative to its pure component vapor pressure:

$$RH \equiv 100 \frac{p_w}{p_w^*} \tag{4.2-24}$$

Combining this equation with Equation 4.2-22, we obtain the percent relative humidity in terms of the S solutes at equilibrium:

$$RH^* \equiv 100\left(1 - \sum_{s=1}^{S} x_s\right) \tag{4.2-25}$$

Example 4.2-2 Water Vapor in Exhaled Air

A human exhales air at a total pressure of 101.3 kPa and a temperature of 37°C. Water transport in the lungs is such an efficient process that exhaled air is essentially in equilibrium with surrounding airway mucus. If the total concentration of all individual ions and molecules in mucus is 0.3 M, then determine p_w and RH of the expired air.

Solution

The molar density of pure water at 37°C is $c_w = \rho_w/M_w = 993/18 = 55.3$ M. Thus, the molar density of mucus is about $55.3 + 0.3 = 55.8$ M, and the mole fraction of solutes it contains is $x_s = 0.3/(55.3 + 0.3) = 0.005$. According to Equation 4.2-25, the percent relative humidity of water vapor in equilibrium with mucus will then be

$$RH^* = 100(1 - 0.005) = 99.5\% \tag{4.2-26}$$

Using the Antoine equation, we determine the equilibrium vapor pressure of pure water at 37°C:

$$\log\left(p_w^*[\text{kPa}]\right) = 7.23245 - \frac{1750.286}{(273 + 37) - 38} = 0.797 \Rightarrow p_w^* = 6.26 \text{ kPa} \tag{4.2-27}$$

From the definition of percent relative humidity (Eq. 4.2-24), we obtain

$$p_w = (RH/100)p_w^* = (99.5/100)(6.26) = 6.23 \text{ kPa} \tag{4.2-28}$$

Because the mole fraction of solute in mucus is so small, p_w is close to the equilibrium vapor pressure of pure water.

4.2.3 Multiphase Equilibrium

Biological tissues frequently contain more than two phases. To illustrate equilibrium among multiple phases, we focus on a system consisting of a homogeneous phase A adjacent to heterogeneous phase B that is a mixture of two homogeneous subphases, B1 and B2. As shown in Figure 4.2-4, the solute concentration C_s^{B1} is in equilibrium with concentration C_s^{B2} as well as with the concentration C_s^{A}. It follows that C_s^{B2} and C_s^{A} must also be in equilibrium with each other.

If there are no electrical effects, we can use the concentration distribution coefficients defined in Equations 4.2-5a,b to describe the mutual equilibria among phases A, B1, and B2:

$$C_s^{B1} = \lambda_s^{B1,A} C_s^{A}, \quad C_s^{B2} = \lambda_s^{B2,A} C_s^{A}, \quad C_s^{B1} = \lambda_s^{B1,B2} C_s^{B2} \tag{4.2-29a-c}$$

These equations reveal that the three equilibrium distribution coefficients are not independent. Comparing the ratio of the first two equations to the third equation indicates that $\lambda_s^{B1,B2} = \lambda_s^{B1,A}/\lambda_s^{B2,A}$.

We often find it more convenient to work with the average concentration of solute in the heterogeneous material B rather than the individual concentrations in the two subphases. If ε represents the volume fraction of subphase B1, then the volume averaged concentration of solute s in the heterogeneous phase B is

$$\bar{C}_s^{B} = \varepsilon C_s^{B1} + (1-\varepsilon) C_s^{B2} \tag{4.2-30}$$

By combining this equation with Equations 4.2-29a,b, we arrive at an equilibrium relationship between C_s^{A} and $\bar{C}_s^{B}$:

$$\bar{C}_s^{B} = \lambda_s^{B,A} C_s^{A}; \quad \lambda_s^{B,A} = \varepsilon \lambda_s^{B1,A} + (1-\varepsilon) \lambda_s^{B2,A} \tag{4.2-31a,b}$$

This equation indicates that $\lambda_s^{B,A}$ is an effective distribution coefficient of solute s between the heterogeneous and homogeneous phases. When the homogeneous phase is a gas mixture containing species s at a partial pressure p_s^{A}, then we can define equilibrium in terms of the

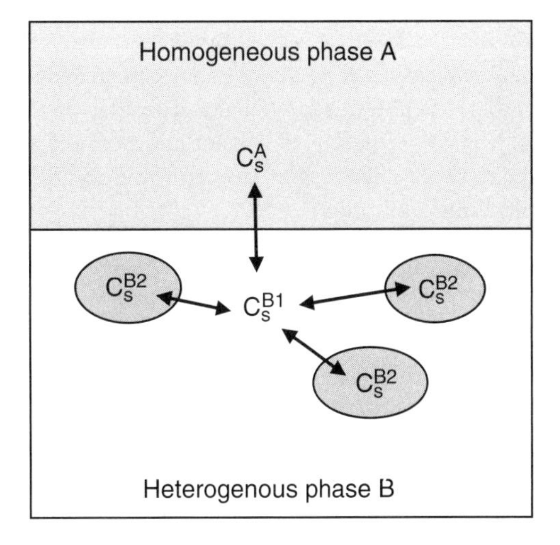

Figure 4.2-4 Solute equilibrium between homogeneous phase A and heterogeneous phase B.

total gas content $\widehat{C}_s$ of dissolved solute in the heterogeneous phase. The equivalent forms of Equation 4.2-31a,b for this situation are as follows:

$$\widehat{C}_s = \alpha_s p_s^A; \quad \alpha_s = \varepsilon \alpha_s^{B1} + (1-\varepsilon)\alpha_s^{B2} \tag{4.2-32a,b}$$

where α_s is an effective Bunsen solubility for gaseous species s in the heterogeneous liquid phase as a whole, while α_s^{B1} and α_s^{B2} are the actual Bunsen solubilities in the two subphases.

Example 4.2-3 Bunsen Solubility of O_2 in Red Blood Cells

According to Table A3-3, the Bunsen solubility of O_2 at body temperature is 0.0211 ml(STP)/dL-kPa in human blood plasma and 0.0219 ml(STP)/dL-kPa in human whole blood. If whole blood has a hematocrit (volume percent of red cells) of 45%, estimate the Bunsen solubility of O_2 within red blood cells.

Solution

Blood (B) is essentially a two-phase mixture of red blood cells (RBCs) suspended in plasma (P). Taking ε_H as the RBC volume fraction, we formulate $\alpha_{O_2}^{RBC}$ in terms of $\alpha_{O_2}^{B}$ and $\alpha_{O_2}^{P}$ by rearranging Equation 4.2-32b:

$$\alpha_{O_2}^{B} = \varepsilon_H \alpha_{O_2}^{RBC} + (1-\varepsilon_H)\alpha_{O_2}^{P} \Rightarrow \alpha_{O_2}^{RBC} = \frac{\alpha_{O_2}^{B} - (1-\varepsilon_H)\alpha_{O_2}^{P}}{\varepsilon_H} \tag{4.2-33a,b}$$

With $\varepsilon_H = 45/100 = 0.45$, the value of $\alpha_{O_2}^{RBC}$ is

$$\alpha_{O_2}^{RBC} = \frac{0.0219-(1-0.45)(0.0211)}{0.45} = 0.0229 \text{ ml(STP)/dL-kPa} \tag{4.2-34}$$

Thus, O_2 is about 10% more soluble in RBCs than in plasma.

4.3 Membrane Equilibria

Suppose two homogeneous solutions containing different concentrations of a particular solute are separated by a membrane. Because a membrane is very thin, we treat it as a two-dimensional dividing surface that can sustain discontinuities in both solute concentration and electrical potential. We also consider the membrane to be ideally semipermeable. That is, it is so permeable to solvent and possibly some solutes that they equilibrate across the membrane, while it is so impermeable to the remaining solutes that they are incapable of reaching equilibrium.

4.3.1 Electrochemical Equilibrium

We first analyze the electrochemical equilibrium established when a membrane semipermeable to water and some ions separates two aqueous electrolyte solutions. As an example, consider a membrane that is permeable to water and to chloride anions Cl^- but not to sodium cations Na^+. As shown in Figure 4.3-1, this membrane separates two NaCl solutions, a phase A containing a high NaCl concentration and a phase B containing a low NaCl concentration. Both phases have constant volumes.

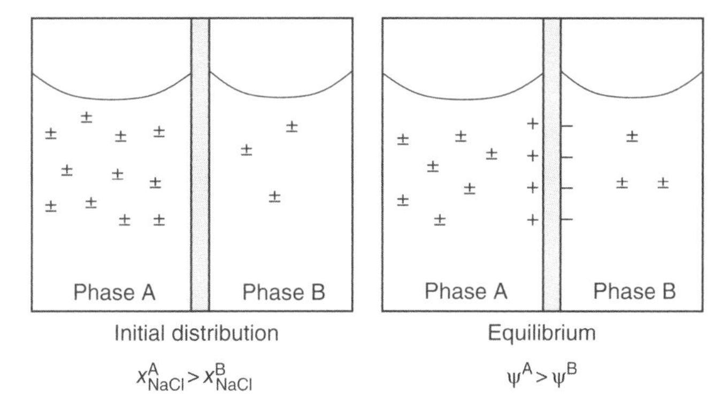

Figure 4.3-1 Charge distribution with electrochemical membrane equilibrium.

As a strong electrolyte, NaCl quantitatively dissociates into Na^+ and Cl^- when it is dissolved in water. Because electrostatic forces only allow these counterions to separate by a very small distance, they exist in aqueous solution as Na^+Cl^- ion pairs. Although the membrane prevents Na^+ transport from phase A to phase B, some Cl^- ions can overcome electrostatic forces and diffuse to the B surface of the membrane. The Na^+ ions that are left behind remain on the A surface. Equilibrium is reached when the difference in electrical potential created by this small charge separation is able to balance the difference in chemical potential of Cl^- between the bulk phases of the two compartments.

The criterion for this equilibrium state is that the electrochemical potential of Cl^- or any other permeable substance i with a charge z_i be equal on the two sides of the membrane. Combining Equation 3.5-15 with Equation 3.6-2 and replacing the mole fractions x_i by C_i/c_L, we formulate the electrochemical potential for species i in phase A as

$$\mu_i^A = z_i\mathcal{F}\psi^A + \mu_i^{*A}(T,P) + \mathcal{R}T\left[\ln\left(\gamma_i^A \frac{C_i^A}{c_L^A}\right)\right] \tag{4.3-1}$$

and in phase B as

$$\mu_i^B = z_i\mathcal{F}\psi^B + \mu_i^{*B}(T,P) + \mathcal{R}T\left[\ln\left(\gamma_i^B \frac{C_i^B}{c_L^B}\right)\right] \tag{4.3-2}$$

After equating μ_i^A and μ_i^B, we solve for the electrical potential that is created by the equilibration of i across an ideal semipermeable membrane:

$$\psi^A - \psi^B = \frac{\mu_i^{*B} - \mu_i^{*A}}{z_i\mathcal{F}} + \frac{\mathcal{R}T}{z_i\mathcal{F}}\ln\left(\frac{c_L^A\gamma_i^B C_i^B}{c_L^B\gamma_i C_i^A}\right) \tag{4.3-3}$$

For the common situation where phases A and B are both dilute aqueous solutions at the same temperature and pressure, $\mu_i^{*A} = \mu_i^{*B}$, $\gamma_i^A \approx \gamma_i^B$ and $c_L^A \approx c_L^B$. Equation 4.3-3 then becomes

$$\psi^A - \psi^B = -\frac{\mathcal{R}T}{z_i\mathcal{F}}\ln\left(\frac{C_i^A}{C_i^B}\right) \tag{4.3-4}$$

Notice that for a negatively charged ion like Cl^-, this Nernst potential difference has the same direction as the solute concentration difference; the opposite is true for a positively charged ion.

Example 4.3-1 Resting Potential of a Squid Giant Axon

Resting potential is the electrical voltage at the interior of a neuron or another excitable cell relative to its surroundings when the cell is not stimulated. Among the ions present in physiological fluids, cell membranes are generally most permeable to K^+. Concentrations of K^+ ions have been measured for the squid giant axon in the resting state: $C_K^A = 344\,\text{mM}$ in intracellular fluid and $C_K^B = 10\,\text{mM}$ in extracellular fluid (Mountcastle, 1974). Estimate the resting potential of the axon at 20°C by assuming that K^+ is in electrochemical equilibrium across the cell membrane.

Solution

Taking into account a charge number of $z_K = +1$ for the K^+ ion, Equation 4.3-4 predicts that

$$\psi^A - \psi^B = -1000\left[\frac{(8.314)(293)}{(+1)(96500)}\ln\left(\frac{344}{10}\right)\right] = -89.3\ \text{mV} \tag{4.3-5}$$

Since the observed potential difference from the inside to the outside of the squid axon at 20°C is −77 mV, the potassium ion is not exactly at equilibrium. We will return to this problem in Chapter 10 where we will predict the resting potential more accurately by accounting for other ions and for limitations in membrane permeability.

4.3.2 Osmotic Pressure

We now consider the equilibrium established for a membrane that is ideally semipermeable to water molecules but impermeable to all solutes. This membrane separates a phase A consisting of an aqueous solution of various solutes s from phase B comprised of pure water (Fig. 4.3-2). Open manometers allow the volumes in the individual phases to change, but the total liquid volume is constant.

As shown on the left side of the figure, the liquid levels in the two manometers are initially the same. Thereafter, water will attempt to equalize the solute concentrations in the two solutions by permeating from phase B to phase A. This causes an increase in volume and dilution

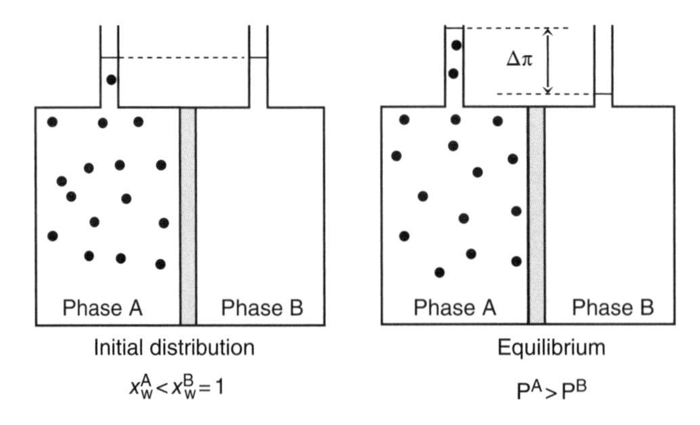

Figure 4.3-2 Transmembrane osmotic pressure balanced by hydrostatic pressure.

of phase A with a simultaneous decrease in volume of phase B. Since each phase is sealed except for its manometer, this volume shift increases the liquid level in phase A relative to that in phase B. To formulate the equilibrium that is eventually reached, we equate the chemical potential of water (Eq. 3.6-2) in the two phases:

$$\mu_w^{*A}(P^A, T) + \mathcal{R}T\ln(\gamma_w^A x_w^A) = \mu_w^{*B}(P^B, T) + \mathcal{R}T\ln(\gamma_w^B x_w^B) \quad (4.3\text{-}6)$$

Here, P^A and P^B are the hydrostatic pressures associated with the manometer levels. Realizing that $\gamma_w^B x_w^B$ must be equal to one since phase B contains pure water, and referring to $\gamma_w^A x_w^A$ simply as $\gamma_w x_w$, Equation 4.3-6 reduces to

$$\mathcal{R}T\ln(\gamma_w x_w) = \mu_w^{*B} - \mu_w^{*A} \quad (4.3\text{-}7)$$

Because we are usually dealing with solutions of essentially constant molar density c_w, Equation 3.6-8 can be integrated to find the difference in μ_w^* between the phases in terms of their pressure difference:

$$\int_{\mu_w^{*A}}^{\mu_w^{*B}} \left(\frac{\partial \mu_w^*}{\partial P}\right)_T dP = \int_{P^A}^{P^B} \frac{1}{c_w} dP \Rightarrow \mu_w^{*B} - \mu_w^{*A} = \frac{1}{c_w}(P^B - P^A) \quad (4.3\text{-}8)$$

Equation 4.3-7 now becomes

$$P^A - P^B = -\mathcal{R}Tc_w\ln(\gamma_w x_w) \quad (4.3\text{-}9)$$

A convenient substitute for γ_w in this equation is the thermodynamic osmotic coefficient Φ_w that we define as

$$\Phi_w \equiv 1 + (\ln\gamma_w/\ln x_w) \quad (4.3\text{-}10)$$

This allows us to write

$$P^A - P^B = -\mathcal{R}Tc_w\Phi_w\ln x_w \quad (4.3\text{-}11)$$

Since the mole fractions of water and all solutes must sum to one,

$$x_w = 1 - \sum_{s=1}^{S} x_s \quad (4.3\text{-}12)$$

and Equation 4.3-11 becomes

$$P^A - P^B = -\mathcal{R}Tc_w\Phi_w\ln\left(1 - \sum_{s=1}^{S} x_s\right) \quad (4.3\text{-}13)$$

In a dilute solution, $\Sigma x_s \ll 1$ and the logarithmic function can be simplified by using the approximation $\ln(1 - \Sigma x_s) \approx -\Sigma x_s$. Also in a dilute solution, $C_s \approx c_w x_s$ and we write Equation 4.3-13 as follows:

$$P^A - P^B = \mathcal{R}T\Phi_w\sum_{s=1}^{S} C_s \quad (4.3\text{-}14)$$

It is customary to define this equilibrium pressure difference as the osmotic pressure π such that

$$\pi \equiv P^A - P^B = \mathcal{R}T\Phi_w \sum_{s=1}^{S} C_s \tag{4.3-15}$$

The value of Φ_w depends on the influence of solutes on the activity of water. Changing either the particular solutes that make up the solution or the mole fractions of a given set of solutes can modify Φ_w. On the contrary, Φ_w is not influenced by membrane properties as long as the membrane is ideally semipermeable to water.

When the solution in phase A is very dilute such that $\gamma_w \to 1$, Equation 4.3-10 indicates that $\Phi_w \to 1$. Equation 4.3-15 then reduces to the Van't Hoff equation for osmotic pressure of an ideal solution:

$$\pi = \mathcal{R}T\sum_{s=1}^{S} C_s \tag{4.3-16}$$

The total molar concentration of all solute molecules $\sum_s C_s[\text{osm/L}]$ is the osmolarity of a solution. In aqueous solutions, this sum includes both positive ions and their negative counterions, undissociated electrolyte molecules, and nonelectrolyte molecules. As is the case for the equilibrium vapor pressure of a liquid solution, osmotic pressure does not depend on the nature of the solutes, but only on their concentration. For this reason, osmotic pressure and vapor pressure are classified as colligative properties.

Example 4.3-2 Osmotic Pressure of Blood Plasma

The most plentiful ions in blood plasma are Na^+(137 mM), K^+(5 mM), Cl^-(103 mM), and HCO_3^- (28 mM). The most abundant uncharged species are glucose (6 mM) and urea (4 mM). The total concentration of other ionic and uncharged chemical species is about 20 mM. Estimate the osmotic pressure of plasma using the Van't Hoff relationship. If the actual osmotic pressure is 730 kPa, determine the osmotic and the activity coefficients of water in plasma.

Solution

The total osmolarity of solute in blood plasma is

$$\sum_s C_s = \frac{137+5+103+28+6+4+20}{1000} = 0.303 \text{ osmol/L} \tag{4.3-17}$$

The corresponding osmotic pressure estimated with the Van't Hoff relationship is

$$\pi = 8.31(310)(0.303) = 781 \text{ kPa} \tag{4.3-18}$$

We determine the osmotic coefficient from the ratio of the actual osmotic pressure to the ideal osmotic pressure:

$$\Phi_w = 730/781 = 0.935 \tag{4.3-19}$$

To find the activity coefficient of water, we note that the molar density of plasma at 37°C is about 56 M so that the mole fraction of water is $x_w = (56 - 0.304)/56 = 0.995$. Rearranging Equation 4.3-10, we get

$$\gamma_w = (x_w)^{(\Phi_w - 1)} = (0.995)^{0.935-1} = 1.0003 \tag{4.3-20}$$

This demonstrates that water molecules in blood plasma essentially behave in an ideal manner. We also see that a small positive deviation of γ_w from one corresponds to a more extreme negative deviation of Φ_w from one.

In biological systems, aqueous solutes are present on both sides of a membrane, and we obtain the osmotic pressure difference after separately applying Equation 4.3-15 to each compartment:

$$\left(\pi^A - \pi^B\right) \equiv \left(P^A - P^B\right) = \mathcal{R}T\left(\Phi_w^A \sum_{s=1}^{S} C_s^A - \Phi_w^B \sum_{s=1}^{S} C_s^B\right) \tag{4.3-21}$$

Intracellular and extracellular fluids have similar osmotic pressures of about 0.3 osmol/L. It is important to minimize variations in this value since imbalances in osmotic pressure between body compartments lead to an undesirable redistribution of water. To quantify the tendency toward such water shifts, an osmotic concentration of 0.3 osmol/L is designated as isotonic, concentrations less than 0.3 osmol/L are hypotonic, and concentrations greater than 0.3 osm/L are hypertonic. In equalizing the osmotic pressure between compartments, water will migrate toward the compartment of higher tonicity. That is, water will shift either from hypotonic compartments toward isotonic compartments or from isotonic compartments toward hypertonic compartments.

Example 4.3-3 Mechanical Strength of a Red Blood Cell Membrane

A red blood cell suspended in isotonic medium is a biconcave disk with a volume of 90 μm^3 and a surface area of 124 μm^2. Only 57% of this volume contains osmotically active material. When resuspended in a hypotonic solution, water diffuses into the cell and causes it to swell (Fig. 4.3-3). In a sufficiently hypotonic solution, the RBC becomes nearly spherical, and the membrane stretches slightly to create a surface tension that is balanced by a residual osmotic pressure difference. When surface tension exceeds the membrane tensile strength, the cell will lyse, spilling its contents into the surrounding suspension medium.

In experiments with RBCs resuspended in a series of hypotonic solutions of decreasing NaCl concentration, the median concentration at which cell lysis occurred was 0.41 g NaCl per 100 ml of water. What is the hydrostatic pressure difference across a cell membrane when it ruptures? Determine the tensile strength of the membrane.

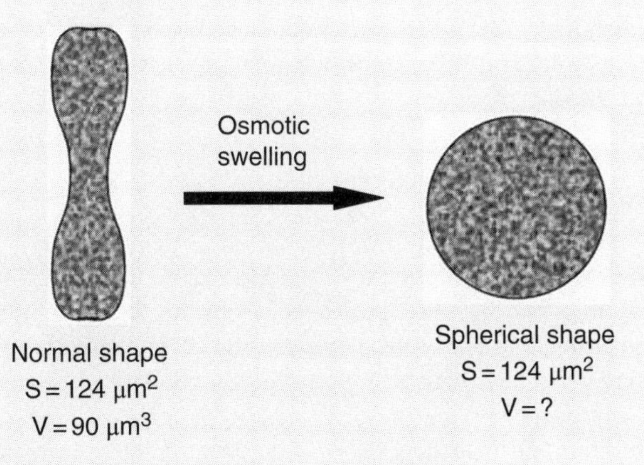

Figure 4.3-3 Swelling of an erythrocyte in hypotonic salt solution.

Solution

Because the plasma membrane of the RBC is quite stiff, we assume that its area is constant. Thus, a spherically swollen cell has an surface area of 124 μm^2, a radius of $a_c = \sqrt{124/4\pi} =$ 3.14 μm, and a total volume of $(4\pi/3)(3.14)^3 = 130\ \mu m^3$. Since the volume of osmotically inactive material in a cell is always the same as its initial value, the non-osmotically active volume is (1 − 0.57)(90) = 39 μm^3. The osmotically active volume of the normal biconcave cell is (90 − 39) = 51 μm^3 and of the spherically swollen cell is (130 − 39) = 91 μm^3.

We expect the intracellular fluid of an unswollen cell to have an isotonic concentration of 0.3 osmol/L, and we assume that solutes do not leak across the cell membrane during osmotic swelling. Under these conditions, we compute the osmolarity in the interior (A) of the swollen cell from the ratio of its initial and final osmotically active volumes:

$$\sum_s C_s^A = (0.3)\frac{51}{91} = 0.17 \text{ osmol/L} \tag{4.3-22}$$

Since NaCl is a strong electrolyte, each molecule dissociates into Na^+ and Cl^- ions, and the osmolarity of an NaCl solution is twice its molarity. We can compute the osmolarity of the suspending medium (B) at cell lysis from the mass concentration and molecular weight (58.5 Da) of NaCl as follows:

$$\sum_s C_s^B = 2\left(\frac{0.41}{58.5}\right) = 0.014 \text{ osmol/100 ml} = 0.14 \text{ osmol/L} \tag{4.3-23}$$

Assuming ideal behavior of water in the osmotically active portion of an RBC, the Van't Hoff equation provides an estimate of the pressure difference necessary to lyse a cell:

$$(P^A - P^B) = (\pi^A - \pi^B) = (8.314)(310)(0.17 - 0.14) = 77 \text{ kPa} \tag{4.3-24}$$

To compute the surface tension of the membrane corresponding to this pressure difference, we employ the Young–Laplace equation (Adamson, 1967; p 4–6) over a spherical membrane shell:

$$\text{Surface tension} = \frac{a_c(P^A - P^B)}{2} = \frac{3.14 \times 10^{-6}(77)}{2} \text{kPa-m} = 0.12 \text{ N/m} \tag{4.3-25}$$

This tensile strength of an RBC membrane is not much larger than the surface tension of an air–water interface, which is 0.07 N/m. It is not surprising that some RBCs lyse when blood comes in contact with air.

4.3.3 Colloid Osmotic Pressure

In biological systems, colloid osmotic pressure (also called oncotic pressure) refers to the chemical equilibrium occurring when protein solutions are separated by a membrane that is impermeable to protein but permeable to water, electrolytes, and other small solute molecules. Oncotic pressures are particularly important in the regulation of water balance between tissue and blood compartments.

With a concentration normally between 0.7 and 0.9 mM, albumin is the most abundant protein in blood plasma. The data points in Figure 4.3-4 measured in isotonic saline solution (Ott, 1956) display an increasingly positive deviation of oncotic pressure from the prediction

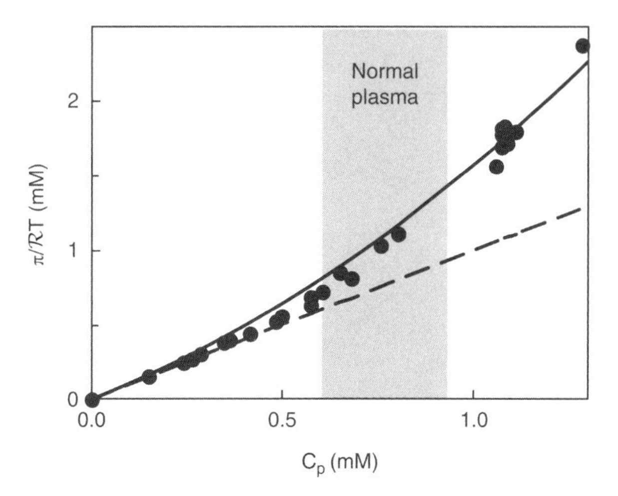

Figure 4.3-4 Gibbs–Donnan effect of albumin in physiological saline solution (Data from Ott (1956)).

of the Van't Hoff equation (dashed line) as albumin concentration C_p increases. The source of this Gibbs–Donnan effect is the nonuniform partitioning of ions across a membrane because of their electrostatic interactions with charged albumin molecules. The osmotic pressure produced by this nonuniform ion distribution adds to the osmotic pressure produced by the protein.

Suppose that a membrane separates an aqueous solution of a charged protein P and ions i in compartment A from a solution of the same ions in compartment B. By employing Equation 4.3-4, we can define an equilibrium distribution coefficient of permeating ions i between the two compartments as

$$\lambda^{A,B} \equiv \left(\frac{C_i^A}{C_i^B}\right)^{-1/z_i} = \exp\left[\frac{\mathcal{F}}{\mathcal{R}T}(\psi^A - \psi^B)\right] \tag{4.3-26}$$

where z_i is the ionic charge and $(\psi^A - \psi^B)$ is the Nernst potential difference induced by the protein. Since $\lambda^{A,B}$ only depends on $(\psi^A - \psi^B)$, it has the same value for all ions that permeate the membrane, no matter what their charge.

We will restrict our attention to solutions containing a strong electrolyte, such as sodium chloride or potassium chloride, that completely dissociates into a univalent cation (i = c) and a univalent anion (i = a). For such a 1–1 electrolyte, $z_c = +1$ and $z_a = -1$, and according to Equation 4.3-26,

$$\lambda^{A,B} \equiv \frac{C_c^B}{C_c^A} = \frac{C_a^A}{C_a^B} \tag{4.3-27}$$

When $\lambda^{A,B} > 1$ (corresponding to $\psi^A > \psi^B$), $C_a^A > C_a^B$ and $C_c^B > C_c^A$, indicating that there are an excess of anions in compartment A and an excess of cations in compartment B. This equation can also be written as the more familiar Gibbs–Donnan concentration relationship:

$$C_c^A C_a^A = C_c^B C_a^B \tag{4.3-28}$$

To formulate the relationship of $\lambda^{A,B}$ to protein concentration C_p and protein charge z_p, we apply the electroneutrality principle. That is, there can be no net charge on an electrolyte solution as a whole. In compartment A, electrical charges on all ions and protein molecules must sum to zero:

$$C_c^A + C_{H^+}^A - C_a^A - C_{OH^-}^A + z_P C_P = 0 \tag{4.3-29}$$

In compartment B, where protein is absent, the charges on all ions must also sum to zero:

$$C_c^B + C_{H^+}^B - C_a^B - C_{OH^-}^B = 0 \tag{4.3-30}$$

These equations include the concentrations of H^+ and OH^- ions that are always present in aqueous solutions. In most solutions, these concentrations can be neglected, and the combination of Equations 4.3-27, 4.3-29, and 4.3-30 yields

$$\lambda^{A,B} = \frac{z_P C_P}{2C_a^B} + \sqrt{1 + \left(\frac{z_P C_P}{2C_a^B}\right)^2} \begin{cases} <1 & \text{when } z_P < 0 \\ =1 & \text{when } z_P = 0 \\ >1 & \text{when } z_P > 0 \end{cases} \tag{4.3-31}$$

When $z_p > 0$ such that $\lambda^{A,B} > 1$, Equation 4.3-27 predicts that anions are in excess on the protein side of the membrane, while cations are in excess on the protein-free side. That is, a positively charged protein attracts anions while repelling cations. The opposite is true for a negatively charged protein.

Assuming that the solution of interest behaves in an ideal manner, we can formulate oncotic pressure by applying the Van't Hoff equation to the cations and anions as well as to the protein. Neglecting the contribution of hydrogen and hydroxyl ions, we obtain the osmotic pressure in compartment A

$$\pi^A = \mathcal{R}T\left(C_c^A + C_a^A + C_P\right) \tag{4.3-32}$$

and in compartment B

$$\pi^B = \mathcal{R}T\left(C_c^B + C_a^B\right) \tag{4.3-33}$$

Combining Equation 4.3-27 with Equations 4.3-31–4.3-33, we arrive at the following expression for colloidal osmotic pressure:

$$\pi_p \equiv \left(\pi^A - \pi^B\right) = \mathcal{R}TC_p + 2\mathcal{R}TC_a^B\left[\sqrt{1 + \left(\frac{z_P C_P}{2C_a^B}\right)^2} - 1\right] \tag{4.3-34}$$

The first term on the right side of this equation is the ideal osmotic pressure of the protein alone. The second term is the Gibbs–Donnan effect that depends on the magnitude but not the sign of the protein charge. It also depends inversely on the ion concentration; when $C_a^B \gg z_p C_p$, the Gibbs–Donnan effect is suppressed and the colloidal osmotic pressure reduces to the Van't Hoff prediction of $\mathcal{R}TC_p$.

Example 4.3-4 Albumin Charge Estimate from Colloid Osmotic Pressure

The data points in Figure 4.3-4 were observed when aqueous solutions of albumin were separated by a collodion membrane from a 150 mM NaCl solution. To minimize variations in protein charge, the albumin solutions were buffered at a physiological pH of 7.3–7.4. The purpose of this example is to simulate these data by accounting for both the osmotic pressure of the protein and the Gibbs–Donnan effect.

Solution

Dividing both sides of Equation 4.3-34 by $\mathcal{R}T$ and substituting $C_a^B = 0.150$ mM, we obtain

$$\frac{\pi_p}{\mathcal{R}T} = C_p + 0.3\left[\sqrt{(z_p C_p)^2 + 0.09} - 0.3\right] \text{ mM} \tag{4.3-35}$$

The solid curve in Figure 4.3-4 is a nonlinear regression of this Gibbs–Donnan equation to the data points with an estimate of $z_p^2 = 344$ that minimized the sum of the squared error in $\pi_p/\mathcal{R}T$. Therefore, the magnitude of z_p is 19, but based on the Gibbs–Donnan effect alone, we cannot tell whether z_p is positive or negative. The albumin molecule has positively charged amino $(-NH_3^+)$ and negatively charged carboxyl ($-COO^-$) side groups. At a pH > 7, carboxyl groups are in excess of amino groups, so we expect that z_p is negative.

4.4 Electrical Double Layer

Biological solids often have a surface charge that affects the spatial distribution of mobile ions in an adjacent electrolyte solution. For example, a negative charge on a cell surface can originate from integral membrane proteins in the plasma membrane. Being fixed in space, this charge attracts cations and repel anions from extracellular solution. The resulting excess of positive charge produces a positive electrical potential in the vicinity of the surface relative to the potential far from the surface.

Our analysis follows the double-layer model of Gouy and Chapman (Adamson, 1967; p 210–214). In this model, a very thin layer of fixed charges is uniformly distributed along a planar surface. Adjoining the surface is a second layer containing mobile cations and anions in aqueous solution (Fig. 4.4-1). In this "diffuse" layer, we represent the molar concentrations of anions (i = a) and cations (i = c) at position y relative to the solid surface as $C_i(y)$. A corresponding distribution of electrical potential is designated as $\psi(y)$.

In general, an electrical potential that varies in space generates an electric field **E** according to the electric field equation $\mathbf{E} = -\nabla\psi$. The divergence of the electric field $\nabla \cdot \mathbf{E}$ equals the ratio of the local charge density σ to electrical permittivity ε_o, a material property representing

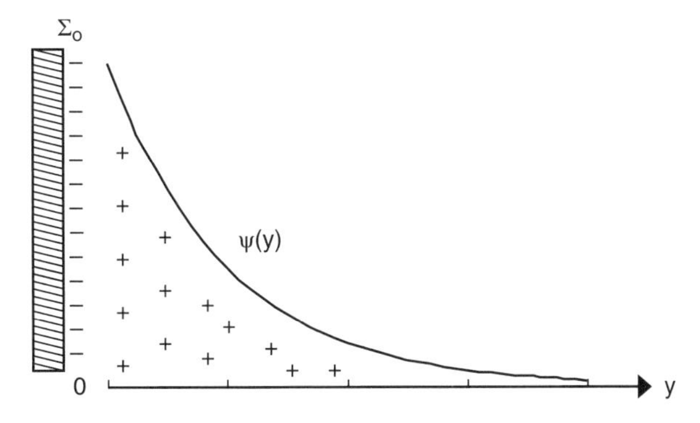

Figure 4.4-1 Electrical double layer at a negatively charged surface.

the inverse of the force produced between two unit charges relative to their squared distance of separation. This leads to the Poisson equation:

$$\nabla \cdot \mathbf{E} = \frac{\sigma}{\varepsilon_o} \Rightarrow -\nabla^2 \psi = \frac{\sigma}{\varepsilon_o} \qquad (4.4\text{-}1a,b)$$

In this application, the electrical potential and therefore the charge density vary only along the y coordinate direction. Consequently, we write the Poisson equation in the diffuse layer as

$$\frac{d^2\psi}{dy^2} = -\frac{1}{\varepsilon_o}\sigma(y) \qquad (4.4\text{-}2)$$

Two boundary conditions are available to solve this equation. Adjacent to the solid surface, we specify a constant value for the electrical potential:

$$y \to 0 \; : \; \psi = \psi_o \qquad (4.4\text{-}3)$$

Far from the surface, where solution is undisturbed by fixed surface charges, spatial changes in ψ are negligible:

$$y \to \infty \; : \; \frac{d\psi}{dy} = 0 \Rightarrow \psi = \psi_\infty \qquad (4.4\text{-}4)$$

We can determine σ at any y by summing the local charge densities of all the ions:

$$\sigma = \mathcal{F} \sum_i z_i C_i \qquad (4.4\text{-}5)$$

In undisturbed solution far from the surface, ion concentrations simultaneously approach constant values of $C_i = C_{i\infty}$, and $\sigma = 0$ because of the electroneutrality principle. Equation 4.4-5 therefore gives rise to

$$y \to \infty \; : \; \sum_i z_i C_{i\infty} = 0 \qquad (4.4\text{-}6)$$

Electroneutrality also applies to the double layer as a whole. This means that the fixed surface charge Σ_o must balance the collection of mobile charges in the diffuse layer:

$$\Sigma_o = -\int_0^\infty \sigma(y)dy \qquad (4.4\text{-}7)$$

Ion concentration can be directly related to electrical potential by asserting that electrochemical equilibrium applies locally. In that case, the Nernst equation (Eq. 4.3-4) predicts that

$$C_i = C_{i\infty} \exp\left[-\frac{\mathcal{F}z_i(\psi - \psi_\infty)}{\mathcal{R}T}\right] \quad (i = a, c) \qquad (4.4\text{-}8)$$

To obtain an analytical solution for $\psi(y)$ and $C_i(y)$, we restrict our attention to a symmetric electrolyte consisting of a cation (i = c) with a charge $z_c = z$ and an anion (i = a) with the opposite charge $z_a = -z$. It follows from Equation 4.4-6 that $C_{c\infty} = C_{a\infty} \equiv C_\infty$. Also, we make the model equations less cumbersome by introducing the dimensionless variables:

$$\mathit{y} \equiv \frac{y}{L_d}, \quad \mathit{C}_c \equiv \frac{C_c}{C_\infty}, \quad \mathit{C}_a \equiv \frac{C_a}{C_\infty} \quad \mathit{\psi} \equiv \frac{\mathcal{F}z}{\mathcal{R}T}(\psi - \psi_\infty), \quad \mathit{\sigma} \equiv \frac{\mathcal{F}zL_d^2}{\mathcal{R}T\varepsilon_o}\sigma \qquad (4.4\text{-}9a\text{-}e)$$

and the dimensionless parameters:

$$\psi_o \equiv \frac{\mathcal{F}z(\Psi_o - \Psi_\infty)}{\mathcal{R}T}, \quad \Sigma_o \equiv \frac{\mathcal{F}zL_d}{\mathcal{R}T\varepsilon_o}\Sigma_o \tag{4.4-10a,b}$$

These equations contain the Debye length L_d, the characteristic thickness of the diffuse layer:

$$L_d \equiv \sqrt{\frac{\varepsilon_o \mathcal{R}T}{2z^2 C_\infty \mathcal{F}^2}} \tag{4.4-11}$$

The model equations (Eqs. 4.4-2, 4.4-5, 4.4-7, and 4.4-8) are now given in dimensionless form as

$$\frac{d^2\psi}{dy^2} = -\sigma \tag{4.4-12}$$

$$\sigma = \frac{1}{2}(C_c - C_a), \quad \Sigma_o = -\int_0^\infty \sigma dy \tag{4.4-13a,b}$$

$$C_c = e^{-\psi}, \quad C_a = e^{+\psi} \tag{4.4-14a,b}$$

The dimensionless forms of the boundary conditions are

$$y = 0 \quad : \quad \psi = \psi_o \tag{4.4-15}$$

$$y \to \infty \quad : \quad \frac{d\psi}{dy} = 0 \Rightarrow \psi = 0 \tag{4.4-16}$$

Combining Equations 4.4-13a and 4.4-14a,b with 4.4-12, we obtain

$$\frac{d^2\psi}{dy^2} = \frac{1}{2}(e^{+\psi} - e^{-\psi}) \tag{4.4-17}$$

This equation and the boundary condition at $y \to \infty$ are satisfied when

$$\frac{d\psi}{dy} = \pm\left(e^{+\psi/2} - e^{-\psi/2}\right) \tag{4.4-18}$$

Only the negative solution is physically reasonable because the magnitude of ψ must continuously decrease toward zero as y increases. To solve this equation, we separate variables and then integrate with the boundary condition at $y = 0$ as a lower limit:

$$\int_{\psi_o}^{\psi} \frac{d\psi}{e^{+\psi/2} - e^{-\psi/2}} = -\int_0^y dy \tag{4.4-19}$$

After the variable transformation $\xi = e^{\psi/2} \Rightarrow d\xi = (e^{\psi/2}/2)d\psi$, this equation becomes

$$2\int_{2\ln\psi_o}^{2\ln\psi} \frac{d\xi}{1-\xi^2} = \int_0^y dy \tag{4.4-20}$$

Completing the integration, we obtain

$$y = \ln\left[\frac{\xi+1}{\xi-1}\right]_{e^{\psi_o/2}}^{e^{\psi/2}} \Rightarrow e^y = \left(\frac{e^{\psi/2}+1}{e^{\psi/2}-1}\right)\left(\frac{e^{\psi_o/2}-1}{e^{\psi_o/2}+1}\right) \tag{4.4-21a,b}$$

Rearranging this equation, we get the final form for the potential distribution:

$$\psi = 2\ln\left[\frac{\left(e^{\psi_o/2}+1\right)+\left(e^{\psi_o/2}-1\right)e^{-y}}{\left(e^{\psi_o/2}+1\right)-\left(e^{\psi_o/2}-1\right)e^{-y}}\right] \tag{4.4-22}$$

To relate surface charge density to surface potential, we substitute Equation 4.4-12 into 4.4-13b and make use of Equations 4.4-15, 4.4-16, and 4.4-18:

$$\Sigma_o = -\int_0^\infty \sigma dy = \int_0^\infty \frac{d^2\psi}{dy^2}dy = \left[\frac{d\psi}{dy}\right]_0^\infty = e^{+\psi_o/2} - e^{-\psi_o/2} \tag{4.4-23}$$

We define the partition coefficient of an ion across the diffusion layer as the concentration ratio $C_{ao} \equiv C_a(0)/C_\infty$ for the anion and $C_{co} \equiv C_c(0)/C_\infty$ for the cation. Utilizing either Equation 4.4-14a or b in Equation 4.4-23, we get

$$\Sigma_o = C_{co}^{-1/2} - C_{co}^{+1/2} \quad \text{or} \quad \Sigma_o = C_{ao}^{+1/2} - C_{ao}^{-1/2} \tag{4.4-24a,b}$$

Solving each of these equations for C_{co} and C_{ao}, we get

$$C_{co} = \left(\sqrt{1+\frac{\Sigma_o^2}{4}} - \frac{\Sigma_o}{2}\right)^2, \quad C_{ao} = \left(\sqrt{1+\frac{\Sigma_o^2}{4}} + \frac{\Sigma_o}{2}\right)^2 \tag{4.4-25a,b}$$

When $\Sigma_o > 0$, Equation 4.4-25a indicates that $C_c(0)/C_\infty < 1$. In other words, a positive fixed charge repels cations from the surface toward the bulk solution. According to Equation 4.4-25b, the surface charge has the opposite effect on anions.

This double-layer model for a symmetric electrolyte can be simplified when $|\psi_o| \ll 1$. With this Debye–Hückel approximation,

$$\psi \approx e^{-y}\Sigma_o, \quad C_c \approx e^{-\Sigma_o}, \quad C_a \approx e^{+\Sigma_o} \tag{4.4-26a-c}$$

These equations indicate that reversing the sign on the surface charge changes the sign but not the magnitude of electrical potential and also interchanges the cation and anion distributions.

Example 4.4-1 Double-Layer Characteristics in Biological Solutions

Compute the Debye length associated with extracellular fluid. Determine the distributions of ions and electrical potential in the diffuse layer adjacent to a negatively charged cell surface.

Solution

Since extracellular fluid mainly contains Na^+, Cl^-, and HCO_3^-, we will consider this solution to be a 1–1 electrolyte ($z = z_c = -z_a = 1$) with ion concentrations of $C_{a\infty} = C_{c\infty} = C_\infty = 0.15$ M. For this relatively dilute solution, the value of $\varepsilon_o = 6.94 \times 10^{-10}$ A-s/(V-m) for water is a reasonable approximation of the electrical permittivity. We compute the Debye thickness (Eq. 4.4-11) at a temperature of 37°C as:

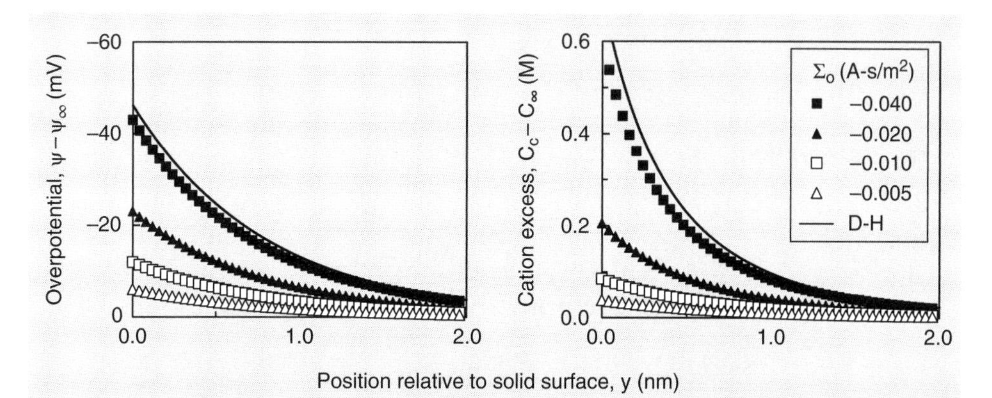

Figure 4.4-2 Double-layer properties in biological solutions.

$$L_d = \sqrt{\frac{6.94\times10^{-10}(8.314)(310)}{(2\times0.15\times1000)(9.65\times10^4)^2}} = 8\times10^{-10}\ \text{m} = 0.8\ \text{nm} \qquad (4.4\text{-}27)$$

Thus, the extracellular (as well as intracellular) diffuse layer is relatively thin, even compared to a typical cell membrane thickness of 10 nm.

For any assumed value of Σ_o, we determine ψ_o from Equation 4.4-23, $\psi(y)$ from Equation 4.4-22, and $C_c(y)$ from Equation 4.4-14a. We then transform these variables into their dimensional forms with Equation 4.4-9a,b,d. Figure 4.4-2 provides the resulting distributions of overpotential $(\psi-\psi_\infty)$ and cation excess $(C_c - C_{c\infty})$ for several negative values of surface charge Σ_o. Both distributions are strongly affected by the magnitude of Σ_o, and they reach undisturbed conditions within three Debye lengths from the solid surface. We also computed these distributions using the Debye–Hückel approximation (curves). When $|\Sigma_o| \leq 0.02\,\text{A-s/m}^2$, the Debye–Hückel results are indistinguishable from the more exact model.

References

Adamson AW. Physical Chemical of Surfaces. 2nd ed. New York: Interscience Publishers; 1967.

Lange AN. Handbook of Chemistry. New York: McGraw-Hill; 1967, p 1436, 1450.

Mountcastle VB. Medical Physiology. 13th ed. St Louis: Mosby; 1974, p 3, Vol 1.

Ott H. Die Errechnung des kolloidosmotischen serumdruckes aus dem eiweiss-sprektrum und das mittlere molekulargewicht der serumeiweissfraktionen. Klinische Wochenschrift. 1956; 34:1079–1083.

Chapter 5

Chemical Reaction Equilibrium

The generation of energy and the synthesis of new substances by chemical reactions are essential for the survival of an organism. In this chapter, we develop the quantitative basis for chemical reaction equilibrium. We consider two types of important reactions: acid dissociation and reversible binding. Applications include the basis of protein charge, ligand binding to receptor proteins, and O_2 and CO_2 content in blood.

5.1 Equilibrium Criterion

The stoichiometry of a reaction between two reactants, A_1 and A_2, that form two products, A_3 and A_4, is represented as

$$\nu_1' A_1 + \nu_2' A_2 \rightleftharpoons \nu_3'' A_3 + \nu_4'' A_4 \tag{5.1-1}$$

where ν_1' and ν_2' represent the stoichiometric coefficients of the reactants and ν_3'' and ν_4'' are the stoichiometric coefficients of the products. These coefficients always have positive values. The double arrow indicates a reversible reaction that is capable of occurring in either the forward or reverse direction, depending on the concentrations of reactants and products relative to their equilibrium values.

Consider a system that contains an equilibrium mixture of m_1 and m_2 moles of reactants and m_3 and m_4 moles of products (Fig. 5.1-1). Suppose reaction equilibrium is disturbed by converting a small amount of reactant A_1, represented by $dm_1 < 0$, into products. Using Equation 5.1-1, we can predict the resulting changes in the amounts of A_2, A_3, and A_4:

$$dm_2 = \left(\frac{\nu_2'}{\nu_1'}\right) dm_1, \quad dm_3 = -\left(\frac{\nu_3''}{\nu_1'}\right) dm_1, \quad dm_4 = -\left(\frac{\nu_4''}{\nu_1'}\right) dm_1 \tag{5.1-2a-c}$$

Biomedical Mass Transport and Chemical Reaction: Physicochemical Principles and Mathematical Modeling, First Edition. James S. Ultman, Harihara Baskaran, and Gerald M. Saidel.

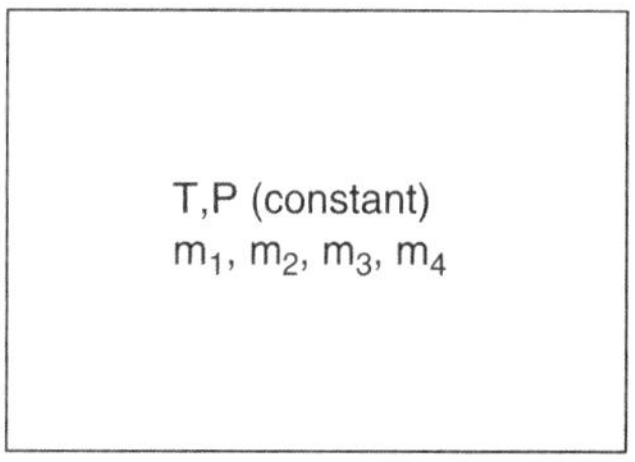

Figure 5.1-1 Equilibrium solution of reactants and products in a well-mixed reactor.

T,P (constant)
m_1, m_2, m_3, m_4

To express these equations in a more general and compact form, we introduce modified stoichiometric coefficients for a reaction of R different reactants that form P different products. The modified coefficients have positive values for reaction products and negative values for reactants:

$$\nu_i = \begin{cases} -\nu_i' & (i = 1,\ 2\ldots,\ R) \\ +\nu_i'' & (i = R+1, R+2\ldots, R+P) \end{cases} \qquad (5.1\text{-}3a,b)$$

For such a reaction Equation 5.1-2a-c can be generalized as follows:

$$\frac{dm_1}{\nu_1} = \frac{dm_2}{\nu_2} = \ldots = \frac{dm_I}{\nu_I} \equiv d\xi_r \Rightarrow dm_i = \nu_i d\xi_r \quad (i = 1, 2, \ldots I) \qquad (5.1\text{-}4a,b)$$

where $I = R + P$ is the number of substances participating in the reaction. The quantity ξ_r, which represents the extent to which reaction occurs, does not depend on the stoichiometric coefficients. By incorporating Equation 5.1-4b into Equation 3.5-21, we establish a criterion for reaction equilibrium at constant T and P:

$$\sum_{i=1}^{I} \mu_i dm_i = \sum_{i=1}^{I} (\nu_i \mu_i) d\xi_r = 0 \qquad (5.1\text{-}5)$$

Since $d\xi_r$ is nonzero, this equation is only satisfied if

$$\sum_{i=1}^{I} \nu_i \mu_i = 0 \qquad (5.1\text{-}6)$$

Using Equation 3.5-15 to replace the electrochemical potential by the sum of its electrical and chemical contributions,

$$\sum_{i=1}^{I} \nu_i \mu_i = F\psi \sum_{i=1}^{I} \nu_i z_i + \sum_{i=1}^{I} \nu_i \breve{\mu}_i = 0 \qquad (5.1\text{-}7)$$

Since a chemical reaction must be electrically balanced, $\sum_i \nu_i z_i = 0$ and the equilibrium criterion reduces to

$$\sum_{i=1}^{I} \nu_i \breve{\mu}_i = 0 \qquad (5.1\text{-}8)$$

We define the summation on the left side of this equation as the Gibbs free energy of reaction, a quantity that represents the abundance in free energy of products $\sum_i \nu_i'' \breve{\mu}_i$ compared to the free energy of reactants $\sum_i \nu_i' \breve{\mu}_i$:

$$\Delta G_r \equiv \sum_{i=1}^{I} \nu_i \breve{\mu}_i = \sum_{i=R+1}^{I} \nu_i'' \breve{\mu}_i - \sum_{i=1}^{R} \nu_i' \breve{\mu}_i \tag{5.1-9}$$

According to Equation 5.1-8, $\Delta G_r = 0$ is the equilibrium condition at which no net reaction occurs. When $\Delta G_r < 0$, the free energy of reactants exceeds the free energy of products, and we expect the reactants to be spontaneously converted to products. Conversely, we expect products to be spontaneously converted to reactants if $\Delta G_r > 0$.

As an alternative thermodynamic driving force, one that is positive for a reaction proceeding in the forward direction, we define the affinity of a reaction as

$$A_r \equiv -\Delta G_r = -\sum_{i=1}^{I} \nu_i \breve{\mu}_i \tag{5.1-10}$$

5.2 Equilibrium Coefficients

5.2.1 Gas Phase

To express the reaction equilibrium criterion in a useful form, we must relate $\breve{\mu}_i$ to measurable quantities. For a chemical reaction occurring in an ideal gas mixture consisting of I reactants and products, this relationship is provided by Equation 3.6-1a such that Equation 5.1-8 becomes

$$\frac{\mu^o(T)}{\mathcal{R}T} \sum_{i=1}^{I} \nu_i + \ln\left(\prod_{i=1}^{I} p_i^{\nu_i} \right) = 0 \Rightarrow \prod_{i=1}^{I} p_i^{\nu_i} = \exp\left(-\frac{\mu^o(T)}{\mathcal{R}T} \sum_{i=1}^{I} \nu_i \right) \tag{5.2-1a,b}$$

Defining the temperature-dependent exponential on the right side of Equation 5.2-1b as the equilibrium coefficient κ_p for a gas-phase reaction, we obtain

$$\kappa_p(T) = \prod_{i=1}^{I} p_i^{\nu_i} = \frac{\prod_{i=R+1}^{P} p_i^{\nu_i''}}{\prod_{i=1}^{R} p_i^{\nu_i'}} \tag{5.2-2}$$

Since chemical reactions rarely occur in the gas phase of biological systems, this relationship will not be particularly useful.

5.2.2 Liquid Phase

For a chemical reaction occurring in a liquid phase, we substitute Equation 3.6-2 into Equation 5.1-8 and get

$$\frac{1}{\mathcal{R}T} \sum_{i=1}^{I} \nu_i \mu_i^* + \ln \prod_{i=1}^{I} (\gamma_i x_i)^{\nu_i} = 0 \tag{5.2-3}$$

Analogous to ΔG_r in Equation 5.1-9, we define a standard free energy of reaction as

$$\Delta G_r^* \equiv \sum_{i=1}^{I} \nu_i \mu_i^* \tag{5.2-4}$$

Since $\mu_i^*(T,P)$ is only weakly related to pressure, ΔG_r^* is primarily a function of T. By expressing mole fraction in terms of molar concentration, $x_i = C_i/c_L$, we can write Equation 5.2-3 as

$$\prod_{i=1}^{I} (\gamma_i C_i)^{\nu_i} = (c_L)^{\nu_T} \exp\left(-\frac{\Delta G_r^*}{\mathcal{R}T}\right) \tag{5.2-5}$$

where $\nu_T \equiv \sum_i \nu_i$ is the sum of all the modified stoichiometric coefficients. We define the left side of this equation as the equilibrium coefficient for chemical reaction in a liquid solution:

$$\kappa_c(T) \equiv \prod_{i=1}^{I} (\gamma_i C_i)^{\nu_i} = (c_L)^{\nu_T} \exp\left(-\frac{\Delta G_r^*}{\mathcal{R}T}\right) \tag{5.2-6}$$

Because activity coefficients in biological systems are typically not known and most body fluids are relatively dilute aqueous solutions, it is common to assume that $\Pi_i(\gamma_i)^{\nu_i} \approx 1$ so that

$$\kappa_c(T) = \prod_{i=1}^{I} C_i^{\nu_i} = \frac{\prod_{i=R+1}^{I} C_i^{\nu_i''}}{\prod_{i=1}^{R} C_i^{\nu_i'}} = (c_L)^{\nu_T} \exp\left(-\frac{\Delta G_r^*}{\mathcal{R}T}\right) \tag{5.2-7}$$

The concentration ratio in this equation is equal to κ_c only when a reacting system is in an equilibrium state. Consider a mixture of reactants and products that are initially in equilibrium in a closed well-mixed container. Suppose we add more reactant to the system so that the concentration ratio decreases to a value below κ_c. A spontaneous reaction would then occur in the forward direction, elevating product concentrations and diminishing reactant concentrations until the concentration ratio again reaches its equilibrium value, κ_c. By the same reasoning, the addition of product to an equilibrium reacting mixture promotes spontaneous reaction in the reverse direction. These responses to perturbations from reaction equilibrium, forward reaction by addition of reactants and backward reaction by addition of product, are the basis of the law of mass action.

Equation 5.2-7 indicates that κ_c is large when equilibrium product concentrations are large relative to equilibrium reactant concentrations. Thus, product formation is favored over reactant formation when $\Delta G_r^* < 0$. If $\Delta G_r^* \ll -\mathcal{R}T$, then product formation is so highly favored that reaction in the forward direction is essentially irreversible.

For a reaction carried out at constant temperature, the definition of Gibbs free energy in Equation 3.5-1 implies that

$$\Delta G_r = \Delta H_r - T\Delta S_r \tag{5.2-8}$$

where ΔH_r and ΔS_r are the enthalpy and entropy changes associated with chemical reaction. A spontaneous reaction will occur when ΔH_r is sufficiently negative and/or ΔS_r is sufficiently positive to produce a value of $\Delta G_r < 0$. Consider the complete combustion of one mole of glucose given in Equation 1.1-1. If this reaction is carried out at $P = 101.3$ kPa and $T = 25°C$, then $\Delta H_r = -2810$ kJ and $T\Delta S_r = 62$ kJ so that $\Delta G_r = -2872$ kJ. Because $-\Delta H_r$ is much greater than $T\Delta S_r$, the exothermic nature of this particular reaction makes a far more important contribution to the lowering of ΔG_r than the disordering of molecules. The highly negative ΔG_r relative to $\mathcal{R}T = 2$ kJ indicates that glucose combustion can be considered to be an irreversible reaction.

5.3 Acid Dissociation

5.3.1 Monovalent Acids

In aqueous solutions, acids ionize into a hydrogen ion and a complementary base, affecting both pH and electrical potential. Consider the dissociation of water into hydrogen ions and hydroxyl ions:

$$H_2O \rightleftharpoons H^+ + OH^- \tag{5.3-1}$$

According to Equation 5.2-7, the equilibrium coefficient for this reaction is given by

$$\kappa_c = \frac{C_{H^+} C_{OH^-}}{C_{HOH}} \tag{5.3-2}$$

Because molar concentrations of hydrogen and hydroxyl ions can change over several orders of magnitude, it is convenient to define logarithmic concentration measures:

$$pH \equiv -\log(C_{H^+}), \quad pOH \equiv -\log(C_{OH^-}) \tag{5.3-3a,b}$$

Consequently, Equation 5.3-2 becomes

$$pH + pOH = -\log(\kappa_c C_{HOH}) \tag{5.3-4}$$

where $\kappa_c C_{HOH}$ is the dissociation constant of water. The ionization of water is so weak that C_{HOH} is essentially constant. Since $\kappa_c C_{HOH}$ has a value of 10^{-14} M,

$$pH + pOH = 14 \tag{5.3-5}$$

A neutral solution is one in which $C_{H^+} = C_{OH^-}$. Thus, in either pure water or neutral solutions, pH = pOH = 7. In acidic solutions, $C_{H^+} > C_{OH^-}$ so that pH < 7 and pOH > 7. In basic solutions, $C_{H^+} < C_{OH^-}$ so that pH > 7 and pOH < 7.

We now examine the dissociation in aqueous solution of a monovalent acid into a hydrogen ion and its counterion A^-:

$$HA \rightleftharpoons H^+ + A^- \tag{5.3-6}$$

The equilibrium constant for this reaction given by

$$\kappa_c = \frac{C_{H^+} C_{A^-}}{C_{HA}} \tag{5.3-7}$$

is a measure of the strength of an acid. Large values of κ_c correspond to a strong acid that readily donates H^+ ions to a solution. Taking the logarithm of both sides of this equation and defining $pK \equiv -\log \kappa_c$, we arrive at

$$pH = pK + \log\left(\frac{C_{A^-}}{C_{HA}}\right) \tag{5.3-8}$$

Thus, we can predict the degree of dissociation of a monovalent acid from the difference between its pK and the pH of a solution. If pH is equal to pK, then $C_{A^-} = C_{HA}$ and the acid is 50% dissociated. If pH is greater than pK, then $C_{A^-} > C_{HA}$ and acid dissociation is more than 50%.

5.3.2 Complex Acids

Carboxyl (–COOH) and amide ($-NH_2$) are monoacid groups that coexist in many amino acids (Table A3-6). They have markedly different pK values of about 2 and 10, respectively. The dissociation of a carboxyl group into $-COO^-$ and H^+ reduces the charge on the amino acid by one unit. The association of the amide group with H^+ to form $-NH_3^+$ increases the amino acid charge by one unit. The net charge on a protein, consisting of a multitude of amino acids, depends on the number of monoacid groups and their degree of acid dissociation. Protein charge can have a strong influence on its transport, particularly in the vicinity of membranes and surfaces.

To illustrate the relationship between solution pH and molecular charge, consider an amino acid such as glycine that consists of one carboxyl group, one amide group, and one nondissociable hydrocarbon group (–R). The two possible acid dissociations of such a molecule are

$$^+NH_3\text{-CRH-COOH} \rightleftharpoons \mathbf{^+NH_3\text{-CRH-COO}^-} + H^+ \quad : \quad pK_1 \approx 2 \tag{5.3-9}$$

$$\mathbf{^+NH_3\text{-CRH-COO}^-} \rightleftharpoons H^+ + NH_2\text{-CRH-COO}^- \quad : \quad pK_2 \approx 10 \tag{5.3-10}$$

The molecule in bold font is a zwitterion that has positive and negative charges on different side groups, but no net charge.

Suppose a solution of this amino acid is adjusted to a low pH by adding a strong acid. According to the law of mass action, the high concentration of H^+ ions would reverse both reactions and favor the formation of $^+NH_3$–CRH–COOH. This would increase the charge on the amino acid. Alternatively, if the solution were adjusted to a high pH by adding a strong base, then the low concentration of H^+ would favor the forward reaction sequence with its formation of NH_2–CRH–COO^-, thereby reducing the amino acid charge.

We can quantify this process by separately applying Equation 5.3-8 to the two dissociation reactions given in Equations 5.3-9 and 5.3-10. Representing the $^+NH_3$-CRH-COOH cation as HAH^+, the $^+NH_3$-CRH-COO^- zwitterion as HA, and the NH_2-CRH-COO^- anion as A^-, we obtain

$$pH = pK_1 + \log\left[\frac{C_{HA}}{C_{HAH^+}}\right] \tag{5.3-11}$$

for the first reaction and

$$pH = pK_2 + \log\left[\frac{C_{A^-}}{C_{HA}}\right] \tag{5.3-12}$$

for the second reaction. By adding these equations, we eliminate the zwitterion concentration and obtain the relative concentrations of anions to cations:

$$\log\left[\frac{C_{A^-}}{C_{HAH^+}}\right]^{1/2} = pH - \left(\frac{pK_1 + pK_2}{2}\right) \tag{5.3-13}$$

When pH is equal to $(pK_1 + pK_2)/2$, this equation indicates that $C_{A^-} = C_{HAH^+}$. This pH at which the positive and negative charges on the collection of amino acid molecules are balanced is defined as the isoelectric point:

$$pI \equiv \frac{pK_1 + pK_2}{2} \tag{5.3-14}$$

When pH is greater than pI, $C_{A^-} > C_{HAH^+}$ and the amino acid molecules have a net negative charge. Conversely, the net charge is positive when pH is less than pI.

Example 5.3-1 Relationship between pH and Glycine Charge

Glycine undergoes the two acid dissociation steps indicated by Equations 5.3-9 and 5.3-10 with a $pK_1 = 2.34$ and a $pK_2 = 9.6$. Suppose that solid glycine is mixed with pure water in a closed container. The pH of this mixture is increased by slowly adding a sodium hydroxide solution to the container. In a second experiment, the pH of the same starting mixture is decreased by slowly adding hydrochloric acid. The purpose of this example is to determine the net electrical charge on the glycine molecules as the pH of the solution changes.

Solution

According to Equation 5.3-14, glycine has an isoelectric point of $pI = (2.34 + 9.6)/2 = 5.97$. If C_o represents the total concentration of glycine in the solution, then by a molar balance on the three forms of glycine

$$C_{HAH^+} + C_{HA} + C_{A^-} = C_o \tag{5.3-15}$$

Rearranging Equations 5.3-11 and 5.3-12 yields two more equations for the three unknown concentrations:

$$C_{HAH^+} = 10^{(2.34-pH)} C_{HA} \tag{5.3-16}$$

$$C_{A^-} = 10^{(pH-9.60)} C_{HA} \tag{5.3-17}$$

The molecular charge on a collection of glycine molecules can be expressed as the net concentration of charge $(C_{HAH^+} - C_{A^-})$ divided by the overall concentration C_o. By combining Equations 5.3-15–5.3-17, we obtain

$$\text{Molecular charge} = \frac{10^{(2.34-pH)} - 10^{(pH-9.60)}}{1 + 10^{(2.34-pH)} + 10^{(pH-9.60)}} \tag{5.3-18}$$

Equation 5.3-18 predicts that molecular charge depends on pH but not on the overall concentration of glycine. As illustrated by the plot of this equation in Figure 5.3-1, there is little change in net charge when $8 > pH > 4$, a range that surrounds the isoelectric point.

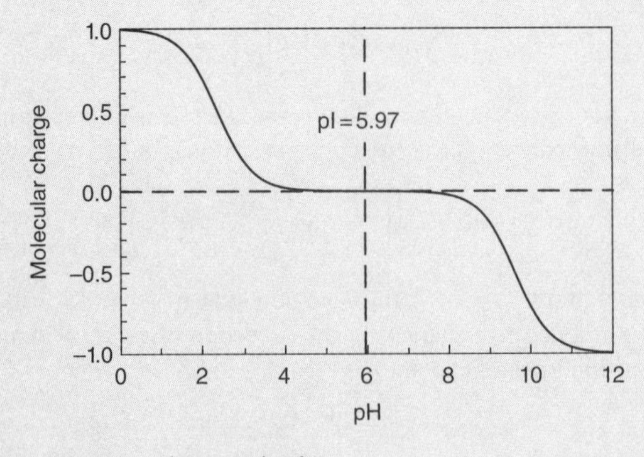

Figure 5.3-1 Electrical charge on glycine molecules.

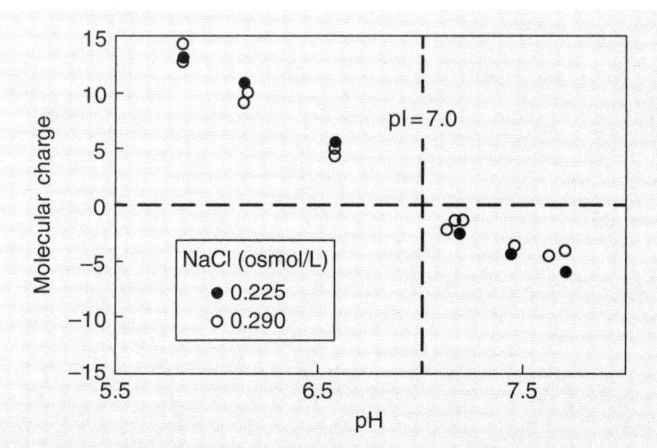

Figure 5.3-2 Electrical charge on hemoglobin in salt solutions (Data from Gary-Bobo and Solomon (1968)).

At pH $\ll$ 2.34, essentially all of the glycine is in the HAH^+ form and the molecular charge approaches +1. At pH $\gg$ 9.2, almost all of the glycine is present as A^- and the molecular charge approaches −1. In the intermediate pH ranges, the molecular charge changes by about −0.5 per unit increase in pH.

Because protein molecules can contain many carboxyl, amide, and other acidic side groups, their charge changes more linearly with respect to changes in pH and is more sensitive to pH. This is illustrated by data for hemoglobin dissolved in saline solutions (Gary-Bobo and Solomon, 1968) shown in Figure 5.3-2.

5.4 Ligand–Receptor Binding

The reversible noncovalent binding of an endogenous molecule or a drug (the ligand) to a protein molecule (the receptor) is an important part of many biological processes. Binding of solutes to proteins is necessary for facilitated and active transport of large molecules across cell membranes. Binding of reactants to active sites on enzyme molecules is required for the catalysis of metabolic reactions. Binding of extracellular signaling molecules to integral membrane proteins is an important step in the regulation of cell function. Binding of a pharmaceutical to the surface of specific cells is a key aspect of targeted drug delivery.

For a receptor containing a single binding site, the nature of the binding process depends on the specificity of the site with respect to different ligands. Monovalent binding occurs when a protein site is highly specific to one type of ligand. Often two or more ligands with similar structures can bind at the same site in a process called competitive binding. It is also possible for a receptor to contain more than one binding site allowing multiple ligand molecules to simultaneously bind to the protein. In such allosteric binding processes, the sites might be fixed in structure and properties, or they might be interconvertible from one form to another.

5.4.1 Monovalent Binding

The reversible monovalent binding of a single ligand L from a liquid solution to a receptor containing a single binding site R is the simplest ligand–receptor binding mechanism. This process, depicted in Figure 5.4-1, has an equilibrium binding coefficient given by

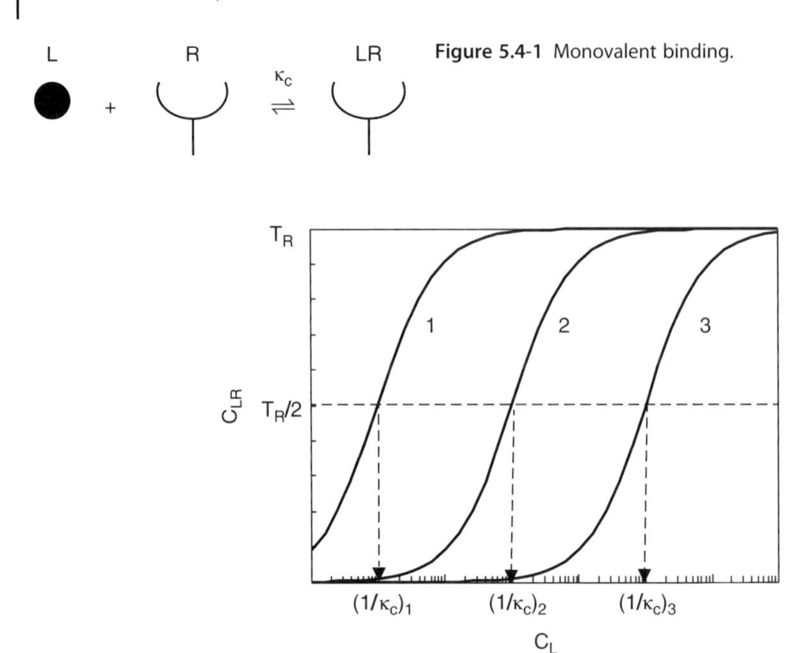

Figure 5.4-1 Monovalent binding.

Figure 5.4-2 Monovalent binding curve.

$$\kappa_c = \frac{C_{LR}}{C_L C_R} \tag{5.4-1}$$

Because this is a reversible process, we also recognize that $1/\kappa_c$ is the equilibrium constant for the dissociation of LR into L and R. Because it is usually not possible to separately measure the concentration of occupied and unoccupied binding sites, it is convenient to introduce the total binding site concentration:

$$T_R = C_R + C_{LR} \tag{5.4-2}$$

Equations 5.4-1 and 5.4-2 can now be solved to show that

$$C_{LR} = \frac{T_R \kappa_c C_L}{1 + \kappa_c C_L} \tag{5.4-3}$$

As illustrated in Figure 5.4-2, this equation is usually graphed with a logarithmic abscissa in order to spread C_L over the wide range of values over which dissociation occurs. According to this equation, C_{LR} saturates at level of T_R when $C_L \gg 1/\kappa_c$ and reaches one-half of this level when $C_L = 1/\kappa_c$. The larger is the dissociation constant, $1/\kappa_c$, the more the binding curves are shifted to the right and the weaker is L–R binding.

Since ligands are soluble molecules, $1/\kappa_c$ and C_L are always given in units of moles per unit volume of solution. Receptor sites can be found on dissolved proteins that are free to be transported in bulk solution, but they can also be associated with membrane proteins that are confined to cell surfaces. For a receptor in solution, C_{LR} and T_R are expressed as moles per unit solution volume. However, if a receptor is fixed at a surface, then it is more logical to specify C_{LR} and T_R as moles per unit area of the surface. Because surface areas are not ordinarily measured, the C_{LR} and T_R values for surface-bound receptors are often normalized by a parameter that is proportional to the surface area. The following example illustrates how total protein mass extracted from a membrane can serve as a surrogate for surface area.

Example 5.4-1 Benzodiazepine Binding to Synaptic Receptors

Benzodiazepine is a psychoactive drug used in treating anxiety, insomnia, agitation, seizures, and muscle spasms. The equilibrium binding of radioactively labeled benzodiazepine to a protein fraction extracted from synaptic membranes of the rat brain has been measured by Bowling and DeLorenzo (1982). As shown in Figure 5.4-3, the resulting data are expressed as picomoles of bound diazepam per milligram of membrane protein. Can these measurements be modeled as a monovalent binding of benzodiazepine to surface receptors within the membrane protein fraction?

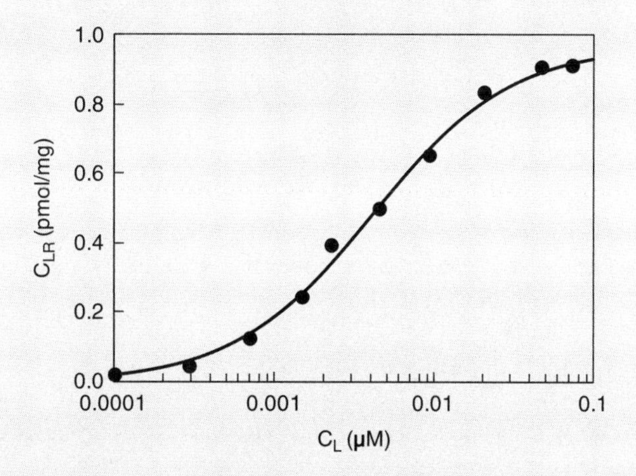

Figure 5.4-3 Benzodiazepine binding to rat brain cell receptors.

Solution

We fit the data points to the monovalent model given by Equation 5.4-3 by a nonlinear regression of C_{LR} versus C_L in which T_R and κ are free parameters. By minimizing the squared error between the measured and predicted values of C_{LR}, we obtained estimates of $T_R = 0.980$ pmol/mg and $\kappa_c = 243\ \mu M^{-1}$. Since the data points lie close to the regression curve, we conclude that the benzodiazepine binding can be modeled as a monovalent process.

5.4.2 Competitive Binding

To examine the nature of the equilibrium that arises from competitive binding, consider the case of two ligands, L_1 and L_2, that are both capable of binding to the same site R (Fig. 5.4-4). The binding coefficients for this process are

$$\kappa_{C_1} = \frac{C_{L_1R}}{C_{L_1}C_R}, \quad \kappa_{C_2} = \frac{C_{L_2R}}{C_{L_2}C_R}, \tag{5.4-4a,b}$$

and the total concentration of binding sites is

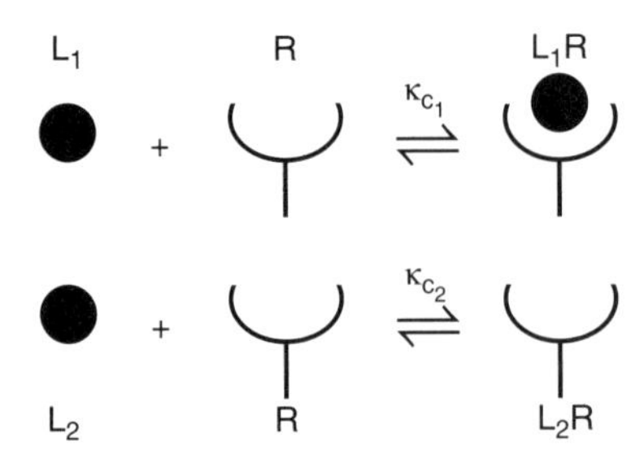

Figure 5.4-4 Competitive binding.

$$T_R = C_R + C_{L_1R} + C_{L_2R} \tag{5.4-5}$$

By combining these equations, we obtain the concentration of the two forms of bound ligand:

$$C_{L_1R} = \frac{T_R \kappa_1 C_{L_1}}{1 + \kappa_{C_1} C_{L_1} + \kappa_{C_2} C_{L_2}}$$

$$C_{L_2R} = \frac{T_R \kappa_2 C_{L_2}}{1 + \kappa_{C_1} C_{L_1} + \kappa_{C_2} C_{L_2}} \tag{5.4-6a,b}$$

When $C_{L_2} = 0$ such that competition by ligand L_2 is eliminated, the equation for C_{L_1R} reduces the same form as Equation 5.4-3 for monovalent binding. When $C_{L_2} > 0$, competition by ligand L_2 reduces C_{L_1R}. Similarly, competition by ligand L_1 reduces the value of C_{L_2R}.

Example 5.4-2 Norephedrine Effect on Pindolol-Receptor Binding

Pindolol, a drug that blocks β-adrenergic receptors, is used in the treatment of angina pectoris, arrhythmias, glaucoma, and hypertension. Salinas *et al.* (2005) observed the effect of norephedrine on the equilibrium binding of pindolol to membrane-bound receptors on cardiac smooth muscle cells.

In each experiment, a known weight of cell membranes isolated from the cardiac tissue of pigs was suspended in Krebs buffer solution containing norephedrine and radioactively labeled pindolol. After allowing a sufficient time to reach equilibrium, the radioactivity of a membrane sample was determined. Figure 5.4-5 shows the data points from experiments repeated at various norephedrine concentrations but a constant pindolol concentration. The relative activity refers to the ratio of the pindolol radioactivity at a particular norephedrine concentration to the radioactivity in the absence of norephedrine. Assuming these data can be explained by competitive binding between pindolol and norephedrine, determine the equilibrium constant for the binding of norephedrine to cardiac β-adrenergic receptors.

Solution

We designate norephedrine as ligand L_1, pindolol as ligand L_2, and the binding site on the β-adrenergic receptor as R. When the norephedrine concentration in the suspending medium is zero, $C_{L_1} = 0$ and Equation 5.4-6b predicts a bound pindolol concentration given by

$$(C_{L_2R})_o = \frac{T_R \kappa_{C_2} C_{L_2}}{1 + \kappa_{C_2} C_{L_2}} \tag{5.4-7}$$

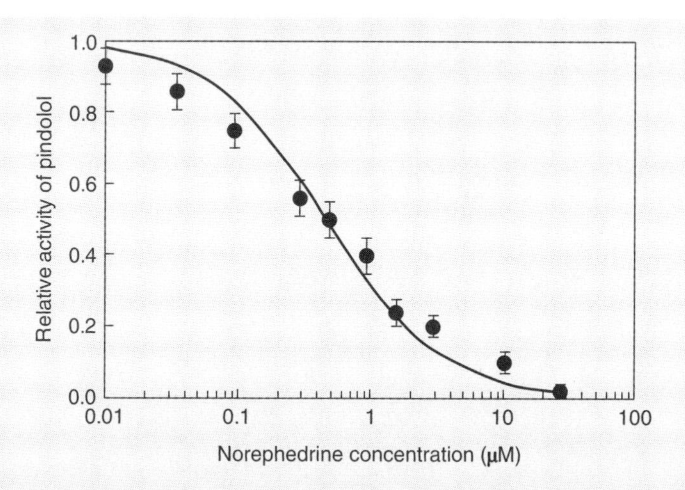

Figure 5.4-5 Displacement of bound pindolol by norephedrine.

The concentration of bound pindolol in the presence of norephedrine relative to that in the absence of norephedrine is given by the ratio of Equations 5.4-6b to 5.4-7:

$$\frac{C_{L_2R}}{(C_{L_2R})_o} = \frac{1}{1 + [\kappa_{C_1}/(1 + \kappa_{C_2} C_{L_2})] C_{L_1}} \qquad (5.4\text{-}8)$$

Because radioactivity is proportional to the amount of bound pindolol in a membrane sample, $C_{L_2R}/(C_{L_2R})_o$ is equal to the relative activity shown on the ordinate of Figure 5.4-5.

A nonlinear regression with $\kappa_{C_1}/(1 + \kappa_{C_2} C_{L_2})$ as a free parameter was used to fit the data points to Equation 5.4-8. The regression curve shown in the figure is based on $\kappa_{C_1}/(1 + \kappa_{C_2} C_{L_2}) = 2.20\ \mu M^{-1}$, the parameter value that produced the minimum squared error between measured and predicted relative activities. Because the amount of radioactive pindolol used in the experiments was exceedingly small, it is reasonable to assume that $\kappa_{C_2} C_{L_2} \ll 1$, and we estimate that the binding coefficient of norephedrine is $\kappa_{C_1} \approx 2.20\ \mu M^{-1}$.

5.4.3 Allosteric Binding

Noninterconverting Receptor Sites

Figure 5.4-6 illustrates a model for allosteric binding of a single ligand L to two receptor sites, R_1 and R_2, to form two different complexes LR_1 and LR_2. It is assumed in this model that R_1 and R_2 are distinct entities that cannot interconvert from one form to the other. Since each of these reactions is an independent monovalent binding process, equilibrium can be described by separately applying Equation 5.4-3 to each of the two reactions:

$$C_{LR_1} = \frac{T_{R_1} \kappa_{C_1} C_L}{1 + \kappa_{C_1} C_L}, \quad C_{LR_2} = \frac{T_{R_2} \kappa_{C_2} C_L}{1 + \kappa_{C_2} C_L} \qquad (5.4\text{-}9a,b)$$

where $T_{R_1} \equiv C_{R_1} + C_{LR_1}$ and $T_{R_2} \equiv C_{R_2} + C_{LR_2}$ are the total concentrations of each type of binding site. Typically, experiments do not reveal the distribution of a ligand among different types of receptor sites. Rather, they measure the total concentration of bound ligand. For

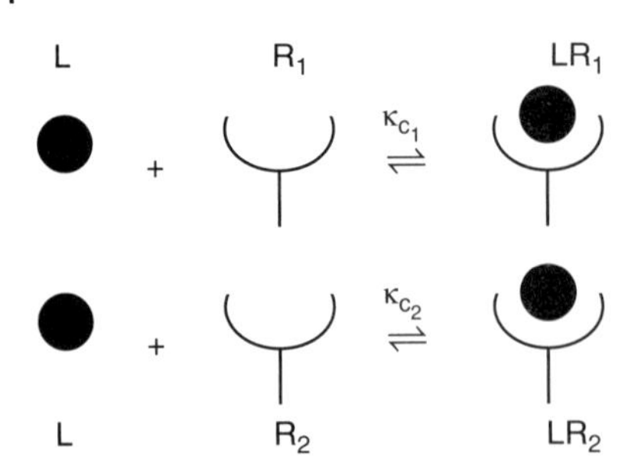

Figure 5.4-6 Allosteric binding of a ligand to noninterconverting receptor sites.

the binding of ligand L to noninterconverting sites R_1 and R_2, we sum Equation 5.4-9a,b to obtain the total bound ligand concentration:

$$C_{LR_1} + C_{LR_2} = \left[\frac{\chi\kappa_{C_1} + (1-\chi)\kappa_{C_2} + \kappa_{C_1}\kappa_{C_2}C_L}{1 + (\kappa_{C_1} + \kappa_{C_2})C_L + \kappa_{C_1}\kappa_{C_2}C_L^2}\right] T_R C_L \tag{5.4-10}$$

where

$$T_R \equiv T_{R_1} + T_{R_2} = C_{R_1} + C_{R_2} + C_{LR_1} + C_{LR_2} \tag{5.4-11}$$

is the total concentration of both types of binding sites,

$$\chi \equiv C_{T_1}/C_T \tag{5.4-12}$$

is the fraction of R_1 binding sites, and $(1-\chi)$ is the fraction of R_2 sites.

Interconverting Receptor Sites

Allosteric binding can also occur at a binding site that interconverts between two different forms, $R_1 \rightleftharpoons R_2$ (Fig. 5.4-7). We define $\kappa_{12} \equiv C_{R_2}/C_{R_1}$ as an equilibrium constant for the interconversion of the receptor sites from the R_1 to the R_2 forms. The concentrations of the individual ligand–receptor complexes are then given by

$$C_{LR_1} = \frac{T_R\kappa_{C_1}C_L}{(1+\kappa_{12}) + \kappa_{C_1}[1 + (\kappa_{12}\kappa_{C_2}/\kappa_{C_1})]C_L}$$

$$C_{LR_2} = \frac{T_R\kappa_{12}\kappa_{C_2}C_L}{(1+\kappa_{12}) + \kappa_{C_1}[1 + (\kappa_{12}\kappa_{C_2}/\kappa_{C_1})]C_L} \tag{5.4-13a,b}$$

As before, T_R represents the total concentration of both types of binding sites, whether they are unoccupied or occupied by ligand. By adding these two equations, the total concentration of bound ligand is obtained as

$$C_{LR_1} + C_{LR_2} = \frac{T_R(\kappa_{C_1} + \kappa_{12}\kappa_{C_2})C_L}{(1+\kappa_{12}) + \kappa_{C_1}[1 + (\kappa_{12}\kappa_{C_2}/\kappa_{C_1})]C_L} \tag{5.4-14}$$

Dimensionless Comparison

In comparing allosteric binding with converting sites to allosteric binding with noninterconverting sites, it is convenient to minimize the number of parameters by introducing the following dimensionless quantities:

$$\mathcal{C}_{LR} \equiv \frac{C_{LR_1} + C_{LR_2}}{T_R}, \quad \mathrm{C}_L = \kappa_{C_1}C_L, \quad \beta \equiv \frac{\kappa_{C_2}}{\kappa_{C_1}} \tag{5.4-15a-c}$$

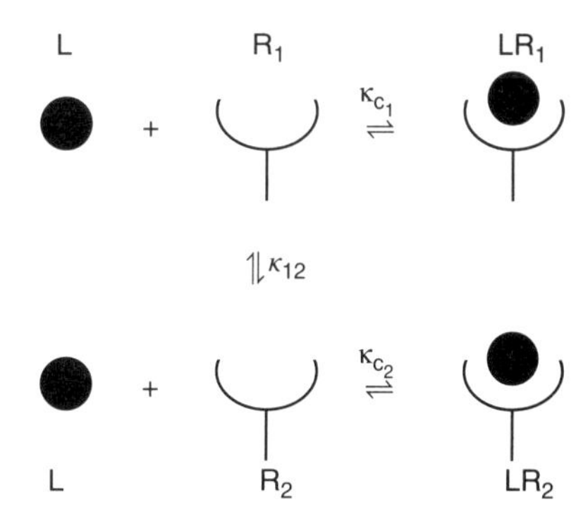

Figure 5.4-7 Allosteric binding of a ligand to interconverting receptor sites.

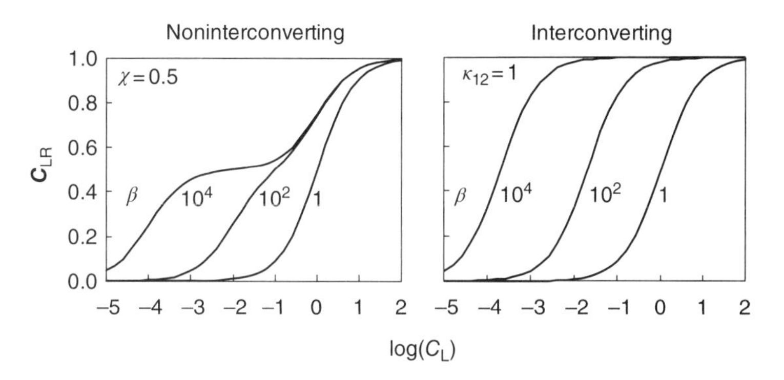

Figure 5.4-8 Allosteric binding curves for noninterconverting and interconverting receptor sites.

For a receptor that has noninterconverting sites, the dimensionless form of Equation 5.4-10 is

$$C_{LR} = \frac{(\chi + \beta - \chi\beta)C_L + \beta C_L^2}{1 + (1 + \beta)C_L + \beta C_L^2} \tag{5.4-16}$$

and for a receptor that has interconverting sites, Equation 5.4-14 becomes

$$C_{LR} = \frac{(1 + \kappa_{12}\beta)C_L}{(1 + \kappa_{12}) + (1 + \kappa_{12}\beta)C_L} \tag{5.4-17}$$

Figure 5.4-8 contains overall binding curves for C_{LR} versus C_L. The left side is a plot of Equation 5.4-16 when $\chi = 0.5$, corresponding to equal concentrations of the R_1 and R_2 sites when they do not interconvert. The right side illustrates the behavior of Equation 5.4-17 when $\kappa_{12} = 1$ such that the concentrations of the interconverting sites are equal.

When R_1 and R_2 have the same binding strength, their separate behaviors should be indistinguishable. Thus, the overall binding curves for noninterconverting and converting sites are both equivalent to the same monovalent binding curve when $\beta = 1$ (This degenerate case is independent of the particular values of χ and κ_{12}).

Because the noninterconverting binding sites are independent, each of their overall binding curves is the sum of two independent monovalent binding curves. As β increases above one so that the binding strengths of the R_1 and R_2 sites diverge, their separate binding behaviors become more apparent. At the extreme value of $\beta = 10^4$, the behavior of the R_1 site is clearly visible when $\log(C_L) < -2$, while the added behavior of the R_2 site is revealed when $\log(C_L) > -1$. Unlike noninterconverting binding sites, the behavior of interconverting R_1 and R_2 is linked. Thus, their overall binding curves have similar shapes at all β.

Almost all of the overall binding curves in Figure 5.4-8 resemble monovalent binding curves. Thus, from total bound ligand measurements alone, it can be difficult to judge whether allosteric binding is occurring.

5.5 Equilibrium Models of Blood Gas Content

5.5.1 Blood Chemistry

Erythrocytes or red blood cells (RBCs) contain hemoglobin, a metalloprotein that binds O_2. As illustrated in Example 4.2-1, the transport of physically dissolved O_2 in blood is insufficient to satisfy a person's metabolic O_2 demand. Rather, it is the oxyhemoglobin complex that is largely responsible for O_2 transport. Similarly, physically dissolved CO_2 plays a minor role in CO_2 transport. It is reversible hydration of CO_2 to form bicarbonate ion HCO_3^- that plays the major role. Figure 5.5-1 illustrates how O_2 and CO_2 in dissolved and chemically bound forms are distributed between RBC, plasma, and a gas phase that is adjacent to blood, as occurs in the lungs.

The total content of either O_2 or CO_2 in blood is the sum of its physically dissolved $\hat{C}_i^d$ and chemically bound $\hat{C}_i^b$ contributions:

$$\hat{C}_i = \hat{C}_i^d + \hat{C}_i^b \quad (i = O_2 \text{ or } CO_2) \tag{5.5-1}$$

A key variable for quantifying both $\hat{C}_i^d$ and $\hat{C}_i^b$ is the partial pressure p_i of species i that would exist in a gas phase that is equilibrated with blood (Table 5.5-1). The content of dissolved gas can be expressed as the product of p_i and a volume average Bunsen solubility α_i so that

$$\hat{C}_i = \alpha_i p_i + \hat{C}_i^b \tag{5.5-2}$$

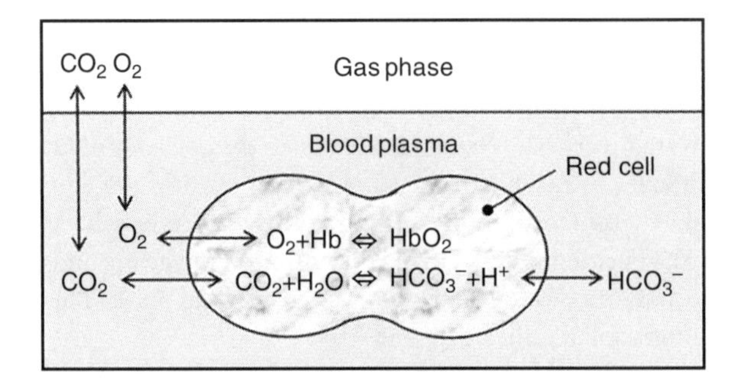

Figure 5.5-1 Distribution of O_2 and CO_2 between a gas phase and blood.

Table 5.5-1 Typical O_2 and CO_2 Parameters in Blood for a Resting Person Breathing Ambient Air

	p_{O_2} (kPa)	Percent O_2 Chemically Combined	p_{CO_2} (kPa)	Percent CO_2 Chemically Combined	pH
Venous	5.3	99	6.2	94	7.36
Arterial	12.7	98	5.3	94	7.40

The bound portion of O_2 content depends on p_{O_2} and to a lesser extent on p_{CO_2}, pH, and T. Likewise, bound CO_2 depends on p_{CO_2} and also varies somewhat with p_{O_2}, pH, and T. These relationships are governed by the following chemical reactions:

$$\begin{aligned}
&\text{Oxyhemoglobin formation:} && O_2 + Hb \rightleftharpoons HbO_2 \\
&\text{Bicarbonate formation:} && CO_2 + H_2O \overset{E}{\rightleftharpoons} H_2CO_3 \rightleftharpoons H^+ + HCO_3^- \\
&\text{Carbamino formation:} && CO_2 + Hb(NH_2) \rightleftharpoons Hb(NHCOO^-) + H^+ \\
&\text{Isohydric shift:} && H^+ + HbO_2 \rightleftharpoons H^+Hb + O_2 \\
&\text{Acid buffering:} && ca + H^+ \rightleftharpoons c^+ + a^- + H^+ \rightleftharpoons Ha + c^+
\end{aligned} \qquad (5.5\text{-3a-e})$$

The first three reactions represent the principal mechanisms by which O_2 and CO_2 are chemically bound in blood:

- A hemoglobin molecule contains four heme binding sites Hb, each with a ferrous iron atom linked to a central globin by a separate polypeptide chain. An oxyhemoglobin group HbO_2 is formed when a molecule of dissolved O_2 binds reversibly to a Hb site. Since the RBC membrane is impermeable to hemoglobin, both Hb and HbO_2 are confined to the RBC interior.
- Bicarbonate formation is a two-step process. Dissolved CO_2 is first hydrated to form carbonic acid H_2CO_3. This inherently slow reaction is enzymatically accelerated by carbonic anhydrase E. Formation of H_2CO_3 is followed by its rapid dissociation into bicarbonate HCO_3^- and H^+ ions. Bicarbonate formation largely occurs within the RBC where E is confined. Since HCO_3^- permeates through the cell membrane, it is able to reach the blood plasma.
- Carbon dioxide can also bind to the amide groups of proteins, forming carbamino groups $-NHCOO^-$. This reversible reaction occurs with hemoglobin inside the RBC and with extracellular albumin and plasma proteins.

The fourth reaction, the isohydric shift, has a secondary influence on both O_2 and CO_2 binding. The Bohr effect refers to a reduction in O_2 content in the RBC due to an increase in p_{CO_2}. According to the law of mass action, an elevation of p_{CO_2} shifts the bicarbonate as well as the carbamino formation reactions in the forward direction, thereby increasing H^+ concentration. This, in turn, enhances the isohydric shift, causing a reduction in the content of bound O_2. The Haldane effect is defined as a decrease in CO_2 content caused by an increase in p_{O_2}. When there is an increase in p_{O_2}, the isohydric shift occurs in reverse so that H^+ increases. By the law of mass action, an increased H^+ level means that the carbamino and bicarbonate formation reactions are also reversed and bound CO_2 content is thereby reduced.

The fifth reaction, acid buffering, occurs by the exchange of H^+ with sodium or potassium cations c^+ that are present in weak electrolytes ca. The anions a^- of these electrolytes are typically charged proteins or phosphate ions. The natural buffering capacity of blood restricts the magnitude of the Bohr and Haldane effects that are both modulated by H^+.

5.5.2 Oxygen Content

Oxygen Saturation

Consider a sample of whole blood in which ρ_{Hb_4} is the hemoglobin mass per unit volume (as typically reported). Since each hemoglobin molecule can bind four O_2 molecules, the molar binding capacity of the sample is $4(\rho_{Hb_4}/M_{Hb_4})$ where M_{Hb4} is the hemoglobin molecular weight. The corresponding binding capacity for O_2 on a volumetric basis is

$$\hat{C}^b_{O_2,max} = \left(\frac{4}{M_{Hb_4} c^o_G}\right) \rho_{Hb_4} \tag{5.5-4}$$

The most common variant of hemoglobin in adults is hemoglobin A, which has a molecular weight of 64,458 Da. In that case, the proportionality factor $(4/M_{Hb_4} c^o_G)$ is equal to 1.39 ml (STP)O_2 per gram of hemoglobin.

The extent of O_2 binding to hemoglobin is customarily expressed in terms of the O_2 saturation fraction that can be defined in terms of either O_2 content or molar O_2 concentration:

$$S_{O_2} \equiv \frac{\hat{C}^b_{O_2}}{\hat{C}^b_{O_2,max}} = \frac{C^b_{O_2}}{C^b_{O_2,max}} \tag{5.5-5}$$

Combining Equations 5.5-2 and 5.5-5 with this equation yields the following expression for total O_2 content:

$$\hat{C}_{O_2} = \alpha_{O_2} p_{O_2} + \hat{C}^b_{O_2,max} S_{O_2} \tag{5.5-6}$$

Hill Model

Because O_2 binding occurs on a time scale that is about 1% of the residence time of blood in a capillary, it is often reasonable to assume that S_{O_2} can be described by reaction equilibrium models. According to the Hill (1910) model, O_2 binds with clusters of n heme sites according to the following reaction stoichiometry:

$$Hb_n + nO_2 \rightleftharpoons (HbO_2)_n \tag{5.5-7}$$

The equilibrium binding constant for this reaction is given by

$$\kappa_c = \frac{C_{(HbO_2)n}}{C_{(Hb)n} C^n_{O_2}} \tag{5.5-8}$$

where $C_{(HbO_2)n}$, $C_{(Hb)n}$, and C_{O_2} are molar concentrations per unit volume of blood. Since the total number of hemoglobin binding sites is the sum of the unoccupied and the occupied sites, the saturation fraction is

$$S_{O_2} = \frac{C_{(HbO_2)n}}{C_{(Hb)n} + C_{(HbO_2)n}} \tag{5.5-9}$$

Eliminating $C_{(HbO_2)n}/C_{(Hb)n}$ between Equations 5.5-8 and 5.5-9, we get

$$S_{O_2} = \frac{\kappa_c C_{O_2}^n}{1 + \kappa_c C_{O_2}^n} = \frac{\left(\kappa_c^{1/n} C_{O_2}\right)^n}{1 + \left(\kappa_c^{1/n} C_{O_2}\right)^n} \tag{5.5-10}$$

By noting that $C_{O_2} = c_G^o \hat{C}_{O_2}^d = c_G^o \alpha_{O_2} p_{O_2}$ and defining $\kappa_p \equiv c_G^o \alpha_{O_2} \kappa_c^{1/n}$, we obtain an alternative formulation of S_{O_2}:

$$S_{O_2} = \frac{\left(\kappa_p p_{O_2}\right)^n}{1 + \left(\kappa_p p_{O_2}\right)^n} \tag{5.5-11}$$

Adair Model

The more realistic Adair model of O_2 saturation is based on the sequential oxidation of the four Hb sites on each hemoglobin molecule (Mountcastle, 1974; p 1403):

$$\begin{aligned}
&O_2 + Hb_4 \rightleftharpoons Hb_4O_2 &&; \ \kappa_{p2} \equiv \frac{C_{Hb_4O_4}}{C_{Hb_4O_2} p_{O_2}} \\
&O_2 + Hb_4O_2 \rightleftharpoons Hb_4O_4 &&; \ \kappa_{p2} \equiv \frac{C_{Hb_4O_4}}{C_{Hb_4O_2} p_{O_2}} \\
&O_2 + Hb_4O_4 \rightleftharpoons Hb_4O_6 &&; \ \kappa_{p3} \equiv \frac{C_{Hb_4O_6}}{C_{Hb_4O_4} p_{O_2}} \\
&O_2 + Hb_4O_6 \rightleftharpoons Hb_4O_8 &&; \ \kappa_{p4} \equiv \frac{C_{Hb_4O_8}}{C_{Hb_4O_6} p_{O_2}}
\end{aligned} \tag{5.5-12a-d}$$

For this model, fractional saturation of hemoglobin binding sites is given by

$$S_{O_2} = \frac{\kappa_{p1} p_{O_2} + 2\kappa_{p1}\kappa_{p2} p_{O_2}^2 + 3\kappa_{p1}\kappa_{p2}\kappa_{p3} p_{O_2}^3 + 4\kappa_{p1}\kappa_{p2}\kappa_{p3}\kappa_{p4} p_{O_2}^4}{4\left(1 + \kappa_{p1} p_{O_2} + \kappa_{p1}\kappa_{p2} p_{O_2}^2 + \kappa_{p1}\kappa_{p2}\kappa_{p3} p_{O_2}^3 + \kappa_{p1}\kappa_{p2}\kappa_{p3}\kappa_{p4} p_{O_2}^4\right)} \tag{5.5-13}$$

Magaria Model

A drawback of the Adair model is the need to specify four separate binding constants. Fortunately, experimental observations indicate that the first three Hb sites on a hemoglobin molecule bind O_2 with equal affinities so that $\kappa_{p1} = \kappa_{p2} = \kappa_{p3} \equiv \kappa_p$. Once the first three Hb sites have been oxidized, the affinity of the fourth Hb site for O_2 increases by two orders of magnitude such that $\kappa_{p4}/\kappa_p = 100$. This allows a simplification of the Adair model known as the Magaria model (Mountcastle, 1974; p 1403):

$$S_{O_2} = \frac{\left(1 + 1/\kappa_p p_{O_2}\right)^3 + 99}{\left(1 + 1/\kappa_p p_{O_2}\right)^4 + 99} \tag{5.5-14}$$

Example 5.5-1 Hill and Magaria Models Compared to O_2 Saturation Data

Estimate the parameters in the Hill and Magaria models that best fit the oxyhemoglobin dissociation data reported by Severinghaus (1966) for blood at $T = 37°C$ and an arterial $pH = 7.4$ (Fig. 5.5-2).

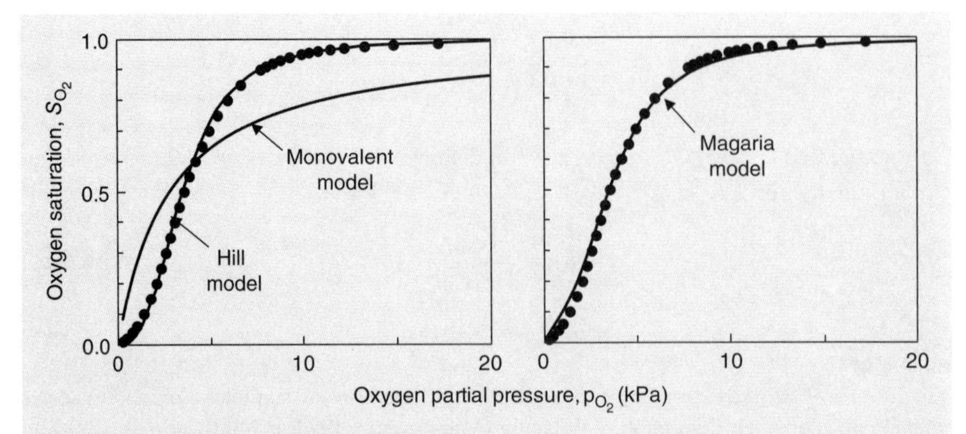

Figure 5.5-2 Oxyhemoglobin dissociation for human blood at 37°C and pH = 7.4.

Solution

Regressing the data points to Equation 5.5-14 by minimizing the sum of the squared error of the model-generated S_{O_2} values, we determine that $\kappa_p = 0.0725\ \text{kPa}^{-1}$ and the Magaria model becomes

$$S_{O_2} = \frac{\left(1 + 13.8/p_{O_2}[\text{kPa}]\right)^3 + 99}{\left(1 + 13.8/p_{O_2}[\text{kPa}]\right)^4 + 99} \tag{5.5-15}$$

A regression of the data to Equation 5.5-11 yields estimates of n = 2.8 and $\kappa_p = 0.283\ \text{kPa}^{-1}$ (equivalent to $\kappa_c = 1.23 \times 10^{-4}\ \mu\text{M}^{-2.8}$), and we can write the Hill model as

$$S_{O_2} = \frac{\left(0.283 p_{O_2}[\text{kPa}]\right)^{2.8}}{1 + \left(0.283 p_{O_2}[\text{kPa}]\right)^{2.8}} \tag{5.5-16}$$

The simplest form of the Hill model occurs when each of the four Hb binding sites on a hemoglobin molecules acts independently of one another such that n = 1. This is equivalent to approximating the allosteric binding of O_2 to the hemoglobin molecule as a monovalent process. A regression of this model to the data results in an estimate of $\kappa_p = 0.358\ \text{kPa}^{-1}$ indicating that

$$S_{O_2} = \frac{0.358 p_{O_2}[\text{kPa}]}{1 + 0.358 p_{O_2}[\text{kPa}]} \tag{5.5-17}$$

The regressed model equations are shown by the curves in Figure 5.5-2. We conclude that monovalent binding is not a reasonable model because it cannot produce the inflection point suggested by the data points. However, the Hill model with n = 2.8 and the Magaria model both fit the data very well. In most applications, it is convenient to use the Hill model because it has the simpler algebraic form.

Virtual Pressure Correction

The values for κ_p found in Example 5.5-1 apply to the p_{CO_2}, pH, and T at which the S_{O_2} data in Figure 5.5-2 were obtained. To extend the models to different conditions, we employ a virtual O_2 partial pressure (Kelman, 1966):

$$p_{O_2}^{virt} = \frac{\left[10^{-0.024(T-37°C)}\right]}{\left[10^{0.4(7.4-pH)}\left(p_{CO_2}/5.3\ kPa\right)^{0.06}\right]} p_{O_2} \tag{5.5-18}$$

This equation reduces to $p_{O_2}^{virt} = p_{O_2}$ at conditions typical of arterial blood: $p_{CO_2} = 5.3$ kPa, pH = 7.4, and T = 37°C. The Hill and Magaria models can now be given in more general forms as

$$S_{O_2}\left(p_{O_2}, p_{CO_2}, pH, T\right) = \frac{\left(0.283 p_{O_2}^{virt}[kPa]\right)^{2.8}}{1 + \left(0.283 p_{O_2}^{virt}[kPa]\right)^{2.8}} \tag{5.5-19}$$

$$S_{O_2}\left(p_{O_2}, p_{CO_2}, pH, T\right) = \frac{\left(1 + 13.8/p_{O_2}^{virt}[kPa]\right)^{3} + 99}{\left(1 + 13.8/p_{O_2}^{virt}[kPa]\right)^{4} + 99} \tag{5.5-20}$$

When computing S_{O_2} in this manner, it is the $\left[10^{0.4(7.4-pH)}\left(p_{CO_2}/5.3\ kPa\right)^{0.06}\right]$ factor appearing in $p_{O_2}^{virt}$ that gives rise to the Bohr effect.

Example 5.5-2 Effect of p_{CO_2} and pH on the Oxyhemoglobin Dissociation Curve

Determine the effect of p_{CO_2} and pH on the oxyhemoglobin dissociation curve using the Hill model with the Bohr correction. Compare the model predictions to the data points (Altman and Dittmer, 1971) shown on the right graph of Figure 5.5-3. These data were obtained at body temperature.

Solution

We compute S_{O_2} from Equation 5.5-19 with $p_{O_2}^{virt}$ determined from Equation 5.5-18 for various values of p_{CO_2} and pH when T = 37°C. Consistent with the Bohr effect, the resulting oxyhemoglobin saturation curves are shifted to the right as p_{CO_2} increases at a fixed pH (Fig. 5.5-3; left graph). Increasing the pH at a fixed p_{CO_2} has the opposite effect, a result that closely matches the data points (Fig. 5.5-3; right graph).

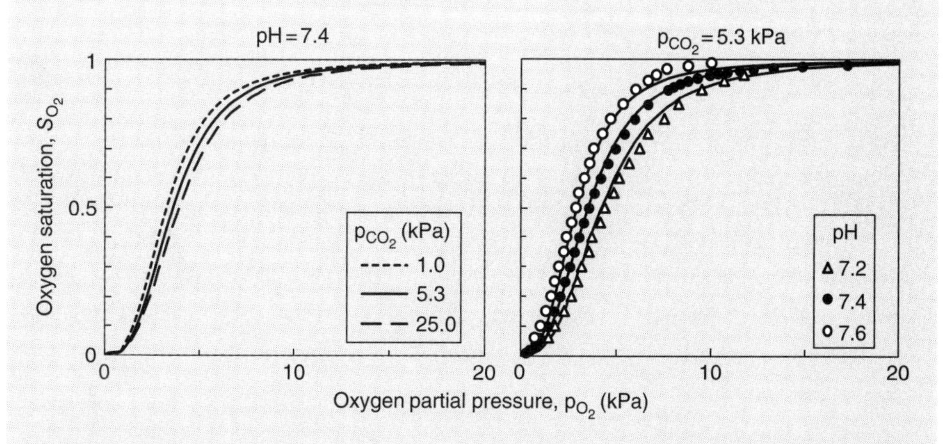

Figure 5.5-3 Oxyhemoglobin dissociation predicted by the Hill model including the Bohr effect (Data from Altman and Dittmer (1971)).

Example 5.5-3 Oxygen Uptake in the Pulmonary Circulation

Using the Hill model, compute the equilibrium content of O_2 in the venous and arterial blood of a resting person. Assuming that reaction equilibrium applies locally at the inlet and the outlet of the pulmonary circulation, compute oxygen uptake $\dot{V}_{O_2}$ through the capillary walls at a typical blood flow of Q = 5 L/min = 50 dL/min. How does this compare to the $\dot{V}_{O_2}$ that would occur if no hemoglobin was present in blood? How important is the Bohr effect in aiding blood oxygenation by the lungs?

The Bunsen solubility of O_2 in blood at a body temperature of T = 37°C and a typical hemoglobin concentration of $\rho_{Hb_4} = 15$ g/dL is $\alpha_{O_2} = 0.0219$ ml(STP)/(dL-kPa) (Table A3-3). Typical values of p_{O_2}, p_{CO_2}, and pH in venous and arterial blood of a resting adult appear in Table 5.5-1.

Solution

The capacity of blood for bound O_2 determined with Equation 5.5-4 is

$$\hat{C}^b_{O_2,max} = 1.39(15) = 20.9 \text{ ml(STP)}O_2/\text{dL blood} \tag{5.5-21}$$

The computations required to obtain O_2 content in venous blood using Equation 5.5-2 are

$$p^{virt}_{O_2,v} = 5.3\left[10^{-0.024(37-37)}\right]/\left[10^{0.4(7.40-7.36)}(6.2/5.3)^{0.06}\right] = 5.06$$

$$S_{O_2,v} = [(0.283)(5.06)]^{2.8}/\left\{1 + [(0.283)(5.06)]^{2.8}\right\} = 0.732$$

$$\alpha_{O_2} p_{O_2,v} = (0.0219)(5.30) = 0.1 \text{ ml(STP)/dL} \tag{5.5-22a-e}$$

$$\hat{C}^b_{O_2,max} S_{O_2,v} = (20.9)(0.732) = 15.3 \text{ ml(STP)/dL}$$

$$\hat{C}_{O_2,v} \equiv 15.3 + 0.1 = 15.4 \text{ ml(STP)/dL}$$

A similar set of computations for arterial blood indicate that $p^{virt}_{O_2,a} = 12.7$ kPa, $S_{O_2,a} = 0.973$, $\alpha_{O_2} p_{O_2,a} = 0.3$ ml(STP)/dL, $\hat{C}^b_{O_2,max} S_{O_2,a} = 20.3$ ml(STP)/dL, and $\hat{C}_{O_2,a} = 20.6$ ml(STP)/dL.

To determine $\dot{V}_{O_2}$, we use the O_2 molar balance that we previously derived for the pulmonary circulation (Eq. 4.2-16):

$$\dot{V}_{O_2} = Q\left(\hat{C}_{O_2,a} - \hat{C}_{O_2,v}\right) = 50(20.6 - 15.4) = 260 \text{ ml(STP)/min} \tag{5.5-23}$$

This $\dot{V}_{O_2}$ falls within the normal physiological range for adults during resting conditions. Without hemoglobin in blood, O_2 would be transported as dissolved O_2 alone, and $\dot{V}_{O_2}$ would only be 8 ml(STP)/min (Example 4.2-1). Thus, about 97% of respiratory O_2 uptake depends on oxyhemoglobin binding.

We can determine O_2 uptake in the absence of the Bohr effect by computing $p^{virt}_{O_2}$ as if $\left[10^{0.4(7.4-pH)}(p_{CO_2}/5.3 \text{ kPa})^{0.06}\right]$ is equal to one in venous as well as arterial blood. With this condition, $\hat{C}_{O_2,v} = 15.9$ ml(STP)/dL, the value of $\hat{C}_{O_2,a}$ is unchanged, and the O_2 uptake is

$$\dot{V}_{O_2} = 50(20.6 - 15.9) = 235 \text{ ml(STP)/min} \tag{5.5-24}$$

Comparing this result to Equation 5.5-23, we see that the Bohr effect enhances pulmonary blood oxygenation by about 10%.

5.5.3 Carbon Dioxide Content

In this section, we determine how the simultaneous reactions involving O_2 and CO_2 (Eq. 5.5-3a-e) affect the CO_2 content of blood. Ultimately, we want to analyze changes in both O_2 and CO_2 content as blood flows through a capillary bed or a man-made device in which gas exchange occurs. As a reasonable approximation, we assume that chemical reactions are sufficiently fast relative to transport rates that the dissolved and bound portions of these gases are in local reaction equilibrium.

Fixed CO_2 Content

We first consider CO_2 content in an isolated blood sample or a fixed point in the circulatory system. Of the total CO_2 content in blood, approximately 5% is physically dissolved, 90% is reversibly bound as HCO_3^-, and 5% is reversibly bound to proteins as carbamino compounds. Since most carbamino formation occurs with hemoglobin inside of RBCs and relatively little occurs with plasma proteins, dissolved CO_2 and bicarbonate ions account for virtually all of the CO_2 content in blood plasma:

$$\widehat{C}_{CO_2}^{plasma} = \alpha_{CO_2}^{plasma} p_{CO_2} + \widehat{C}_{HCO_3^-}^{plasma} \tag{5.5-25}$$

We can relate the molar concentrations of the species that participate in the bicarbonate reaction (Eq. 5.5-3b) from its equilibrium constant:

$$\kappa_C = \frac{C_{H^+}^{plasma} C_{HCO_3^-}^{plasma}}{C_{CO_2}^{plasma} C_{H_2O}^{plasma}} \tag{5.5-26}$$

With $C_{CO_2}^{plasma} = c_G^o \left(\alpha_{CO_2}^{plasma} p_{CO_2} \right)$, $pH = -\log C_{H^+}^{plasma}$, and $pK_{CO_2} \equiv -\log(\kappa_c C_{H_2O})$, we solve Equation 5.5-26 for the plasma bicarbonate content:

$$\hat{C}_{HCO_3^-}^{plasma} = \frac{\hat{C}_{HCO_3^-}^{plasma}}{c_G^o} = \alpha_{CO_2}^{plasma} p_{CO_2} 10^{pH - pK_{CO_2}} \tag{5.5-27}$$

Combining this result with Equation 5.5-25, we get

$$\widehat{C}_{CO_2}^{plasma} = \alpha_{CO_2}^{plasma} \left(1 + 10^{pH - pK_{CO_2}}\right) p_{CO_2} \tag{5.5-28}$$

The ratio of the CO_2 content in whole blood to that of plasma is represented by following empirical function (McHardy, 1967; Eq. 2):

$$\frac{\widehat{C}_{CO_2}}{\widehat{C}_{CO_2}^{plasma}} = 1 - \frac{0.0215 \hat{C}_{O_2,max}^{b} [ml(STP)/dL]}{(2.244 - 0.422 S_{O_2})(8.740 - pH)} \tag{5.5-29}$$

For a hemoglobin binding capacity of $\hat{C}_{O_2,max}^{b} = 1.39 \rho_{Hb_4}$, the combination of Equations 5.5-28 and 5.5-29 yields the CO_2 content in whole blood:

$$\widehat{C}_{CO_2} = \left\{ 1 - \frac{0.0299 \rho_{Hb_4} [g/dL]}{(2.244 - 0.422 S_{O_2})(8.740 - pH)} \right\} \alpha_{CO_2}^{plasma} \left(1 + 10^{pH - pK_{CO_2}}\right) p_{CO_2} \tag{5.5-30}$$

Thus, a blood sample will have a CO_2 content that depends on the constant parameters ρ_{Hb_4}, $\alpha_{CO_2}^{plasma}$, and pK_{CO_2} and the variables S_{O_2}, pH, and p_{CO_2}.

Example 5.5-4 CO_2 Excretion from the Pulmonary Circulation

Compute the arterial and venous CO_2 contents in blood with a hemoglobin content of ρ_{Hb_4} = 15 gm/dL. Also determine the CO_2 excretion rate from the lungs at a pulmonary blood flow of Q = 5 L/min. Use the O_2 and CO_2 partial pressures in Table 5.5-1 and the hemoglobin saturation values of $S_{O_2,v} = 0.732$ and $S_{O_2,a} = 0.973$ found in Example 5.5-3. At a body temperature of 37°C, $pK_{CO_2} = 6.1$, and $\alpha_{CO_2}^{plasma} = 0.508$ ml(STP)/(dL-kPa).

Solution

Employing Equation 5.5-29, we find that

$$\left(\frac{\hat{C}_{CO_2,v}}{\hat{C}_{CO_2,v}^{plasma}}\right) = 1 - \frac{0.0299(15)}{[2.244 - 0.422(0.732)](8.74 - 7.36)} = 0.832 \quad (5.5\text{-}31)$$

$$\left(\frac{\hat{C}_{CO_2,a}}{\hat{C}_{CO_2,a}^{plasma}}\right) = 1 - \frac{0.0299(15)}{[2.244 - 0.422(0.973)](8.74 - 7.40)} = 0.817 \quad (5.5\text{-}32)$$

By multiplying these ratios by $\hat{C}_{CO_2,a}^{plasma}$ (Eq. 5.5-28), we arrive at CO_2 content in whole blood:

$$\hat{C}_{CO_2,v} = 0.832(0.508)\left(1 + 10^{7.36-6.1}\right)(6.2) = 50.3 \text{ ml(STP)/dL} \quad (5.5\text{-}33)$$

$$\hat{C}_{CO_2,a} = 0.817(0.508)\left(1 + 10^{7.40-6.1}\right)(5.3) = 46.1 \text{ ml(STP)/dL} \quad (5.5\text{-}34)$$

To calculate the steady-state CO_2 excretion rate $\dot{V}_{CO_2}$, we apply the material balance given by Equation 5.5-23 to CO_2:

$$\dot{V}_{CO_2} = Q\left(\hat{C}_{CO_2,v} - \hat{C}_{CO_2,a}\right) = 50(50.3 - 46.1) = 210 \text{ ml(STP)/dL} \quad (5.5\text{-}35)$$

Notice that the CO_2 contents in arterial and venous blood are two to three times larger than the corresponding O_2 contents found in Example 5.5-3. Even so, the arterial-to-venous difference in content of the two gases is similar so that their molar exchange ratio is close to one.

Changing CO_2 Content

We now analyze the change of CO_2 content in blood that accompanies an addition or loss of O_2 and CO_2 relative to some reference condition. For example, consider a blood element transversing a muscle capillary with the entering arterial blood taken as the reference condition. Oxygen is transported out of the capillary blood, while CO_2 is absorbed into the blood. This results in an increase in CO_2 content caused by two factors: a forward shift in the bicarbonate and carbamino formation reactions following an increase in p_{CO_2} and a reverse Haldane effect associated with a decrease in S_{O_2}. For small changes of p_{CO_2} and S_{O_2}, we approximate CO_2 content as a linear function of these variables:

$$\hat{C}_{CO_2} = \beta_0 + \beta_1 p_{CO_2} + \beta_2 S_{O_2} \quad (5.5\text{-}36)$$

Selecting a reference set of conditions for p_{CO_2} and S_{O_2}, we can also write that

$$\hat{C}_{CO_2,ref} = \beta_0 + \beta_1 p_{CO_2,ref} + \beta_2 S_{O_2,ref} \quad (5.5\text{-}37)$$

Taking the difference between the last two equations, we get

$$\hat{C}_{CO_2} = \hat{C}_{CO_2,ref} + \beta_1\left(p_{CO_2} - p_{CO_2,ref}\right) + \beta_2\left(S_{O_2} - S_{O_2,ref}\right) \quad (5.5\text{-}38)$$

We evaluate β_1 using an empirical equation from McHardy (1967; p 300):

$$\beta_1[\text{ml(STP)CO}_2/(\text{dL-kPa})] = 0.0835\hat{C}^{b}_{O_2,max}[\text{ml(STP)O}_2/\text{dL}] + 1.58 \quad (5.5\text{-}39)$$

and we deduce the Haldane factor β_2 from Loeppky and coworkers (1983; p 173):

$$\beta_2[\text{ml(STP)CO}_2/\text{dL}] = -0.257\hat{C}^{b}_{O_2,max}[\text{ml(STP)O}_2/\text{dL}] \quad (5.5\text{-}40)$$

With $\hat{C}^{b}_{O_2,max} = 1.39\rho_{Hb_4}$, Equation 5.5-38 can now be written as

$$\widehat{C}_{CO_2} = \widehat{C}_{CO_2,ref} + (0.116\rho_{Hb_4} + 1.58)(p_{CO_2} - p_{CO_2,ref}) - 0.357\rho_{Hb_4}(S_{O_2} - S_{O_2,ref}) \quad (5.5\text{-}41)$$

This equation requires the following units: $\widehat{C}_{CO_2}$[ml(STP)/dL], ρ_{Hb_4}[g/dL], and p_{CO_2}[kPa]. Since Equations 5.5-30 and 5.5-41 must both lead to the same value of $\widehat{C}_{CO_2}$, they can be equated to find pH as a function of p_{CO_2} and S_{O_2}, once values for $\widehat{C}_{CO_2,ref}$, $p_{CO_2,ref}$, and $S_{O_2,ref}$ are provided.

Example 5.5-5 Validation of the Capillary Equilibrium Model

In this example, we examine the ability of Equation 5.5-41 to reproduce the $\widehat{C}_{CO_2}$ versus p_{CO_2} data points shown in the left graph of Figure 5.5-4 (Comroe *et al.*, 1962). These data, collected at a fixed S_{O_2} of either 0.7 or 0.975, were obtained by varying p_{O_2} as well as p_{CO_2} in a blood sample. We will also determine the effect of p_{CO_2} on pH of the blood sample.

Solution

Selecting the triangular data point as a reference condition ($p_{CO_2,ref} = 5.33$ kPa, $S_{O_2,ref} = 0.975$, $\widehat{C}_{CO_2,ref} = 48.4$ ml(STP)/dL), Equation 5.5-41 becomes

$$\widehat{C}_{CO_2}[\text{ml(STP)/dL}] = 48.4 + (0.116\rho_{Hb}[\text{g/dL}] + 1.58)(p_{CO_2}[\text{kPa}]\text{-}5.33) - 0.357\rho_{Hb}(S_{O_2} - 0.975) \quad (5.5\text{-}42)$$

We fit this equation to the $\widehat{C}_{CO_2}(p_{CO_2})$ data points by using ρ_{Hb_4} (not given for these measurements) as an adjustable parameter. By minimizing the summed squared error, we arrive at an estimate of $\rho_{Hb} = 14.5$ g/dL. Substituting this value back into Equation 5.5-42 produces the lines shown in the left graph of Figure 5.5-4 for the two different oxyhemoglobin saturations.

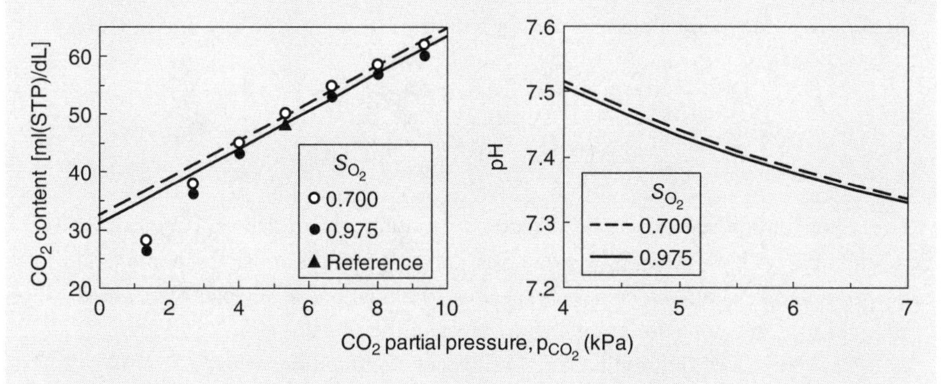

Figure 5.5-4 CO_2 dissociation curves and associated pH changes (Data from Comroe *et al.* (1962)).

These lines follow the data quite well when p_{CO_2} is in a typical arterial–venous range of 5.3–6.2 kPa characteristic of resting conditions. Our use of a linear model is a satisfactory approximation even at higher p_{CO_2} levels that occur during exercise or because of pulmonary disease.

To simulate pH behavior, we substitute $\rho_{Hb} = 14.5$ g/dL, $pK_{CO_2} = 7.1$ and $\alpha_{CO_2}^{plasma} = 0.508$ ml(STP)/(dL-kPa) into Equation 5.5-30:

$$\widehat{C}_{CO_2} = \left[1 - \frac{0.0299(14.5)}{(2.244 - 0.422S_{O_2})(8.740 - pH)}\right](0.508)\left(1 + 10^{pH-6.1}\right)p_{CO_2} \tag{5.5-43}$$

We then solve this equation for pH using various pairs of p_{CO_2} and $\widehat{C}_{CO_2}$ values that satisfy Equation 5.5-42 when S_{O_2} is either 0.975 or 0.700. This procedure results in the two curves in the right graph of the figure. As we expect, pH at a fixed S_{O_2} decreases as p_{CO_2} increases because of the elevation in hydrogen ions that accompanies increased bicarbonate formation. We also see that pH at a fixed p_{CO_2} decreases as S_{O_2} increases.

Simultaneous Changes in O_2 and CO_2 Contents

To model O_2 and CO_2 transport in blood flowing through a capillary, we must simultaneously predict $\widehat{C}_{O_2}$ and $\widehat{C}_{CO_2}$ from the reaction equilibria associated with local p_{O_2} and p_{CO_2} conditions. To illustrate an approach to this problem, we use typical conditions in arterial blood ($p_{O_2,a} = 12.7$ kPa, $p_{CO_2,a} = 5.3$ kPa, $S_{O_2} = 0.973$) as reference values in Equation 5.5-41 to provide one prediction of CO_2 content:

$$\widehat{C}_{CO_2} = 46.1 + \left(0.116\rho_{Hb_4} + 1.58\right)\left(p_{CO_2} - 5.3\right) - 0.357\rho_{Hb_4}\left(S_{O_2} - 0.973\right) \tag{5.5-44}$$

After setting $pK_{CO_2} = 6.1$ and $\alpha_{CO_2}^{plasma} = 0.508$ ml(STP)/(dL-kPa), we obtain another prediction of CO_2 content from Equation 5.5-30:

$$\widehat{C}_{CO_2} = \left[1 - \frac{0.0299\rho_{Hb_4}}{(2.244 - 0.422S_{O_2})(8.740 - pH)}\right](0.503)\left(1 + 10^{pH-6.1}\right)p_{CO_2} \tag{5.5-45}$$

To formulate O_2 content, we use Equation 5.5-6 with $\alpha_{O_2} = 0.0219$ ml/(dL-kPa):

$$\widehat{C}_{O_2} = 0.0219p_{O_2} + 1.39\rho_{Hb_4}S_{O_2} \tag{5.5-46}$$

Finally, we model hemoglobin saturation with the Hill equation (Eqs. 5.5-18 and 5.5-19):

$$S_{O_2} = \left[\frac{\left(0.313 \times 10^{0.4(pH-7.4)}p_{O_2}/p_{CO_2}^{0.06}\right)^{2.8}}{1 + \left(0.313 \times 10^{0.4(pH-7.4)}p_{O_2}/p_{CO_2}^{0.06}\right)^{2.8}}\right] \tag{5.5-47}$$

These four nonlinear algebraic equations contain six variables ($\widehat{C}_{O_2}$[ml(STP)/dL], $\widehat{C}_{CO_2}$[ml(STP)/dL], p_{O_2}[kPa], p_{CO_2}[kPa], S_{O_2}, and pH) in addition to the hemoglobin mass concentration ρ_{Hb_4}[g/dL]. Once ρ_{Hb_4} is specified, we can choose any pair of arbitrary values of p_{O_2} and p_{CO_2} and solve for equilibrium values of the remaining four variables.

Figure 5.5-5 presents the results of several sets of computations when $\rho_{Hb_4} = 15$ g/dL. The two graphs show the $\widehat{C}_{O_2}$ and $\widehat{C}_{CO_2}$ values that would occur in a blood sample as a result of

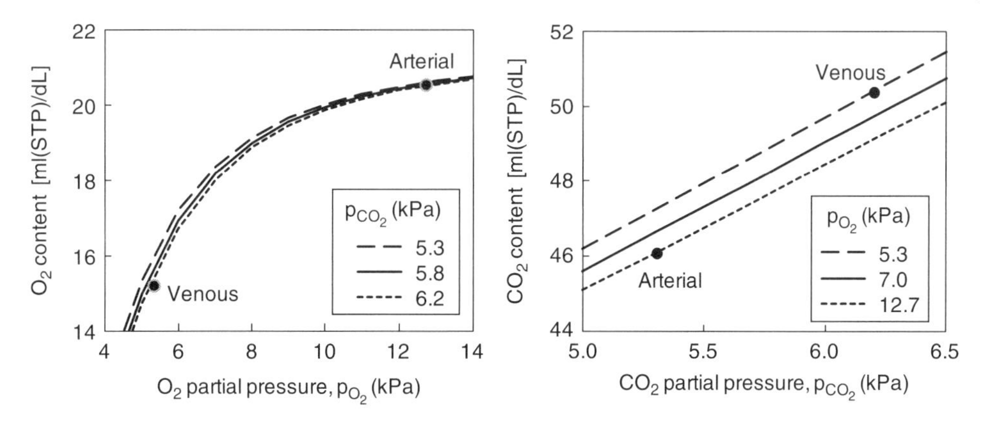

Figure 5.5-5 O_2 and CO_2 changes along a pulmonary capillary.

various combinations of p_{O_2} and p_{CO_2} values. The Bohr effect is indicated in the left graph by the small downward shift of the $\hat{C}_{O_2}$ versus p_{O_2} curves occurring when p_{CO_2} increases. A much larger Haldane effect is seen in the right graph by the downward shift of the $\hat{C}_{CO_2}$ versus p_{CO_2} lines when p_{O_2} increases.

Specific points have been placed in this figure to indicate p_{O_2} and p_{CO_2} conditions typically present at the venous and arterial (i.e., reference) ends of a capillary. The composition of a blood element as it moves through a capillary should follow some trajectory between this pair of points. To determine this trajectory, we would need to know the rates of O_2 and CO_2 transfer across the capillary walls.

References

Altman PL, Dittmer DS. Respiration and Circulation. Bethesda: Federation of American Societies for Experimental Biology; 1971, p 197.

Bowling AC, DeLorenzo RJ. Micromolar affinity benzodiazepine receptors: identification and characterization in central nervous system. Science. 1982; 216:1247–1250.

Comroe JH, Forster RE, Dubois AB, Briscoe WA, Carsen E. The Lung-Clinical Physiology and Pulmonary Function Tests. Chicago: Year Book Medical Publishers; 1962, p 154.

Gary-Bobo CM, Solomon AK. Properties of hemoglobin solutions in RBCs. J Gen Physiol. 1968; 52:825–853.

Hill AV. The possible effects of the aggregation of the molecules of haemoglobin on its dissociation curves. Proc Physiol Soc. 1910; 40:4–7.

Kelman GR. Digital computer subroutine for the conversion of oxygen tension into saturation. J Appl Physiol. 1966; 21:1375–1376.

Loeppky JA, Luft UC, Fletcher ER. Quantitative description of whole blood CO_2 dissociation curve and Haldane effect. Respir Physiol. 1983; 51:167–181.

McHardy GJ. The relationship between the differences in pressure and content of carbon dioxide in arterial and venous blood. Clin Sci. 1967; 32:299–309.

Mountcastle VB. Medical Physiology. 13th ed. St Louis: Mosby; 1974, Vol. 2.

Salinas C, Muzic RF, Jr, Berridge M, Ernsberger P. PET imaging of myocardial β-adrenergic receptors with fluorocarazolol. J Cardiovasc Pharmacol. 2005; 46:222–231.

Severinghaus JW. Blood gas calculator. J Appl Physiol. 1966; 21:1108–1116.

Part III
Fundamentals of Rate Processes

Chapter 6

Nonequilibrium Thermodynamics and Transport Rates

Whereas classical thermodynamics can only predict the direction of spontaneous transport and reaction processes, nonequilibrium thermodynamics provides a framework for developing rate equations for such processes. Nonequilibrium thermodynamics presumes that the material points of a continuum are each in a different equilibrium state. Relationships between the fluxes and driving forces of irreversible processes occurring between material points can then be deduced from local entropy generation rates (Appendix B2). As is also true for classical thermodynamics, nonequilibrium thermodynamics does not elucidate the mechanisms by which processes occur. Rather, it produces phenomenological equations that must be validated by experimental observation.

In this chapter, we will demonstrate how the formalisms of nonequilibrium thermodynamics are used to derive rate equations for binary diffusion, multicomponent diffusion, and membrane transport. But first, we will define the variables necessary to specify transport rates.

6.1 Transport Velocities and Fluxes

6.1.1 Molar and Mass Average Velocity

As discussed in Chapter 2, properties at a material point of a continuum are determined by averaging the underlying molecular contributions. Consider a fluid of I different chemical species. By counting the molecules of a particular chemical species i at a material point of known volume, we arrive at a local molar concentration C_i. Similarly by averaging the velocities of these molecules, we obtain the local velocity $\mathbf{u}_i$ of species i.

Biomedical Mass Transport and Chemical Reaction: Physicochemical Principles and Mathematical Modeling,
First Edition. James S. Ultman, Harihara Baskaran, and Gerald M. Saidel.

We define fluid velocity as the sum of $\mathbf{u_i}$, each weighted by the fraction of a solution occupied by species i. Weighting by mole fraction $x_i = C_i/c$ leads to the molar average velocity:

$$\mathbf{u}^* \equiv \sum_{i=1}^{I} x_i \mathbf{u_i} = \frac{1}{c}\sum_{i=1}^{I} C_i \mathbf{u_i} \tag{6.1-1}$$

whereas weighting by mass fraction $\omega_i = \rho_i/\rho$ provides a mass average velocity:

$$\mathbf{u} \equiv \sum_{i=1}^{I} \omega_i \mathbf{u_i} = \frac{1}{\rho}\sum_{i=1}^{I} \rho_i \mathbf{u_i} \tag{6.1-2}$$

In general, the numerical value of an average fluid velocity depends on the choice of the weighting factor. Under some circumstances, however, different types of average velocities will have similar values. Consider the difference between molar and mass average velocity:

$$\mathbf{u}^* - \mathbf{u} = \sum_{i=1}^{I} (x_i - \omega_i)\mathbf{u_i} \tag{6.1-3}$$

For a species i with a molecular weight M_i, its mass per mole of solution is $x_i M_i$, and its mass fraction is

$$\omega_i = M_i x_i / M \tag{6.1-4}$$

where M is the mean molecular weight of the solution. Therefore, the velocity difference in Equation 6.1-3 is equivalent to

$$\mathbf{u}^* - \mathbf{u} = \sum_{i=1}^{I} \left(1 - \frac{M_i}{M}\right) x_i \mathbf{u_i} \tag{6.1-5}$$

The relation between mean molecular weight and species mole fractions is $M = \sum_{i=1}^{I} M_i x_i$ (Eq. 2.1-2). When dealing with a dilute aqueous solution, all solutes (i = 1,2,...I − 1) have a mole fractions that are very small compared to water (i = I) so that $x_i \approx 0$ (i = 1,2,...I − 1), $x_I \approx 1$ and $M \approx M_I$. In that situation, Equation 6.1-5 indicates that there is little difference between mass average and mole average velocity. Similarly, the dominant components of dry air, N_2 and O_2, have comparable molecular weights, so that $M \approx M_{N_2} \approx M_{O_2}$ and there is only a small difference between mass average and mole average velocity.

6.1.2 Convective Flux

The product $C_i\mathbf{u_i}$[mol/(s-m^2)] corresponds to the molar flux at which a species i crosses a stationary plane that is perpendicular to the direction of $\mathbf{u_i}$:

$$\mathbf{N_i} \equiv C_i \mathbf{u_i} = c x_i \mathbf{u_i} \tag{6.1-6}$$

Similarly, the product $\rho_i\mathbf{u_i}$[kg/(s-m^2)] corresponds to the mass flux at which a substance i crosses a stationary plane that is perpendicular to the direction of $\mathbf{u_i}$:

$$\mathbf{n_i} \equiv \rho_i \mathbf{u_i} = \rho \omega_i \mathbf{u_i} \tag{6.1-7}$$

Transport occurs by pure convection when all I substances in a fluid have the same species velocity, that is, $\mathbf{u}_1 = \mathbf{u}_2 = \ldots = \mathbf{u}_I$. According to Equations 6.1-1 and 6.1-2, this means that $\mathbf{u}_i = \mathbf{u} = \mathbf{u}^*$ for all $i = 1,2,\ldots I$. Thus, for transport by convection, the fluxes of species i are

$$[\mathbf{N}_i]_{conv} = C_i \mathbf{u}^*, \quad [\mathbf{n}_i]_{conv} = \rho_i \mathbf{u} \tag{6.1-8a,b}$$

6.1.3 Diffusive Flux

Diffusion is the movement of a species i relative to an average movement of all species.

In other words, when diffusion of a substance i occurs, its flux is different than its flux due to pure convection. We define the molar diffusion flux of substance i relative to its molar convective flux as

$$\mathbf{J}_i \equiv \mathbf{N}_i - [\mathbf{N}_i]_{conv} = C_i(\mathbf{u}_i - \mathbf{u}^*) \tag{6.1-9}$$

The mass diffusion flux relative to the mass convective flux is defined in an analogous fashion:

$$\mathbf{j}_i \equiv \mathbf{n}_i - [\mathbf{n}_i]_{conv} = \rho_i(\mathbf{u}_i - \mathbf{u}) \tag{6.1-10}$$

Here, the relative velocities, $(\mathbf{u}_i - \mathbf{u}^*)$ and $(\mathbf{u}_i - \mathbf{u})$, are alternative ways of expressing a diffusion velocity of component i. The diffusive fluxes of the components in a solution are not all independent since, by summing Equations 6.1-9 or 6.1-10 over all I substances, we obtain

$$\sum_{i=1}^{I} \mathbf{J}_i = \sum_{i=1}^{I} \mathbf{u}_i C_i - \mathbf{u}^* \sum_{i=1}^{I} C_i = 0 \tag{6.1-11}$$

$$\sum_{i=1}^{I} \mathbf{j}_i = \sum_{i=1}^{I} \mathbf{u}_i \rho_i - \mathbf{u} \sum_{i=1}^{I} \rho_i = 0 \tag{6.1-12}$$

The combination of Equations 6.1-1, 6.1-8a, and 6.1-9 or Equations 6.1-2, 6.1-8b, and 6.1-10 reveals that

$$\mathbf{N}_i = \mathbf{u}^* C_i + \mathbf{J}_i$$
$$\mathbf{n}_i = \mathbf{u}\rho_i + \mathbf{j}_i \tag{6.1-13a,b}$$

Since $\mathbf{u}^* = \sum_i C_i \mathbf{u}_i^* / c = \sum_i \mathbf{N}_i / c$ and $\mathbf{u} = \sum_i \rho_i \mathbf{u}_i / \rho = \sum_i \mathbf{n}_i / \rho$, these equations can also be written as

$$\mathbf{N}_i = x_i \sum_{j=1}^{I} \mathbf{N}_j + \mathbf{J}_i$$
$$\mathbf{n}_i = \omega_i \sum_{j=1}^{I} \mathbf{n}_j + \mathbf{j}_i \tag{6.1-14a,b}$$

Example 6.1-1 Convection and Diffusion in a Stefan Tube

A Stefan tube is an apparatus for measuring the diffusion coefficient in a vapor–gas mixture (Fig. 6.1-1). For example, to measure the diffusion coefficient of water vapor in air, pure liquid

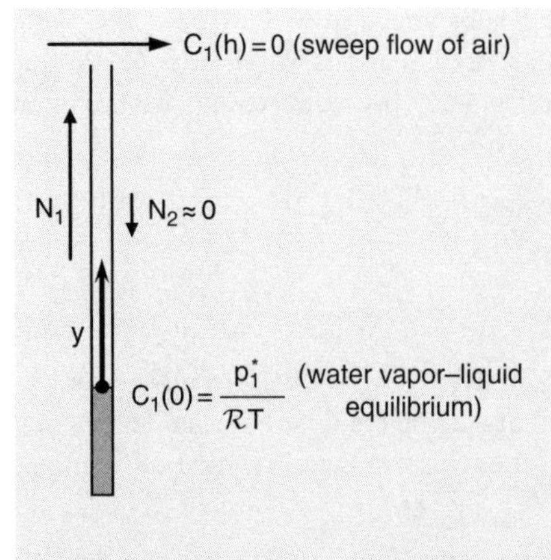

Figure 6.1-1 Vapor diffusion through a Stefan tube.

water is placed at the bottom of a vertical tube, and dry air is blown across the top of the tube. Water vapor formed by evaporation from the gas–liquid interface at $y = 0$ is transported up the gas column and out of the tube at $y = h$. The evaporation rate is determined from the height of the slowly receding liquid level.

Suppose that water is evaporating at a pseudo-steady rate of 42.4 nmol/s in a 5 mm-diameter Stefan tube containing a $h = 0.2$ m long air column. Determine the convective, diffusive, and overall molar fluxes of water vapor and air at the gas–liquid interface. Is there a substantial difference between the molar and mass average gas phase velocities at the interface? At the operating conditions of $T = 70°C$ and $P = 101.3$ kPa, the molar density an ideal gas is $c_G = 35.5\ \text{mol/m}^3$, and the equilibrium vapor pressure of water is $p_1^* = 31.2\text{kPa}$.

Solution

Dry air is a mixture of about 78 mol%N_2 and 21 mol%O_2. The remaining 1 mol% consists of small amounts of several inert gases. Because N_2(28Da) and O_2(36Da) have similar molecular weights, we will treat air as a single-component gas. Then, the gas in the Stephan tube can be considered to be a binary mixture of water vapor (component 1) and air (component 2).

For a cylindrical Stefan tube with a 5 mm diameter, the cross-sectional area available for axial transport is $1.96 \times 10^{-5}\ \text{m}^2$, and the molar flux of water vapor from the gas-liquid interface is given by

$$N_1 = \frac{42.4 \times 10^{-9}}{1.96 \times 10^{-5}} = 0.00216\ \text{mol/s-m}^2 \tag{6.1-15}$$

Because air has such a low solubility in liquid water, the flux of air at the gas–liquid interface can be neglected:

$$N_2 = 0 \tag{6.1-16}$$

If the interface is receding very slowly, then N_1 and N_2 are essentially fluxes relative to fixed coordinates, and the molar velocity can be computed by applying Equations 6.1-1 and 6.1-6.

$$u^* = \frac{C_1 u_1 + C_2 u_2}{C_G} = \frac{N_1 + N_2}{C_G} = \frac{2.16 \times 10^{-3} + 0}{35.5} = 6.08 \times 10^{-5}\ \text{m/s} \tag{6.1-17}$$

The convective fluxes of air and water vapor at the interface can be expressed in terms of their partial pressures by applying the ideal gas law.

$$[N_i]_{conv} = C_i u^* = \left(\frac{p_i}{\mathcal{R}T}\right) u^* \quad (i = 1,2) \tag{6.1-18}$$

where the partial pressure p_i varies along the z direction. Assuming local equilibrium at the interface, $p_1 = p_1^*$ and $p_2 = P - p_1^*$ such that

$$\begin{aligned} [N_1]_{conv} &= \frac{(31200)(6.08\times10^{-5})}{(8.31)(343)} = 6.66\times10^{-4}\ \text{mol/s-m}^2 \\ [N_2]_{conv} &= \frac{(101300-31200)(6.08\times10^{-5})}{(8.31)(343)} = 1.50\times10^{-3}\ \text{mol/s-m}^2 \end{aligned} \tag{6.1-19a, b}$$

We can compute the diffusive fluxes at the interface by taking the difference between the overall and convective fluxes:

$$\begin{aligned} J_1 &= 2.16\times10^{-3} - 6.66\times10^{-4} = 1.50\times10^{-3}\ \text{mol/s-m}^2 \\ J_2 &= 0 - 1.50\times10^{-3} = -1.50\times10^{-3}\ \text{mol/s-m}^2 \end{aligned} \tag{6.1-20a, b}$$

The difference between the mole average and mass average velocities can be found from Equations 6.1-5 and 6.1-6:

$$u^* - u = \left[1 - \frac{M_1}{M_1x_1 + M_2(1-x_1)}\right]\frac{N_1}{c_G} + \left[1 - \frac{M_2}{M_1x_1 + M_2(1-x_1)}\right]\frac{N_2}{c_G} \tag{6.1-21a, b}$$

With $N_2 = 0$ and $x_1 = p^*/P = 31{,}200/101300 = 0.308$, we compute this velocity difference as

$$u^* - u = \left[1 - \frac{18}{18(0.308) + 30(1-0.308)}\right]\frac{0.00216}{35.5} = 1.92\times10^{-5}\ \text{m/s} \tag{6.1-22}$$

Thus, because of the considerable water vapor concentration and the substantially lower molecular weight of water compared to air, the difference between molar and mass average gas velocities is almost one-third of the molar velocity computed in Equation 6.1-17.

6.2 Stefan–Maxwell Equation

According to the second law of thermodynamics, reversible processes occurring in a closed system result in a dissipation of heat equal to TdS. In nonequilibrium thermodynamics, we apply this concept to irreversible processes by defining an entropy generation function as the rate of heat dissipation per unit volume:

$$\vartheta \equiv \frac{T}{V}\frac{dS}{dt} \tag{6.2-1}$$

For electrochemical diffusion in a fluid composed of I nonreacting species, the entropy generation function at constant T and P is given by Equation B2-13:

$$\vartheta = -\sum_{i=1}^{I} \mathbf{N}_i \cdot \nabla\mu_i = -\sum_{i=1}^{I} (cx_i\mathbf{u}_i)\cdot\nabla\mu_i = -\sum_{i=1}^{I} c\mathbf{u}_i \cdot (x_i\nabla\mu_i) \tag{6.2-2}$$

Because the summation of any constant between 1.0 and I is simply I times that constant, an alternative form of the entropy generation function is

$$\vartheta = -\frac{1}{I}\sum_{j=1}^{I}\left[\sum_{i=1}^{I} c\mathbf{u}_i \cdot (x_i \nabla \mu_i)\right] \tag{6.2-3}$$

The arithmetic average of the molar fluxes of all I species in a solution is $\sum_j \mathbf{N}_j/I = \sum_j c\mathbf{u}_j/I$. Taking the dot product of this quantity with the Gibbs–Duhem equation for a nonequilibrium process (Eq. B2-25) yields

$$\sum_{j=1}^{I} \frac{c\mathbf{u}_j}{I} \cdot \sum_{i=1}^{I} x_i \nabla \mu_i = \frac{1}{I}\sum_{j=1}^{I}\sum_{i=1}^{I} c\mathbf{u}_j \cdot (x_i \nabla \mu_i) = 0 \tag{6.2-4}$$

Adding this zero-valued term to the entropy production rate given by Equation 6.2-3, we obtain

$$\vartheta = -\frac{1}{I}\sum_{j=1}^{I}\sum_{i=1}^{I} c\mathbf{u}_i \cdot (x_i \nabla \mu_i) + \frac{1}{I}\sum_{j=1}^{I}\sum_{i=1}^{I} c\mathbf{u}_j \cdot (x_i \nabla \mu_i) = 0 \tag{6.2-5}$$

The entropy production rate can now be expressed in terms of the relative species velocity $(\mathbf{u}_i - \mathbf{u}_j)$:

$$\vartheta = -\frac{1}{I}\sum_{i=1}^{I}\sum_{j=1}^{I} c\left(\mathbf{u}_i - \mathbf{u}_j\right) \cdot (x_i \nabla \mu_i) \tag{6.2-6}$$

In general, the entropy generation function is a sum of the products of fluxes with their driving forces. For electrochemical diffusion processes governed by Equation 6.2-6, $c(\mathbf{u}_i - \mathbf{u}_j)$ is a form of the molar flux of a species i, and $x_i \nabla \mu_i$ is its "conjugate" driving force. A principal postulate of nonequilibrium thermodynamics states that fluxes are a linear sum of all the driving forces or, alternatively, that driving forces are a linear sum of all the fluxes. We choose the latter option:

$$-x_i \nabla \mu_i = \sum_{j=1}^{I} R_{ij} c\left(\mathbf{u}_i - \mathbf{u}_j\right) \quad (i = 1, 2 \ldots I) \tag{6.2-7}$$

The phenomenological coefficient R_{ij} represents the resistance to movement of molecules i exerted by molecules of species j. In terms of molar fluxes, $\mathbf{u}_i = \mathbf{N}_i/cx_i$ and $\mathbf{u}_j = \mathbf{N}_j/cx_j$ so that

$$-x_i \nabla \mu_i = \sum_{j=1}^{I} R_{ij}\left(\frac{\mathbf{N}_i}{x_i} - \frac{\mathbf{N}_j}{x_j}\right) \quad (i = 1, 2 \ldots I) \tag{6.2-8}$$

To express R_{ij} in terms of the interaction between individual i and j molecules, we scale it by the mole fractions of the two species (viz., $R_{ij}/x_i x_j$). As a phenomenological coefficient that represents mass conductance rather than resistance, we define a Stefan–Maxwell diffusion coefficient D_{ij} that is proportional to the reciprocal of $R_{ij}/x_i x_j$:

$$D_{ij} \equiv \left(\frac{\mathcal{R}T}{c}\right)\frac{1}{\left(R_{ij}/x_i x_j\right)} \tag{6.2-9}$$

According to the Onsager reciprocal relation between phenomenological coefficients, $R_{ij} = R_{ji}$ so that $D_{ij} = D_{ji}$.

Combining this relation with Equation 6.2-8, we arrive at the generalized Stefan–Maxwell equation that can be applied to electrochemical diffusion in a gas mixture as well as in a liquid solution:

$$-\frac{x_i \nabla \mu_i}{\mathcal{R}T} = \sum_{j=1}^{I} \frac{(x_j \mathbf{N}_i - x_i \mathbf{N}_j)}{cD_{ij}} \quad (i = 1, 2, \ldots I) \tag{6.2-10}$$

When chemical reaction occurs simultaneous to diffusion, an additional term consisting of the reaction rate as a flux and the reaction affinity as its conjugate driving force must be added to the entropy generation rate (Eq. B2-20). When transport properties are isotropic (i.e., independent of direction), however, Curie's theorem stipulates that this extra term does not affect the electrochemical flux–force equation since the orders of molar flux (a vector of order one) and reaction affinity (a scalar of order zero) differ by an odd number. Therefore, Equation 6.2-10 is valid in an isotropic medium, whether or not chemical reaction of the diffusing species is occurring.

6.3 Diffusion of Uncharged Substances

6.3.1 Binary Diffusion

Consider the diffusion of one component (i = 1) through a second component (i = 2) in a binary mixture. The Stefan–Maxwell equation for species 1 is given by

$$-\frac{x_1 \nabla \mu_1}{\mathcal{R}T} = \sum_{j=1}^{2} \frac{(x_j \mathbf{N}_1 - x_1 \mathbf{N}_j)}{cD_{1j}} = \frac{x_2 \mathbf{N}_1 - x_1 \mathbf{N}_2}{cD_{12}} \tag{6.3-1}$$

From Equation 6.1-14a, $\mathbf{N}_1 = x_1(\mathbf{N}_1 + \mathbf{N}_2) + \mathbf{J}_1$ so that

$$-\frac{x_1 \nabla \mu_1}{\mathcal{R}T} = \frac{1}{cD_{12}} \{x_2 \mathbf{N}_1 - [\mathbf{N}_1(1 - x_1) - \mathbf{J}_1]\} \tag{6.3-2}$$

With $x_1 + x_2 = 1$, this equation reduces to

$$\mathbf{J}_1 = -\left(\frac{c x_1 D_{12}}{\mathcal{R}T}\right) \nabla \mu_1 \tag{6.3-3}$$

To relate this diffusion flux to composition, we first combine Equation 3.5-15 with Equation 3.6-2 to obtain the electrochemical potential of species i in a multicomponent mixture:

$$\mu_i(T, P, x_1) = \mu_i^*(T, P) + \mathcal{R}T \ln(\gamma_i x_i) + z_i \mathcal{F} \psi \tag{6.3-4}$$

When T and P are constant, this equation provides a direct relationship between the chemical potential gradient and the gradients of mole fraction and electrical potential:

$$\nabla \mu_i = \mathcal{R}T \nabla \ln(\gamma_i x_i) + z_i \mathcal{F} \nabla \psi = \mathcal{R}T \left(\frac{1}{x_i} + \frac{1}{\gamma_i} \frac{d\gamma_i}{dx_i}\right) \nabla x_i + z_i \mathcal{F} \nabla \psi \tag{6.3-5}$$

Normally, all species in a gas mixture are uncharged. In a binary liquid solution, the two components must be uncharged since maintaining electroneutrality would require at least three distinct components—a cation, an anion, and a solvent. In either case, $z_i = 0$ and with some manipulation, we can write Equation 6.3-5 for species 1 as:

$$\nabla\mu_1 = \frac{\mathcal{R}T\Gamma}{x_1}\nabla x_1 \tag{6.3-6}$$

where Γ is a thermodynamic factor that we have defined as

$$\Gamma \equiv 1 + \frac{d(\ln\gamma_1)}{d(\ln x_1)} = 1 + \frac{d(\ln\gamma_2)}{d(\ln x_2)} \tag{6.3-7}$$

The second equality in this equation arises from Equation 3.6-14. For an ideal binary solution, $\gamma_1 = \gamma_2 = 1$ so that $\Gamma = 1$. In general, however, substitution of Equation 6.3-6 into Equation 6.3-3 yields

$$\mathbf{J}_1 = -c\Gamma D_{12}\nabla x_1 \tag{6.3-8}$$

According to Equation 6.1-11, the diffusion flux of component 2 must be the negative of the flux of component 1 so that

$$\mathbf{J}_2 = c\Gamma D_{12}\nabla x_1 = -c\Gamma D_{12}\nabla x_2 \tag{6.3-9}$$

We can consolidate the flux equations of the two components into a single equation:

$$\mathbf{J}_i = -c\Gamma D_{12}\nabla x_i \quad (i = 1,2) \tag{6.3-10}$$

Since Γ and D_{12} are positive quantities, the negative sign in Equation 6.3-10 indicates that the diffusion vectors ($\mathbf{J}_1$,$\mathbf{J}_2$) point in a direction that is opposite to their mole fraction gradients (∇x_1,∇x_2). In most situations, the individual values of Γ and D_{12} are unknown, and ΓD_{12} is expressed as a single lumped parameter $\mathcal{D}_{12}$ called the binary diffusion coefficient. The resulting flux expression for binary diffusion is known as Fick's first law:

$$\mathbf{J}_i = -c\mathcal{D}_{12}\nabla x_i \quad (i = 1,2) \tag{6.3-11}$$

Following from the Onsager reciprocal relation, $D_{ij} = D_{ji}$ so that $\mathcal{D}_{12} = \mathcal{D}_{21}$.

6.3.2 Multicomponent Diffusion

We now turn our attention to diffusion when more than two species are present. We restrict this analysis to gas mixtures and to liquid solutions consisting exclusively of uncharged species. We also consider that all species behave in a thermodynamically ideal manner so that $\nabla\mu_i = (\mathcal{R}T/x_i)\nabla x_i$ and $D_{ij} = \mathcal{D}_{ij}$. The Stefan–Maxwell equation can then be written as

$$-\nabla x_i = \sum_{j=1}^{I}\frac{(x_j\mathbf{N}_i - x_i\mathbf{N}_j)}{c\mathcal{D}_{ij}} \quad (i = 1,2,\ldots,I) \tag{6.3-12}$$

Employing Equations 6.1-11 and 6.1-14a and after considerable algebraic manipulation, we can rewrite the Stefan–Maxwell equation in terms of the diffusion fluxes (Taylor and Krishna, 1993):

$$c\nabla x_i = -\sum_{j=1}^{I-1} B_{ij}\mathbf{J}_j \quad (i = 1,2,\ldots,I-1) \tag{6.3-13}$$

When $i = j$, the B_{ij} are given by

$$B_{ii} = \frac{x_i}{\mathcal{D}_{iI}} + \sum_{\substack{k=1 \\ (k \neq i)}}^{I} \frac{x_k}{\mathcal{D}_{ik}} \quad (i = 1, 2, \ldots, I-1) \tag{6.3-14}$$

and when $i \neq j$, the B_{ij} are

$$B_{ij} = -x_i \left(\frac{1}{\mathcal{D}_{ij}} - \frac{1}{\mathcal{D}_{iI}} \right) \quad \begin{matrix} (i = 1, 2, \ldots, I-1) \\ (j = 1, 2, \ldots, I-1) \end{matrix} \tag{6.3-15}$$

Equation 6.3-13 is equivalent to the linear matrix equation

$$c \begin{bmatrix} \nabla x_1 \\ \nabla x_2 \\ \vdots \\ \nabla x_{I-1} \end{bmatrix} = - \begin{bmatrix} B_{11} & B_{12} & \cdots & B_{1,I-1} \\ B_{21} & B_{22} & \cdots & B_{2,I-1} \\ \vdots & \vdots & \ddots & \vdots \\ B_{I-1,1} & B_{I-1,2} & \cdots & B_{I-1,I-1} \end{bmatrix} \begin{bmatrix} J_1 \\ J_2 \\ \vdots \\ J_{I-1} \end{bmatrix} \Rightarrow c[\nabla x] = -[B][J] \tag{6.3-16a, b}$$

where the [B] matrix has diagonal elements given by Equation 6.3-14 and off-diagonal elements given by Equation 6.3-15. We can solve this equation for the flux matrix by premultiplying both sides of Equation 6.3-16 by the inverse of the [B] matrix (Appendix C2). Defining this inverse as

$$\begin{bmatrix} \mathcal{D}'_{11} & \mathcal{D}'_{12} & \cdots & \mathcal{D}'_{1,I-1} \\ \mathcal{D}'_{21} & \mathcal{D}'_{22} & \cdots & \mathcal{D}'_{2,I-1} \\ \vdots & \vdots & \ddots & \vdots \\ \mathcal{D}'_{I-1,1} & \mathcal{D}'_{I-1,2} & \cdots & \mathcal{D}'_{I-1,I-1} \end{bmatrix} \equiv \begin{bmatrix} B_{11} & B_{12} & \cdots & B_{1,I-1} \\ B_{21} & B_{22} & \cdots & B_{2,I-1} \\ \vdots & \vdots & \ddots & \vdots \\ B_{I-1,1} & B_{I-1,2} & \cdots & B_{I-1,I-1} \end{bmatrix}^{-1} \Rightarrow [\mathcal{D}'] \equiv [B]^{-1} \tag{6.3-17a, b}$$

we obtain

$$\begin{bmatrix} J_1 \\ J_2 \\ \vdots \\ J_{I-1} \end{bmatrix} = -c \begin{bmatrix} \mathcal{D}'_{11} & \mathcal{D}'_{12} & \cdots & \mathcal{D}'_{1,I-1} \\ \mathcal{D}'_{21} & \mathcal{D}'_{22} & \cdots & \mathcal{D}'_{2,I-1} \\ \vdots & \vdots & \ddots & \vdots \\ \mathcal{D}'_{I-1,1} & \mathcal{D}'_{I-1,2} & \cdots & \mathcal{D}'_{I-1,I-1} \end{bmatrix} \begin{bmatrix} \nabla x_1 \\ \nabla x_2 \\ \vdots \\ \nabla x_{I-1} \end{bmatrix} \Rightarrow [J] = c[\mathcal{D}'][\nabla x] \tag{6.3-18a, b}$$

The matrix elements $\mathcal{D}'_{ij}$ represent the multicomponent diffusion coefficient of species i through species j. The $\mathcal{D}'_{ij}$ are not necessarily symmetric, although the $\mathcal{D}_{ij}$ must be. Also, the $\mathcal{D}'_{ij}$ can depend on composition even when the $\mathcal{D}_{ij}$ do not.

Equation 6.3-18 differs from Fick's law by the appearance of terms in which the diffusion of a species i depends on the mole fraction gradient of other species j. For example, the flux of component 1 as given by Equation 6.3-18 is

$$J_1 = -c\mathcal{D}'_{11} \nabla x_1 - c\mathcal{D}'_{12} \nabla x_2 - \ldots - c\mathcal{D}'_{1,I-1} \nabla x_{I-1} \tag{6.3-19}$$

The lead term on the right side of this equation corresponds to the Fickian diffusion of species 1 with a diffusion coefficient $\mathcal{D}'_{11}$. The additional terms are due to the coupled diffusion of species 1 with the other species in the mixture.

Example 6.3-1 Ternary Diffusion of Respiratory Gases

In the smallest airways of the lung, diffusion is the dominant transport process. Ignoring water vapor, the gas mixture in those airways contains about 15 volume percent O_2 (component 1), 5 volume percent CO_2 (component 2), and 80 volume percent N_2 (component 3). The binary diffusion coefficients of these gases at a body temperature of 37°C and pressure of 101.3 kPa are $\mathcal{D}_{12} = \mathcal{D}_{21} = 1.66 \times 10^{-5}$ m²/s, $\mathcal{D}_{13} = \mathcal{D}_{31} = 2.19 \times 10^{-5}$ m²/s, and $\mathcal{D}_{23} = \mathcal{D}_{32} = 1.77 \times 10^{-5}$ m²/s. Assuming that the gas mixture is ideal, formulate the diffusional flux of the three gas components in terms of their mole fraction gradients.

Solution

The elements of the [B] matrix are computed from Equations 6.3-14 and 6.3-15 by taking I = 3:

$$[B] = \begin{bmatrix} 4.64 \times 10^4 & -2.19 \times 10^3 \\ -1.87 \times 10^2 & 5.71 \times 10^4 \end{bmatrix} \text{ s/m}^2 \tag{6.3-20}$$

Using the specific relationships for the adjoint and determinant of a 2 × 2 matrix (Eqs. C2-7 and C2-8), we obtain

$$\text{adj}[B] = \begin{bmatrix} 5.71 \times 10^4 & 2.19 \times 10^3 \\ 1.87 \times 10^2 & 4.64 \times 10^4 \end{bmatrix} \text{ s/m}^2$$

$$\det[B] = (4.64 \times 10^4)(5.71 \times 10^4) - (-2.19 \times 10^3)(-1.87 \times 10^2) = 2.65 \times 10^9 \text{ s}^2/\text{m}^4 \tag{6.3-21a,b}$$

The matrix of multicomponent diffusion coefficients is given by the ratio of the adjoint matrix to the determinant:

$$[D'] = [B]^{-1} = \frac{1}{2.65 \times 10^9} \begin{bmatrix} 5.71 \times 10^4 & 2.19 \times 10^3 \\ 1.87 \times 10^2 & 4.64 \times 10^4 \end{bmatrix} = \begin{bmatrix} 2.16 \times 10^{-5} & 8.26 \times 10^{-7} \\ 7.06 \times 10^{-8} & 1.75 \times 10^{-5} \end{bmatrix} \text{ m}^2/\text{s} \tag{6.3-22}$$

With $c_G = 39.3$ mol/m³ when T = 37°C and P = 101.3 kPa, we can express the fluxes of O_2 and CO_2 in terms of their mole fraction gradients ∇y_i according to Equation 6.3-18a:

$$\begin{bmatrix} \mathbf{J}_1 \\ \mathbf{J}_2 \end{bmatrix} = -39.3 \begin{bmatrix} 2.16 \times 10^{-5} & 8.26 \times 10^{-7} \\ 7.06 \times 10^{-8} & 1.75 \times 10^{-5} \end{bmatrix} \cdot \begin{bmatrix} \nabla y_1 \\ \nabla y_2 \end{bmatrix} \tag{6.3-23}$$

so that

$$\begin{aligned} \mathbf{J}_1 &= -(8.49 \times 10^{-4} \text{ [mol/s-m}^2])(\nabla y_1 + 0.0382 \nabla y_2) \\ \mathbf{J}_2 &= -(6.88 \times 10^{-4} \text{ [mol/s-m}^2])(\nabla y_2 + 0.00403 \nabla y_1) \end{aligned} \tag{6.3-24a,b}$$

The J_3 flux can be found by using Equation 6.1-11 and introducing the J_3 mole fraction gradient as $\nabla y_3 = -(\nabla y_1 + \nabla y_2)$:

$$\mathbf{J}_3 = -(\mathbf{J}_1 + \mathbf{J}_2) = -(8.52 \times 10^{-4} \text{ [mol/s-m}^2])(\nabla y_3 + 0.155 \nabla y_2) \tag{6.3-25}$$

Because uptake rates of O_2(component 1) and CO_2(component 2) in the respiratory airspaces are similar, we expect the values of ∇y_1 and ∇y_2 to be comparable. In that case, Equation 6.3-24a,b indicates that coupled diffusion for these two gases is not nearly as important as Fickian diffusion. The uptake rate of nitrogen (component 3) in the lungs is very small so that ∇y_3 is probably less than ∇y_2. According to Equation 6.3-25, this means that coupled diffusion can be important in determining the nitrogen flux.

6.3.3 Pseudo-binary Diffusion

Example 6.3-1 problem illustrates a specific situation where we can ignore coupled diffusion effects. We can specify three general cases where this is true, allowing us to treat multicomponent diffusion as a pseudo-binary process.

Case 1
All species are sufficiently similar in molecular weight that their binary diffusion coefficients are approximately equal. This is often a reasonable approximation for diffusion in respiratory gas mixtures. In this case, $\mathcal{D}_{ij} = \mathcal{D}_m$ (all $i \neq j$) so that $B_{ii} = 1/\mathcal{D}_m$ (Eq. 6.3-14) and $B_{ij} = 0$ (Eq. 6.3-15). The multicomponent diffusion coefficients (Eq. 6.3-17) then reduce to $\mathcal{D}'_{ii} = \mathcal{D}_m$ and $\mathcal{D}'_{ij} = 0$, and the diffusion fluxes (Eq. 6.3-16) can be written for each of the I components as

$$J_i = -c\mathcal{D}_m \nabla x_i \quad (i = 1,2,\ldots,I) \tag{6.3-26}$$

Case 2
All solutes ($i = 1, 2,\ldots I - 1$) *are present at very small mole fractions in a solvent ($i = I$).* This is a reasonable approximation for the dilute aqueous solutions that usually occur in biological systems. In this case, $x_I \rightarrow 1$ and $x_i \rightarrow 0$ ($i = 1,\ldots,I - 1$) so that $B_{ii} = 1/\mathcal{D}_{iI}$ (Eq. 6.3-14) and $B_{ij} = 0$ (Eq. 6.3-15). The multicomponent diffusion coefficients (Eq. 6.3-17) then reduce to $\mathcal{D}'_{ii} = \mathcal{D}_{i,I}$ and $\mathcal{D}'_{ij} = 0$, and the diffusion fluxes (Eq. 6.3-16) for each of the solutes are

$$J_i = -c\mathcal{D}_{i,I} \nabla x_i \quad i = 1,2,\ldots(I-1) \tag{6.3-27}$$

Case 3
There are $i = 1,\ldots,K$ *trace components present at very small mole fractions, and other components,* $i = K + 1,\ldots,I$, *have binary diffusion coefficients that are equal to one another.* According to *case 1*, the collection of components $i = K + 1,\ldots,I$ can be approximated as a single component that we view as pure solvent w. With a separate binary diffusion coefficient $\mathcal{D}_{i,w}$ between each trace component and the solvent, the diffusion flux of the trace components is then given by *case 2* as

$$J_i = -c\mathcal{D}_{i,w} \nabla x_i \quad (i = 1,2,\ldots,K) \tag{6.3-28}$$

These three cases encompass most diffusion processes in biomedical systems. Thus, with few exceptions, we model the diffusion of component i through a multicomponent gas mixture or liquid solution with a Fickian diffusion equation:

$$\mathbf{J}_i = -c\mathcal{D}_i \nabla x_i \tag{6.3-29}$$

where $\mathcal{D}_i$ represents a pseudo-binary diffusion coefficient of species i through the mixture or solution.

6.4 Diffusion of Electrolytes

In a dilute electrolyte solution, we can treat the electrochemical diffusion of each ion through the solvent as a pseudo-binary process with a diffusion coefficient $\mathcal{D}_i$. With ions i that are thermodynamically ideal, the chemical potential gradient is obtained by setting $\gamma_i = 1$ in Equation 6.3-5:

$$\nabla \mu_i = \frac{\mathcal{R}T}{x_i} \nabla x_i + z_i \mathcal{F} \nabla \psi \tag{6.4-1}$$

and the binary diffusion rate equation (Eq. 6.3-3) for a solution of molar density c_L becomes

$$\mathbf{J}_i = -c_L \mathcal{D}_i \left[\nabla x_i + \left(\frac{z_i \mathcal{F}}{\mathcal{R}T} \right) x_i \nabla \psi \right] \tag{6.4-2}$$

This Nernst–Planck equation indicates that an ion flux $\mathbf{J}_i$ is the net result of random movement by a concentration gradient $c_L \nabla x_i$ and directed drift by an electric field $-\nabla \psi$. An interesting implication is that an ion concentration gradient can generate an electric field. To demonstrate this phenomenon in concrete terms, consider the free diffusion of a strong 1–1 electrolyte such as NaCl in the absence of an externally applied electric field.

When a strong 1–1 electrolyte is dissolved in water, it completely dissociates to form a cation with a charge of $z_c = +1$ and an anion with a charge of $z_a = -1$. To preserve electroneutrality in a solution of this electrolyte, the local mole fractions of the cations x_c and anions x_a must be equal to each other so that

$$x_c = x_a \equiv x_{ca} \Rightarrow \nabla x_c = \nabla x_a \equiv \nabla x_{ca} \tag{6.4-3a, b}$$

where the subscript ca indicates a quantity associated with the salt as a whole. Applying the electrochemical flux equation to the separate ions, we obtain

$$\mathbf{J}_c = -c_L \mathcal{D}_c \left[\nabla x_c + \left(\frac{\mathcal{F} x_c}{\mathcal{R}T} \right) \nabla \psi \right] = -c_L \mathcal{D}_c \left[\nabla x_{ca} + \left(\frac{\mathcal{F} x_c}{\mathcal{R}T} \right) \nabla \psi \right] \tag{6.4-4}$$

$$\mathbf{J}_a = -c_L \mathcal{D}_a \left[\nabla x_a - \left(\frac{\mathcal{F} x_a}{\mathcal{R}T} \right) \nabla \psi \right] = -c_L \mathcal{D}_a \left[\nabla x_{ca} - \left(\frac{\mathcal{F} x_a}{\mathcal{R}T} \right) \nabla \psi \right] \tag{6.4-5}$$

Absent a closed circuit in which electrical current can flow, the transport rates of positive and negative ion charges must be equal. For a 1–1 electrolyte solution, in particular, the total convective and diffusive flux of the anion is equal to that of the cation:

$$(c_L x_a)\mathbf{u}^* + \mathbf{J}_a = (c_L x_c)\mathbf{u}^* + \mathbf{J}_c \tag{6.4-6}$$

With $x_c = x_a$, the diffusion fluxes of the anion $\mathbf{J}_a$ and the cation $\mathbf{J}_c$ must be equal to each other:

$$\mathbf{J}_a = \mathbf{J}_c \equiv \mathbf{J}_{ca} \tag{6.4-7}$$

Equating Equation 6.4-4 to Equation 6.4-5, we get

$$\mathcal{D}_c\left[\nabla x_{ca} + \left(\frac{\mathcal{F}x_{ca}}{\mathcal{R}T}\right)\nabla\psi\right] = \mathcal{D}_a\left[\nabla x_{ca} - \left(\frac{\mathcal{F}x_{ca}}{\mathcal{R}T}\right)\nabla\psi\right] \tag{6.4-8}$$

Rearranging this equation, we find that

$$\nabla\psi = -\left(\frac{\mathcal{D}_c - \mathcal{D}_a}{\mathcal{D}_c + \mathcal{D}_a}\right)\left(\frac{\mathcal{R}T}{x_s\mathcal{F}}\right)\nabla x_{ca} \tag{6.4-9}$$

This equation can be substituted into either Equation 6.4-4 or 6.4-5 to obtain the molar diffusion flux of salt:

$$\mathbf{J}_{ca} = -c_L\mathcal{D}_{ca}\nabla x_{ca} \tag{6.4-10}$$

where

$$\mathcal{D}_{ca} \equiv \left(\frac{2\mathcal{D}_c\mathcal{D}_a}{\mathcal{D}_c + \mathcal{D}_a}\right) \tag{6.4-11}$$

Equations 6.4-10 and 6.4-11 indicate that free diffusion of a completely dissociated 1–1 electrolyte taken as a whole follows Fick's law with an overall diffusion coefficient that is an average of the individual cation and anion diffusion coefficients (viz., $\mathcal{D}_{ca}$ is the ratio of the squared geometric mean $\mathcal{D}_c\mathcal{D}_a$ to the arithmetic mean $(\mathcal{D}_c + \mathcal{D}_a)/2$).

Equation 6.4-9 demonstrates that electrolyte diffusion induces a gradient of electric potential that is proportional to the mole fraction gradient of the electrolyte and also to the diffusion coefficient of an individual cation relative to an individual anion. When $\mathcal{D}_c$ is greater than $\mathcal{D}_a$, the vectors $\nabla\psi$ and $\mathbf{J}_{ca}$ both point in the same direction as $-\nabla x_{ca}$. This occurs because the positive-charged cations are able to diffuse slightly faster than the negative-charged anions, creating an increase in electrical potential in the direction of the ion fluxes. If the individual diffusion coefficients are identical, then no potential is generated.

6.5 Transport across Membranes

6.5.1 Entropy Generation Function for Uncharged Solutes

We now apply the concepts of nonequilibrium thermodynamics to formulate the transport rates of an uncharged solute s and water w across a planar membrane (Katchalsky and Curran, 1965). The membrane has a thickness h_m and separates two well-mixed, dilute solutions with different solute concentrations (Fig. 6.5-1). We suppose that both species behave in a thermodynamically ideal manner. The system is at a uniform temperature, but the two solutions exist at different hydrostatic pressures.

For a binary nonelectrolyte solution, the electrochemical potentials of s and w reduce to their chemical potentials, $\breve{\mu}_s$ and $\breve{\mu}_w$, and the entropy generation function (Eq. 6.2-2) for the solution in the membrane is given by

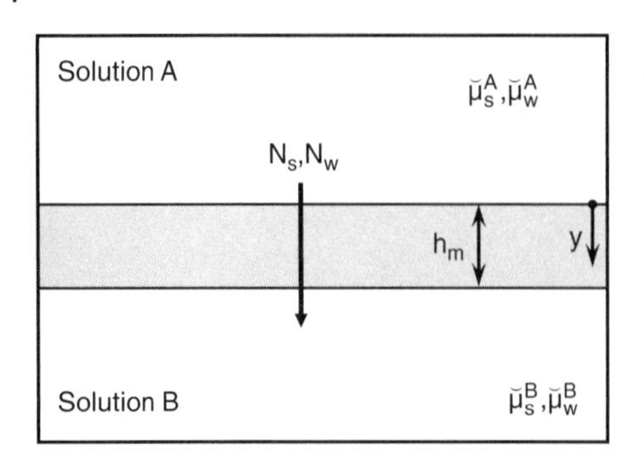

Figure 6.5-1 Transport across a membrane dividing two well-mixed solutions.

$$\vartheta = -\mathbf{N}_s \cdot \nabla \breve{\mu}_s - \mathbf{N}_w \cdot \breve{\nabla} \mu_w \qquad (6.5\text{-}1)$$

It is not necessary to include a term for entropy production of the membrane material itself since it is stationary. As shown in Figure 6.5-1, the fluxes through a membrane are aligned in the y direction perpendicular to the membrane surface. The dot products in Equation 6.5-1 are then equivalent to the products of flux components in the y direction, N_s and N_w, with the scalar chemical potential derivatives, $d\breve{\mu}_s/dy$ and $d\breve{\mu}_w/dy$, respectively:

$$\vartheta = -\sum_{i=1}^{I} N_i \frac{d\breve{\mu}_i}{dy} = -N_s \frac{d\breve{\mu}_s}{dy} - N_w \frac{d\breve{\mu}_w}{dy} \qquad (6.5\text{-}2)$$

Fluxes and compositions cannot be measured inside a membrane, so this equation is of no practical use. To circumvent this limitation, we integrate the entropy generation function over the thickness of the membrane:

$$\vartheta_m \equiv \int_0^{h_m} \vartheta dy = -\int_0^{h_m} \left(N_s \frac{d\breve{\mu}_s}{dy} + N_w \frac{d\breve{\mu}_w}{dy} \right) dy \qquad (6.5\text{-}3)$$

Given that ϑ is the entropy production rate per unit volume, ϑ_m represents the entropy production rate per unit surface of the membrane. Because a membrane is thin, transients in membrane transport processes die out very rapidly, and N_s and N_w will not vary appreciably with y. Thus, the integration can be completed to yield

$$\vartheta_m = N_s[\breve{\mu}_s(0) - \breve{\mu}_s(h_m)] + N_w[\breve{\mu}_w(0) - \breve{\mu}_w(h_m)] \qquad (6.5\text{-}4)$$

where the chemical potentials are evaluated just inside the two surfaces of the membrane. Assuming that interfacial equilibrium occurs at these two surfaces, we equate the electrochemical potentials of s and w just inside the membrane with their corresponding chemical potentials, $\breve{\mu}_s^A$ and $\breve{\mu}_s^B$, in external solutions A and B:

$$\breve{\mu}_s(0) = \breve{\mu}_s^A, \quad \breve{\mu}_s(h_m) = \breve{\mu}_s^B, \quad \breve{\mu}_w(0) = \breve{\mu}_w^A, \quad \breve{\mu}_s(h_m) = \breve{\mu}_w^B \qquad (6.5\text{-}5a,d)$$

so that

$$\vartheta_m = N_s(\breve{\mu}_s^A - \breve{\mu}_s^B) + N_w(\breve{\mu}_w^A - \breve{\mu}_w^B) \qquad (6.5\text{-}6)$$

6.5.2 Chemical Potential Driving Forces

The constitutive relationships between chemical potentials and mole fractions of the uncharged ideal species s and w in the external solutions A and B are given by

$$\begin{aligned} \breve{\mu}_s^k &= \breve{\mu}_s^{*k}(T,P) + \mathcal{R}T\ln x_s^k \\ \breve{\mu}_w^k &= \breve{\mu}_w^{*k}(T,P) + \mathcal{R}T\ln x_w^k \end{aligned} \quad (k = A, B) \qquad (6.5\text{-}7a,b)$$

The difference in the chemical potential for solute between the two solutions is

$$\left(\breve{\mu}_s^A - \breve{\mu}_s^B\right) = \left(\mu_s^{*A} - \mu_s^{*B}\right) + \mathcal{R}T\ln\left(\frac{x_s^A}{x_s^B}\right) \qquad (6.5\text{-}8)$$

and similarly for water

$$\left(\breve{\mu}_w^A - \breve{\mu}_w^B\right) = \left(\mu_w^{*A} - \mu_w^{*B}\right) + \mathcal{R}T\ln\left(\frac{x_w^A}{x_w^B}\right) \qquad (6.5\text{-}9)$$

The partial molar volumes of the solution with respect to solute $\hat{V}_s$ and to water $\hat{V}_w$ are essentially independent of pressure, allowing us to integrate Equation 3.6-5 for each of the two species:

$$\left(\mu_s^{*A} - \mu_s^{*B}\right) = \hat{V}_s\left(P^A - P^B\right), \quad \left(\mu_w^{*A} - \mu_w^{*B}\right) = \hat{V}_w\left(P^A - P^B\right) \qquad (6.5\text{-}10a,b)$$

Noting also that $x_s + x_w = 1$ in each solution, Equations 6.5-8 and 6.5-9 become

$$\left(\breve{\mu}_s^A - \breve{\mu}_s^B\right) = \hat{V}_s\left(P^A - P^B\right) + \mathcal{R}T\ln\left(\frac{x_s^A}{x_s^B}\right) \qquad (6.5\text{-}11)$$

$$\left(\breve{\mu}_w^A - \breve{\mu}_w^B\right) = \hat{V}_w\left(P^A - P^B\right) + \mathcal{R}T\ln\left(\frac{1 - x_s^A}{1 - x_s^B}\right) \qquad (6.5\text{-}12)$$

Because nonequilibrium thermodynamics applies to small departures from equilibrium, x_s^A and x_s^B must be similar in value. Thus, $\left|\left(x_s^A/x_s^B\right) - 1\right| \ll 1$ and we can utilize the approximation (Abramowitz and Stegun, 1965) that

$$\ln\left(\frac{x_s^A}{x_s^B}\right) \approx 2\frac{\left(x_s^A/x_s^B - 1\right)}{\left(x_s^A/x_s^B + 1\right)} \qquad (6.5\text{-}13)$$

Also, $x_s^A \ll 1$ and $x_s^B \ll 1$ for dilute solutions so that

$$\ln\left(\frac{1 - x_s^A}{1 - x_s^B}\right) = \ln\left(1 - x_s^A\right) - \ln\left(1 - x_s^B\right) \approx -x_s^A + x_s^B \qquad (6.5\text{-}14)$$

The molar density of a dilute aqueous solution is approximated by the inverse of the partial molar volume of water. Thus, the solute mole fractions in solutions A and B can be expressed in terms of their molar concentrations as $x_s^A \approx \hat{V}_w C_s^A$ and $x_s^B \approx \hat{V}_w C_s^B$. Combining Equations 6.5-13 and 6.5-14 with Equations 6.5-11 and 6.5-12, we obtain

$$\left(\breve{\mu}_s^A - \breve{\mu}_s^B\right) = \hat{V}_s\Delta P + \frac{\mathcal{R}T}{C_s^{ave}}\Delta C_s \qquad (6.5\text{-}15)$$

$$\left(\breve{\mu}_w^A - \breve{\mu}_w^B\right) = \hat{V}_w\Delta P - \mathcal{R}T\hat{V}_w\Delta C_s \qquad (6.5\text{-}16)$$

where we have defined $\Delta P = P^A - P^B$, $\Delta C_s \equiv \left(C_s^A - C_s^B\right)$, and $C_s^{ave} \equiv \left(C_s^A + C_s^B\right)/2$.

6.5.3 Kedem–Katchalsky Equations

Substituting Equations 6.5-15 and 6.5-16 into Equation 6.5-6 results in a entropy generation equation:

$$\vartheta_m = \left(N_s\hat{V}_s + N_w\hat{V}_w\right)\Delta P + \left(\frac{N_s}{C_s^{ave}} - \hat{V}_w N_w\right)\mathcal{R}T\Delta C_s \tag{6.5-17}$$

Recognizing that the mean molar velocity of solution through the membrane is

$$u^* \equiv \hat{V}_s N_s + \hat{V}_w N_w \tag{6.5-18}$$

and the diffusion velocity of solute relative to the velocity of water is

$$u_{s-w} \equiv \frac{N_s}{C_s^{ave}} - \hat{V}_w N_w \tag{6.5-19}$$

we can rewrite the entropy generation rate as

$$\vartheta_m = u^*\Delta P + u_{s-w}\mathcal{R}T\Delta C_s \tag{6.5-20}$$

In the nomenclature of nonequilibrium thermodynamics, each term in this equation represents flux–force pair. Expressing the fluxes u^* and u_{s-w} as linear homogeneous functions of the driving forces gives rise to the Kedem–Katchalsky equations for binary transport across a membrane:

$$u^* = L_{PP}\Delta P + L_{PC}\mathcal{R}T\Delta C_s \tag{6.5-21}$$

$$u_{s-w} = L_{CP}\Delta P + L_{CC}\mathcal{R}T\Delta C_s \tag{6.5-22}$$

Here, L_{PP} and L_{CC} are the phenomenological coefficients for the principal flux–force pairs, and $L_{PC} = L_{CP}$ are the coefficients for the coupled forces and fluxes.

6.5.4 Starling Equations

It is desirable to modify these transport rate equations so that they can be directly related to measurements. Solving Equations 6.5-18 and 6.5-19 for N_s, we obtain

$$N_s = \frac{C_s^{ave}(u^* + u_{s-w})}{1 + C_s^{ave}\hat{V}_s} \tag{6.5-23}$$

In dilute solutions, the volume fraction of solute $C_s^{ave}\hat{V}_s$ is much less than one so that

$$N_s \approx C_s^{ave}(u^* + u_{s-w}) = C_s^{ave}[u^* + (L_{PC}\Delta P + L_{CC}\mathcal{R}T\Delta C_s)] \tag{6.5-24}$$

By solving Equation 6.5-21 for ΔP and substituting the result into Equation 6.5-24, we get

$$N_s = u^* C_s^{ave}\left(1 + \frac{L_{PC}}{L_{PP}}\right) + \mathcal{R}TC_s^{ave}\left(L_{CC} - \frac{L_{PC}^2}{L_{PP}}\right)\Delta C_s \tag{6.5-25}$$

Because we assumed that water in the solutions external to the membrane behaves in a thermodynamically ideal manner, the osmotic pressure difference across the membrane is given by the Van't Hoff equation ($\Delta\pi = \mathcal{R}T\Delta C_s$), and we can write Equation 6.5-21 as

$$u^* = L_{PP}\Delta P + L_{PC}\Delta\pi \tag{6.5-26}$$

Based on Equations 6.5-25 and 6.5-26, we define three positive-valued coefficients that have a direct physical meaning:

$$L_P \equiv L_{PP} = \left(\frac{u^*}{\Delta P}\right)_{\Delta C_s = 0} \quad \text{hydraulic permeability} \tag{6.5-27}$$

$$\sigma_S \equiv -\frac{L_{PC}}{L_{PP}} = \left(\frac{u^* C_s^{ave} - N_s}{u^* C_s^{ave}}\right)_{\Delta C_s = 0} \quad \text{Staverman reflection coefficient} \tag{6.5-28}$$

$$P_s \equiv L_{CC} - \frac{L_{PC}^2}{L_{PP}} = \left(\frac{N_s}{\Delta C_S}\right)_{u^* = 0} \quad \text{solute permeability} \tag{6.5-29}$$

Equations 6.5-25 and 6.5-26 now become

$$N_s = u^* C_s^{ave}(1 - \sigma_S) + P_s \Delta C_s \tag{6.5-30}$$

$$u^* = L_P(\Delta P - \sigma_S \Delta \pi) \tag{6.5-31}$$

Equations 6.5-30 and 6.5-31 are known as the Starling equations. Equation 6.5-30 shows that the solute flux N_s results from a combination of convective transport by the fluid velocity $u*$ and diffusion by the concentration driving force ΔC_S. From the convection term, we deduce that $(1 - \sigma_S)$ is the fraction of solute molecules capable of being dragged across the membrane by the solvent flow; equivalently, σ_S is the fraction of solute molecules that are rejected by the membrane. For an ideal semipermeable membrane, $\sigma_S = 1$ so that convection does not contribute to solute transport.

Equation 6.5-31 indicates that fluid velocity through a membrane is driven by the hydraulic pressure difference and is opposed by the osmotic pressure difference multiplied by the fraction of the molecules rejected by the membrane. For a solute that freely permeates a membrane, $\sigma_S = 0$ and $\Delta\pi$ has no effect on u^*. For an impermeable solute, $\sigma_S=1$ and $\Delta\pi$ has the same importance as the hydraulic pressure difference.

In realistic situations, a membrane is surrounded by solutions containing many solutes. Assuming that each solute acts independently of the others and has its own permeability and reflection coefficient, Equation 6.5-30 still applies and Equation 6.5-31 can be generalized as follows:

$$u^* = L_P\left(\Delta P - \sum_{s=1}^{S} \sigma_S \Delta \pi_S\right) \tag{6.5-32}$$

Example 6.5-1 Determination of L_p and $\sigma_{albumin}$ for a Capillary Wall

Starling equations are frequently used to analyze transport through the capillary wall by treating it as a membrane. Efflux of liquid through the wall of a single capillary from a frog mesentery has been measured at transcapillary pressure differences from 1.18 to 4.90 kPa (Curry *et al.*, 1976). At each pressure difference, data were obtained in the same capillary in two separate experiments: (i) in control experiments, the osmotic pressure was balanced across the capillary wall; and (ii) in other experiments, the osmotic pressure was increased to 5.59 kPa

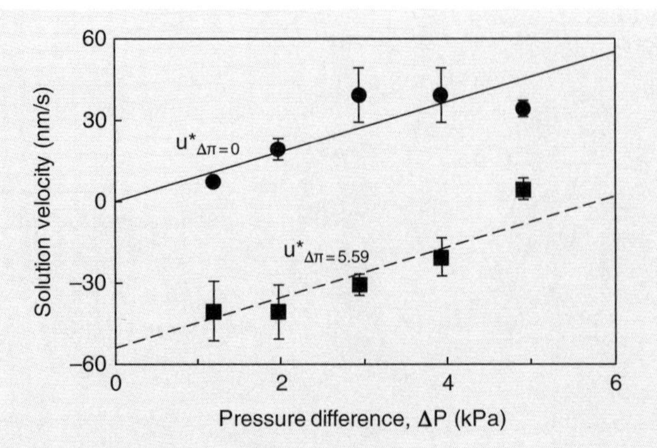

Figure 6.5-2 Convection through the wall of a single capillary.

in the interior of the capillary by the addition of albumin to the perfusate. Figure 6.5-2 illustrates the mean ± standard deviation of replicated measurements of fluid velocity u^* in the two experiments at each of the transcapillary pressures ΔP. From these data, determine the values of L_p and $\sigma_{albumin}$ for the capillary wall and evaluate its permeability to albumin.

Solution

For the control experiment in which there is no osmotic pressure difference across the capillary wall, Equation 6.5-32 indicates that

$$u^*_{\Delta\pi=0} = L_p \Delta P \tag{6.5-33}$$

From the slope of a linear least squares regression of the $u^*_{\Delta\pi=0} - \Delta P$ data shown by the solid line in Figure 6.5-2, we obtain a value of $L_p = 9.29$ nm/(s-kPa).

We can also use Equation 6.5-32 to estimate $\sigma_{albumin}$ from the data of the other experiments. Since albumin is only contributing solute to the osmotic differences, this equation reduces to

$$u^*_{\Delta\pi=5.59} = L_P(\Delta P - \sigma_{albumin}\Delta\pi_{albumin}) \Rightarrow \sigma_{albumin} = \frac{u^*_{\Delta\pi=0} - u^*_{\Delta\pi=5.59}}{9.29(5.59)} \tag{6.5-34a, b}$$

Thus, we can compute a separate value of $\sigma_{albumin}$ at each ΔP from the corresponding difference between $u^*_{\Delta\pi=0}$ and $u^*_{\Delta\pi=5.59}$ measurements. The mean ± standard deviation of the five computed $\sigma_{albumin}$ values was 1.03 ± 0.14. The dashed line in Figure 6.5-2 was drawn by inserting values of $L_p = 9.29$ nm/(s-kPa), $\sigma_{albumin} = 1.03$, and $\Delta\pi_{albumin} = 5.59$ kPa into Equation 6.5-34a.

The mean value of $\sigma_{albumin}$ is not significantly different from one. This indicates that the capillary wall in the frog mesentery is virtually impermeable to albumin. In addition, the data points in Figure 6.5-2 do not deviate systematically from the predicted lines. This supports the conclusion that neither the hydraulic conductivity nor the reflection coefficient depends on the mechanical stresses associated with the hydrodynamic pressure difference.

References

Abramowitz M, Stegun IA. Handbook of Mathematical Functions. Washington: US Government Printing Office; 1965, Eq. 4.1.27.

Curry FE, Mason JC, Michel CC. Osmotic reflexion coefficients of capillary walls to low molecular weight hydrophilic solutes measured in single perfused capillaries of frog mesentery. J Physiol Lond. 1976; 261:319–336.

Katchalsky A, Curran PE. Nonequilibrium Thermodynamics in Biophysics. Cambridge: Harvard University Press; 1965, ch. 10.

Taylor R, Krishna R. Multicomponent Mass Transfer. New York: John Wiley & Sons, Inc.; 1993, ch. 1-2.

Chapter 7

Mechanisms and Models of Diffusion

Mass conservation equations in combination with the transport rate equations discussed in Chapter 6 are the primary tools for analyzing mass transport in cells and tissues as well as in medical mass transfer devices. Whereas mass conservation is a universal principle that holds for all substances, rate equations depend on the structure and composition of a particular material and the molecular mechanism by which a process occurs. For an ordinary diffusion process, the order-of-magnitude of diffusion coefficients $\mathcal{D}_i$ are

- Component i in a gas mixture at ambient pressure.....10^{-1} cm^2/s = 10^{-5} m^2/s
- Small solute molecule i in a liquid solution................10^{-5} cm^2/s = 10^{-9} m^2/s
- Large solute molecule i in a liquid solution................10^{-6} cm^2/s = 10^{-10} m^2/s

Gas mixtures exist in an expanded state with weak intermolecular forces, explaining why the $\mathcal{D}_i$ of components in a gas phase is relatively large. Liquids are in a compressed state with relatively large intermolecular forces such that the $\mathcal{D}_i$ of solutes are much smaller. In general, $\mathcal{D}_i$ is inversely related to molecular size and directly related to temperature whether diffusion is occurring in a gas or liquid phase.

Although diffusion coefficients of simple solutes in homogeneous phases are frequently available (e.g., Tables A4-1 and A4-2), $\mathcal{D}_i$ values for biochemical molecules in biological materials are limited (e.g., Table A4-3). The primary goal of this chapter is to present some practical methods of estimating diffusion coefficients. In the first portion of the chapter, we discuss simple molecular models of diffusion in homogeneous media. Because of the multiphase nature of most biological tissues, the final sections of the chapter focus on mass transport in heterogeneous materials.

Biomedical Mass Transport and Chemical Reaction: Physicochemical Principles and Mathematical Modeling, First Edition. James S. Ultman, Harihara Baskaran, and Gerald M. Saidel.

7.1 Transport Rates in Homogeneous Materials

Following Equation 6.1-13, we write the overall molar flux $\mathbf{N}_i$ and overall mass flux $\mathbf{n}_i$ of a substance i in a multicomponent solution as the sum of convective and diffusive components:

$$\mathbf{N}_i \equiv c\mathbf{u}^* x_i + \mathbf{J}_i$$
$$\mathbf{n}_i = \rho\mathbf{u}\omega_i + \mathbf{j}_i \quad (7.1\text{-}1a,b)$$

where x_i and ω_i are mole and mass fractions, $\mathbf{u}^*$ and $\mathbf{u}$ represent molar and mass average velocities, and $\mathbf{J}_i$ and $\mathbf{j}_i$ are the molar and mass diffusion fluxes. For a binary or pseudo-binary diffusion process, nonequilibrium thermodynamics leads to the following relationship between molar diffusion flux and mole fraction:

$$\mathbf{J}_i = -c\mathcal{D}_i \nabla x_i \quad (6.3\text{-}29)$$

where $\mathcal{D}_i$ is the pseudo-binary diffusion coefficient of substance i through the solution and c is the molar density. Using relationships between molar and mass quantities (Bird *et al.*, 2002), an equivalent relationship between mass diffusion flux and mass fraction is

$$\mathbf{j}_i = -\rho\mathcal{D}_i \nabla \omega_i \quad (7.1\text{-}2)$$

The overall molar and mass fluxes are

$$\mathbf{N}_i = c\mathbf{u}^* x_i - c\mathcal{D}_i \nabla x_i \quad (7.1\text{-}3)$$
$$\mathbf{n}_i = \rho\mathbf{u}\omega_i - \rho\mathcal{D}_i \nabla \omega_i \quad (7.1\text{-}4)$$

When molar density is constant, Equation 7.1-3 becomes

$$\mathbf{N}_i = c\mathbf{u}^* x_i - \mathcal{D}_i \nabla (c x_i) = \mathbf{u}^* C_i - \mathcal{D}_i \nabla C_i \quad (7.1\text{-}5)$$

Similarly, if the mass density ρ is constant, then Equation 7.1-4 becomes

$$\mathbf{n}_i = \rho\mathbf{u}\omega_i - \mathcal{D}_i \nabla (\rho\omega_i) = \mathbf{u}\rho_i - \mathcal{D}_i \nabla \rho_i \quad (7.1\text{-}6)$$

Dividing this overall mass flux by the molecular weight M_i and noting that $\mathbf{N}_i = \mathbf{n}_i/M_i$ and $C_i = \rho_i/M_i$, we get

$$\mathbf{N}_i = \mathbf{u}C_i - \mathcal{D}_i \nabla C_i \quad (7.1\text{-}7)$$

While the forms of the molar flux equations for constant molar density c (Eq. 7.1-5) and constant mass density ρ (Eq. 7.1-7) materials are the same, the convective portion of the flux is based on the molar-averaged velocity $\mathbf{u}^*$ when c is constant and on massaveraged velocity $\mathbf{u}$ when ρ is constant. It is only when the difference between these velocities is very small that the flux equations are essentially equivalent. This will be the case for dilute liquid solutions and for gas mixtures consisting primarily of components with similar molecular weights (Eq. 6.1-5).

7.2 Diffusion Coefficients in Gases

7.2.1 Kinetic Theory

The molecules in a gaseous medium are separated by large distances in comparison to the diameter of the molecules themselves. According to the kinetic theory of gases, each gas

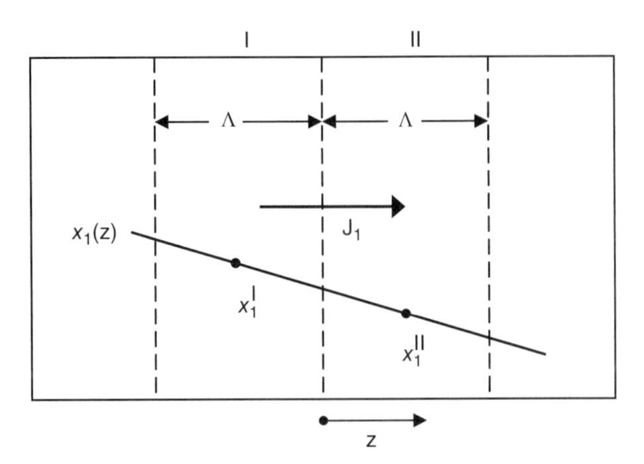

Figure 7.2-1 Gas-phase diffusion.

molecule is in a random motion that is unaffected by the presence of other molecules except when two molecules collide. Estimates of diffusion coefficients using kinetic theory incorporate two properties of a gas: the frequency f of molecular collisions with a stationary plane (either real or imagined) of unit area and the mean distance Λ that a molecule travels between collision with another molecule.

Consider one-dimensional transport in a binary mixture of gas components 1 and 2 in a closed container (Fig. 7.2-1). The container is rigid and the gas is at a constant T and P such that molar gas density c_G is constant and there is no convective transport. We examine the z-directed diffusion flux J_1 of component 1 by constructing a reference plane at z = 0 that separates a region I at $0 > z > -\Lambda$ from a second region II at $+\Lambda > z > 0$.

Since Λ is the mean distance between molecular collisions, any molecule of gas component 1 that reaches the left side of the reference plane must originate in region I. Of the frequency f of all molecular collisions on the left side of the reference plane, the fraction of these due to component 1 is equal to its average mole fraction x_1^{I} in region I. Thus, the collision frequency of component 1 on the left side of the reference plane is fx_1^{I}. Similarly, the collision frequency of component 1 on the right side of the reference plane is fx_1^{II}, and the net collision rate on the two sides of the reference plane is $\left(fx_1^{\mathrm{I}} - fx_1^{\mathrm{II}}\right)$. The corresponding molar flux in the z direction is

$$[J_1]_{z=0} = \frac{1}{\mathcal{A}} f\left(x_1^{\mathrm{I}} - x_1^{\mathrm{II}}\right) \tag{7.2-1}$$

where $\mathcal{A}$ is Avogadro's number. When Λ is small relative to the length scale over which we apply the diffusion equation, the mole fraction distribution $x_1(z)$ can be approximated as a linear function of z. Then,

$$x_1^{\mathrm{I}} = [x_1]_{z=0} - \frac{\Lambda}{2}\left[\frac{dx_1}{dz}\right]_{z=0} \tag{7.2-2}$$

$$x_1^{\mathrm{II}} = [x_1]_{z=0} + \frac{\Lambda}{2}\left[\frac{dx_1}{dz}\right]_{z=0} \tag{7.2-3}$$

Substituting Equations 7.2-2 and 7.2-3 into Equation 7.2-1 results in

$$[J_1]_{z=0} = -\frac{f\Lambda}{\mathcal{A}}\left[\frac{dx_1}{dz}\right]_{z=0} \tag{7.2-4}$$

This equation is equivalent to the z-component of Fick's law as stated in Equation 6.3-11 when the binary diffusion coefficient of component 1 is given by

$$\mathcal{D}_{12} \equiv \left(\frac{f\Lambda}{c_G \mathcal{A}}\right) \tag{7.2-5}$$

7.2.2 Ideal Gas Model

An ideal gas consists of rigid molecules that undergo elastic collisions. For a pure ideal gas 1 of molecular weight M_1 and molecular cross section A_1, the collision frequency and the mean free path are (Bird *et al.*, 2002, Eqs. 1.4-1–1.4-3)

$$f \propto c_G\sqrt{\frac{T}{M_1}}, \quad \Lambda \propto \frac{1}{c_G A_1} \tag{7.2-6a, b}$$

Substituting these relations into Equation 7.2-5 with $c_G = P/\mathcal{R}T$, we determine the self-diffusivity of gas 1:

$$\mathcal{D}_{11} \propto \frac{1}{c_G A_1}\sqrt{\frac{T}{M_1}} \propto \frac{T^{3/2}}{PA_1}\sqrt{\frac{1}{M_1}} \tag{7.2-7}$$

This result can be generalized for the diffusion coefficient of a binary gas mixture of rigid spherical molecules by defining proper averages of M and A for the two gas components:

$$M_{12} = \left(\frac{1}{M_1} + \frac{1}{M_2}\right)^{-1}, \quad A_{12} = \pi\left(\frac{a_1 + a_2}{2}\right)^2 \tag{7.2-8}$$

such that

$$\mathcal{D}_{12} \propto \frac{T^{3/2}}{P(a_1 + a_2)^2}\sqrt{\frac{1}{M_1} + \frac{1}{M_2}} \tag{7.2-9}$$

Here, a_1 and a_2 represent the molecular radii of the two components. A particularly useful semiempirical equation based on Equation 7.2-9 was developed by Fuller *et al.* (1966):

$$\mathcal{D}_{12} = \frac{101.3 \times 10^{-7} T^{1.75}}{P\left[(\sum \hat{\upsilon})_1^{1/3} + (\sum \hat{\upsilon})_2^{1/3}\right]^2}\sqrt{\frac{1}{M_1} + \frac{1}{M_2}} \tag{7.2-10}$$

where a_1 and a_2 have been replaced by the cube root of the corresponding molecular volumes, $(\sum \hat{\upsilon})_1$ and $(\sum \hat{\upsilon})_2$. When using this equation, the molecular volumes are computed as the sum of their constituent atomic volumes (Table 7.2-1), and specific units of $\mathcal{D}_{12}[m^2/s]$, $T[°K]$, and $P[kPa]$ must be employed. The $\mathcal{D}_{12}$ values predicted in this way are usually within 10% of measured values for both polar and nonpolar gases at ambient pressures.

Table 7.2-1 Atomic and Molecular Diffusion Volumes for the Fuller Equation[a]

Atoms and Atomic Increments: $\hat{\upsilon}(10^{-3}\ m^3/kg\,atom)$			
C	16.5	(Cl)	19.5
H	1.98	(S)	17.0
O	5.48	Aromatic ring	−20.2
(N)	5.69	Heterocyclic ring	−20.2
Simple Molecules: $\hat{\upsilon}(10^{-3}\ m^3/kg\,atom)$			
H_2	7.07	CO	18.9
D_2	6.70	CO_2	26.9
He	2.88	N_2O	35.9
N_2	17.9	NH_3	14.9
O_2	16.6	H_2O	12.7
Air	20.1	$(CC1_2F_2)$	114.8
Ar	16.1	(SF_6)	69.7
Kr	22.8	$(C1_2)$	37.7
(Xe)	37.9	(Br_2)	67.2
Ne	5.59	(SO_2)	41.1

[a] Parentheses indicate that the listed value is based on only a few data points.

Example 7.2-1 Diffusion Coefficient of Ethanol in Air

Estimate the binary diffusion coefficient of ethanol in air at 313°K and 101 kPa.

Solution

Let component 1 be ethanol (CH_3CH_2OH, $M_1 = 46$ Da) and component 2 be air that we will treat as a pure substance ($M_2 = 29$ Da). With the molecular volumes determined from Table 7.2-1

$$\left(\sum \hat{\upsilon}\right)_1 = 2(16.5) + 6(1.98) + 1(5.48) = 50.4, \quad \left(\sum \hat{\upsilon}\right)_2 = 20.1 \tag{7.2-11a,b}$$

we compute a diffusion coefficient of

$$\mathcal{D}_{12} = \frac{101.3 \times 10^{-7}(313)^{1.75}}{(101.3)\left[(50.4)^{1/3} + (20.1)^{1/3}\right]^2}\sqrt{\frac{1}{46} + \frac{1}{29}} = 1.34 \times 10^{-5}\ m^2/s \tag{7.2-12}$$

The measured value listed in Table A4-1 is 1.45×10^{-5} m^2/s, so there is a 7.5% error in the prediction.

7.3 Diffusion Coefficients in Liquids

7.3.1 Einstein Model

In liquids, where intermolecular forces are much greater than in gases, a diffusion model of a solute s through a liquid solvent w can be developed by combining concepts from

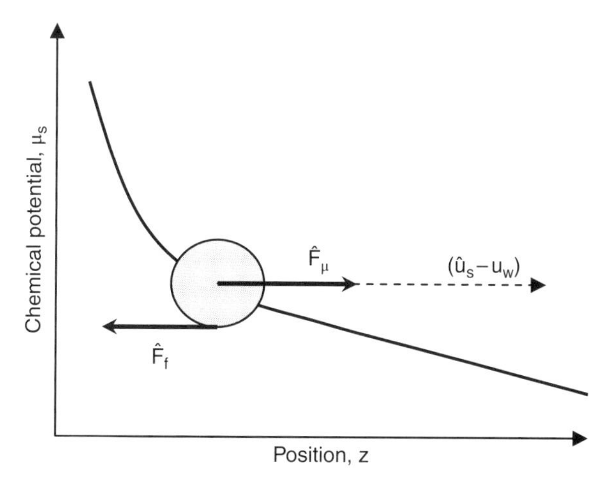

Figure 7.3-1 Diffusion of a solute molecule s through a liquid solvent w.

thermodynamics and hydrodynamics. Provided that solute molecules are much larger than solvent molecules, we visualize each solute molecule as a solid particle moving at a velocity $\hat{u}_s$ through a continuum of solvent moving at a velocity u_w. The difference between these velocities $(\hat{u}_s - u_w)$ is established by a balance between a thermodynamic body force $\hat{F}_\mu$ that propels a solute molecule and the frictional surface force $\hat{F}_f$ that retards the molecule as it moves through surrounding solvent (Fig. 7.3-1). The caret above these symbols indicates that they refer to quantities associated with a single molecule.

The mechanical work performed by the force $\hat{F}_\mu$ to move a solute molecule from position z to z + dz is

$$d\hat{w}_s = \hat{F}_\mu dz \tag{7.3-1}$$

Now suppose we add dm_s moles of solute to a solution held at constant T and P. If this process occurs in a reversible manner, then Equation 3.5-14 specifies the change in added work per mole:

$$\left(\frac{\partial w'_{rev}}{\partial m_s}\right)_{T,P,m_i} = -\mu_s \tag{7.3-2}$$

and the added work per added molecule is

$$\hat{w}_{s,rev} \equiv \frac{1}{\mathcal{A}}\left(\frac{\partial w'_{s,\,rev}}{\partial m_s}\right)_{T,P,m_i} = -\frac{\mu_s}{\mathcal{A}} \tag{7.3-3}$$

With the assumption that the diffusion process is occurring locally in a reversible manner, $\hat{w}_{s,rev} \approx \hat{w}_s$ and we combine Equations 7.3-1 and 7.3-3 to obtain

$$\hat{F}_\mu = \frac{d\hat{w}_{s,\,rev}}{dz} = -\frac{1}{\mathcal{A}}\frac{d\mu_s}{dz} \tag{7.3-4}$$

Excluding electrical effects, $\mu_s = \breve{\mu}_s$ and we utilize Equation 3.6-2 to get

$$\hat{F}_\mu = -\frac{\mathcal{R}T}{\mathcal{A}}\frac{d(\ln \gamma_s x_s)}{dz} \tag{7.3-5}$$

Hydrodynamic considerations reveal that the frictional force on a single sphere or ellipsoid moving through an incompressible fluid is proportional to its velocity relative to that of the solvent phase:

$$\hat{F}_f = -\frac{1}{\delta_s^\infty}(\hat{u}_s - u_w) \tag{7.3-6}$$

where δ_s^∞ is a mobility constant. The negative sign is necessary in this equation since $\hat{F}_f$ is opposite in direction to $(\hat{u}_s - u_w)$. The superscript "∞" on the mobility emphasizes the fact that the solution is sufficiently dilute that solute molecules act independently of one another. Equating the frictional force to the thermodynamic force, we get

$$\hat{F}_f = \hat{F}_\mu \Rightarrow \frac{(\hat{u}_s - u_w)}{\delta_s^\infty} = -\frac{\mathcal{R}T}{\mathcal{A}}\frac{d(\ln\gamma_s x_s)}{dz} \tag{7.3-7a,b}$$

In a dilute solution, we can approximate u_w by the molar average fluid velocity u^*, and the solute activity coefficient γ_s is close to one. The solute diffusion flux is then given by

$$J_s \equiv c_L x_s(\hat{u}_s - u^*) = -\mathcal{K}T\delta_s^\infty c_L \frac{dx_s}{dz} \tag{7.3-8}$$

where we have used the relationship between the gas constant and Boltzmann's constant, $\mathcal{R} = \mathcal{K}\mathcal{A}$. Comparing this result to the z-component of Fick's diffusion flux (Eq. 6.3-11), which does not account for electrical effects, we find that the diffusivity of a solute in dilute solution is directly related to the hydrodynamic mobility as

$$\mathcal{D}_s^\infty = \mathcal{K}T\delta_s^\infty \tag{7.3-9}$$

7.3.2 Diffusion Coefficients of Nonelectrolytes

The Stokes solution (Bird *et al.*, 2002, Eq. 2.6-15) for steady-state low Reynolds number flow of an unbounded, incompressible Newtonian fluid around a single sphere predicts a hydrodynamic mobility of

$$\delta_s^\infty = \frac{1}{6\pi a_s \mu} \tag{7.3-10}$$

where a_s is the sphere radius and μ is the fluid viscosity. Combining this equation with Equation 7.3-9 results in the Stokes–Einstein equation for the diffusion coefficient of solute s in a dilute solution:

$$\mathcal{D}_s^\infty = \frac{\mathcal{K}T}{6\pi a_s \mu} \tag{7.3-11}$$

It is not practical to use this equation to predict a binary diffusion coefficient since the diffusion radius a_s is usually not known. Instead, the Stokes–Einstein equation provides a way of estimating the size of a solute for which a diffusion coefficient has been measured. Equation 7.3-11 can also be used as a basis for extrapolating diffusion coefficients from one temperature to another since it predicts that $\mathcal{D}_s^\infty \mu/T$ is a constant for a particular solute–solvent pair.

Example 7.3-1 Diffusion Radius of Albumin in Water

Determine the diffusion radius of albumin at infinite dilution in water at 20°C.

Solution

According to Table A4-2, $\mathcal{D}_s^\infty$ (20°C) is 6.1 × 10^{-11} m^2/s. The viscosity of water at 20°C is 1.01 × 10^{-3} kg/(m-s). Rearranging the Stokes–Einstein equation, we compute a diffusion radius of albumin as

$$a_s = \frac{\mathcal{K}T}{6\pi\mu\mathcal{D}_s^\infty} = \frac{(1.38\times10^{-23})(298)}{6\pi(1.01\times10^{-3})(6.1\times10^{-11})} = 3.54\times10^{-9}\,\text{m} = 3.54\text{ nm} \quad (7.3\text{-}12)$$

This is considerably larger than the diameter of a water molecule whose radius is on the order of 0.1 nm.

In a semiempirical extension of the Stokes–Einstein equation (Wilke and Chang, 1955), the size of a solute molecule is estimated from the sum of its constituent atomic volumes $(\Sigma\hat{\upsilon})_s$ (Table 7.3-1), and $\mathcal{D}_s^\infty$ is computed as

$$\mathcal{D}_s^\infty = 7.4\times10^{-15}\frac{T\sqrt{\phi_w M_w}}{\left(\sum\hat{\upsilon}\right)_s^{0.6}\mu} \quad (7.3\text{-}13)$$

Table 7.3-1 Atomic and Molecular Diffusion Volumes for the Wilke–Chang Equation

Atoms and Atomic Increments: $\hat{\upsilon}(10^{-3}\text{m}^3/\text{kg atom})$			
Bromine	27.0	Oxygen in methyl esters	9.1
Carbon	14.8	Oxygen in higher esters	11.0
Chlorine	24.6	Oxygen in acids	12.0
Hydrogen	3.7	Oxygen in methyl ethers	9.9
Iodine	37.0	Oxygen in higher ethers	11.0
Nitrogen	15.6	Sulfur	25.6
Nitrogen in primary amines	10.5	Benzene ring	−15.0
Nitrogen in secondary amines	12.0	Naphthalene ring	−30.0
Oxygen	7.4		
Simple Molecules: $\hat{\upsilon}(10^{-3}\text{m}^3/\text{kg atom})$			
Air	29.9	H_2S	32.9
Br_2	53.2	I_2	71.5
Cl_2	48.4	N_2	31.2
CO	30.7	NH_3	25.8
CO_2	34.0	NO	23.6
COS	51.5	N_2O	36.4
H_2	14.3	O_2	25.6
H_2O	18.9	SO_2	44.8

Here M_w is solvent molecular weight and ϕ_w is a solvent association factor that is 1.0 for nonpolar solvents like benzene or ethers, 1.5 for ethanol, 1.9 for methanol, and 2.26 for water. Computations with Equation 7.3-13 require specific units for $\mathcal{D}_s^\infty$ [m²/s], μ[Pa-s], T[°K], and $(\Sigma\hat{\upsilon})_s[10^{-3}\ m^3/kg\ atom]$. Unlike the Stokes–Einstein equation, the Wilke–Chang equation is not restricted to solute molecules that are large compared to the solvent or to nonpolar solvents. The predictions of this equation are within ±10% of measured values.

When a liquid solution is not dilute, $\mathcal{D}_s$ will depend on solute concentration. For example, in an aqueous solution of sucrose or urea, $\mathcal{D}_s$ decreases as solute concentration increases (Table A4-2). At a higher concentration, a solution becomes more viscous, thereby diminishing the mobility of the solute molecule. We can estimate this effect by interpreting μ in the Wilke–Chang equation as the viscosity of the solution rather than that of the solvent.

Example 7.3-2 Diffusion Coefficient of Sucrose in Water

Estimate the diffusion coefficient of sucrose in water at 25°C.

Solution

Water has a molecular weight of 18 Da and its viscosity at 25°C is 9.02×10^{-4} kg/(m-s). The diffusion volume of sucrose ($C_{12}H_{22}O_{11}$) can be obtained from adding its atomic contributions from Table 7.3-1:

$$\left(\sum\hat{\upsilon}\right)_s = 12(14.8)+22(3.7)+11(7.4)=340.4 \tag{7.3-14}$$

so that

$$\mathcal{D}_s^\infty = \frac{7.40\times10^{-15}(298)\sqrt{(2.26)(18)}}{(340.4)^{0.6}(9.02\times10^{-4})} = 4.72\times10^{-10}\ m^2/s \tag{7.3-15}$$

This value is about 10% lower than the measured value of $\mathcal{D}_s^\infty = 0.523\times10^{-9}\ m^2/s$ given in Table A4-2 when $C_s \to 0$.

7.3.3 Diffusion Coefficients of Electrolytes

Ions in aqueous solution are surrounded by hydration shells that increase their diffusion radii above the values that are implied by the atomic volumes given in Table 7.3-1. Thus, it is not practical to use the Wilke–Chang equation to estimate the diffusion coefficient of electrolytes. Rather, these diffusion coefficients are obtained from electrical conductance data (Table 7.3-2).

The electrical conductance Λ_i of ion i in an aqueous solution is the current-to-voltage ratio that results when an electrical potential difference is applied between an anode and cathode submerged in the fluid. For a dilute solution, the limiting electrical conductance Λ_i^∞ [m²/(ohm·eq)] of an ion with a charge z_i is related to its hydrodynamic mobility (Snell *et al.*, 1965):

$$\delta_i^\infty = \left(\frac{\mathcal{A}}{\mathcal{F}^2}\right)\frac{\Lambda_i^\infty}{|z_i|} \tag{7.3-16}$$

Table 7.3-2 Ionic Conductances at Infinite Dilution in Water at 25°C

Cation (c)	Λ_c^∞ [cm²/(ohm-eq)]	Anion (a)	Λ_a^∞ [cm²/(ohm-eq)]
Ag^+	61.9	Br^-	78.4
Ba^{2+}	63.6	Cl^-	76.35
H^+	349.8	ClO_3^-	64.6
Li^+	38.7	ClO_4^-	67.4
Na^+	50.10	F^-	55.4
K^+	73.5	I^-	76.8
NH_4^+	73.6	NO_3^-	71.46
Ca^{2+}	59.5	OH^-	198.6
Cu^{2+}	56.6	CO_3^{2-}	69.3
Mg^{2+}	53.0	SO_4^{2-}	80.0
Zn^{2+}	52.8	Acetate$^-$	40.9

Abstracted from Perry *et al.* (1963).

Combining this equation with Equation 7.3-9, the diffusion coefficient of an ion can be evaluated from Λ_i^∞ as

$$\mathcal{D}_i^\infty = \left(\frac{\mathcal{R}T}{\mathcal{F}^2}\right)\frac{\Lambda_i^\infty}{|z_i|} \tag{7.3-17}$$

To predict the diffusion coefficient of a particular electrolyte, the diffusion coefficients of its constituent cations and anions must be combined in an appropriate manner. For example, during free diffusion of a strong 1–1 electrolyte "ca" composed of a cation "c" and an anion "a," we obtain the diffusion coefficient from Equations 6.4-11 and 7.3-17 as

$$\mathcal{D}_{ca}^\infty \equiv \left(\frac{2\mathcal{D}_c^\infty \mathcal{D}_a^\infty}{\mathcal{D}_c^\infty + \mathcal{D}_a^\infty}\right) = \left(\frac{2\mathcal{R}T}{\mathcal{F}^2}\right)\left(\frac{\Lambda_c^\infty \Lambda_a^\infty}{\Lambda_c^\infty + \Lambda_a^\infty}\right) \tag{7.3-18}$$

Example 7.3-3 Diffusion Coefficient of Sodium Chloride in Water

Estimate the diffusion coefficient of sodium chloride in dilute aqueous solution at 18°C.

Solution

Finding the limiting conductances of Na^+ and Cl^- at 25°C in Table 7.3-2 and correcting to a temperature of 18°C using Table 7.3-3, we determine that

$$\begin{aligned}\Lambda_{Na}^\infty &= 50.10 + 1.092(18-25) + 4.72\times10^{-3}(18-25)^2 - 1.15\times10^{-5}(18-25)^3\\ &= 43.08\ \text{cm}^2/(\text{ohm·eq}) = 0.004308\ \text{m}^2/(\text{ohm·eq})\\ \Lambda_{Cl}^\infty &= 76.35 + 1.540(18-25) + 4.65\times10^{-3}(18-25)^2 - 1.28\times10^{-5}(18-25)^3\\ &= 65.80\ \text{cm}^2/(\text{ohm·eq}) = 0.00658\ \text{m}^2/(\text{ohm·eq})\end{aligned} \tag{7.3-19a,b}$$

Table 7.3-3 Effect of Temperature on Limiting Ionic Conductance[a]

Ion	A_1	$A_2 \times 10^2$	$A_2 \times 10^4$
H^+	4.816	−1.031	−0.767
Li^+	0.890	0.441	−0.204
Na^+	1.092	0.472	−0.115
K^+	1.433	0.406	−0.318
Cl^-	1.540	0.465	−0.128
Br^-	1.544	0.447	−0.230
I^-	1.509	0.438	−0.217

Abstracted from Perry *et al.* (1963).
[a] $\Lambda_i^o(T) = \Lambda_i^o(25°C) + A_1[T(°C) - 25] + A_2[T(°C) - 25]^2 + A_3[T(°C) - 25]^3$.

Substituting these values into Equation 7.3-18, we predict the diffusion coefficient for NaCl at infinite dilution:

$$\mathcal{D}_{NaCl}^{\infty} \equiv \left[\frac{2(8.31)(291)}{(96500)^2}\right]\left[\frac{(0.00431)(0.00658)}{0.00431 + 0.00658}\right] = 1.35 \times 10^{-9}\ m^2/s \tag{7.3-20}$$

This is 7% larger than the diffusion coefficient of 1.26×10^{-9} m^2/s that has been measured for NaCl in a 0.05 molar solution at 18°C.

7.4 Transport in Porous Media Models of Tissue

Biological tissues are ordinarily heterogeneous, consisting of two or more immiscible phases. Some tissues, such as blood, are a suspension of particles in a continuous medium. Other materials, such as cell membranes, may be more suitably modeled as a porous solid matrix containing fluid-filled voids. In this section, we will discuss porous media models of tissue, whereas Section **7.5** is devoted to suspension models.

7.4.1 Representative Volume Element and Volume Averaging

Often it is not feasible to use a flux equation such as Equation 7.1-7 to analyze transport in the different phases of a heterogeneous material. Using this microscopic equation would require a detailed description of internal interfaces and the discontinuities in transport properties that they produce. By spatial averaging a microscopic equation, however, we can arrive at a macroscopic description that treats tissue as if it is a single homogeneous phase.

Consider a porous medium consisting of a solid matrix with a constant volume fraction of pores that are filled with a single interstitial fluid. There are two progressively larger scales at which we can view the spatial distribution of properties in this material (Fig. 7.4-1). The *microscale* refers to a collection of material points, each containing an array of atoms and molecules; the *macroscale* is a collection of representative elemental volumes (REV), each containing intermingled solid and fluid phases. A transport property at the microscale is obtained by averaging over the many atoms and molecules associated with a material point.

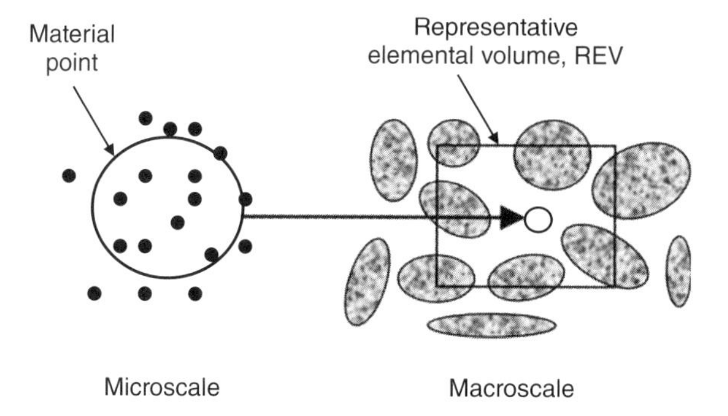

Figure 7.4-1 Size scales associated with a heterogeneous medium.

This point is contained within one phase only, either the solid or fluid phase of an REV. To determine a macroscale property, we average a microscale property over all the material points in both phases included in an REV. This effectively eliminates internal interfaces and allows us to arrive at a single value for a macroscale property. Because we want a macroscale quantity to be representative of the combined solid and interstitial liquid, an REV must be large compared to the characteristic graininess of the porous medium. At the same time, an REV must be small enough compared to the transport domain that the spatial distribution of macroscale properties is mathematically differentiable.

Suppose ψ is a microscopic scalar or vector quantity such as concentration or velocity that varies within a phase as well as from phase to phase. We define the corresponding macroscopic quantity $\bar{\psi}$ as a volume average of ψ over the REV domain ΔV^{REV}:

$$\bar{\psi} \equiv \frac{1}{\Delta V^{REV}} \int_{\Delta V^{REV}} \psi dV \tag{7.4-1}$$

With ε representing the volume fraction of the material occupied by fluid-filled pores, we also define a volume average quantity that is only associated with the pore domain $\varepsilon \Delta V^{REV}$. This is often called an intrinsic average quantity:

$$\bar{\psi}^{P} \equiv \frac{1}{\varepsilon \Delta V^{REV}} \int_{\varepsilon \Delta V^{REV}} \psi dV \tag{7.4-2}$$

In the following discussion, we consider microscopic properties that are associated with the interstitial liquid but are absent from the solid domain $(1-\varepsilon)\Delta V^{REV}$. For such a property, subdivision of the integral in Equation 7.4-1 leads to

$$\bar{\psi} = \frac{1}{\Delta V^{REV}} \left(\int_{\varepsilon \Delta V^{REV}} \psi dV + \cancel{\int_{(1-\varepsilon)\Delta V^{REV}} \psi dV} \right) \tag{7.4-3}$$

allowing us to conclude that

$$\bar{\psi}^{P} = \left(\frac{1}{\varepsilon}\right) \bar{\psi} \tag{7.4-4}$$

In common applications, local velocity is zero in a stationary solid matrix of a porous medium. Equation 7.4-4 can thus be used to relate intrinsic velocity in a pore $\bar{\mathbf{u}}^p$ to macroscopic velocity associated with the material as a whole $\bar{\mathbf{u}}$:

$$\bar{\mathbf{u}}^p = \left(\frac{1}{\varepsilon}\right)\bar{\mathbf{u}} \tag{7.4-5}$$

This Dupuit–Forchheimer relationship predicts that small values of ε lead to intrinsic velocities in the pores that are considerably greater than macroscopic velocities. For the interstitial fluid of brain tissues, for example, $\varepsilon \approx 0.1$ so that $\bar{\mathbf{u}}^p = 10\bar{\mathbf{u}}$.

For a solute s confined to pores, we can also use Equation 7.4-4 to relate intrinsic solute concentration $\bar{C}_s^p$ and flux $\bar{\mathbf{N}}_s^p$ to their corresponding macroscopic quantities:

$$\bar{C}_s^p = \left(\frac{1}{\varepsilon}\right)\bar{C}_s, \quad \bar{\mathbf{N}}_s^p = \left(\frac{1}{\varepsilon}\right)\bar{\mathbf{N}}_s \tag{7.4-6a, b}$$

7.4.2 Hydrodynamic Model of a Porous Medium

Figure 7.4-2 illustrates the axial transport of a single molecule of radius a_s along a straight cylindrical pore of radius a_p. The figure also shows a molecule of the same radius that cannot be transported because it is in contact with the pore wall. To avoid the latter situation, the center point of a solute molecule can only reside in the central region of the pore where $a_p - a_s > r > 0$. We define an accessibility factor ϖ_s as the ratio of the pore cross section that is accessible to the solute, $\pi(a_p - a_s)^2$, to the total pore cross section πa_p^2:

$$\varpi_s = \left(1 - \frac{a_s}{a_p}\right)^2 \tag{7.4-7}$$

For large solutes such as proteins and polysaccharides, ϖ_s can be substantially less than one. Because of their small molecular size, however, solvents can access virtually the entire pore cross section.

In the Einstein model of diffusion in an unconfined fluid, the friction force $\hat{F}_f$ on a solute molecule is directly proportional to its velocity relative to the solvent $(\hat{u}_s - u_w)$ and inversely proportional to its hydrodynamic mobility δ_s^∞ (Eq. 7.3-6). For diffusion in the confines of a

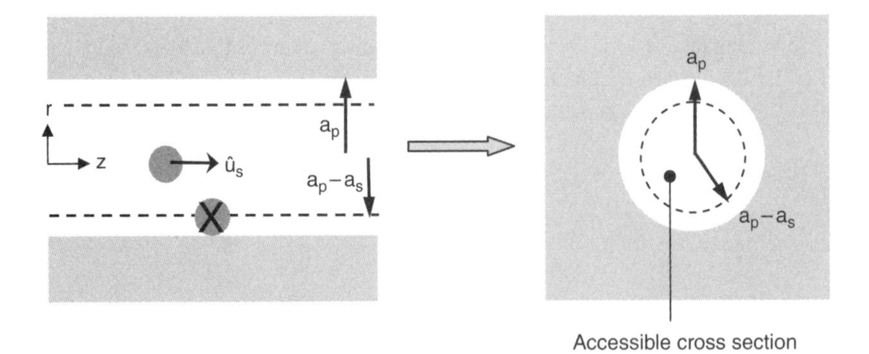

Figure 7.4-2 Transport of a molecule along a straight pore.

pore, we use a similar relationship that includes two additional hydrodynamic parameters, A and B (Anderson and Quinn, 1974):

$$\left(\hat{u}_s^p - A u_w^p\right) = -B\delta_s^\infty \hat{F}_f \tag{7.4-8}$$

Here, $\hat{u}_s^p$ is the local velocity of a solute molecule along a longitudinal flow streamline in the pore, and u_w^p is the local velocity of the solvent that surrounds the solute molecule. In an unconfined medium, A = 1 and B = 1. Within a pore, however, A and B deviate from unity as a consequence of retarded solvent velocity and diminished solute mobility, respectively. After solving this equation for $\hat{u}_s^p$, we can express the local solute flux along a flow streamline as:

$$N_s^p \equiv \left(c_L x_s^p\right)\hat{u}_s^p = A c_L x_s^p u_w^p - B c_L x_s^p \delta_s^\infty \hat{F}_f \tag{7.4-9a, b}$$

where x_s^p is a local value of solute mole fraction. Since the friction force $\hat{F}_f$ is equal to the same thermodynamic force F_μ as in the Stokes–Einstein model (Eq. 7.3-7b), we can further write that

$$N_s^p = A c_L x_s^p u_w^p - B\delta_s^\infty \mathcal{K} T c_L \frac{d\left(\gamma_s x_s^p\right)}{dz} \tag{7.4-10}$$

In a dilute solution, c_L is constant and $\gamma_s \to 1$. Also, we identify the solute concentration in a pore as $C_s^p = c_L x_s^p$ and the diffusion coefficient in an unconfined fluid as $\mathcal{D}_s^\infty = \mathcal{K}T\delta_s^\infty$ (Eq. 7.3-11). Thus,

$$N_s^p = A u_w^p C_s^p - B\mathcal{D}_s^\infty \frac{dC_s^p}{dz} \tag{7.4-11}$$

Assuming steady, laminar flow of an incompressible Newtonian solvent through a long cylindrical pore and neglecting the effect of solute molecules, the radial distribution of axial velocity is given by the Hagen–Poiseuille solution (Bird *et al.*, 2002; eqs. 2.3-18 and 2.3-20):

$$u_w^p = 2\bar{u}^p\left(1 - \frac{r^2}{a_p^2}\right) \tag{7.4-12}$$

where $\bar{u}^p$ is the average velocity over a pore cross section. The solute flux along a flow streamline can now be written as

$$N_s^p = 2A(r)\bar{u}^p\left(1 - \frac{r^2}{a_p^2}\right)C_s^p - \mathcal{D}_s^\infty B(r)\frac{dC_s^p}{dz} \tag{7.4-13}$$

We expect $N_s^p(r,z)$ to be a cylindrically symmetric quantity. Averaging N_s^p over all streamlines that occupy the pore cross section, we obtain

$$\bar{N}_s^p(z) \equiv \frac{1}{\pi a_p^2}\int_0^{2\pi}\int_0^{a_p} N_s^p(r,z)\,r\,dr\,d\theta = \frac{2}{a_p^2}\int_0^{a_p} N_s^p\,r\,dr \tag{7.4-14}$$

Combining this relation with Equation 7.4-13, we get

$$\bar{N}_s^p = \frac{4\bar{u}^p}{a_p^2}\int_0^{a_p} A\left(1 - \frac{r^2}{a_p^2}\right)C_s^p\,r\,dr - \frac{2\mathcal{D}_s^\infty}{a_p^2}\frac{d}{dz}\left[\int_0^{a_p} B C_s^p\,r\,dr\right] \tag{7.4-15}$$

Following the suggestion of Anderson and Quinn (1974), we assume that $C_s^p(r,z)$ is radially uniform at a concentration $C_s^{pa}(z)$ in the accessible pore cross section and is zero in the inaccessible cross section. Then, C_s^p is a continuous function of z but a discontinuous function of r:

$$C_s^p(z) = \begin{cases} C_s^{pa}(z) & \text{when } (a_p - a_s) > r \geq 0 \\ 0 & \text{when } r \geq (a_p - a_s) \end{cases} \tag{7.4-16}$$

Inserting this expression into Equation 7.4-15, we get

$$\bar{N}_s^p = \frac{4\bar{u}^p C_s^{pa}}{a_p^2} \int_0^{(a_p - a_s)} A\left(1 - \frac{r^2}{a_p^2}\right) r\mathrm{d}r - \frac{2\mathcal{D}_s^\infty C_s^{pa}}{a_p^2} \frac{\mathrm{d}}{\mathrm{d}z}\left[\int_0^{(a_p - a_s)} B r\mathrm{d}r\right] \tag{7.4-17}$$

The average solute concentration across the entire pore cross section can be formulated as

$$\bar{C}_s^p(z) \equiv \frac{1}{\pi a_p^2} \int_0^{2\pi} \int_0^{a_p} C_s^p(z) r\mathrm{d}r\mathrm{d}\theta = \frac{2C_s^{pa}}{a_p^2} \int_0^{(a_p - a_s)} r\mathrm{d}r\mathrm{d}\theta \tag{7.4-18}$$

so that

$$\bar{C}_s^p(z) = \frac{(a_p - a_s)^2}{a_p^2} C_s^{pa}(z) = \varpi_s C_s^{pa}(z) \tag{7.4-19}$$

Combining this relation with Equation 7.4-17, we get

$$\bar{N}_s^p = \frac{4\bar{u}^p \bar{C}_s^p}{(a_p - a_s)^2} \int_0^{a_p - a_s} A\left(1 - \frac{r^2}{a_p^2}\right) r\mathrm{d}r - \frac{2\mathcal{D}_s^\infty}{(a_p - a_s)^2} \frac{\mathrm{d}\bar{C}_s^p}{\mathrm{d}z} \int_0^{a_p - a_s} B r\mathrm{d}r \tag{7.4-20}$$

After some algebraic manipulations, this equation can be written in an equivalent form:

$$\bar{N}_s^p = (2 - \varpi_s)\bar{A}\bar{u}^p \bar{C}_s^p - \bar{B}\mathcal{D}_s^\infty \frac{\mathrm{d}\bar{C}_s^p}{\mathrm{d}z} \tag{7.4-21}$$

where we have defined average hydrodynamic parameters as

$$\bar{A} = \frac{\int_0^{a_p - a_s} A(r)\left(a_p^2 - r^2\right) r\mathrm{d}r}{\int_0^{a_p - a_s} \left(a_p^2 - r^2\right) r\mathrm{d}r}, \quad \bar{B} = \frac{\int_0^{a_p - a_s} B(r) r\mathrm{d}r}{\int_0^{a_p - a_s} r\mathrm{d}r} \tag{7.4-22a,b}$$

To put Equation 7.4-21 in a form that is comparable to a microscopic flux equation, we define a convection correction factor β_s and a hindered diffusion coefficient $\bar{\mathcal{D}}_s$ that account for the interaction of solvent and solute with the pore walls:

$$\beta_s \equiv (2 - \varpi_s)\bar{A}, \quad \bar{\mathcal{D}}_s \equiv \mathcal{D}_s^\infty \bar{B} \tag{7.4-23a,b}$$

The intrinsic flux is then given in terms of intrinsic concentration and velocity as

$$\bar{N}_s^p = \beta_s \bar{u}^p \bar{C}_s^p - \bar{\mathcal{D}}_s \frac{d\bar{C}_s^p}{dz} \tag{7.4-24}$$

Usually, we are interested in how the macroscopic flux through the entire surface of a porous material depends on the macroscopic concentration and macroscopic velocity across that surface. Using Equations 7.4-5 and 7.4-6, we transform Equation 7.4-24 to a macroscopic form:

$$\bar{N}_s = \left(\frac{\beta_s}{\varepsilon}\right) \bar{u}\bar{C}_s - \bar{\mathcal{D}}_s \frac{d\bar{C}_s}{dz} \tag{7.4-25}$$

We can extend this equation to three dimensions by replacing $\bar{N}_s$ with the flux vector $\bar{\mathbf{N}}_s$ and $d\bar{C}_s/dz$ with the vector gradient $\nabla\bar{C}_s$:

$$\bar{\mathbf{N}}_s = \left(\frac{\beta_s}{\varepsilon}\right) \bar{\mathbf{u}}\bar{C}_s - \bar{\mathcal{D}}_s \nabla\bar{C}_s \tag{7.4-26}$$

This equation is similar to Equation 7.1-7 for a homogeneous material. The primary difference is the need to use β_s and ε to correct the convection term.

With rare exception, heterogeneous materials do not contain straight parallel pores. Figure 7.4-3 shows an example of a porous medium containing crooked undivided pores and another material with a brick-and-mortar arrangement of symmetrically dividing pores. In such materials, the value of $\bar{\mathcal{D}}_s$ is smaller than predicted by Equation 7.4-23b because the pores are not aligned with the macroscopic diffusion direction z. Defining a tortuosity factor $\tau_p > 1$ as the actual length of the diffusion path relative to the distance a molecule transverses in the z direction, we correct the hindered diffusion coefficient as follows:

$$\bar{\mathcal{D}}_s = \frac{D_s^\infty \bar{B}}{\tau_p} \tag{7.4-27}$$

When a tissue has a regular structure, it is possible to make a priori estimates of τ_p. For example, the pore structure of the skin's stratum corneum can be idealized by the equally spaced brick-and-mortar geometry in Figure 7.4-3 (Barbero and Frasch, 2006). The tortuosity is then given by (Nielsen, 1967)

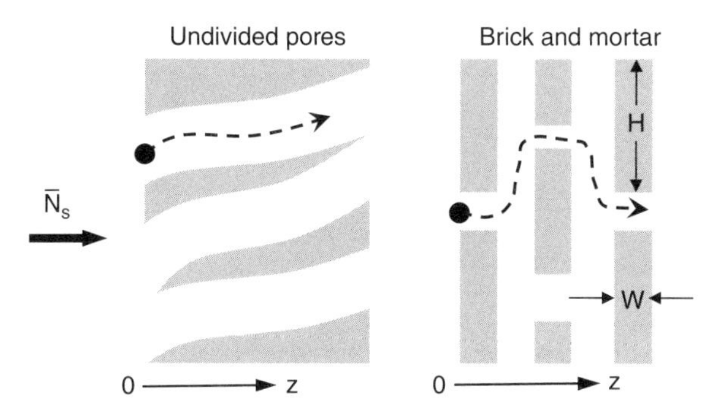

Figure 7.4-3 Examples of tortuous porous materials.

$$\tau_p = 1 + \left(\frac{H}{2W}\right)(1-\varepsilon) \tag{7.4-28}$$

where H and W are the height and width of a brick, respectively, and ε is the volume fraction of mortar. Because pore structures are usually so irregular, however, the value of τ_p for a particular material cannot be predicted. Rather, it can be inferred from transport data, as in the following example problem.

7.4.3 Renkin Model of Solute Diffusion

As an approximation to the parameters $\bar{A}$ and $\bar{B}$ that are defined by Equation 7.4-22a,b, we apply the hydrodynamic result for a sphere that is confined to the centerline of a straight cylindrical tube (Anderson and Quinn, 1974):

$$\bar{A} \approx A(0) = \left[1 - \frac{2}{3}\left(\frac{a_s}{a_p}\right)^2 - 0.163\left(\frac{a_s}{a_p}\right)^3\right] \tag{7.4-29}$$

$$\bar{B} \approx B(0) = \left[1 - 2.1044\left(\frac{a_s}{a_p}\right) + 2.089\left(\frac{a_s}{a_p}\right)^3 - 0.948\left(\frac{a_s}{a_p}\right)^5\right] \tag{7.4-30}$$

Combining Equation 7.4-23b with Equations 7.4-7 and 7.4-29, we obtain the convection correction coefficient

$$\beta_s = \left[1 + 2\frac{a_s}{a_p} - \left(\frac{a_s}{a_p}\right)^2\right]\left[1 - \frac{2}{3}\left(\frac{a_s}{a_p}\right)^2 - 0.163\left(\frac{a_s}{a_p}\right)^3\right] \tag{7.4-31}$$

and substituting Equation 7.4-30 into Equation 7.4-27, we get the hindered diffusion coefficient

$$\bar{\mathcal{D}}_s = \frac{\mathcal{D}_s^\infty}{\tau_p}\left[1 - 2.1044\left(\frac{a_s}{a_p}\right) + 2.089\left(\frac{a_s}{a_p}\right)^3 - 0.948\left(\frac{a_s}{a_p}\right)^5\right] \tag{7.4-32}$$

For very small solute molecules, $\beta_s \approx 1$ and $\bar{\mathcal{D}}_s \approx \mathcal{D}_s^\infty/\tau_p$, and the macroscopic flux equation reduces to

$$\bar{N}_s = \frac{\bar{u}}{\varepsilon}\bar{C}_s - \frac{\mathcal{D}_s^\infty}{\tau_p}\frac{d\bar{C}_s}{dz} \tag{7.4-33}$$

Example 7.4-1 Pore Model of the Extracellular Matrix

The interstitial space in tissue is a heterogeneous medium containing a mixture of liquid and insoluble structural proteins. Nugent and Jain (1984) studied the *in vivo* diffusion of dextrans in the interstitial space of normal rat tissue and neoplastic rat tissue (i.e., a tumor formed by an abnormal proliferation of cells). The dextran molecules were tagged with fluorescein isothiocyanate such that $\bar{\mathcal{D}}_s$ could be inferred from the spatial distribution of extravascular

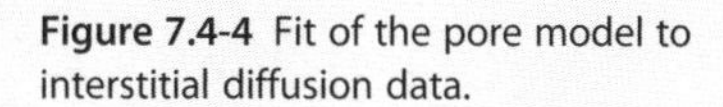
Figure 7.4-4 Fit of the pore model to interstitial diffusion data.

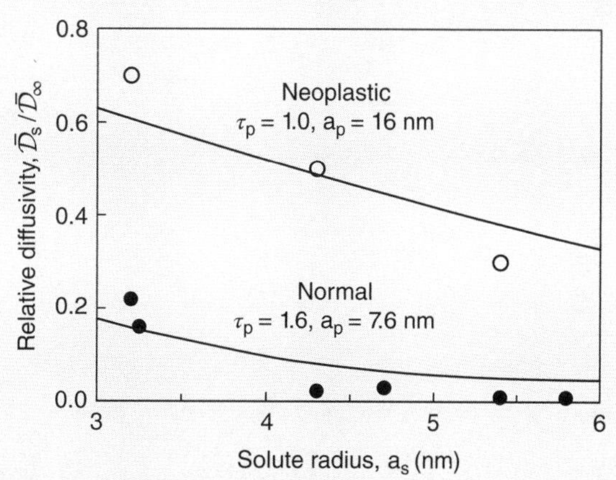

fluorescence emission. The dextrans, with molecular weights ranging from 19,400 to 71,800 Da, can be idealized as spherical molecules with radii from 3.1 to 5.8 nm.

What can we learn about the comparative geometry of these tissues by applying the Renkin pore model to the data in Figure 7.4-4?

Solution

If the Renkin model is capable of representing the diffusivity data, then according to Equation 7.4-32

$$\frac{\bar{\mathcal{D}}_s}{\mathcal{D}_s^\infty} = \frac{1}{\tau_p}\left[1-2.0144\left(\frac{a_s}{a_p}\right)+2.089\left(\frac{a_s}{a_p}\right)^3-0.948\left(\frac{a_s}{a_p}\right)^5\right] \tag{7.4-34}$$

The curves in Figure 7.4-4 represent nonlinear least squares regressions of the data to this equation with both τ_p and a_p employed as adjustable parameters and with the constraint that $\tau_p \geq 1$. Although we do not expect pores in extracapillary tissue to contain an array of identical cylindrical passages, the τ_p and a_p estimates shown in the figure can be interpreted as "effective values" of these geometric quantities. Judging from these results, the larger $\bar{\mathcal{D}}_s/\mathcal{D}_s^\infty$ in neoplastic tissue occurs because its pore tortuosity is less and its pore size is greater than in normal tissue.

7.4.4 Hydraulic and Solute Permeabilities

Hydraulic and solute permeabilities directly relate transport rates to pressure and concentration driving forces, respectively. To formulate permeabilities in terms of more fundamental material properties and geometry, we separately consider flow and diffusion through a planar slab of a porous material (Fig. 7.4-5). The slab has a thickness h and cross-sectional area A. It contains pores of radius a_p, tortuosity τ_p, and length $\tau_p h$. Homogeneous fluids A and B contact the slab at its boundaries where z = 0 and z = h.

Hydraulic Permeability

To analyze hydraulic permeability, we impose pressures P^A and P^B in the external fluids. At the boundaries of the slab, the pressures in the porous medium, P(0) and P(h), are equal to P^A

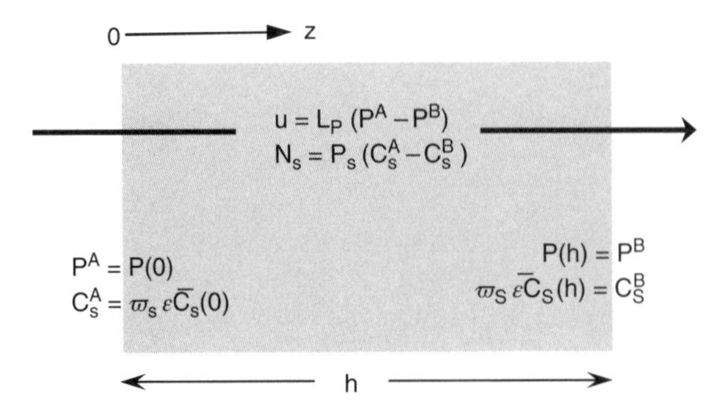

Figure 7.4-5 Transport through a porous medium in contact with homogeneous media.

and P^B. At steady state, the macroscopic velocity $\bar{u}$ produced by these pressures will be constant and equal to the velocity u of the external solutions.

Within liquid-filled pores of heterogeneous materials, flow is typically laminar with viscous forces being much greater than inertial forces. Also, the pore length is much greater than its diameter. It is therefore appropriate to use the Hagen–Poiseuille equation to relate the cross-sectional average velocity $\bar{u}^P$ through a pore to the pressure drop between the pore openings (Bird *et al.*, 2002; Eq. 2.3-20):

$$\bar{u}^P = \frac{a_p^2}{8\mu\tau_p h}[\bar{P}(0) - \bar{P}(h)] \tag{7.4-35}$$

where μ is the fluid viscosity. From the one-dimensional form of Equation 7.4-5, we see that the average pore velocity is related to the macroscopic velocity by $\bar{u}^P = \bar{u}/\varepsilon = u/\varepsilon$. In addition, $P(0) = P^A$ and $P(h) = P^B$ so that Equation 7.4-35 can be written as

$$u = \frac{a_p^2 \varepsilon}{8\mu\tau_p h}\left(P^A - P^B\right) \tag{7.4-36}$$

The hydraulic permeability L_P is defined as the ratio of velocity to the pressure drop (Eq. 6.5-27):

$$L_P \equiv \frac{u}{P^A - P^B} = \frac{a_p^2 \varepsilon}{8\mu\tau_p h} \tag{7.4-37}$$

We can generalize Equation 7.4-37 to a heterogeneous solid consisting of packed particles. Suppose n is the number of pores of mean radius a_p in a solid of volume Ah, tortuosity τ_p, and porosity ε. Then the total volume of pores is $\varepsilon Ah = n\left(\pi a_p^2 \tau_p h\right)$, the total particle volume is $(1-\varepsilon)(Ah) = (1-\varepsilon)\left(n\pi a_p^2 \tau_p h/\varepsilon\right)$, and the total pore surface is $n(2\pi a_p \tau_p h)$. Since the surface-to-volume ratio of an individual particle ϕ among a group of packed particles is equivalent to the ratio of total particle surface to total particle volume,

$$\phi = \frac{n(2\pi a_p \tau_p h)}{(1-\varepsilon)n\pi a_p^2 \tau_p h/\varepsilon} = \frac{2\varepsilon}{a_p(1-\varepsilon)} \tag{7.4-38}$$

Substituting ϕ into Equation 7.4-36, we get

$$u = \frac{\varepsilon^3}{2\mu\phi^2\tau_p h(1-\varepsilon)^2}\left(p^A - p^B\right) \Rightarrow L_p = \frac{\varepsilon^3}{2\mu\phi^2\tau_p h(1-\varepsilon)^2} \qquad (7.4\text{-}39a,b)$$

For spherical particles of diameter d, the particle surface-to-volume ratio is 6/d so that

$$u = \frac{d^2\varepsilon^3}{72\mu\tau_p h(1-\varepsilon)^2}\left(p^A - p^B\right) \Rightarrow L_p = \frac{d^2\varepsilon^3}{72\mu\tau_p h(1-\varepsilon)^2} \qquad (7.4\text{-}40a,b)$$

This equation has the same form as the Carman–Kozeny equation (Nield and Bejan, 1992) that applies to heterogeneous solids composed of packed fibers or nearly spherical particles; in that case, $72\tau_p$ has an empirical value of 180.

Solute Permeability

To analyze solute permeability, we impose solute concentrations C_s^A and C_s^B in the external homogeneous fluids. At steady state, the concentration driving force $\left(C_s^A - C_s^B\right)$ will produce a constant macroscopic solute flux $\bar{N}_s$ that is equal to the solute flux N_s in the external solutions.

An important aspect of mass transport through a porous medium is the solute boundary conditions between the pore mouths and the two external solutions. Solute can only occupy the accessible region of the pore entrances where its concentrations are $C_s^{pa}(0)$ and $C_s^{pa}(h)$. These concentrations must equal those of the adjoining homogeneous solutions so that $C_s^{pa}(0) = C_s^A$ and $C_s^{pa}(h) = C_s^B$. According to Equation 7.4-19, $\bar{C}_s^p(z) = \varpi_s C_s^{pa}(z)$ where $\bar{C}_s^p(z)$ is solute concentration averaged across an entire pore cross section. Therefore, relationships between $\bar{C}_s^p(0)$ and $\bar{C}_s^p(h)$ and the external solute concentrations are given by

$$\bar{C}_s^p(0) = \varpi_s C_s^A, \quad \bar{C}_s^p(h) = \varpi_s C_s^B \qquad (7.4\text{-}41a,b)$$

Another form of these boundary conditions can be obtained by noting that $\bar{C}_s^p(z) = \bar{C}_s(z)/\varepsilon$ (Eq. 7.4-6a):

$$\bar{C}_s(0) = \varpi_s \varepsilon C_s^A, \quad \bar{C}_s(h) = \varpi_s \varepsilon C_s^B \qquad (7.4\text{-}42a,b)$$

The factor $\varpi_s\varepsilon$ serves as a partition coefficient between the macroscopic solute concentration of the porous media and the external solute concentration. Unlike a thermodynamic partition coefficient that is based on continuity of chemical potential at an interface, $\varpi_s\varepsilon$ is based on continuity of concentration and derives its value from geometric considerations.

For diffusion in the absence of convection, the one-dimensional form of the macroscopic solute flux equation (Eq. 7.4-25) becomes

$$\bar{N}_s = -\bar{\mathcal{D}}_s \frac{d\bar{C}_s}{dz} \qquad (7.4\text{-}43)$$

Provided that $\bar{\mathcal{D}}_s$ is constant, we can integrate this equation between the slab surfaces:

$$\bar{N}_s = \frac{\bar{\mathcal{D}}_s}{h}\left[\bar{C}_s(0) - \bar{C}_s(h)\right] \qquad (7.4\text{-}44)$$

With $\bar{N}_s = N_s$, $\bar{C}_s(0) = \varpi_s \varepsilon C_s^A$, and $\bar{C}_s(h) = \varpi_s \varepsilon C_s^B$, this equation becomes

$$N_s = \frac{\varpi_s \varepsilon \bar{\mathcal{D}}_s}{h}(C_s^A - C_s^B) \tag{7.4-45}$$

The solute permeability P_s is defined by Equation 6.5-29 as the ratio of the solute flux to the external concentration difference such that

$$P_s \equiv \frac{N_s}{C_s^A - C_s^B} = \frac{\varpi_s \varepsilon \bar{\mathcal{D}}_s}{h} \tag{7.4-46}$$

7.5 Transport in Suspension Models of Tissue

7.5.1 Fiber Matrix Model

A matrix of insoluble macromolecular fibers suspended in a homogeneous solution (Fig. 7.5-1; left) is a useful representation of some biological media. Using a probabilistic analysis, Curry and Michel (1980) modeled diffusion (without convection) of spherical solute molecules of radius a_s through a suspension of randomly oriented impermeable fibers of radius a_f. The length-to-diameter ratio of the fibers was assumed to be so large that end effects could be ignored. By considering the likelihood of tangential contacts of solute molecules with surrounding fibers, they determined an accessibility factor given by

$$\varpi_s = \frac{1}{\varepsilon}\exp\left[-(1-\varepsilon)\left(\frac{2a_s}{a_f} + \frac{a_s^2}{a_f^2}\right)\right]; \quad \varepsilon = 1 - \pi a_f^2 L_f \tag{7.5-1a,b}$$

where ε is the void fraction occupied by suspending medium and L_f is the summed length of all the fibers in a unit volume of suspension.

From the viewpoint of kinetic theory, the presence of fibers in a suspension affects the diffusion coefficient by reducing the mean free path of solute motion. In the fiber matrix model, the mean free path was deduced from the distribution of distances that a solute molecule travels between fiber contacts. This resulted in a hindered diffusion coefficient given by

$$\bar{\mathcal{D}}_s = \mathcal{D}_s^\infty \exp\left[-\sqrt{1-\varepsilon}\left(1 + \frac{a_s}{a_f}\right)\right] \tag{7.5-2}$$

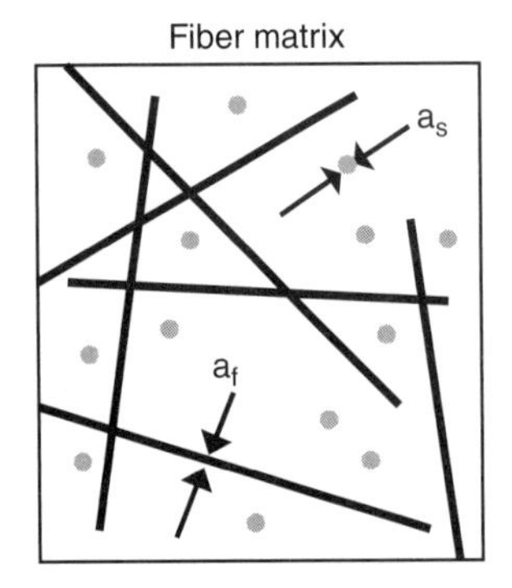

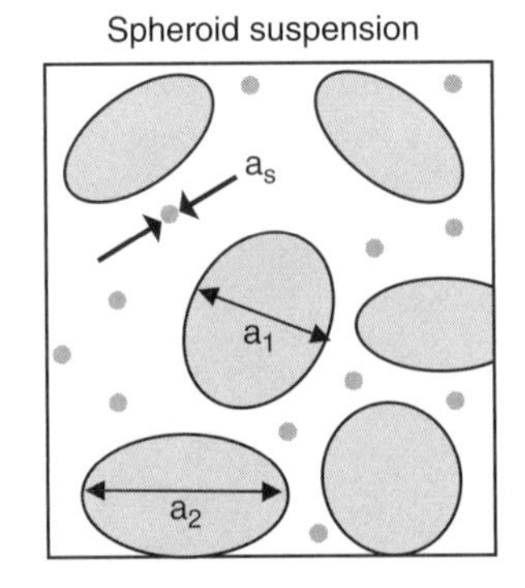

Figure 7.5-1 Suspension models of tissue.

where $\mathcal{D}_s^\infty$ is the diffusion coefficient of the solute in the suspending medium alone. In the absence of fibers, $\varepsilon \rightarrow 1$ and these equations reduce to the expected result that $\varpi_s \rightarrow 1$ and $\bar{\mathcal{D}}_s \rightarrow \mathcal{D}_s^\infty$. If the entire material is filled with fibers, then $\varepsilon \rightarrow 0$ and the equations do not exhibit the expected result that $\varpi_s \rightarrow 0$ and $\bar{\mathcal{D}}_s \rightarrow 0$. Therefore, this model is not appropriate for high fiber loadings.

Equations 7.4-46, 7.5-1a,b, and 7.5-2 give rise to the permeability of solute through a fiber matrix of thickness h:

$$P_s = \frac{\varpi_s \varepsilon \bar{\mathcal{D}}_s}{h} = \frac{\mathcal{D}_s^\infty}{h} \exp\left[-(1-\varepsilon)\left(\frac{2a_s}{a_f} + \frac{a_s^2}{a_f^2}\right)\right] \exp\left[-\sqrt{1-\varepsilon}\left(1 + \frac{a_s}{a_f}\right)\right] \tag{7.5-3}$$

Example 7.5-1 Fiber Matrix Model of the Extracellular Matrix

Determine whether the $\bar{\mathcal{D}}_s/\mathcal{D}_s$ data presented in Example 7.4-1 can be better explained by the fiber matrix model than by the Renkin pore model.

Solution

If the fiber matrix model is appropriate, then the data should follow the relationship

$$\frac{\bar{\mathcal{D}}_s}{\mathcal{D}_s^\infty} = \exp\left[-\sqrt{1-\varepsilon}\left(1 + \frac{a_s}{a_f}\right)\right] \tag{7.5-4}$$

Hyaluronate, an abundant fibrous constituent of the extracellular matrix, has a radius of 0.54 nm. A nonlinear least squares analysis of the data to Equation 7.5-4 was performed with $a_f = 0.54$ nm and ε treated as an adjustable parameter. The results are represented by the two solid curves in Figure 7.5-2. Using the corresponding ε estimates, we can compute the fiber length per unit tissue volume from the relation $L_f = (1-\varepsilon)/(\pi a_f^2)$. For normal tissue,

$$L_f = \frac{(1-0.92)(10^6)}{\pi(0.54)^2} = 87{,}000\ \mu m/\mu m^3 \tag{7.5-5}$$

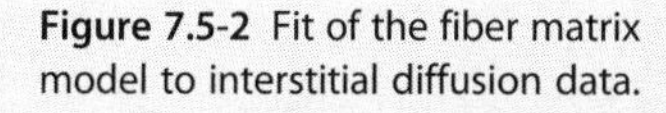

Figure 7.5-2 Fit of the fiber matrix model to interstitial diffusion data.

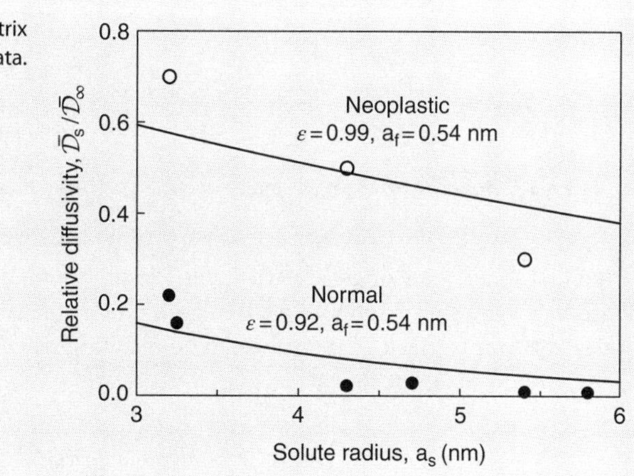

and for neoplastic tissue

$$L_f = \frac{(1-0.99)(10^6)}{\pi(0.54)^2} = 11{,}000\ \mu m/\mu m^3 \qquad (7.5\text{-}6)$$

In other words, if all the fibers in a 1 μm^3 cube of normal tissue were connected end to end, they could crisscross the cube about 87,000 times. In the same-size cube of neoplastic tissue, the fibers could crisscross only 11,000 times. Thus, according to the fiber matrix model, the larger $\bar{\mathcal{D}}_s/\mathcal{D}_s^\infty$ of the neoplastic tissue is due to a smaller packing density of fibers than in normal tissue.

From a comparison of Figures 7.4-4 and 7.5-2, we see that the fiber matrix and pore models provide virtually the same fit to the data. Thus, based on these data alone, we cannot tell which of the two models is a more accurate representation of reality.

7.5.2 Spheroidal Suspension Models

Diffusion of Inert Solute through a Suspension of Spherical Particles

The hindered diffusion coefficient for a solute in a particulate suspension can be determined by summing the contributions of diffusion in the vicinity of individual particles. To illustrate this, we describe ordinary diffusion through a suspension of identical spherical particles using the approach of Stroeve and coworkers (1976a). We consider the special case in which solute is soluble but nonreactive in both the suspending medium and the particles.

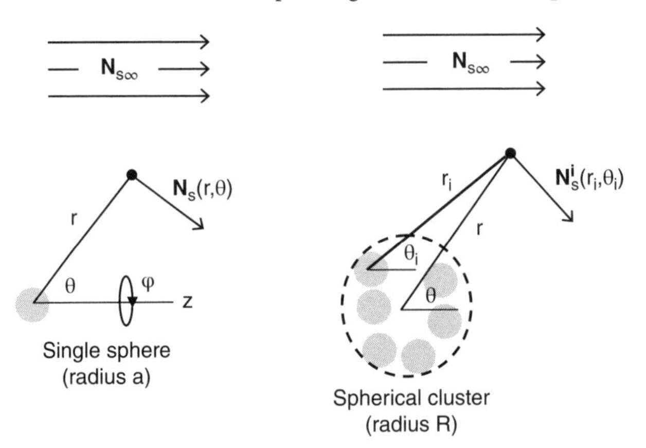

Figure 7.5-3 Diffusion surrounding a single spherical particle and a particle cluster.

We first focus on a single spherical particle of radius "a" surrounded by an infinite fluid phase (Fig. 7.5-3; left). Suppose we impose a uniform rectilinear gradient of solute concentration far from surface of the sphere. The resulting deviation of a local solute flux $\mathbf{N}_s(r,\theta)$ from flux $\mathbf{N}_{s\infty}$ far from the particle is

$$\mathbf{N}_s - \mathbf{N}_{s\infty} = |N_{s\infty}|\nabla\left[\left(\frac{1-\chi_s}{2+\chi_s}\right)\frac{a^3}{r^2}\cos\theta\right] \quad (r \geq a) \qquad (7.5\text{-}7)$$

where $|N_{s\infty}|$ is the magnitude of $\mathbf{N}_{s\infty}$ and χ_s is the permeability of solute through the sphere relative to the suspending medium:

$$\chi_s \equiv \frac{\lambda_s^{d,c} \mathcal{D}_s^d}{\mathcal{D}_s^c} \tag{7.5-8}$$

Here, $\mathcal{D}_s^d$ and $\mathcal{D}_s^c$ are the diffusion coefficients of solute s in the dispersed particle phase (d) and the continuous suspending medium (c), and $\lambda_s^{d,c}$ is the thermodynamic partition coefficient of solute between the two phases.

For a cluster of n noninteracting spheres confined to a spherical region of radius R (Fig. 7.5-3; right), the deviation of the external flux field from the undisturbed flux can be obtained by summing the individual contributions $\mathbf{N}_s^i$ of the spheres:

$$\mathbf{N}_s^{\text{cluster}} - \mathbf{N}_{s\infty} = \sum_{i=1}^{n} \left(\mathbf{N}_s^i - \mathbf{N}_{s\infty}\right) \tag{7.5-9}$$

Taking r_i and θ_i as spherical coordinates relative to the center of the ith sphere, we formulate $\left(\mathbf{N}_s^i - \mathbf{N}_{s\infty}\right)$ with Equation 7.5-7 such that Equation 7.5-9 becomes

$$\mathbf{N}_s^{\text{cluster}} - \mathbf{N}_{s\infty} = |N_{s\infty}|\nabla\left[\sum_{i=1}^{n}\left(\frac{1-\chi_s}{2+\chi_s}\right)\frac{a^3}{r_i^2}\cos\theta_i\right] \tag{7.5-10}$$

If we observe the effect of the cluster at a faraway point, then r_i and θ_i can be approximated by the spherical coordinates r and θ relative to the center of cluster. Equation 7.5-10 then becomes

$$\mathbf{N}_s^{\text{cluster}} - \mathbf{N}_{s\infty} = N_{s\infty}\nabla\left[\left(\frac{1-\chi_s}{2+\chi_s}\right)\frac{na^3}{r^2}\cos\theta\right] \quad (r \gg R) \tag{7.5-11}$$

To obtain the hindered diffusion coefficient $\bar{\mathcal{D}}_s$ of solute through the particle cluster and its associated suspending medium, we treat the region $R > r \geq 0$ as a hypothetical single phase with a macroscopic solute permeability ratio given by

$$\bar{\chi}_s \equiv \frac{\lambda_s^{d,c} \bar{\mathcal{D}}_s}{\mathcal{D}_s^c} \tag{7.5-12}$$

After replacing "a" by R and χ_s by $\bar{\chi}_s$, we write Equation 7.5-7 for the particle cluster as

$$\mathbf{N}_s^{\text{cluster}} - \mathbf{N}_{s\infty} = |N_{s\infty}|\nabla\left[\left(\frac{1-\bar{\chi}_s}{2+\bar{\chi}_s}\right)\frac{R^3}{r^2}\cos\theta\right] \quad (r \geq R) \tag{7.5-13}$$

Since the external flux deviation caused by the actual cluster and its single-phase representation are equivalent, Equation 7.5-11 must be equal to Equation 7.5-13. It follows that

$$na^3\left(\frac{1-\chi_s}{2+\chi_s}\right) = R^3\left(\frac{1-\bar{\chi}_s}{2+\bar{\chi}_s}\right) \tag{7.5-14}$$

Realizing that $(1 - na^3/R^3) \equiv \varepsilon$ is the volume fraction of the suspending medium associated with the cluster, we can solve this equation for $\bar{\mathcal{D}}_s$:

$$\bar{\mathcal{D}}_s = \mathcal{D}_s^c\left\{1-(1-\chi_s)\left[\frac{3(1-\varepsilon)}{3-\varepsilon(1-\chi_s)}\right]\right\} \tag{7.5-15}$$

Because we have assumed that the spheres do not interact, this result is limited to dilute suspensions, explaining why the predicted $\bar{\mathcal{D}}_s$ does not depend on the particle size.

Diffusion of Inert Solute through a Suspension of Spheroidal Particles

Fricke (1924) modeled the electrical conductance of a suspension of arbitrarily oriented spheroids, each with a minor axis a_1 and major axis a_2 (Fig. 7.5-1; right). His approach was mathematically analogous to a transport analysis for solute diffusion through a suspension of noninteracting spheroids. From his result, the hindered diffusion coefficient is predicted to be

$$\bar{\mathcal{D}}_s = \mathcal{D}_s^c \left[1 - \frac{(1-\varepsilon)}{1/(1-\chi_s) - \varepsilon/(1+\eta_s)} \right] \tag{7.5-16}$$

where ε is the volume fraction of suspending medium and η_s is a shape factor that depends on χ_s and a_1/a_2. For a sphere, $a_1/a_2 = 1$ and $\eta_s = 2$, no matter what the value of χ_s. In that case, Equation 7.5-16 reduces to Equation 7.5-15. Figure 7.5-4 shows the η_s values for oblate spheroids ($a_1 < a_2$). When the solute permeability of the dispersed phase is less than the continuous phase ($1 > \chi_s > 0$), the left graph indicates that $\eta_s < 2$. In that case, Equation 7.5-16 predicts that diffusion through a suspension of oblate spheroids is less than diffusion through a suspension of spherical particles. This relationship changes when the permeability of the dispersed phase is greater than the continuous phase ($1 > 1/\chi_s > 0$). In that case, the right graph predicts that $\eta_s > 2$, indicating that a suspension of oblate spheroid promotes more solute diffusion than a suspension of spherical particles.

Diffusion of Reactive Solute through a Suspension of Spherical Particles

Stroeve and coworkers (1976a) analyzed solute diffusion through a suspension of spheres in which reversible monovalent binding of solute s to a carrier molecule p occurs in the dispersed phase. This binding reaction is given by

$$s + p \rightleftharpoons sp; \quad \kappa_c = \left[\frac{C_{sp}^d}{C_p^d C_s^d} \right]_{\text{equilibrium}} \tag{7.5-17a,b}$$

where κ_c is the equilibrium binding coefficient of solute to carrier and C_s^d and C_{sp}^d are the respective concentrations of free solute s and sp solute–protein complex in the spheres. Because transport occurs by both diffusion of s and sp, an effective diffusion coefficient of solute s in the spheres may be expressed as

$$\mathcal{D}_{s,\text{eff}}^d = \mathcal{D}_s^d (1 + F) \tag{7.5-18}$$

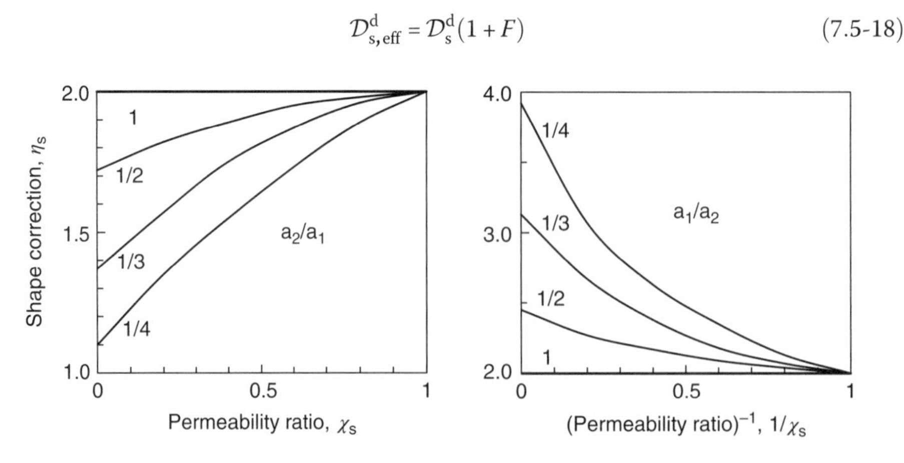

Figure 7.5-4 Shape correction function for oblate spheroids.

Here, $\mathcal{D}_s^d$ is the diffusion coefficient within the spheres when no carrier is present, and F is an enhancement factor that depends on the diffusion rate of solute relative to the binding rate of solute to the carrier. When solute diffusion in the spheres is very fast compared to the solute binding rate, then $F \to 0$. At the other extreme when the binding reaction is very fast, the concentrations of s, p, and sp are related by the binding equilibrium given in Equation 7.5-17b, and F reaches its maximum value of

$$F = \frac{\mathcal{D}_{sp}^d \kappa_c T_p^d}{\mathcal{D}_s^d} \frac{1}{\left(1 + \kappa_c C_s^d\right)^2} \tag{7.5-19}$$

In this equation, $\mathcal{D}_{sp}^d$ is the diffusion coefficient of the sp complex, and $T_p^d \equiv C_p^d + C_{sp}^d$ is the total concentration of carrier in its p and sp forms. The degree of diffusion enhancement predicted by Equation 7.5-19 depends on the magnitude of κ_c. For very small κ_c, little sp complex is formed and Equation 7.5-19 indicates that $F \to 0$. For very large κ_c, Equation 7.5-19 again indicates that $F \to 0$, this time because carrier is saturated with solute so that diffusion gradients in C_{sp}^d cannot exist. Thus, F is greatest at an intermediate value of κ_c. Because C_s^d typically varies with spatial position in the diffusion field, F generally depends on position.

Although derived for spherical suspensions, it is reasonable to apply Stroeve's model to nonspherical suspensions by using Fricke's equation with χ_s redefined as

$$\chi_s \equiv \frac{\lambda_s^{d,c} \mathcal{D}_{s,\text{eff}}^d}{\mathcal{D}_s^c} = \frac{\lambda_s^{d,c} \mathcal{D}_s^d}{\mathcal{D}_s^c}(1 + F) \tag{7.5-20}$$

Example 7.5-2 Diffusion of Oxygen through Red Cell Suspensions

Stroeve and coworkers (1976b) measured the permeability of O_2 through red blood cell suspensions. Red blood cells in isotonic saline solution were placed in an apparatus where they formed a stagnant planar film. The ratio of the observed steady-state O_2 flux to the O_2 partial pressure difference measured across the entire film was equivalent to a spatially averaged diffusion coefficient $\langle \bar{\mathcal{D}}_{O_2} \rangle$. These data were reported as $\langle \bar{\mathcal{D}}_{O_2} \rangle$ relative to the O_2 diffusion coefficient $\mathcal{D}_{O_2}^c$ in cell-free saline solution. Results were obtained over a wide range of hematocrit when partial pressures at the two film surfaces were fixed at 3.24 and 0.267 kPa as well as when they were 12.0 and 0.64 kPa (Fig. 7.5-5). Compare the data reported by these investigators to the prediction of the Fricke suspension model corrected for oxyhemoglobin binding.

Solution

We will approximate oxyhemoglobin formation in the red blood cells as a rapid monovalent binding reaction so that F is given by Equation 7.5-19. Associating subscript s with dissolved O_2 and subscript p with the heme binding site Hb, the combination of Equations 7.5-19 and 7.5-20 yields

$$\chi_{O_2} = \frac{\lambda_{O_2}^{d,c} \mathcal{D}_{O_2}^d}{\mathcal{D}_{O_2}^c} \left[1 + \frac{\mathcal{D}_{HbO_2}^d \kappa_c T_{Hb}^d}{\mathcal{D}_{O_2}^d} \frac{1}{\left(1 + \kappa_c C_{O_2}^d\right)^2} \right] \tag{7.5-21}$$

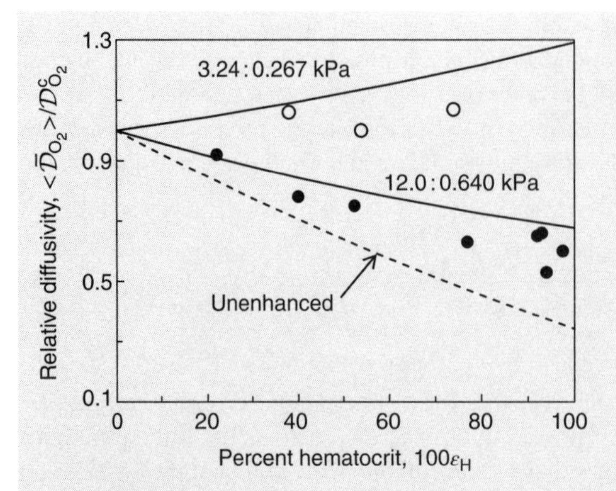

Figure 7.5-5 Diffusion of oxygen through red cell suspensions at pH = 7 and 25°C.

Since O_2 is a gaseous solute, it is natural to use partial pressure as a variable. By making use of the following relations:

$$c^d_{O_2} = \left(\alpha^d_{O_2} c^o_G\right) p_{O_2}, \quad \lambda^{d,c}_{O_2} = \frac{\alpha^d_{O_2}}{\alpha^c_{O_2}}, \quad \kappa_c = \left(\frac{1}{\alpha^d_{O_2} c^o_G}\right)\kappa_p \tag{7.5-22a-c}$$

Equation 7.5-21 becomes

$$\chi_{O_2} = \frac{\alpha^d_{O_2}\mathcal{D}^d_{O_2}}{\alpha^c_{O_2}\mathcal{D}^c_{O_2}}\left[1 + \kappa_p\left(\frac{\mathcal{D}^d_{HbO_2} T^d_{Hb}}{c^o_G \alpha^d_{O_2}\mathcal{D}^d_{O_2}}\right)\frac{1}{\left(1+\kappa_p p_{O_2}\right)^2}\right] \tag{7.5-23}$$

From parameter values in Stroeve *et al.* (1976b, table 3), we compute grouped parameter values of $\left(\alpha^d_{O_2} D^d_{so}/\alpha^c_{O_2} D^c_s\right) = 0.342$, $\kappa_p = 0.533\ \text{kPa}^{-1}$ and $\mathcal{D}^d_{HbO_2} T^d_{Hb}/c^o_G \alpha^d_{O_2}\mathcal{D}^d_{O_2} = 15.9$ kPa. Equation 7.5-23 now becomes

$$\chi_{O_2} = 0.342\left\{1 + \frac{8.48}{\left[1 + (0.533\ \text{kPa}^{-1})p_{O_2}\right]^2}\right\} \tag{7.5-24}$$

Since $\bar{\mathcal{D}}_{O_2}$ depends on χ_s, it varies with p_{O_2} within the film. By integrating Equation 7.5-16 between $p_{O_2,1}$:$p_{O_2,2}$ boundary values, we obtain $\langle\bar{\mathcal{D}}_{O_2}\rangle/\mathcal{D}^c_{O_2}$ across the entire film:

$$\frac{\langle\bar{\mathcal{D}}_{O_2}\rangle}{\mathcal{D}^c_{O_2}} = \frac{1}{p_{O_2,1} - p_{O_2,2}}\int_{p_{O_2,1}}^{p_{O_2,2}} \frac{\bar{\mathcal{D}}_{O_2}}{\mathcal{D}^c_{O_2}} dp_{O_2} = 1 - \frac{1}{p_{O_2,1} - p_{O_2,2}}\int_{p_{O_2,1}}^{p_{O_2,2}}\left[\frac{\varepsilon_H}{1/(1-\chi_{O_2}) - (1-\varepsilon_H)/(1+\eta_{O_2})}\right] dp_{O_2} \tag{7.5-25}$$

where $\varepsilon_H = (1-\varepsilon)$ is the volume fraction of red cells or fractional hematocrit. Values of $\langle\bar{\mathcal{D}}_{O_2}\rangle/\mathcal{D}^c_{O_2}$ for comparison to the data are obtained by numerical integration of Equation 7.5-25 with $\chi_{O_2}(p_{O_2})$ evaluated from Equation 7.5-24 and $\eta_{O_2}(\chi_{O_2})$ determined from Figure 7.5-4 using a cell shape factor of $a_1/a_2 = 1/4$ suggested by Fricke (1924).

The $\langle\bar{\mathcal{D}}_{O_2}\rangle/\mathcal{D}^c_{O_2}$ computed for the two $p_s(0) : p_s(L)$ conditions at which data are available appear as the solid curves in Figure 7.5-5. Also presented is an "unenhanced" curve for carrier-free spheroids. This curve was obtained by fixing $a_1/a_2 = 1/4$, $\chi_{O_2} = 0.342$, and $\eta_{O_2} = 2.73$. At a fixed hematocrit, enhancement of $\langle\bar{\mathcal{D}}_{O2}\rangle/\mathcal{D}^c_{O2}$ due to oxyhemoglobin binding is less for the experiments carried out at the higher p_{O_2} levels. This occurs because saturation of hemoglobin binding sites is more extensive than at the lower p_{O_2} levels, thereby restricting the oxyhemoglobin concentration gradient within the spheroids.

We also observe that the 12.0 : 0.64 kPa curve has a negative slope, whereas the 3.24 : 0.267 kPa curve has a positive slope. For the 12.0 : 0.64 kPa curve, the effective diffusion coefficient of O_2 within the spheroids is less than its diffusion coefficient in the suspending medium. Therefore, $\langle \bar{\mathcal{D}}_{O_2} \rangle / \mathcal{D}^c_{O_2}$ falls as more cells are added to the suspension. For the 3.24 : 0.267 kPa curve, diffusion enhancement by oxyhemoglobin binding is so great that the effective diffusion coefficient within the spheroids is larger than in the suspending medium, and $\langle \bar{\mathcal{D}}_{O_2} \rangle / \mathcal{D}^c_{O_2}$ rises with the addition of cells to the suspension.

At a normal percent hematocrit of 45%, the data in Figure 7.5-5 indicate that the O_2 diffusion coefficient in a red cell suspension is 0.8–1.1 times than in its suspending medium alone. In general, the diffusion coefficients predicted by the model overestimate the data to a small degree. This may arise from the assumption that the O_2-hemoglobin binding reaction is sufficiently rapid to reach equilibrium. It may also result from the approximation of oxyhemoglobin binding as a monovalent process.

In addition to ordinary diffusion by random molecular motion, the convection of blood cells within a continuous liquid phase can cause local mixing effects. For example, the fluid shear stresses present as blood flows along a tube impart a local rotation of blood cells that enhances transport (Keller, 1971). The effective diffusion coefficient of a stationary suspension predicted by Equation 7.5-25 does not account for this potentially important phenomenon.

References

Anderson JL, Quinn JA. Restricted transport in small pores. Biophys J. 1974; 14:130–150.

Barbero AM, Frasch HF. Transcellular route of diffusion through stratum corneum: results from finite element models. J Pharm Sci. 2006; 95: 2186–2194.

Bird RB, Stewart WE, Lightfoot EN. Transport Phenomena. 2nd ed. New York: John Wiley & Sons, Inc.; 2002, p 535.

Curry FE, Michel CC. A fiber-matrix model of capillary permeability. Microvasc Res. 1980; 20:96–66.

Fricke H. A mathematical treatment of the electric conductivity and capacity of disperse systems. I. The electrical conductivity of a suspension of homogeneous spheroids. Phys Rev. 1924; 24:575–587.

Fuller EN, Schettler PD, Giddings JC. A new method for prediction of binary gas-phase diffusion coefficients. Ind Eng Chem. 1966; 58:19–27.

Keller KH. Effect of fluid shear on mass transport in flowing blood. Fed Proc. 1971; 30:1591–1599.

Nield DA, Bejan A. Convection in Porous Media. New York: Springer Verlag; 1992, p 5–7.

Nielsen LE. Models for the permeability of filled polymer systems. J Macromol Sci (Chem). 1967; A1:929–942.

Nugent LJ, Jain RK. Pore and fiber-matrix models for diffusive transport in normal and neoplastic tissues. Microvasc Res. 1984; 28:270–274.

Perry JH, Chilton CH, Kirkpatrick SD. Chemical Engineers' Handbook. 4th ed. New York: McGraw-Hill; 1963, p 24.

Snell FM, Shulman S, Spencer RP, Moos C. Biophysical Principles of Structure and Function. Reading: Addison-Wesley; 1965, Eq. 19-30.

Stroeve P, Smith KA, Colton CK. An analysis of carrier facilitated transport in heterogeneous media. AIChE J. 1976a; 22:1125–1132.

Stroeve P, Colton CK, Smith KA. Steady state diffusion of oxygen in red blood cell and model suspensions. AIChE J. 1976b; 22:1133–1142.

Wilke CR, Chang P. Correlation of diffusion coefficients in dilute solutions. AIChE J 1955; 1:264–270.

Chapter 8

Chemical Reaction Rates

More often than not, mass transport in biological systems serves to shuttle reactants and products to and from local sites of chemical reactions. This coupling between transport and reaction is taken into account in mass balance equations by specifying the reaction rate in terms of reactant and product concentrations. The purpose of this chapter is to describe how these rate equations can be formulated for reactions with known mechanisms. Of particular interest are ligand–receptor interactions, individual enzyme reactions, and reaction networks that form the basis for cellular metabolism and regulation.

8.1 General Kinetic Models

8.1.1 Reaction Rates in a Closed System

It is convenient to express reaction rate with reference to a well-mixed system in which there are no inputs or outputs. A species balance on this closed system requires that the molar rate of accumulation of a substance i be equal to its molar formation rate R_i. With $m_i(t)$ moles of substance i in the system, its accumulation rate is dm_i/dt and the species material balance is

$$\dot{R}_i = \left[\frac{dm_i}{dt}\right]_{\substack{\text{closed}\\ \text{system}}} \tag{8.1-1}$$

The reaction rate $\dot{R}$ is an extensive quantity that depends on the size of the reaction vessel. Normalizing this by the reactor volume V produces an intensive reaction rate that is applicable to a reactor of any size.

Biomedical Mass Transport and Chemical Reaction: Physicochemical Principles and Mathematical Modeling, First Edition. James S. Ultman, Harihara Baskaran, and Gerald M. Saidel.

$$R_i = \frac{1}{V}\left[\frac{dm_i}{dt}\right]_{\substack{\text{closed}\\ \text{system}}} \tag{8.1-2}$$

For the particular reaction:

$$\nu_1{}'A_1 + \nu_2{}'A_2 \rightleftharpoons \nu_3{}''A_3 + \nu_4{}''A_4 \tag{5.1-1}$$

we have previously shown that a small departure from reaction equilibrium of one reactant or product, causes a change in the remaining substances given by

$$dm_i = \nu_i d\xi_r \tag{5.1-4}$$

where ξ_r is the extent of reaction, and ν_i is a modified stoichiometric coefficient: $\nu_i = -\nu_i{}' < 0$ for a reactant and $\nu_i = +\nu_i{}'' > 0$ for a product. With this relation, the intensive reaction rate of species i can be written as

$$R_i = \nu_i\left[\frac{1}{V}\frac{d\xi_r}{dt}\right]_{\substack{\text{closed}\\ \text{system}}} \tag{8.1-3}$$

No matter how many reactant and product species i are associated with a reaction, ξ_r does not depend on ν_i. Therefore, we can define an intrinsic reaction rate $\bar{r}$ that is independent of ν_i.

$$\bar{r} \equiv \left[\frac{1}{V}\frac{d\xi_r}{dt}\right]_{\substack{\text{closed}\\ \text{system}}} = \frac{R_i}{\nu_i} \tag{8.1-4}$$

In Equations 8.1-2 and 8.1-4, we defined R_i and $\bar{r}$ on the basis of a closed well-mixed system. Since these reaction rates depend on reactant and product concentrations in a unique manner, they apply equally well to a spatially distributed system. In that case, local values of R_i and $\bar{r}$ are related to local reactant and product concentrations.

The molecularity of a reaction is the number of different reactant molecules that must collide to produce the reaction products. The order with respect to a chemical species i is the exponent on its concentration C_i as it appears in the reaction rate equation. Overall order is the sum of the exponents on all reactant concentrations in the rate equation. Although all chemical reactions are reversible in theory, reactions with a very large equilibrium constant convert essentially all of their reactants to products. Such reactions that tend to occur in the forward direction only and are considered to be irreversible reactions.

8.1.2 Single-Step Reactions

Consider the following irreversible reaction in which a molecule of C is converted to a molecule of A. The probability that this reaction will occur is directly proportional to the concentration of C molecules so that

$$C \xrightarrow{k_r} A \ : \ \bar{r}[\text{mol/L-s}] = k_r\left[s^{-1}\right]C_C[\text{mol/L}] \tag{8.1-5}$$

With a molecularity of one, this reaction is first order with respect to C and also has an overall order of one. The proportionality constant k_r is a first-order reaction rate coefficient that generally increases with temperature. When the temperature is constant as is common in mammals, the reaction-rate coefficient is constant also. We will usually consider this coefficient to be a constant.

A more complex example of an irreversible reaction involves the collision of a molecule of A with two molecules of B to produce a molecule of C. The probability of molecular collisions of A and B leading to the formation C is proportional C_A and to $C_B \times C_B$.

$$A + 2B \xrightarrow{k_r} C \ : \ \bar{r}[\text{mol/L-s}] = k_r\left[L^2/\text{mol}^2\text{-s}\right]C_A[\text{mol/L}]C_B^2\left[\text{mol}^2/L^2\right] \tag{8.1-6}$$

The molecularity of this reaction is three. It is first order with respect to A, second order with respect to B, and third order overall.

Finally, consider a reversible reaction in which one molecule of A collides with two molecules of B to produce a molecule of C which, in turn, can decompose into A and B. This reaction has a net rate that can be formulated from the difference between forward and reverse rates given in Equations 8.1-5 and 8.1-6.

$$A + 2B \underset{k_{-r}}{\overset{k_r}{\rightleftharpoons}} C \ : \ \bar{r} = k_r C_A C_B^2 - k_{-r} C_C \tag{8.1-7}$$

Here, the reaction rate constants for the forward reaction k_r and the backward reaction k_{-r} usually have different numerical values. The molecularity and overall order of this reaction is three in the forward direction and one in the reverse direction. The net rate $\bar{r}$ adheres to the principle of mass action. That is, relatively large concentrations of reactants A and B tend to drive the reaction in the forward direction leading to $\bar{r} > 0$, whereas large concentrations of product C drive the reaction in the reverse direction resulting in $\bar{r} < 0$.

8.2 Basis of Reaction Rate Equations

8.2.1 Equilibrium Constraint on Reaction Rate Expressions

The specific form of a reaction rate equation can be inferred from equilibrium considerations. For example, consider the reaction given by Equation 5.1-1 which has an equilibrium constant (Eq. 5.2-7) given by

$$\kappa_C(T) = \left(\frac{C_3^{\nu_3''} C_4^{\nu_4''}}{C_1^{\nu_1'} C_2^{\nu_2'}}\right)_{eq} \tag{8.2-1}$$

The overall intrinsic rate $\bar{r}$ of this reaction is the net result of a forward rate $\bar{r}_{forward}$ and a reverse rate $\bar{r}_{reverse}$. When equilibrium occurs, there is no net reaction rate.

$$\bar{r} = \bar{r}_{forward} - \bar{r}_{reverse} = 0 \tag{8.2-2}$$

We can satisfy both of these equations by asserting that the forward reaction rate is proportional to the numerator of κ_C and the reverse reaction rate is proportional to the denominator. By symbolizing the associated proportionality constants as k_r and k_{-r}, we get

$$\bar{r}_{forward} = k_r C_1^{\nu_1'} C_2^{\nu_2'}, \quad \bar{r}_{reverse} = k_{-r} C_3^{\nu_3''} C_4^{\nu_4''} \tag{8.2-3a,b}$$

and Equations 8.2-1 and 8.2-2 indicate that

$$\frac{k_r}{k_{-r}} = \left(\frac{C_3^{\nu_3''} C_4^{\nu_4''}}{C_1^{\nu_1'} C_2^{\nu_2'}}\right)_{eq} = \kappa \tag{8.2-4}$$

The overall rate is then given by

$$\bar{r} = k_r C_1^{\nu_1'} C_2^{\nu_2'} - k_{-r} C_3^{\nu_3''} C_4^{\nu_4''} = k_r \left(C_1^{\nu_1'} C_2^{\nu_2'} - \frac{1}{\kappa} C_3^{\nu_3''} C_4 \nu^{\nu_4''} \right) \tag{8.2-5}$$

When κ is so large that the reverse reaction rate is negligible relative to the forward reaction, we can consider this to be an irreversible reaction with the kinetics

$$\nu_1' A_1 + \nu_2' A_2 \rightarrow \nu_3'' A_3 + \nu_4'' A_4; \quad \bar{r} = k_r C_1^{\nu_1'} C_2^{\nu_2'} \tag{8.2-6a, b}$$

Example 8.2-1 A Model of Oxygen–Hemoglobin Binding Kinetics

Using dynamic measurements of light absorption, Gibson (1970) determined the time-varying increase in oxyhemoglobin saturation in a closed reactor after a deoxygenated hemoglobin solution was rapidly mixed with an oxygenated hemoglobin-free solution (Fig. 8.2-1). The mixture contained a 41.5 μM concentration of heme groups and was maintained at pH = 7 and T = 21.5°C. As shown in Table 8.2-1, experiments were terminated at different times t_f depending on the initial oxygen concentration $C_{O_2}(0)$ of the mixture.

Using Gibson's data, we will test whether the binding reaction in the Hill model of oxyhemoglobin equilibrium can be extended to simulate dynamic, nonequilibrium behavior.

$$Hb_n + nO_2 \underset{}{\overset{k_r}{\rightleftharpoons}} (HbO_2)_n \tag{5.5-7}$$

Here, n is the number of heme groups that simultaneously bind with oxygen.

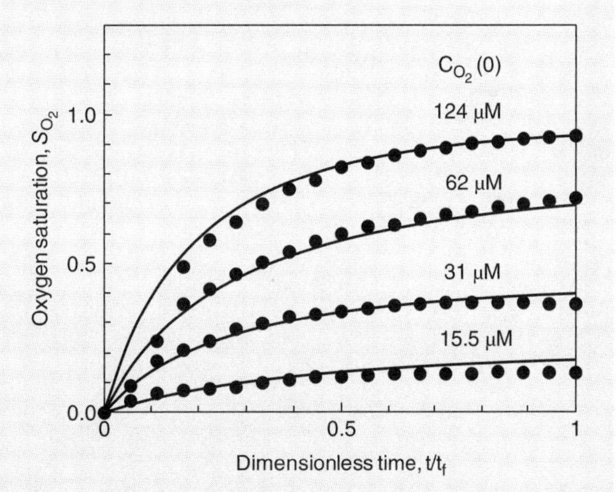

Figure 8.2-1 Oxygen–hemoglobin reaction dynamics in a closed well-mixed reactor.

Table 8.2-1 Length of Experiments t_f at the Initial O_2 Concentrations $C_{O_2}(0)$

$C_{O_2}(0)$ (μM)	15.5	31	62	124
t_f (ms)	36	36	18	9

Solution

For the reaction given in Equation 5.5-7, the equilibrium constant is

$$\kappa_C = \left[\frac{C_{(HbO_2)n}}{C_{(Hb)n}C_{O_2}^n}\right]_{eq} \tag{8.2-7}$$

and inherent reaction rate predicted by Equation 8.2-5 is given by

$$\bar{r} = k_r\left(C_{O_2}^n C_{Hbn} - \frac{1}{\kappa_C}C_{(HbO_2)n}\right) \tag{8.2-8}$$

The intensive reaction rates of O_2 and $(HbO_2)_n$ are prescribed by Equation 8.1-4 as

$$R_{O_2} = \nu_{O_2}\bar{r} = -nk_r\left(C_{O_2}^n C_{Hbn} - \frac{1}{\kappa_C}C_{(HbO_2)n}\right)$$
$$R_{(HbO_2)n} = \nu_{(HbO_2)n}\bar{r} = +k_r\left(C_{O_2}^n C_{Hbn} - \frac{1}{\kappa_C}C_{(HbO_2)n}\right) \tag{8.2-9a,b}$$

Since the experiments were carried out in a closed reactor, we can apply Equation 8.1-2 to express the material balance equations for O_2 and for (HbO_2).

$$\frac{dm_{O_2}}{dt} = VR_{O_2}, \quad \frac{dm_{(HbO_2)n}}{dt} = VR_{(HbO_2)n} \tag{8.2-10a,b}$$

The number of moles of these components in the reactor is $m_{O_2} = VC_{O_2}$ and $m_{(HbO_2)n} = VC_{(HbO_2)n}$, and the liquid volume V contained in the reactor is constant. With the reaction rates specified in Equation 8.2-9a,b, we can rewrite the material balances as

$$\frac{dC_{O_2}}{dt} = -nk_r\left(C_{O_2}^n C_{Hbn} - \frac{1}{\kappa_C}C_{(HbO_2)n}\right) \tag{8.2-11}$$

$$\frac{dC_{(HbO_2)n}}{dt} = k_r\left(C_{O_2}^n C_{Hbn} - \frac{1}{\kappa_C}C_{(HbO_2)n}\right) \tag{8.2-12}$$

Taking the ratio of these two equations and integrating subject to the initial concentrations $C_{O_2} = C_{O_2}(0)$ and $C_{(HbO_2)n} = 0$, we obtain

$$\frac{dC_{O_2}}{dC_{(HbO_2)n}} = -n \Rightarrow C_{O_2} = C_{O_2}(0) - nC_{(HbO_2)n} \tag{8.2-13}$$

Combining this equation with the material balance on $(HbO_2)_n$, we get

$$\frac{dC_{(HbO_2)n}}{dt} = k_r\left[\left(C_{O_2}(0) - nC_{(HbO_2)n}\right)^n C_{Hbn} - \frac{1}{\kappa_C}C_{(HbO_2)n}\right] \tag{8.2-14}$$

At any time, the total concentration of heme groups Hb in the closed reactor must be a constant equal to the sum of its unoxidized and oxidized forms.

$$T_{Hb} = n\left(C_{(HbO_2)n} + C_{Hbn}\right) \tag{8.2-15}$$

Solving this equation for C_{Hbn} and substituting into Equation 8.2-14 results in

$$\frac{dC_{(HbO_2)n}}{dt} = k_r\left[\left(C_{O_2}(0) - nC_{(HbO_2)n}\right)^n\left(\frac{T_{Hb}}{n} - C_{(HbO_2)n}\right) - \frac{1}{\kappa_C}C_{(HbO_2)n}\right] \tag{8.2-16}$$

By introducing oxyhemoglobin saturation as the ratio of the oxidized heme groups to the total heme groups

$$S_{O_2} \equiv \frac{nC_{(HbO_2)n}}{T_{Hb}} \tag{8.2-17}$$

Equation 8.2-16 becomes

$$\frac{dS_{O_2}}{dt} = k_r\left[(C_{O_2}(0) - T_{Hb}S_{O_2})^n(1 - S_{O_2}) - \frac{1}{\kappa}S_{O_2}\right] \tag{8.2-18}$$

We simulated each set of $S_{O_2}(t)$ data at a given $C_{O_2}(0)$ by numerical solution of Equation 8.2-18. Simulations were carried out from $t = 0$ to $t = t_f$ with a fixed value of $T_{Hb} = 41.5\ \mu M$. Best estimates of $n = 1.81$, $k_r = 1.01 \times 10^{-4} s^{-1} (\mu M)^{-1.81}$, and $\kappa_C = 0.00667 (\mu M)^{-1.81}$ were determined by minimizing the squared error between measured and simulated S_{O_2} values summed over all data points. The curves in Figure 8.2-1 show the simulations using these parameter estimates compared to the data points.

In Chapter 5, we found different saturation parameter values of $n = 2.8$ and $\kappa_C = 0.000123\ \mu M^{-2.8}$ based on equilibrium S_{O_2} data from whole blood (Example 5.5-1). This may reflect a difference between oxyhemoglobin binding in dilute hemoglobin solutions and that occurring in blood cells whose hemoglobin content is about 5000 μM.

8.2.2 Transition State Theory

For an understanding of the molecular origin of the rate equation, consider the simple irreversible reaction

$$A + BC \xrightarrow{k_r} AB + C \tag{8.2-19}$$

that has an inherent rate given by

$$\bar{r} = k_r C_A C_{BC} \tag{8.2-20}$$

The reaction of a molecule of A with a molecule of BC requires that the two molecules become close enough that the necessary rearrangement of chemical bonds can take place. According to transition state theory, this reaction will be successful only if the reactant molecules acquire sufficient potential energy to overcome repulsive forces and release this energy as product molecules are formed.

As shown in Figure 8.2-2, this process can be visualized by graphing the energy of the molecules as a function of their relative position or reaction coordinate. The molecular configuration that has the highest energy, corresponding to the closest proximity of the A and BC molecules, is the activated complex AB^*C. While the free energy ΔG_r^* of the activated complex relative to reactants A and BC is positive, the free energy change ΔG_r of the reaction given by Eq. 8.2-19 must be negative if it is to be spontaneous.

Assuming that the formation of the activated complex is reversible, the mechanism of the overall reaction given in Equation 8.2-19 can be represented by the following two separate reaction steps:

$$\begin{gathered} A + BC \underset{k_{-1}}{\overset{k_1}{\rightleftharpoons}} A^*BC \\ A^*BC \xrightarrow{k_2} AB + C \end{gathered} \tag{8.2-21a,b}$$

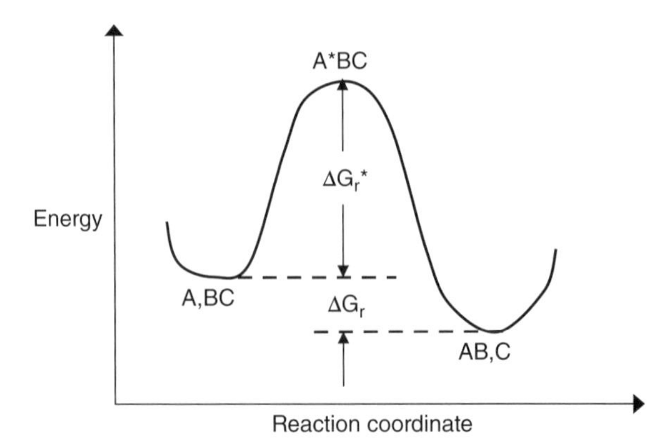

Figure 8.2-2 Reaction path in the transition-state theory.

We assume that the forward and reverse rates of the reversible step are fast compared to the irreversible step. This quasi-equilibrium hypothesis allows us to formulate the concentration of activated complex in terms of an equilibrium constant (Appendix C3).

$$\kappa_c^* = \frac{k_1}{k_{-1}} = \left(\frac{C_{A^*BC}}{C_A C_{BC}}\right)_{eq} \Rightarrow C_{A^*BC} = \kappa_c^* C_A C_{BC} \tag{8.2-22a,b}$$

After applying Equation 5.2-6 to express κ_c^* in terms of ΔG_r^*, we get

$$C_{A^*BC} = \frac{1}{c_L} \exp\left(-\frac{\Delta G_r^*}{\mathcal{R}T}\right) C_A C_{BC} \tag{8.2-23}$$

where c_L is molar density of the solution. Because the rate-limiting step in the formation of products is the rate at which A^*BC dissociates in the second reaction step, we can formulate the inherent rate of the overall reaction as

$$\bar{r} = k_2 C_{A^*BC} = \frac{k_2}{c_L} \exp\left(-\frac{\Delta G_r^*}{\mathcal{R}T}\right) C_A C_{BC} \tag{8.2-24}$$

Comparing this to Equation 8.2-20, we see that

$$k_r(T) = \frac{k_2(T)}{c_L} \exp\left(-\frac{\Delta G_r^*}{\mathcal{R}T}\right) \tag{8.2-25}$$

As a practical matter, reaction rate constants frequently follow a simplified form of this equation known as the Arrhenius relationship:

$$k_r(T) = k_o \exp\left(-\frac{E_a}{\mathcal{R}T}\right) \tag{8.2-26}$$

where the frequency factor k_o and the activation energy E_a are constants. It is clear from this equation that k_r increases with temperature. As a rule of thumb, k_r approximately doubles for every temperature increase of 10°K.

8.3 Multi-Step Reactions

In most of the examples we have considered so far, chemical reaction has been viewed as a single-step process in which one set of products is formed from one set of reactants. In such elementary reactions, the overall stoichiometric equation reflects the actual mechanism of the reaction. In most situations of practical interest, such as the reaction scheme we treated in transition state theory, reactions occur in multiple steps, and the overall rate of formation of any substance is the sum of its formation rates in the individual steps. In a reaction consisting of J steps, we represent the jth step as:

$$\nu_{1j}{}'A_1 + \nu_{2j}{}'A_2 \rightleftharpoons \nu_{3j}{}''A_3 + \nu_{4j}{}''A_4 \qquad (8.3\text{-}1)$$

Following from Equation 8.1-4, the intensive rate R_{ij} of species i in this reaction step is related to the inherent rate $\bar{r}_j$ of the step by

$$R_{ij} = \nu_{ij}\bar{r}_j \qquad (8.3\text{-}2)$$

Here, $\nu_{ij} = -\nu_{ij}{}'$ for reactants, $\nu_{ij} = +\nu_{ij}{}''$ for products, and $\nu_{ij} = 0$ when substance i does not participate in reaction step j.

Since the overall formation rate of substance i is the sum of its formation rates in the J reaction steps,

$$R_i = \sum_{j=1}^{J} R_{ij} = \sum_{j-1}^{J} \nu_{ij}\bar{r}_j \qquad (8.3\text{-}3)$$

For a reaction in which there is a total of I substances (both reactants and products) and J reaction steps, the relation between $R_i(i = 1.2.I)$ and $\bar{r}_j(j = 1,2,\ldots J)$ is written in matrix form as follows:

$$\begin{bmatrix} R_1 \\ R_2 \\ \vdots \\ R_I \end{bmatrix} = \begin{bmatrix} \nu_{1,1} & \nu_{1,2} & \cdots & \nu_{1,J} \\ \nu_{2,1} & \nu_{2,2} & \cdots & \nu_{2,J} \\ \vdots & \vdots & & \vdots \\ \nu_{I,1} & \nu_{I,2} & \cdots & \nu_{I,J} \end{bmatrix} \begin{bmatrix} \bar{r}_1 \\ \bar{r}_2 \\ \vdots \\ \bar{r}_J \end{bmatrix} \Rightarrow [R] = [\nu][\bar{r}] \qquad (8.3\text{-}4a,b)$$

where $[\nu]$ is the stoichiometric coefficient matrix.

Example 8.3-1 Rate Equations for a Reaction with Parallel Steps

As an example of a nonelementary reaction, we consider

$$\begin{aligned} &A \xrightarrow{k_1} B \\ &A \xrightarrow{k_2} C \end{aligned} \qquad (8.3\text{-}5a,b)$$

What are the intensive rates of reaction of A, B, and C for these two parallel first-order irreversible reactions?

Solution

This reaction scheme consists of three substances (i = A,B,C) and two reaction steps (j = 1,2). The inherent reaction rates for the reaction steps are related to the reactant concentration by

$$\bar{r}_1 = k_1C_A, \quad \bar{r}_2 = k_2C_A \qquad (8.3\text{-}6a,b)$$

The reaction rate matrices and the stoichiometric coefficient matrix are given by

$$[R]=\begin{bmatrix}R_A\\R_B\\R_C\end{bmatrix},\quad [\bar{r}]=\begin{bmatrix}\bar{r}_1\\\bar{r}_2\end{bmatrix},\quad [\nu]=\begin{bmatrix}\nu_{A,1} & \nu_{A,2}\\\nu_{B,1} & \nu_{B,2}\\\nu_{C,1} & \nu_{C,2}\end{bmatrix}=\begin{bmatrix}-1 & -1\\+1 & 0\\0 & +1\end{bmatrix} \qquad (8.3\text{-}7a\text{-}c)$$

Substituting these matrices into Equation 8.3-4, the intensive reaction rates are

$$\begin{bmatrix}R_A\\R_B\\R_C\end{bmatrix}=\begin{bmatrix}-1 & -1\\+1 & 0\\0 & +1\end{bmatrix}\begin{bmatrix}\bar{r}_1\\\bar{r}_2\end{bmatrix}=\begin{bmatrix}-\bar{r}_1-\bar{r}_2\\+\bar{r}_1\\+\bar{r}_2\end{bmatrix} \qquad (8.3\text{-}8)$$

The final equations for R_A, R_B, and R_C are obtained by substituting Equation 8.3-6a,b into Equation 8.3-8.

$$\begin{aligned}R_A &= -k_1C_A-k_2C_A\\R_B &= k_1C_A\\R_C &= k_2C_A\end{aligned} \qquad (8.3\text{-}9a\text{-}c)$$

Example 8.3-2 Rate Equations for a Reaction with Series Steps

As an another example of a nonelementary reaction, consider the series reactions

$$A+2B\xrightarrow{k_1}C\xrightarrow{k_2}D \qquad (8.3\text{-}10)$$

that can also be written in terms of two individual reaction steps.

$$\begin{aligned}A+2B &\xrightarrow{k_1}C\\C &\xrightarrow{k_2}D\end{aligned} \qquad (8.3\text{-}11a,b)$$

What are the intensive rates of reaction of A,B, C, and D?

Solution

For this reaction scheme, there are four substances (i = A,B,C,D) and two reaction steps (j = 1,2). The first reaction step is an irreversible third-order reaction, and the second step is an irreversible first-order reaction. The inherent reaction rates for individual reaction steps are

$$\bar{r}_1=k_1C_AC_B^2,\quad \bar{r}_2=k_2C_C \qquad (8.3\text{-}12a,b)$$

The reaction rate matrices and the stoichiometric coefficient matrix for these reaction steps are

$$[R]=\begin{bmatrix}R_A\\R_B\\R_C\\R_D\end{bmatrix},\quad [\bar{r}]=\begin{bmatrix}\bar{r}_1\\\bar{r}_2\end{bmatrix},\quad [\nu]=\begin{bmatrix}\nu_{A,1} & \nu_{A,2}\\\nu_{B,1} & \nu_{B,2}\\\nu_{C,1} & \nu_{C,2}\\\nu_{D,1} & \nu_{D,2}\end{bmatrix}=\begin{bmatrix}-1 & 0\\-2 & 0\\+1 & -1\\0 & +1\end{bmatrix} \qquad (8.3\text{-}13a\text{-}c)$$

The matrix relationship between the intensive and inherent reaction rates is therefore

$$\begin{bmatrix}R_A\\R_B\\R_C\\R_D\end{bmatrix}=\begin{bmatrix}-1 & 0\\-2 & 0\\+1 & -1\\0 & +1\end{bmatrix}\begin{bmatrix}\bar{r}_1\\\bar{r}_2\end{bmatrix}=\begin{bmatrix}-\bar{r}_1\\-2\bar{r}_1\\\bar{r}_1-\bar{r}_2\\+\bar{r}_2\end{bmatrix} \qquad (8.3\text{-}14)$$

and final equations for R_A, R_B, R_C, and R_D are

$$R_A = R_B/2 = -k_1 C_A C_B^2$$
$$R_C = k_1 C_A C_B^2 - k_2 C_C \qquad (8.3\text{-}15a\text{-}c)$$
$$R_D = k_2 C_C$$

8.4 Ligand–Receptor Kinetics

Chapter 5 presented the equilibrium relationships of a few ligand–receptor binding processes including monovalent binding of one ligand to a receptor site and competitive binding of two ligands to a single receptor site. In this section, we examine the dynamics of such processes when they deviate from their equilibrium state.

8.4.1 Monovalent Binding

Consider the reversible monovalent binding of a ligand L to the single binding site R of a receptor according to the following reaction:

$$L + R \underset{k_{-r}}{\overset{k_r}{\rightleftharpoons}} LR \qquad (8.4\text{-}1)$$

Referring to Equation 8.2-5, the inherent rate associated with this reaction mechanism is

$$\bar{r} = k_r\left(C_L C_R - \frac{1}{\kappa_c} C_{LR}\right); \quad \kappa_c = \frac{k_r}{k_{-r}} = \frac{C_{LR}}{C_L C_R} \qquad (8.4\text{-}2a,b)$$

In this application, the equilibrium constant κ_c reflects the binding strength between L and R. By employing Equation 8.3-4, the intensive reaction rates of L, R, and LR are related to the inherent reaction rate by

$$\begin{bmatrix} R_L \\ R_R \\ R_{LR} \end{bmatrix} = \begin{bmatrix} -1 \\ -1 \\ +1 \end{bmatrix} \bar{r} \qquad (8.4\text{-}3)$$

so that

$$R_L = R_R = -R_{LR} = -\bar{r} = k_r\left(\frac{1}{\kappa} C_{LR} - C_L C_R\right) \qquad (8.4\text{-}4)$$

To determine the concentration dynamics of a particular process, these rates must be incorporated into material balance equations. For example, consider a well-mixed system of volume V in which substances L, R, and LR can accumulate and react, but are not transported across its boundaries. The molar balances on these substances are given by

$$\frac{d(C_R V)}{dt} = R_R V, \quad \frac{d(C_{LR} V)}{dt} = R_{LR} V, \quad \frac{d(C_L V)}{dt} = R_L V \qquad (8.4\text{-}5a\text{-}c)$$

If V is constant, these equations after substitution of Equation 8.4-4 become

$$-\frac{dC_{LR}}{dt}=\frac{dC_L}{dt}=\frac{dC_R}{dt}=k_r\left(\frac{1}{\kappa_c}C_{LR}-C_LC_R\right) \tag{8.4-6}$$

The sum of the R and LR concentration derivatives yields

$$\frac{d}{dt}(C_R+C_{LR})=0 \tag{8.4-7}$$

Thus, the overall receptor concentration, $T_R = C_R + C_{LR}$, is a constant that is always equal to the sum of the initial concentrations $C_R(0)$ and $C_{RL}(0)$.

$$T_R \equiv C_R + C_{LR} = C_R(0) + C_{LR}(0) \tag{8.4-8}$$

Similarly, the overall ligand concentration, $T_L = C_L + C_{LR}$ is the sum of the initial concentrations $C_L(0)$ and $C_{LR}(0)$.

$$T_L = C_L(0) + C_{LR}(0) \tag{8.4-9}$$

Combining Equations 8.4-6, 8.4-8, and 8.4-9, we arrive at a single equation for $C_{LR}(t)$.

$$\frac{1}{k_r}\frac{dC_{LR}}{dt}=(T_L-C_{LR})(T_R-C_{LR})-\frac{1}{\kappa_c}C_{LR} \tag{8.4-10}$$

Although an analytical solution to this nonlinear equation exists, a special linear approximation is often valid. If the initial concentration of unbound ligand is much greater than the total concentration of binding sites, then the total ligand concentration at any time is approximately equal to the initial ligand concentration. Also, the bound ligand concentration is always much less than the initial ligand concentration (i.e., $T_L \approx C_L(0) \gg C_{LR}$). This allows us to approximate Equation 8.4-10 as

$$\frac{1}{k_r}\frac{dC_{LR}}{dt}\approx C_L(0)(T_R-C_{LR})-\frac{1}{\kappa_c}C_{LR} \tag{8.4-11}$$

We find the steady-state solution to this equation at $t \rightarrow \infty$ by setting the time derivative equal to zero.

$$(C_{LR})_\infty \equiv \frac{T_R\kappa_cC_L(0)}{1+\kappa_cC_L(0)} \tag{8.4-12}$$

Because this model of monovalent binding is restricted to a closed system, we expect the process to reach equilibrium when this steady state is reached. The formula for $(C_{LR})_\infty$ is, in fact, identical to the monovalent binding equilibrium that we previously derived (Eq. 5.4-3). Combining Equations 8.4-11 and 8.4-12 to eliminate T_R, we get

$$\frac{\kappa_c}{k_r}\frac{dC_{LR}}{dt}=[1+\kappa_cC_L(0)]\left[(C_{LR})_\infty-C_{LR}\right] \tag{8.4-13}$$

It is convenient to transform this model equation in terms of the dimensionless quantities:

$$t\equiv\frac{k_r}{\kappa_c}\mathrm{t},\quad F\equiv\frac{C_{LR}}{(C_{LR})_\infty},\quad \kappa\equiv\kappa_cC_L(0) \tag{8.4-14a-c}$$

where F is the fraction of occupied binding sites at dimensionless time t relative to the number of sites occupied at steady state. Equation 8.4-13 then becomes

$$\frac{dF}{dt}=(1+\kappa)(1-F) \tag{8.4-15}$$

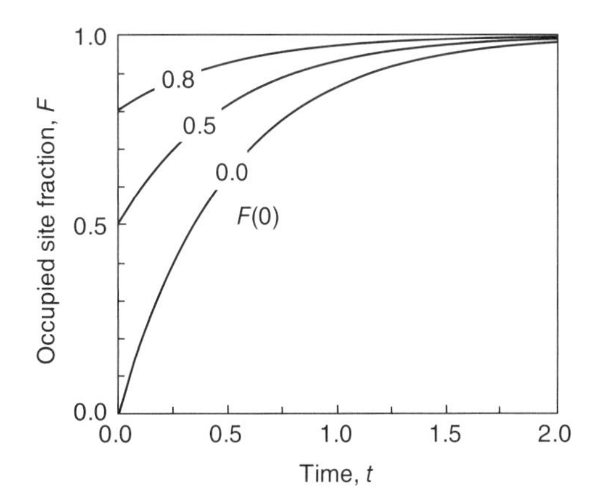

Figure 8.4-1 Monovalent binding dynamics in a closed well-mixed reactor.

Solving this equation with the dimensionless initial condition

$$F(0) \equiv \frac{C_{LR}(0)}{(C_{LR})_\infty} \tag{8.4-16}$$

we obtain

$$F = 1 + [F(0) - 1]\exp[-(1+\kappa)t] \tag{8.4-17}$$

Figure 8.4-1 shows the dynamic behavior of F for different values of the initial values when $\kappa = 1$. As $F(0)$ increases, the occupied sites more rapidly approach their steady-state occupancy. If the κ value was increased above one, then the approach to steady state at any $F(0)$ would be even more rapid.

8.4.2 Competitive Binding

We now analyze the competition between ligands L_1 and L_2 for the same binding site R on a receptor. This process has two reaction steps

$$\begin{aligned} L_1 + R &\underset{k_{-1}}{\overset{k_1}{\rightleftharpoons}} L_1R \\ L_2 + R &\underset{k_{-2}}{\overset{k_2}{\rightleftharpoons}} L_2R \end{aligned} \tag{8.4-18a,b}$$

with inherent rates and equilibrium constants given by

$$\begin{aligned} \bar{r}_1 &= k_1\left(C_{L_1}C_R - \frac{1}{\kappa_{C_1}}C_{L_1R}\right); \quad \kappa_{C_1} = \frac{k_1}{k_{-1}} \\ \bar{r}_2 &= k_2\left(C_{L_2}C_R - \frac{1}{\kappa_{C_2}}C_{L_2R}\right); \quad \kappa_{C_2} = \frac{k_2}{k_{-2}} \end{aligned} \tag{8.4-19a-d}$$

Based on Equation 8.3-4, the intensive rate equations in matrix form are

$$\begin{bmatrix} R_{L_1} \\ R_R \\ R_{L_1R} \\ R_{L_2} \\ R_{L_2R} \end{bmatrix} = \begin{bmatrix} -1 & 0 \\ -1 & -1 \\ +1 & 0 \\ 0 & -1 \\ 0 & +1 \end{bmatrix} \begin{bmatrix} \bar{r}_1 \\ \bar{r}_2 \end{bmatrix} \tag{8.4-20}$$

or equivalently

$$\begin{aligned} R_{L_1} &= -R_{L_1R} = -\bar{r}_1 = k_{c_1}\left(\frac{1}{\kappa_{c_1}}C_{L_1R} - C_{L_1}C_R\right) \\ R_{L_2} &= -R_{L_2R} = -\bar{r}_2 = k_{c_2}\left(\frac{1}{\kappa_{c_2}}C_{L_2R} - C_{L_2}C_R\right) \\ R_R &= -(\bar{r}_1 + \bar{r}_2) = k_1\left(\frac{1}{\kappa_{c_1}}C_{L_1R} - C_{L_1}C_R\right) + k_2\left(\frac{1}{\kappa_{c_2}}C_{L_2R} - C_{L_2}C_R\right) \end{aligned} \tag{8.4-21a-c}$$

In a closed, well-mixed system of volume V, the molar balances on R, L_1, L_1R, L_2, and L_2R have the same form as Equation 8.4-5a-c.

$$\frac{d(C_iV)}{dt} = R_iV \quad (i = R, L_1, L_1R, L_2, L_2R) \tag{8.4-22}$$

As was the case for monovalent binding, the concentrations of the five species are not independent. Rather, the total concentration of receptor sites as well as the total concentration of ligands of each type must be constant.

$$T_R = C_R + (C_{L_1R} + C_{L_2R}), \quad T_{L_1} = C_{L_1} + C_{L_1R}, \quad T_{L_2} = C_{L_2} + C_{L_2R} \tag{8.4-23a-c}$$

After some manipulation, Equations 8.4-21a-c, 8.4-22, and 8.4-23a-c reduce to two nonlinear, coupled, model equations.

$$\frac{1}{k_1}\frac{dC_{L_1R}}{dt} = C_{L_1}C_R - \frac{1}{\kappa_{c_1}}C_{L_1R} = [T_{L_1} - C_{L_1R}][T_R - (C_{L_1R} + C_{L_2R})] - \frac{1}{\kappa_{c_1}}C_{L_1R} \tag{8.4-24}$$

$$\frac{1}{k_2}\frac{dC_{L_2R}}{dt} = C_{L_2}C_R - \frac{1}{\kappa_{c_2}}C_{L_2R} = [T_{L_2} - C_{L_2R}][T_R - (C_{L_1R} + C_{L_2R})] - \frac{1}{\kappa_{c_2}}C_{L_2R} \tag{8.4-25}$$

Suppose that the initial concentrations of unbound ligand, $C_{L_1}(0)$ and $C_{L_2}(0)$, are much greater than the total concentration of receptor sites. Then, $T_{L_1} \approx C_{L_1}(0) \gg C_{L_1R}$ and $T_{L_2} \approx C_{L_2}(0) \gg C_{L_2R}$, and we can approximate the model by two linear equations.

$$\begin{aligned} \frac{\kappa_{c_1}}{k_1}\frac{dC_{L_1R}}{dt} + [1 + \kappa_{c_1}C_{L_1}(0)]C_{L_1R} &= [1 + \kappa_{c_1}C_{L_1}(0) + \kappa_{c_2}C_{L_2}(0)](C_{L_1R})_\infty \\ &\quad - \kappa_{c_1}C_{L_1}(0)C_{L_2R} \end{aligned} \tag{8.4-26}$$

$$\begin{aligned} \frac{\kappa_{c_2}}{k_2}\frac{dC_{L_2R}}{dt} + [1 + \kappa_{c_2}C_{L_2}(0)]C_{L_2R} &= [1 + \kappa_{c_1}C_{L_1}(0) + \kappa_{c_2}C_{L_2}(0)](C_{L_2R})_\infty \\ &\quad - \kappa_{c_2}C_{L_2}(0)C_{L_1R} \end{aligned} \tag{8.4-27}$$

where

$$(C_{L_1R})_\infty = \frac{T_R\kappa_{c_1}C_{L_1}(0)}{1 + \kappa_{c_1}C_{L_1}(0) + \kappa_{c_2}C_{L_2}(0)}, \quad (C_{L_2R})_\infty = \frac{T_R\kappa_{c_2}C_{L_2}(0)}{1 + \kappa_{c_1}C_{L_1}(0) + \kappa_{c_2}C_{L_2}(0)} \tag{8.4-28a,b}$$

These steady-state solutions are equivalent to the equations that we previously derived for competitive binding equilibrium (Eq. 5.4-6a,b). By introducing the dimensionless variables,

$$t = \frac{k_1}{\kappa_{C_1}} t, \quad F_{L_1R} = \frac{C_{L_1R}}{(C_{L_1R})_\infty + (C_{L_2R})_\infty}, \quad F_{L_2R} = \frac{C_{L_2R}}{(C_{L_1R})_\infty + (C_{L_2R})_\infty} \tag{8.4-29a-c}$$

we arrive at dimensionless model equations:

$$\frac{dF_{L_1R}}{dt} + (1 + \kappa_{C_1}) F_{L_1R} = \kappa_{C_1} \left[\left(\frac{1 + \kappa_{C_1} + \kappa_{C_2}}{\kappa_{C_1} + \kappa_{C_2}} \right) - F_{L_2R} \right] \tag{8.4-30}$$

$$\frac{1}{\xi} \frac{dF_{L_2R}}{dt} + (1 + \kappa_{C_2}) F_{L_2R} = \kappa_{C_2} \left[\left(\frac{1 + \kappa_{C_1} + \kappa_{C_2}}{\kappa_{C_1} + \kappa_{C_2}} \right) - F_{L_1R} \right] \tag{8.4-31}$$

with dimensionless parameters given by:

$$\xi \equiv \frac{k_2 \,\kappa_{C_1}}{k_1 \,\kappa_{C_2}}, \quad \kappa_{C_i} \equiv \kappa_{C_i} C_{L_i}(0) \quad (i = 1,2) \tag{8.4-32a,b}$$

and dimensionless initial conditions:

$$F_{L_1R}(0) = \frac{C_{L_1R}(0)}{(C_{L_1R})_\infty + (C_{L_2R})_\infty}, \quad F_{L_2R}(0) = \frac{C_{L_2R}(0)}{(C_{L_1R})_\infty + (C_{L_2R})_\infty} \tag{8.4-33a,b}$$

The use of dimensionless variables reduces the number of independent parameters from five (i.e., k_1, k_2, κ_1, κ_2, T_R) to three (i.e., ξ, κ_{C_1}, κ_{C_2}).

Figure 8.4-2 illustrates the system behavior obtained from an analytical solution to Equations 8.4-30, 8.4-31, and 8.4-33a,b when $F_{L_1R}(0) = F_{L_2R}(0) = 0$ (i.e., all the receptor binding sites are initially unoccupied). In all three panels of this figure, we have set $\kappa_{C_1} = \kappa_{C_2} = 1$ so there is no thermodynamic advantage to the formation of L_1R compared to L_2R. Consequently, the fraction of binding sites occupied by L_1 and by L_2 each approach 0.5 at large *t*. Prior to this steady state, however, the dynamics of F_{L_1R} and F_{L_2R} differ in a manner that depends on ξ.

With $\kappa_{C_1} = \kappa_{C_2} = 1$, the ξ parameter reduces to k_2/k_1, the characteristic rate of L_2R formation relative to the rate of L_1R formation. In the center graph, $\xi = 1$ and the behaviors of F_{L_1R} and F_{L_2R} are identical because the two characteristic formation rates are equal. In the left graph,

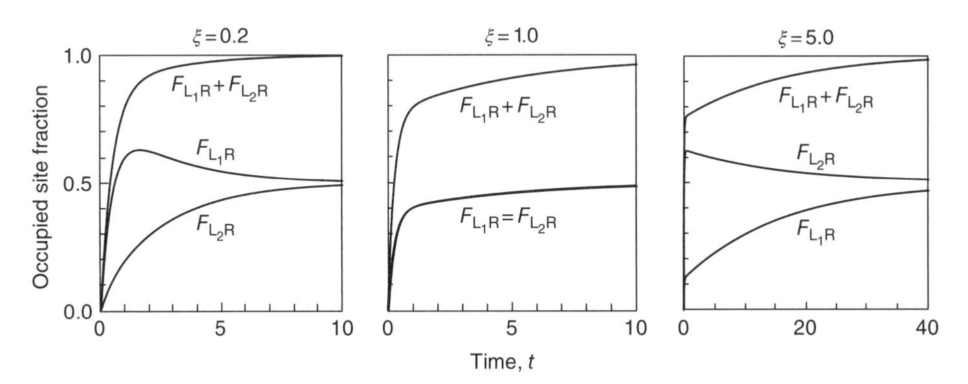

Figure 8.4-2 Competitive binding dynamics in a closed well-mixed reactor.

$\xi < 1$ and F_{L_1R} initially overshoots its steady-state value because L_1R formation occurs at a faster characteristic rate than L_2R formation. In other words, L_1 is able to overpopulate the receptor sites in the short term before ligand L_2 can catch up. The reverse is true in the right graph where $\xi > 1$ such that the characteristic rate of L_2R dominates; F_{L_2R} rises very quickly and initially overshoots its steady-state level, while the formation of F_{L_1R} is simultaneously suppressed below its steady-state level.

8.5 Enzyme Kinetics

8.5.1 Enzyme Behavior

An enzyme is a protein molecule with one or more active sites that speed up the conversion of specific biochemical substrates to products without a net change in the enzyme itself. As a simple example, consider the hydrolysis of urea into ammonia and carbon dioxide, a reaction that is catalyzed by the urease enzyme. As illustrated in Figure 8.5-1 (Snell *et al.*, 1965, p 351), water and urea are adsorbed at an active site of urease (a) where they form a urea–water–urease complex (b). The chemical bonds in urea are thereby distorted, making the urea more vulnerable to bond breakage and rearrangements. Once carbon dioxide and ammonia products are formed, they desorb from the active site, and the enzyme relaxes to its original configuration (c).

In a closed system, we expect the binding of substrate to enzyme sites to increase with substrate concentration C_S. When substrate concentration C_S is very high, however, the active sites on an entire collection of enzyme molecules become occupied and the rate of product formation R_P reaches a maximum level V_m. Thus, at a fixed total enzyme concentration T_E, enzyme reactions exhibit the saturation kinetics shown in Figure 8.5-2. This behavior is characterized by the following three parameters:

$$\begin{aligned} V_m &\equiv \lim_{C_S \to \infty} R_P && \text{Rate at saturation} \\ k_{cat} &\equiv V_m / T_E && \text{Turnover number} \\ K_m &\equiv [C_S]_{R_p = V_m/2} && \text{Michaelis constant} \end{aligned} \tag{8.5-1a-c}$$

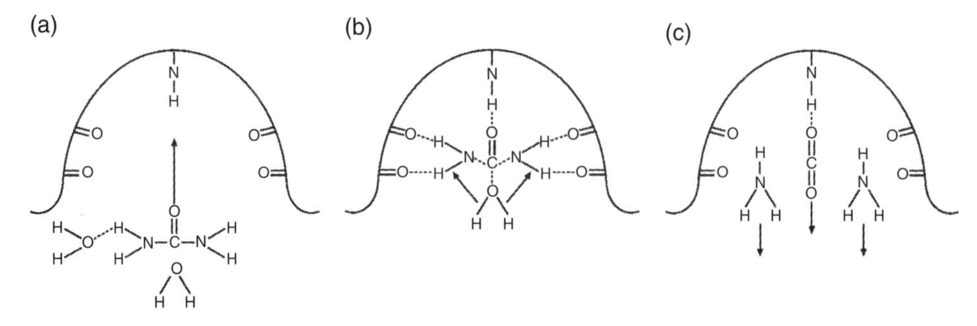

Figure 8.5-1 Hydrolysis of urea into ammonia and carbon dioxide at the active site of urease. (a) Absorption of urea and water. (b) Formation of urea–water complex. (c) Desorption of urea. (Reproduced from figure 23.1; Snell *et al.*, 1965. Courtesy of Joanna Ramage).

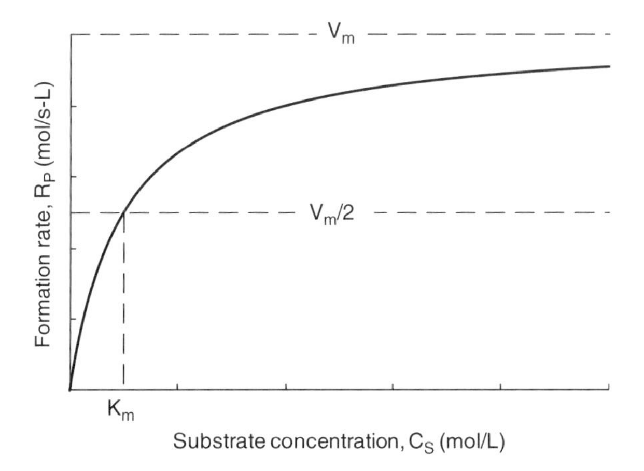

Figure 8.5-2 Saturation kinetics of a single substrate enzyme reaction.

Table 8.5-1 Enzyme Reaction Rate Parameters

Enzyme	Substrate	k_{cat} (s^{-1})	K_m (mol/L)
Acetylcholinesterase	Acetylcholine	1.4×10^4	9×10^{-5}
Carbonic anhydrase	CO_2	1×10^6	0.012
Carbonic anhydrase	HCO_3^-	4×10^5	0.026
Catalase	H_2O_2	4×10^7	1.1
Crotonase	Crotonyl-CoA	5.7×10^3	2×10^{-5}
Fumarase	Funarate	800	5×10^{-6}
Fumarase	Malate	900	2.5×10^{-5}
β–Lactamase	Benzylpenicillin	2.0×10^3	2×10^{-5}

From Zubay *et al.* (1993).

Table 8.5-2 Activation Energies (E_a) of Some Catalytic Reactions

Reaction	Catalyst	Act. Energy (kJ)
Sucrose → Glucose + Fructose	Acid (pH ≪ 7)	108.8
	Invertase	48.1
	None	75.3
$2H_2O_2 \rightarrow 2H_2O + O_2$	Colloidal platinum	49.0
	Liver catalase	20.9

From Snell *et al.* (1965, p 352).

Values of these parameters for various enzymes are given in Table 8.5-1. Table 8.5-2 compares the activation energies of uncatalyzed, synthetically catalyzed, and enzymatically catalyzed reactions. It is clear that enzymes have the extraordinary ability to lower activation energy. According to Equation 8.2-26, this not only increases the reaction rate constant but also diminishes the sensitivity of the rate constant to changes in temperature. Perhaps the most important feature of an active enzyme site is its specificity for a particular substrate. This allows for a robust regulation of metabolic reactions by the cellular control of enzyme expression.

8.5.2 Michaelis–Menten Kinetics

In the Michaelis–Menten model of enzyme kinetics, a single substrate S reversibly binds at the active site ♥ of an enzyme E to form an intermediate complex E♥S. The complex then decomposes irreversibly into a product P and the original enzyme.

$$\mathrm{E + S \underset{k_{-1}}{\overset{k_1}{\rightleftharpoons}} ES}$$
$$\mathrm{ES \overset{k_2}{\rightarrow} E + P} \qquad (8.5\text{-}2a, b)$$

To simplify analysis of these kinetics, we assume that substrate–enzyme binding is very fast compared to product formation so that the first reaction step is always close to equilibrium. This is equivalent to the quasi-equilibrium assumption that we applied in the transition-state state theory. The concentrations of E, S, and ES can then be related by an equilibrium binding constant.

$$\kappa_{ES} \equiv \frac{k_1}{k_{-1}} = \left(\frac{C_{ES}}{C_E C_S}\right) \qquad (8.5\text{-}3)$$

The stoichiometric reaction indicates that rate of utilization of S equals the rate of formation of P.

$$-R_S = R_P = k_2 C_{ES} \qquad (8.5\text{-}4)$$

Because individual values of C_E and C_{ES} are usually difficult to measure, it is useful to introduce the total enzyme concentration.

$$T_E = C_E + C_{ES} \qquad (8.5\text{-}5)$$

Combining this with Equation 8.5-3 to eliminate C_E, we find

$$C_{ES} = \frac{T_E C_S}{(1/\kappa_{ES}) + C_S} \qquad (8.5\text{-}6)$$

so that the reaction rates become

$$-R_S = R_P = \frac{T_E k_2 C_S}{(1/\kappa_{ES}) + C_S} \qquad (8.5\text{-}7)$$

For this rate expression, the enzyme parameters defined in Equation 8.5-1a-c correspond to

$$V_m = k_2 T_E, \quad k_{cat} = k_2, \quad K_m = 1/\kappa_{ES} \qquad (8.5\text{-}8a\text{-}c)$$

and Equation 8.5-7 can be written in the conventional form

$$R_P = -R_S = \frac{V_m C_S}{K_m + C_S} \qquad (8.5\text{-}9)$$

Depending on the magnitude of C_S relative to the K_m value, this Michaelis–Menten equation exhibits two limiting cases.

$$R_P = -R_S = \begin{cases} (V_m/K_m)C_S & \text{when } C_S/K_m \to 0 \\ V_m & \text{when } C_S/K_m \to \infty \end{cases} \tag{8.5-10}$$

Thus, the rate of product formation can be approximated as a constant when substrate concentration is high and as a first-order, irreversible reaction with a rate constant V_m/K_m when substrate concentration is low.

Specific values of K_m and V_m can be inferred from measurements of R_P or R_S at various C_S by non-linear regression of the data to the Michaelis-Menten equation. Alternatively, the data can be transformed in a manner that allows the rate parameters to be estimated as the slope and intercept of a linear regression.

$$\text{Lineweaver-Burk}: \quad \left(\frac{1}{R_P}\right) = \frac{K_m}{V_m}\left(\frac{1}{C_S}\right) + \frac{1}{V_m}$$

$$\text{Eadie-Hofstee}: \quad (R_P) = -K_m\left(\frac{R_P}{C_S}\right) + V_m \tag{8.5-11a-c}$$

$$\text{Hanes-Woolf}: \quad \left(\frac{C_S}{R_P}\right) = \frac{1}{V_m}(C_S) + \frac{K_m}{V_m}$$

Once V_m is known, it can be used to obtain k_{cat} from Equation 8.5-8a,b provided that the molar concentration of the enzyme is known. This can be problematic if the molecular weight of the enzyme or its purity is unknown.

Example 8.5-1 Cellular Detoxification of Hydrogen Peroxide

A cell initially contains a concentration of 2 mM hydrogen peroxide (H_2O_2), a potent oxidizing agent. The cell also contains a concentration of 1.0 μM catalase, an enzyme that promotes the decomposition of hydrogen peroxide according to overall reaction

$$2H_2O_2 \xrightarrow{\text{catalase}} 2H_2O + O_2 \tag{8.5-12}$$

How long will it take for 90% of the initial hydrogen peroxide to be decomposed?

Solution

Assuming that the cell is a closed well-mixed system of constant volume, a material balance on the H_2O_2 is given by

$$\frac{dC_S}{dt} = R_S \tag{8.5-13}$$

If this reaction follows simple Michaelis–Menten kinetics, then Equations 8.5-9 and 8.5-13 indicate that

$$\frac{dC_S}{dt} = -\frac{V_m C_S}{K_m + C_S} \tag{8.5-14}$$

This equation can be solved by separation of variables and application of the initial condition Cs(0).

$$t = \frac{1}{V_m}\left[C_S(0) - C_S + K_m \ln\frac{C_S(0)}{C_S}\right] = \frac{1}{k_{cat}T_E}\left[C_S(0) - C_S + K_m \ln\frac{C_S(0)}{C_S}\right] \tag{8.5-15}$$

From the problem statement and the parameters in Table 8.5-1: $k_{cat} = 4 \times 10^7\ s^{-1}$, $T_E = 1 \times 10^{-6}$ M, $K_m = 1.1$ M, $C_S(0) = 0.002$ M, and $C_S/C_S(0) = 0.1$. The reaction time can thus be computed from Equation 8.5-15.

$$t = \frac{1}{(4\times 10^7)(10^{-6})}\left[\left(0.002 - \frac{0.002}{10}\right) + (1.1)\ln(10)\right] = 0.0634\ s = 70\ ms \tag{8.5-16}$$

Because of its large turnover number $k_{cat,}$ catalase is extremely effective in decomposing H_2O_2 even though the enzyme concentration T_E is very small. In addition, because K_m is much larger than the peroxide level, the reaction kinetics are approximately first order, and the $[C_s(0) - C_S]$ term in Equation 8.5-15 contributes very little compared to $K_m \ln C_S(0)/C_S$.

8.5.3 Enzyme Inhibition

The reaction of a desired substrate at an active enzyme site can be inhibited by substances that either compete for occupancy of the site or somehow lower the activity of the active site. Some inhibitors form covalent bonds with an enzyme causing irreversible inhibition, whereas others bind to an enzyme in a reversible manner. In illustrating enzyme inhibition, we will restrict our attention to the mechanisms of reversible inhibition in Figure 8.5-3.

Single-Site Competitive Inhibition

When an inhibitor I has characteristics that are similar to a substrate S, it is possible for S and I to compete for the same active site ♥. This is the situation during competitive inhibition shown on the left side of the figure where both E♥S and E♥I complexes are both formed, but only the E♥S complex can produce product P; the enzyme is inactivated when the E♥I complex is formed. The kinetics of competitive inhibition originate from the following three reaction steps:

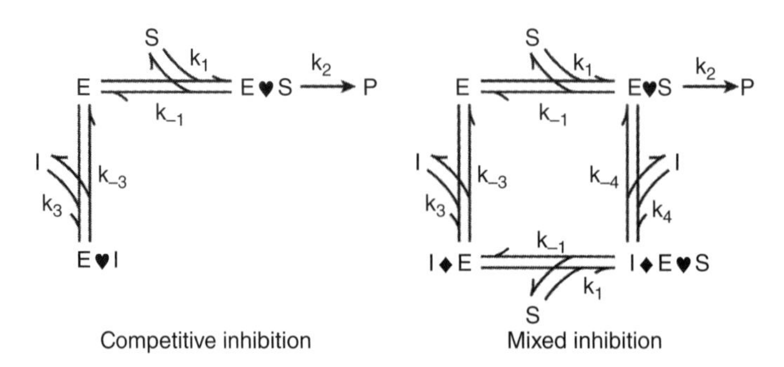

Figure 8.5-3 Reaction schemes for enzyme inhibition (Courtesy of Joanna Ramage).

$$\begin{aligned}&E + S \underset{k_{-1}}{\overset{k_1}{\rightleftharpoons}} ES\\&ES \xrightarrow{k_2} E + P\\&E + I \underset{k_{-3}}{\overset{k_3}{\rightleftharpoons}} EI\end{aligned} \tag{8.5-17a-c}$$

We apply the quasi-equilibrium hypothesis to both of the reversible reaction steps by defining the binding constants:

$$\kappa_{ES} \equiv \frac{k_1}{k_{-1}} = \frac{C_{ES}}{C_E C_S}, \quad \kappa_{EI} \equiv \frac{k_3}{k_{-3}} = \frac{C_{EI}}{C_E C_I} \tag{8.5-18a,b}$$

Introducing the total enzyme concentration:

$$T_E \equiv C_E + C_{ES} + C_{EI} \tag{8.5-19}$$

we arrive at the concentration of bound substrate.

$$C_{ES} = \frac{T_E C_S}{(1 + \kappa_{EI} C_I)/\kappa_{ES} + C_S} \tag{8.5-20}$$

The rate of product formation is

$$R_P = k_2 C_{ES} = \frac{k_2 T_E C_S}{(1 + \kappa_{EI} C_I)/\kappa_{ES} + C_S} \tag{8.5-21}$$

In terms of the dimensionless quantities,

$$\mathcal{C}_S \equiv \kappa_{ES} C_S, \quad \mathcal{C}_I \equiv \kappa_{EI} C_I, \quad \mathcal{R}_P = \frac{1}{k_2 T_E} R_P \tag{8.5-22a-d}$$

the dimensionless rate of product formation is

$$\mathcal{R}_P = \frac{\mathcal{C}_S}{1 + \mathcal{C}_I + \mathcal{C}_S} \tag{8.5-23}$$

and its derivative is

$$\frac{d\mathcal{R}_P}{d\mathcal{C}_S} = \frac{1}{1 + \mathcal{C}_I + \mathcal{C}_S} - \frac{\mathcal{C}_S}{(1 + \mathcal{C}_I + \mathcal{C}_S)^2} = + \frac{1 + \mathcal{C}_I}{(1 + \mathcal{C}_I + \mathcal{C}_S)^2} \tag{8.5-24}$$

The dimensionless reaction rate $\mathcal{R}_p$ is plotted in Figure 8.5-4 for various values of dimensionless substrate concentration. When $\mathcal{C}_I = 0$, Equation 8.5-23 reduces to Equation 8.5-9 for an uninhibited substrate reaction. In that case, $\mathcal{R}_p$ asymptotically approaches one as $\mathcal{C}_S \to \infty$ and the slope $d\mathcal{R}_p/d\mathcal{C}_S$ approaches one as $\mathcal{C}_S \to 0$. At values of $\mathcal{C}_I > 0$, this initial slope decreases by the factor $1/(1 + \mathcal{C}_I)$, but the asymptotic maximum is still equal to one.

Two-Site Mixed Inhibition

Sometimes an inhibitor I has characteristics so different from the substrate S that it cannot bind to active site ♥. It is still possible for I to interfere with product formation by allosteric binding to a second site ♦. For example, in the mechanism shown on the right side of Figure 8.5-3, an enzyme molecule can contain either complex E♥S or I♦E. It is also possible for these two complexes to coexist on the same enzyme molecule, a situation that we denote as I♦E♥S. However, only an enzyme molecule containing the E♥S complex alone is capable of producing product. The kinetics of this mixed inhibition process are represented by the following reaction steps:

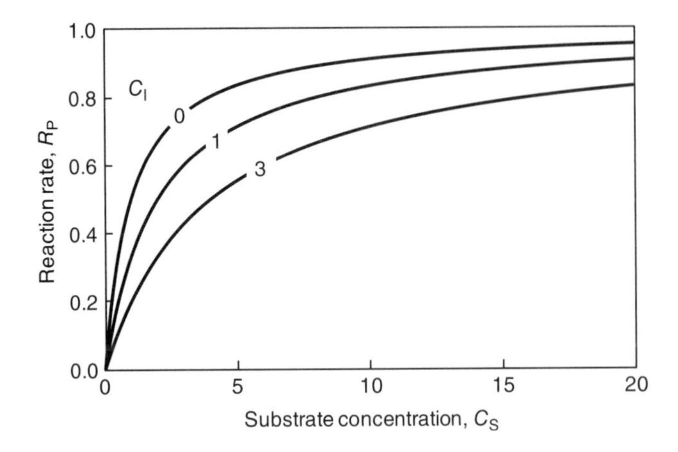

Figure 8.5-4 Competitive inhibition of substrate S by inhibitor I at a single active site.

$$\begin{aligned} &E + S \underset{k_{-1}}{\overset{k_1}{\rightleftharpoons}} ES \\ &ES \overset{k_2}{\rightarrow} E + P \\ &E + I \underset{k_{-3}}{\overset{k_3}{\rightleftharpoons}} IE \\ &ES + I \underset{k_{-4}}{\overset{k_4}{\rightleftharpoons}} IES \\ &IE + S \underset{k_{-1}}{\overset{k_1}{\rightleftharpoons}} IES \end{aligned} \qquad (8.5\text{-}25a\text{-}e)$$

Notice that the presence of bound inhibitor at its site ♦ does not interfere with the binding kinetics of substrate at its site ♥ so that the rate constants are the same in reaction steps 8.5-25a and e.

Using the quasi-equilibrium hypothesis for all four reversible binding reactions, the final equation for the formation rate of product P is found by the same approach as that for single-site inhibition.

$$R_P = \frac{k_2 T_E C_S}{K_m(1 + \kappa_{IE} C_I) + C_S(1 + \kappa_{IES} C_I)} \qquad (8.5\text{-}26)$$

where $T_E \equiv (C_E + C_{ES} + C_{IE} + C_{IES})$. The binding constants in this equation are defined as:

$$\kappa_{ES} \equiv \frac{k_1}{k_{-1}} = \frac{C_{ES}}{C_E C_S} = \frac{C_{IES}}{C_{IE} C_S}, \quad \kappa_{IE} \equiv \frac{k_3}{k_{-3}} = \frac{C_{IE}}{C_E C_I}, \quad \kappa_{IES} \equiv \frac{k_4}{k_{-4}} = \frac{C_{IES}}{C_{ES} C_I} \qquad (8.5\text{-}27a\text{-}c)$$

and the Michaelis constant is:

$$K_m \equiv \frac{1 + k_2/k_{-1}}{\kappa_{ES}} \qquad (8.5\text{-}28)$$

In terms of the dimensionless quantities,

$$\mathcal{C}_S \equiv \frac{C_S}{K_m}, \quad \mathcal{C}_I \equiv \kappa_{IE} C_I, \quad \mathcal{R}_P = \frac{R_P}{k_2 T_E}, \quad \kappa \equiv \frac{\kappa_{IES}}{\kappa_{IE}} \qquad (8.5\text{-}29a\text{-}e)$$

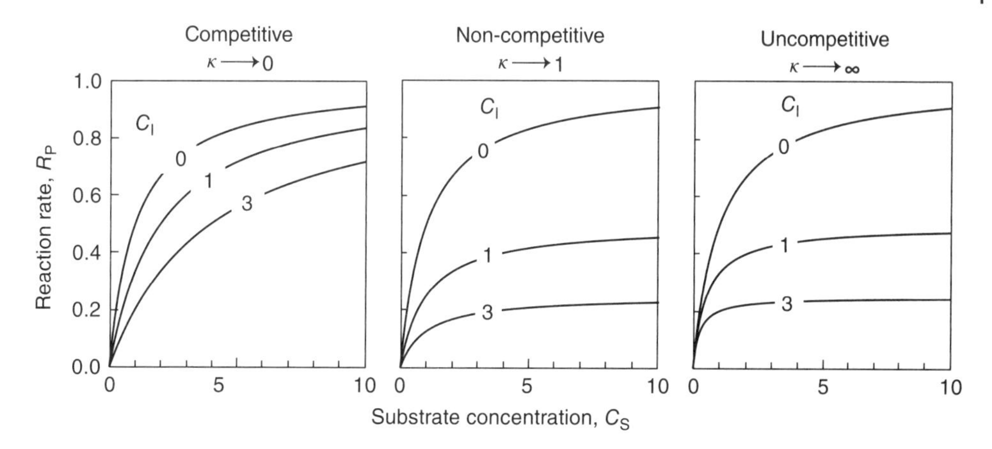

Figure 8.5-5 Modes of mixed inhibition.

The rate of product formation is

$$R_P = \frac{C_S}{1 + C_I + (1 + \kappa C_I)C_S} \tag{8.5-30}$$

From this equation, we can identify the following three modes of mixed inhibition

$$R_P = \begin{cases} \dfrac{C_S}{1 + C_I + C_S} & \text{when } \kappa \to 0 \quad \text{Competitive} \\ \dfrac{C_S}{(1 + C_I)(1 + C_S)} & \text{when } \kappa \to 1 \quad \text{Non-competitive} \\ \dfrac{C_S}{1 + (1 + \kappa C_I)C_S} & \text{when } \kappa \to \infty \quad \text{Uncompetitive} \end{cases} \tag{8.5-31}$$

When $\kappa \ll 1$ ($\kappa_{IES} \ll \kappa_{IE}$), the binding strength of I to an enzyme molecule already containing an E♥S complex is very weak, and the rate of product formation is equivalent to single-site competitive binding (Eq. 8.5-23). During non-competitive binding, $\kappa \approx 1$ ($\kappa_{IES} \approx \kappa_{IE}$), indicating that the binding strength of I to an enzyme molecule is the same, whether or not a E♥S complex is present. Uncompetitive inhibition occurs when $\kappa \gg 1$ ($\kappa_{IES} \gg \kappa_{IE}$), such that the binding strength of I to an enzyme molecule free of the E♥S complex is weak. Figure 8.5-5 contains dimensionless plots of the product formation rate predicted under these three conditions.

8.6 Urea Cycle as a Reaction Network

8.6.1 Reaction Rate Equations

The kinetics in the previous section were restricted to a single enzyme that converts a single substrate to a single product. In reality, the reaction pathways in metabolic processes can involve multiple enzymes acting on multiple substrates. For example, consider the urea cycle

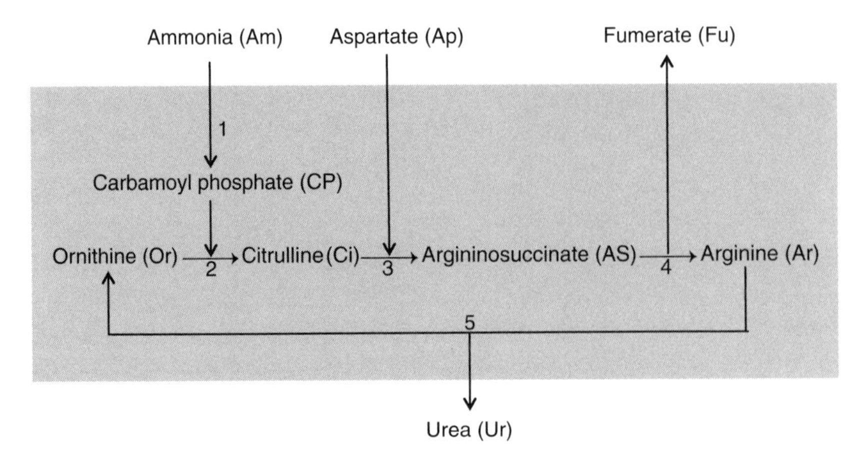

Figure 8.6-1 Urea cycle.

that occurs within a hepatocyte. The goal of this process is to transform ammonia, a toxic byproduct of protein catabolism, into urea that can be excreted by the kidneys.

The gray rectangle in Figure 8.6-1 represents a cell into which two key reactants, ammonia and aspartate, enter at molar rates I_{Am} and I_{Ap} per unit cell volume. Simultaneously, two key reaction products, urea and fumerate, exit the cell at molar rates O_{Ur} and O_{Fu} per unit volume. Within the cell, the urea cycle consists of the following enzymatic reactions:

$$\begin{aligned} &Am \rightarrow CP \quad \bar{r}_1 \\ &CP + Or \rightarrow Ci \quad \bar{r}_2 \\ &Ci + Ap \rightarrow AS \quad \bar{r}_3 \\ &AS \rightarrow Fu + Ar \quad \bar{r}_4 \\ &Ar \rightarrow Or + Ur \quad \bar{r}_5 \end{aligned} \tag{8.6-1a-e}$$

Not shown in figure is the required coupling of the first and third reactions to ATP hydrolysis resulting in the formation of ADP, AMP, and inorganic phosphate ions.

The intensive reaction rates of the nine different chemical species in the urea cycle are given in terms of inherent reaction rates as

$$\begin{aligned} &R_{Am} = -\bar{r}_1, \quad R_{CP} = \bar{r}_1 - \bar{r}_2, \quad R_{Ci} = \bar{r}_2 - \bar{r} \\ &R_{Ap} = -\bar{r}_3, \quad R_{AS} = \bar{r}_4 - \bar{r}_3, \quad R_{Fu} = \bar{r}_4 \\ &R_{Or} = \bar{r}_5 - \bar{r}_2, \quad R_{Ar} = \bar{r}_4 - \bar{r}_5, \quad R_{Ur} = \bar{r} \end{aligned} \tag{8.6-2a-i}$$

Because they are nonelementary, the five reactions in Equation 8.6-1 each consist of a number of reaction steps. For example, the conversion of arginine into ornithine and urea in reaction 5 is catalyzed at an active site of the enzyme arginase (E). As shown in Figure 8.6-2, the steps that constitute this reaction are

$$\begin{aligned} &E + Ar \underset{k_{-1}}{\overset{k_1}{\rightleftharpoons}} E{\bullet}Ar \\ &E{\bullet}Ar \overset{k_2}{\rightarrow} E{\bullet}Or + Ur \\ &E + Or \underset{k_{-3}}{\overset{k_3}{\rightleftharpoons}} E{\bullet}Or \end{aligned} \tag{8.6-3a-c}$$

These kinetics are a variation of single-site competitive inhibition (Eq. 8.5-17). By applying the quasi-equilibrium hypothesis to both of the reversible reaction steps, we arrive at an

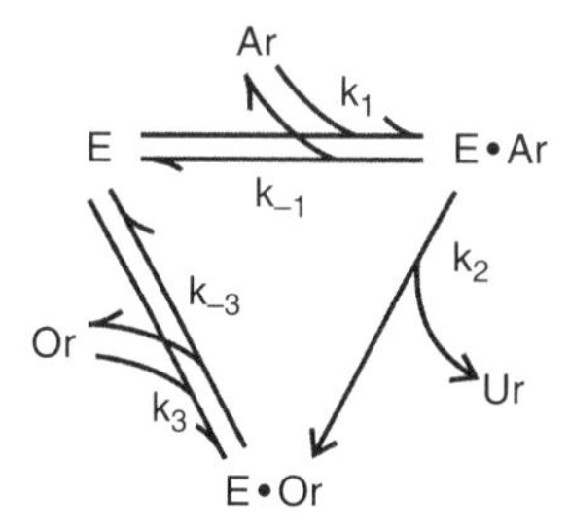

Figure 8.6-2 Hydrolysis of arginine to ornithine and urea by arginase (Adapted from Mulquiney and Kuchel (2003). Courtesy of Joanna Ramage).

equation for E • Ar concentration in terms of Ar concentration, Or concentration and total enzyme concentration.

$$C_{E\bullet Ar} = \frac{T_E C_{Ar}}{(1 + \kappa_{E\bullet Or} C_{Or})/\kappa_{E\bullet Ar} + C_{Ar}} \tag{8.6-4}$$

where

$$\kappa_{E\bullet Ar} \equiv \frac{k_1}{k_{-1}} = \frac{C_{E\bullet Ar}}{C_E C_{Ar}}, \quad \kappa_{E\bullet Or} \equiv \frac{k_3}{k_{-3}} = \frac{C_{E\bullet Or}}{C_E C_{Or}}, \quad T_E \equiv C_E + C_{E\bullet Ar} + C_{E\bullet Or} \tag{8.6-5a-c}$$

The rate of urea production by the arginine reaction can now be written as

$$R_{Ur} = k_2 C_{E\bullet Ar} = \frac{k_2 T_E \kappa_{E\bullet Ar} C_{Ar}}{1 + \kappa_{E\bullet Or} C_{Or} + \kappa_{E\bullet Ar} C_{Ar}} \tag{8.6-6}$$

8.6.2 Material Balances

Assuming that the urea cycle occurs in a well-mixed cell of constant volume, molar balances on the five species confined to the cell yield

$$\frac{dC_{CP}}{dt} = R_{CP}, \quad \frac{dC_{Ci}}{dt} = R_{Ci}, \quad \frac{dC_{AS}}{dt} = R_{AS}, \quad \frac{dC_{Or}}{dt} = R_{Or}, \quad \frac{dC_{Ar}}{dt} = R_{Ar} \tag{8.6-7a-e}$$

For the four species that are transferred across the cell boundaries,

$$\begin{aligned} \frac{dC_{Am}}{dt} &= R_{Am} + I_{Am}, \quad \frac{dC_{Ap}}{dt} = R_{Ap} + I_{Ap} \\ \frac{dC_{Fu}}{dt} &= R_{Fu} - O_{Fu}, \quad \frac{dC_{Ur}}{dt} = R_{Ur} - O_{Ur} \end{aligned} \tag{8.6-8a-d}$$

To simplify the solution of these equations, we will see what happens when one step of the reaction series represented by Equation 8.6-1a-e responds very slowly to an initial change in ammonia input compared to the other four steps. For example, suppose that arginine-to-ornithine conversion is this rate-limiting reaction step. In that case, the $\bar{r}_4$ reaction rate determines the urea production rate, while the intrinsic rates of the fast reaction steps are essentially equal to the ammonia input rate.

$$\bar{r}_1 = \bar{r}_2 = \bar{r}_3 = \bar{r}_4 = I_{Am} \tag{8.6-9}$$

Making use of Equation 8.6-2g-i, this equation leads to

$$R_{Ar} = -R_{Or}, \quad R_{Ar} = I_{Am} - R_{Ur} \tag{8.6-10a,b}$$

Equation 8.6-10a allows us to combine Equation 8.6-7d,e and obtain

$$\frac{dC_{Or}}{dt} + \frac{dC_{Ar}}{dt} = 0 \Rightarrow C_T \equiv C_{Or} + C_{Ar} = \text{constant} \tag{8.6-11a,b}$$

Substituting Equation 8.6-11b into Equation 8.6-6, we get the urea formation rate written exclusively in terms of arginine concentration.

$$R_{Ur} = \frac{k_2 T_E \kappa_{E\bullet Ar} C_{Ar}}{1 + \kappa_{E\bullet Or} C_T + (\kappa_{E\bullet Ar} - \kappa_{E\bullet Or}) C_{Ar}} \tag{8.6-12}$$

Combining this equation with Equations 8.6-7e and 8.6-10b, we obtain the arginine molar balance in terms of C_{Ar}.

$$\frac{dC_{Ar}}{dt} = I_{Am} - \frac{k_2 T_E \kappa_{E\bullet Ar} C_{Ar}}{1 + \kappa_{E\bullet Or} C_T + (\kappa_{E\bullet Ar} - \kappa_{E\bullet Or}) C_{Ar}} \tag{8.6-13}$$

We can formulate C_T in terms of initial conditions by assuming that the reaction cycle is initially at steady state such that $(dC_{Ar}/dt)_{t=0} = 0$. After specifying $C_{Ar}(0) = C_o$ and $I_{Am} = I_o$, Equation 8.6-13 indicates that

$$C_T = \frac{(k_2 T_E \kappa_{E\bullet Ar} - \kappa_{E\bullet Ar} I_o + \kappa_{E\bullet Or} I_o) C_o - I_o}{I_o \kappa_{E\bullet Or}} \tag{8.6-14}$$

We now consider the response of the urea cycle to a step increase in the rate of ammonia input from $I_{Am} = I_o$ at $t \le 0$ to $I_{Am} = I_\infty$ at $t > 0$. After inserting $I_{Am} = I_\infty$ in Equation 8.6-13 and using Equation 8.6-14 to eliminate C_T, we get

$$\frac{dC_{Ar}}{dt} = I_\infty - \frac{C_{Ar}}{C_o/I_o + (1 - \kappa_{E\bullet Or}/\kappa_{E\bullet Ar})(C_{Ar} - C_o)/k_2 T_E} \tag{8.6-15}$$

After this equation is solved for $C_{Ar}(t)$, we can determine $R_{Ur}(t)$ from a combination of Equations 8.6-12 and 8.6-14.

$$R_{Ur} = \frac{C_{Ar}}{C_o/I_o + (1 - \kappa_{E\bullet Or}/\kappa_{E\bullet Ar})(C_{Ar} - C_o)/k_2 T_E} \tag{8.6-16}$$

8.6.3 Dimensional Analysis and Simulations

We simplify the form of the model equations, by defining the dimensionless variables

$$t \equiv \frac{I_\infty}{C_o} t, \quad C_{Ar} \equiv \frac{C_{Ar}}{C_o}, \quad R_{Ur} \equiv \frac{R_{Ur}}{I_o} \tag{8.6-17a-c}$$

Using the dimensionless initial condition $C_{Ar}(0) = 1$, Equation 8.6-15 can be written in dimensionless form

$$\frac{dC_{Ar}}{dt} = 1 - \frac{C_{Ar}}{\beta_1 + \beta_2(1 - C_{Ar})} \quad (t > 0) \tag{8.6-18}$$

and Equation 8.6-16 becomes

$$R_{Ur} = \frac{\beta_1 C_{Ar}}{\beta_1 + \beta_2(1 - C_{Ar})} \tag{8.6-19}$$

The two dimensionless parameters in these equations are defined as

$$\beta_1 \equiv \frac{I_\infty}{I_o}, \quad \beta_2 \equiv \frac{I_\infty}{k_2 T_E}\left(\frac{\kappa_{E\bullet Or}}{\kappa_{E\bullet Ar}} - 1\right) \tag{8.6-20a,b}$$

Since I_∞/k_2T_E is a positive quantity, β_2 could have a positive or negative value, depending on the relative strengths of arginine and ornithine binding to arginase. Based on the ratios of rate coefficients reported for these binding reactions (Mulquiney and Kuchel, 2003), we infer that the arginine binding coefficient $\kappa_{E\bullet Ar}$ is less than the ornithine binding coefficient $\kappa_{E\bullet Or}$. Thus, realistic values of β_2 are greater than zero.

This model exhibits some interesting limiting behaviors at very small times and very large times. The initial slope of $C_{Ar}(t)$ is obtained from Equation 8.6-18 by setting $C_{Ar}(0) = 1$.

$$\lim_{t\to 0}\left(\frac{dC_{Ar}}{dt}\right) = 1 - \frac{1}{\beta_1} \tag{8.6-21}$$

Its ultimate steady-state concentration is obtained by setting $(dC_{Ar}/dt) = 0$.

$$\lim_{t\to\infty} C_{Ar} = \frac{\beta_1 + \beta_2}{1 + \beta_2} \tag{8.6-22}$$

Combining these limiting behaviors of $C_{Ar}(t)$ with Equation 8.6-19, we obtain corresponding formulas for $R_{Ur}(t)$.

$$\lim_{t\to 0}\left(\frac{dR_{Ur}}{dt}\right) = \lim_{t\to 0}\left(\frac{dR_{Ur}}{dC_{Ar}}\frac{dC_{Ar}}{dt}\right) = \left(1 + \frac{\beta_2}{\beta_1}\right)\left(1 - \frac{1}{\beta_1}\right) \tag{8.6-23}$$

$$\lim_{t\to\infty} R_{Ur} = \beta_1 \tag{8.6-24}$$

In performing numerical simulations of Equations 8.6-18 and 8.6-19 with various positive β_2 values, we set $\beta_1 = 2$, corresponding to a doubling of the ammonia step input I_∞. According to the results in Figure 8.6-3, the arginine concentration C_{Ar} at any β_2 value increases with time,

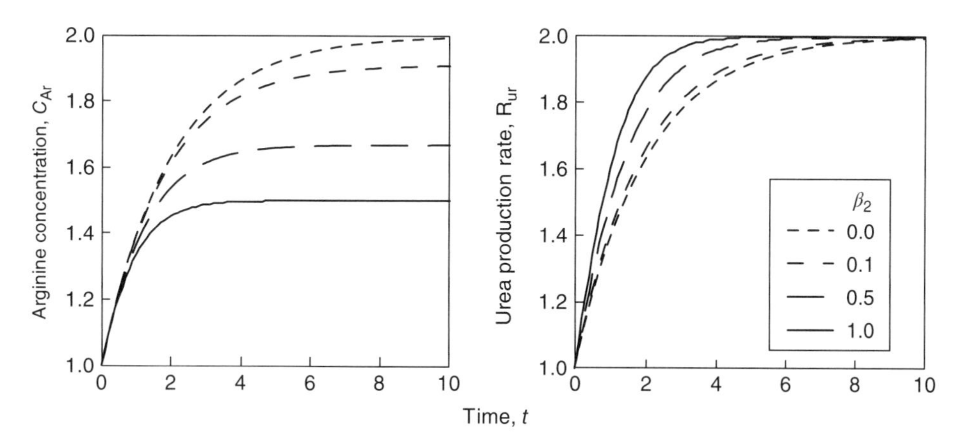

Figure 8.6-3 Response of a urea cycle to a doubling of the ammonia input ($\beta_1 = 2$).

and this causes R_{ur} to also increase with time. However, at any given time, increasing β_2 values produce increases in R_{ur} and decreases in C_{Ar}. Notice that both $C_{Ar}(t)$ and $R_{Ur}(t)$ exhibit the initial slopes and steady-state limits of the model specified by Equations 8.6-21–8.6-24. For R_{Ur}, in particular, the doubling of the ammonia input rate eventually causes a doubling of the urea production rate.

References

Gibson, QH. The reaction of oxygen with hemoglobin and the kinetic basis of the effect of salt on binding of oxygen. J Biol Chem. 1970; 245:3285–3288.

Mulquiney PJ, Kuchel PW. Modelling Metabolism with Mathematica. Boca Raton: CRC; 2003, p 69.

Snell FM, Shulman S, Spencer RP, Moos C. Biophysical Principles of Structure and Function. Reading: Addison-Wesley; 1965, p 352.

Zubay G. Biochemistry. Dubuque: Wm. C. Brown; 1993, p 209.

Part IV

Transport Models in Fluids and Membranes

Chapter 9

Unidirectional Transport

In this chapter, we discuss mass transfer in homogeneous fluids when velocity and composition vary in only one spatial direction. This focus on unidirectional transport simplifies mathematical analysis, while allowing us to discuss important concepts that are still applicable to multidimensional problems. The material in this chapter is also the foundation for later developments of transport rates across membranes and interfaces. A detailed presentation of three-dimensional, unsteady-state equations of mass and momentum transport will follow in the next part of the book.

We begin the chapter with the development of the differential species and solution mass balances for simultaneous convection and diffusion in a single spatial direction. These unidirectional equations will then be used in illustrations of transport under steady- and unsteady-state conditions in the absence and the presence of chemical reaction.

9.1 Unidirectional Transport Equations

9.1.1 Species Fluxes

The molar flux equation for ordinary diffusion in a homogeneous solution was given in Chapter 7 for a solution with constant molar density as

$$\mathbf{N}_i = \mathbf{u}^* C_i - \mathcal{D}_i \nabla C_i \tag{7.1-5}$$

and for a solution of constant mass density as

$$\mathbf{N}_i = \mathbf{u} C_i - \mathcal{D}_i \nabla C_i \tag{7.1-7}$$

Biomedical Mass Transport and Chemical Reaction: Physicochemical Principles and Mathematical Modeling, First Edition. James S. Ultman, Harihara Baskaran, and Gerald M. Saidel.

Table 9.1-1 Components of the Molar Flux Vector: Constant Mass Density

Rectangular coordinates:

$$N_{i,x} = u_x C_i - \mathcal{D}_i \frac{\partial C_i}{\partial x}, \quad N_{i,y} = u_y C_i - \mathcal{D}_i \frac{\partial C_i}{\partial y}, \quad N_{i,z} = u_z C_i - \mathcal{D}_i \frac{\partial C_i}{\partial z}$$

Cylindrical coordinates:

$$N_{i,r} = u_r C_i - \mathcal{D}_i \frac{\partial C_i}{\partial r}, \quad N_{i,\theta} = u_\theta C_i - \mathcal{D}_i \frac{1}{r}\frac{\partial C_i}{\partial \theta}, \quad N_{i,z} = u_z C_i - \mathcal{D}_i \frac{\partial C_i}{\partial z}$$

Spherical coordinates:

$$N_{i,r} = u_r C_i - \mathcal{D}_i \frac{\partial C_i}{dr}, \quad N_{i,\theta} = u_\theta C_i - \mathcal{D}_i \frac{1}{r}\frac{\partial C_i}{d\theta}, \quad N_{i,\varphi} = u_\varphi C_i - \mathcal{D}_i \frac{1}{r\sin\theta}\frac{\partial C_i}{\partial \varphi}$$

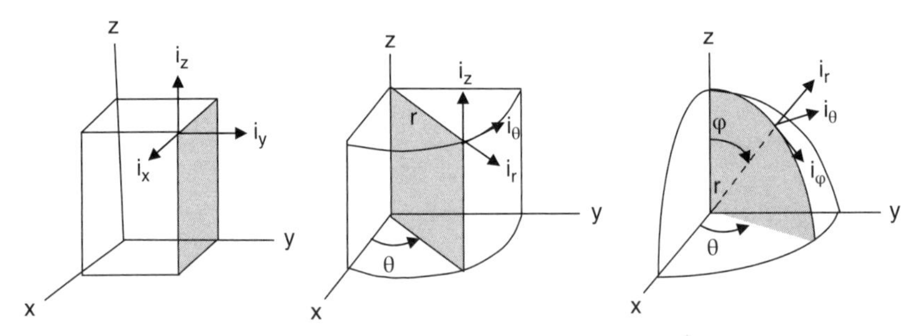

Figure 9.1-1 Rectangular (left), cylindrical (center), and spherical coordinates (right). The unit base vectors (i_j) indicate the direction in which the scalar components of vectors are resolved.

These vector equations can be represented by three scalar components. Table 9.1-1 lists the $\mathbf{N_i}$ components in rectangular, cylindrical, and spherical coordinates (Fig. 9.1-1) for a solution of constant mass density. The $\mathbf{N_i}$ components for a fluid of constant molar density are obtained by replacing the components of the mass-average velocity vector in all table entries by the corresponding components of the molar-average velocity.

9.1.2 Rectilinear Transport

There are a few special cases of transport occurring in a single spatial direction, along either a rectilinear coordinate y or a radial coordinate r (Fig. 9.1-2). In this section, we develop the solution and species material balances for rectilinear transport in either a fluid of constant mass density or a constant molar density. We will present the analogous equations for radial transport in the next section.

Consider a flat slab bounded by the $x - z$ planes of a rectangular coordinate system. These planes have equal areas S_y and are located between coordinate positions y and $y + \Delta y$. When this slab is sufficiently long in the x and z directions relative to its thickness Δy, we can neglect edge effects. Spatial changes in mass-averaged velocity u_y and species molar concentration C_i are then restricted to the y direction so that the molar flux vector reduces to its y component $N_{i,y}$. Corresponding to Equation 2.3-4, the molar balance for species i in this control volume is

$$\frac{\partial(C_i \Delta V)}{\partial t} = \left\{ [\dot{N}_{i,y}]_y - [\dot{N}_{i,y}]_{y+\Delta y} \right\} + R_i \Delta V \tag{9.1-1}$$

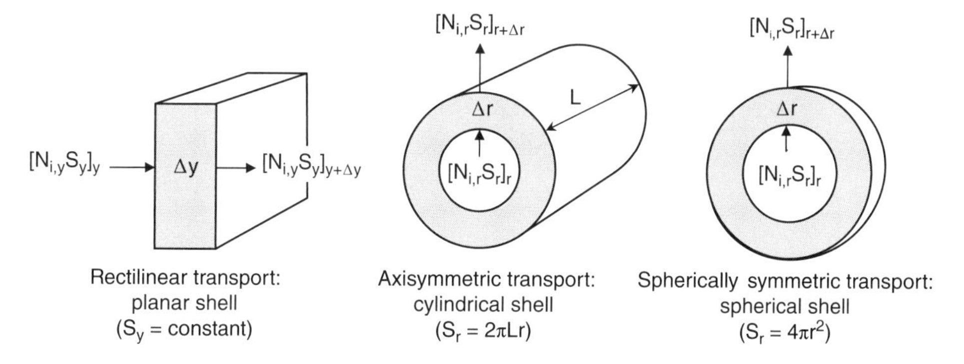

Figure 9.1-2 Control volumes for unidirectional transport.

where ΔV is the size of the control volume; $[\dot{N}_{i,y}]_y$ and $[\dot{N}_{i,y}]_{y+\Delta y}$ are the molar rates at which component i enters and exits the control volume; and R_i is the molar rate per unit volume at which component i is produced by chemical reaction. Noting that $\dot{N}_{i,y} = N_{i,y}S_y$ and $\Delta V = S_y \Delta y$, which is constant, Equation 9.1-1 becomes

$$\frac{\partial C_i}{\partial t} = \frac{1}{S_y \Delta y}\left\{[N_{i,y}S_y]_y - [N_{i,y}S_y]_{y+\Delta y}\right\} + R_i \tag{9.1-2}$$

Taking the limit of this equation as Δy approaches zero,

$$\frac{\partial C_i}{\partial t} + \lim_{\Delta y \to 0}\frac{1}{S_y}\left\{\frac{[N_{i,y}S_y]_{y+\Delta y} - [N_{i,y}S_y]_y}{\Delta y}\right\} = R_i \tag{9.1-3}$$

Realizing that the second term defines a the y derivative of $N_{i,y}S_y$, this equation becomes

$$\frac{\partial C_i}{\partial t} + \frac{1}{S_y}\frac{\partial (N_{i,y}S_y)}{\partial y} = R_i \tag{9.1-4}$$

For a flat slab, the transport cross section S_y is independent of y so that

$$\frac{\partial C_i}{\partial t} + \frac{\partial N_{i,y}}{\partial y} = R_i \tag{9.1-5}$$

Constant Mass Density

We derive an overall mass balance by multiplying Equation 9.1-5 by molecular weight M_i and summing over all I substances in the solution:

$$\frac{\partial}{\partial t}\sum_{i=1}^{I} M_i C_i + \frac{\partial}{\partial y}\left(\sum_{i=1}^{I} M_i N_{i,y}\right) = \sum_{i=1}^{I} M_i R_i \tag{9.1-6}$$

The summation on the right side of this equation is zero since no net mass can be produced by chemical reactions. Identifying the other two summations as $\sum_i M_i C_i = \rho$ and $\sum_i M_i N_{i,y} = \sum_i n_{i,y} = \rho u_y$, we write the overall mass balance as

$$\frac{\partial \rho}{\partial t} + \frac{\partial (\rho u_y)}{\partial y} = 0 \tag{9.1-7}$$

When ρ is constant, a reasonable assumption for biological liquids, the y component of molar flux in rectangular coordinates is given in Table 9.1-1 as

$$N_{i,y} = u_y C_i - \mathcal{D}_i \frac{\partial C_i}{\partial y} \tag{9.1-8}$$

For such a constant mass density fluid, Equations 9.1-5, 9.1-7, and 9.1-8 lead to the transport equations:

$$\begin{aligned} &\frac{\partial u_y}{\partial y} = 0 \\ &\frac{\partial C_i}{\partial t} + u_y \frac{\partial C_i}{\partial y} = \frac{\partial}{\partial y}\left(\mathcal{D}_i \frac{\partial C_i}{\partial y}\right) + R_i \end{aligned} \tag{9.1-9a, b}$$

Constant Molar Density

The transport equations for fluids of constant c, often a realistic idealization for gas mixtures, are derived in a similar manner. When summed over all species and noting that $\sum_i C_i = c$ and $\sum_i N_{i,y} = c u_y^*$, Equation 9.1-5 leads to an overall molar balance:

$$\frac{\partial c}{\partial t} + \frac{\partial \left(c u_y^*\right)}{\partial y} = \sum_{i=1}^{I} R_i \tag{9.1-10}$$

When c is constant, the rectilinear flux expression based on molar-average velocity is given by

$$N_{i,y} = u_y^* C_i - \mathcal{D}_i \frac{\partial C_i}{\partial y} \tag{9.1-11}$$

and Equations 9.1-5, 9.1-10, and 9.1-11 lead to the transport equations for a solution of constant molar density:

$$\begin{aligned} &\frac{\partial u_y^*}{\partial y} = \frac{1}{c}\sum_{i=1}^{I} R_i \\ &\frac{\partial C_i}{\partial t} + u_y^* \frac{\partial C_i}{\partial y} = \frac{\partial}{\partial y}\left(\mathcal{D}_i \frac{\partial C_i}{\partial y}\right) + R_i - \frac{C_i}{c}\sum_{i=1}^{I} R_i \end{aligned} \tag{9.1-12a, b}$$

The additional reaction term $\sum_i R_i$ that appears in these equations compared to those for a constant density fluid represents the rate at which the total moles of solution increase because of chemical reactions. Although there is no change of mass because of chemical reaction, $\sum_i R_i$ will only disappear when, according to the stoichiometric reaction, the total moles of all reactants are equal to the total moles of all products.

9.1.3 Radial Transport

For an axisymmetric concentration field in a cylindrical shell that is long enough to neglect end effects, C_i only depends on r and t. Similarly, for a concentration field with symmetry about the center point of a spherical shell, C_i only depends on r and t. Whereas the cross section S_y for rectilinear transport is a constant, the cross section $S_r(r)$ for radial transport depends on radial position.

The overall mass balance for radial transport will be the same as Equation 9.1-4 with the rectilinear coordinate y replaced by the radial coordinate r:

$$\frac{\partial C_i}{\partial t} + \frac{1}{S_r}\frac{\partial (N_{i,r} S_r)}{\partial r} = R_i \tag{9.1-13}$$

To obtain an overall mass balance, we multiply this equation by molecular weight M_i and sum over the I substances in the solution:

$$\frac{\partial}{\partial t}\sum_{i=1}^{I} M_i C_i + \frac{1}{S_r}\frac{\partial}{\partial r}\left(S_r \sum_{i=1}^{I} M_i N_{i,r}\right) = \sum_{i=1}^{I} M_i R_i \tag{9.1-14}$$

Since $\sum_i M_i R_i = 0$, $\sum_i M_i C_i = \rho$, and $\sum_i M_i N_{i,r} = \sum_i n_{i,r} = \rho u_r$, we can write this equation as

$$\frac{\partial \rho}{\partial t} + \frac{1}{S_r}\frac{\partial (\rho u_r S_r)}{\partial r} = 0 \tag{9.1-15}$$

In a constant mass density fluid, the radial flux in axisymmetric and spherically symmetric concentration fields is given in Table 9.1-1 as

$$N_{i,r} = u_r C_i - \mathcal{D}_i \frac{\partial C_i}{\partial r} \tag{9.1-16}$$

and Equations 9.1-13, 9.1-15, and 9.1-16 lead to

Table 9.1-2 Unidirectional Radial Transport

Constant Mass Density, ρ	
Axisymmetric ($S_r \propto r$)	$\frac{\partial (r u_r)}{\partial r} = 0$
	$\frac{\partial C_i}{\partial t} + u_r \frac{\partial C_i}{\partial r} = \frac{1}{r}\frac{\partial}{\partial r}\left(\mathcal{D}_i r \frac{\partial C_i}{\partial r}\right) + R_i$
Spherically symmetric ($S_r \propto r^2$)	$\frac{\partial (r^2 u_r)}{\partial r} = 0$
	$\frac{\partial C_i}{\partial t} + u_r \frac{\partial C_i}{\partial r} = \frac{1}{r^2}\frac{\partial}{\partial r}\left(\mathcal{D}_i r^2 \frac{\partial C_i}{\partial r}\right) + R_i$
Constant Molar Density c	
Axisymmetric ($S_r \propto r$)	$\frac{\partial (r u_r^*)}{\partial r} = \frac{r}{c}\sum_{k=1}^{I} R_k$
	$\frac{\partial C_i}{\partial t} + u_r^* \frac{\partial C_i}{\partial r} = \frac{1}{r}\frac{\partial}{\partial r}\left(\mathcal{D}_i r \frac{\partial C_i}{\partial r}\right) + R_i - \frac{C_i}{c}\sum_{k=1}^{I} R_k$
Spherically symmetric ($S_r \propto r^2$)	$\frac{\partial (r^2 u_r^*)}{\partial r} = \frac{r^2}{c}\sum_{k=1}^{I} R_k$
	$\frac{\partial C_i}{\partial t} + u_r^* \frac{\partial C_i}{\partial r} = \frac{1}{r^2}\frac{\partial}{\partial r}\left(\mathcal{D}_i r^2 \frac{\partial C_i}{\partial r}\right) + R_i - \frac{C_i}{c}\sum_{k=1}^{I} R_k$

$$\frac{\partial(u_r S_r)}{\partial r} = 0$$
$$\frac{\partial C_i}{\partial t} + u_r \frac{\partial C_i}{\partial r} = \frac{1}{S_r}\frac{\partial}{\partial r}\left(\mathcal{D}_i S_r \frac{\partial C_i}{\partial r}\right) + R_i \quad (9.1\text{-}17a,b)$$

By accounting for the dependence of transport cross section on radial position, $S_r \propto r$ for a cylindrical shell and $S_r \propto r^2$ for a spherical shell, Table 9.1-2 lists the final transport equations for radial transport in fluids of constant ρ. The radial transport equations for a fluid of constant c (derived in a similar manner) are also listed in Table 9.1-2

9.2 Steady-State Diffusion

9.2.1 Rectilinear Diffusion

We first discuss the simple problem of unidirectional, steady-state diffusion of a non-reacting substance s in the y direction through a slab bounded by planes located at $y = y_1$ and $y = y_2$. With $dC_s/dt = 0$ and $R_s = 0$, a molar balance (Eq. 9.1-4) indicates that

$$\frac{d(N_{s,y}S_y)}{dy} = \frac{d\dot{N}_{s,y}}{dy} = 0 \quad (9.2\text{-}1)$$

Thus, the transport rate $\dot{N}_{s,y}$ is constant. Since S_y is constant, the species flux $N_{s,y}$ is also constant. For now, we ignore convective transport relative to diffusive transport. The diffusion flux equation for a solute s in fluids with constant mass density and constant molar density is then the same:

$$N_{s,y} = -\mathcal{D}_s \frac{dC_s}{dy} \quad (9.2\text{-}2)$$

We integrate this equation subject to a boundary condition $C_s(y = y_1) = C_{s1}$:

$$\int_{y_1}^{y} N_{s,y}dy = -\int_{C_{s1}}^{C_s} \mathcal{D}_s dC_s \quad (9.2\text{-}3)$$

Assuming that $\mathcal{D}_s$ is constant, we can complete this integration:

$$C_s(y) = C_{s1} - \left(\frac{N_{s,z}}{\mathcal{D}_s}\right)(y - y_1) \quad (9.2\text{-}4)$$

As we expect from physical observations, C_s decreases with y when $N_{s,y}$ is a positive quantity. By evaluating Equation 9.2-4 at the second boundary concentration, $C_s(y_2) = C_{s2}$, we can express the species flux and transport rates in terms of an overall concentration difference between two boundaries of the slab:

$$N_{s,y} = \frac{\mathcal{D}_s}{y_2 - y_1}(C_{s1} - C_{s2}) \quad (9.2\text{-}5)$$

or equivalently

$$\dot{N}_{s,y} = \frac{\mathcal{D}_s S_y}{y_2 - y_1}(C_{s1} - C_{s2}) \tag{9.2-6}$$

9.2.2 Radial Diffusion

During steady-state, radial diffusion of a non-reacting solute s through a cylindrical or spherical shell, the molar balance equation (Eq. 9.1-17b) reduces to

$$\frac{d(N_{s,r}S_r)}{dr} = \frac{d\dot{N}_{s,r}}{dr} = 0 \tag{9.2-7}$$

Therefore, $\dot{N}_{s,r}$ is constant, whereas $N_{s,r}$ is inversely proportional to $S_r(r)$, which depends on r. In a fluid with either a constant mass density or a constant molar density, the radial transport rate in the absence of convection is

$$\dot{N}_{s,r} = -\mathcal{D}_s S_r \frac{dC_s}{dr} \tag{9.2-8}$$

When $\mathcal{D}_s$ is constant, we can integrate this equation with the concentration limits of C_{s1} at $r = r_1$ and C_{s2} at $r = r_2$, corresponding at the two surfaces of either a cylindrical or a spherical shell:

$$\dot{N}_{s,r} = \mathcal{D}_s \left(\int_{r_1}^{r_2} \frac{dr}{S_r} \right)^{-1} (C_{s1} - C_{s2}) \tag{9.2-9}$$

Example 9.2-1 Polarographic Oxygen Electrode in Quiescent Blood

Oxygen partial pressure in blood and other fluids is frequently monitored by observing the electrical current i passing between a gold electrode biased at a negative voltage and a silver reference electrode coated with silver chloride. When these electrodes are immersed in blood, the following reactions occur at the surfaces of the electrodes:

$$\begin{aligned} &O_2 + 2H_2O + 4e^- \rightarrow 4OH^- \quad &\text{Gold cathode} \\ &4Ag^o + 4Cl^- \rightarrow 4AgCl + 4e^- \quad &\text{Silver anode} \end{aligned} \tag{9.2-10a, b}$$

As shown in Figure 9.2-1, electrons in the external circuit migrate through a voltage source in the direction of decreasing electrical potential, that is, away from the anode and toward the cathode. In the blood, chloride anions diffuse to the silver anode where they are stripped of electrons while forming AgCl. Simultaneously, O_2 and water diffuse to the inert cathode where they combine with electrons to form hydroxyl anions. In this manner, a complete circuit is formed by the conductance of electrons through the external circuit and the transport of negatively charged ions through the solution.

Electroneutrality is maintained in the solution because each chloride anion incorporated into the anode as silver chloride is replaced by a hydroxyl anion that is released into solution at the cathode. Because chloride ions are present in blood at high concentrations relative to

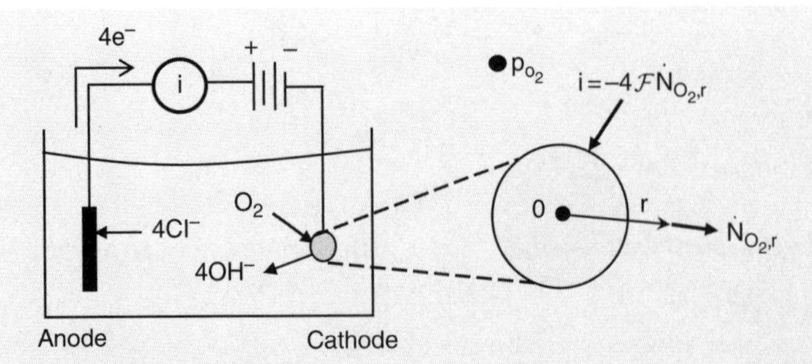

Figure 9.2-1 Polarographic electrode system for monitoring O_2 With electrochemical circuit (left) and O_2 diffusion conditions at the cathode (right).

O_2, the electrical current through the solution is limited by O_2 diffusion to the cathode surface. And with proper selection of the bias voltage (approximately −0.9 V), the cathode reaction is so fast that O_2 concentration at the cathode surface is close to zero.

Consider a polarographic electrode system with spherical cathode of diameter d = 1 mm placed in a container of stagnant blood at a temperature of 37°C. We will develop a model to estimate the electrical current i generated by this electrode relative to the O_2 partial pressure p_{O_2} in the blood.

Solution

Placing the origin of a radial coordinate at the center of the cathode, the diffusion surface at any r position is $S_r = 4\pi r^2$. If the cathode diameter is small compared to the container size, steady-state diffusion will result in an O_2 partial pressure distribution in the blood that is spherically symmetric and depends only on the radial position. On the cathode surface, O_2 concentration is suppressed by the rapid polarographic reaction so that $C_{s1} \simeq 0$ at $r_1 = d/2$. Far from the surface, O_2 partial pressure remains at its undisturbed value p_{O_2} so that $C_{s2} = \alpha_{O_2} c_G^o p_{O_2}$ at $r_2 \to \infty$. Employing Equation 9.2-9, we obtain the O_2 transport rate in the r direction:

$$\dot{N}_{O_2,r} = \mathcal{D}_{O_2}\left(\int_{d/2}^{\infty} \frac{1}{4\pi r^2} dr\right)^{-1} \left(0 - \alpha_{O_2} c_G^o p_{O_2}\right) = -2\pi d \alpha_{O_2} c_G^o p_{O_2} \tag{9.2-11}$$

where $\mathcal{D}_{O_2}$ is the diffusion coefficient of O_2 through blood. With four moles of electrons required for each mole of O_2 that reacts at the cathode surface, the corresponding electrical current is

$$i = -4\mathcal{F}\dot{N}_{O_2,r} \tag{9.2-12}$$

where $\mathcal{F}$ is Faraday's constant. The negative sign in this equation is necessary because $\dot{N}_{O_2,r}$ is positive when O_2 movement is away from the cathode, whereas a positive current i (i.e., external electron flow toward the cathode) requires O_2 diffusion toward the cathode. Combining Equation 9.2-11 with Equation 9.2-12, we get an electrode sensitivity of

$$\frac{i}{p_{O_2}} = 8\pi\mathcal{F} c_G^o \alpha_G \mathcal{D}_{O_2} d \tag{9.2-13}$$

Neglecting the effect of hemoglobin, we approximate the diffusion coefficient of O_2 in whole blood by that in serum, $\mathcal{D}_{O_2}(37°C) \approx 1.87 \times 10^{-5}$ cm^2/s (Table A4-3). The Bunsen solubility coefficient of O_2 in whole blood is $\alpha_{O_2}(37\,°C) = 2.19 \times 10^{-7}$ ml(STP)/ml-Pa (Table A3-3). With remaining parameter values of $c_G^o = 4.46 \times 10^{-5}$ mol/ml(STP), d = 0.1 cm, and $\mathcal{F} = 96{,}500$ A-s/mol, we find

$$\frac{i}{p_{O_2}} = 8\pi(96{,}500)(4.46\times10^{-5})(2.19\times10^{-7})(1.87\times10^{-5})(0.1) = 4.43\times10^{-2}\ \mu A/kPa \quad (9.2\text{-}14)$$

That is, a 1 mm polarographic O_2 electrode generates a current of 44 nanoamps for each kilopascal of O_2 partial pressure in stagnant blood.

Integrating Equation 9.2-9 for the radial diffusion rate of species s through a cylindrical shell of length L and diffusion surface $S_r = 2\pi rL$, we obtain

$$\dot{N}_{s,r} = \frac{2\pi L \mathcal{D}_s}{\ln(r_2/r_1)}(C_{s1} - C_{s2}) \quad (9.2\text{-}15)$$

Similarly, the diffusion rate of species s through a spherical shell with diffusion surface $S_r = 4\pi r^2$ is

$$\dot{N}_{s,r} = \frac{4\pi r_1 r_2 \mathcal{D}_s}{r_2 - r_1}(C_{s1} - C_{s2}) \quad (9.2\text{-}16)$$

We can consolidate these two equations and the corresponding equation for rectilinear diffusion (Eq. 9.2-6) into a single transport rate equation:

$$\dot{N}_s = \frac{\mathcal{D}_s \tilde{S}}{h}(C_{s1} - C_{s2}) \quad (9.2\text{-}17)$$

where we have defined the generalized diffusion path length h and mean surface area $\tilde{S}$ as

$$h \equiv \begin{cases} y_2 - y_1 \\ r_2 - r_1 \\ r_2 - r_1 \end{cases} \quad \tilde{S} \equiv \begin{cases} S_y & \text{Planar slab} \\ \dfrac{2\pi L(r_2 - r_1)}{\ln(r_2/r_1)} & \text{Cylindrical shell} \\ 4\pi r_1 r_2 & \text{Spherical shell} \end{cases} \quad (9.2\text{-}18a\text{-}c)$$

Whereas $\tilde{S}$ has a constant value S_y for a planar slab, the effect of surface curvature on radial diffusion is accounted for by the dependence of $\tilde{S}$ on r_1 and r_2.

Example 9.2-2 Pulmonary Oxygen Uptake

Oxygen uptake in the lungs results from transport across alveolar–capillary membranes that consist of adjacent layers of surfactant, epithelium, interstitial space, and endothelium. Roughly speaking, the alveolar–capillary membrane of each alveolus is a spherical shell, about 2 μm in thickness (Fig. 9.2-2). This membrane separates an inner gas space of approximate radius 100 μm from an outer capillary blood sheath. With the origin of a radial coordinate at the center of the gas space, the gas side of the membrane is located at $r_1 = 100$ μm and the blood side is at $r_2 = 102$ μm.

There are about 300 million alveoli that contribute to the O_2 uptake rate $\dot{V}_{O_2}$ in the adult human lung. During maximal aerobic exercise, essentially all of the alveolar capillaries are perfused with venous blood at a $p_{O_2}^{ven} \approx 2$ kPa and the alveolar gas space contains a $p_{O_2}^{alv} = 13$ kPa. Using this information, we will estimate $\dot{V}_{O_2}$.

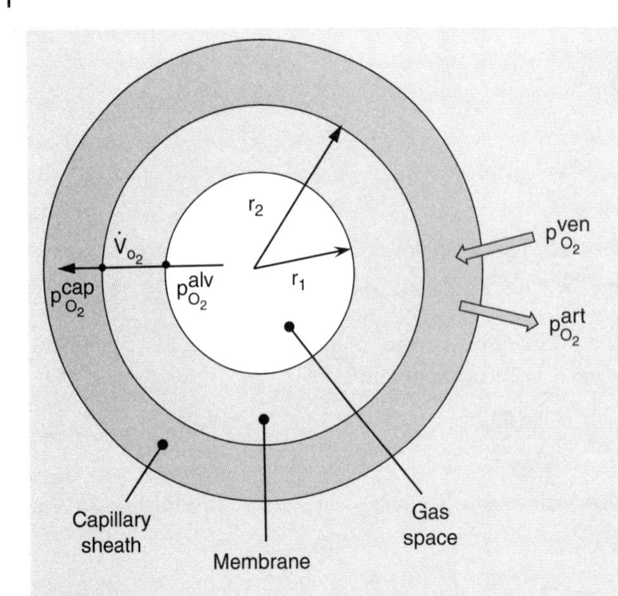

Figure 9.2-2 Oxygen transport across the alveolar–capillary membrane.

Solution

Assume that (i) transport through the alveolar gas and through the capillary sheath is so rapid that $\dot{V}_{O_2}$ is limited by transport through the membrane; (ii) net convection through the membrane is negligible; (iii) transmembrane diffusion is so rapid that O_2 partial pressure in end-capillary blood, $p_{O_2}^{art}$, is equilibrated with $p_{O_2}^{alv}$; and (iv) O_2 partial pressure within the capillaries, $p_{O_2}^{cap}$, can be approximated as a constant equal to the arithmetic average of $p_{O_2}^{ven}$ and $p_{O_2}^{art}$.

At the surfaces of the membrane, we can express the boundary concentrations, C_{s1} and C_{s2}, in terms of the O_2 partial pressures in the alveolar gas and the capillaries, respectively:

$$C_{s1} = c_G^o \alpha_{O_2} p_{O_2}^{alv}, \quad C_{s2} = c_G^o \alpha_{O_2} p_{O_2}^{cap} \tag{9.2-19a,b}$$

where α_{O_2} is the Bunsen solubility of O_2 in the membrane. Equation 9.2-16 can now be written in terms of the volumetric O_2 uptake rate as

$$\dot{V}_{O_2} \equiv \frac{\dot{N}_{O_2}}{c_G^o} = \left(\frac{4\pi\alpha_{O_2}\mathcal{D}_{O_2}}{1/r_1 - 1/r_2}\right)\left(p_{O_2}^{alv} - p_{O_2}^{cap}\right) \tag{9.2-20}$$

If $p_{O_2}^{art}$ is the same as $p_{O_2}^{alv}$, then $p_{O_2}^{cap} = (2.0 + 13.0)/2 = 7.5\,\text{kPa}$. For lung tissue at $T = 37°C$, Table A4-3 indicates that $\mathcal{D}_{O_2} = 2.30 \times 10^{-5}\,\text{cm}^2/\text{s}$ and $\alpha_{O_2} = 1.8 \times 10^{-4}$ ml(STP)/ml-kPa. The O_2 uptake of each alveolus is then

$$\dot{V}_{O_2} = \frac{4\pi(1.8\times10^{-4})(2.30\times10^{-5})}{(1/100\times10^{-4}) - (1/102\times10^{-4})}(13.0 - 7.5) = 1.5\times10^{-7}\ \text{ml(STP)/s} \tag{9.2-21}$$

The O_2 uptake for the entire lung is

$$(\dot{V}_{O_2})_{lung} = 1.94\times10^{-7}(300\times10^6)(60) = 2700\ \text{ml(STP)/min} \tag{9.2-22}$$

This is comparable to the values of $(\dot{V}_{O_2})_{lung}$ measured during strenuous exercise. During rest, $(\dot{V}_{O_2})_{lung}$ is about one tenth of this value. Two processes are responsible for this large difference. First, less O_2 is taken up by systemic tissues during rest so that $p_{O_2}^{ven}$ is larger than during exercise. This leads to a smaller $p_{O_2}^{cap}$ and therefore a smaller $\left(p_{O_2}^{alv} - p_{O_2}^{cap}\right)$ driving force for O_2 uptake. Second, with a lower blood pressure during rest, fewer pulmonary capillaries are open. This reduces the effective surface area for transferring O_2.

9.3 Diffusion with Parallel Convection

Formulation in Terms of Volumetric Flow Rate

We now consider unidirectional transport in a mixture of I different chemical species in which convection occurs in parallel with diffusion. To generalize our analysis to both rectilinear and radial transport, we drop the directional subscript on all vector components that are now understood to refer to either r or y directions (e.g., $u_y \& u_r \rightarrow u$, $S_y \& S_r \rightarrow S$, $N_{i,y} \& N_{i,r} \rightarrow N_i$). By also defining a coordinate l that can refer to either y or r direction, we can unify the rectilinear and radial flux equations:

$$N_s = \begin{cases} u^* C_s - \mathcal{D}_s \dfrac{dC_s}{dl} & \text{(constant c)} \\ u C_s - \mathcal{D}_s \dfrac{dC_s}{dl} & \text{(constant } \rho) \end{cases} \tag{9.3-1}$$

We can further unify the transport rate equations by introducing the volumetric flow Q. In a fluid of constant molar density, the molar transport rate of solution is Qc, whereas with constant mass density, the mass transport rate of solution is Qρ. This allows us to write Q in two different forms, depending on the nature of the solution:

$$Q \equiv \begin{cases} \dfrac{1}{c}\displaystyle\sum_{i=1}^{I} \dot{N}_i = \dfrac{1}{c}\sum_{i=1}^{I} (C_i u_i) S = u^* S & \text{(constant c)} \\ \dfrac{1}{\rho}\displaystyle\sum_{i=1}^{I} \dot{n}_i = \dfrac{1}{\rho}\sum_{i=1}^{I} (\rho_i u_i) S = u S & \text{(constant } \rho) \end{cases} \tag{9.3-2}$$

The molar rate equations for rectilinear or radial diffusion of a substance s in either a constant molar or a mass density fluid can now be represented by a single equation:

$$\dot{N}_s = N_s S = \left\{ \begin{matrix} \mathbf{u}^* S C_s - \mathcal{D}_s S \dfrac{dC_s}{dl} & \text{(constant c)} \\ \mathbf{u} S C_s - \mathcal{D}_s S \dfrac{dC_s}{dl} & \text{(constant } \rho) \end{matrix} \right\} = Q C_s - \mathcal{D}_s S \frac{dC_s}{dl} \tag{9.3-3}$$

Integrating this equation between the bounding surfaces at $l = l_1$ and $l = l_2$, we obtain

$$\int_{C_{s1}}^{C_{s2}} \frac{\mathcal{D}_s dC_s}{QC_s - \dot{N}_s} = \int_{l_1}^{l_2} \frac{dl}{S} = \frac{h}{\tilde{S}} \tag{9.3-4}$$

where h is the diffusion path length, and $\tilde{S}$ is the mean diffusion cross section that is defined in Equation 9.2-18a-c. Under steady-state conditions and in the absence of chemical reactions, the $\dot{N}_i$ in a constant molar density fluid as well as the $\dot{n}_i$ in a constant mass density fluid are constant. Thus, Q has a constant value for these two types of fluids. Assuming that $\mathcal{D}_s$ is also constant, we can complete the integration in Equation 9.3-4:

$$\ln\frac{\dot{N}_s - QC_{s2}}{\dot{N}_s - QC_{s1}} = \frac{hQ}{\mathcal{D}_s\tilde{S}} \tag{9.3-5}$$

Solving this equation for $\dot{N}_s$, we obtain

$$\dot{N}_s = Q\left(\frac{C_{s1}e^{Pe_h} - C_{s2}}{e^{Pe_h} - 1}\right) \tag{9.3-6}$$

Here, we have introduced a dimensionless Peclet number that reflects the importance of convection relative to diffusion:

$$Pe_h \equiv \frac{Qh}{\mathcal{D}_s\tilde{S}} \tag{9.3-7}$$

Equation 9.3-6 can also be expressed as a sum of convective and diffusive transport rates:

$$\dot{N}_s = Q\tilde{C}_s + \frac{\mathcal{D}_s\tilde{S}}{h}(C_{s1} - C_{s2}) \tag{9.3-8}$$

where we have defined a mean concentration of the solute as

$$\tilde{C}_s \equiv \left[C_{s1} - (C_{s1} - C_{s2})\left(\frac{1}{Pe_h} - \frac{1}{e^{Pe_h} - 1}\right)\right] \tag{9.3-9}$$

Formulation in Terms of a Flux Ratio

We now consider an alternative form of the transport rate equation for non-reacting species in a solution of I species with a constant molar density c. By replacing Q with $\Sigma_i\dot{N}_i/c$, Equation 9.3-5 becomes

$$\ln\frac{c\dot{N}_s - C_{s2}\sum_{i=1}^{I}\dot{N}_i}{c\dot{N}_s - C_{s1}\sum_{i=1}^{I}\dot{N}_i} = \frac{h\sum_{i=1}^{I}\dot{N}_i}{c\mathcal{D}_s\tilde{S}} \tag{9.3-10}$$

At steady state, the transport rates of all species are constant, and thus the transport rate of the entire solution $\Sigma_i\dot{N}_i$ is also constant. After defining the dimensionless ratio of the solution flux to the species flux as

$$\phi_s \equiv \frac{\sum_i \dot{N}_i}{\dot{N}_s} \tag{9.3-11}$$

we solve Equation 9.3-10 for the transport rate and obtain

$$\dot{N}_s = \frac{c\mathcal{D}_s\widetilde{S}}{\phi_s h}\ln\frac{1-\phi_s C_{s2}/c}{1-\phi_s C_{s1}/c} \quad (9.3\text{-}12)$$

This equation can also be written as

$$\dot{N}_s = \left(\frac{\mathcal{D}_s\widetilde{S}}{h}\right)\frac{C_{s1}-C_{s2}}{(1-\phi_s C_s/c)_{LM}} \quad (9.3\text{-}13)$$

where $(1-\phi_s C_s/c)_{LM}$ is a log-mean correction to the driving force of species s.

$$(1-\phi_s C_s/c)_{LM} \equiv \frac{(1-\phi_s C_{s2}/c)-(1-\phi_s C_{s1}/c)}{\ln\frac{(1-\phi_s C_{s2}/c)}{(1-\phi_s C_{s1}/c)}} \quad (9.3\text{-}14)$$

The LM factor corrects for the effect of convection on solute transport. When $\phi_s C_{s1}/c$ and $\phi_s C_{s2}/c$ are less than 0.1, the LM factor is close to one. For this special case, Equation 9.3-13 approaches Equation 9.2-17 for steady-state diffusion alone. This is often true of biological liquids but not necessarily of gas mixtures.

Example 9.3-1 Diffusion Coefficient in a Stefan Tube

In this example, we will compute the pseudo-binary diffusion coefficient of water vapor (component 1) in air (component 2) for the Stefan tube (Fig. 6.1-1) described in Example 6.1-1. In that problem, water evaporated at a molar flux $N_1 = 2.16\times10^{-3}$ mol/(s-m^2) into a gas column where it diffused over a path length of h = 0.2 m. The mean molar velocity of the air and water vapor mixture in the gas column was $u^* = 6.08\times10^{-5}$ m/s. At the experimental conditions of T = 70°C and P = 101.3 kPa, the molar density of liquid water is $c_L = 54{,}300$ mol/m^3, the molar density in the gas column is $c_G = 35.5$ mol/m^3, and the vapor pressure of water is $p_1^* = 31{,}200$ Pa.

Solution

Using the water flux and molar density of liquid water, we find that the air–water interface at the bottom of the gas column recedes at a velocity

$$u_{interface} = \frac{N_1}{c_L} = \frac{2.16\times10^{-3}}{54{,}300} = 3.98\times10^{-8}\,\text{m/s} = 3.44\,\text{mm/day} \quad (9.3\text{-}15)$$

Since this is less than one thousandth the magnitude of u^*, transport in the gas column will rapidly and continually adjust to a new steady state, while the gas–liquid interface moves. With such pseudo-steady behavior, the water vapor transport rate is approximated by Equation 9.3-13.

Because O_2 and N_2 have limited solubilities in liquid water, the flux of air across the air–water interface is negligible and $\phi_1 = (\dot{N}_1 + \dot{N}_2)/\dot{N}_1 = 1$. Water vapor is transported up the gas column in a rectilinear fashion so that its diffusion cross section, $\widetilde{S} = S_y$, is also a constant. Thus, Equation 9.3-13 reduces to

$$\dot{N}_1 = \left(\frac{\mathcal{D}_1 S_y}{h}\right)\frac{C_1(0)-C_1(h)}{(1-C_1/c_G)_{LM}} \quad (9.3\text{-}16)$$

where $C_1(0)$ and $C_1(h)$ are the water vapor concentrations at the two ends of the gas column. In terms of its flux, $N_1 = \dot{N}_1/S_y$, the diffusion coefficient of water vapor in the Stefan tube is given by

$$\mathcal{D}_1 = N_1 h \frac{(1 - C_1/c_G)_{LM}}{C_1(0) - C_1(h)} \tag{9.3-17}$$

The sweep flow of dry air over the open end of the gas column at $y = h$ is sufficiently fast that $C_1(h) \approx 0$. When vapor–liquid equilibrium occurs at the other end of the column where $y = 0$, the partial pressure of water is equal to its vapor pressure. The corresponding vapor concentration is

$$C_1(0) = \frac{p^*}{\mathcal{R}T} = \frac{31200}{8.31(343)} = 11.0\,\text{mol/m}^3 \tag{9.3-18}$$

The log-mean factor corresponding to these end concentrations is

$$(1 - C_1/c)_{LM} = \frac{(1-0) - (1 - 11.0/35.5)}{\ln\dfrac{(1-0)}{(1 - 11.0/35.5)}} = 0.837 \tag{9.3-19}$$

This result indicates that diffusion in the Stefan tube produces convection that increases overall transport by a factor of $1/(1 - C_1/c)_{LM} \approx 1.2$. Substituting numerical values into Equation 9.3-16, we compute the diffusion coefficient of water vapor in air at 70°C as

$$\mathcal{D}_1 = \frac{(2.16 \times 10^{-3})(0.2)(0.837)}{(11.0 - 0)} = 3.26 \times 10^{-5}\ \text{m}^2/\text{s} \tag{9.3-20}$$

Although the alternative rate expressions for $\dot{N}_s$ given by Equations 9.3-6 and 9.3-13 are mathematically equivalent, they include the effect of convection in different ways. To utilize Equation 9.3-6, we must specify volumetric flow rate. When there is forced convection and Q can be measured or inferred from fluid pressure differences, it is convenient to use Equation 9.3-6. In other situations where a relationship between species transport rates is known, Equation 9.3-13 can be used. This is the case in Example 9.3-1 where we could specify that $\phi_i = 1$ for water vapor.

9.4 Diffusion with Chemical Reaction

9.4.1 Metabolic Demand of a Cell

To illustrate the effect of chemical reaction on transport, consider the diffusion of a substrate s that is undergoing metabolism within a cell according to the Michaelis–Menten kinetics with the following intensive rate:

$$R_s = -\left(\frac{V_m C_s}{K_m + C_s}\right) \tag{9.4-1}$$

where the negative sign is necessary since substrate s is depleted. For simplicity, we model the cell as a homogeneous spherical body of radius "a". If substrate concentration C_s at the cell surface has a uniform value C_{so}, then the spatial distribution of C_s will be spherically symmetric. Because the cell is primarily composed of water, its mass density is approximately that of water. Therefore, the solute concentration equation (using Table 9.1-2) is

$$\frac{\partial C_s}{\partial t} + u_r \frac{\partial C_s}{\partial r} = \frac{1}{r^2}\frac{\partial}{\partial r}\left(\mathcal{D}_s r^2 \frac{\partial C_s}{dr}\right) - \left(\frac{V_m C_s}{K_m + C_s}\right) \tag{9.4-2}$$

Assuming further that the process is in steady state, convection is negligible, and $\mathcal{D}_s$ is constant, this equation becomes

$$\frac{\mathcal{D}_s}{r^2}\frac{d}{dr}\left(r^2 \frac{dC_s}{dr}\right) = \left(\frac{V_m C_s}{K_m + C_s}\right) \tag{9.4-3}$$

The following boundary conditions apply on the surface and at the center of the cell:

$$\begin{aligned} r = a &: C_s = C_{so} \\ r = 0 &: C_s \text{ is finite} \end{aligned} \tag{9.4-4a, b}$$

By defining the dimensionless quantities

$$C \equiv \frac{C_s}{C_{so}}, \quad r \equiv \frac{r}{a}, \quad Da \equiv \frac{a^2 V_m}{\mathcal{D}_s C_{so}}, \quad K \equiv \frac{K_m}{C_{so}} \tag{9.4-5a-d}$$

the differential equation and its boundary conditions become

$$\frac{1}{r^2}\frac{d}{dr}\left(r^2 \frac{dC}{dr}\right) = Da\left(\frac{C}{K + C}\right) \tag{9.4-6}$$

and

$$\begin{aligned} r = 1 &: C = 1 \\ r = 0 &: C = \text{finite} \end{aligned} \tag{9.4-7a, b}$$

where the Damkohler number Da is the ratio of a characteristic reaction rate $a^3 V_m$ to a characteristic diffusion rate $a\mathcal{D}_s C_{so}$. Because the reaction rate term is nonlinear in C, this equation does not have an exact analytical solution. One special case for which there is an analytical solution occurs when $K_m \ll C_S$ or equivalently when $K \ll C$ throughout the cell. Equation 9.4-6 then becomes

$$\frac{1}{r^2}\frac{d}{dr}\left(r^2 \frac{dC}{dr}\right) = Da \tag{9.4-8}$$

By integrating with respect to r, dividing the result by r^2 and integrating again, we obtain the solution

$$C = \frac{Da}{6} r^2 - \frac{k_1}{r} + k_2 \tag{9.4-9}$$

Evaluating the constants of integration, k_1 and k_2, with the two boundary conditions, we obtain a final expression for the concentration distribution:

$$C = 1 - \frac{Da}{6}\left(1 - r^2\right) \tag{9.4-10}$$

When diffusion takes place very rapidly compared to the metabolic reaction, $Da \to 0$ and this equation predicts that C is one throughout the intracellular fluid. For nonzero values of Da, C continuously decreases between the cell surface and cell center where its value is $C=(1-Da/6)$. When $Da>6$, this solution is invalid since it predicts negative C values.

Since we have assumed that $K_m \ll C_S$, substrate is consumed at its maximum rate per unit volume V_m, and its depletion rate for an individual cell of volume $(4/3)\pi a^3$ is

$$-\dot{R}_s = \left(\frac{4}{3}\pi a^3\right)V_m \tag{9.4-11}$$

With this relation, the concentration distribution written in dimensional form is

$$C_s = C_{so} - \frac{(-\dot{R}_s)}{8\pi\mathcal{D}_s a}\left(1-\frac{r^2}{a^2}\right) \tag{9.4-12}$$

Example 9.4-1 Diffusion-Reaction of Oxygen in a Cell

Hepatocytes are being cultured in a well-mixed suspending medium. The cell surfaces are kept at a constant O_2 partial pressure by continuous aeration of the medium. Human hepatocyte volume is 4900 μm^3, and O_2 consumption is fairly constant at 280 μm^3(STP)/min per cell. Assuming that aeration supplies sufficient O_2 to maintain saturation kinetics (i.e., $C_{O_2} \gg K_m$) throughout the cell interior, how does O_2 partial pressure at the cell center compare to that at the cell surface?

Solution

The following relationships express O_2 concentration in terms of partial pressure p_{O_2} and O_2 molar depletion rate in terms of volumetric utilization $\dot{V}_{O_2}$:

$$C_{O_2} = \alpha_{O_2} c_G^o p_{O_2}, \quad -\dot{R}_{O_2} = c_G^o \dot{V}_{O_2} \tag{9.4-13a,b}$$

With these relations, Equation 9.4-12 becomes

$$p_{O_2} = (p_{O_2})_{r=a} - \frac{\dot{V}_{O_2}}{8\pi\alpha_{O_2}\mathcal{D}_s a}\left(1-\frac{r^2}{a^2}\right) \tag{9.4-14}$$

Aeration in a well-mixed medium will provide the hepatocytes with a surface partial pressure of $(p_{O_2})_{r=a} = 0.21(101) = 21.2$ kPa. The effective spherical radius of a cell is $a = [(3/4\pi)\,4900]^{1/3} = 10\ \mu m$. From Table A4-3, we estimate that $\alpha_{O_2} \approx 2\times10^{-4}\ kPa^{-1}$ and $\mathcal{D}_{O_2} \approx 1000\ \mu m^2/(s)$. Given these parameter values, the O_2 partial pressure at $r=0$ is

$$(p_{O_2})_{r=0} = 21.2 - \frac{280/60}{8\pi(2\times10^{-4})(1000)(10)} = 21.1\ \text{kPa} \tag{9.4-15}$$

Thus, diffusion processes are so rapid compared to reaction that the O_2 distribution is essentially uniform throughout a hepatocyte. This result is subject to the assumption that O_2 consumption occurs at its maximum rate throughout the cell. If this was not the case, then the O_2 distribution would be even more uniform.

9.4.2 Augmented Diffusion by Protein Binding

Revisiting the earlier problem of unidirectional, steady-state diffusion through a planar slab, let us consider what happens when there is reversible binding between solute s and a soluble protein p to form an sp complex.

In the schematic of this process shown in Figure 9.4-1, the surfaces of the slab are maintained at constant solute concentrations such that $C_s(0) > C_s(h)$, and s diffuses from $y = 0$ to $y = h$. Since the formation of sp depends directly on C_s, we expect $C_{sp}(0) > C_{sp}(h)$ and sp will diffuse in the same direction as s. For a protein that cannot cross the surfaces of the slab, p will accumulate at $y = h$. At steady state, $C_p(h) > C_p(0)$ and p will diffuse from $y = h$ to $y = 0$ at the same rate that sp diffuses in the opposite direction.

To quantify this process, we start with the molar balances for the rectilinear transport of s, p, and sp as prescribed by Equation 9.1-4. Under steady-state conditions and with constant S_y,

$$\frac{dN_s}{dy} = R_s, \quad \frac{dN_p}{dy} = R_p, \quad \frac{dN_{sp}}{dy} = R_{sp} \qquad (9.4\text{-}16\text{a-c})$$

where R_s, R_p, and R_{sp} are formation rates per unit volume in the solute–protein binding reaction. Because of the stoichiometry of this reaction, $R_s = R_p = -R_{sp}$. Thus, we can eliminate reaction rates by adding Equation 9.4-16a,c and also by adding Equation 9.4-16b,c:

$$\frac{d}{dy}\left(N_s + N_{sp}\right) = 0, \quad \frac{d}{dy}\left(N_p + N_{sp}\right) = 0 \qquad (9.4\text{-}17\text{a,b})$$

Equation 9.4-17a indicates that the total flux of s is spatially uniform:

$$(N_s)_T \equiv N_s + N_{sp} = \text{constant} \qquad (9.4\text{-}18)$$

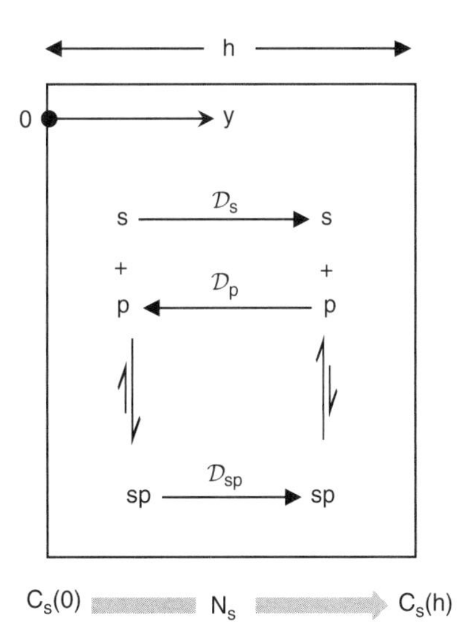

Figure 9.4-1 Augmented diffusion of solute s when $C_s(0) > C_s(h)$.

Similarly, Equation 9.4-17b indicates that the total flux of p is spatially uniform. But neither p nor sp can leave the slab, so that the total flux of p is zero at $y = 0$ and $y = h$ and everywhere else:

$$(N_p)_T \equiv N_p + N_{sp} = 0 \tag{9.4-19}$$

In the absence of convection, the one-dimensional molar fluxes in the y direction are described by Fick's law:

$$N_s = -\mathcal{D}_s \frac{dC_s}{dy}, \quad N_p = -\mathcal{D}_p \frac{dC_p}{dy}, \quad N_{sp} = -\mathcal{D}_{sp} \frac{dC_{sp}}{dy} \tag{9.4-20a-c}$$

and Equations 9.4-18 and 9.4-19 become

$$(N_s)_T = -\mathcal{D}_s \frac{dC_s}{dy} - \mathcal{D}_{sp} \frac{dC_{sp}}{dy} = \text{constant} \tag{9.4-21}$$

$$(N_p)_T = -\mathcal{D}_p \frac{dC_p}{dy} - \mathcal{D}_{sp} \frac{dC_{sp}}{dy} = 0 \tag{9.4-22}$$

Because a protein molecule is ordinarily much larger than the ligand that binds to it, $\mathcal{D}_p$ and $\mathcal{D}_{sp}$ are approximately equal. From Equation 9.4-22, we then deduce that

$$\mathcal{D}_p \left(\frac{dC_p}{dy} + \frac{dC_{sp}}{dy} \right) = 0 \Rightarrow T_p \equiv C_p + C_{sp} = \text{constant} \tag{9.4-23a,b}$$

That is, the total concentration of protein binding sites T_p is independent of y. If the binding rate of s to p is sufficiently rapid relative to the individual diffusion rates of s and p, then the binding reaction is in local equilibrium at any y. In that case, C_s is an algebraic function of C_{sp}. We define this function as the saturation fraction S_s:

$$S_s(C_s) \equiv \frac{C_{sp}}{T_p} \tag{9.4-24}$$

When this relation is used in Equation 9.4-21, we obtain

$$(N_s)_T = -\mathcal{D}_s \frac{dC_s}{dy} - \mathcal{D}_{sp} T_p \frac{dS_s}{dy} = -\mathcal{D}_s \frac{dC_s}{dy} - \mathcal{D}_p T_p \left(\frac{dS_s}{dC_s} \frac{dC_s}{dy} \right) \tag{9.4-25}$$

or in a more compact form

$$(N_s)_T = -\mathcal{D}_{s,\text{eff}} \frac{dC_s}{dy} \tag{9.4-26}$$

where $\mathcal{D}_{s,\text{eff}}$ is a local effective diffusion coefficient given by

$$\mathcal{D}_{s,\text{eff}} \equiv \mathcal{D}_s \left(1 + T_p \frac{\mathcal{D}_p}{\mathcal{D}_s} \frac{dS_s}{dC_s} \right) \tag{9.4-27}$$

This expression indicates that the diffusion of s can be augmented by a diffusible protein that reversibly binds with s. The augmentation depends on the protein concentration, the ratio of the protein and solute diffusion coefficients, and the slope of the binding fraction function dS_s/dC_s. Typically, this slope is steepest at low solute concentrations and approaches zero at solute concentrations that are so high that the carrier protein is saturated with s.

Therefore, we expect augmentation along the direction of transport to increase as C_s decreases. We can formulate the spatial average $\mathcal{D}_{s,eff}$ between y = 0 and y = h by integrating Equation 9.4-27:

$$\langle \mathcal{D}_{s,eff} \rangle \equiv \frac{1}{C_s(0) - C_s(h)} \int_{C_s(0)}^{C_s(h)} \mathcal{D}_{s,eff} dC_s = \mathcal{D}_s + \mathcal{D}_p T_p \left[\frac{S_s(0) - S_s(h)}{C_s(0) - C_s(h)} \right] \tag{9.4-28}$$

where $S_s(0)$ and $S_s(h)$ represent the fraction of p bound to s at y = 0 and y = h, respectively.

Example 9.4-2 Oxygen Diffusion Through Hemoglobin Solutions

Keller and Friedlander (1966) measured the total O_2 flux $(N_{O_2})_T$ through h = 1.2 mm thick films of stagnant hemoglobin solution with different pairs of O_2 partial pressures, $p^A_{O_2}$ and $p^B_{O_2}$, on the film surfaces. The corresponding $\langle \mathcal{D}_{O_2,eff} \rangle$ values were computed by taking the ratio of $(N_{O_2})_T/h$ to the concentration driving force $\alpha_{O_2} c^o_G \left(p^A_{O_2} - p^B_{O_2} \right)$. The purpose of this example problem is to simulate the $\langle \mathcal{D}_{O_2,eff} \rangle$ data points shown in the lower graph of Figure 9.4-2.

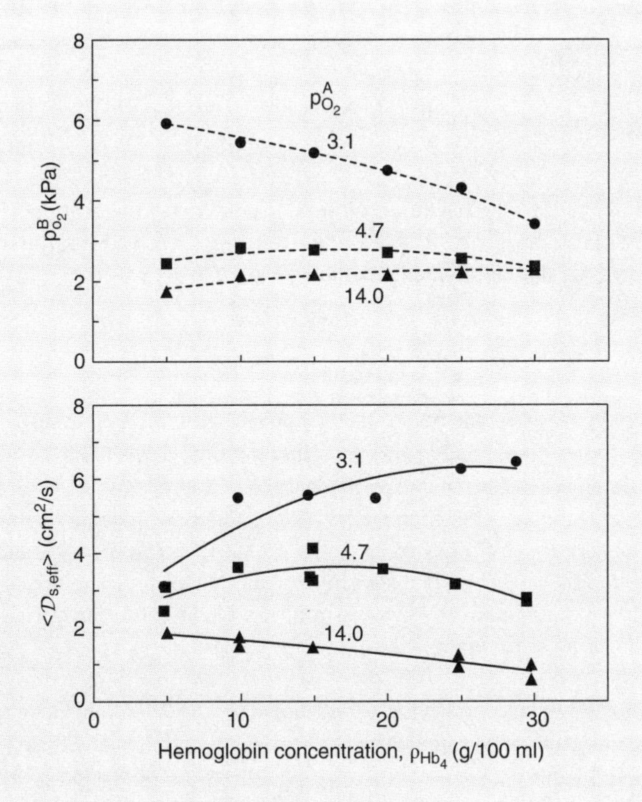

Figure 9.4-2 Augmented diffusion through hemoglobin solutions.

Table 9.4-1 Properties of the Hemoglobin Test Solutions[a]

ρ_{Hb_4} (gm/100 ml)	5	10	15	20	25	30
$\mathcal{D}_{O_2}$ (10^{-5} cm^2/s)	1.84	1.64	1.43	1.23	1.02	0.82
$\mathcal{D}_{Hb_4}$ (10^{-7} cm^2/s)	6.87	4.85	3.04	2.06	1.70	1.53
pH	7.10	7.00	6.90	6.80	6.72	6.65

[a] All data collected at a temperature of 25°C.

The individual points in the upper graph of Figure 9.4-2 show the combinations of $p_{O_2}^A$ and $p_{O_2}^B$ in experiments performed on six aqueous solutions with different hemoglobin mass concentrations ρ_{Hb_4} (Table 9.4-1). The dashed curves are quadratic least squares regressions that we used to construct intervening $p_{O_2}^A : p_{O_2}^B : \rho_{Hb_4}$ conditions.

Solution

With interfacial equilibrium at the film surfaces, the bounding O_2 concentrations are $C_{O_2}(0) = \alpha_{O_2} c_G^o p_{O_2}^A$ and $C_{O_2}(h) = \alpha_{O_2} c_G^o p_{O_2}^B$. Since a hemoglobin molecule contains four heme groups (Hb), each capable of binding one O_2 molecule, the molar concentration of binding sites in the film (T_{Hb}) is related to ρ_{Hb_4} and to the molecular weight of hemoglobin (M_{Hb_4} = 66,800 Da): $T_{Hb} = 4\rho_{Hb_4}/M_{Hb_4}$. With these relations, Equation 9.4-28 can be written as

$$\langle \mathcal{D}_{O_2,eff} \rangle = \mathcal{D}_{O_2} + \frac{4\rho_{Hb_4}}{M_{Hb_4}} \left[\frac{S_{O_2}\left(p_{O_2}^A, pH\right) - S_{O_2}\left(p_{O_2}^B, pH\right)}{\alpha_{O_2} c_G^o \left(p_{O_2}^A - p_{O_2}^B\right)} \right] \mathcal{D}_{Hb_4} \tag{9.4-29}$$

We will represent S_{O_2} by the Hill model of reaction equilibrium in Equation 5.5-11, accounting for different pH by the factor $10^{m(pH-7.4)}$ appearing in Equation 5.5-18. The resulting equation for the fractional saturation of a hemoglobin molecule with O_2 is

$$S_{O_2}(p_{O_2}, pH) = \frac{\left(10^{m(pH-7.4)} \kappa_p p_{O_2}\right)^n}{1 + \left(10^{m(pH-7.4)} \kappa_p p_{O_2}\right)^n} \tag{9.4-30}$$

We regressed the $\langle \mathcal{D}_{O_2,eff} \rangle$ data points to Equations 9.4-29 and 9.4-30 using a constant value of $\alpha_{O_2} c_G^o(25^o) = 1 \times 10^{-5}$ mol/(L-kPa) and the solution properties in Table 9.4-1. The values of the unknown parameters that minimized the least square error of this regression were $m = 0.46$, $\kappa_p = 1.07$, and $n = 2.25$. We found other values for these parameters in whole blood (Section 5.5), namely, $m = 0.40$, $\kappa_p = 0.283$, and $n = 2.85$. This discrepancy is due in part to a difference in chemical properties between hemoglobin dissolved in homogeneous solution and hemoglobin segregated in the cytoplasm of red blood cells. It might also result from the assumption of local oxyhemoglobin binding equilibrium in our augmented diffusion model.

We constructed the solid curves in the lower graph of Figure 9.4-2 from Equations 9.4-29 and 9.4-30 by using our parameter estimates for the hemoglobin solution and the $p_{O_2}^A : p_{O_2}^B : \rho_{Hb_4}$ conditions along the dashed curves in the upper graph. These model predictions of $\langle \mathcal{D}_{O_2,eff} \rangle$ are in reasonable agreement with the data points. When ρ_{Hb_4} is fixed, $\langle \mathcal{D}_{O_2,eff} \rangle$ decreases with increases in $p^A_{O_2}$. This occurs because higher $p^A_{O_2}$ levels (and the correspondingly higher $p^B_{O_2}$ levels) lead to more oxyhemoglobin binding at each film surface but smaller differences in S_{O_2} between the two surfaces.

9.5 Unsteady-State Diffusion

We will now analyze unsteady-state diffusion of an inert solute s from a planar surface into an adjacent, semi-infinite, stagnant phase. When convection and reaction are eliminated, the solute balances for rectilinear transport through a solution with constant $\mathcal{D}_s$ and either constant ρ (Eq. 9.1-9b) or constant c (Eq. 9.1-12b) reduce to

$$\frac{\partial C_s}{\partial t} = \mathcal{D}_s \frac{\partial^2 C_s}{\partial y^2} \tag{9.5-1}$$

where y is the direction perpendicular to the surface. This is commonly referred to as Fick's Second Law of diffusion. To complete the formulation of this problem, we specify an initial condition and two boundary conditions:

$$\begin{aligned} t = 0 \quad &: C_s = C_{s\infty} \\ y = 0 \quad &: C_s = C_{s0} \\ y \to \infty \quad &: C_s = C_{s\infty} \end{aligned} \tag{9.5-2a-c}$$

where C_{s0} and $C_{s\infty}$ are constants. According to the first condition, the concentration of solute s is initially $C_{s\infty}$ throughout the diffusion domain. The second condition indicates that the surface concentration is held at C_{s0}. The third condition stipulates that far from the surface, the solute concentration is undisturbed by diffusion and thus remains at its initial value.

As described in Appendix C.6, this is one of the general classes of problems without characteristic bounds on the independent variables. Consequently, the partial differential equation can be transformed into an ordinary differential equation by combining the two independent variables y and t into a single similarity variable η. If we define the dimensionless variables

$$C = \frac{C_s - C_{s\infty}}{C_{s0} - C_{s\infty}}, \eta \equiv \frac{y}{2\sqrt{\mathcal{D}_s t}} \tag{9.5-3a,b}$$

then the problem is transformed to

$$\frac{d^2 C}{d\eta^2} + \eta \frac{dC}{d\eta} = 0 \tag{9.5-4}$$

with the conditions

$$\begin{aligned} \eta = 0 \quad &: C = 1 \\ \eta \to \infty \quad &: C = 0 \end{aligned} \tag{9.5-5a,b}$$

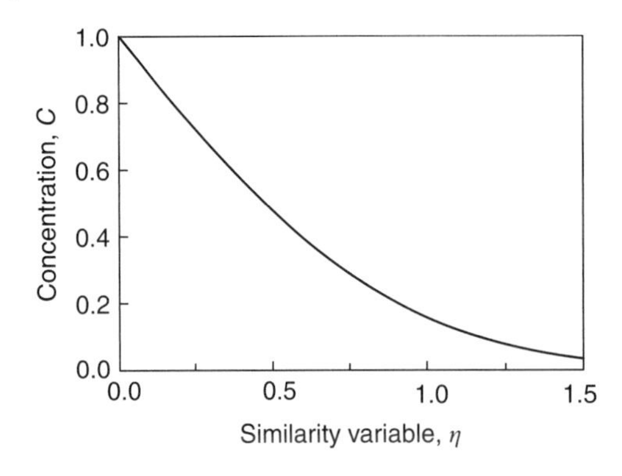

Figure 9.5-1 Unsteady diffusion through a semi-infinite domain.

After integrating twice and applying the boundary conditions, we obtain

$$C(\eta) = 1 - \frac{\int_0^\eta \exp(-\xi^2)d\xi}{\int_0^\infty \exp(-\xi^2)d\xi} = 1 - \frac{2}{\sqrt{\pi}}\int_0^\eta \exp(-\xi^2)d\xi = \text{erfc}(\eta) \tag{9.5-6}$$

where we have introduced the complementary error function erfc(η). As illustrated in Figure 9.5-1, the dimensionless concentration decreases monotonically from erfc(0) = 1.0 to erfc(∞) = 0.

We define the penetration distance y_p of the solute from the surface as that position where C_s falls 99% of the way from C_{s0} to $C_{s\infty}$. This corresponds to a dimensionless concentration of $C = 0.01$ that, according to Equation 9.5-6, is reached at $\eta = \text{erfc}^{-1}(0.01)$, which has a tabulated value of 1.83 (Abramowitz and Stegun, 1965). Thus, the penetration distance varies with time according to

$$y_p = 2\eta\sqrt{\mathcal{D}_s t} = 3.7\sqrt{\mathcal{D}_s t} \tag{9.5-7}$$

Although the 99% criterion is arbitrary, the important point is that the penetration distance of an inert substance by diffusion is proportional to $\sqrt{\mathcal{D}_s t}$. For respiratory gases in air, $\mathcal{D}_s$ is on the order of 0.1 cm^2/s so that diffusion in the lung occurs over a distance of about 2 cm during the 2 s of a normal inhalation. Thus, diffusion makes an important contribution to gas transport in the lung airways, particularly in the alveoli whose diameters are about 0.02 cm. The diffusion of respiratory gases in blood $\mathcal{D}_s$ is on the order of 10^{-5} cm^2/s, which leads to a penetration distance of 0.02 cm during the 1 s of blood flow through a capillary. This still constitutes an important source of O_2 and CO_2 transport since capillaries are only about 0.001 cm in diameter.

The diffusion flux of solute at the y = 0 surface is given by

$$[N_s]_{y=0} = -\mathcal{D}_s\left[\frac{\partial C_s}{\partial y}\right]_{y=0} = \sqrt{\frac{D_s}{\pi t}}(C_{s0} - C_{s\infty}) \tag{9.5-8}$$

Initially, this flux is infinite because the solute concentration undergoes a discontinuous jump from C_{s0} to $C_{s\infty}$. As time progresses and solute diffuses further and further beyond y = 0, the flux at the surface decreases.

References

Abramowitz M, Stegun IA. Handbook of Mathematical Functions. Washington: US Government Printing Office; 1965, p. 311.

Keller KH, Friedlander SK. The steady-state transport of oxygen through hemoglobin solutions. J Gen Physiol. 1966; 49:663–679.

Chapter 10

Membrane Transport I

Convection and Diffusion Processes

The selective separation of solutes in biological systems is regulated by thin transport barriers such as cell membranes and capillary walls. In medical devices, synthetic membranes are often the means by which bodily functions such as blood oxygenation by the lungs or urea removal by the kidneys are artificially provided to a patient.

Depending on their properties, passive transport of chemical species across biological membranes occurs along different paths. Small hydrophobic molecules can diffuse through the phospholipid bilayer of plasma membranes, while ions and small hydrophilic molecules move through ion channels by parallel convection and diffusion. Large hydrophilic molecules that cannot cross a biological membrane by either of these routes are often transported by special molecules called carrier or transporter proteins. Carrier molecules also participate in the active transport of molecules in opposition to their electrochemical gradients.

In this chapter, we present rate equations for passive transport across biological as well as synthetic membranes. We begin with a discussion of ordinary diffusion and then consider the combination of forced convection occurring in parallel with diffusion. Finally, we examine electrochemical processes by which charged species are transported through transmembrane channels.

10.1 Ordinary Diffusion

Let us first focus on ordinary diffusion of an uncharged solute across a membrane separating two solutions containing different solute concentrations. Typical examples of this are the

Biomedical Mass Transport and Chemical Reaction: Physicochemical Principles and Mathematical Modeling,
First Edition. James S. Ultman, Harihara Baskaran, and Gerald M. Saidel.

transport of hydrophobic molecules across cell membranes and the transport of respired gases through the alveolar–capillary membranes of the lungs. Transport by diffusion alone is common for membranes that lack the pores through which convection can occur. Ordinary diffusion can also dominate transport through porous membranes when there are no net fluid pressure and osmotic pressure differences to support convection.

10.1.1 Nonequilibrium Thermodynamics

According to nonequilibrium thermodynamics, the molar flux of a nonreactive, uncharged solute s across a membrane by a combination of diffusion and convection is given by the Starling equation:

$$N_s = u^* C_s^{ave}(1 - \sigma_s) + P_s \Delta C_s \tag{6.5-30}$$

where u^* is the magnitude of the molar-averaged solution velocity, ΔC_s is the difference between solute concentrations in the external solutions, and C_s^{ave} is their arithmetic average. The phenomenological parameters associated with this equation are the hydraulic permeability L_p, the solute permeability P_s, and the reflection coefficient σ_s.

The derivation of Equation 6.5-30 incorporated four important restrictions: (i) The membrane must be sufficiently thin that accumulation rates are negligible and transport occurs in a one-dimensional pseudo-steady-state manner; (ii) both external solutions are dilute so that concentration driving forces can be approximated by linear concentration differences; (iii) equilibrium applies at the two membrane–solution interfaces; and (iv) external solutions are of a similar nature so that partition coefficients are equal at the two membrane surfaces.

In the absence of convection, the first term in this equation is zero, and the molar transport rate through a membrane surface S_m is

$$\dot{N}_s = N_s S_m = P_s S_m \left(C_s^A - C_s^B\right) \tag{10.1-1}$$

Although nonequilibrium thermodynamics provides a framework for analyzing transport through membranes, it does not explain the molecular mechanisms underlying these flux equations or provide a means of estimating the associated phenomenological parameters. For this, we must rely on more detailed models from hydrodynamics and diffusion theory.

10.1.2 Mechanistic Models

Consider the diffusion of a solute s across a homogeneous planar membrane that separates well-mixed solutions A and B that contain solute concentrations C_s^A and C_s^B (Fig. 10.1-1). We solved the diffusion model for such a process previously (Eq. 9.2-6). For a membrane of surface area S_m and thickness h_m, we obtain the molar transport rate

$$\dot{N}_s = \frac{\mathcal{D}_s S_m}{h_m}[C_s(0) - C_s(h_m)] \tag{10.1-2}$$

where $\mathcal{D}_s$ is the diffusion coefficient of solute through the membrane. The solute concentrations $C_s(0)$ and $C_s(h_m)$ are located immediately within the membrane surfaces at positions $y = 0$ and $y = h_m$. If interfacial equilibrium applies at the membrane surfaces, then $C_s(0)/C_s^A$ and $C_s(h_m)/C_s^B$ are equivalent to concentration partition coefficients between the membrane material and external solution A and B, respectively:

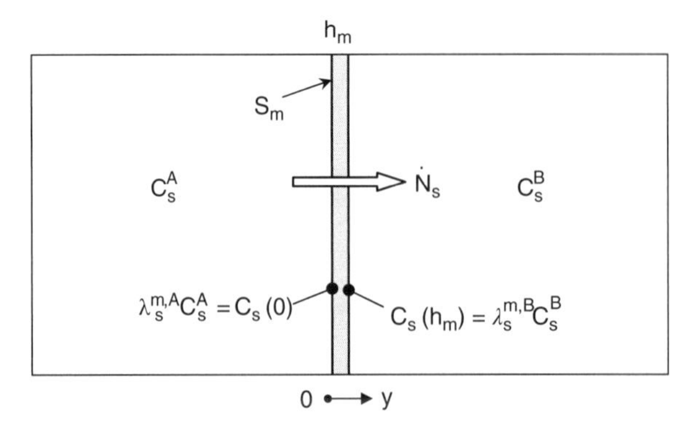

Figure 10.1-1 Ordinary diffusion through a membrane.

$$\lambda_s^{m,A} = \frac{C_s(0)}{C_s^A}, \quad \lambda_s^{m,B} = \frac{C_s(h_m)}{C_s^B}, \quad \lambda_s^{A,B} = \frac{C_s^A}{C_s^B} = \frac{\lambda_s^{m,B}}{\lambda_s^{m,A}} \tag{10.1-3a-c}$$

Here, $\lambda_s^{A,B}$ represents the partition coefficient between the two external solutions if they were in direct contact. This allows us to rewrite Equation 10.1-2 as

$$\dot{N}_s = P_s S_m \left(C_s^A - \lambda_s^{A,B} C_s^B\right); \quad P_s = \frac{\lambda_s^{m,A} \mathcal{D}_s}{h_m} \tag{10.1-4a,b}$$

Since P_s is proportional to $\mathcal{D}_s$, we expect it to be inversely related to solute size. For external solutions that are similar in nature such as two aqueous solutions, their partition coefficients with the membrane material are essentially equal. In that case, $\lambda_s^{A,B} = \lambda_s^{m,B}/\lambda_s^{m,A} = 1$, and Equation 10.1-4a reduces to Equation 10.1-1 derived from nonequilibrium thermodynamics. For a membrane separating two dissimilar phases, however, a value of $\lambda_s^{A,B} \neq 1$ must be included in the transport rate equation.

We sometimes find it useful to utilize a specific solute permeability $\hat{P}_s$ that is independent of membrane thickness. We can then rewrite Equation 10.1-4 in an alternative manner:

$$\dot{N}_s = \frac{\hat{P}_s S_m}{h_m} \left(C_s^A - \lambda_s^{A,B} C_s^B\right); \quad \hat{P}_s = \lambda_s^{m,A} \mathcal{D}_s \tag{10.1-5a,b}$$

Example 10.1-1 Homogeneous Model of the Red Cell Membrane

Most small uncharged hydrophobic molecules are thought to enter cells by diffusion through the lipid bilayer portion of the plasma membrane. Leib and Stein (1986) reported P_s values in the range of 0.00007–90 μm/s for the transport of solutes of various sizes through the red cell membrane (Fig. 10.1-2). Before making these measurements, specific protein transporters that might augment the diffusion process were disabled by the use of the appropriate inhibitors. The purpose of this problem is to determine how well these solute permeability data can be explained by the diffusion model described by Eq. 10.1-4.

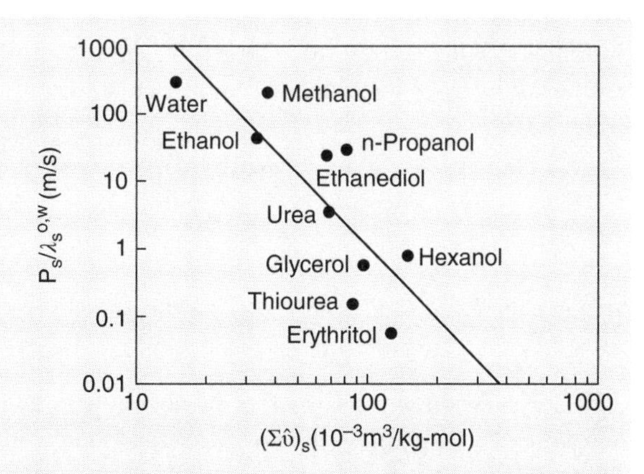

Figure 10.1-2 Permeability of various organic solutes through the red cell membrane.

Solution

Equation 10.1-4b indicates that

$$\frac{P_s}{\lambda_s^{m,A}} \equiv \frac{\mathcal{D}_s}{h_m} \tag{10.1-6}$$

The exact values of $\lambda_s^{m,A}$ and $\mathcal{D}_s$ necessary to validate this equation are not available. Instead, we will assume that $\lambda_s^{m,A}$ for the lipid bilayer of the plasma membrane is proportional to the tabulated values of the octanol–water partition coefficient $\lambda_s^{o,w}$ (Table A3-1). We will also assume that the $\mathcal{D}_s$ values are inversely related to the molecular volumes of the solutes $(\Sigma\hat{\upsilon})_s$ as prescribed by the Wilke-Chang equation (Eq. 7.3-13). Equation 10.1-6 can then be approximated as

$$\frac{P_s}{\lambda_s^{o,w}} = a\left[(\Sigma\hat{\upsilon})_s\right]^{-b} \tag{10.1-7}$$

where a and b are empirical constants and $(\Sigma\hat{\upsilon})_s$ is determined from by summing the atomic volumes $\hat{\upsilon}$ that constitute the solute molecule (Table 7.3-1).

A linear least-squares regression of $\log(P_s/\lambda_s^{m,o})$ versus $\log(\Sigma\hat{\upsilon})_s$ using "a" and "b" as adjustable parameters is indicated by the solid line in Figure 10.1-2. On average, the model produces a reasonable representation of the data with parameter estimates of $a = 7.14$ and $b = 3.55$. For most of the solutes, however, the random error between predicted and measured $P_s/\lambda_s^{m,o}$ values is substantial. This might be due to the use of octanol–water partition coefficient as a surrogate for the true membrane partition coefficient or to the assumption that $\mathcal{D}_s$ of all solutes share the same proportionality constant "a" with their estimated molecular volumes.

The value of $b = 3.55$ is much larger than the value of 0.6 that appears in the Wilke-Chang equation. In other words, transport is more sensitive to molecular size than one would expect from diffusion through a homogeneous material. Possibly, there is some molecular sieving of solute through gaps between the aligned phospholipid molecules. Such an effect would be sensitive to the shape as well as the size of a solute. We conclude that the diffusion model is insufficient to completely explain the reported P_s values of this group of solutes through the red cell membrane.

The molar transport rate of gaseous solutes can be formulated in terms of a partial pressure driving force by expressing equilibrium partitioning at the membrane–solution interfaces in terms of the Bunsen solubility coefficient of the gas in the membrane, α_s^m:

$$C_s(0) = c_G^o \alpha_s^m p_s^A, \quad C_s(h_m) = c_G^o \alpha_s^m p_s^B \qquad (10.1\text{-}8a,b)$$

where p_s^A and p_s^B are the solute partial pressures in the external solutions. Equation 10.1-2 can then be rewritten as

$$\dot{N}_s = \frac{\alpha_s^m c_G^o \mathcal{D}_s S_m}{h_m}\left(p_s^A - p_s^B\right) \qquad (10.1\text{-}9)$$

This expression is valid whether or not the two external solutions are similar in nature because the use of partial pressures accounts for any differences. The equation is equally useful in modeling transfer of a gaseous solute across a membrane that separates two liquid solutions, two gas mixtures or a liquid solution and a gas mixture.

It is often convenient to specify the transport of gaseous solutes in terms of a volumetric rate $\dot{V}_s = \dot{N}_s / c_G^o$ that ordinarily has units of STP gas volume per unit time. Equation 10.1-9 then becomes

$$\dot{V}_s = P_s^G S_m \left(p_s^A - p_s^B\right) \quad \text{or} \quad \dot{V}_s = \frac{\hat{P}_s^G S_m}{h_m}\left(p_s^A - p_s^B\right) \qquad (10.1\text{-}10a,b)$$

where we have defined the permeability and the specific permeability relative to a partial pressure driving force as

$$P_s^G \equiv \frac{\alpha_s^m \mathcal{D}_s}{h_m} \quad \text{and} \quad \hat{P}_s^G \equiv \alpha_s^m \mathcal{D}_s \qquad (10.1\text{-}11a,b)$$

Since $\lambda_s^{m,A} = \alpha_s^m / \alpha_s^A$, we can relate these permeabilities to those we previously defined for a concentration driving force: $P_s^G = \alpha_s^A P_s$ and $\hat{P}_s^G = \alpha_s^A \hat{P}_s$.

Table 10.1-1 Specific Permeability $\hat{P}_s^G$ [10^{-12} ml(STP)/(s-m-Pa)] of Gases Through Homogeneous Synthetic Membranes at 37°C

Material \ Gas, s	Oxygen	Carbon Dioxide	Nitrogen	Helium
Silicone rubber	4,910	24,600	2,450	3,200
Teflon	126	304	63.8	748

Adapted from Galletti (1968).

Recall that during one-dimensional, rectilinear or radial, steady-state transport of a non-reactive solute, the product of solute flux and surface area available for diffusion is dependent on neither time nor position. For rectilinear transport through a flat membrane, the flux and surface area are each constant. For radial transport through a curved membrane with a non-negligible thickness, however, the diffusion surface increases with radial position. Therefore, the solute flux must decrease in a complementary manner.

By analogy to Equation 9.2-18a-c, we account for this effect by replacing S_m in Equation 10.1-5 or Equation 10.1-10 with the mean surface area for diffusion $\tilde{S}_m$:

$$\tilde{S}_m \equiv \begin{cases} S_m & \text{for flat membranes} \\ \dfrac{2\pi L(r_B - r_A)}{\ln(r_B/r_A)} & \text{for cylindrical membranes of length L} \\ 4\pi r_A r_B & \text{for spherical membranes} \end{cases} \tag{10.1-12}$$

Here, r_A and r_B are the radii of curvature of the membrane surfaces.

Example 10.1-2 Maximum O_2/CO_2 Transfer in a Blood Gas Exchanger

We wish to design a membrane gas exchanger to be used during open-heart surgery. The device will contain hollow cylindrical fibers fabricated from silicone rubber with an outside diameter of 400 μm and a wall thickness of 50 μm (Fig. 10.1-3). Oxygen-enriched air, free of CO_2, will enter the device on the outside of the fibers. Venous blood will enter on the inside of the fibers with $p_{O_2} = 5.3$ kPa and $p_{CO_2} = 6.2$ kPa.

Determine the minimum length L of fiber necessary to transfer 200 ml(STP)/min of CO_2 from the blood to the gas phase, as is typically required for a normal adult at rest. How much oxygen enrichment is necessary to achieve a respiratory quotient of 0.8 at this fiber length?

Solution

The minimum fiber length occurs when the p_{CO_2} driving force for diffusion remains at its maximum (i.e., inlet) value everywhere along the fiber wall. Replacing S_m in Equation 10.1-10b by $\tilde{S}_m$ from Equation 10.1-12b, we obtain

$$\dot{V}_s = \frac{\hat{P}_s^G \tilde{S}_m}{h_m}(p_s^A - p_s^B) = \frac{2\pi L(r_B - r_A)}{h_m \ln(r_B/r_A)} \hat{P}_s^G (p_s^A - p_s^B) \tag{10.1-13}$$

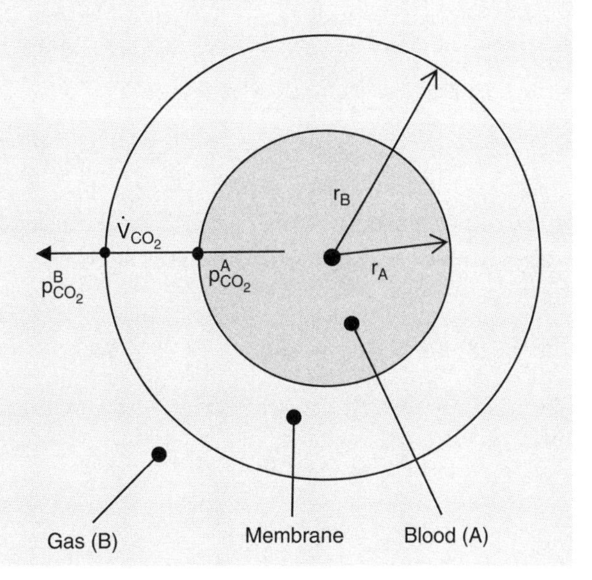

Figure 10.1-3 Carbon dioxide transport through a hollow fiber.

Recognizing that the membrane thickness is given by $h_m = (r_B - r_A)$, the total length of fiber required when $s = CO_2$ is

$$L = \frac{\dot{V}_{CO_2} \ln(r_B/r_A)}{2\pi \hat{P}^G_{CO_2}\left(p^A_{CO_2} - p^B_{CO_2}\right)} \tag{10.1-14}$$

For CO_2 diffusion through a silicone membrane, Table 10.1-1 indicates that $\hat{P}^G_{CO_2} = 2.46 \times 10^{-5}$ ml(STP)/(s-m-kPa). The other parameter values required to evaluate L are $\dot{V}_{CO_2} = 200$ ml(STP)/min, $p^A_{CO_2} = 6.2$ kPa, $p^B_{CO_2} = 0$, $r^B = 400/2 = 200$ μm, and $r^A = 200 - 50 = 150$ μm. Thus,

$$L = \frac{(200/60)\ln(0.020/0.015)}{2\pi(2.46 \times 10^{-5})(6.2 - 0)} = 1000 \text{ m} \approx 0.6 \text{ miles} \tag{10.1-15}$$

If the gas exchanger is to have a reasonable size, the entering blood must be split among many fibers. The 1000 m fiber length could, for example, be accommodated in a 1-m-long device by incorporating 1000 parallel fibers. The total outer surface of the fibers in such a device would be

$$A^B_m = 2\pi(0.020/100)(1000) = 1.26 \text{ m}^2 \tag{10.1-16}$$

This is almost two orders of magnitude smaller than 50–100 m^2 alveolar surface of the adult human lung. There are three reasons for this: (i) The human lung needs to have a large reserve to excrete the large amounts of CO_2 produced during heavy exercise; (ii) radial diffusion resistance in blood substantially impedes CO_2 excretion; and (iii) a progressive loss of driving force occurs along a fiber as CO_2 is removed from blood.

The O_2 enrichment necessary to achieve the desired O_2 transfer can be found by solving for $p^B_{O_2}$ with Equation 10.1-13 when $s = O_2$:

$$p^B_{O_2} = p^A_{O_2} - \frac{\dot{V}_{O_2} \ln(r_B/r_A)}{2\pi \hat{P}^G_{O_2} L} \tag{10.1-17}$$

The appropriate parameter values to complete the computation are $\hat{P}^{O_2}_G = 4.91 \times 10^{-6}$ ml(STP)/(s-m-kPa), $\dot{V}_{O_2} = -\dot{V}_{CO_2}/RQ = -200/0.8 = -250$ ml(STP)/min, $p^A_{O_2} = 5.3$ kPa and $L = 1000$ m. We compute a $p^B_{O_2}$ value of

$$p^B_{O_2} = 5.3 - \frac{(-250/60)\ln(0.020/0.015)}{2\pi(4.91 \times 10^{-6})(1000)} = 44 \text{ kPa} \tag{10.1-18}$$

Since the normal p_{O_2} in room air is 21 kPa, this means that the air entering the device must be enriched with O_2 by at least a factor of about two.

10.1.3 Selectivity

To quantify the relative ease with which two different solutes diffuse through a homogeneous membrane, we can use the ratio of either their permeabilities or specific permeabilities. The resulting selectivity factor is

$$S_{1,2} \equiv \frac{P_1}{P_2} = \frac{\hat{P}_1}{\hat{P}_2} = \frac{\lambda_1^{m,A}\mathcal{D}_1}{\lambda_2^{m,A}\mathcal{D}_2} \quad \text{or} \quad S^G_{1,2} \equiv \frac{P^G_1}{P^G_2} = \frac{\hat{P}^G_1}{\hat{P}^G_2} = \frac{\alpha_1^m \mathcal{D}_1}{\alpha_2^m \mathcal{D}_2} \tag{10.1-19a,b}$$

Equation 10.1-19a, applicable to any pair of solutes, expresses selectivity in terms of relative molar transport rates at the same concentration driving force; Equation 10.1-19b, applicable to gaseous solutes only, characterizes selectivity by the relative volumetric transport rates at the same partial pressure driving force. A comparison of CO_2 and O_2 permeability data (Table 10.1-1), for example, indicates that $S^G_{CO_2,O_2}$ values are larger than 1.0 for both silicone and Teflon membranes. Thus, given equal partial pressure driving forces, volumetric CO_2 transport is greater than the volumetric O_2 transport. Or stated in a different way, a small CO_2 driving force can provide the same volumetric transport as a larger O_2 driving force. This property is useful in the design of artificial lungs for which volumetric transport of these two gases must be similar, but the available partial pressure driving force for CO_2 is much smaller than the driving force that can be provided for O_2.

10.2 Diffusion with Parallel Convection

The convection of liquid solutions across heterogeneous membranes by fluid and osmotic pressure driving forces is critical for regulating water balance in physiological systems. Convection through cell membranes mainly occurs through pores or channels, while convection across capillary walls occurs through intercellular gaps. Hollow fibers used in hemodialysis devices have microporous structures through which convection can occur. It is important to consider how diffusion is influenced by parallel convection in such porous media.

10.2.1 Nonequilibrium Thermodynamics

Consider a planar membrane with surface area S_m separating well-mixed aqueous solutions A and B (or any other two solutions that are similar in nature) containing a single solute s. In general, the solute concentrations, C_s^A and C_s^B, and the hydrostatic pressures, P^A and P^B, are different in the two solutions (Fig. 10.2-1). The transmembrane solute flux N_s and velocity u^* predicted by nonequilibrium thermodynamics for simultaneous diffusion and convection across the membrane are

$$N_s = u^* C_s^{ave}(1 - \sigma_s) + P_s \Delta C_s \tag{6.5-30}$$

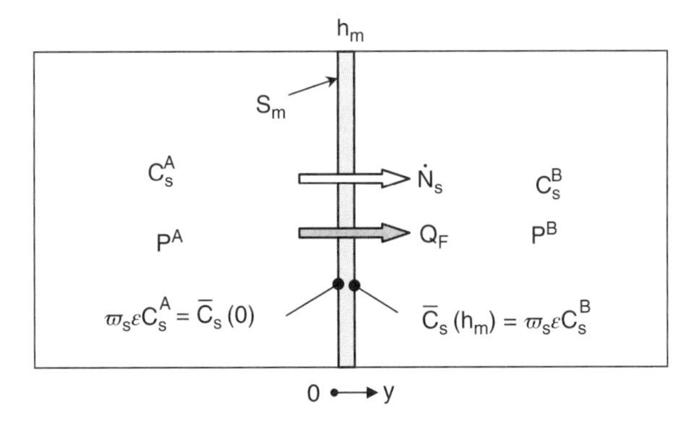

Figure 10.2-1 Parallel diffusion and convection through a membrane.

$$u^* = L_P(\Delta P - \sigma_s \mathcal{R}T\Delta C_s) \tag{6.5-31}$$

Multiplying both sides of these equations by S_m, we obtain the solute transport rate $\dot{N}_s$ and the volumetric rate of solution filtration Q_F:

$$\dot{N}_s \equiv N_s S_m = Q_F C_s^{ave}(1-\sigma_s) + P_s S_m (C_s^A - C_s^B) \tag{10.2-1}$$

$$Q_F \equiv u^* S_m = L_P S_m \left[(P^A - P^B) - \mathcal{R}T\sigma_s (C_s^A - C_s^B)\right] \tag{10.2-2}$$

10.2.2 Mechanistic Models

In Chapter 7, we formulated the phenomenological coefficient L_p for a heterogeneous medium of arbitrary thickness h and containing a volume fraction ε of cylindrical pores, each with a radius a_p and tortuosity τ_p. For a porous membrane in particular, $h \to h_m$ and $\varepsilon \to \varepsilon_m$ so that Equations 7.4-37 becomes

$$L_P = \frac{a_p^2 \varepsilon_m}{8\mu\tau_p h_m} \tag{10.2-3}$$

To determine how the other two phenomenological coefficients, P_s and σ_s, depend on microscopic membrane properties, we apply the one-dimensional form of the macroscopic flux equation for simultaneous convection and diffusion in heterogeneous media. Replacing $\varepsilon \to \varepsilon_m$ in Equation 7.4-26,

$$\bar{N}_s = \frac{\beta_s}{\varepsilon_m}\bar{u}\bar{C}_s - \bar{\mathcal{D}}_s \frac{d\bar{C}_s}{dy} \tag{10.2-4}$$

where β_s is the hydrodynamic correction factor, and $\bar{\mathcal{D}}_s$ is the hindered diffusion coefficient. Multiplying both sides of this equation by S_m and noting that $Q_F = \bar{u}S_m$, we arrive at the solute transport rate:

$$\dot{N}_s = \frac{\beta_s}{\varepsilon_m} Q_F \bar{C}_s - \bar{\mathcal{D}}_s S_m \frac{d\bar{C}_s}{dy} \tag{10.2-5}$$

Provided that $\dot{N}_s$, $\beta_s Q_F/\varepsilon_m$, and $\bar{\mathcal{D}}_s S_m$ are constant, the integral of this equation between the membrane surfaces at $y = 0$ and $y = h_m$ leads to

$$\dot{N}_s = \frac{\beta_s Q_F}{\varepsilon_m}\left[\frac{\bar{C}_s(0)e^{Pe_m} - \bar{C}_s(h_m)}{e^{Pe_m} - 1}\right] \tag{10.2-6}$$

where

$$Pe_m \equiv \frac{\beta_s Q_F h_m}{\varepsilon_m \bar{\mathcal{D}}_s S_m} \tag{10.2-7}$$

is the Peclet number associated with the membrane. This result is analogous to parallel convection and diffusion through a homogeneous fluid phase (Eqs. 9.3-6 and 9.3-7). Applying the interfacial condition given in Equation 7.4-42a,b to rewrite Equation 10.2-6 in terms of external concentrations C_s^A and C_s^B, we get

$$\dot{N}_s = \varpi_s \beta_s Q_F \left(\frac{e^{Pe_m} C_s^A - C_s^B}{e^{Pe_m} - 1}\right) \tag{10.2-8}$$

Here, ϖ_s is the accessibility of membrane pores to solute. By rearranging this equation, the transport rate can also be expressed as the sum of convective and diffusive contributions:

$$\dot{N}_s = \varpi_s \beta_s Q_F \widetilde{C}_s + \frac{\varepsilon_m \varpi \bar{\mathcal{D}}_s S_m}{h_m}\left(C_s - C_s^B\right) \tag{10.2-9}$$

where

$$\widetilde{C}_s \equiv \left[C_s^A - \left(C_s^A - C_s^B\right)\left(\frac{1}{Pe_m} - \frac{1}{e^{Pe_m} - 1}\right)\right] \tag{10.2-10}$$

When convection dominates diffusion so that Pe_m is large, the transport rate has the limiting form

$$\lim_{Pe_m \to \infty} \dot{N}_s = \varpi_s \beta_s Q_F C_s^A \tag{10.2-11}$$

On the contrary, when diffusion dominates convection so that Pe_m is small,

$$\lim_{Pe_m \to 0} \dot{N}_s = \varpi_s \beta_s Q_F C_s^{ave} + \frac{\varepsilon_m \varpi_s \bar{D}_s S_m}{h_m}\left(C_s^A - C_s^B\right) \tag{10.2-12}$$

Equation 10.2-12, having the same form as Equation 10.2-1, leads us to conclude that the nonequilibrium formulation is limited to processes in which diffusion dominates convection. A comparison of the two equations also provides the following parameter relationships:

$$P_s = \frac{\varpi_s \varepsilon_m \bar{\mathcal{D}}_s}{h_m}, \quad \sigma_s = 1 - \varpi \beta_s \tag{10.2-13a,b}$$

We can now rewrite the general expression for $\dot{N}_s$ (Eq. 10.2-8) in terms of the phenomenological coefficients of nonequilibrium thermodynamics:

$$\dot{N}_s = (1 - \sigma_s) Q_F \left(\frac{e^{Pe_m} C_s^A - C_s^B}{e^{Pe_m} - 1}\right); \quad Pe_m = \frac{(1 - \sigma_s) Q_F}{P_s S_m} \tag{10.2-14a,b}$$

This equation applies to membranes with very little curvature. For a spherical or cylindrical membrane, we can include curvature effects by replacing S_m appearing in Equation 10.2-14b by the mean surface area $\widetilde{S}_m$ defined in Equations 10.1-12a-c.

By utilizing ϖ_s (Eq. 7.4-7), β_s (Eq. 7.4-31), and $\bar{\mathcal{D}}_s$ (Eq. 7.4-32) given by the Renkin model, we can write P_s and σ_s in terms of solute and pore sizes:

$$\sigma_s = 1 - \left(1 - \frac{a_s}{a_p}\right)^2 \left[1 + 2\frac{a_s}{a_p} - \left(\frac{a_s}{a_p}\right)^2\right]\left[1 - \frac{2}{3}\left(\frac{a_s}{a_p}\right)^2 - 0.163\left(\frac{a_s}{a_p}\right)^3\right] \tag{10.2-15}$$

$$P_s = \frac{\varepsilon_m \mathcal{D}_s^\infty}{\tau_p h_m}\left(1 - \frac{a_s}{a_p}\right)^2 \left[1 - 2.1044\left(\frac{a_s}{a_p}\right) + 2.089\left(\frac{a_s}{a_p}\right)^3 - 0.948\left(\frac{a_s}{a_p}\right)^5\right] \tag{10.2-16}$$

10.2.3 Selectivity and Sieving

In the absence of convection, the selectivity factor of a heterogeneous membrane for a solute 1 relative to a solute 2 can be expressed in terms of their permeabilities as

$$S_{12} \equiv \frac{P_1}{P_2} = \frac{\varpi_1 \bar{\mathcal{D}}_1}{\varpi_2 \bar{\mathcal{D}}_2} \tag{10.2-17}$$

When convection occurs in parallel with diffusion, it is useful to characterize the membrane as a molecular sieve and the separation of solutes is referred to as ultrafiltration. During ultrafiltration, solution in the upstream compartment A is called the retentate, while solution that reaches the downstream compartment B is the filtrate. The sieving coefficient for this process is the ratio of the filtrate concentration C_s^B to the retentate concentration C_s^A:

$$\Sigma_s \equiv \frac{C_s^B}{C_s^A} \tag{10.2-18}$$

Recognizing that the molar rate at which solute appears in the filtrate, $\dot{N}_s = Q_F C_s^B$, must be the same as the transport rate through the membrane given by Equation 10.2-14a, we obtain

$$\Sigma_s = \frac{(1-\sigma_s)}{1-\sigma_s \exp(-Pe_m)} \tag{10.2-19}$$

The sieving coefficient is a measure of the loss of a solute to the filtrate relative to the remaining solute in the retentate. The greatest loss occurs when diffusion dominates convection such that $Pe_m \to 0$ and $\Sigma_s = 1$. In that limit, it is preferable to use the selectivity factor given by Equation 10.2-17 to characterize membrane performance. The least loss occurs when convection dominates diffusion such that $Pe_m \to \infty$ and $R_s \to (1 - \sigma_s)$.

A different but complementary parameter for characterizing the selectivity of a porous membrane is the retention coefficient:

$$R_s \equiv 1 - \frac{C_s^B}{C_s^A} = 1 - \Sigma_s = \frac{\sigma_s[1-\exp(-Pe_m)]}{1-\sigma_s \exp(-Pe_m)} \tag{10.2-20}$$

The retention coefficient indicates how efficiently the membrane prevents a solute from being lost from the retentate to the filtrate. The maximum retention $R_s \to \sigma_s$ occurs when $Pe_m \to \infty$, whereas $R_s \to 0$ is predicted when $Pe_m \to 0$.

Example 10.2-1 Diffusion and Ultrafiltration Through a Membrane

Cellophane membranes were used in the first practical hemodialysis devices. When immersed in aqueous solution, these membranes become hydrated, forming a porous heterogeneous material. Renkin (1954) carried out both diffusion and ultrafiltration measurements on Visking cellophane membranes mounted in the apparatus depicted in Figure 10.2-2. A thickness

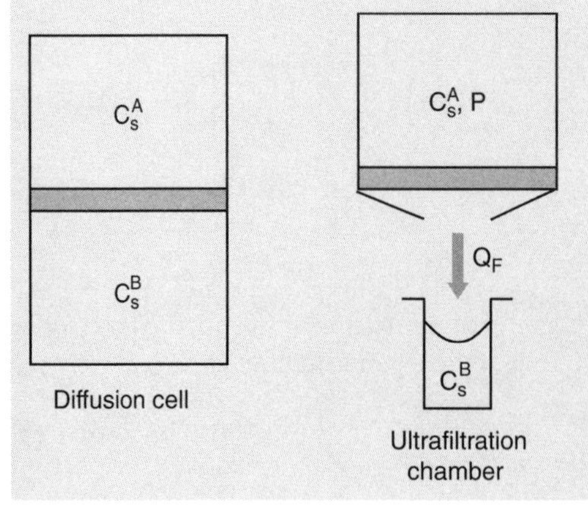

Figure 10.2-2 Apparatus for determining membrane performance.

$h_m = 54.7$ μm of the hydrated membrane was measured with calipers, and a pore fraction $\varepsilon_m = 0.664$ was inferred from water content.

Diffusion experiments were performed on solutes of different molecular sizes (Table 10.2-1) using a diffusion cell with two well-mixed compartments separated by the membrane. Initially, compartment B contained water, and compartment A contained an aqueous solution with known solute concentration C_s^A. Values of $P_s/\mathcal{D}_s^\infty$ were determined from measurements of $C_s^B(t)$.

Ultrafiltration experiments were performed on several of the same solutes using an apparatus consisting of a closed retentate compartment A and an open filtrate compartment that drained into a sampling vial B. During each experiment, a constant filtration flow was maintained by a pressure P in the retentate compartment. A corresponding value of R_s was determined from initial and final retentate concentrations and concentrations measured in the sampling vial. These observations of R_s were repeated at different filtrate fluxes Q_F/S_m of 0.2 and 0.6 μm/s-cm^2 of membrane surface. A hydraulic permeability of $L_P = 9.5 \times 10^{-7}$ μm/(Pa-s) was determined from Q_F-Δp data obtained in separate experiments with pure water.

The purpose of this example is to see how well the data points in Figure 10.2-3 can be simulated by the Renkin pore model and to determine the pore radius predicted by the model.

Solution

The model of the $P_s/\mathcal{D}_s$ measurements is given by Equation 10.2-16, and the R_s measurements are modeled by Equations 10.2-14b, 10.2-15, and 10.2-20. All together, these equations contain parameters h_m, ε_m, a_s, $\mathcal{D}_s$, and Q_F/S_m that have known values in this problem. They also contain parameters a_p and τ_p that are not known individually but are related by Equation 10.2-3:

Table 10.2-1 Solute Properties

Solute, s	H^3HO	Urea	Glucose	Antipyrine	Sucrose	Raffinose
a_s (nm)	0.197	0.27	0.357	0.396	0.44	0.564
$\mathcal{D}_s$ (10^{-5} cm^2/s)	2.36	1.45	0.68	0.65	0.55	0.42

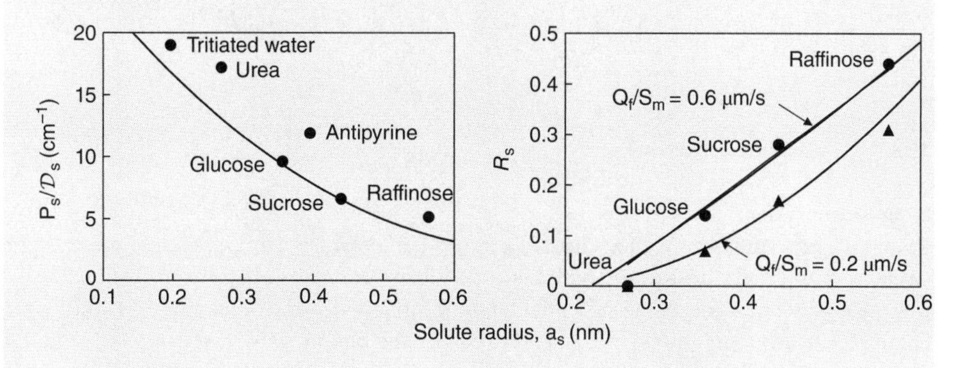

Figure 10.2-3 Comparison of diffusion and ultrafiltration data points to model simulations.

$$\tau_p = \frac{a_p^2 \varepsilon_m}{8\mu L_p h_m} = \frac{0.664\left(10^{-3}\right)^2}{8\left(9.0\times10^{-4}\right)\left(9.5\times10^{-7}\right)(54.7)} a_p^2 = 1.78 a_p^2 \left(\text{nm}^2\right) \quad (10.2\text{-}21)$$

where $\mu = 9.0 \times 10^{-4}$ Pa-s is the viscosity of water at the temperature 25°C of the experiments. Thus, in fitting the Renkin model equations to the data, a_p is the only unknown parameter.

Treating a_p as an adjustable parameter, we find that a value of $a_p = 1.49$ nm minimizes the summed squared error between all the $P_s/\mathcal{D}_s$ and R_s measurements and their model predictions. Inserting this value of a_p in Equation 10.2-15, we compute a range of σ_s from 0.129 for urea to 0.442 for raffinose. The Pe_m calculated from Equation 10.2-14 covered ranges of 0.139–1.09 at the lower filtration flux and 0.417–3.26 at the higher filtration flux.

Judging from the agreement between the data points and the regression curves in Figure 10.2-3, we conclude that the Renkin model provides a reasonable simulation of solute and solvent transport through a Visking membrane.

10.3 Cell Membrane Channels

A traffic of ions across cell membranes is necessary for physiological processes as diverse as neural transmission, muscle contraction, and cell volume control. Because of their low solubility in organic solvents, ions are not transported across the phospholipid bilayer of the plasma membrane. Instead, ions move through fluid-filled channels formed by integral proteins in the phospholipid bilayer.

Although the channels select for particular ions, ion transport through a channel does not require an energy source as with active transport. Rather, this is a passive process occurring along an electrochemical gradient. Excitable cells such as neurons contain "gated" channels through which ion transport is greatly enhanced by chemical, electrical, or mechanical activation. When in an activated or "open" state, a single gated channel is capable of transporting on the order of one-million ions per second.

Carboxyl, amine, and other dissociable groups on a channel protein can impart a fixed charge that has a strong influence on ion selectivity. Consider, for example, the negatively charged end-plate channel associated with a synapse between two neurons. This chemically gated channel is activated by the binding of acetylcholine to a specific receptor on the channel protein. At acetylcholine concentrations above a threshold level, positively charged cations permeate the end-plate channel at a rate that is inversely related to their size. Yet, because of their negative charge, electrical repulsion prevents anions from crossing the channel.

10.3.1 Electrodiffusion Model

Model Formulation

From an electrical point of view, the collection of ion channels in a cell membrane acts as a variable conductor with the surrounding lipid bilayer acting as a parallel capacitor. In the presence of a transmembrane potential, a continuous current of ions can flow through a channel. However, an ion current into (or out of) the bilayer only occurs when a dynamic change in potential is available to charge (or discharge) this capacitor. We will focus on situations in which membrane potential does not vary with time. The bilayer then acts as an insulator, and ion current is only determined by transport through channels.

In the absence of convection, the electrochemical flux for each ion "i" that passes through a channel "c" can be represented by the y-component of the Nernst–Planck equation (Eq. 6.4-2). For an ion with a charge number z_i in a fluid of constant molar density,

$$J_i^c(r,y) = -\mathcal{D}_i\left[\frac{\partial C_i^c(r,y)}{\partial y} + \left(\frac{z_i\mathcal{F}}{\mathcal{R}T}\right)C_i^c(r,y)\frac{\partial\psi(r,y)}{\partial y}\right] \tag{10.3-1}$$

Similar to the procedure followed for the pore model (Eq. 7.4-14), we will volume average the local flux $J_i^c(y,r)$ over the channel cross section. Assuming that $\psi(r,y)$ is uniform over the entire cross section, we obtain

$$\bar{J}_i^c(y) = -\bar{\mathcal{D}}_i\left[\frac{d\bar{C}_i^c(y)}{dy} + \left(\frac{z_i\mathcal{F}}{\mathcal{R}T}\right)\bar{C}_i^c\frac{d\psi(y)}{dy}\right] \tag{10.3-2}$$

where $\bar{\mathcal{D}}_i$ is a hindered diffusion coefficient whose relationship to the local diffusion coefficient is given by Equation 7.4-27. Equation 10.3-2 states that the overall transport of ion i is the sum of two components: (i) pure diffusion driven by the concentration gradient $d\bar{C}_i^c/dy$ and (ii) drift imposed by the electric field, $E = -d\psi/dy$, acting on the ion charge.

To complete this electrodiffusion model of ion transport, we introduce the Poisson equation, an electrostatic relationship between the spatial variation of voltage and the density of electrical charge. The local charge density $\sigma(y)$ associated with a channel is composed of: (i) the charge density of each of the I different ion species i that can simultaneously pass through the channel, $\mathcal{F}z_i\bar{C}_i^c(y)$; and (ii) the charge density of the stationary channel protein, $\sigma_p(y)$. The one-dimensional form of Poisson's equation (Eq. 4.4-2) is thus written as

$$-\varepsilon_o\frac{d^2\psi}{dy^2} = \sigma(y) = \mathcal{F}\sum_{i=1}^{I} z_i\bar{C}_i^c(y) + \sigma_p(y) \tag{10.3-3}$$

where ε_o is the electrical permittivity parameter.

Boundary Conditions

Transport through ion channels has been studied by observing the electrical current in response to known transmembrane potential and concentration differences. These experiments were first performed on large single cells, typically the squid giant axon, into which miniature electrodes, chemical sensors and fluid transfer pipettes could be directly inserted. Studies today are more often conducted on a microscopic membrane patch isolated from a cell membrane with a micropipette. In either case, we idealize the membrane as a planar structure whose surfaces, A and B, are exposed to fluids containing constant ion concentrations, C_i^A and C_i^B, and uniform electric potentials, ψ^A and ψ^B (Fig. 10.3-1). Consistent with the usual convention, we associate A with intracellular fluid and B with extracellular fluid, and we consider $\bar{J}_i^c$ to be an outward directed ion flux, while i_i is the corresponding inward directed current.

The electric potential is continuous between a membrane surface and its external solution. Identifying $y = 0$ and $y = h_m$ as the intracellular and extracellular surfaces, respectively, the boundary conditions on the electrical potential are:

$$\psi(0) = \psi^A, \quad \psi(h_m) = \psi^B \tag{10.3-4a,b}$$

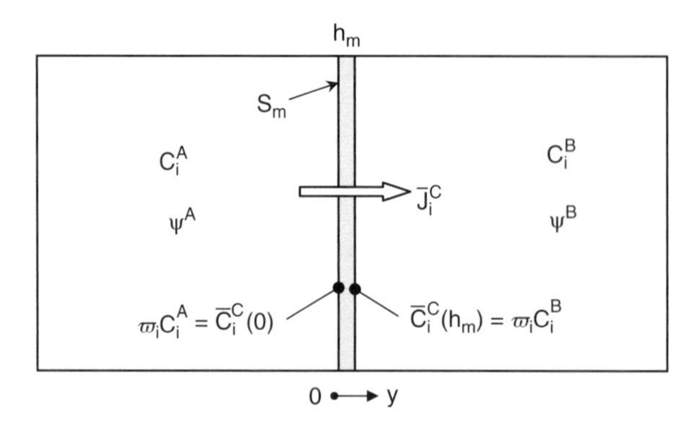

Figure 10.3-1 Electrochemical transport through an ion channel.

Consider the openings of a channel located at $y = 0$ and $y = h_m$. At these positions, the concentrations of ion i averaged over the channel cross section are $\bar{C}_i^c(0)$ and $\bar{C}_i^c(h_m)$, and the ion concentrations in the adjoining homogeneous solutions are C_i^A and C_i^B. The requirement that solute concentrations be continuous between the accessible portion of the channel openings and the external solutions leads to the boundary conditions prescribed by Equation 7.4-41a,b.

$$\bar{C}_i^c(0) = \varpi_i C_i^A, \quad \bar{C}_i^c(h_m) = \varpi_i C_i^B \tag{10.3-5a,b}$$

Here, ϖ_i, the fraction of the channel cross section accessible to ion i, depends on the radius of a transported solute relative to the channel radius. Unlike an uncharged molecule, the radius of an ion includes the hydration layers formed by electrostatic attraction of water molecules. In a relatively large channel, an ion is transported with a full complement of hydration layers, whereas in a small channel, the ions may shed hydration layers in order to squeeze through.

Goldman Assumption

To examine the relative magnitude of terms in the Poisson equation, we introduce the dimensionless variables:

$$y = \frac{y}{h_m}, \quad C = \frac{\bar{C}_i^P}{(C_i^A + C_i^B)}, \quad \psi = \frac{\mathcal{F}}{\mathcal{R}T}\Psi, \quad \sigma \equiv \frac{\sigma_p}{\sigma_{p,max}} \tag{10.3-6a-d}$$

where $\sigma_{p,max}$ is the maximum protein charge density along the channel. In terms of these quantities, Equation 10.3-3 becomes

$$\frac{d^2\psi}{dy^2} = -\frac{\mathcal{F}^2 h_m^2 \sum_i z_i (C_i^A + C_i^B)}{\mathcal{R}T\varepsilon_o}\left[C + \frac{\sigma_{p,max}}{\mathcal{F}\sum_i z_i (C_i^A + C_i^B)}\sigma\right] \tag{10.3-7}$$

All the dimensionless variables in this equation except ψ are scaled to be of order-of-magnitude one, whereas ψ is not constrained to a specific magnitude. We are particularly interested in analyzing ion transport when the right side of this equation is very small. This will occur when (i) the channel is sufficiently short that $h_m \ll \sqrt{\mathcal{R}T\varepsilon_o / \mathcal{F}^2 \Sigma_i z_i (C_i^A + C_i^B)}$ and

(ii) the protein charge density does not exceed the ion charge density such that $\left|\sigma_{p,max}\right| \leq \mathcal{F}\Sigma_i\left(C_i^A + C_i^B\right)|z_i|$. In that case,

$$\frac{d^2\psi}{dy^2} \approx 0 \Rightarrow \frac{d^2\psi}{dy^2} = 0 \qquad (10.3\text{-}8a,b)$$

Integrating twice and applying Equations 10.3-4a,b to evaluate the two constants of integration leads to

$$\psi - \psi^B = \psi_m\left(1 - \frac{y}{h_m}\right) \Rightarrow \frac{d\psi}{dy} = -\frac{\psi_m}{h_m} \qquad (10.3\text{-}9a,b)$$

where $\psi_m \equiv \psi^A - \psi^B$.

This approximation of a constant electric potential gradient $d\psi/dy$, which is equivalent to a constant electric field, is referred to as the Goldman assumption. Although the Goldman assumption greatly simplifies mathematical solution of the electrodiffusion equations, it excludes the possible contribution of internal protein charges to the electric field.

Steady-State Single Ion Current

Fixing $d\psi/dy$ by the Goldman assumption allows us to solve for the ion flux along a single channel without having to simultaneously solve the Poisson equation. Substituting Equation 10.3-9b into Equation 10.3-2, we have

$$\frac{d\bar{C}_i^c(y)}{dy} - \left(\frac{z_i\mathcal{F}\psi_m}{\mathcal{R}Th_m}\right)\bar{C}_i^c - \frac{\bar{J}_i^c}{\bar{\mathcal{D}}_i} = 0 \qquad (10.3\text{-}10)$$

If ion transport is in a steady or pseudo-steady state and there is no convective transport, then the electrodiffusion flux $\bar{J}_i^c$ must be constant by material balance considerations. Provided that $\bar{\mathcal{D}}_i$ is also a constant, this equation can be integrated subject to the first boundary condition (Eq. 10.3-5a) to obtain the concentration distribution:

$$\bar{C}_i^c(y) = \left[\varpi_i C_i^A + \left(\frac{h_m\bar{J}_i^c}{z_i\psi_m\bar{\mathcal{D}}_i}\right)\right]\exp\left[\left(\frac{z_i\psi_m}{h_m}\right)y\right] - \left(\frac{h_m\bar{J}_i^c}{z_i\psi_m\bar{\mathcal{D}}_i}\right) \qquad (10.3\text{-}11)$$

where $\psi_m \equiv (\mathcal{F}/\mathcal{R}T)\psi_m$. By applying the second boundary condition (Eq. 10.3-5b), we solve for the outward directed ion flux through a single channel:

$$\bar{J}_i^c = \frac{\varpi_i\bar{\mathcal{D}}_i z_i\psi_m}{h_m}\left(\frac{C_i^A - C_i^B e^{-z_i\psi_m}}{1 - e^{-z_i\psi_m}}\right) \qquad (10.3\text{-}12)$$

Consider a membrane with a volume fraction ε_i of open channels that are transporting ion i. In the absence of current flow through the lipid bilayer, Equation 7.4-6b indicates that the overall flux of ion i through such a membrane will be $\bar{J}_i = \varepsilon_i\bar{J}_i^c$ such that

$$\bar{J}_i = \frac{\varpi_i\varepsilon_i\bar{\mathcal{D}}_i z_i\psi_m}{h_m}\left(\frac{C_i^A - C_i^B e^{-z_i\psi_m}}{1 - e^{-z_i\psi_m}}\right) \qquad (10.3\text{-}13)$$

Since $\bar{J}_i$ is an outward flux of an ion i of charge z_1, it corresponds to an electric current per unit surface, $\mathcal{F}z_i\bar{J}_i$, flowing inward through a membrane. For a membrane of surface area S_m, the inward electrical current is

$$i_i = \mathcal{F}z_i\bar{J}_i S_m = \mathcal{F}z_i^2\psi_m S_m\left(\frac{\varpi_i\varepsilon_i\bar{\mathcal{D}}_i}{h_m}\right)\left(\frac{C_i^A - C_i^B e^{-z_i\psi_m}}{1-e^{-z_i\psi_m}}\right) \quad (10.3\text{-}14)$$

Equation 10.3-14 can be written in a more compact form by introducing the permeability of the membrane to ion i, $P_i = \varpi_i\varepsilon_i\bar{\mathcal{D}}_i/h_m$ (Eq. 10.2-13a):

$$i_i = \mathcal{F}z_i^2\psi_m P_i S_m\left(\frac{C_i^A - C_i^B e^{-z_i\psi_m}}{1-e^{-z_i\psi_m}}\right) \quad (10.3\text{-}15)$$

For gated channels, ε_i and therefore P_i will increase with the degree of electrical or chemical activation.

10.3.2 Resting Potential

When a cell is in its resting (i.e., unexcited) state, channel-mediated ion transport is balanced by active transport processes such as carrier-mediated transport by the Na^+–K^+ pump. This results in a steady-state difference between individual ion concentrations across the cell membrane. With no external means to close the electrical circuit, the ion currents generated by these concentration differences must sum to zero:

$$\sum_{i=1}^{I} i_i = 0 \quad (10.3\text{-}16)$$

Assuming that the Goldman assumption holds, we combine this with Equation 10.3-15 to obtain an implicit relation for the membrane potential generated by all I permeating ions:

$$\sum_{i=1}^{I} P_i z_i^2\left(\frac{C_i^A - C_i^B e^{-z_i\psi_m}}{1-e^{-z_i\psi_m}}\right) = 0 \quad (10.3\text{-}17)$$

If an ion j has a permeability that is very large compared to the other ions, then the membrane potential will closely approach the Nernst equilibrium potential of that ion:

$$P_j z_j^2\left(\frac{C_i^A - C_i^B e^{-z_i\psi_m}}{1-e^{-z_i\psi_m}}\right) = 0 \Rightarrow \psi_m = -\frac{\mathcal{R}T}{z_i\mathcal{F}}\ln\left(\frac{C_j^A}{C_j^B}\right) \quad (10.3\text{-}18a,b)$$

Since the prevalent ions in physiological solutions are univalent, $|z_i| = 1$ and an explicit formulation for the transmembrane potential can be found from Equation 10.3-17:

$$\psi_m = \left(\frac{\mathcal{R}T}{\mathcal{F}}\right)\ln\left(\frac{\sum_i^{cations} P_i C_i^B + \sum_i^{anions} P_i C_i^A}{\sum_i^{cations} P_i C_i^A + \sum_i^{anions} P_i C_i^B}\right) \quad (10.3\text{-}19)$$

This result is known as the Goldman–Hodgkin–Katz equation.

Example 10.3-1 Resting Potential of a Squid Giant Axon

The principal species that contribute to the electrical activity of a squid giant axon are potassium, sodium, and chloride ions. In the resting state, the permeability of K^+ through the axon membrane is 75 times that of Na^+ and 60 times that of Cl^-. The ionic compositions on the two sides of the axon membrane are given in Table 10.3-1.

Determine the transmembrane potential of the neuron at 37°C.

Solution

We will utilize Equation 10.3-19 in the following form:

$$\psi_m = \left(\frac{\mathcal{R}T}{\mathcal{F}}\right)\ln\left[\frac{C_K^B + (P_{Na}/P_K)C_{Na}^B + C_{Cl}^A(P_{Cl}/P_K)}{C_K^A + (P_{Na}/P_K)C_{Na}^A + C_{Cl}^B(P_{Cl}/P_K)}\right] \tag{10.3-20}$$

Inserting the values given in the problem statement:

$$\psi_m = 1000\left[\frac{(8.314)(310)}{(96{,}500)}\right]\ln\left[\frac{(10/1)+(460/75)+(80/60)}{(344/1)+(65/75)+(540/60)}\right] = -80.4\text{ mV} \tag{10.3-21}$$

This result is closer to the observed resting potential of $\psi_m = -77$ mV than the Nernst potential of −89.3 we computed for K^+, the ion with the highest permeability (Example 4.3-1). Thus, the GHK model, which includes electrical interactions between ions as well as to their individual diffusion characteristics, is more realistic than an equilibrium model of resting potential.

Table 10.3-1 Intracellular(A) and Extracellular(B) Ion Concentrations

Compartment	C_{Na} (mM)	C_K (mM)	C_{Cl} (mM)
A	65	344	80
B	460	10	540

From Mountcastle (1974).

10.3.3 Voltage Clamp Measurements

In a closed circuit, voltage clamp experiment, dynamic changes in current are measured after the membrane potential of an excitable cell is suddenly "clamped" to a more negative (hyperpolarized) or less negative (depolarized) level relative to the resting potential. In addition to modifying the rate of ion migration through individual channels, membrane polarization is capable of opening additional, voltage-gated channels.

For the giant squid axon, in particular, the current changes observed during a voltage clamp are governing by the transport of Na^+ and K^+ through their respective ion channels. During membrane hyperpolarization, Na^+ flux first increases above its resting level and then decreases back to its resting level. Presumably, this is due to a transient opening of Na^+ channels. Simultaneously, K^+ flux continuously increases toward a constant value, indicating a sustained opening of K^+ channels. Once a hyperpolarized steady state is reached, the electrical current through the membrane is overwhelmingly due to the K^+ current flux predicted by Equation 10.3-15.

Example 10.3-2 Voltage Clamp Experiments on the Squid Giant Axon

Figure 10.3-2 shows steady-state K^+ currents measured during three voltage clamp experiments performed on a squid giant axon (Rojas and Ehrenstein, 1965). Each experiment began a few minutes after intracellular perfusion with 600 mM KCl solution. In the first experiment, a series of voltage clamps was performed at successively higher levels of hyperpolarization, while the cell was externally perfused with a 10 mM KCl solution(unfilled circles). In the second experiment, a series of voltage clamps was carried out with the 10 mM extracellular solution replaced by a 600 mM KCl solution (filled triangles). During the third experiment, the voltage clamps were repeated with an extracellular solution again containing 10 mM KCl (filled circles). The purpose of this example is to simulate these experiments using the electrodiffusion model of ion transport.

Solution

The first and third experiments are modeled by substituting $\mathcal{F} = 9.65 \times 10^7$ mA-s/mol, $C_K^A = 6.00 \times 10^{-4}$ mol/ml, $C_K^B = 0.10 \times 10^{-4}$ mol/ml, and $z_K = +1$ into Equation 10.3-15:

$$\frac{i_K}{S_m}[\text{mA/cm}^2] = 10^4\left(\frac{5.79 - 0.0965e^{-\psi_m}}{1 - e^{-\psi_m}}\right)\psi_m P_K[\text{cm/s}] \tag{10.3-22}$$

A least-squares error regression of this current flux equation to the i_K/S_m versus ψ_m data resulted in an estimate of $P_K = 0.0107$ cm/s, the only adjustable parameter. The second experiment is similarly modeled by substituting $\mathcal{F} = 9.65 \times 10^7$ mA-s/mol, $C_K^A = C_K^B = 6.00 \times 10^{-4}$ mol/ml, and $z_K = +1$ into Equation 10.3-15:

$$\frac{i_K}{S_m}[\text{mA/cm}^2] = 5.79 \times 10^4 \psi_m P_K[\text{cm/s}] \tag{10.3-23}$$

A least-squares error regression of this equation to the data resulted a permeability estimate of $P_K = 0.00331$ cm/s.

As shown by the upper curve in Figure 10.3-2, the electrodiffusion model with the Goldman assumption is able to simulate the curvilinear behavior of the data obtained with an extracellular KCl concentration of 10 mM, a value that is substantially below

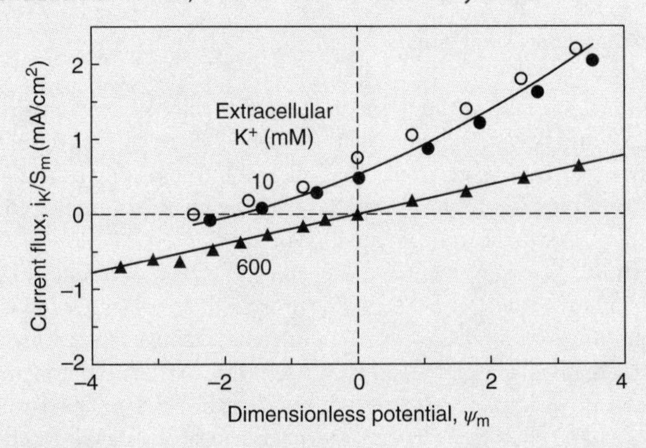

Figure 10.3-2 Steady-state K^+ current during voltage clamp with 600 mM intracellular KCl (Data from Rojas and Ehrenstein (1965)).

the 600 mM intracellular concentration. As indicated by the lower curve, the model was also able to simulate the linear form of the data observed when the extracellular and intracellular KCl concentrations were both 600 mM. A similar fit to these data has been obtained with a more exact solution to the electrodiffusion equations that does not include the Goldman assumption (Leuchtag, 2008). In both cases, the P_K estimated from the 10 mM KCl data is about three times larger than the P_K estimated from the 600 mM KCl data. In terms of the mathematical model, this may reflect a difference in either the fraction of a channel cross section ϖ_i that is accessible to transport or the fraction of open channels ε_i that are present at the two KCl concentration conditions.

References

Galletti PM. Advances in heart-lung machines. In: Levine SN, editor. Advances in Biomedical Engineering and Biomedical Physics. Vol 2. New York: Wiley-Interscience; 1968, pp 121–167.

Leib WR, Stein WD. Non-Stokesian nature of transverse diffusion within human red cell membranes. J Membrane Biol. 1986; 92:114–119.

Leuchtag HR. Voltage-Sensitive Ion Channels: Biophysics of Molecular Excitability. Netherlands: Springer; 2008, p. 151–162.

Mountcastle VB. Medical Physiology. 13th ed. St Louis: Mosby; 1974, Vol 1, p 3.

Renkin EM. Filtration, diffusion and molecular sieving through porous cellulose membranes. J Gen Physiol. 1954; 38:225–243.

Rojas E, Ehrenstein G. Voltage clamp experiments on axons with potassium as the only internal and external cation. J Cell Comp Physiol. 1965; 66:71–78.

Chapter 11

Membrane Transport II

Carrier-Mediated Processes

As discussed in Chapter 10, channel proteins provide fluid-filled pores through which solutes cross biological membranes by electrochemical diffusion. In contrast, carrier or transporter proteins are integral membrane components that are intimately involved in solute transport. After attachment of a solute molecule to a carrier binding site on one surface of a membrane, the protein translocates the bound solute to the other surface of the membrane. By releasing the solute from this downstream surface, the carrier protein assumes its original state with the unoccupied binding site repositioned at the upstream membrane surface. The carrier is then ready to transport another solute molecule.

The primary goal of this chapter is to develop models of both passive and active modes of carrier-mediated membrane transport. As is frequently done (Schultz, 1980), we will model the kinetics of carrier translocation by first-order reaction rate equations. Though these empirical equations lack the theoretical underpinnings of the diffusion equations we used to describe membrane transport in Chapter 10, they are able to provide valuable insights and reasonable simulations of experimental observations.

11.1 Facilitated Transport of a Single Substance

The simplest carrier-mediated process, facilitated transport, occurs when a single solute S crosses a cell membrane surface by monovalent binding to a specific site on a transporter T to form a solute–transporter complex ST according to the reaction:

Biomedical Mass Transport and Chemical Reaction: Physicochemical Principles and Mathematical Modeling, First Edition. James S. Ultman, Harihara Baskaran, and Gerald M. Saidel.

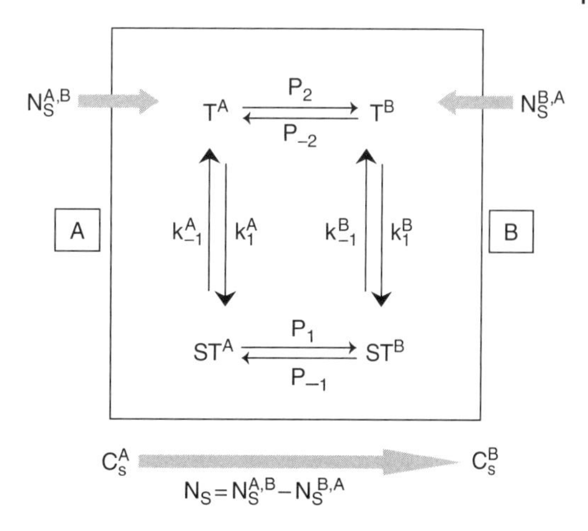

Figure 11.1-1 Facilitated transport of solute S with monovalent binding to transporter T.

$$S(\text{solution}) + T(\text{surface}) \rightleftharpoons ST(\text{surface}) \tag{11.1-1}$$

Because it is a passive process, facilitated transport can only proceed in the direction of decreasing electrochemical potential. In this section, we restrict ourselves to uncharged solutes for which facilitated transport is in the direction of decreasing concentration.

A kinetic model of facilitated transport between external solutions A and B is shown in Figure 11.1-1. Rather than our usual treatment of transport as a sum of convective and diffusive contributions, we attribute the net solute flux N_S to one-way fluxes, $N_S^{A,B}$ and $N_S^{B,A}$, that cross the membrane between the external fluid phases. Fluxes N_S and $N_S^{A,B}$ are positive when they occur in the $A \rightarrow B$ direction, while $N_S^{B,A}$ is positive when it occurs in the $B \rightarrow A$ direction. Thus,

$$N_S = N_S^{A,B} - N_S^{B,A} \tag{11.1-2}$$

Since a transporter protein is embedded in the membrane, it cannot move across the membrane. Rather, it is reversible conformational changes of the transporter molecule that translocates solute binding sites between the membrane surfaces. Hereafter, we will refer to the translocation of a binding site occupied by solute as ST translocation. Similarly, we will refer to the translocation of a binding site unoccupied by solute as T translocation.

The one-way fluxes of S are equal to the corresponding rates of ST translocation between the two membrane surfaces. Employing rate expressions that are first order with respect to the concentrations, C_{ST}^A and C_{ST}^B, at the membrane surfaces, the one-way fluxes are given by

$$N_S^{A,B} = P_1 C_{ST}^A, \quad N_S^{B,A} = P_{-1} C_{ST}^B \tag{11.1-3a,b}$$

and the net solute flux is

$$N_S = P_1 C_{ST}^A - P_{-1} C_{ST}^B \tag{11.1-4}$$

We use analogous rate expressions for T translocation, so that the net flux of free transporter is

$$N_T = P_2 C_T^A - P_{-2} C_T^B \tag{11.1-5}$$

Since no transporter molecules can enter or leave the membrane, $N_S = -N_T$ at steady state:

$$\left(P_1 C_{ST}^A - P_{-1} C_{ST}^B\right) = \left(P_{-2} C_T^B - P_2 C_T^A\right) \tag{11.1-6}$$

In addition, the total concentration of occupied and unoccupied sites on the two membrane surfaces must be constant:

$$T_T = \left(C_{ST}^A + C_T^A\right) + \left(C_{ST}^B + C_T^B\right) \tag{11.1-7}$$

To complete the model, we need to specify the binding and dissociation rates of S to T on the two membrane surfaces. We represent the monovalent binding rate on surface j = A,B as $k_1^j C_S^j C_T^j$ and the corresponding rate of dissociation as $k_{-1}^j C_S^j C_T^j$. Although an analytical solution to the model equations for such kinetics exists (Stein, 1990), we will consider a simplified case in which the influence of solute concentration on solute flux will be more obvious. Suppose that the rate constants k_i^j for the binding reaction are all so large relative to translocation rate constants P_i that surface binding reactions are in equilibrium. The surface concentrations are then related by equilibrium dissociation constants:

$$\kappa^A = \frac{k_{-1}^A}{k_1^A} = \frac{C_S^A C_T^A}{C_{ST}^A}, \quad \kappa^B = \frac{k_{-1}^B}{k_1^B} = \frac{C_S^B C_T^B}{C_{ST}^B} \tag{11.1-8a,b}$$

When the dissociation constants are the same on both membrane surfaces, $\kappa^A = \kappa^B \equiv \kappa$ and we combine Equations 11.1-6, 11.1-7, and 11.1-8 with Equations 11.1-3a,b to obtain

$$N_S^{A,B} = \frac{P_1 T_T \left(P_{-1} C_S^B + P_{-2}\kappa\right) C_S^A}{C_S^A \left[(P_1 + P_{-1}) C_S^B + (P_1 + P_{-2})\kappa\right] + \kappa\left[(P_{-1} + P_2) C_S^B + (P_2 + P_{-2})\kappa\right]} \tag{11.1-9}$$

$$N_S^{B,A} = \frac{P_{-1} T_T \left(P_1 C_S^A + P_2\kappa\right) C_S^B}{C_S^B \left[(P_1 + P_{-1}) C_S^A + (P_{-1} + P_2)\kappa\right] + \kappa\left[(P_1 + P_{-2}) C_S^A + (P_2 + P_{-2})\kappa\right]} \tag{11.1-10}$$

To further simplify our discussion, we assume that the translocation rate constants are the same in both directions across the membrane so that $P_1 = P_{-1}$ and $P_2 = P_{-2}$. The net flux of S across the membrane is then given by

$$N_S = N_S^{A,B} - N_S^{B,A} = \frac{T_T P_1 P_2 \kappa \left(C_S^A - C_S^B\right)}{\kappa(P_1 + P_2)\left(C_S^A + C_S^B\right) + 2\left(P_1 C_S^A C_S^B + P_2 \kappa^2\right)} \tag{11.1-11}$$

The static head condition refers to the relationship between C_S^A and C_S^B that is in force when $N_S = 0$. For this single-solute model, Equation 11.1-11 indicates that the static head condition is $C_S^A = C_S^B$. To produce a sustained flux, $C_S^A > C_S^B$ and a positive driving force $\left(C_S^A - C_S^B\right)$ must be maintained.

Equation 11.1-11 is symmetric in the sense that interchanging C_S^A and C_S^B reverses the sign of N_S without altering its magnitude. Symmetry is also apparent from the one-way fluxes; N_S^B can be obtained by interchanging A and B in the equation for N_S^A. This symmetry is a consequence of assuming that the dissociation constant κ is the same on both sides of the membrane and that $P_1 = P_{-1}$ and $P_2 = P_{-2}$.

Depending on the relative value of P_1 to P_2 and of C_S^B to κ, the one-way flux equations reduce to three limiting cases. Equation 11.1-9 indicates that

$$N_S^{A,B} = \begin{cases} \dfrac{P_1 T_T C_S^A}{C_S^A\left(2 + \kappa/C_S^B\right) + \kappa} & \left(P_1 \gg P_2 \text{ and } C_S^B \geq \kappa\right) \\ \dfrac{(T_T P_1/2) C_S^A}{C_S^A + \kappa} & (P_1 = P_2) \\ \dfrac{P_1 T_T C_S^A}{C_S^A\left(1 + C_S^B/C_S^A\right) + 2\kappa} & \left(P_1 \ll P_2 \text{ and } C_S^B \leq \kappa\right) \end{cases} \qquad (11.1\text{-}12)$$

When $P_1 = P_2$, this equation is identical in form to a Michaelis–Menten rate equation (Eq. 8.5-9) in which the maximum flux is $V_m = T_T P_1/2$ and the half-saturation constant is $K_m = \kappa$. In this case, the structure of the kinetic model of one-way transport is analogous to that of single-substrate enzyme kinetics.

Often C_S^A is referred to as a cis-concentration since it exists in the external solution A from which $N_S^{A,B}$ originates. Conversely, C_S^B is a trans-concentration that refers to the external solution B toward which $N_S^{A,B}$ is directed. When $P_1 = P_2$, $N_S^{A,B}$ is affected by the cis-concentration but not the trans-concentrations. When P_1 and P_2 are unequal, $N_{ST}^{A,B}$ can be influenced by both the cis- and the trans-concentrations of solute. If $P_1 \gg P_2$ and $C_S^B \geq \kappa$, then an increase C_S^B causes $N_{ST}^{A,B}$ to increase. This is referred to as trans-stimulation. However, when $P_1 \ll P_2$ and $C_S^B \leq \kappa$, an increase C_S^B causes $N_{ST}^{A,B}$ to decrease, which is known as trans-inhibition. Neither cis-stimulation nor trans-inhibition of one-way fluxes can occur during ordinary diffusion through a homogeneous membrane or transport through ion channels.

11.2 Cotransport of Two Substrates

The previous section dealt with transport of just a single uncharged solute on a single site of a carrier protein, a process often referred to as uniport. In that case, the solute flux moves in the direction of decreasing concentration. When two different solutes are cotransported by the same carrier protein, it is possible for one of the solutes to be transported in a direction opposite to its concentration driving force. We will consider a model of such secondary active transport in which two solutes, S_1 and S_2, cross a membrane by competitive binding to a single site on a transporter T.

This transporter can exist in one of three states: T bound to S_1 as S_1T; T bound to S_2 as S_2T; or free T bound to neither S_1 nor S_2 (Fig. 11.2-1). In formulating a transport model, we make the following simplifying assumptions: the solute–transporter reactions are so rapid that the dissociation of S_1T and S_2T on both sides of the membrane are at equilibrium; the dissociation constants associated with each of these reactions, κ_1 and κ_2, are the same on the two sides of the membrane; and the translocation rate constant P has the same value for all three states of the transporter and in both directions of carrier protein translocation.

In this case, the equilibrium constants of S_1T and S_2T are given by

$$\kappa_1 = \frac{C_{S_1}^A C_T^A}{C_{S_1T}^A} = \frac{C_{S_1}^B C_T^B}{C_{S_1T}^B}, \quad \kappa_2 = \frac{C_{S_2}^A C_T^A}{C_{S_2T}^A} = \frac{C_{S_2}^B C_T^B}{C_{S_2T}^B} \qquad (11.2\text{-}1a,b)$$

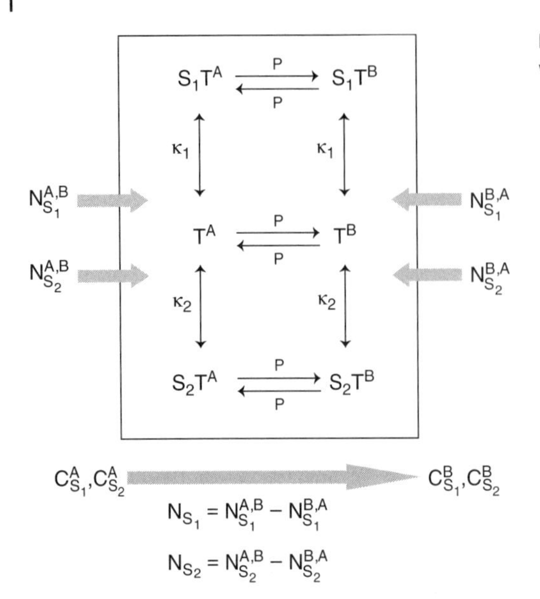

Figure 11.2-1 Cotransport of solutes S_1 and S_2 with competitive binding to transporter T.

Similar to facilitated transport, the net rate of S_1T and S_2T translocations from surface A to surface B must be equal to the net rate of T translocation in the opposite direction:

$$P\left(C_{S_1T}^A - C_{S_1T}^B\right) + P\left(C_{S_2T}^A - C_{S_2T}^B\right) = P\left(C_T^B - C_T^A\right) \tag{11.2-2}$$

and the total concentration of solute binding sites is constant:

$$T_T = \left(C_{S_1T}^A + C_{S_2T}^A + C_T^A\right) + \left(C_{S_1T}^B + C_{S_2T}^B + C_T^B\right) \tag{11.2-3}$$

Solving the preceding equations for the one-way fluxes of S_1, in terms of the surface concentrations of S_1T and S_2T, we obtain

$$N_{S_1}^{A,B} \equiv PC_{S_1T}^A = \frac{(T_TP/2)C_{S_1}^A}{\kappa_1\left[1 + \left(C_{S_2}^A/\kappa_2\right)\right] + C_{S_1}^A} \tag{11.2-4}$$

$$N_{S_1}^{B,A} \equiv PC_{S_1T}^B = \frac{(T_TP/2)C_{S_1}^B}{\kappa_1\left[1 + \left(C_{S_2}^B/\kappa_2\right)\right] + C_{S_1}^B} \tag{11.2-5}$$

These fluxes are identical in form to the Michaelis–Menten rate equations for an enzyme reaction with competitive inhibition of S_1 by S_2 (Eq. 8.5-21). Notice that each of these one-way fluxes of S_1 is inhibited by a cis-concentration of solute S_2.

The net flux of S_1 is the difference between the one-way fluxes:

$$N_{S_1} = \frac{T_TP}{2}\left\{\frac{C_{S_1}^A}{\kappa_1\left[1 + \left(C_{S_2}^A/\kappa_2\right)\right] + C_{S_1}^A} - \frac{C_{S_1}^B}{\kappa_1\left[1 + \left(C_{S_2}^B/\kappa_2\right)\right] + C_{S_1}^B}\right\} \tag{11.2-6}$$

The direction of N_{S_1} relative to its concentration driving force $\left(C_{S_1}^A - C_{S_1}^B\right)$ is inferred from the static head condition obtained from Equation 11.2-6 by setting $N_{S_1} = 0$:

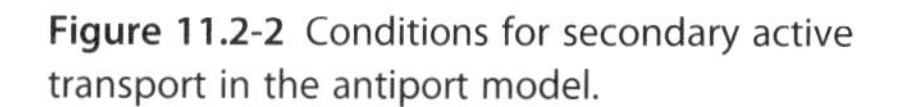

Figure 11.2-2 Conditions for secondary active transport in the antiport model.

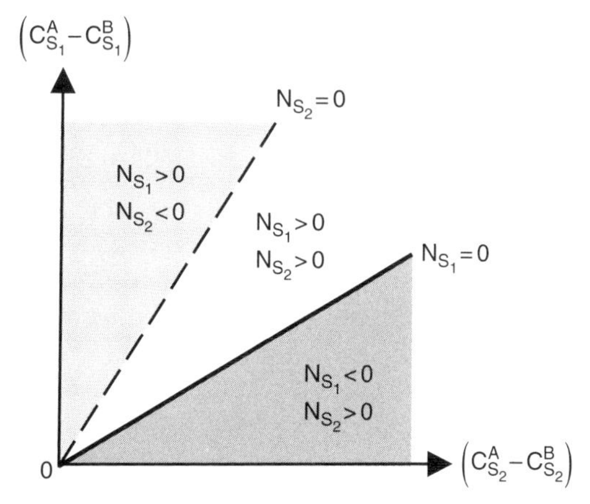

$$\left(C^A_{S_1}-C^B_{S_1}\right)=\frac{C^A_{S_1}}{\kappa_2+C^A_{S_2}}\left(C^A_{S_2}-C^B_{S_2}\right) \tag{11.2-7}$$

Provided that $C^A_{S_1}$ and $C^A_{S_2}$ are assigned constant values, a graph of this equation with $\left(C^A_{S_2}-C^B_{S_2}\right)$ on the abscissa and $\left(C^A_{S_1}-C^B_{S_1}\right)$ on the ordinate is a straight line with a positive slope $C^A_{S_1}/\left(\kappa_2+C^A_{S_2}\right)$. This is shown by the solid line in Figure 11.2-2 that focuses on the quadrant of the graph where the concentration driving forces $\left(C^A_{S_2}-C^B_{S_2}\right)$ and $\left(C^A_{S_1}-C^B_{S_1}\right)$ are both positive. Anywhere above the solid line, $\left(C^A_{S_1}-C^B_{S_1}\right)$ is larger than the static head condition for zero flux so that N_{S_1} must be positive. Below the line, $\left(C^A_{S_1}-C^B_{S_1}\right)$ is smaller than the static head so that N_{S_1} is negative. Thus, in the dark gray region of the graph, $\left(C^A_{S_1}-C^B_{S_1}\right)>0$ while $N_{S_1}<0$.

Because there is symmetry between the transport of the two solutes, N_{S_2} can be obtained from Equation 11.2-6 by interchanging the subscripts 1 and 2 on all the variables and parameters in the equation:

$$N_{S_2}=\frac{T_T P}{2}\left\{\frac{C^A_{S_2}}{\kappa_2\left[1+\left(C^A_{S_1}/\kappa_1\right)\right]+C^A_{S_2}}-\frac{C^B_{S_2}}{\kappa_2\left[1+\left(C^B_{S_1}/\kappa_1\right)\right]+C^B_{S_2}}\right\} \tag{11.2-8}$$

Setting $N_{S_2}=0$, the static head condition for S_2 is

$$\left(C^A_{S_1}-C^B_{S_1}\right)=\frac{\kappa_1+C^A_{S_1}}{C^A_{S_2}}\left(C^A_{S_2}-C^B_{S_2}\right) \tag{11.2-9}$$

This equation is graphed as the dashed line in Figure 11.2-2, again for the case where $C^A_{S_1}$ and $C^A_{S_2}$ are assigned constant values. Given that κ_1 and κ_2 are both positive quantities, this line has a positive slope of $\left(\kappa_1+C^A_{S_1}\right)/C^A_{S_2}$ which is greater than the slope $C^A_{S_1}/\left(\kappa_2+C^A_{S_2}\right)$ of the static head condition for N_{S_1}. Anywhere in the space to the right of the dashed line,

$\left(C_{S_2}^A - C_{S_2}^B\right)$ is greater than its static head value so that N_{S_2} is positive. To the left of the dashed line, N_{S_2} must be negative because $\left(C_{S_2}^A - C_{S_2}^B\right)$ is below its static head value. Thus, in the light gray region of the graph, $\left(C_{S_2}^A - C_{S_2}^B\right) > 0$ while $N_{S_2} < 0$.

To summarize, in the light gray region of the graph, S_1 is transported from surface A to surface B, which is in the direction of its concentration difference. Conversely, S_2 transport occurs from surface B to surface A, opposite to its concentration difference. This uphill movement of S_2 requiring energy provided by the downhill movement of S_1 is a form of active transport. In the dark gray region of the graph, it is the downhill S_2 transport from surface A to surface B that actively transports S_1 in opposite direction. The processes in both the dark gray and light gray regions are known as antiport since the fluxes of the two solutes are in opposite directions. The uphill transport of a solute during antiport is classified as secondary active transport to distinguish it from primary active transport that derives its energy from a stored chemical source.

Between the dark and light gray regions in the graph, there is neither antiport nor active transport since the fluxes of S_1 and S_2 both occur from surface A to surface B, in the same direction as their concentration driving forces.

11.3 Simulation of Tracer Experiments

11.3.1 Cotransport of a Labeled and Unlabeled Solute

Experiments that use a labeled form of a solute as a tracer can help us understand mechanisms as well test mathematical models of its transport dynamics. A tracer formed by attaching a radioactive atom or small fluorescent group to a solute molecule has physical–chemical properties that are virtually the same as its unlabeled counterpart. In addition, highly sensitive methods are available for monitoring concentrations of such tracers.

Figure 11.3-1 illustrates the simultaneous transport of unlabeled solute S and labeled solute S^* that compete for a single binding site on a transporter T at the two surfaces of a cell membrane. This model is similar to our previous description of antiport (Fig. 11.2-1) except that we now assume that ST and S^*T molecules are so similar that they both have the same dissociation constant κ and we allow the translocation constant P_2 for the free transporter molecule T to be different from P_1 for the solute–transporter complex.

The model is then governed by the following equations:

$$\kappa = \frac{C_S^A C_T^A}{C_{ST}^A} = \frac{C_{S^*}^A C_T^A}{C_{S^*T}^A} = \frac{C_S^B C_T^B}{C_{ST}^B} = \frac{C_{S^*}^B C_T^B}{C_{S^*T}^B} \tag{11.3-1}$$

$$P_1\left(C_{ST}^A - C_{ST}^B\right) + P_1\left(C_{S^*T}^A - C_{S^*T}^B\right) = P_2\left(C_T^B - C_T^A\right) \tag{11.3-2}$$

$$T_T = \left(C_{ST}^A + C_{S^*T}^A + C_T^A\right) + \left(C_{ST}^B + C_{S^*T}^B + C_T^B\right) \tag{11.3-3}$$

that lead to a net efflux of S^* and S are given by

$$N_{S^*} = P_1\left(C_{S^*T}^A - C_{S^*T}^B\right) = \frac{T_T P_1\left[\left(P_1 \underset{\sim}{C}_S^B + P_2\kappa\right)C_{S^*}^A - \left(P_1 \underset{\sim}{C}_S^A + P_2\kappa\right)C_{S^*}^B\right]}{\underset{\sim}{C}_S^A\left[2P_1 \underset{\sim}{C}_S^B + (P_1 + P_2)\kappa\right] + \kappa\left[(P_1 + P_2)\underset{\sim}{C}_S^B + 2P_2\kappa\right]} \tag{11.3-4}$$

Figure 11.3-1 Cotransport of labeled solute S* and unlabeled solute S.

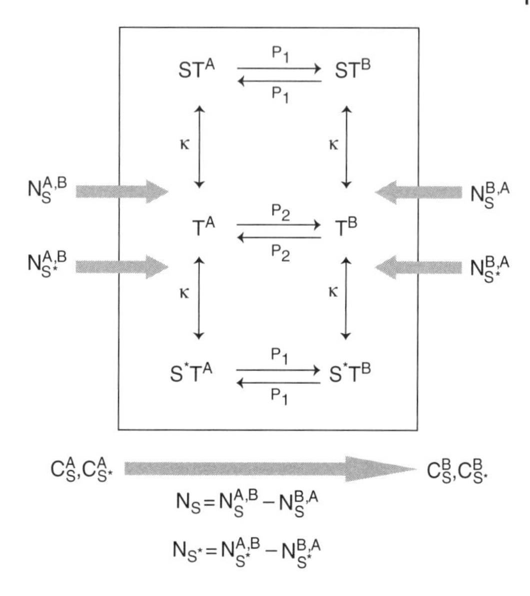

$$N_S = P_1\left(C_{ST}^A - C_{ST}^B\right) = \frac{T_T P_1\left[\left(P_1 \underset{\sim}{C}_S^B + P_2\kappa\right)C_S^A - \left(P_1 \underset{\sim}{C}_S^A + P_2\kappa\right)C_S^B\right]}{\underset{\sim}{C}_S^A\left[2P_1\underset{\sim}{C}_S^B + (P_1+P_2)\kappa\right] + \kappa\left[(P_1+P_2)\underset{\sim}{C}_S^B + 2P_2\kappa\right]} \tag{11.3-5}$$

Here, $\underset{\sim}{C}_S^A \equiv C_S^A + C_{S^*}^A$ and $\underset{\sim}{C}_S^B \equiv C_S^B + C_{S^*}^B$ represent the summed concentrations of labeled and unlabeled solute molecules on the two membrane surfaces. Frequently, the concentration of tagged solute is much less than untagged solute ($\underset{\sim}{C}_S^A \approx C_S^A$ and $\underset{\sim}{C}_S^B \approx C_S^B$), and the translocation rate constants of the unoccupied and occupied receptors sites are almost the same ($P_1 \approx P_2 \equiv P$). The overall flux equations then reduce to the simpler forms:

$$N_{S^*} = \frac{T_T P}{2}\left(\frac{C_{S^*}^A}{C_S^A + \kappa} - \frac{C_{S^*}^B}{C_S^B + \kappa}\right) \tag{11.3-6}$$

$$N_S = \frac{T_T P}{2}\left(\frac{C_S^A}{C_S^A + \kappa} - \frac{C_S^B}{C_S^B + \kappa}\right) \tag{11.3-7}$$

Example 11.3-1 Trans-Stimulation of Glucose Across Red Cells

Studies of glucose transport across the human red cell membrane have used a combination of radiolabeled glucose S* and unlabeled glucose S. Because of their almost identical molecular structure, it is reasonable to assume that S* and S compete for the same binding site on their hexose transporter T.

Miller and coworkers (1968a, b) studied the initial trans-stimulation of glucose efflux from red cells. The cells were preloaded with a known S* concentration and no S. They were then resuspended in a medium containing either no S* and no S (series I) or no S and an S* concentration equal to

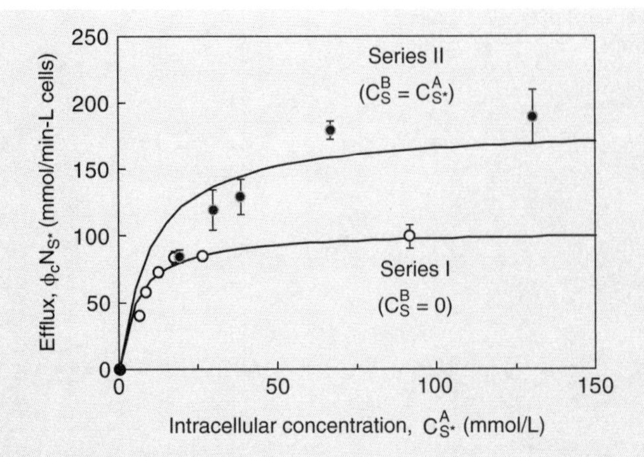

Figure 11.3-2 Trans-stimulation of labeled glucose transport.

the intracellular S^* concentration (series II). Using superscripts A and B to designate intracellular and extracellular membrane surfaces, respectively, this means that initially $C_{S^*}^A > 0$, $C_S^A = 0$, and $C_S^B = C_{S^*}^B = 0$ in series I experiments. In series II experiments, $C_{S^*}^A > 0$, $C_S^A = 0$, $C_S^B = 0$, and $C_{S^*}^B = C_{S^*}^A$.

In both series of experiments, the initial efflux of label from the cells was deduced from the drop in their radioactivity immediately following resuspension in the test medium. The resulting data were reported as the tracer efflux per liter of the packed cells (Fig. 11.3-2). The purpose of this problem is to see how well our competitive inhibition model simulates these data.

Solution

For the series I experiments, the summed solute concentrations are $\underset{\sim}{C}_S^A = C_{S^*}^A$ and $\underset{\sim}{C}_S^B = 0$, and the initial efflux of S^* according to Equation 11.3-4 is

$$[N_{S^*}]_I = \frac{(T_T P_1 P_2) C_{S^*}^A}{(P_1 + P_2) C_{S^*}^A + 2\kappa P_2} \tag{11.3-8}$$

Representing the surface-to-volume ratio of the cells by ϕ_c allows us to express the initial efflux of labeled solute per unit cell volume as $\phi_c N_{S^*}$. Equation 11.3-8 can now be rewritten as

$$[\phi_c N_{S^*}]_I = \left[\left(\frac{P_1 + P_2}{\phi_c T_T P_1 P_2} \right) + \frac{1}{C_{S^*}^A} \left(\frac{2\kappa}{\phi_c T_T P_1} \right) \right]^{-1} \tag{11.3-9}$$

where the $\phi_c T_T$ product represents the transporter concentration per unit cell volume. Performing a nonlinear least-squares regression of the series I data to this equation, we obtain the two parameter estimates $(P_1 + P_2)/\phi_c T_T P_1 P_2 = 0.00964\ \text{min/mM}$ and $2\kappa/(\phi_c T_T P_1) = 0.0544\ \text{min}$.

For the series II experiments, the summed solute concentrations are $\underset{\sim}{C}_S^A = \underset{\sim}{C}_S^B = C_{S^*}^A$, and the initial efflux is

$$[N_{S^*}]_{II} = \frac{(T_T P_1/2)\left(P_2 + P_1 \underset{\sim}{C}_{S^*}^A/\kappa \right) C_{S^*}^A}{\left[(P_1 + P_2) + P_1 \underset{\sim}{C}_{S^*}^A/\kappa \right] C_{S^*}^A + \kappa P_2} \tag{11.3-10}$$

Restating this equation in terms of the efflux per unit cell volume and substituting the parameter values found from the series I data, we get

$$[\phi_c N_{S^*}]_{II} = \frac{1+\left(\frac{P_1}{P_2\kappa}\right)C^A_{S^*}}{0.0193+\left[0.0193\left(\frac{P_1}{P_2\kappa}\right)-0.0544\left(\frac{P_1}{P_2\kappa}\right)^2\right]C^A_{S^*}+\frac{0.0544}{C^A_{S^*}}} \qquad (11.3\text{-}11)$$

A nonlinear least-squares regression of the series II data to this equation results in a parameter estimate of $P_1/P_2\kappa = 0.254\ \text{mM}^{-1}$.

The curves in Figure 11.3-2 are the flux behaviors predicted by the two regressions. The model simulation of the series I experiments that used two adjustable parameters agrees quite well with the data. The simulation of the series II experiments properly predicts a trans-stimulation of S^* efflux by the presence of extracellular S (i.e., the series II data is above the series I data). However, using only one extra adjustable model parameter, the series II simulation does not fit the data as well as the series I simulations.

Combining the three parameter estimates from the two regressions, we further get

$$\begin{aligned} \kappa &= \frac{(0.0544)}{2(0.00964)-0.254(0.0544)} = 9.95\ \text{mM} \\ P_1/P_2 &= 0.254(9.95) = 2.53 \\ \phi_c T_T P_1 &= 2(9.95)/0.0544 = 366\ \text{mM/min} \end{aligned} \qquad (11.3\text{-}12\text{a-c})$$

With $P_1/P_2 > 1$, this result indicates that the inherent translocation rate of the occupied transporter site is much greater than that of the unoccupied site. In other words, translocation of the hexose transporter is more likely to occur once its binding site becomes occupied.

Example 11.3-2 Antiport of Glucose Across Red Cells

Baker and Widdas (1973) observed the dynamic trans-stimulation of radiolabeled glucose influx across the red cell membrane by unlabeled glucose. After preloading cells with intracellular concentrations of $C^A_S(0) = 76\text{mM}$ unlabeled glucose and $C^A_{S^*}(0) = 0$ labeled glucose, the cells were resuspended in an isotonic solution that contained $C^B_{S^*}(0) = 4\text{mM}$ labeled glucose and $C^B_S(0) = 0$ unlabeled glucose. Thereafter, the concentration $C^A_{S^*}(t)$ of labeled glucose in the intracellular fluid was measured at several time points.

As shown by the data points in the left graph of Figure 11.3-3, labeled glucose accumulated in the cells during the first minute of the experiment, reaching a peak concentration that was ten times its extracellular concentration. Thus, during this initial phase of the experiment, labeled glucose was actively transported against its concentration driving force. As the experiment proceeded, the intracellular concentration of label continually decreased to reach equilibrium with its extracellular concentration. The purpose of this example is to determine whether our cotransport model of S^* and S exchange can anticipate this transient antiport behavior.

Solution

We assume the cell contents are well mixed, and we take N_{S^*} and N_S to represent effluxes from the cell interior A to the extracellular fluid B. Material balances on the intracellular concentrations of S and S^* are given by

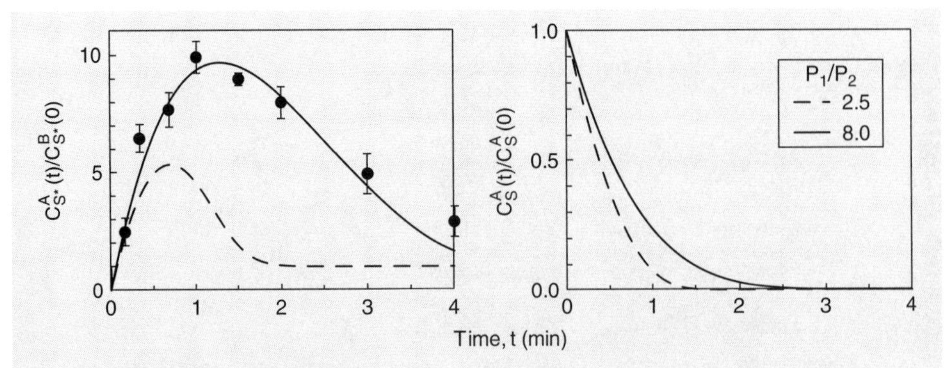

Figure 11.3-3 Antiport of labeled glucose.

$$\frac{d(V_c C^A_{S^*})}{dt} = -S_c N_{S^*}, \quad \frac{d(V_c C^A_S)}{dt} = -S_c N_S \tag{11.3-13a, b}$$

If the volume V_c and surface area S_c of the individual cells are constant, then their surface-to-volume ratio ϕ_c is constant and these equations become

$$\frac{dC^A_{S^*}}{dt} = -\phi_c N_{S^*}, \quad \frac{dC^A_S}{dt} = -\phi_c N_S \tag{11.3-14a, b}$$

To complete the model, we need to relate N_{S^*} and N_S to their concentration driving forces. Because these experiments were carried out on suspensions with a small volume fraction of cells, the composition of extracellular fluid B remained essentially constant at its initial condition. Consequently, $C^B_{S^*}(t) \approx C^B_{S^*}(0) = 4$ mM and $C^B_S(t) \approx C^B_S(0) = 0$. This allows us to reduce Equations 11.3-4 and 11.3-5 to

$$N_{S^*} = \frac{T_T P_1 \left[C^A_{S^*} - (P_1/P_2\kappa) C^B_{S^*} C^A_S - C^B_{S^*}\right]}{\left[1 + (P_1/P_2) + 2(P_1/P_2\kappa) C^B_{S^*}\right] \left(C^A_{S^*} + C^A_S\right) + \left[2\kappa + (P_1/P_2) C^B_{S^*} + C^B_{S^*}\right]} \tag{11.3-15}$$

$$N_S = \frac{T_T P_1 \left[1 + (P_1/P_2\kappa) C^B_{S^*}\right] C^A_S}{\left[1 + (P_1/P_2) + 2(P_1/P_2\kappa) C^B_{S^*}\right] \left(C^A_{S^*} + C^A_S\right) + \left[2\kappa + (P_1/P_2) C^B_{S^*} + C^B_{S^*}\right]} \tag{11.3-16}$$

Combining Equations 11.3-14, 11.3-15, and 11.3-16, we obtain

$$\frac{dC^A_{S^*}}{dt} = \frac{\phi_c T_T P_1 \left[(P_1/P_2\kappa) C^B_{S^*} C^A_S + C^B_{S^*} - C^A_{S^*}\right]}{\left[1 + (P_1/P_2) + 2(P_1/P_2\kappa) C^B_{S^*}\right] \left(C^A_{S^*} + C^A_S\right) + \left[2\kappa + (P_1/P_2) C^B_{S^*} + C^B_{S^*}\right]} \tag{11.3-17}$$

$$\frac{dC^A_S}{dt} = \frac{-\phi_c T_T P_1 \left[1 + (P_1/P_2\kappa) C^B_{S^*}\right] C^A_S}{\left[1 + (P_1/P_2) + 2(P_1/P_2\kappa) C^B_{S^*}\right] \left(C^A_{S^*} + C^A_S\right) + \left[2\kappa + (P_1/P_2) C^B_{S^*} + C^B_{S^*}\right]} \tag{11.3-18}$$

that have initial conditions $C^A_{S^*}(0) = 0$ and $C^A_S(0) = 76$ mM. The dashed curves in Figure 11.3-3 were generated by a numerical solution to these equations with the parameter values that we found in Example 11.3-1: $\kappa = 9.95$ mM/min, $P_1/P_2 = 2.53$, and $\phi_c T_T P_1 = 366$ mM/(L-min). These results are expressed as intracellular concentrations, $C^A_{S^*}(t)$ and $C^A_S(t)$ relative to their respective starting values, $C^B_{S^*}(0) = 4$ mM and $C^A_S(0) = 76$ mM.

Although the model simulation exhibits the dynamic trends of the $C^A_{S^*}(t)/C^B_{S^*}(0)$ data, it underestimates the magnitude and the appearance time of the peak value. This could be due to an osmotic shrinking of red cell volume occurring with the loss of intracellular glucose. It is also possible that the parameters deduced in Example 11.3-1 are not appropriate for the cell population used in the present example problem. By increasing P_1/P_2 from a value of 2.53–8.00, the simulation shown by the solid curve in the left graph matches the data quite well.

11.3.2 Inhibition of Carrier-Mediated Transport

One tool for investigating the mechanism of carrier-mediated transport is to observe the effect of a molecule that reversibly inhibits the binding of a solute to the transporter protein. The analysis of such experiments depends on whether (i) the inhibition is competitive or noncompetitive, (ii) the binding kinetics are the same on both membrane surfaces, and (iii) the inhibitor itself is translocated by the transporter protein. We will consider a class of tracer experiments in which the transport of labeled and unlabeled solute is affected by a nontranslocated, competitive inhibitor "I" that is fixed to the inside surface of cells and therefore is not transported to the outside surface of the cells.

Such experiments can be analyzed with an extension of the cotransport tracer model (Fig. 11.3-1) that accounts for the binding of I to T to form complex IT on an intracellular surface A. We assume that I, T, and IT are in equilibrium with a dissociation constant given by

$$\kappa_I = \frac{C_I C_T^A}{C_{IT}} \tag{11.3-19}$$

where C_I and C_{IT} are the molar concentrations of I and IT associated with surface A. We also assume that I has no influence on the translocation constant P and the dissociation constant κ. Although Equations 11.3-1 and 11.3-2 are still applicable to the new model, the total transporter concentration defined by Equation 11.3-3 must be modified to include C_{IT}:

$$T_T = \left(C_{ST}^A + C_{S^*T}^A + C_T^A + C_{IT}\right) + \left(C_{ST}^B + C_{S^*T}^B + C_T^B\right) \tag{11.3-20}$$

Solving Equations 11.3-1, 11.3-2, 11.3-19, and 11.3-20 for $C_{S^*T}^A$ and $C_{S^*T}^B$ allows us to express the efflux equation for S^* in terms of $C_{S^*}^A$ and $C_{S^*}^B$:

$$N_{S^*} = \frac{T_T P_1 \left[\left(P_1 \underset{\sim}{C}_S^B + P_2\kappa\right) C_{S^*}^A - \left(P_1 \underset{\sim}{C}_S^A + P_2\kappa\right) C_{S^*}^B \right]}{\underset{\sim}{C}_S^A \left[2P_1 \underset{\sim}{C}_S^B + (P_1 + P_2)\kappa\right] + \kappa\left[(P_1 + P_2)\underset{\sim}{C}_S^B + 2P_2\kappa\right] + (C_I\kappa/\kappa_I)\left(P_1 \underset{\sim}{C}_S^B + P_2\kappa\right)} \tag{11.3-21}$$

When little inhibitor is present or the inhibitor has a large dissociation constant, $C_I/\kappa \to 0$ and this equation reduces to Equation 11.3-4. Otherwise, the inhibitor increases the denominator of this expression so that tracer transport is reduced.

A useful way of estimating parameters in this model is the equilibrium exchange experiment in which unlabeled solute is maintained at the same concentration in the intracellular and extracellular fluids, while a small concentration of labeled solute is used to observe the uniport of tracer. We can simplify Equation 11.3-21 for equilibrium exchange by defining $\underset{\sim}{C}_S^A = \underset{\sim}{C}_S^B = C_S$ and noting that $C_{S^*}^A \ll C_S^A$ and $C_{S^*}^B \ll C_S^B$:

$$N_{S^*} = \frac{T_T P_1/2}{\dfrac{(P_1/P_2\kappa)C_S^2 + (1 + P_1/P_2)C_S + \kappa}{(P_1/P_2\kappa)C_S + 1} + \left(\dfrac{\kappa}{2\kappa_I}\right)C_I}\left(C_{S^*}^A - C_{S^*}^B\right) \tag{11.3-22}$$

For experiments in a dilute cell suspension, $C_{S^*}^B$ is essentially constant.

Example 11.3-3 Inhibition of Glucose Uniport Across Red Cells

Baker and Widdas (1973) also studied the effect of ethylidene glucose, a competitive nontranslocating inhibitor, on the equilibrium exchange of radiolabeled glucose across the human red cell membrane. In experiments designed to restrict inhibitor to the inside membrane surface, red cells were first equilibrated with a solution containing a concentration C_S of unlabeled glucose, concentration C_I of inhibitor, and a very small concentration C_{S^*} of radiolabeled glucose. The cells were then resuspended in a medium containing the same concentration C_S of unlabeled glucose in the absence of S^* and I. This suspension contained less than one volume percent of cells.

Due to the loss of intracellular tracer to the suspending medium, the cell radioactivity was observed to decrease in an exponential manner equivalent to a diminishing S^* concentration of

$$C_{S^*}^A = C_{S^*}^A(0)\exp(-P^* t) \tag{11.3-23}$$

The $1/P^*$ data points in Figure 11.3-4 were obtained in experiments carried out at different C_I, while C_S was fixed at either 10 or 20 mM. We want to determine whether our inhibition model is consistent with these equilibrium exchange data and, if so, we wish to estimate κ_I.

Solution

Assuming that the cell contents are well mixed, we perform a material balance on intracellular S^* and get

$$\frac{d(V_c C_{S^*}^A)}{dt} = -S_c N_{S^*} \tag{11.3-24}$$

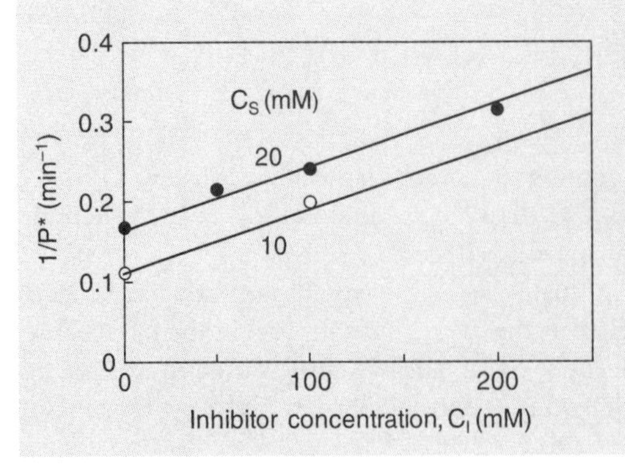

Figure 11.3-4 Inhibition of glucose uniport through red cells by ethylidene glucose.

where V_c is the cell volume, and S_c is the cell surface area. Representing the surface-to-volume ratio of a cell as ϕ_c, Equation 11.3-24 becomes

$$\frac{dC_{S^*}^A}{dt} = -\phi_c N_{S^*} \tag{11.3-25}$$

Because the cell suspension is so dilute, $C_{S^*}^B$ remains close to its initial value of zero, and Equation 11.3-22 can be approximated as

$$N_{S^*} = \frac{T_T P_1}{2}\left[\frac{(P_1/P_2\kappa)C_S^2 + (1 + P_1/P_2)C_S + \kappa}{(P_1/P_2\kappa)C_S + 1} + \left(\frac{\kappa}{2\kappa_I}\right)C_I\right]^{-1} C_{S^*}^A \tag{11.3-26}$$

Given the initial concentration $C_{S^*}^A(0)$, Equations 11.3-25 and 11.3-26 combined have an exponential solution given by Equation 11.3-23 where

$$P^* = \frac{\phi_c T_T P_1}{2}\left[\frac{(P_1/P_2\kappa)C_S^2 + (1 + P_1/P_2)C_S + \kappa}{(P_1/P_2\kappa)C_S + 1} + \left(\frac{\kappa}{2\kappa_I}\right)C_I\right]^{-1} \tag{11.3-27}$$

By inverting this equation, we obtain a linear relation between the inverse of P^* and the solute and inhibitor concentrations:

$$\frac{1}{P^*} = \left(\frac{\kappa}{\phi_c T_T P_1 \kappa_I}\right)C_I + \left(\frac{2}{\phi_c T_T P_1}\right)\left[\frac{(P_1/P_2\kappa)C_S^2 + (1 + P_1/P_2)C_S + \kappa}{(P_1/P_2\kappa)C_S + 1}\right] \tag{11.3-28}$$

Employing the values of $\kappa = 9.95$ mM, $P_1/P_2 = 2.53$, and $\phi_c T_T P_1 = 366$ mM/min from Example 11.3-1, we performed a least-squares regression of this relation to the P^* data by employing κ_I as an adjustable parameter. The results shown by the lines in Figure 11.3-3 provide an good simulation of the data points when $\kappa_I = 34.0$ mM. This dissociation constant is three times the value of κ, indicating that ethylidene glucose binds to the glucose transporter more weakly than glucose itself.

11.4 Primary Active Transport

11.4.1 A Model of Primary Active Transport

We can modify our model of facilitated transport to illustrate how primary active transport of a single-solute S can occur by coupling translocation of a carrier-solute complex to chemical reaction with a high-energy substrate such as ATP. Consider a carrier protein that, in addition to its binding site for solute, incorporates an active site for ATP hydrolysis and an amino acid side group that is simultaneously phosphorylated.

The carrier exists in three states: free carrier protein T; a complex ST consisting of S bound to the carrier; and a solute–transporter complex STP in which the amino acid side group has been phosphorylated (Fig. 11.4-1). Although T spontaneously translocates across the membrane, ST does not translocate until STP is formed. For simplicity, we let the translocation rate constant P be the same for both T and STP. As a consequence of fast reaction processes, we assume that T and S are in equilibrium with ST with a dissociation constant κ_1 that is the same on the two membrane surfaces. We also assume that the relative concentrations of ST

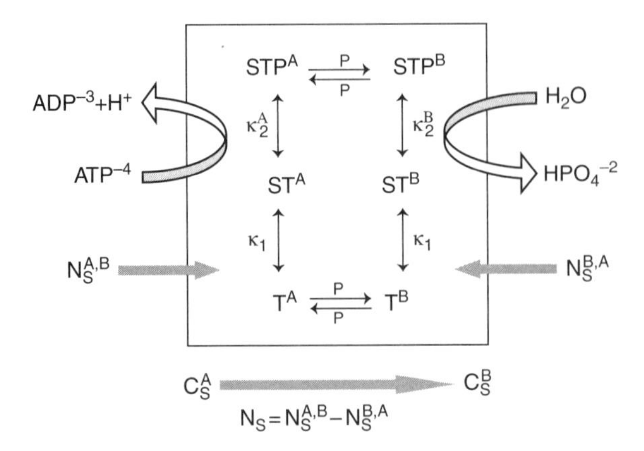

Figure 11.4-1 Active transport by coupling to an exergonic chemical reaction (Rx).

and STP can be expressed by equilibrium constants κ_2^A and κ_2^B, which, as we will show in the succeeding text, must have different values on the two membrane surfaces to produce active transport.

The reaction equilibria for the three forms of the transporter present in this process are given by

$$\kappa_1 = \frac{C_S^A C_T^A}{C_{ST}^A} = \frac{C_S^B C_T^B}{C_{ST}^B}, \quad \kappa_2^A = \frac{C_{STP}^A}{C_{ST}^A}, \quad \kappa_2^B = \frac{C_{STP}^B}{C_{ST}^B} \tag{11.4-1a-c}$$

With transporter confined to the membrane, we also know that

$$P\left(C_T^A - C_T^B\right) = P\left(C_{STP}^B - C_{STP}^A\right) \tag{11.4-2}$$

$$T_T = \left(C_T^A + C_{ST}^A + C_{STP}^A\right) + \left(C_T^B + C_{ST}^B + C_{STP}^B\right) \tag{11.4-3}$$

Solving these equations for the one-way fluxes of S, we obtain

$$N_S^{A,B} = \frac{T_T P \kappa_2^A \left(\kappa_2^B C_S^B + \kappa_1\right) C_S^A}{\left(\kappa_2^A + \kappa_2^B + 2\kappa_2^A \kappa_2^B\right) C_S^A C_S^B + \kappa_1\left(1 + 2\kappa_2^A\right) C_S^A + \kappa_1\left(1 + 2\kappa_2^B\right) C_S^B + 2\kappa_1^2} \tag{11.4-4}$$

$$N_S^{B,A} = \frac{T_T P \kappa_2^B \left(\kappa_2^A C_S^A + \kappa_1\right) C_S^B}{\left(\kappa_2^A + \kappa_2^B + 2\kappa_2^A \kappa_2^B\right) C_S^A C_S^B + \kappa_1\left(1 + 2\kappa_2^A\right) C_S^A + \kappa_1\left(1 + 2\kappa_2^B\right) C_S^B + 2\kappa_1^2} \tag{11.4-5}$$

The net flux of solute S is then given by

$$N_S = \frac{T_T P \kappa_1 \left(\kappa_2^A C_S^A - \kappa_2^B C_S^B\right)}{\left(\kappa_2^A + \kappa_2^B + 2\kappa_2^A \kappa_2^B\right) C_S^A C_S^B + \kappa_1\left(1 + 2\kappa_2^A\right) C_S^A + \kappa_1\left(1 + 2\kappa_2^B\right) C_S^B + 2\kappa_1^2} \tag{11.4-6}$$

To determine the conditions under which primary active transport of a solute can occur, we set $N_S = 0$ to obtain the static head condition:

$$\left(C_S^A - C_S^B\right) = \left(\frac{\kappa_2^B - \kappa_2^A}{\kappa_2^B}\right) C_S^A \tag{11.4-7}$$

This equation is plotted in Figure 11.4-2 where the solid line with a negative slope applies when $\kappa_2^B < \kappa_2^A$ and the dashed line with a positive slope holds when $\kappa_2^B > \kappa_2^A$. Above the solid

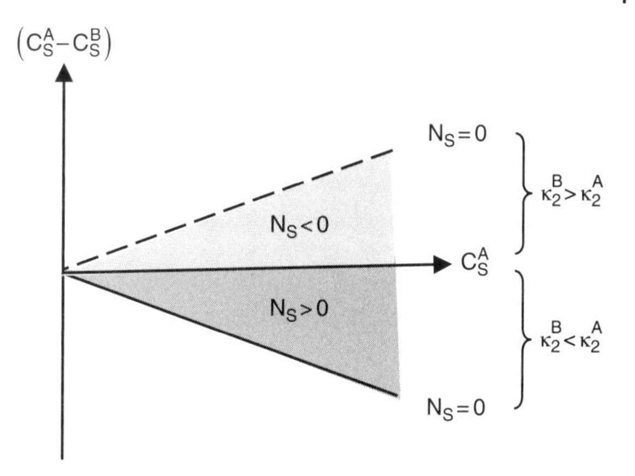

Figure 11.4-2 Conditions for primary active transport.

line, $(C_S^A - C_S^B)$ is greater than its static head value so that $N_S > 0$. Between the solid line and the abscissa, $(C_S^A - C_S^B) < 0$ while $N_S > 0$. Thus, in the dark gray region of the graph where $\kappa_2^B < \kappa_2^A$, solute S is actively transported from surface A to surface B. By similar reasoning, active transport of S from surface B to surface A occurs in the light gray region of the graph when $\kappa_2^B > \kappa_2^A$.

11.4.2 ATP Concentration Constraint

Referring again to Figure 11.4-1, the phosphorylation of ST at surface A by ATP entering from external solution is given by the reaction:

$$ST + ATP^{-4} \rightleftharpoons STP^{-2} + ADP^{-3} + H^{+}; \quad \kappa_{ST} \equiv \left(\frac{C_{STP}^A C_{ADP}^A C_H^A}{C_{ST}^A C_{ATP}^A} \right) \tag{11.4-8a,b}$$

After translocation of STP to surface B, dephosphorylation of STP occurs according to the reaction:

$$STP^{-2} + H_2O \rightleftharpoons ST + HPO_4^{-2}; \quad \kappa_{STP} \equiv \left(\frac{C_{ST}^B C_{HPO_4}^B}{C_{STP}^B} \right) \tag{11.4-9a,b}$$

Comparing Equations 11.4-1b,c–11.4-8b, and 11.4-9b, we see that the equilibrium coefficients are related:

$$\kappa_2^A = \left(\frac{C_{ATP}^A}{C_{ADP}^A C_H^A} \right) \kappa_{ST}, \quad \kappa_2^B = \left(C_{HPO_4}^B \right) \frac{1}{\kappa_{STP}} \tag{11.4-10a,b}$$

Together with these relationships for κ_2^A and κ_2^B, Equation 11.4-6 represents a model for primary active transport of solute S from surface A to surface B that depends directly on the concentrations of ATP and ADP in contact with membrane surface A.

Consider the case of a positive solute flux that occurs during active transport when $\kappa_2^B < \kappa_2^A$. Equations 11.4-10a,b then lead to the inequality:

$$\left(C_{HPO_4}^B \right) \frac{1}{\kappa_{STP}} < \left(\frac{C_{ATP}^A}{C_{ADP}^A C_H^A} \right) \kappa_{ST} \Rightarrow \frac{C_{ATP}^A}{C_{ADP}^A C_H^A C_{HPO_4}^B} > \frac{1}{\kappa_{ST} \kappa_{STP}} \tag{11.4-11a,b}$$

For each solute molecule that is transported between the two external solutions, one molecule of transporter undergoes the sum of the reactions given by Equations 11.4-8a and 11.4-9a:

$$ATP^{-4} + H_2O \rightleftharpoons ADP^{-3} + HPO_4^{-2} + H^+ \tag{11.4-12}$$

Thus, the combined phosphorylation and dephosphorylation of ST is equivalent to the complete hydrolysis of one molecule of ATP. The equilibrium constant κ_{ATP} for this overall reaction is equivalent to the product of the equilibrium constants κ_{ST} and κ_{STP} of the two individual reactions. The condition for active transport given by Equation 11.4-11b can now be written as

$$\frac{C_{ATP}^{A}}{C_{ADP}^{A} C_{HPO_4}^{B}} > \frac{C_{H}^{A}}{\kappa_{ATP}} \tag{11.4-13}$$

Under normal physiological conditions, the right side of this equation is essentially constant. This condition, equivalent to $\kappa_2^B < \kappa_2^A$, indicates that ATP concentration in excess of ADP concentration on membrane surface A must exceed HPO_4^- concentration on membrane surface B by a critical value.

11.4.3 Limiting Solute Flux

Corresponding to the schematic representation in Figure 11.4-1, Figure 11.4-3 shows the fluxes and chemical potentials associated with each step of our active transport model. By applying some concepts from irreversible thermodynamics, we will develop a constraint on the maximum transmembrane solute flux that can occur in this process.

Since we have assumed that reaction processes at the membrane surfaces are at equilibrium, the entropy generation function of the overall transport process depends only on the irreversible translocation steps occurring within the membrane. The net translocation flux of STP from surface A to surface B, N_{STP}, is driven by the chemical potential driving force $(\breve{\mu}_{STP}^{A} - \breve{\mu}_{STP}^{B})$, while the net translocational flux of T from surface B to surface A has an associated driving force $(\breve{\mu}_{T}^{B} - \breve{\mu}_{T}^{A})$. By analogy with Equation 6.5-4, the entropy generation function of carrier protein transport is given by sum of the flux-force products of T and STP:

$$\vartheta_m = N_{STP}\left(\breve{\mu}_{STP}^{A} - \breve{\mu}_{STP}^{B}\right) + N_T\left(\breve{\mu}_{T}^{B} - \breve{\mu}_{T}^{A}\right) \tag{11.4-14}$$

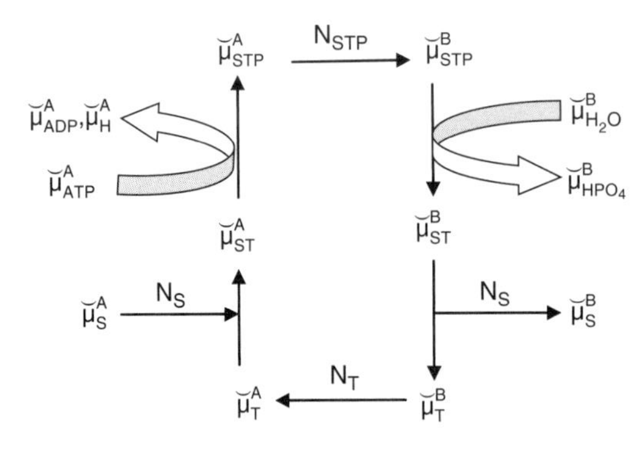

Figure 11.4-3 Fluxes and chemical potentials in a primary active transport model.

For the binding of S to T (Eq. 11.1-1), we infer statements of chemical equilibrium on each of the two membrane surfaces from Equation 5.1-8:

$$\breve{\mu}_{ST}^{A}-\breve{\mu}_{S}^{A}-\breve{\mu}_{T}^{A}=0,\quad \breve{\mu}_{ST}^{B}-\breve{\mu}_{S}^{B}-\breve{\mu}_{T}^{B}=0 \tag{11.4-15a,b}$$

For the phosphorylation of ST by ATP on surface A (Eq. 11.4-8a) and the dephosphorylation of STP on surface B (Eq. 11.4-9a), the equilibrium relations are

$$\begin{aligned}(\mu_{STP}^{A}+\mu_{ADP}^{A}+\mu_{H}^{A})-(\breve{\mu}_{ST}^{A}+\breve{\mu}_{ATP}^{A})&=0\\ \left(\breve{\mu}_{ST}^{B}+\breve{\mu}_{HPO_4}^{B}\right)-\left(\breve{\mu}_{STP}^{B}+\breve{\mu}_{H_2O}^{B}\right)&=0\end{aligned} \tag{11.4-16a,b}$$

Upon substitution of the four equilibrium equations into Equation 11.4-14, we obtain

$$\begin{aligned}\vartheta_m = {}&N_{STP}\left(\breve{\mu}_{ST}^{A}-\breve{\mu}_{ST}^{B}+\breve{\mu}_{ATP}^{A}-\breve{\mu}_{ADP}^{A}-\mu_{H}^{A}-\breve{\mu}_{HPO_4}^{B}+\breve{\mu}_{H_2O}^{B}\right)\\ &+N_T\left(\breve{\mu}_{S}^{A}-\breve{\mu}_{S}^{B}+\breve{\mu}_{ST}^{B}-\breve{\mu}_{ST}^{A}\right)\end{aligned} \tag{11.4-17}$$

With the transporter protein confined to but not accumulating in the membrane, $N_T = N_{STP}$ and this equation becomes

$$\vartheta_m = N_{STP}\left(\breve{\mu}_{ATP}^{A}-\breve{\mu}_{ADP}^{A}-\mu_{H}^{A}-\breve{\mu}_{HPO_4}^{B}+\breve{\mu}_{H_2O}^{B}\right)+N_{STP}\left(\breve{\mu}_{S}^{A}-\breve{\mu}_{S}^{B}\right) \tag{11.4-18}$$

Furthermore, S crosses the membrane when the STP complex breaks down to reform T. Thus, $N_{STP} = N_S$ and

$$\vartheta_m = N_S\left(\breve{\mu}_{S}^{A}-\breve{\mu}_{S}^{B}\right)+N_{STP}\left(\breve{\mu}_{ATP}^{A}-\breve{\mu}_{ADP}^{A}-\breve{\mu}_{H}^{A}-\breve{\mu}_{HPO_4}^{B}+\breve{\mu}_{H_2O}^{B}\right) \tag{11.4-19}$$

The N_{STP} flux must be equal to the molar ATP depletion per time per unit surface. We will represent this rate as $-R_{ATP}$. Following Equation 5.1-10, we identify the reaction affinity for ATP hydrolysis as $A_{ATP} \equiv -\left(\breve{\mu}_{ADP}^{B}+\breve{\mu}_{H}^{B}+\breve{\mu}_{HPO_4}^{A}-\breve{\mu}_{ATP}^{B}-\breve{\mu}_{H_2O}^{B}\right)$. Equation 11.4-19 can now be written as

$$\vartheta_m = N_S\left(\breve{\mu}_{S}^{A}-\breve{\mu}_{S}^{B}\right)-R_{ATP}A_{ATP} \tag{11.4-20}$$

This form of ϑ_m is more useful than Equation 11.4-14 because it incorporates only those variables that can, in principle, be observed in the fluids external to the membrane material. For a spontaneous process, $\vartheta_m \geq 0$ and Equation 11.4-20 can be restated as an inequality:

$$N_S\left(\breve{\mu}_{S}^{A}-\breve{\mu}_{S}^{B}\right) > R_{ATP}A_{ATP} \tag{11.4-21}$$

Since N_s, $-R_{ATP}$, A_{ATP}, and $\left(\breve{\mu}_{S}^{B}-\breve{\mu}_{S}^{A}\right)$ are all positive quantities during active transport from surface A to surface B, it is useful to rewrite Equation 11.4-21 as

$$N_S < \frac{(-R_{ATP})A_{ATP}}{\breve{\mu}_{S}^{B}-\breve{\mu}_{S}^{A}} \tag{11.4-22}$$

This result indicates that the maximum flux possible in this model of primary active transport is the ratio of the entropy generation rate by ATP hydrolysis, $-R_{ATP}A_{ATP}$, to the chemical potential difference that opposes the solute flux, $\left(\breve{\mu}_{S}^{B}-\breve{\mu}_{S}^{A}\right)$.

11.5 Electrical Effects on Ion Transport

In this section, we discuss the effect of electrical potential difference on ion transport across a membrane. In order to conserve charge, chemical reactions must be electrically balanced. Thus, the presence of charge on a transporter or solute–transporter complex does not affect reaction equilibrium coefficients (see Eqs. 5.1-6–5.1-8). Electrical charges can, however, affect the translocation rate constants by a process that we will model with transition state theory.

Figure 11.5-1 illustrates the energy-reaction coordinate diagram for the translocation of a transporter (or solute–transporter complex) X from surface A to surface B of a membrane. In completing this process, the transporter undergoes a change from structure X^A on surface A to structure X^B on surface B by passing through an activated state X^*. This reaction pathway, in the absence of electrical effects, is shown by the dashed curve in the figure. With no difference in electrical potential between surfaces A and B, the translocation rate factor can be expressed by Equation 8.2-25:

$$P_+^o = k_+ \exp\left(-\frac{\Delta G_+^*}{\mathcal{R}T}\right) \tag{11.5-1}$$

where the subscript "+" indicates that the direction of translocation is from surface A to surface B; k_+ is a translocation coefficient that depends on temperature; and ΔG_+^* is the standard free energy of the $X^A \rightarrow X^*$ transition.

We now impose an electrical potential difference $\Delta\psi$ across a membrane with a potential $+\Delta\psi/2$ on surface A and $-\Delta\psi/2$ on surface B. The alteration this imposes on the reaction path is shown by the solid curve in the figure. If X carries a charge z_x, the relative energy of X^A will increase by $z_x\mathcal{F}(\Delta\psi/2)$, thereby reducing the free energy for the $X^A \rightarrow X^*$ transition from ΔG_+^* to $\Delta G_+^* - \mathcal{F}z_X\Delta\psi/2$. The translocation rate constant is now given by

$$P_+ = k_+ \exp\left[-\frac{\left(\Delta G_+^* - \mathcal{F}z_X\Delta\psi/2\right)}{\mathcal{R}T}\right] \tag{11.5-2}$$

Thus, the relationship between the translocation rate constants in the presence and absence of a change in electrical potential is

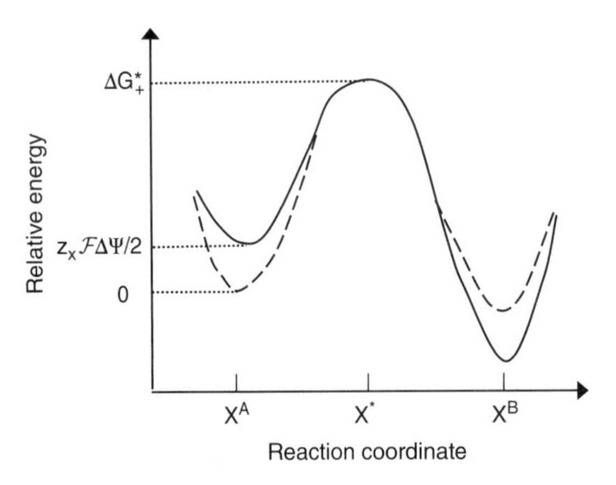

Figure 11.5-1 Translocation of transporter X from surface A to surface B of a membrane.

$$P_+ = \xi P_+^o \tag{11.5-3}$$

where ξ is given as

$$\xi = \exp\left(\frac{\mathcal{F} z_X \Delta\psi}{2\mathcal{R}T}\right) \tag{11.5-4}$$

When transporter X translocates in the opposite direction, between surface B and surface A, the imposition of the same potential drop between surface A and surface B will increase the activation energy of the $X^B \rightarrow X^*$ transition from ΔG_-^* to $\Delta G_-^* + \mathcal{F} z_X \Delta\psi/2$. In that case, the relationship between the translocation rate constant in the presence and absence of an electrical potential drop is

$$P_- = \xi^{-1} P_-^o \tag{11.5-5}$$

When z_X and $\Delta\psi$ have the same sign, then $\xi > 1$ so that P_+ is enhanced, whereas P_- is diminished relative to their electrically neutral values. When z_X and $\Delta\psi$ do not have the same sign, the opposite is true.

As illustrated in the following example, electrical effects can be incorporated into one of our previous carrier-mediated models by including the factor ξ.

Example 11.5-1 Electrical Effects During Uniport

An ionic solute S crosses a cell membrane from an intracellular space A to an extracellular space B by monovalent binding with a transporter T. Suppose that cells containing S at an intracellular concentration C_S^A are placed in a suspending medium that is free of S and that an intracellular-to-extracellular potential of $\psi_m = -100$ mV exists across the cell membranes.

Compare the initial efflux of S out of the cells for three different cases: (a) S and T are both uncharged; (b) S has a charge of +1 and T is uncharged; (c) S is uncharged and T has a charge of +1. In the absence of electrical effects, we will assume that the translocation constants of ST and T are equal and the same in both translocation directions.

Solution

This situation can be modeled by facilitated transport with the one-way fluxes given by Equations 11.1-9 and 11.1-10. Taking the difference between these equations to get the net flux, and then setting $C_S^B = 0$:

$$N_S = \frac{(P_1 P_{-2} T_T) C_S^A}{C_S^A (P_1 + P_{-2}) + \kappa (P_2 + P_{-2})} \tag{11.5-6}$$

where P_1 and P_2 are the translocation constants for ST and T, respectively, in the A → B direction, and P_{-2} is the translocation constant for T in the opposite direction (Fig. 11.1-1). To account for the effect of a charged solute or transporter, we need only correct these three translocation constants using

$$\xi = \exp\left[\frac{\mathcal{F} z_X \psi_m}{2\mathcal{R}T}\right] = \left\{\exp\left[\frac{(96500)(-0.100)}{2(8.31)(310)}\right]\right\}^{z_X} = (0.154)^{z_x} \tag{11.5-7}$$

For the translocation rate factors in the forward direction, we use Equation 11.5-3:

$$P_1 = 0.154^{+z_{ST}} P_1^o, \quad P_2 = 0.154^{+z_T} P_2^o \tag{11.5-8a,b}$$

and in the reverse direction, we apply Equation 11.5-5:

$$P_{-2} = 0.154^{-z_T} P_{-2}^o \tag{11.5-9}$$

Here, z_{ST} is the charge on ST, and z_T is the charge on T. Since we are assuming in the absence of charge that the translocation constants of ST and T are equal and the same in both directions, $P_1^o = P_2^o = P_{-2}^o \equiv P^o$. Combining Equations 11.5-8a,b, 11.5-9, and 11.5-6, we obtain

$$N_S = \frac{(T_T P^o) C_S^A}{(0.154^{+z_T} + 0.154^{-z_{ST}}) C_S^A + \left(0.154^{2z_T - z_{ST}} + 0.154^{-z_{ST}}\right)\kappa} \tag{11.5-10}$$

With the charge on ST equal to the sum of the charges on S and T, the fluxes for the three cases are

(a) $z_{ST} = z_T = 0$

$$N_S = \frac{0.500(T_T P^o) C_S^A}{C_S^A + \kappa} \tag{11.5-11}$$

(b) $z_{ST} = +1 + 0 = +1,\ z_T = 0$

$$N_S = \frac{0.133(T_T P^o) C_S^A}{C_S^A + 1.73\kappa} \tag{11.5-12}$$

(c) $z_{ST} = 0 + 1 = +1,\ z_T = +1$

$$N_S = \frac{0.150(T_T P^o) C_S^A}{C_S^A + \kappa} \tag{11.5-13}$$

Among these three cases, N_S is greatest for case (a) when both S and T are uncharged. The flux in case (b) is smaller than in case (a) because the rate constant P_1 governing the translocation of the ST complex toward surface B is reduced. Whereas the flux in case (c) is also diminished by the electrostatic reduction in P_1, there is an increase in P_{-2} relative to P_2 that enhances the recycling of positively charged T toward surface A. Because of this compensatory effect, the flux in case (c) is somewhat larger than in case (b).

References

Baker GF, Widdas WF. The asymmetry of the facilitated transfer system for hexoses in human red cells and the simple kinetics of a two component model. J Physiol. 1973; 231:143–165.

Miller DM. The kinetics of selective biological transport. III. Erythrocyte-monosaccharide transport data. Biophysical J. 1968a; 8:1329–1338.

Miller DM. The kinetics of selective biological transport. IV. Assessment of three carrier systems using the erythrocyte-monosaccharide transport data. Biophysical J. 1968b; 8:1339–1352.

Schultz SG. Basic Principles of Membrane Transport. Cambridge: Cambridge University Press; 1980, p 95–115, ch 6.

Stein WD. Channels, Carriers, and Pumps: An Introduction to Membrane Transport. New York: Academic Press; 1990, p 167.

Chapter 12

Mass Transfer Coefficients and Chemical Separation Devices

In this chapter, we discuss mass transport across the membranes and interfaces that separate fluid-filled compartments in biological systems and medical devices. To account for geometry and physical–chemical properties at interfaces, we introduce solute fluxes that incorporate an individual mass transfer coefficient. For interfacial and membrane transport occurring simultaneously, we develop a solute flux equation that combines individual mass transfer coefficients of the solute with its membrane permeability to arrive at an overall mass transfer coefficient. The use of mass transfer coefficients with material balances in the design and evaluation of blood oxygenators and hemodialyzers is illustrated at the end of the chapter.

12.1 Transport Through a Single Phase

12.1.1 Individual Mass Transfer Coefficient

In practical applications, it is essential to determine the rate at which a substance diffuses between the bulk phase of a fluid and its bounding surface. Commonly, forced convection primarily occurs parallel to the surface and transverse to diffusion, which mainly occurs perpendicular to the surface. Examples of such convection–diffusion processes include nutrient transport from flowing blood toward a capillary wall and O_2 transfer from a flowing gas phase toward a membrane surface in a blood oxygenator. In these situations, a solute concentration gradient that drives mass transport is compressed into a thin boundary layer or "unstirred layer" adjacent to the surface. Outside this layer in the bulk solution, the solute concentration is approximately uniform (Fig. 12.1-1). Thus, the major resistance to diffusion is governed by characteristics of the unstirred layer.

Biomedical Mass Transport and Chemical Reaction: Physicochemical Principles and Mathematical Modeling,
First Edition. James S. Ultman, Harihara Baskaran, and Gerald M. Saidel.

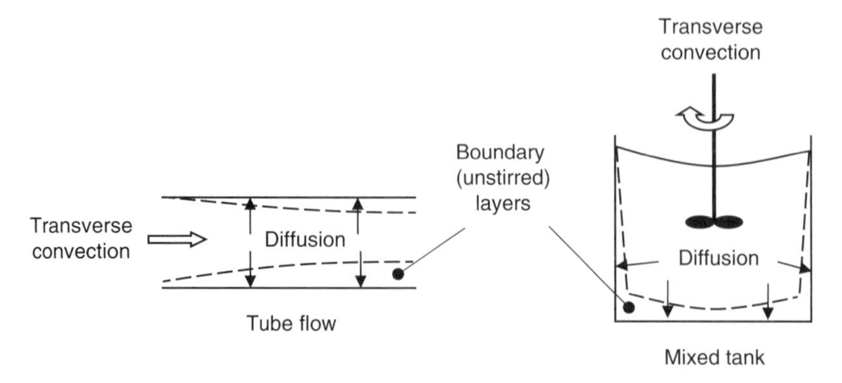

Figure 12.1-1 Simultaneous diffusion and transverse convection near a solid wall.

Based on empirical evidence, the local flux at which a solute s is transported from a point on a surface to the bulk solution can be expressed as

$$N_{s,loc} = k_{s,loc}(C_{so} - C_{sb}) \tag{12.1-1}$$

where C_{so} is the solute concentration at the surface, C_{sb} is the solute concentration in bulk solution, S is the area of the surface, and $k_{s,loc}$ is an individual mass transfer coefficient representing the conductance of mass through the unstirred layer. To compute the molar transport rate over a finite surface S, this expression must be integrated over S:

$$\dot{N}_s = \int_S N_{s,loc} dS = \int_S k_{s,loc}(C_{so} - C_{sb}) dS \tag{12.1-2}$$

Provided that the driving force for transport $(C_{so} - C_{sb})$ is uniform over the surface, this leads to the definition of a surface-averaged mass transfer coefficient k_s:

$$\dot{N}_s = k_s S(C_{so} - C_{sb}); \quad k_s \equiv \frac{1}{S}\int_S k_{s,loc} dS \tag{12.1-3a,b}$$

When dealing with transport of a gas dissolved in a liquid, it is useful to use a partial pressure driving force that is related to concentration driving force by the equilibrium relationship: $C_s = \alpha_s c_G^o p_s$. In addition, we express Equation 12.1-3a in terms of standard gas volumes transferred per unit time:

$$\dot{V}_s \equiv \frac{\dot{N}_s}{c_G^o} = \alpha_s k_s S(p_{so} - p_{sb}) \tag{12.1-4}$$

This equation is also valid for transport in a gas phase, in which case $\alpha_s = 1/\mathcal{R}Tc_G^o$.

As we illustrate in Chapter 16, k_s can be determined for any geometry and flow field of interest by solving multidimensional diffusion equations. For most practical problems, however, mass transfer coefficients based on more simple models can capture the essence of mass transfer rate across a boundary layer. Two steady-state models of this type are the stagnant film and penetration models.

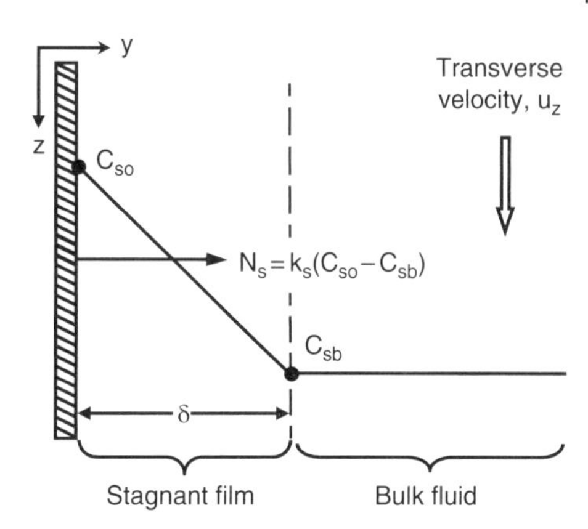

Figure 12.1-2 Stagnant film model of solute transport.

12.1.2 Stagnant Film Model

Figure 12.1-2 illustrates the diffusion of solute s at a flux N_s in the y direction perpendicular to a surface, while fluid is flowing at a velocity u_z in the z direction parallel to the surface. In addition to applying a steady (or pseudo-steady) state analysis, the stagnant film model assumes as follows: (i) fluid friction forces create a stagnant film of thickness δ adjacent to the surface, (ii) the bulk fluid external to the stagnant film is well mixed by convection so that transport is limited by diffusion through the stagnant film, and (iii) the film is sufficiently thin that curvature effects can be neglected.

Under these conditions, the solute flux for steady-state, one-dimensional diffusion through a stagnant film is given by an equation analogous to Equation 9.2-5. In terms of the variables in Figure 12.1-2, $N_{s,y} \to N_s$, $(y_2 - y_1) \to \delta$, $C_{s_1} \to C_{so}$, and $C_{s_2} \to C_{sb}$ so that the molar rate of solute transport across the film is

$$N_s = \left(\frac{\mathcal{D}_s}{\delta}\right)(C_{so} - C_{sb}) \tag{12.1-5}$$

A comparison of this equation to Equation 12.1-3a indicates how an individual mass transfer coefficient is related to the diffusion coefficient and the thickness of the stagnant film:

$$k_s \equiv \frac{\mathcal{D}_s}{\delta} \tag{12.1-6}$$

Since we do not know how δ depends on fluid properties or system geometry, the stagnant film model alone is not applicable for quantitative prediction, but can be used for empirical characterization.

12.1.3 Penetration Model

The penetration model explicitly accounts for a convective solute flux $N_{s,z}$ in the z direction in addition to a diffusive flux $N_{s,y}$ in the y direction. The material balance for this two-dimensional problem can be obtained by an extension of Equation 9.1-5, a relationship that applies to unsteady-state transport and reaction in the y direction only. For a nonreacting species s, this unidirectional equation reduces to

$$\frac{\partial C_s(y,t)}{\partial t} + \frac{\partial N_{s,y}(y,t)}{\partial y} = 0 \tag{12.1-7}$$

To account for transport also occurring in the z direction, we need only add the gradient of the flux component $N_{s,z}$ in that direction:

$$\frac{\partial C_s(y,z,t)}{\partial t} + \frac{\partial N_{s,y}(y,z,t)}{\partial y} + \frac{\partial N_{s,z}(y,z,t)}{\partial z} = 0 \tag{12.1-8}$$

If transient processes are sufficiently fast, then the system reaches a pseudo-steady state, and $\partial C_s/\partial t$ can be omitted from this equation:

$$\frac{\partial N_{s,y}}{\partial y} + \frac{\partial N_{s,z}}{\partial z} = 0 \tag{12.1-9}$$

With diffusion dominating in the y direction perpendicular to the fluid surface and convection dominating in the x direction parallel to the surface, the corresponding solute fluxes in a solution of constant mass density are

$$N_{s,y} = u_y C_s - \mathcal{D}_s \frac{\partial C_s}{\partial y} \approx -\mathcal{D}_s \frac{\partial C_s}{\partial y} \tag{12.1-10}$$

$$N_{s,z} = u_z C_s - \mathcal{D}_s \frac{\partial C_s}{\partial z} \approx u_z C_s \tag{12.1-11}$$

Substituting these simplified flux equations into Equation 12.1-9, we obtain

$$u_z \frac{\partial C_s}{\partial z} = \mathcal{D}_s \frac{\partial^2 C_s}{\partial y^2} \tag{12.1-12}$$

Specifying solute concentrations of C_{so} at the surface where $y = 0$ and C_{sb} in the entering flow stream where $z = 0$, the boundary conditions for this equation are

$$\begin{aligned} z = 0 \quad &: \quad C_s = C_{sb} \\ y = 0 \quad &: \quad C_s = C_{so} \\ y \to \infty &: \quad C_s = C_{sb} \end{aligned} \tag{12.1-13a-c}$$

The last condition states that far away from the surface, the solute concentration is not disturbed from its entering value at $z = 0$. Although this model can be solved for any steady-state velocity distribution $u_z(y,z)$, we consider the simplest case, namely, a constant velocity. This idealization is the most appropriate when friction at the fluid surface is very small as occurs when a gas flows over a liquid or solid surface.

We define a transit or contact time of an individual fluid element as

$$\theta \equiv \frac{z}{u_z} \tag{12.1-14}$$

As illustrated in Figure 12.1-3, θ increases while the element moves along the surface, allowing solute to diffuse farther into surrounding fluid. The locus of points at which solute concentration approaches its bulk solution value C_{sb} traces the boundary of the "penetration layer."

Converting Equation 12.1-12 and its boundary conditions from (y,z) variables to (θ,y) variables, we obtain

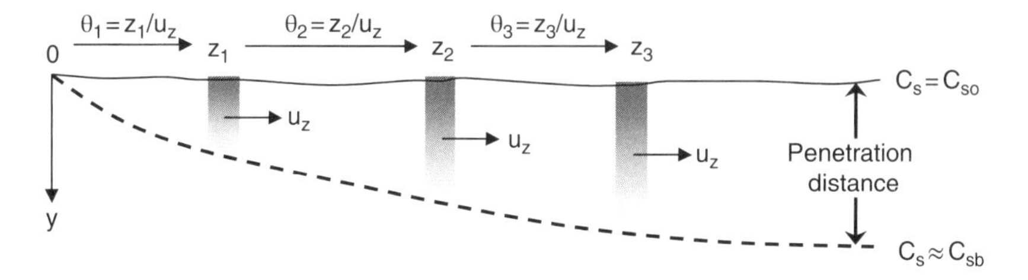

Figure 12.1-3 Penetration distance with increasing contact time (θ).

$$\frac{\partial C_s}{\partial \theta} = \mathcal{D}_s \frac{\partial^2 C_s}{\partial y^2} \tag{12.1-15}$$

and

$$\begin{aligned} \theta = 0 \quad &: \quad C_s = C_{sb} \\ y = 0 \quad &: \quad C_s = C_{so} \\ y \to \infty &: \quad C_s = C_{sb} \end{aligned} \tag{12.1-16a-c}$$

These mathematical relations are the same as Equations 9.5-1 and 9.5-2a-c that govern unsteady one-dimensional diffusion through a semi-infinite stagnant medium except that clock time t has been replaced by the residence time θ, and the undisturbed solute concentration $C_{s\infty}$ is now written as C_{sb}. Making these changes in Equation 9.5-8, we obtain the diffusion flux for the penetration model:

$$\left[N_{s,y}\right]_{y=0} = \sqrt{\frac{\mathcal{D}_s}{\pi\theta}}(C_{so} - C_{sb}) = \sqrt{\frac{\mathcal{D}_s u_z}{\pi z}}(C_{so} - C_{sb}) \tag{12.1-17}$$

This equation indicates that diffusion flux occurring in the y direction perpendicular to the surface decreases in the z direction along the surface. According to Equation 12.1-1, the local mass transfer coefficient is the ratio of this flux to the concentration driving force:

$$k_{s,loc} = \frac{\left[N_{s,y}\right]_{y=0}}{(C_{so} - C_{sb})} = \sqrt{\frac{\mathcal{D}_s u_z}{\pi z}} \tag{12.1-18}$$

Based on Equation 12.1-6, we can define a local penetration distance that is equivalent to a stagnant film thickness:

$$\delta_{loc} \equiv \frac{\mathcal{D}_s}{k_{s,loc}} = \sqrt{\frac{\pi \mathcal{D}_s z}{u_z}} \tag{12.1-19}$$

Thus, δ_{loc} increases with diffusion coefficient and downstream distance but decreases when the transverse velocity increases. The progressive increase in δ_{loc} with downstream distance z is an inherent property of a concentration boundary layer that is developing along a surface.

For a surface of length L, the spatially averaged mass transfer coefficient and its corresponding penetration distance are

$$k_s \equiv \frac{1}{L}\int_0^L \sqrt{\frac{\mathcal{D}_s u_z}{\pi z}} dz = \sqrt{\frac{2\mathcal{D}_s u_z}{\pi L}}, \quad \delta \equiv \frac{\mathcal{D}_s}{k_s} = \sqrt{\frac{\pi \mathcal{D}_s L}{2u_z}} \tag{12.1-20a,b}$$

12.1.4 Dimensional Analysis

Dimensionless equations are the most efficient means of expressing the relationship between k_s and the various material properties and geometric parameters of a particular transport process. For example, consider radial transport of a solute from the inner wall of a cylindrical tube into a flowing liquid as shown on the left side of Figure 12.1-1. There are seven parameters of the system that we expect to govern this process:

k_s [m/s]	Axially averaged mass transfer coefficient
$\mathcal{D}_s$ [m^2/s]	Diffusion coefficient for solute s
ρ [kg/m^3]	Mass density of the solution
μ [kg/(s-m)]	Shear viscosity of the solution
u [m/s]	Cross-sectionally averaged fluid velocity
d [m]	Tube diameter
L [m]	Tube length

In the absence of a mathematical model, we can utilize the Buckingham Pi method to determine the number of dimensionless groups necessary to describe this process. According to this method, a functional relationship can be written among *d* quantities that have *q* fundamental units by utilizing *d–q* dimensionless groups. The seven quantities that we have listed have three fundamental units (mass, length, and time), so we only require (7 – 3) = 4 dimensionless groups to form a unique dimensionless relationship. Among the different sets of four dimensionless groups into which the dimensional quantities could be arranged, the set that is customarily adopted is as follows:

$Sh_d = \dfrac{k_s d}{\mathcal{D}_s}$	*Sherwood number,* the ratio of the mass transfer coefficient (k_s) to a diffusion velocity ($\mathcal{D}_s/d$) in the radial direction
$Re_d = \dfrac{\rho u d}{\mu}$	*Reynolds number,* the ratio of convection parallel to the surface (u) to the radial propagation velocity of friction ($\mu/\rho d$)
$Sc = \dfrac{\mu}{\rho \mathcal{D}_s}$	*Schmidt number,* the ratio of the radial propagation velocity of friction ($\mu/\rho d$) to the radial diffusion velocity ($\mathcal{D}_s/d$)
$\dfrac{L}{d}$	*Aspect ratio* of tube dimensions

Thus, for radial transport of solute simultaneous to flow through a tube, we conclude that the dimensionless mass transfer coefficient is a function of the three other dimensionless groups:

$$Sh_d = f\left(Re_d, Sc, \frac{L}{d}\right) \tag{12.1-21}$$

Table 12.1-1 contains Sherwood number equations for liquid flows in several commonly encountered geometries. For some entries, such as "flow parallel to a flat plate" and "mass transfer from a spinning disk," the solute fluxes were obtain by solving more general model equations (as presented in Section 14.1). For more complex systems such as spheres randomly packed in a tube, the equations are based on experimental data. In most cases, the dimensionless equations follow a form given by

Table 12.1-1 Dimensionless, Surface-Averaged, Mass Transfer Coefficients

1. Fully developed flow through circular tubes (Middleman, 1998)

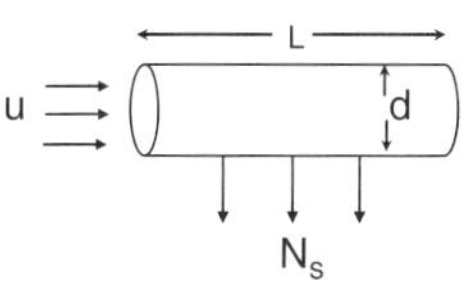

(a) Laminar flow with developing concentration profile

$Sh_d = 1.62[Re_d Sc(d/L)]^{1/3} : Re_d < 2100,\ L/d < 0.025 Re_d Sc$

(b) Laminar flow with fully developing concentration profile

$Sh_d = 3.66 :\ Re_d < 2100,\ L/d > 0.025 Re_d Sc$

(c) Turbulent flow

$Sh_d = 0.023\ Re_d^{0.83} Sc^{0.44} :\ Re_d > 4000$

2. Flow parallel to a flat plate (Treybal, 1980)

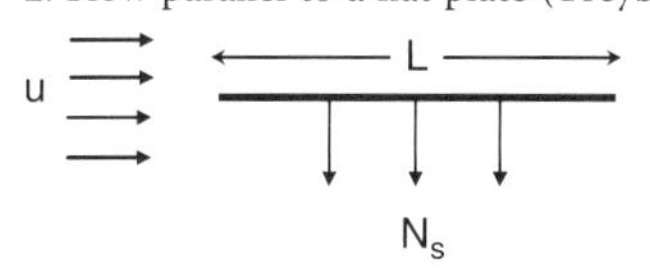

$Sh_L = 0.664\ Re_L^{1/2} Sc^{1/3}$

3. Flow perpendicular to cylinders

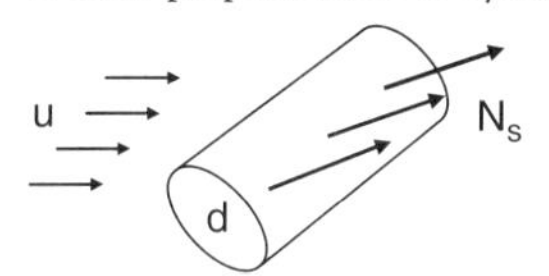

(a) Single cylinder (Treybal, 1980)

$Sh_d = \left(0.35 + 0.34 Re_d^{0.5} + 0.15 Re_d^{0.58}\right) Sc^{0.3}$

(b) Uniform array of cylinders (Cussler, 1997, p 226–227)

$Sh_d = 0.80\ Re_d^{0.47} Sc^{1/3}$

4. Flow parallel to uniform arrays of cylinders (Yang and Cussler, 1986)

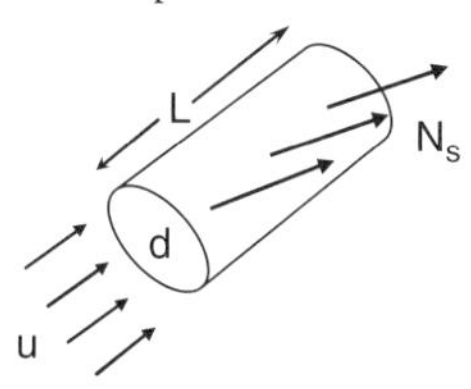

(a) $Sh_d = 1.25[Re_d(d/L)]^{0.93} Sc^{0.33}$: $\varepsilon = 0.97$

(b) $Sh_d = 0.022 Re_d^{0.60} Sc^{0.33}$: $\varepsilon = 0.74$

(c) $Sh_d = 0.24$: $\varepsilon = 0.60$

5. Flow past spheres

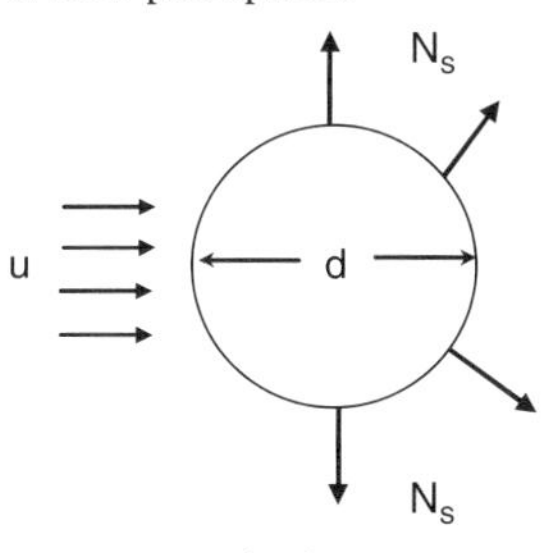

(a) Single sphere (Cussler, 1997, p 226–227)

$Sh_d = 2.0 + 0.6 Re_d^{1/2} Sc^{1/3}$

(b) Spheres randomly packed in a tube (Treybal, 1980)

$Sh_d = (0.25/\varepsilon) Re_d^{0.69} Sc^{1/3}$

6. Mass transfer from a spinning circular disk (Cussler, 1997, p 226–227)

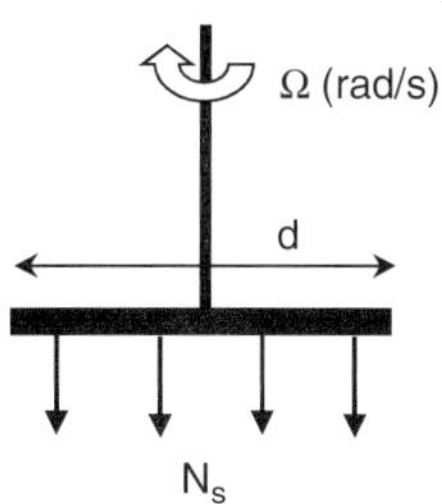

$Sh_d = 1.24 Re_\Omega^{1/2} Sc^{1/3} \left(Re_\Omega \equiv \rho d^2 \Omega / 4\mu\right)$

$(Re_d = \rho u d/\mu : Re_L = \rho u L/\mu : Sh_d = k_s d/\mathcal{D}_s : Sh_L = k_s L/\mathcal{D}_s : Sc = \mu/\rho\mathcal{D}_s)$

u, superficial velocity through packed tube or across tube bank

ε, void fraction between particle packing or tube banks

$$\mathrm{Sh_l} = a Re_l^b Sc^c \left(\frac{l'}{l}\right)^g \tag{12.1-22}$$

where Sh_l and Re_l are based on a characteristic length l. The aspect ratio l'/l is only necessary when there is a second characteristic length l' associated with the geometry of interest. In addition to shape of the interface, the values of the a, b, c, and g constants depend on the nature of the velocity and concentration fields. For example, the parameters values appropriate for mass transfer in a tube depend on whether the flow is laminar ($Re_d < 2100$) or turbulent ($Re_d > 4000$) and whether a concentration boundary layer is developing ($L/d < 0.05 Re_d Sc$) or fully developed ($L/d > 0.2 Re_d Sc$).

When mass transfer occurs through the walls of a tube or channel with a noncircular cross section, then the Reynolds and Sherwood numbers are frequently based on the hydraulic diameter:

$$d_H \equiv 4\frac{\text{Cross-sectional area}}{\text{Wetted perimeter}} \tag{12.1-23}$$

For a circular tube of diameter d that is filled with fluid, the cross-sectional area is $\pi d^2/4$ and the wetted perimeter is πd so that its hydraulic diameter is the same as its actual diameter.

Example 12.1-1 Hydraulic Diameters in an Artificial Kidney

An artificial kidney contains a parallel array of $n_f = 17{,}000$ uniformly spaced, thin-walled, cylindrical hollow fibers of diameter $d_f = 0.025$ cm through which blood flow is uniformly distributed. This array is contained within a cylindrical shell of diameter $d = 4.7$ cm through which physiological saline flows. Determine the hydraulic radius of the fiber array and of the shell that surrounds the fibers.

Solution

The cross-sectional areas available for blood and saline flows are

$$\text{Fiber array}: \pi(d_f/2)^2 n_f = 8.35\,\text{cm}^2 \quad \text{Shell}: \pi(d/2)^2 - \pi(d_f/2)^2 n_f = 9.01\,\text{cm}^2$$

The wetted perimeters are

$$\text{Fiber array}: (\pi d_f) n_f = 1340\,\text{cm} \quad \text{Shell}: \pi d + (\pi d_f) n_f = 1350\,\text{cm}$$

The hydraulic diameters are

$$\text{Fiber array}: 4(8.35/1340) = 0.025\,\text{cm}, \quad \text{Shell}: 4(9.01/1350) = 0.027\,\text{cm}$$

Notice that the hydraulic diameter of the parallel fiber array is the same as the diameter of an individual fiber.

The penetration model of k_s in Equation 12.1-20a can be expressed in the form of Equation 12.1-22 as

$$Sh_L = \sqrt{\frac{2}{\pi}} Re_L^{1/2} Sc_L^{1/2} \tag{12.1-24}$$

The $b = 1/2$ power predicted for the Reynolds number is consistent with most of the entries in Table 12.1-1. However, the power of $c = 1/2$ for the Schmit number is somewhat larger than the actual values in the table, which are usually closer to $c = 1/3$.

Because of the robustness of the dependence of Sh_L on $Sc^{1/3}$, dimensionless correlations of mass transfer coefficient data are often expressed in terms of the Chilton–Colburn j factor:

$$j_D(Re_L) \equiv \frac{Sh_L}{Re_L Sc^{1/3}} \tag{12.1-25}$$

Entry 2 in the table, for example, would then be written as

$$j_D = 0.664 Re_L^{1/2} \quad (12.1\text{-}26)$$

Example 12.1-2 Bare Oxygen Electrode in Flowing Blood

In Example 9.2-1, we determined that the O_2 sensitivity of a 1 mm polarographic electrode placed in a stagnant blood is 0.044 μA/kPa. Suppose the same electrode is implanted in a blood vessel with an average longitudinal velocity u (Fig. 12.1-4). Determine how the sensitivity is affected by u.

Solution

Referring to Figure 12.1-4, $\dot{V}_{O_2}$ represents the volumetric O_2 transport rate through an unstirred blood layer from the cathode surface to the bulk solution. That is, $\dot{N}_{O_2,r} = c_G^o \dot{V}_{O_2}$ and Equation 9.2-12 for the electric current becomes

$$i = -4\mathcal{F} c_G^o \dot{V}_{O_2} \quad (12.1\text{-}27)$$

We use Equation 12.1-4 to formulate $\dot{V}_{O_2}$ for a spherical cathode of diameter d and surface area $S = \pi d^2$ by setting $p_{so} = 0$ on the cathode surface and $p_{sb} = p_{O_2}$ in the bulk blood phase:

$$\dot{V}_{O_2} = \alpha_{O_2} k_{O_2} (\pi d^2)(0 - p_{O_2}) \quad (12.1\text{-}28)$$

With this expression, the electrical current is given by

$$i = (4\pi d^2 \mathcal{F} c_G^o \alpha_{O_2} k_{O_2}) p_{O_2} \quad (12.1\text{-}29)$$

The properties for blood at 37°C necessary to complete this problem are $\alpha_{O_2} = 2.2 \times 10^{-7}$ ml(STP)/ml-Pa, $\mu/\rho = 0.017$ cm²/s, and $\mathcal{D}_{O_2}(37°C) = 1.87 \times 10^{-5}$ cm²/s. With u given in centimeter per second, other required parameters are d = 0.1 cm, $\mathcal{F} = 96{,}500$ A-s/mol, and $c_G^o = 4.46 \times 10^{-5}$ mol/ml(STP). The values of the dimensionless groups necessary to determine k_{O_2} are

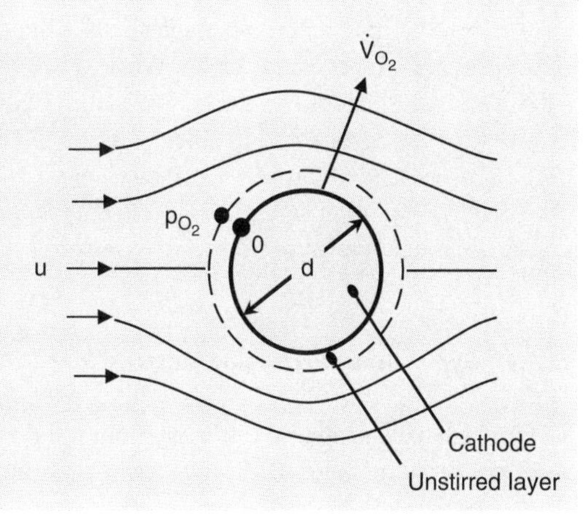

Figure 12.1-4 Oxygen transport during flow around an electrode.

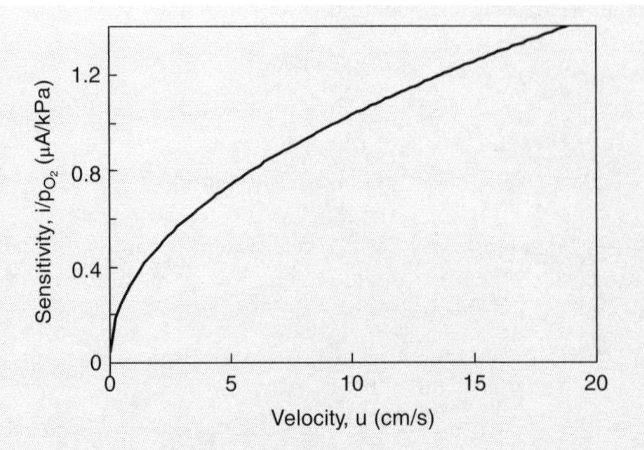

Figure 12.1-5 Current output sensitivity of a bare oxygen electrode.

$$\mathrm{Re_d} = \frac{0.1\mathrm{u}}{0.017} = 5.88\mathrm{u[cm/s]}, \quad \mathrm{Sc} = \frac{0.017}{1.87 \times 10^{-5}} = 909 \qquad (12.1\text{-}30\mathrm{a,b})$$

Assuming the diameter of the blood vessel is much larger than the diameter of the O_2 electrode, we apply entry 5a of Table 12.1-1 to determine the individual mass transfer coefficient adjacent to the electrode surface:

$$\begin{aligned} k_{O_2}[\mathrm{cm/s}] &= \frac{\mathcal{D}_{O_2}}{d} Sh_d = \frac{1.87 \times 10^{-5}}{0.1}\left[2.0 + 0.6(5.88)^{1/2}(909)^{1/3}\sqrt{u}\right] \\ &= \left(3.74 + 26.3\sqrt{u[\mathrm{cm/s}]}\right) \times 10^{-4} \end{aligned} \qquad (12.1\text{-}31)$$

Substituting this result into Equation 12.1-29, we predict the sensitivity of the cathode output:

$$\begin{aligned} \frac{i[\mu A]}{p_{O_2}[\mathrm{kPa}]} &= 4\pi(96{,}500)\left(2.2 \times 10^{-7}\right)\left(4.46 \times 10^{-5}\right)(0.1)^2\left(3.74 + 26.3\sqrt{u[\mathrm{cm/s}]}\right) \times 10^{+5} \\ &= \left(4.43 + 31.3\sqrt{u[\mathrm{cm/s}]}\right) \times 10^{-2} \end{aligned} \qquad (12.1\text{-}32)$$

When $u = 0$, this equation predicts the same electrode sensitivity as we computed in Example 9.2-1 for stagnant blood. When $u > 0$, blood flow creates a thin unstirred layer at the electrode surface. The shortened path length for diffusion through this layer increases the electrode current above its value when $u = 0$. For example, at a value of $u = 10$ cm/s, typical for a large vein, Equation 12.1-32 predicts an electrode sensitivity of $i/p_{O2} = 1.01$ μA/kPa, which is about twenty times larger than it is in stagnant blood. As u becomes greater, the unstirred layer becomes thinner and the electrode sensitivity increases even more (Fig. 12.1-5).

12.1.5 Hydraulically Permeable Surfaces

So far, we have considered diffusion from a fluid to a solid surface in the presence of transverse convection alone. At the surface of a porous material, convection driven by hydraulic

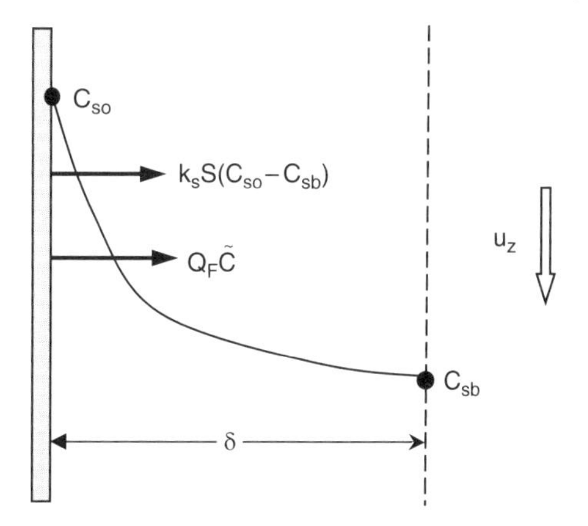

Figure 12.1-6 Film model of solute convection and diffusion in parallel.

and osmotic pressure forces can also occur toward the surface, in parallel with diffusion (Fig. 12.1-6). Similar to stagnant film theory, we assume there is a thin layer adjacent to the surface in which transverse convection has a minimal influence. A transport rate equation can then be obtained from our previous result for one-dimensional parallel convection and diffusion (Eq. 9.3-6):

$$\dot{N}_s = Q_F\left(\frac{C_{so}\,e^{Pe} - C_{sb}}{e^{Pe} - 1}\right) \tag{12.1-33}$$

where C_{so} and C_{sb} are the bounding solute concentrations between the surface and the bulk solution, Q_F is a constant volumetric filtration flow through the surface, and Pe is the Peclet number inferred from Equation 9.3-7 for a film layer of thickness δ:

$$Pe \equiv \frac{Q_F\delta}{\mathcal{D}_s\tilde{S}} \tag{12.1-34}$$

As in the stagnant film model, we define $k_s = \mathcal{D}_s/\delta$ as an individual mass transfer coefficient. We also assume that the surface film is so thin that it can be approximated by a planar slab such that $\tilde{S} \rightarrow S$. Therefore, the Peclet number can be determined as

$$Pe = \frac{Q_F}{k_sS} \tag{12.1-35}$$

The value of k_s is generally affected by Q_F. However, when convective transport toward a surface is small relative to diffusion (i.e., $Pe < 1$), we can approximate k_s by dimensionless correlations such as those in Table 12.1-1.

With some algebraic manipulation, Equation 12.1-33 can be also written as

$$\dot{N}_s = k_sS(C_{so} - C_{sb}) + Q_F\tilde{C};\quad \tilde{C} \equiv \left[C_{so} - (C_{so} - C_{sb})\left(\frac{1}{Pe} - \frac{1}{e^{Pe} - 1}\right)\right] \tag{12.1-36a, b}$$

Comparing this to Equation 12.1-3a, we see that the effect of the filtration flow is accounted for by a convection term $Q_F\tilde{C}$ in which $\tilde{C}$ is an intermediate concentration between C_{so} and C_{sb}.

12.2 Transport Through Multiple Phases

In this section, we will analyze transport in multiphase systems where diffusion must overcome a series of film resistances. Building on the formalism of individual mass transfer coefficients introduced in the previous section, we will see that the solute flux is now proportional to an overall mass transfer coefficient.

12.2.1 Diffusion at a Two-Phase Interface

Some problems include the transport of a solute s across an interface that separates two immiscible fluids. As shown in Figure 12.2-1, we allow for imperfect mixing in both fluids by placing an unstirred layer adjacent to both sides of the interface. While C_{sb}^{A} and C_{sb}^{B} represent solute concentrations in the bulk solutions, C_{so}^{A} and C_{so}^{B} are solute concentrations immediately adjacent to the interface. As is customary, we assume that the interfacial concentrations are in thermodynamic equilibrium. For a thermodynamically ideal solution, this means that

$$C_{so}^{A} = \lambda_{s}^{A,B} C_{so}^{B} \tag{12.2-1}$$

where $\lambda_{s}^{A,B}$ is a temperature-dependent partition coefficient. More generally, equilibrium is nonideal, following a curvilinear relationship illustrated in Figure 12.2-1. Point "a" in the right graph represents the equilibrium concentrations at an interface, and point "b" indicates the solute concentrations in the two bulk phases.

If solute s does not accumulate at the interface, its transport rate toward the interface from solution A must match its transport rate away from the interface into solution B. Employing Equation 12.1-3a, the transport rate through the interface can be written in two ways:

$$\dot{N}_{s} = k_{s}^{A} S\left(C_{sb}^{A} - C_{so}^{A}\right) = k_{s}^{B} S\left(C_{so}^{B} - C_{sb}^{B}\right) \tag{12.2-2}$$

where k_{s}^{A} and k_{s}^{B} are the mass transfer coefficients associated with the unstirred layers on the A and B sides of the interface. Since interfacial solute concentrations are difficult if not

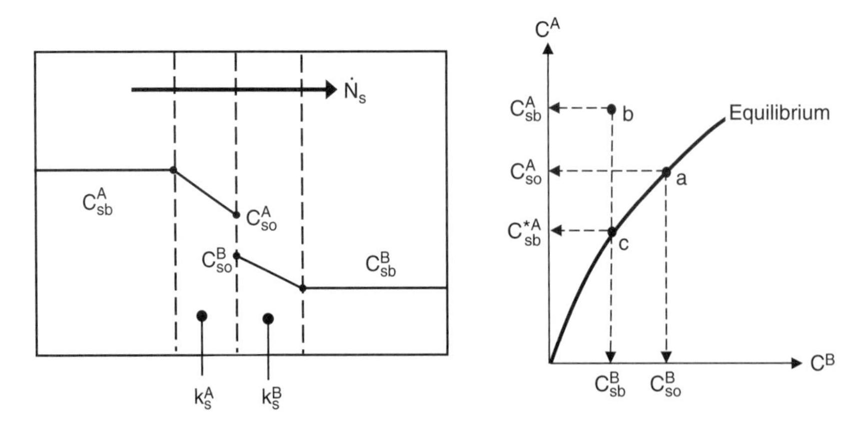

Figure 12.2-1 Diffusion between two nonideal immiscible solutions.

impossible to measure, these equations are insufficient for computing the transport rate. With equilibrium at the interface between the two fluids, we can write another relationship between the solute concentrations:

$$\Lambda_s^{A,B} \equiv \frac{C_{so}^A - C_{sb}^{*A}}{C_{so}^B - C_{sb}^B} \tag{12.2-3}$$

where C_{sb}^{*A} is the concentration of A that would (hypothetically) be in equilibrium with C_{sb}^B, and $\Lambda_s^{A,B}$ is the slope of a cord connecting points "a" and "c" along the equilibrium curve in Figure 12.2-1. By eliminating C_{so}^A and C_{so}^B from Equations 12.2-2 and 12.2-3, we express the transport rate in a form that only requires solute concentrations within the fluids:

$$\dot{N}_s = K_s S\left(C_{sb}^A - C_{sb}^{*A}\right) \tag{12.2-4}$$

Here, K_s is the overall mass transfer coefficient given by

$$\frac{1}{K_s} \equiv \frac{1}{k_s^A} + \frac{\Lambda_s^{A,B}}{k_s^B} \tag{12.2-5}$$

For nonideal solutions, $\Lambda_s^{A,B}$ will vary with solute concentration because of the nonlinearity of the equilibrium relationship. However, for liquid–liquid or gas–liquid systems that are ideal, the equilibrium plot has a constant slope $\lambda_s^{A,B}$ so that $\Lambda_s^{A,B} = \lambda_s^{A,B}$ and $C_{sb}^{*A} = \lambda_s^{A,B} C_b^B$. We can then rewrite Equations 12.2-4 and 12.2-5 as

$$\dot{N}_s = K_s S\left(C_{sb}^A - \lambda_s^{A,B} C_{sb}^B\right); \quad \frac{1}{K_s} \equiv \frac{1}{k_s^A} + \frac{\lambda_s^{A,B}}{k_s^B} \tag{12.2-6a,b}$$

Even for nonideal systems, we can use this rate expression provided that the equilibrium curve is reasonably straight over the region of interest.

Equation 12.2-6b states that the overall diffusion resistance $1/K_s$, is the sum of the individual diffusion resistances, $1/k_s^A$ and $\lambda_i^{A,B}/k_s^B$. Often, one of these two individual resistances is much larger than the other, and transport is limited or controlled by that film. For example, suppose that $1/k_s^A \gg \lambda_i^{A,B}/k_s^B$. In that case, the flux reduces to $\dot{N}_s = k_s S\left(C_{sb}^A - \lambda_s^{A,B} C_{sb}^B\right)$.

12.2.2 Diffusion Through a Membrane

We now consider solute transport across a planar membrane that separates two fluid-filled compartments. When we discussed this process in Chapter 10, the membrane was the limiting resistance to transport. This assumption is only valid when the contents of the compartments are perfectly mixed, right up to the membrane surfaces. More often than not, however, unstirred layers present on both membrane surfaces reduce the transport rate (Fig. 12.2-2). In the following analysis, we will restrict our attention to membranes through which convection can be neglected.

Recall that the transport rate through a homogeneous planar membrane in contact with two thermodynamically ideal solutions is given by Equation 10.1-4a. For membrane surface concentrations of C_{so}^A and C_{so}^B, this equation becomes

$$\dot{N}_s = P_s S_m\left(C_{so}^A - \lambda_s^{A,B} C_{so}^B\right) \tag{12.2-7}$$

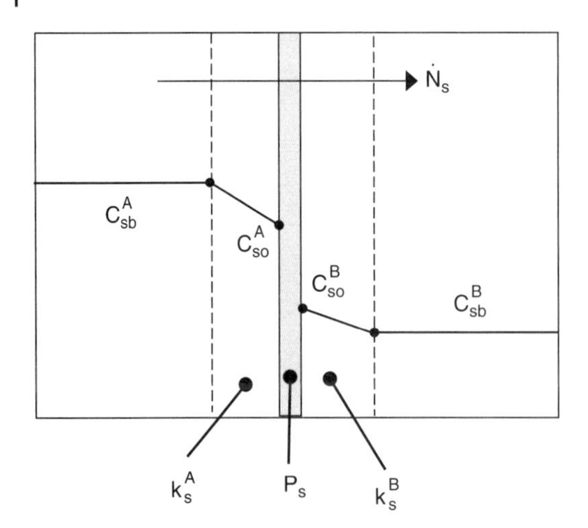

Figure 12.2-2 Diffusion through a membrane and adjacent unstirred layers.

where P_s is the solute permeability through the membrane. When this equation is combined with Equations 12.2-2 to eliminate C_{so}^A and C_{so}^B, we obtain

$$\dot{N}_s = K_s S_m \left(C_{sb}^A - \lambda_s^{A,B} C_{sb}^B\right); \quad \frac{1}{K_s} \equiv \frac{1}{k_s^A} + \frac{1}{P_s} + \frac{\lambda_s^{A,B}}{k_s^B} \tag{12.2-8a,b}$$

Equation 12.2-8a,b is the same as Equation 12.2-6a,b except for the addition of the membrane resistance $1/P_s$ to the overall mass transfer coefficient. Because k_s^A and k_s^B depend directly on the velocities acting along the membrane surfaces, the associated film resistances can be diminished by increasing the fluid flows over the membrane. At very high flows over both sides of the membrane, $1/k_s^A \to 0$, $1/k_s^B \to 0$, and K_s reaches its maximum value of P_s.

When a membrane is curved, its interfaces with the adjacent unstirred layers have different surface areas, S^A and S^B. In that case, Equation 12.2-8a,b should be modified as follows:

$$\dot{N}_s = (K_s S)\left(C_{sb}^A - \lambda_s^{A,B} C_{sb}^B\right); \quad \frac{1}{K_s S} \equiv \frac{1}{k_s^A S^A} + \frac{1}{P_s \widetilde{S}_m} + \frac{\lambda_s^{A,B}}{k_s^B S^B} \tag{12.2-9a,b}$$

where $\widetilde{S}_m$ is a mean surface area that has been defined in Equation 10.1-12 for cylindrical and spherical membranes. For membranes that are very thin compared to their radius of curvature, $S^A \approx S^B \approx \widetilde{S}_m$ and Equation 12.2-9a,b can be approximated by Equation 12.2-8a,b for a planar membrane.

Whereas interfacial equilibrium usually requires a discontinuity in species concentration between adjoining phases, the partial pressure of a gas species is continuous. Thus, in analyzing transport of a gas across an interface, it is convenient to use a partial pressure driving force instead of a concentration driving force. By replacing C_{sb}^A and C_{sb}^B with their equilibrium equivalents, $\alpha_s^A c_G^o p_{sb}^A$ and $\alpha_s^B c_G^o p_{sb}^B$, and also replacing the equilibrium partition coefficient with $\lambda_s^{A,B} = \alpha_s^A/\alpha_s^B$, we can express the volumetric rate of gas transport, $\dot{V}_s \equiv \dot{N}_s/c_G^o$, as

$$\dot{V}_s = K_s^G S\left(p_{sb}^A - p_{sb}^B\right); \quad \frac{1}{K_s^G S} = \frac{1}{\alpha_s^A k_c^A S^A} + \frac{1}{P_s^G \widetilde{S}_m} + \frac{1}{\alpha_s^B k_c^B S^B} \tag{12.2-10a,b}$$

Here, $P_s^G \equiv \alpha_s^A P_s$ is membrane permeability defined for a gaseous solute (Eq. 10.1-11a), and $K_s^G \equiv \alpha_s^A K_s$ is an overall mass transfer coefficient based on a partial pressure driving force. This equation is applicable to the transport of a gas component through a membrane separating any two fluid-filled compartments, whether they contain liquid or gas phases. It is only necessary to know the Bunsen solubility of component s in each phase. For example, compartment A could contain an aqueous solution and compartment B could contain an organic solution. Alternatively, compartment A could contain a gas mixture, while compartment B contains a liquid solution. In that special case, $\alpha_s^A = 1/\mathcal{R}Tc_G^o$ so that

$$\dot{V}_s \equiv \frac{\dot{N}_s}{c_G^o} = K_s^G S\left(p_{sb}^A - p_{sb}^B\right); \quad \frac{1}{K_s^G S} \equiv \frac{RTc_G^o}{k_c^A S^A} + \frac{1}{P_s^G \tilde{S}_m} + \frac{1}{\alpha_s^B k_c^B S^B} \tag{12.2-11a,b}$$

Example 12.2-1 Membrane-Covered Oxygen Electrode in Flowing Blood

Consider again Example 12.1-1 in which a 1 mm diameter O_2 electrode is used to measure p_{O2} in a large blood vessel. To reduce fouling of the electrode surface by formation of blood clots and deposition of plasma proteins, the electrode is now covered by a Teflon coating with a thickness of 1 mil = 0.0254 mm (Fig. 12.2-3). How does the O_2 sensitivity of this membrane-covered electrode compare to the sensitivity of the bare electrode found in the previous example?

Solution

As in the previous example, we designate the volumetric transport rate of O_2 away from the cathode as $\dot{V}_{O_2}$. Therefore, we identify phase A with the cathode side of the membrane and phase B with the blood side of the membrane. By letting $p_{sb}^A = 0$ on the cathode surface and $p_{sb}^B = p_{O_2}$ in the bulk blood phase, Equation 12.2-10a becomes

$$\dot{V}_{O_2} = K_{O_2}^G S\left(0 - p_{O_2}\right) \tag{12.2-12}$$

Combining this with Equation 12.1-27, the electrode sensitivity is given by

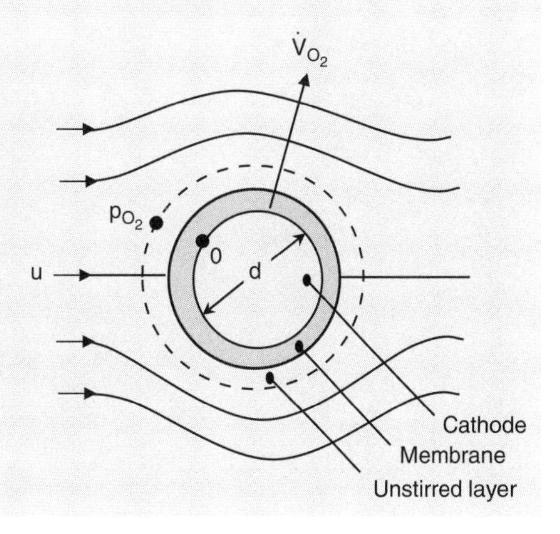

Figure 12.2-3 Oxygen transport during flow around a membrane-coated electrode.

$$\frac{i}{p_{O_2}} = 4\mathcal{F}c_G^o K_{O_2}^G S \tag{12.2-13}$$

To formulate the overall transport resistance through blood and the Teflon coating, we utilize the first two terms in Equation 12.2-10b:

$$\frac{1}{K_{O_2}^G S} \equiv \frac{1}{\alpha_{O_2}^A k_{O_2}^A S^A} + \frac{1}{P_{O_2}^G \widetilde{S}_m} \tag{12.2-14}$$

The surface area S^A on the outside of the coating and the mean surface area $\widetilde{S}_m$ of a spherical shell (Eq. 10.1-12) are given in terms of cathode diameter d and coating thickness h_m as

$$S^A = \pi(d + 2h_m)^2, \quad \widetilde{S}_m = \pi d(d + 2h_m) \tag{12.2-15a,b}$$

Combining Equations 12.2-12–12.2-15 and replacing the permeability $P_{O_2}^G$ by the specific permeability $\hat{P}_{O_2}^G \equiv P_{O_2}^G h_m$, we get

$$\frac{i}{p_{O_2}} = 4\pi\mathcal{F}c_G^o \alpha_{O_2}^A d^2 k_{O_2}^A \frac{(1 + 2h_m/d)^2}{\left[1 + (1 + 2h_m/d)\left(\alpha_{O_2}^A k_{O_2}^A h_m / \hat{P}_{O_2}^G\right)\right]} \tag{12.2-16}$$

In computing the electrode sensitivity with this equation, we set $\hat{P}_{O_2}^G = 1.26 \times 10^{-12}$ ml(STP)/(s-cm-Pa) as listed in Table 10.1-1. We obtain k_{O_2} as a function of u as we did in Example 12.1-2; in this case, Re_d and Sh_d are based on the outside diameter of the coating, $d = 1 + 2(0.0254) = 1.05$ mm rather than the bare electrode diameter d.

In Figure 12.2-4, the sensitivity of the coated electrode predicted by Equation 12.2-16 is shown by the solid curve. The behavior of the cathode without its membrane covering shown by the dashed curve is duplicated from Figure 12.1-5.

Compared to the bare cathode, a Teflon coating has the disadvantage of reducing the sensitivity of the electrical output to O_2. However, it also has the advantage of reducing the dependence on blood velocity. In fact, the sensitivity approaches a constant value at large velocities. Noting that $k_{O_2} \to \infty$ as $u \to \infty$, Equation 12.2-16 predicts that this upper limit is

$$\frac{i_\infty}{p_{O_2}} = 4\pi\mathcal{F}c_G^o d^2 \left(1 + \frac{2h_m}{d}\right) \frac{\hat{P}_{O_2}^G}{h_m} \tag{12.2-17}$$

This limiting sensitivity is shown by the dotted line in the figure.

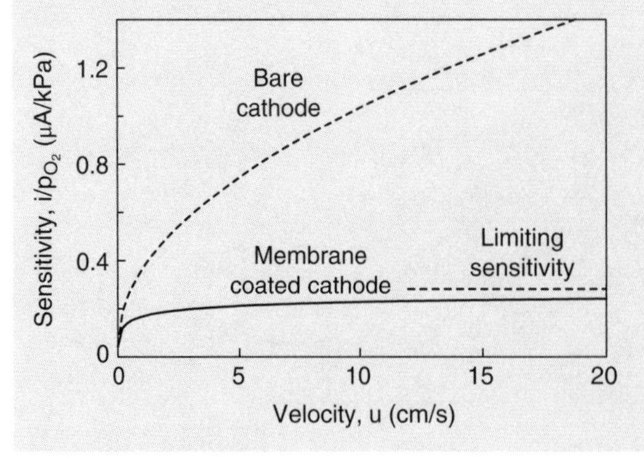

Figure 12.2-4 Current output sensitivity of bare and membrane-coated oxygen electrodes.

12.2.3 Parallel Convection and Diffusion Through a Membrane

Solute transport through a membrane that allows filtration as well as diffusion to occur (Fig. 12.2-5) can be analyzed by extending the approach we used when filtration is absent (Zydney, 1993). At steady state, the molar rate of solute transport $\dot{N}_s$ must be the same across each of the two membrane surfaces. With solution entering the membrane at the same mass density as solution leaving the membrane, the volumetric flow Q_F will also be the same at the surfaces. We can write the solute transport rates through the unstirred layers adjacent to the membrane by employing Equation 12.1-35:

$$\dot{N}_s = Q_F\left(\frac{C_{sb}^A e^{Pe^A} - C_{so}^A}{e^{Pe^A} - 1}\right); \quad Pe^A \equiv \frac{Q_F}{k_s^A S^A} \qquad (12.2\text{-}18a,b)$$

$$\dot{N}_s = Q_F\left(\frac{C_{so}^B e^{Pe^B} - C_{sb}^B}{e^{Pe^B} - 1}\right); \quad Pe^B \equiv \frac{Q_F}{k_s^B S^B} \qquad (12.2\text{-}19a,b)$$

The molar transport rate of solute through a planar heterogeneous membrane in contact with solute concentrations C_{so}^A and C_{so}^B is given by Equation 10.2-14a,b. To allow for membrane curvature, we replace S_m by the mean surface $\widetilde{S}_m$ and obtain

$$\dot{N}_s = (1-\sigma_s)Q_F\left(\frac{C_{so}^A e^{Pe_m} - C_{so}^B}{e^{Pe_m} - 1}\right); \quad Pe_m = \frac{(1-\sigma_s)Q_F}{P_s \widetilde{S}_m} \qquad (12.2\text{-}20a,b)$$

where σ_s is the reflection coefficient (Section 6.5). Since the interfacial concentrations C_{so}^A and C_{so}^B are usually unknown, we would like to solve these equations for $\dot{N}_s$ in terms of C_{sb}^A and C_{sb}^B alone. In general, this is difficult to accomplish since Q_F is affected by the osmotic pressure difference across the membrane, which itself depends on C_{so}^A and C_{so}^B. To avoid this problem,

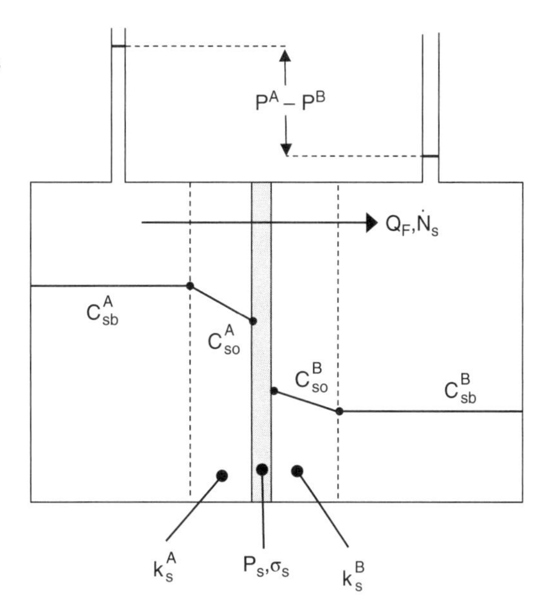

Figure 12.2-5 Parallel diffusion and convection through a membrane and its unstirred layers.

we assume that Q_F is independent of C_{so}^A and C_{so}^B. In that case, the combination of Equations 12.2-18–12.2-20 yields

$$\dot{N}_s = Q_F\left(f_s^A C_{sb}^A - f_s^B C_{sb}^B\right) \quad (12.2\text{-}21)$$

where

$$f_s^A = \frac{(1-\sigma_s)\exp\left(Pe^A + Pe_m + Pe^B\right)}{\sigma_s \exp(Pe^B)\left[\exp(Pe_m) - 1\right] + (1-\sigma_s)\left[\exp(Pe^A + Pe_m + Pe^B) - 1\right]} \quad (12.2\text{-}22)$$

and

$$f_s^B = \frac{(1-\sigma_s)}{\sigma_s \exp(Pe^B)\left[\exp(Pe_m) - 1\right] + (1-\sigma_s)\left[\exp(Pe^A + Pe_m + Pe^B) - 1\right]} \quad (12.2\text{-}23)$$

Although f_s^A and f_s^B are rather complex functions of the system parameters, considerable simplification results when Q_F is so small that all three Peclet numbers Pe^A, Pe_m, and Pe^B are also small. We can then approximate the exponentials in Equations 12.2-22 and 12.2-23 by a third-order series expansion, and Equation 12.2-21 becomes (Jaffrin, 1995)

$$\dot{N}_s = (K_s S)\left(C_{sb}^A - C_{sb}^B\right) + Q_F \widetilde{C}_s \quad (12.2\text{-}24)$$

where

$$\frac{1}{K_s S} = \frac{1}{S^A k_s^A} + \frac{1}{P_s \widetilde{S}_m} + \frac{1}{k_s^B S^B} \quad (12.2\text{-}25)$$

and

$$\widetilde{C}_s = \frac{K_s S}{Q_F}\left[\left(Pe^A + \frac{Pe_m}{2}\right)C_{sb}^A + \left(Pe^B + \frac{Pe_m}{2}\right)C_{sb}^B\right] \quad (12.2\text{-}26)$$

Equation 12.2-24 reveals that the transport rate can be subdivided into diffusive and convective contributions in a manner analogous to Equation 12.1-36a for a single boundary layer. In this case, the overall transport resistance $1/K_s S$ is the sum total of the diffusion resistances of the membrane and the two unstirred layers, and $\widetilde{C}_s$ is the effective concentration at which the solute from the two compartments is convected.

When transport occurs by diffusion alone and there is no transmembrane convection at all, the Peclet numbers are all zero and the flux equation reduces to

$$\dot{N}_s = (K_s S)\left(C_{sb}^A - C_{sb}^B\right); \quad \frac{1}{K_s S} = \frac{1}{S^A k_s^A} + \frac{1}{P_s \widetilde{S}_m} + \frac{1}{k_s^B S^B} \quad (12.2\text{-}27a,b)$$

This result is similar to Equation 12.2-9a,b that we derived for pure diffusion through a homogeneous membrane. For a porous membrane, however, $\lambda_s^{A,B}$ is unity since the fluids within the voids and external to the membrane are assumed to be miscible.

12.2.4 Concentration Polarization

The term "ultrafiltration" refers to the separation of solutes achieved when convection occurs in parallel with diffusion through a membrane. During ultrafiltration, solute rejected by a

Figure 12.2-6 Solute concentration polarization across a membrane.

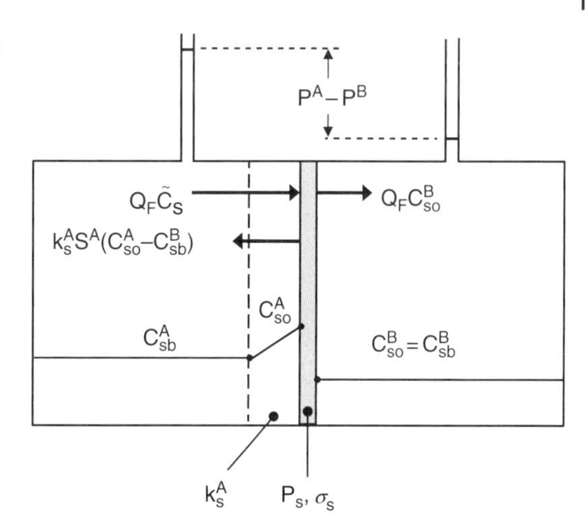

membrane builds up at its upstream surface, thereby generating a surface concentration C_{so}^{A} greater than the bulk concentration C_{sb}^{A}. This concentration polarization results in a back-diffusion of solute from the upstream surface that is in balance with solute convection toward the surface. Concentration polarization is important because elevated solute concentrations at a membrane surface can cause detrimental effects such as blood coagulation and protein deposition. Concentration polarization also modifies the performance of a membrane since back-diffusion affects the retention and sieving of solutes.

To model this process, consider the membrane shown in Figure 12.2-6. Surface A, the retentate side of the membrane, is in contact with an unstirred layer. At surface B, the filtrate side of the membrane, fluid is well mixed so that there is no unstirred layer and $C_{so}^{B} = C_{sb}^{B}$. Solving Equations 12.2-18a–12.2-20a, we obtain the concentration polarization ratio, which is independent of filtrate composition:

$$\frac{C_{so}^{A}}{C_{sb}^{A}} = \frac{1-\sigma_s e^{-Pe_m}}{(1-\sigma_s)+\sigma_s e^{-Pe^A}(1-e^{-Pe_m})} \tag{12.2-28}$$

Figure 12.2-7 shows the effect of $Pe_m/(1-\sigma_s) = Q_F/P_s\tilde{S}_m$ and $Pe^A = Q_F/S^A k_s^A$ on C_{so}^{A}/C_{sb}^{A} as predicted by this equation. The fixed value of $Pe_m/(1-\sigma_s) = 1$ in the left graph means that the characteristic convection and diffusion rates are equal in the membrane, whereas the value of $Pe^A = 1$ in the right graph indicates that these rates are equal in the retinate. When $\sigma_s = 0$, solute freely passes through the membrane, and concentration polarization is absent, no matter what the value of Pe^A or $Pe_m/(1-\sigma_s)$. However, when $\sigma_s > 0$ the rejection of solute molecules from the membrane surface leads to $C_{so}^{A}/C_{sb}^{A} > 0$. The larger σ_s, the greater is this concentration polarization. When σ_s is fixed, either a larger Pe^A in the retinate or a larger $Pe_m/(1-\sigma_s)$ in the membrane results in an increase in concentration polarization.

During ultrafiltration, solute convection through the membrane often dominates diffusion such that Pe_m is very large. Equation 12.2-28 then reaches the following limits:

$$\lim_{Pe_m\to\infty}\frac{C_{so}^{A}}{C_{sb}^{A}} = \frac{1}{1-\sigma_s\left[1-\exp\left(-Q_F/S^A k_s^A\right)\right]} = \begin{cases} 1 & \text{when } Pe^A \to 0 \\ 1/(1-\sigma_s) & \text{when } Pe^A \to \infty \end{cases} \tag{12.2-29}$$

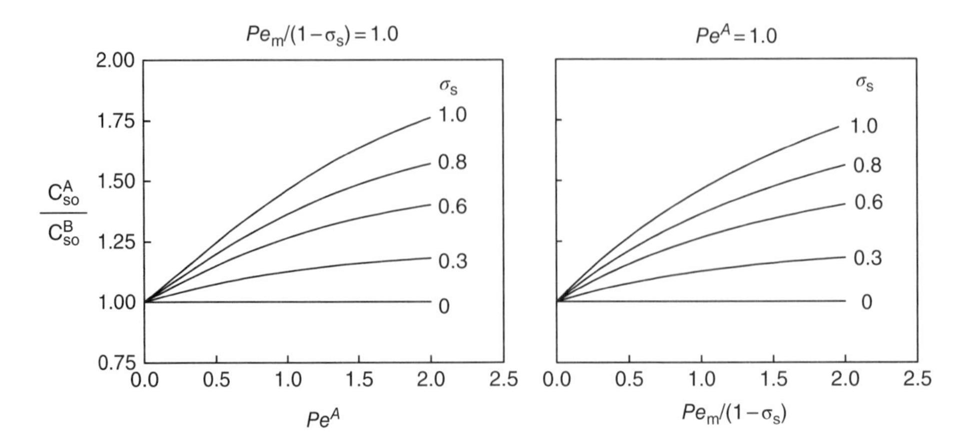

Figure 12.2-7 Solute buildup at a membrane surface during concentration polarization.

The $Pe^A \rightarrow 0$ limit meets our expectation that concentration polarization is not possible when diffusion in the retinate is so great that there is no unstirred layer. The $Pe^A \rightarrow \infty$ limit indicates that C^A_{so}/C^A_{sb} can become exceedingly large when there is very little diffusion through retinate and the membrane is ideally semi-permeable such that $\sigma_s \rightarrow 1$.

To characterize solute separation when there is concentration polarization, we define an overall sieving coefficient as

$$\Sigma_{s0} \equiv \frac{C^B_{sb}}{C^A_{sb}} \tag{12.2-30}$$

By simultaneously solving Equations 12.2-18–12.2-20 with the constraint that $C^B_{so} = C^B_{sb}$, we formulate Σ_{s0} in terms of membrane properties and unstirred layer properties in compartment A:

$$\Sigma_{s0} = \frac{\Sigma_{sm}}{\Sigma_{sm} + (1 - \Sigma_{sm}) \exp(-Pe^A)} \tag{12.2-31}$$

where

$$\Sigma_{sm} \equiv \frac{C^B_o}{C^A_o} = \frac{(1-\sigma_s)}{1 - \sigma_s \exp(-Pe_m)} \tag{12.2-32}$$

is the sieving coefficient of the membrane alone as previously formulated in Equation 10.2-19. Alternatively, the overall retention coefficient is given by

$$R_{s0} \equiv 1 - \Sigma_{s0} = \frac{R_{sm} \exp(-Pe^A)}{(1 - R_{sm}) + R_{sm} \exp(-Pe^A)} \tag{12.2-33}$$

where

$$R_{sm} \equiv 1 - \Sigma_{sm} = \frac{\sigma_s[1 - \exp(-Pe_m)]}{1 - \sigma_s \exp(-Pe_m)} \tag{12.2-34}$$

The value of Σ_{s0} is less than Σ_{sm}, and R_{s0} is greater than R_{sm}. Thus, when compared to transport in the absence of an unstirred layer, concentration polarization decreases molecular sieving while increasing retention.

12.3 Design and Performance of Separation Devices

Mass transfer coefficients are particularly useful in the engineering of chemical separation devices for organ replacement and treatment of life-threatening diseases. We will focus on extracorporeal devices for which blood is continuously removed from the patient, passed through the device, and returned to the patient. Within the device, blood contacts another phase that acts either as a source of a desired nutrient or therapeutic drug, or as a sink for an undesirable metabolite or toxic chemical.

In the following sections, we will discuss two types of commonly used extracorporeal devices, namely, the blood oxygenator (artificial lung) and the blood dialyzer (artificial kidney). Rather than developing general models that require extensive numerical analysis, we will focus on special cases with analytical solutions that provide insights into device performance.

12.3.1 Blood Oxygenation by Membrane Devices

Blood oxygenators are used to provide gas exchange during cardiopulmonary bypass surgery and can be used to support patients with lung failure. Early disk and bubble oxygenators required direct contact of blood with oxygen-enriched air, causing severe red cell hemolysis at the blood–air interface. Later devices greatly reduced blood damage by segregating blood and gas phases with a gas permeable membrane. As in the lungs, efficiency of O_2 transport requires that blood-membrane contact surface and O_2 permeability of the membrane both be large. In current devices, this is often achieved by using hollow fibers with inner diameters of 100–300 μm and wall thicknesses of 10–30 μm. Hollow fibers can be fabricated from a variety of microporous materials that are highly permeable to gases but essentially impermeability to liquids. Special treatments of a hollow fiber surface are sometimes used to minimize blood damage in a device.

In a typical hollow fiber blood oxygenator (ECMO), the outside surfaces of a fiber array are contacted with a patient's entire blood flow, while O_2-enriched air is provided to the inside of the fibers. The fiber wall acts as a membrane, allowing O_2 and CO_2 to be exchanged between the flowing blood and the flowing air. Achieving the required blood oxygenation requires several thousand hollow fibers with a total surface area of 1–3 m^2 enclosed in a plastic shell containing about 300 ml of blood. The particular arrangement of the hollow fibers and the blood path around the fibers differs among commercial devices. The fibers can be laid out in a straight parallel bundle, or they can be wound in a circular pattern. The blood flow can follow a straight path parallel to the fibers or can be guided in a curved path that runs perpendicular to the fiber orientation.

To illustrate a general approach to quantifying performance, we consider a generic oxygenator configuration in which a parallel array of gas-filled fibers and the blood path outside the fibers are both aligned in a longitudinal z direction (Fig. 12.3-1). Because the rate at which O_2 is supplied to the device is usually high, the gas-side O_2 partial pressure is essentially constant at its input value p_{gas} for all z. Thus, although this is a two-phase device, it is only necessary to formulate an O_2 molar balance in the blood phase, and the relative direction of the gas and blood flows (concurrent or countercurrent) does not matter.

The operating parameters on the blood side of this device are: Q, the constant volumetric flow rate of blood; p_{in} and p_{out}, the input and output O_2 partial pressures in the blood space.

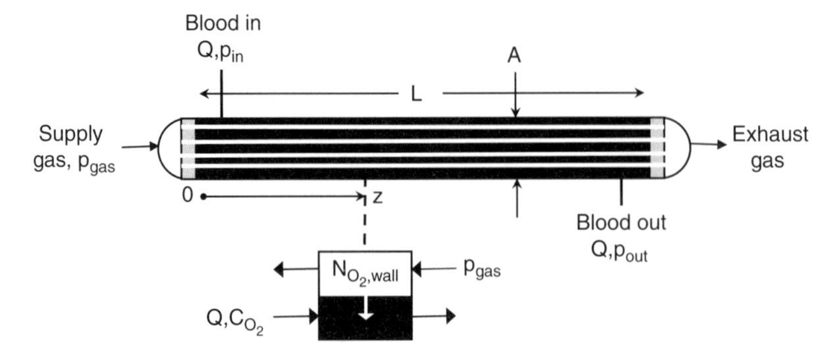

Figure 12.3-1 Hollow fiber blood oxygenator.

The device geometry is characterized by the following constant parameters: V, device volume; A, cross-sectional area of the device in the direction of blood flow; S, total outside surface of fibers; ε, volume fraction of the device that contains blood; a_f, outside fiber radius; and $\phi = 2/a_f$, surface-to-volume ratio of the fibers. The volume of blood in the device is given by either εV or $(V - S/\phi)$. Equating these quantities, we obtain the geometric relationship:

$$\varepsilon = 1 - \frac{S}{V\phi} \tag{12.3-1}$$

We visualize the space outside of the fibers as a single conduit of blood. This conduit is bounded by the outside surfaces of all the fibers so that its cross-sectional area is $A\varepsilon$. With the ratio of the fiber volume to blood volume given by $(1 - \varepsilon)/\varepsilon$, the blood conduit has a surface-to-volume ratio of $\phi(1 - \varepsilon)/\varepsilon$ and a hydraulic diameter of

$$d_H \equiv 4\frac{(\text{Cross-sectional area})}{(\text{Wetted perimeter})} = 4\frac{(\text{Volume})}{(\text{Surface})} = \frac{4\varepsilon}{\phi(1-\varepsilon)} \tag{12.3-2}$$

A solution molar balance on blood is not needed since its flow is constant in this application. By setting $A_t = A\varepsilon$ and $\phi_t = \phi(1 - \varepsilon)/\varepsilon$ in Equation 2.4-25, we arrive at one-dimensional molar balances on O_2 and oxyheme groups HbO_2 in blood:

$$\frac{\partial C_{O_2}}{\partial t} + \frac{Q}{A\varepsilon}\frac{\partial C_{O_2}}{\partial z} = -\frac{\phi(1-\varepsilon)}{\varepsilon}N_{O_2,\text{wall}} + R_{O_2} \tag{12.3-3}$$

$$\frac{\partial C_{HbO_2}}{\partial t} + \frac{Q}{A\varepsilon}\frac{\partial C_{HbO_2}}{\partial z} = -\frac{\phi(1-\varepsilon)}{\varepsilon}N_{HbO_2,\text{wall}} + R_{HbO_2} \tag{12.3-4}$$

For steady-state operation of the oxygenator, the time derivatives are zero. Upon adding the steady-state equations, we obtain

$$\frac{d(C_{O_2} + C_{HbO_2})}{dz} = -\frac{\phi(1-\varepsilon)A}{Q}(N_{O_2,\text{wall}} + N_{HbO_2,\text{wall}}) + \frac{A\varepsilon}{Q}(R_{O_2} + R_{HbO_2}) \tag{12.3-5}$$

Since HbO_2 is not transported across the fiber walls, $N_{HbO_2,\text{wall}} = 0$. Also, the reaction rates of O_2 and HbO_2 are equal and opposite so that $R_{O_2} = -R_{HbO_2}$. The summed solute balance then reduces to

$$\frac{d(C_{O_2} + C_{HbO_2})}{dz} = \frac{\phi(1-\varepsilon)A}{Q}N_{O_2,\text{wall}} \tag{12.3-6}$$

Employing Equation 5.5-6, the sum of O_2 and HbO_2 concentrations can be written in terms of O_2 partial pressure $p_{O_2}(z)$ and oxyhemoglobin saturation fraction $S_{O_2}(z)$:

$$(C_{O_2} + C_{HbO_2}) = c_G^o \left(\alpha_{O_2} p_{O_2} + \widehat{C}^b_{O_2,max} S_{O_2} \right) \tag{12.3-7}$$

and Equation 12.3-6 becomes

$$\alpha_{O_2} c_G^o \left(\frac{dp_{O_2}}{dz} + \frac{\widehat{C}^b_{O_2,max}}{\alpha_{O_2}} \frac{dS_{O_2}}{dz} \right) = \frac{\phi(1-\varepsilon)A}{Q} N_{O_2,wall} \tag{12.3-8}$$

Typically, the oxyhemoglobin reaction is so rapid relative to O_2 transport that chemical equilibrium exists at all z positions. Ignoring changes in pH and p_{CO_2} that somewhat modify oxyhemoglobin binding equilibrium (the Bohr effect), S_{O_2} will only be a function of p_{O_2}. Then, $dS_{O_2}/dz = \left(dS_{O_2}/dp_{O_2}\right) dp_{O_2}/dz$ and we rewrite Equation 12.3-8 as

$$\left(1 + \frac{\widehat{C}^b_{O_2,max}}{\alpha_{O_2}} \frac{dS_{O_2}}{dp_{O_2}} \right) \frac{dp_{O_2}}{dz} = \frac{\phi(1-\varepsilon)A}{\alpha_{O_2} c_G^o Q} N_{O_2,wall} \tag{12.3-9}$$

The $N_{O_2,wall}$ flux must overcome a series of transport resistances in the gas phase, fiber walls, and blood phase. However, because the O_2 diffusivity in a gas phase is so large and microporous fiber walls have such a high O_2 permeability, $N_{O_2,wall}$ is limited by the transport resistance of the blood phase. With an O_2 partial pressures p_{gas} at the outer surface of a fiber and $p_{O_2}(z)$ in the bulk blood phase, Equation 12.1-4 leads to the following relationship for the wall flux:

$$N_{O_2,wall} \approx \alpha_{O_2} c_G^o k_{O_2,loc} \left(p_{gas} - p_{O_2} \right) \tag{12.3-10}$$

In this flux relationship, the molar concentration of dissolved O_2 at the fiber surface in contact with blood is $\alpha_{O_2} c_G^o p_{gas}$, and the dissolved O_2 concentration in the bulk phase of the flowing blood is $\alpha_{O_2} c_G^o p_{O_2}(z)$. Assuming that the mass transfer coefficient $k_{O_2,loc}$ does not markedly change with position along the surface, we can substitute its spatially averaged value k_{O_2}, and Equation 12.3-9 becomes

$$\left(1 + \frac{\widehat{C}^b_{O_2,max}}{\alpha_{O_2}} \frac{dS_{O_2}}{dp_{O_2}} \right) \frac{dp_{O_2}}{dz} = \frac{\phi(1-\varepsilon)A k_{O_2}}{Q} \left(p_{gas} - p_{O_2} \right) \tag{12.3-11}$$

In a conduit of hydraulic diameter d_H and cross-sectional area εA available for blood flow, Equation 12.1-25 leads to a relationship between k_{O_2}, Reynolds number, Schmidt number, and the Chilton–Colburn j factor:

$$k_{O_2} = \left(\frac{\mathcal{D}_{O_2}}{d_H} \right) Sh_d \Rightarrow k_{O_2} = \left(\frac{\mathcal{D}_{O_2}}{d_H} \right) Re_d Sc^{1/3} j_D(Re_d) \tag{12.3-12a,b}$$

where

$$Re_d \equiv \left(\frac{Q}{\varepsilon A} \right) \frac{\rho d_H}{\mu}, \quad Sc \equiv \frac{\mu}{\rho \mathcal{D}_{O_2}} \tag{12.3-13a,b}$$

After substituting Equations 12.3-12 and 12.3-13 into Equation 12.3-11, the governing equation of O_2 transport in the device is written as

$$\left(1+\frac{\widehat{C}^{b}_{O_2,max}}{\alpha_{O_2}}\frac{dS_{O_2}}{dp_{O_2}}\right)\frac{dp_{O_2}}{dz}=\frac{j_D\phi(1-\varepsilon)}{\varepsilon Sc^{2/3}}\left(p_{gas}-p_{O_2}\right) \tag{12.3-14}$$

Taking the integral of this equation over the flow path from z = 0, where blood enters the device with $p_{O_2} = p_{in}$, to z = L where blood exits the device with $p_{O_2} = p_{out}$, we get

$$\int_{p_{in}}^{p_{out}}\left[1+\frac{\widehat{C}^{b}_{O_2,max}}{\alpha_{O_2}}\left(\frac{dS_{O_2}}{dp_{O_2}}\right)\right]\frac{dp_{O_2}}{p_{gas}-p_{O_2}}=\frac{j_D\phi(1-\varepsilon)L}{\varepsilon Sc^{2/3}} \tag{12.3-15}$$

For any device, the length L of the flow path is equivalent to V/A. But for a particular device, the relationship between j_D and Re_d depends on the fiber arrangement. Oxygenator evaluation with an erythrocyte-free aqueous phase in place of blood is convenient for determining this relationship. In that case, $\widehat{C}^{b}_{O_2,max} = 0$ and we solve Equation 12.3-15 with the result that

$$j_D=\frac{\varepsilon Sc^{2/3}}{\phi(1-\varepsilon)L}\ln\left(\frac{p_{gas}-p_{in}}{p_{gas}-p_{out}}\right) \tag{12.3-16}$$

Suppose that p_{out} is measured at various oxygenator settings of p_{in} and Q. The corresponding j_D and Re_d values can then be computed from Equations 12.3-16 and 12.3-13a, and an empirical expression for $j_D(Re_d)$ can be developed.

We obtain a molar balance for O_2 over the entire blood phase in a device by integrating Equation 12.3-8 from z = 0 to z = L:

$$\alpha_{O_2}Q\left(\int_{p_{in}}^{p_{out}}dp_{O_2}+\frac{\widehat{C}^{b}_{O_2,max}}{\alpha_{O_2}}\int_{S_{in}}^{S_{out}}dS_{O_2}\right)=\frac{\phi(1-\varepsilon)A}{c^{o}_{G}}\int_{0}^{L}N_{O_2,wall}dz \tag{12.3-17}$$

where $S_{in} = S_{O_2}(p_{in})$ and $S_{out} = S_{O_2}(p_{out})$. Since $\phi(1-\varepsilon)AN_{O_2,wall}$ is equivalent to the molar oxygenation rate per unit length of fiber bed, the right side of this equation is the volumetric rate of O_2 transfer, $\dot{V}_{O_2}$, at STP conditions. Completing the integration on the left side of the equation, we get

$$\dot{V}_{O_2}=Q\left[\alpha_{O_2}(p_{out}-p_{in})+\widehat{C}^{b}_{O_2,max}(S_{out}-S_{in})\right] \tag{12.3-18}$$

Example 12.3-1 Performance of a Sarns Turbo 440 Blood Oxygenator in an Adult Patient

The Sarns Turbo 440 blood oxygenator (Fig. 12.3-2) consists of an annular bed of microporous hollow fibers that each have an outside radius $a_f = 190$ μm. While O_2-enriched gas flows inside the fibers, blood flows outside of the fibers. The total outside fiber surface is $S = 1.6$ m^2, and the total device volume is $V = 386$ cm^3.

When used in cardiac surgery, blood bypasses the heart and lungs and is pumped through the oxygenator. Within the device, the incoming blood flow Q is split equally between two identical, semicircular paths. The Q/2 flow through each path is distributed over a frontal areaof $A/2 = 20.5$ cm^2. Thus, blood flows over a path length of $L = V/A = 386/(2 \times 20.5) = 9.42$ cm. Water tests of O_2 absorption in this oxygenator (Dierickx *et al.*, 2001) have established that

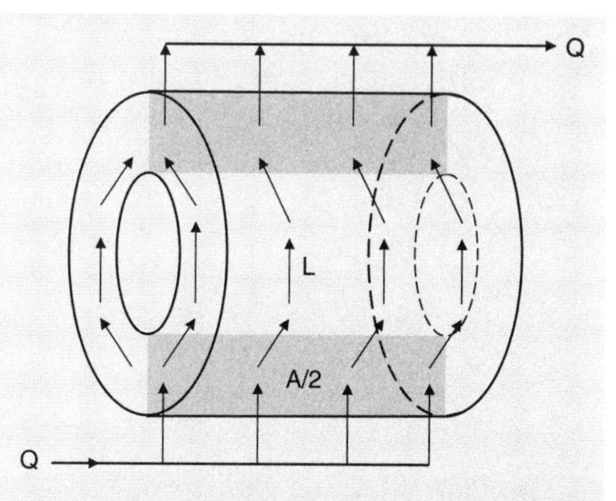

Figure 12.3-2 Blood flow pattern in the fiber bed of the Sarns Turbo oxygenator.

$$j_D = \frac{0.60}{Re_d^{0.31}} \quad \text{when } Re_d Sc > 3200 \tag{12.3-19}$$

Suppose that pure O_2 flows through the hollow fibers at a high flow rate and a pressure close to atmospheric. The venous blood entering the oxygenator typically has an O_2 partial pressure of 5.3 kPa. We will determine how the O_2 partial pressure in exiting blood and the overall uptake rate of O_2 vary with Q.

Solution

The equation governing the performance of this oxygenator is obtained by substituting Equation 12.3-19 into Equation 12.3-15 and noting that $p_{in} = 5.3$ kPa and $p_{gas} = 101$ kPa:

$$\int_{5.3}^{p_{out}} \left[1 + \frac{\hat{C}^b_{O_2,max}}{\alpha_{O_2}} \left(\frac{dS_{O_2}}{dp_{O_2}}\right)\right] \frac{dp_{O_2}}{101 - p_{O_2}} = 0.60 \frac{\phi(1-\varepsilon)L}{\varepsilon Re_d^{0.31} Sc^{2/3}} \tag{12.3-20}$$

The values of the geometric parameters for this device are

$$\phi = \frac{2}{190} = 0.0105\,\mu\text{m}^{-1} = 105\,\text{cm}^{-1}$$
$$\varepsilon = 1 - \frac{1.6 \times 10^4}{386(105)} = 0.605 \tag{12.3-21a-c}$$
$$d_H = \frac{4(0.606)}{41.5} = 0.0584\,\text{cm}$$

Given additional parameter values of $\alpha_{O_2} = 2.2 \times 10^{-4}$ ml(STP)/(ml-kPa), $\mathcal{D}_{O_2} = 1.87 \times 10^{-5}$ cm^2/s, and $\mu/\rho = 0.017$ cm^2/s, the key dimensionless parameters are

$$Re_d = \frac{(Q/60)}{0.605(41.0)}\left(\frac{0.0584}{0.017}\right) = 0.00230Q\ [\text{ml/min}]$$
$$Sc = \frac{0.017}{1.87 \times 10^{-5}} = 909 \tag{12.3-22a,b}$$

These equations indicate that $Re_d Sc = 2.09Q$, and the computation of j_D from Equation 12.3-19 is only valid when $Q > 3200/2.09 = 1530$ ml/min. For a typical value of $\hat{C}^b_{O_2,max} = 0.2\,\text{ml(STP)/ml}$, Equation 12.3-20 can be written as

$$\int_{5.3}^{p_{out}} \left[1+\frac{0.2}{2.2\times10^{-4}}\left(\frac{dS_{O_2}}{dp_{O_2}}\right)\right]\frac{dp_{O_2}}{101-p_{O_2}} = \frac{0.6(105)(1-0.605)(9.42)}{(0.605)(0.00230Q)^{0.31}(909)^{2/3}} \tag{12.3-23}$$

or

$$\int_{5.3}^{p_{out}} \left[1+909\left(\frac{dS_{O_2}}{dp_{O_2}}\right)\right]\frac{dp_{O_2}}{101-p_{O_2}} = \frac{27.1}{Q^{0.31}} \tag{12.3-24}$$

and Equation 12.3-18 becomes

$$\dot{V}_{O_2} = 2.2\times10^{-4}\,Q[(p_{out}-5.3)+909(S_{out}-S_{O_2}(5.3))] \tag{12.3-25}$$

Employing the Hill model of oxyhemoglobin saturation excluding the Bohr effect (Eq. 5.5-16), we obtain

$$S_{O_2} = \frac{(0.283p_{O_2})^{2.8}}{1+(0.283p_{O_2})^{2.8}} \Rightarrow \frac{dS_{O_2}}{dp_{O_2}} = \frac{0.0817p_{O_2}^{1.8}}{\left(1+0.0292p_{O_2}^{2.8}\right)^2} \tag{12.3-26a,b}$$

Equations 12.3-24 and 12.3-25 are now given by

$$Q^{0.31} = 4.24\times10^4\left\{\int_{5.3}^{p_{out}}\left[1+\frac{74.3p_{O_2}^{1.8}}{\left(1+0.0292p_{O_2}^{2.8}\right)^2}\right]\frac{dp_{O_2}}{101-p_{O_2}}\right\}^{-3.23} \tag{12.3-27}$$

$$\dot{V}_{O_2} = 2.2\times10^{-4}Q\left[p_{out}+\frac{900(0.283p_{out})^{2.8}}{1+(0.283p_{out})^{2.8}}-693\right] \tag{12.3-28}$$

where the required units are p_{out}[kPa] and Q[ml/min]. We numerically evaluate the integral in Equation 12.3-27 at successively higher values of p_{out}. At each p_{out} value, we solve Equation 12.3-27 for the corresponding Q value and determine $\dot{V}_{O_2}$ from Equation 12.3-28. The results of these computations are shown by the curves in Figure 12.3-3.

At a typical blood flow of 5 L/min, the Sarns Turbo oxygenator increases p_{O_2} from 5.3 to 10.3 kPa (left graph) by absorbing 201 ml/min of O_2 (right graph). Thus, this device can meet the oxygenation requirements of an adult patient.

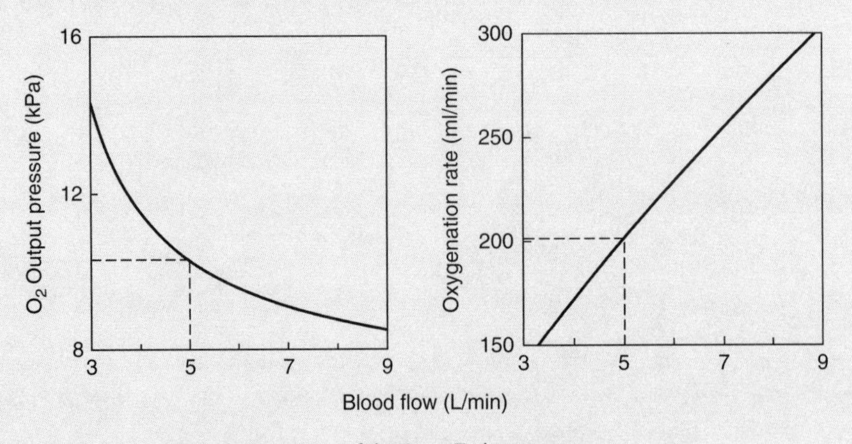

Figure 12.3-3 Performance characteristics of the Sarns Turbo oxygenator.

12.3.2 Blood Purification by Hemodialysis

The removal of potentially toxic metabolites by an extracorporeal device can be used to treat renal insufficiency or to rapidly counteract a drug overdose. In replacing renal function, it is most important to clear urea and creatinine from circulating blood while minimizing the depletion of formed elements and plasma proteins. These objectives can be accomplished by hemodialysis in which solutes passively diffuse from blood into a flowing dialysate liquid through an intervening membrane.

Hemodialyzers frequently use hollow fiber membranes that permit a high flux of urea and creatinine with some plasma filtration to transfer excess water from the patient's blood into the dialysate. We will model an artificial kidney consisting of a parallel bundle of n_f hollow fibers, each of length L (Fig. 12.3-4). The fiber mouths are joined to thin manifolds near the ends of a tubular shell of diameter d. For simplicity, we consider thin-walled cylindrical fibers of approximately equal inner and outer radii a_f. Thus, the total inner and outer surface areas of the fibers, $S = 2\pi a_f L n_f$, are also equal.

There are three important differences between this hemodialyzer and the oxygenator shown in Figure 12.3-1: (i) Only a small portion rather than the entire blood flow of the patient is diverted to a hemodialyzer; (ii) blood flow in a hemodialyzer is usually inside rather than outside the fibers, and dialysate flow is outside the fibers; and (iii) whereas urea and creatinine concentrations substantially change in blood as well as dialysate, the gas phase in an oxygenator has a fairly constant O_2 concentration. To account for composition changes in a hemodialyzer, molar balances must be formulated for the dialysate as well as for blood. Also because of these composition changes, the relative flow directions of blood and dialysate can have a profound effect on device performance. A countercurrent flow of blood and dialysate generally improves separation efficiency compared to a concurrent flow.

For a countercurrent hemodialyzer with the geometry shown in Figure 12.3-4, the volumetric blood inflow Q^B_{in} and dialysate outflow Q^D_{out} occur at z = 0, while the blood outflow Q^B_{out} and dialysate inflow Q^D_{in} occur at z = L. The inlet and outlet concentrations of solute s in blood are $C^B_{s,in}$ and $C^B_{s,out}$, and the corresponding concentrations in dialysate are $C^D_{s,in}$ and $C^D_{s,out}$. For a solute s that is removed from blood, $C^B_{s,in} > C^B_{s,out}$ and $C^D_{s,out} > C^D_{s,in}$. Thus, for this countercurrent arrangement of blood and dialysate flows, solute concentrations in blood and dialysate both decrease with distance z from the blood inlet. We assume that the molar densities of

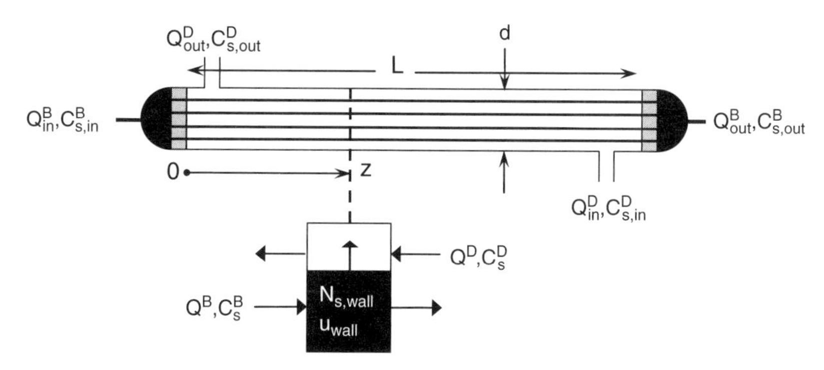

Figure 12.3-4 Hollow fiber dialyzer with countercurrent blood and dialysate flows.

blood and dialysate are constant. Because of water filtration through the fiber walls, however, the flow of blood and of dialysate can change with z.

Balance Equations

We visualize the space inside of all the fibers in the fiber bundle as a single conduit of blood with a cross-sectional area A^B and a surface-to-volume ratio ϕ^B. Similarly, the space outside of the fibers is a conduit of dialysate with a cross-sectional area A^D and a surface-to-volume ratio ϕ^D. Assuming that fluid density is constant throughout the system, we apply Equations 2.4-25 and 2.4-26 to obtain one-dimensional molar balances in both conduits. For a nonreactive species s, the balance equations in blood are given by

$$\frac{\partial C_s^B}{\partial t} + \frac{1}{A^B}\frac{\partial (Q^B C_s^B)}{\partial z} = -\phi^B N_{s,wall}^B \qquad \frac{1}{A^B}\frac{\partial Q^B}{\partial z} = -\phi^B u_{wall}^B \tag{12.3-29a,b}$$

and the dialysate balances are

$$\frac{\partial C_s^D}{\partial t} + \frac{1}{A^D}\frac{\partial (-Q^D C_s^D)}{\partial z} = \phi^D N_{s,wall}^D \qquad \frac{1}{A^D}\frac{\partial (-Q^D)}{\partial z} = \phi^D u_{wall}^D \tag{12.3-30a,b}$$

Here, Q^j are the volumetric flows along the conduits, in the positive z direction in the blood (j = B) and negative direction in the dialysate (j = D); $N_{s,wall}^j$ and u_{wall}^j are the local solute flux and local solution velocity crossing the inner wall (j = B) and the outer wall (j = D) of a fiber. Both of these variables are positive when material is transported from blood to dialysate.

The product of the cross-sectional area of a conduit with its surface-to-volume ratio is equivalent to the total surface area per unit length. For thin-walled fibers, this parameter is the same on the blood and dialysate sides of the fibers: $A^B\phi_B = A^D\phi_D = S/L$. Also for thin-walled fibers, solute flux and solution velocity across the fiber wall are the same on the blood side as they are on the dialysate side: $N_{s,wall}^B = N_{s,wall}^D \equiv N_{s,wall}$ and $u_{wall}^B = u_{wall}^D \equiv u_{wall}$. Thus, the flow equations become

$$\frac{dQ^D}{dz} = \frac{dQ^B}{dz}, \quad \frac{dQ^B}{dz} = -\frac{S}{L}u_{wall} \tag{12.3-31a,b}$$

and the solute concentration equations are

$$A^B\frac{\partial C_s^B}{\partial t} + \frac{\partial (Q^B C_s^B)}{\partial z} = -\frac{S}{L}N_{s,wall} \qquad A^D\frac{\partial C_s^D}{\partial t} - \frac{\partial (Q^D C_s^D)}{\partial z} = \frac{S}{L}N_{s,wall} \tag{12.3-32a,b}$$

In this model, we are interested in steady-state operation for which these equations reduce to

$$\frac{d(Q^D C_s^D)}{dz} = \frac{d(Q^B C_s^B)}{dz}, \quad \frac{d(Q^B C_s^B)}{dz} = -\frac{S}{L}N_{s,wall} \tag{12.3-33a,b}$$

We obtain a relationship between Q^D and Q^B by integrating Equation 12.3-31a from $z = 0$ to any arbitrary z:

$$Q^D(z) = Q^B(z) + \left(Q^D_{out} - Q^B_{in}\right) \quad (12.3\text{-}34)$$

We obtain a relationship between C^D and C^B by integrating Equation 12.3-33a between an arbitrary z and $z = L$. Typically, the substance of interest is not present in the dialysate supply, so that $C^D_{s,in} = 0$ and this integration yields

$$C^D_s(z)Q^D(z) = Q^B(z)C^B_s(z) - Q^B_{out}C^B_{s,out} \quad (12.3\text{-}35)$$

Flux Equations

Equation 12.2-21 provides a relationship between the solute transport rate $\dot{N}_s$ and the volumetric filtration rate Q_F across a membrane of surface area S. By dividing by S and identifying $\dot{N}_s/S \rightarrow N_{s,wall}$ and $Q_F/S \rightarrow u_{wall}$, we convert this rate equation to a relationship between local solute flux and solution filtration velocity:

$$N_{s,wall} = u_{wall}\left(f^B_s C^B_s - f^D_s C^D_s\right) \quad (12.3\text{-}36)$$

For this particular illustration, we have identified (phase A) → (blood B) and (phase B) → (dialysate D), and we have dropped the subscript "b" from the concentration variables. The *f* parameters in this equation are obtained by corresponding modifications to Equations 12.2-22–12.2-23:

$$f^B_s = \frac{(1-\sigma_s)\exp(Pe^B + Pe_m + Pe^D)}{\sigma_s \exp(Pe^B)[\exp(Pe_m) - 1] + (1-\sigma_s)[\exp(Pe^B + Pe_m + Pe^D) - 1]} \quad (12.3\text{-}37)$$

and

$$f^D_s = \frac{(1-\sigma_s)}{\sigma_s \exp(Pe^D)[\exp(Pe_m) - 1] + (1-\sigma_s)[\exp(Pe^B + Pe_m + Pe^D) - 1]} \quad (12.3\text{-}38)$$

We modify the Peclet numbers *Pe* from their definitions (Eqs. 12.2-18b–12.2-20b) by recognizing that $S = S^B = S^D = \tilde{S}$ for a thin-walled fiber:

$$Pe^B \equiv \frac{Q_F}{k^B_s S^B} \rightarrow \frac{u_{wall}}{k^B_s}, \quad Pe^D \equiv \frac{Q_F}{k^D_s S^D} \rightarrow \frac{u_{wall}}{k^D_s}$$
$$Pe_m \equiv \frac{(1-\sigma_s)Q_F}{P_s \tilde{S}} \rightarrow \frac{(1-\sigma_s)u_{wall}}{P_s} \quad (12.3\text{-}39a\text{-}c)$$

The flux equation can be simplified for a solute that encounters little rejection at a membrane surface. In that case, the reflection coefficient σ_s is negligible so that

$$\lim_{\sigma_s \to 0}\left(Pe^B + Pe_f + Pe^D\right) = \frac{u_{wall}}{K_s}; \quad \frac{1}{K_s} = \left(\frac{1}{k^B_s} + \frac{1}{P_s} + \frac{1}{k^D_s}\right) \quad (12.3\text{-}40a,b)$$

and Equations 12.3-37 and 12.3-38 reduce to

$$f^B_s = \frac{1}{1 - \exp(-u_{wall}/K_s)}, \quad f^D_s = \frac{\exp(-u_{wall}/K_s)}{1 - \exp(-u_{wall}/K_s)} \quad (12.3\text{-}41a,b)$$

With these approximations, the flux equation becomes

$$N_{s,wall} = \frac{u_{wall}}{1 - \exp(-u_{wall}/K_s)}\left[C_s^B - C_s^D \exp(-u_{wall}/K_s)\right] \tag{12.3-42}$$

The flux equation can also be simplified in the limit of negligible convection. For a thin-walled fiber, in particular, Equation 12.2-27a,b indicate that

$$N_{s,wall} = K_s\left(C_s^B - C_s^D\right); \quad \frac{1}{K_s} \equiv \frac{1}{k_s^B} + \frac{1}{P_s} + \frac{1}{k_s^D} \tag{12.3-43a,b}$$

Removal Efficiency

The efficiency of a hemodialyzer (or the natural kidney itself) can be expressed as the ratio of the removal rate $\dot{N}_s$ of a species from blood to its input rate in blood:

$$\zeta_s \equiv \frac{\dot{N}_s}{Q_{in}^B C_{s,in}^B} \tag{12.3-44}$$

Instead of this dimensionless quantity, the physiology and medical literature use a performance index called the clearance, which is equal to the removal efficiency multiplied by Q_{in}^B:

$$Q_s^C \equiv \zeta_s Q_{in}^B = \frac{\dot{N}_s}{C_{s,in}^B} \tag{12.3-45}$$

This performance parameter represents that portion of the incoming blood flow that is (hypothetically) cleared of all molecules of species s. Since $\dot{N}_s$ is the difference between the input rate and output rate of a species in blood, $\left(Q_{in}^B C_{s,in}^B - Q_{out}^B C_{s,out}^B\right)$, the removal efficiency and clearance can be expressed as

$$\zeta_s = 1 - \frac{Q_{out}^B C_{s,out}^B}{Q_{in}^B C_{s,in}^B}, \quad Q_s^C = Q_{in}^B - \frac{C_{s,out}^B}{C_{s,in}^B} Q_{out}^B \tag{12.3-46a,b}$$

In principle, we could incorporate the relation of u_{wall} to osmotic and hydrodynamic effects in our hemodialyzer model, but this would result in a complex set of nonlinear governing equations. A less rigorous but simpler approach is to specify u_{wall} as a known function of z (Jaffrin, 1995). In the following two sections, we find solutions to the model equations for two special cases: when u_{wall} is zero and when u_{wall} is constant throughout the device.

12.3.3 Hemodialysis with Negligible Plasma Filtration

When plasma filtration is negligible, $u_w = 0$, and Equation 12.3-31a,b leads to constant blood and dialysate flows of

$$Q^B = Q_{in}^B = Q_{out}^B, \quad Q^D = Q_{in}^D = Q_{out}^D \tag{12.3-47a,b}$$

By combining the species balance in blood (Eq. 12.3-33b) with the transmembrane flux equation (Eq. 12.3-43a), we obtain

$$\frac{d(Q^B C_s^B)}{dz} = -\frac{K_s S}{L}\left(C_s^B - C_s^D\right) \tag{12.3-48}$$

Eliminating C_s^D with Equation 12.3-35 and noting that Q^B is constant yields the final molar balance for a solute s in blood:

$$\frac{dC_s^B}{dz} = \frac{K_s S}{L}\left[\left(\frac{1}{Q^D} - \frac{1}{Q^B}\right)C_s^B - \frac{1}{Q^D}C_{s,out}^B\right] \tag{12.3-49}$$

After separating variables and integrating between z = 0 and z = L, we get

$$\frac{C_{s,out}^B}{C_{s,in}^B} = \frac{1 - Q^D/Q^B}{1 - (Q^D/Q^B)\exp[K_s S(1/Q^B - 1/Q^D)]} \tag{12.3-50}$$

Substituting Equations 12.3-47a and 12.3-50 into Equation 12.3-46a, we predict a removal efficiency of

$$\zeta_s = \frac{1 - \exp[K_s S(1/Q^B - 1/Q^D)]}{(Q^B/Q^D) - \exp[K_s S(1/Q^B - 1/Q^D)]} \tag{12.3-51}$$

and with Equation 12.3-46b, we obtain a clearance of

$$Q_s^C = \frac{1 - \exp[K_s S(1/Q^B - 1/Q^D)]}{1/Q^D - 1/Q^B \exp[K_s S(1/Q^B - 1/Q^D)]} \tag{12.3-52}$$

Although we envisaged the hemodialyzer as a device containing a parallel array of cylindrical hollow fibers enclosed in a cylindrical shell, neither fiber radius and fiber number nor shell diameter appear explicitly in the clearance equation. Rather, such geometric details affect device performance because of their influence on the overall mass transfer coefficient K_s. Because of this, Equation 12.3-52 can be applied to other membrane configurations so long as blood and dialysate are in parallel, countercurrent flow.

Example 12.3-2 Urea Transfer Resistance in a Fresenius Hollow Fiber Hemodialyzer

Mandolfo and coworkers (2003) evaluated the clinical performance of Fresenius dialyzers containing polysulfone hollow fibers with a total fiber surface S of either 1.4 or 2.2 m^2. Table 12.3-1 presents mean urea clearance data for eight patients measured after three hours of dialysis at various blood and dialysate flows. To account for somewhat different urea solubilities in blood and dialysate, blood water (i.e., blood free of solids and solutes) was used as a basis of calculating Q^B and Q_s^C (Clark *et al.*, 2007).

Table 12.3-1 Urea Clearance Q_s^C Measured After 120 min[a]

Q^B (ml blood water/min)[a]	253	253	324	324
Q^D (ml/min)	300	500	300	500
S (m^2)	1.4/2.2	1.4/2.2	1.4/2.2	1.4/2.2
Q_s^C (ml blood water/min)[a]	175/218	203/239	223/226	268/306

[a] Blood water content was backcalculated from whole blood measurements by assuming that 80% of red cell volume and 93% of plasma volume consist of water.

Assuming that plasma filtration does not have a strong impact on urea clearance, determine the relative contribution of the fiber wall and the unstirred blood and dialysate layers to device performance.

Solution

Rearranging Equation 12.3-52, we obtain an explicit formula for K_s:

$$K_s = \frac{1}{S}\left(\frac{1}{Q^D} - \frac{1}{Q^B}\right)^{-1} \ln\left(\frac{1 - Q_s^C/Q^B}{1 - Q_s^C/Q^D}\right) \tag{12.3-53}$$

With this equation and the Q_s^C values in Table 12.3-1, we compute the eight K_s data points shown in Figure 12.3-5. Generally speaking, K_s increases with Q^B when Q^D is fixed and also increases with Q^D when Q^B is fixed. This is expected since a higher flow of either fluid leads to a larger value of their individual mass transfer coefficient.

To account for this flow dependence, we assume that individual mass transfer coefficients for dialysate and blood have the form suggested by Equation 12.1-22: $k_s^D \propto (Re_d^D)^{b_1} \propto (Q^D)^{b_1}$ and $k_s^B \propto (Re_d^B)^{b_2} \propto (Q^B)^{b_2}$. We can then express K_s according to Equation 12.3-43b:

$$\frac{1}{K_s} = \frac{a_1}{(Q^D)^{b_1}} + \frac{a_2}{(Q^B)^{b_2}} + \frac{1}{P_s} \tag{12.3-54}$$

where a_1, b_1, a_2, and b_2 are constants and $1/P_s$ is the permeability of the fiber wall. To estimate a_1, a_2, b_1, b_2, and $1/P_s$, we performed a nonlinear least-squares regression of the data to Equation 12.3-54. The results, shown by the curves in Figure 12.3-5, produced parameter estimates of $a_1 = 97.4 \times 10^5$, $a_2 = 8.92 \times 10^5$, $b_1 = 2.02$, $b_2 = 1.97$, and $1/P_s = 0.101$ with K_s expressed in centimeter per minute, and Q^D and Q^B expressed in milliliter per minute. Because the K_s we computed from the data are highly scattered about the regression curves, we cannot be confident that our model provides a unique simulation of the experimental observations.

Nevertheless, it is useful to see what the model predicts about diffusion resistances in a typical hemodialysis treatment at a blood flow of 300 ml/min and a dialysate of 300 ml/min. The corresponding diffusion resistance of dialysate is

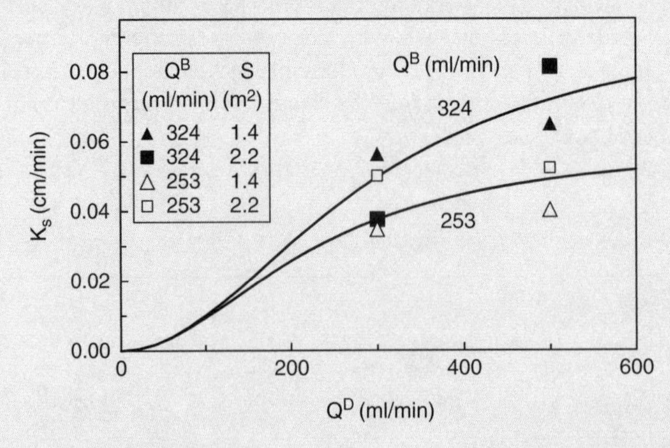

Figure 12.3-5 Effect of dialysate flow on mass transfer coefficient in a Fresenius hemodialyzer without filtration.

$$a_1/(Q^D)^{b_1} = 9.74\times 10^5 (300)^{-2.02} = 9.65 \text{ min/cm} \tag{12.3-55}$$

of the blood is

$$a_2/(Q^B)^{b_2} = 8.92\times 10^5 (300)^{-1.97} = 11.8 \text{ min/cm} \tag{12.3-56}$$

and of the fiber wall is

$$1/P_s = 0.101 \text{ min/cm} \tag{12.3-57}$$

In theory, it is possible to improve hemodialyzer clearance by reducing any of these three diffusion resistances. There is little benefit of using a more permeable membrane to reduce $1/P_m$ which is already relatively small. Reducing the diffusion resistance of the blood boundary layer by substantially increasing the blood flow taken from the patient is medically inadvisable. The most effective way to improve clearance is by operating at a higher dialysate flow.

12.3.4 Hemodialysis with Uniform Filtration

When filtration occurs at a constant transmembrane velocity u_{wall}, then the volumetric filtration rate through the entire fiber bundle is $Q_F = Su_{wall}$. Integrating Equation 12.3-31b between $z = 0$ and any arbitrary z and utilizing Equation 12.3-34, we get the equations for the oppositely directed blood and dialysate flows.

$$Q^B = Q^B_{in} - Q_F(z/L), \quad Q^D = Q^D_{out} - Q_F(z/L) \tag{12.3-58a,b}$$

At $z = L$, these flows are

$$Q^B_{out} = Q^B_{in} - Q_F, \quad Q^D_{in} = Q^D_{out} - Q_F \tag{12.3-59a,b}$$

Employing Equation 12.3-35 with Equation 12.3-58b yields an expression for C^D_s in terms of $(Q^B C^B_s)$ and z:

$$C^D_s = \frac{Q^B C^B_s - Q^B_{out} C^B_{s,out}}{Q^D_{out} - Q_F z/L} \tag{12.3-60}$$

Substituting the general flux equation (Eq. 12.3-36) and the C^D_s equation (Eq. 12.3-60) into the species balance on blood (Eq. 12.3-33b),

$$\frac{d(Q^B C^B_s)}{dz} = -\frac{S}{L} u_{wall}\left(f^B_s C^B_s - f^D_s \frac{Q^B C^B_s - Q^B_{out} C^B_{s,out}}{Q^D_{out} - Q_F z/L}\right) \tag{12.3-61}$$

Utilizing the flow equation (Eq. 12.3-58a) and recalling that $Q_F = Su_{wall}$, we arrive at a final differential equation for the $Q^B C^B_s$ product:

$$\frac{d(Q^B C^B_s)}{d(z/L_f)} + \left(\frac{f^B_s}{Q^B_{in}/Q_F - z/L} - \frac{f^D_s}{Q^D_{out}/Q_F - z/L}\right)(Q^B C^B_s) = -\frac{Q^B_{out} C^B_{s,out} f^D_s}{Q^D_{out}/Q_F - z/L} \tag{12.3-62}$$

Since we are considering the special case of constant u_{wall}, the Peclet numbers in Equation 12.3-39a-c are constant so that f_s^B and f_s^D are also constants.

With its boundary condition at the blood inlet $Q^B(0)C_s^B(0) = Q_{in}^B C_{in}^B$, Equation 12.3-62 has the same structure as a first-order, inhomogeneous, initial value problem that can be solved (Appendix C3) to obtain

$$\frac{Q^B(z)C_s^B(z)}{Q_{out}^B C_{s,out}^B} = \frac{(Q_{in}^B - Q_F z/L)^{f_s^B}}{(Q_{out}^D - Q_F z/L)^{f_s^D}} \left\{ \frac{Q_{in}^B C_{in}^B}{Q_{out}^B C_{s,out}^B} \frac{(Q_{out}^D)^{f_s^D}}{(Q_{in}^B)^{f_s^B}} - Q_F f_s^D \int_0^1 \frac{(Q_{out}^D - Q_F \xi)^{f_s^D - 1}}{(Q_{in}^B - Q_F \xi)^{f_s^B}} d\xi \right\} \quad (12.3\text{-}63)$$

To formulate solute clearance, we evaluate this equation at the blood outlet z = L where $Q^B(L)C_s^B(L) = Q_{out}^B C_{s,out}^B$ and combine the result with Equation 12.3-59a,b:

$$\frac{Q_{out}^B C_{s,out}^B}{Q_{in}^B C_{s,in}^B} = \frac{(Q_{out}^D)^{f_s^D}}{(Q_{in}^B)^{f_s^B}} \left[\frac{(Q_{in}^D)^{f_s^D}}{(Q_{out}^B)^{f_s^B}} + Q_F f_s^D \int_0^1 \frac{(Q_{out}^D - Q_F \xi)^{f_s^D - 1}}{(Q_{in}^B - Q_F \xi)^{f_s^B}} d\xi \right]^{-1} \quad (12.3\text{-}64)$$

Further combining this equation with Equation 12.3-46b, we predict that

$$Q_s^C = Q_{in}^B \left\{ 1 - \frac{(Q_{out}^D)^{f_s^D}}{(Q_{in}^B)^{f_s^B}} \left[\frac{(Q_{in}^D)^{f_s^D}}{(Q_{out}^B)^{f_s^B}} + Q_F f_s^D \int_0^1 \frac{(Q_{out}^D - Q_F \xi)^{f_s^D - 1}}{(Q_{in}^B - Q_F \xi)^{f_s^B}} d\xi \right]^{-1} \right\} \quad (12.3\text{-}65)$$

To complete this integration and determine the clearance of a hemodialyzer, the *f* values must first be established. In evaluating the performance of an existing device, we would know the membrane surface area S and the geometric dimensions on the blood and dialysate sides of the device. This would enable us to estimate individual mass transfer coefficients (and therefore the *f* values) at any specified blood, dialysate, and filtrate flows. Designing a new device is more challenging because the *f* values would not be known *a priori*.

Example 12.3-3 Filtration Effects on Urea Clearance from a PUREMA Hemodialyzer

Leypoldt and associates (2006) evaluated clearance during single-pass, countercurrent blood and dialysate flows in a hemodialyzer containing high-flux polyethersulfone hollow fibers with a total fiber surface of S = 1.46 m^2. Using bovine blood with a fractional hematocrit of 0.33, Q_{in}^B was maintained at 400 ml of whole blood per minute, whereas Q_{in}^D and Q_F varied between experiments.

We wish to see whether the urea clearance data in Table 12.3-2 are consistent with the clearance predicted by Equation 12.3-65. We will assume that urea is such a small molecule that its reflection coefficient σ_s is essentially zero.

Table 12.3-2 Urea Clearance Q_s^C from a PUREMA Hollow Fiber Hemodialyzer[a]

Q_s^C (ml blood water/min)	138	156	163	164	173	190
Q_{in}^B (ml blood water/min)[a]	360	360	360	360	360	360
Q_{in}^D (ml/min)	160	160	160	200	200	200
Q_F (ml plasma/min)	0	16.7	33.3	0	16.7	33.3

[a] Blood water content was backcalculated from whole blood measurements by assuming that 80% of red cell volume and 93% of plasma volume consist of water.

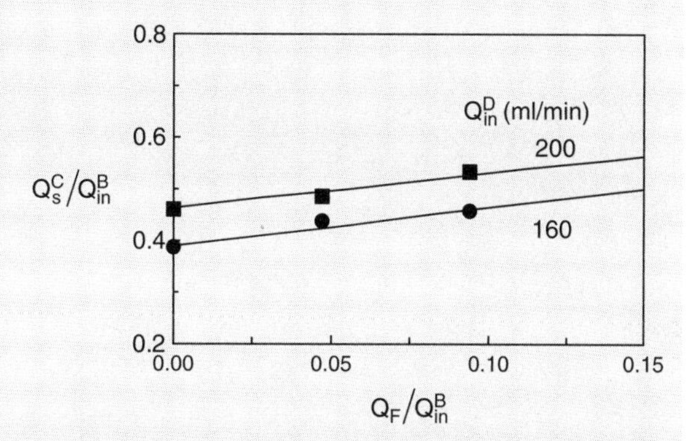

Figure 12.3-6 Urea clearance in a PUREMA hemodialyzer with filtration when $Q_{in}^B = 360\,\text{ml/min}$.

Solution

In this example problem, $\sigma_s \to 0$ so that f_s^B and f_s^D can be determined from Equation 12.3-41a,b. Identifying $u_{wall} \to Q_F/S$, these equations lead to

$$f_s^B = \frac{\exp(Q_F/K_sS)}{\exp(Q_F/K_sS)-1}, \quad f_s^D = \frac{1}{\exp(Q_F/K_sS)-1} \tag{12.3-66a, b}$$

We evaluate K_sS from clearance measurements taken when $Q_F = 0$ by substituting the first and the fourth columns of data from Table 12.3-2 into Equation 12.3-53. We find that $K_sS = 435$ ml/min when $Q^D = 160$ and $Q^B = 360$ ml/min, and $K_sS = 502$ ml/min when $Q^D = 200$ and $Q^B = 360$ ml/min. Values of Q_s^C at these two combinations of Q^D and Q^B can then be determined at any nonzero Q_F by numerical integration of Equation 12.3-65 using Equation 12.3-66a,b to calculate the corresponding f_i^B and f_i^D values.

The results of these computations are shown by the lines in Figure 12.3-6. Over the small range of Q_F/Q_{in}^B in this figure, Q_s^C improves with increasing Q_F in approximately a 1 : 1 manner. This prediction compares favorably with the experimental data points also shown in the figure.

References

Clark WR, Rocha E, Ronco CR. Solute removal by hollow-fiber dialyzers. Contrib Nephrol. 2007; 158:20–33.

Cussler EL. Diffusion: Mass Transfer in Fluid Systems. Cambridge: Cambridge University Press; 1997.

Dierickx PW, De Wachter DS, De Somer F, Van Nooten G, Verdonck PR. Mass transfer characteristics of artificial lungs. Am Soc Artif Intern Organs J. 2001; 47:628–633.

Jaffrin MY. Convective mass transfer in hemodialysis. Artif Organs. 1995; 19:1162–1171.

Leypoldt JK, Kamerath CD, Gilson JF, Friederichs G. Dialyzer clearances and mass transfer-area coefficients for small solutes at low dialysate flow rates. Am Soc Artif Intern Org J. 2006; 52:404–409.

Mandolfo S, Malberti F, Imbasciati E, Cogliati P, Gauly A. Impact of blood and dialysate flow and surface on performance of new polysulfone hemodialysis dialyzers. Int J Artif Organs. 2003; 26:113–120.

Middleman S. An Introduction to Mass and Heat Transfer. New York: John Wiley & Sons, Inc.; 1998, p 294–295.

Treybal RE. Mass-Transfer Operations. 3rd ed. New York: McGraw-Hill; 1980. p 74–75.

Yang M-C, Cussler EL. Designing hollow-fiber contactors. Am Inst Chem Eng J. 1986; 32:1910–1916.

Zydney AL. Bulk mass transport limitations during high-flux hemodialysis. Int Soc Artif Organs. 1993; 17:919–924.

Part V
Multidimensional Processes of Molecules and Cells

Chapter 13

Fluid Mechanics I

Basic Concepts

Mass transport by convection plays an important role in biological systems. Convective processes often result from mechanical pressure induced by muscular contraction. For example, contraction of the heart myocardium produces a hydrodynamic pressure that drives blood flow and contraction of the diaphragm generates an aerodynamic pressure that causes inspiration of air. Convection across porous membranes or cell assemblies such as capillary walls can be driven by osmotic as well as hydraulic pressure.

This chapter presents mass and momentum conservation principles that govern the convection of fluids. Fluid motion also depends on the relationships between frictional force and velocity for the particular material in question. The chapter discusses such material relationships for Newtonian fluids and some non-Newtonian fluids. After summarizing the steps necessary to formulate and solve convection problems, these concepts are applied to examples of fluid flow through tubes.

13.1 Application of Conservation Principles

13.1.1 Mass Conservation in a Flowing System

In performing an overall mass balance, we consider a body that moves through space and deforms in such a manner that it always contains the same atoms and molecules (Fig. 13.1-1; left). From a continuum point of view, this moving or Lagrangian control volume

Biomedical Mass Transport and Chemical Reaction: Physicochemical Principles and Mathematical Modeling, First Edition. James S. Ultman, Harihara Baskaran, and Gerald M. Saidel.

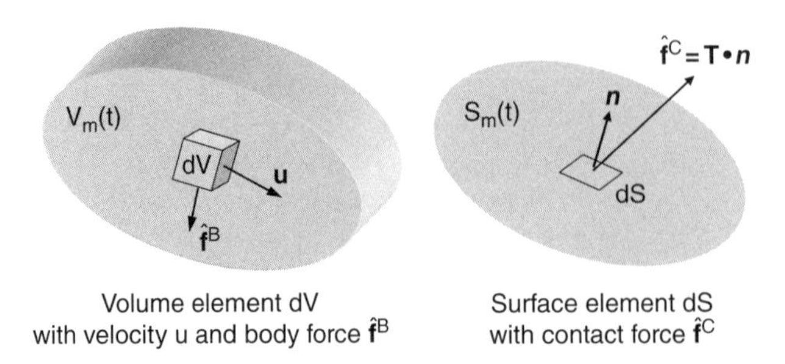

Figure 13.1-1 Moving body of volume V_m and surface S_m enclosing a fixed mass.

contains an unchanging collection of material elements, each of volume dV and mass density ρ. The total mass of the control volume is

$$\text{Mass} = \int_{V_m} \rho dV \tag{13.1-1}$$

Although the control volume $V_m(t)$ can vary with time, its mass is independent of time. Thus, the total or material derivative of the mass associated with the body as it moves through space is zero:

$$\frac{d(\text{Mass})}{dt} = \frac{d}{dt} \int_{V_m(t)} \rho dV = 0 \tag{13.1-2}$$

We can interchange the operations of differentiation and integration according to the Reynolds transport theorem (Eq. B1-19) such that

$$\int_{V_m(t)} \left[\frac{\partial \rho}{\partial t} + \nabla \cdot (\rho \mathbf{u}) \right] dV = 0 \tag{13.1-3}$$

where the partial derivative is taken at a fixed point in space. Since our choice of the body is arbitrary, $V_m(t)$ is also arbitrary and this equation is only satisfied if

$$\frac{\partial \rho}{\partial t} + \nabla \cdot (\rho \mathbf{u}) = 0 \tag{13.1-4}$$

This local mass density relation or "continuity equation" is expressed for rectangular, cylindrical, and spherical coordinates (Fig. 9.1-1) in Table B4-1. When a fluid is incompressible, ρ is constant and the continuity equation simplifies to

$$\nabla \cdot \mathbf{u} = 0 \tag{13.1-5}$$

Scalar components of this equation in rectangular, cylindrical and spherical coordinates are listed in Table 13.1-1.

The continuity equation alone cannot uniquely define a velocity field. In general, this also requires a momentum balance, a material relationship, and suitable boundary and initial conditions. The continuity equation, however, can be used to check the consistency of velocity profiles obtained from experimental observations or theoretical predictions.

Table 13.1-1 Continuity Equation for an Incompressible Fluid

Rectangular coordinates:

$$\frac{\partial u_x}{\partial x}+\frac{\partial u_y}{\partial y}+\frac{\partial u_z}{\partial z}=0$$

Cylindrical coordinates:

$$\frac{1}{r}\frac{\partial}{\partial r}(ru_r)+\frac{1}{r}\frac{\partial u_\theta}{\partial \theta}+\frac{\partial u_z}{\partial z}=0$$

Spherical coordinates:

$$\frac{1}{r^2}\frac{\partial}{\partial r}(r^2u_r)+\frac{1}{r\sin\theta}\frac{\partial}{\partial\theta}(u_\theta\sin\theta)+\frac{1}{r\sin\theta}\frac{\partial u_\phi}{\partial\phi}=0$$

Example 13.1-1 Consistency Check on a Velocity Field

A laboratory instructor constructs an experimental apparatus to measure the velocity field of water flowing around a stationary tube. Initial data indicates that the cylindrical velocity components are

$$u_r=U\cos^2\theta\left[\frac{R}{r}-\frac{1}{2r^3}\right],\quad u_\theta=\frac{U}{r^3}\left[\frac{\sin 2\theta}{4}-\frac{\theta}{2}\right],\quad u_z=-\frac{Uz\cos 2\theta}{r^4} \tag{13.1-6a-c}$$

where R is the tube radius and U is constant. A group of students report measurements of velocity components that differ from those of the instructor:

$$u_r=V\left[1-\left(\frac{R}{r}\right)^3\right]\cos\theta,\quad u_\theta=-V\left[1+\left(\frac{R}{r}\right)^3\right]\sin\theta,\quad u_z=0 \tag{13.1-7a-c}$$

where V is another constant. Are the results of the instructor or the students correct?

Solution

For most practical purposes, water is incompressible, and the continuity equation in cylindrical coordinates (Table 13.1-1) is

$$\frac{1}{r}\frac{\partial}{\partial r}(ru_r)+\frac{1}{r}\frac{\partial u_\theta}{\partial \theta}+\frac{\partial u_z}{\partial z}=0 \tag{13.1-8}$$

Substituting the velocity components from Equation 13.1-6, we get

$$\frac{1}{r}\frac{\partial}{\partial r}\left[r(U\cos^2\theta)\left(\frac{R}{r}-\frac{1}{2r^3}\right)\right]+\frac{1}{r}\frac{\partial}{\partial\theta}\left[\frac{U}{r^3}\left(\frac{\sin 2\theta}{4}-\frac{\theta}{2}\right)\right]-\frac{\partial}{\partial z}\left[\frac{Uz\cos 2\theta}{r^4}\right]=0 \tag{13.1-9}$$

After differentiating and simplifying, what remains is the identity $0\equiv 0$. The same procedure with the velocity components from Equation 13.1-7 yields

$$V\left[1+\left(\frac{R^3}{r^3}-1\right)\cos\theta\right]=0 \tag{13.1-10}$$

This equation cannot be true except for the trivial case where V is zero. Thus, the initial velocity components reported by the instructor are consistent with the continuity equation, whereas the data reported by the students are incorrect.

13.1.2 Momentum Balance in a Flowing System

For fluid flow in the Lagrangian control volume V_m, the rate of change of linear momentum must be balanced by the sum of forces $\mathbf{f}^i$ exerted on that body:

$$\frac{d(\mathbf{Momentum})}{dt} = \sum_i \mathbf{f}^i \tag{13.1-11}$$

Linear momentum of the body is the product of mass and velocity integrated over its volume.

$$\mathbf{Momentum} = \int_{V_m} \rho \mathbf{u} dV \tag{13.1-12}$$

Applying the Reynolds transport theorem (Eq. B1-21), we express the rate of change of linear momentum as

$$\frac{d(\mathbf{Momentum})}{dt} = \frac{d}{dt}\left[\int_{V_m(t)} \rho \mathbf{u} dV\right] = \int_{V_m(t)} \left[\frac{\partial(\rho \mathbf{u})}{\partial t} + \nabla \cdot (\rho \mathbf{u}\mathbf{u})\right] dV \tag{13.1-13}$$

By expanding the time derivative and utilizing the vector identity for $\nabla \cdot (\rho \mathbf{u}\mathbf{u})$ given by Equation B1-26, the integrand on the right side of this equation can be written as

$$\frac{\partial(\rho \mathbf{u})}{\partial t} + \nabla \cdot (\rho \mathbf{u}\mathbf{u}) = \left[\mathbf{u}\frac{\partial \rho}{\partial t} + \mathbf{u}\nabla \cdot (\rho \mathbf{u})\right] + \left[\rho \frac{\partial \mathbf{u}}{\partial t} + \rho \mathbf{u} \cdot \nabla \mathbf{u}\right] \tag{13.1-14}$$

According to the continuity equation, the first term on the right side of this equation is zero. Thus, the combination of Equations 13.1-11 and 13.1-13 leads to

$$\int_{V_m(t)} \left(\rho \frac{\partial \mathbf{u}}{\partial t} + \rho \mathbf{u} \cdot \nabla \mathbf{u}\right) dV = \sum_i \mathbf{f}^i \tag{13.1-15}$$

We classify $\mathbf{f}^i$ as either a body force or a contact force. Body forces such as gravity are external forces that act on the mass within a body. Total body force is obtained by integrating the local force per unit mass $\hat{\mathbf{f}}^B$ over the entire volume of a body:

$$\mathbf{f}^B = \int_{V_m} \rho \hat{\mathbf{f}}^B dV \tag{13.1-16}$$

Contact forces such as pressure and friction act on the surface of a body. Total force on a surface is determined by integrating the local contact force per unit surface $\hat{\mathbf{f}}^C$ over the entire surface S_m of the body:

$$\mathbf{f}^C = \int_{S_m} \hat{\mathbf{f}}^C dS \tag{13.1-17}$$

A local contact force can be expressed as the inner product of a second-order stress tensor $\mathbf{T}$ and the unit normal vector $\boldsymbol{n}$ acting outward from a surface element (Fig. 13.1-1; right):

$$\hat{\mathbf{f}}^C = \mathbf{T} \cdot \boldsymbol{n} \Rightarrow \mathbf{f}^C = \int_{S_m} \mathbf{T} \cdot \boldsymbol{n} dS \tag{13.1-18a,b}$$

By applying the Gauss theorem (Eq. B1-17a), we obtain the total contact force as a volume integral:

$$\mathbf{f}^C = \int_{S_m} (\mathbf{T} \cdot \boldsymbol{n}) dS = \int_{V_m} \nabla \cdot \mathbf{T}^T dV \tag{13.1-19}$$

where the superscript T indicates the transpose of the second-order tensor $\mathbf{T}$. With Equations 13.1-15, 13.1-16, and 13.1-19, the linear momentum balance on the entire body requires that

$$\int_{V(t)} \left(\rho \frac{\partial \mathbf{u}}{\partial t} + \rho \mathbf{u} \cdot \nabla \mathbf{u} - \nabla \cdot \mathbf{T}^T - \rho \hat{\mathbf{f}}^B \right) dV = 0 \tag{13.1-20}$$

For any $V_m(t)$, this integral is zero and we obtain

$$\rho \left(\frac{\partial \mathbf{u}}{\partial t} + \mathbf{u} \cdot \nabla \mathbf{u} \right) = \nabla \cdot \mathbf{T}^T + \rho \hat{\mathbf{f}}^B = 0 \tag{13.1-21}$$

which is the equation of motion. Often, the only body force is due to gravity. In that case, the body force per unit mass is equal to the gravitational acceleration vector

$$\hat{\mathbf{f}}^B = \mathcal{G} \tag{13.1-22}$$

and we can write the equation of motion as

$$\rho \left(\frac{\partial \mathbf{u}}{\partial t} + \mathbf{u} \cdot \nabla \mathbf{u} \right) = \nabla \cdot \mathbf{T}^T + \rho \mathcal{G} \tag{13.1-23}$$

Angular momentum is the vector cross-product of position and velocity vectors. The time derivative of angular momentum in a Lagrangian control volume is balanced by the sum of the moments of all the forces acting on the body. In fluids without an internal source, an angular momentum balance indicates that the stress tensor is symmetric. That is, $\mathbf{T}=\mathbf{T}^T$ so that the equation of motion can also be written as

$$\rho \left(\frac{\partial \mathbf{u}}{\partial t} + \mathbf{u} \cdot \nabla \mathbf{u} \right) = \nabla \cdot \mathbf{T} + \rho \mathcal{G} \tag{13.1-24}$$

13.1.3 Relation of Contact Forces to the Stress Tensor

The contact force vector $\hat{\mathbf{f}}^C$ has three scalar components and the total stress tensor $\mathbf{T}$ has nine scalar stress components. These components can be organized in matrix arrays:

$$\left[\hat{\mathbf{f}}^C\right] = \begin{bmatrix} \hat{f}_1^C \\ \hat{f}_2^C \\ \hat{f}_3^C \end{bmatrix}, \quad [\mathbf{T}] = \begin{bmatrix} T_{11} & T_{12} & T_{13} \\ T_{21} & T_{22} & T_{23} \\ T_{31} & T_{32} & T_{33} \end{bmatrix} \tag{13.1-25}$$

Figure 13.1-2 shows a surface element in the x_2–x_3 plane of a rectangular coordinate system with a unit outward normal $\boldsymbol{n} = [1,0,0]$ pointing in the positive x_1 direction. We will refer to this as the x_1 surface. When written as a matrix operation on its components, Equation 13.1-18a then becomes

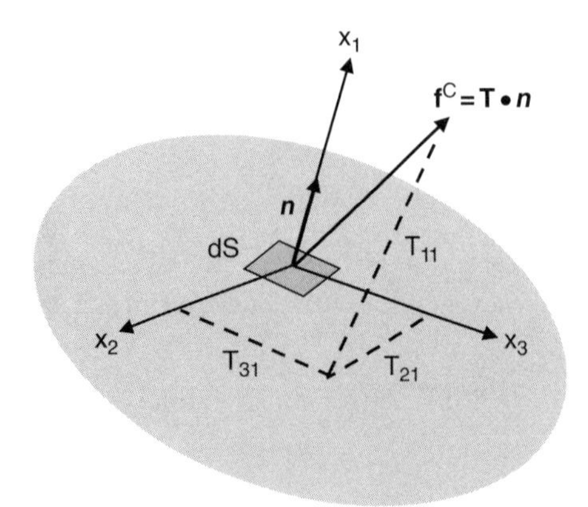

Figure 13.1-2 Decomposition of contact force as stress tensor components.

$$\begin{bmatrix} \hat{f}_1^C \\ \hat{f}_2^C \\ \hat{f}_3^C \end{bmatrix} = \begin{bmatrix} T_{11} & T_{12} & T_{13} \\ T_{21} & T_{22} & T_{23} \\ T_{31} & T_{32} & T_{33} \end{bmatrix} \begin{bmatrix} 1 \\ 0 \\ 0 \end{bmatrix} = \begin{bmatrix} T_{11} \\ T_{21} \\ T_{31} \end{bmatrix} \tag{13.1-26}$$

This equation indicates that the contact force on the x_1 surface has three components: a normal stress T_{11} acting in the x_1 direction perpendicular to the surface and shear stresses T_{21} and T_{31} acting parallel to the surface in the x_2 and x_3 directions, respectively. By also considering surfaces with outward normals in the x_2 and in the x_3 directions, one can infer that the contact force on each of the $x_j(j = 1,2,3)$ surfaces has one normal stress component T_{jj} and two shear stress components $T_{ij}(i \neq j)$. For the collection of all three surfaces, there are three outward normal forces that form the diagonal elements in the [**T**] matrix, and six shear stress components corresponding to the off-diagonal elements in the matrix. For the fluids considered here, the total stress tensor and thus its matrix are symmetric. Therefore, $T_{ij} = T_{ji}$ and only three of the six possible shear stress components have unique values.

We consider two contributions to contact forces: fluid pressure and viscous stress. Fluid pressure acts perpendicular to surfaces with the same magnitude on all surfaces regardless of their orientation. Viscous stress results from fluid friction and can act normal as well as parallel to surfaces. The total stress tensor **T** can be decomposed into a pressure tensor $-P\mathbf{I}$ and a viscous stress tensor $\boldsymbol{\tau}$:

$$\mathbf{T} = -P\mathbf{I} + \boldsymbol{\tau} \Leftrightarrow \begin{bmatrix} T_{11} & T_{12} & T_{13} \\ T_{21} & T_{22} & T_{23} \\ T_{31} & T_{32} & T_{33} \end{bmatrix} = -P \begin{bmatrix} 1 & 0 & 0 \\ 0 & 1 & 0 \\ 0 & 0 & 1 \end{bmatrix} + \begin{bmatrix} \tau_{11} & \tau_{12} & \tau_{13} \\ \tau_{21} & \tau_{22} & \tau_{23} \\ \tau_{31} & \tau_{32} & \tau_{33} \end{bmatrix} \tag{13.1-27a,b}$$

where **I** is the identity tensor. The three diagonal elements of $-P\mathbf{I}$ represent the outward pressure exerted perpendicular to the x_1, x_2, and x_3 surfaces. The negative sign indicates that P is taken to be positive when it acts inward on a surface. A fluid, by its definition, does not support viscous stresses when it is static so that $\boldsymbol{\tau} = 0$ and $\mathbf{T} = -P\mathbf{I}$. For a compressible fluid, P is a material property whose value is related to density and temperature by a thermodynamic equation of state such as the ideal gas law. For an incompressible fluid, P is found by applying the trace operation (i.e., the sum the diagonal components of a square matrix) to Equation 13.1-27b:

$$\sum_{i=1}^{3} T_{ii} = -\sum_{i=1}^{3} I_{ii}P + \sum_{i=1}^{3} \tau_{ii} \Rightarrow P = -\frac{1}{3}\sum_{i=1}^{3} (T_{ii} - \tau_{ii}) \tag{13.1-28a,b}$$

Thus, pressure is the average inward normal stress in excess of the frictional contribution. Like the total stress tensor **T**, the viscous stress tensor $\boldsymbol{\tau}$ has three normal stress components corresponding to the diagonal elements of its matrix representation and six shear stress components represented by the off-diagonal elements. Because **T** is symmetric, $\boldsymbol{\tau}$ is symmetric and only three of its six shear stress components have unique values.

It often simplifies problems involving gravity to define a modified pressure whose gradient is the sum of the actual pressure gradient and a pressure-like contribution due to gravity:

$$\nabla \mathcal{P} = \nabla P - \rho \mathcal{G} \tag{13.1-29}$$

Substituting Equations 13.1-27a and 13.1-29 into Equation 13.1-24, we express the local momentum balance as

$$\rho \left[\frac{\partial \mathbf{u}}{\partial t} + \mathbf{u} \cdot \nabla \mathbf{u} \right] = -\nabla \mathcal{P} + \nabla \cdot \boldsymbol{\tau} \tag{13.1-30}$$

The scalar components of this vector equation in rectangular, cylindrical, and spherical coordinates are given in Tables B4-2, B4-4, and B4-6. This Euler equation of motion can be solved with the continuity equation to determine a velocity field, once we specify a rheological relation between $\boldsymbol{\tau}$ and **u**. Such a rheological relation depends on the mechanical properties of the material being analyzed.

13.2 Mechanical Properties and Rheology of Fluids

13.2.1 Fluid Deformation

A deformable body subjected to an external force will change its shape in a manner that depends on the nature of the material. For a solid, an applied force leads to constant deformation. As an example, placing a weight on the free end of a rubber band stretches the band until it reaches a new length. For a fluid, an applied force results in an ongoing deformation. As an example, imposing a pressure difference across a tube causes a continuous shearing of a fluid as it flows through the tube.

To quantify the rate of deformation of a material element, we introduce the symmetric tensor:

$$\boldsymbol{\gamma} = \frac{1}{2}\left[\nabla \mathbf{u} + (\nabla \mathbf{u})^{T}\right] \tag{13.2-1}$$

where the superscript T represents the transpose of the second-order tensor quantity $\nabla \mathbf{u}$. In a rectangular coordinate system, we can write this deformation rate tensor in matrix form (Table B4-8):

$$\begin{bmatrix} \gamma_{xx} & \gamma_{xy} & \gamma_{xz} \\ \gamma_{yx} & \gamma_{yy} & \gamma_{yz} \\ \gamma_{zx} & \gamma_{zy} & \gamma_{zz} \end{bmatrix} = \begin{bmatrix} \dfrac{\partial u_x}{\partial x} & \dfrac{1}{2}\left(\dfrac{\partial u_x}{\partial y} + \dfrac{\partial u_y}{\partial x}\right) & \dfrac{1}{2}\left(\dfrac{\partial u_x}{\partial z} + \dfrac{\partial u_z}{\partial x}\right) \\ \dfrac{1}{2}\left(\dfrac{\partial u_y}{\partial x} + \dfrac{\partial u_x}{\partial y}\right) & \dfrac{\partial u_y}{\partial y} & \dfrac{1}{2}\left(\dfrac{\partial u_y}{\partial z} + \dfrac{\partial u_z}{\partial y}\right) \\ \dfrac{1}{2}\left(\dfrac{\partial u_z}{\partial x} + \dfrac{\partial u_x}{\partial z}\right) & \dfrac{1}{2}\left(\dfrac{\partial u_z}{\partial y} + \dfrac{\partial u_y}{\partial z}\right) & \dfrac{\partial u_z}{\partial z} \end{bmatrix} \tag{13.2-2}$$

The velocity gradient $\partial u_i/\partial x_j$ represents the rate of movement per unit length of an x_j surface of a material element in the x_i direction. Therefore, a diagonal component γ_{ii} of the deformation rate tensor corresponds to a surface movement perpendicular to the surface itself, indicating that the material element is being stretched in the x_i direction. An off-diagonal component $\gamma_{ij}(i \neq j)$ is the average of surface movements parallel to surfaces in the x_i and the x_j directions. Such movements correspond to a shearing of the material element.

Fluid rheology is often studied in steady laminar flows that produce a single shear deformation without any fluid stretching. For example, consider a liquid between two parallel flat plates separated by a constant spacing h (Fig. 13.2-1). A force F is applied to the upper plate, moving it at constant velocity U parallel to the lower plate that is held stationary. Strong contact forces prevent fluid slippage along the solid surfaces. Therefore, fluid velocity is zero at the lower plate where $y = 0$ and is U at the upper plate where $y = h$. In the fluid-filled gap between the plates, parallel velocity vectors in the x direction have magnitudes that increase in a linear fashion with y. The velocity components are then given by:

$$u_x = \frac{U}{h}y, \quad u_y = u_z = 0 \tag{13.2-3}$$

Substituting these kinematics into Equation 13.2-2 demonstrates that the only nonzero contributions to $\boldsymbol{\gamma}$ are a single pair of equal diagonal components:

$$\gamma \equiv \gamma_{xy} = \gamma_{yx} = \frac{1}{2}\frac{du_x}{dy} = \frac{U}{2h} \tag{13.2-4}$$

Since such a plane shearing flow between parallel plates is inconvenient to produce in a laboratory, other types of simple shearing flows are normally used. Examples of this are shearing of fluid between a rotating outer cylinder and a stationary inner cylinder, shearing of fluid between a rotating circular cone and a stationary parallel plate, and fluid flow through a cylindrical tube. In each case, one pair of equal diagonal components is the only nonzero contribution to the deformation rate tensor.

13.2.2 Newtonian Fluids

We will restrict our discussion of rheological fluid models to fluids of constant density. The most common model of incompressible fluid behavior assumes that the viscous stress tensor is proportional to the deformation rate tensor:

$$\boldsymbol{\tau} = 2\mu\boldsymbol{\gamma} \tag{13.2-5}$$

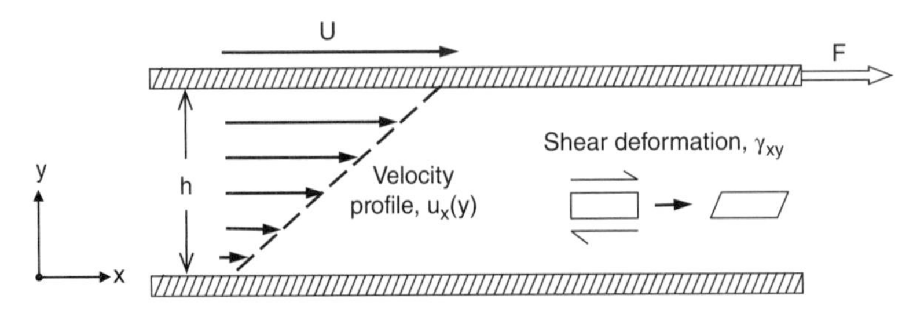

Figure 13.2-1 Plane shearing flow between flat plates.

Table 13.2-1 Properties of Some Liquids at 37°C

Fluid	Shear Viscosity μ (Pa-s)	Density ρ (kg/m³)	Kinematic Viscosity υ ≡ μ/ρ (m²/s)
Air	1.90×10^{-5}	1.14	1.67×10^{-5}
Water	6.93×10^{-4}	9.94×10^{2}	6.98×10^{-7}
Blood plasma	1.2×10^{-3}	1.02×10^{3}	1.18×10^{-6}
Blood (high shear rates)	3×10^{-3}	1.05×10^{3}	2.86×10^{-6}
Glycerin	0.313	1.25×10^{3}	2.49×10^{-4}

where μ is a scalar property called the shear viscosity or simply the viscosity. Viscosity increases with temperature for gases, decreases with temperature for liquids, and is virtually independent of pressure for all liquids. Values of μ for several fluids are listed in Table 13.2-1.

In a simple shearing flow, the components of $\boldsymbol{\gamma}$ are $\gamma_{12} = \gamma_{21} \equiv \gamma$, and the corresponding components of $\boldsymbol{\tau}$ are the viscous shear stresses $\tau_{12} = \tau_{21} \equiv \tau$. In this situation, we can write the Newtonian model as a scalar equation:

$$\tau = 2\mu\gamma \tag{13.2-6}$$

For a Newtonian fluid of constant viscosity and density, the divergence of the viscous stress tensor appearing the Euler equation of motion (Eq. 13.1-30) is

$$\nabla \cdot \boldsymbol{\tau} = \nabla \cdot (2\mu\boldsymbol{\gamma}) = \mu\nabla \cdot \left[\nabla\mathbf{u} + (\nabla\mathbf{u})^{\mathrm{T}}\right] \tag{13.2-7}$$

Substituting the vector identity from Equation B1-29, this relation becomes

$$\nabla \cdot \boldsymbol{\tau} = \mu\left[\nabla^2\mathbf{u} + \nabla(\nabla \cdot \mathbf{u})\right] \tag{13.2-8}$$

With a constant mass density, $\nabla \cdot \mathbf{u} = 0$ and substitution of Equation 13.2-8 into Equation 13.1-30 yields the Navier–Stokes equation:

$$\rho\left(\frac{\partial\mathbf{u}}{\partial t} + \mathbf{u} \cdot \nabla\mathbf{u}\right) = -\nabla\mathcal{P} + \mu\nabla^2\mathbf{u} \tag{13.2-9}$$

The three components of this vector equation for common coordinate systems are given in Tables B4-3, B4-5, and B4-7.

13.2.3 Non-Newtonian Fluids

Many ordinary fluids such as air or physiological saline solution behave as Newtonian fluids. In biological fluids, however, the presence of large molecules and suspended cells can lead to non-Newtonian behavior. These fluids can be classified as either purely viscous or viscoelastic. For example, blood in small-diameter vessels acts as a non-Newtonian, purely viscous fluid, whereas respiratory mucus is a viscoelastic fluid.

Purely viscous and viscoelastic fluids can be distinguished by their responses in a simple shearing flow after the force on the movable surface is suddenly terminated. If a fluid were purely viscous, then frictional forces would slow the motion of the surface to a stop. However, if a fluid were viscoelastic, then the motion of the surface would reverse direction for a short period of time. This occurs because a viscoelastic fluid stores a portion of the mechanical work supplied by an applied force as elastic energy. When the force is removed, the stored energy is transformed into an elastic recoil of the fluid.

We can also distinguish between purely viscous and viscoelastic fluids by the nature of the contact forces developed during a simple shearing flow. A purely viscous fluid will only produce a shear stress, whereas a viscoelastic fluid can exhibit viscous normal stresses in addition to a shear stress.

Purely Viscous Models

Many models have been developed to describe rheological behavior of purely viscous and viscoelastic fluids. We will consider only purely viscous fluids described by a generalized Newtonian equation:

$$\boldsymbol{\tau} = 2\mu_{app}\boldsymbol{\gamma} \tag{13.2-10}$$

Here, μ_{app} is an apparent viscosity, a scalar quantity that depends on an apparent deformation rate γ_{app} that is also a scalar:

$$\gamma_{app} \equiv \left| \sqrt{\frac{\text{tr}(\boldsymbol{\gamma}\cdot\boldsymbol{\gamma})}{2}} \right| \tag{13.2-11}$$

Here, "tr" represents the trace operation. In a simple shearing flow, for example,

$$\gamma_{app} = \left| \sqrt{\frac{1}{2}\text{tr}\left[\begin{bmatrix} 0 & \gamma & 0 \\ \gamma & 0 & 0 \\ 0 & 0 & 0 \end{bmatrix} \bullet \begin{bmatrix} 0 & \gamma & 0 \\ \gamma & 0 & 0 \\ 0 & 0 & 0 \end{bmatrix}\right]} \right| = \left| \sqrt{\frac{1}{2}\text{tr}\begin{bmatrix} \gamma^2 & 0 & 0 \\ 0 & \gamma^2 & 0 \\ 0 & 0 & 0 \end{bmatrix}} \right| = |\gamma| \tag{13.2-12a-c}$$

Table B4-9 gives the forms of γ_{app}^2 in commonly used coordinate systems.

The mechanical behavior of many homogeneous, non-Newtonian viscous fluids can be approximated over a limited range of deformation rates by the Ostwald-de Waele or power-law model. The apparent viscosity for this model is

$$\mu_{app} = m\left(2\gamma_{app}\right)^{n-1} \tag{13.2-13}$$

where m and n are positive constants. The corresponding viscous stress tensor is

$$\boldsymbol{\tau} = 2m\left(2\gamma_{app}\right)^{n-1}\boldsymbol{\gamma} \xrightarrow{\text{Simple shearing flow}} \tau = 2m|2\gamma|^{n-1}\gamma \tag{13.2-14a,b}$$

For the special case of n = 1, the power-law fluid reduces to a Newtonian fluid with a constant value of μ_{app} equal to m. When n < 1, μ_{app} decreases with increasing shear rate. The fluid is then said to be shear thinning. When n > 1, μ_{app} increases with increasing shear rate, that is, the fluid is shear thickening.

The rheology of some suspensions can be described by a Casson model whose apparent viscosity is

$$\mu_{app} = \left[\sqrt{\tau_o/2\gamma_{app}} + \sqrt{\mu_\infty}\right]^2 \tag{13.2-15}$$

The viscous stress tensor is then given by

$$\boldsymbol{\tau} = 2\left[\sqrt{\tau_o/2\gamma_{app}} + \sqrt{\mu_\infty}\right]^2\boldsymbol{\gamma} \xrightarrow{\text{Simple shearing flow}} \sqrt{|\tau|} = \sqrt{\tau_o} + \sqrt{2\mu_\infty|\gamma|} \tag{13.2-16a,b}$$

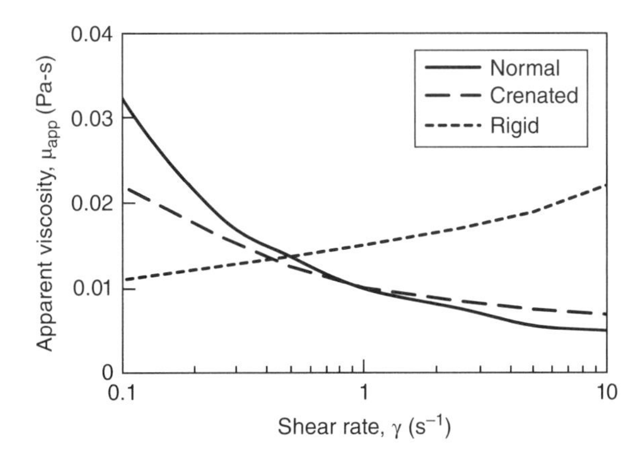

Figure 13.2-2 Red cell deformability and rheology of blood with a 40% hematocrit.

where μ_∞ and τ_o have positive values. Since $|\tau| \to \tau_o$ at low shear rates, we consider τ_o to be a yield stress that must be overcome before fluid motion can occur. At high shear rates, $|\tau| \to 2\mu_\infty|\gamma|$ and the Casson fluid in a shear flow behaves like a Newtonian fluid with a shear viscosity μ_∞.

Rheological Model of Blood

The rheology of normal blood can be modeled by the Casson equation. At very low shear rates, the disk-like erythrocytes aggregate into stacks or rouleaux. The resistance of rouleaux to motion creates very large frictional forces that are the source of τ_o. As shear rate increases, the rouleaux break down into individual cells. This reduces the frictional forces, thereby decreasing the apparent viscosity to its lower limit of μ_∞. The numerical values of τ_o and μ_∞ depend on temperature, hematocrit, and the content of plasma proteins in a particular blood sample (Merrill *et al.*, 1965).

As shown by Wells and Goldstone (1973), the rheology of blood also depends on the state of the erythrocytes (Fig. 13.2-2). Normal blood is a shear-thinning fluid with erythrocytes that are quite deformable. If erythrocytes are pretreated with hypotonic solution, they become more spherical with many small protrusions on the cell membrane. This structural change (or crenation) reduces shear thinning by impeding the rouleaux formation at low shear rates. Making the erythrocytes more rigid by pretreatment with a fixative such as glutaraldehyde causes blood to become shear thickening.

13.3 Model Formulation and Scaling of Fluid Flow

13.3.1 Elements of Model Formulation

The continuity equation and scalar components of the equation of motion with an associated rheological model comprise a nonlinear system of four partial differential equations that describe the three velocity components and pressure. For special cases, these equations can be solved analytically, but numerical solutions are needed for more general cases.

In practice, complex problems such as air flow through the lung airways or blood flow through a capillary bed require some simplifying assumptions even for numerical solution. The following elements are typical in modeling a fluid flow system.

Material Properties

The thermodynamic and rheological properties of the fluid must be specified. For many biomedical applications, fluids can be considered incompressible and Newtonian.

Coordinate System

It is desirable to select a coordinate system with coordinate surfaces that align with boundaries of the flow domain. This reduces the number of independent variables and simplifies the specification of boundary conditions. For example, flow of a fluid through a rigid tube with a circular cross section is conveniently represented by a cylindrical coordinate system. Because of circumferential symmetry, only two coordinates are needed, an axial coordinate z coincident with the centerline of the tube and a radial coordinate r. The boundary between the fluid and the tube wall is specified by assigning a constant value of r, and the flow inlet and outlet boundaries correspond to constant values of z.

Structural and Kinematic Simplification

We can often simplify fluid convection models by idealizing the system structure and the kinematics of the fluid motion. When analyzing air flow through the trachea, for example, we could approximate the geometry by a straight cylindrical tube and assume that the velocity field is one dimensional in the flow direction. Consequently, terms in the governing equations that do not contain the axial velocity component can be neglected. In many problems, an order-of-magnitude analysis of the dimensionless governing equations can lead to further simplification.

Boundary Conditions

The constraints of the physical system are imposed on the transport equations by boundary conditions. On the imaginary boundary through which flow enters a system, the velocity distribution is usually specified. At a flow exit, normal viscous stresses are often so small that the normal velocity gradient can be set to zero. In some systems, pressure rather than velocity and velocity gradients are given at a flow inlet and flow outlet. Within the transport domain, there can be a point or line of geometric symmetry where a velocity component or its spatial derivative vanishes. As discussed below, velocity and pressure conditions at phase interfaces are based on special continuity and momentum balance equations (Slattery, 1999).

13.3.2 Interface Relationships

Conservation of Mass

Consider an interface that separates two immiscible solutions designated as + and – phases in Figure 13.3-1. In general, discontinuities in velocity ($\mathbf{u}^+$, $\mathbf{u}^-$) and density (ρ^+, ρ^-) exist at an interface that moves at a velocity $\mathbf{u}^S$. Excluding surface active materials, an interface is a two-dimensional entity that has no mass of its own, but mass can be transferred across it. Thus, overall mass is conserved when the mass transfer rate toward an interface from

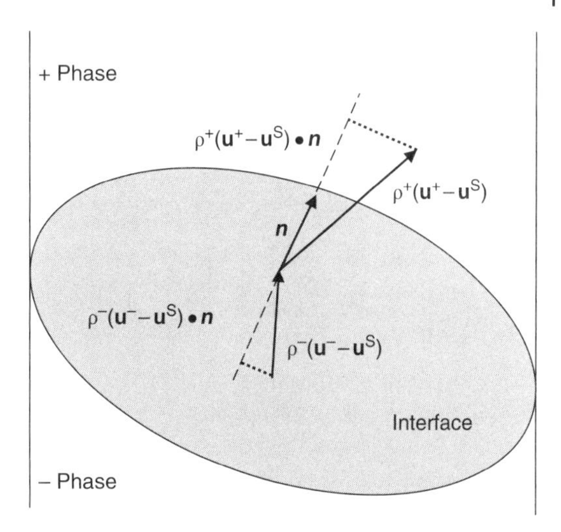

Figure 13.3-1 Convection of mass through an interface moving at a velocity $\mathbf{u}^S$.

the −phase is equal to the mass transfer rate from an interface toward the +phase. This leads to the following jump condition at a point on the surface:

$$\rho^+\left(\mathbf{u}^+-\mathbf{u}^S\right)\cdot\boldsymbol{n}=\rho^-\left(\mathbf{u}^--\mathbf{u}^S\right)\cdot\boldsymbol{n} \tag{13.3-1}$$

where the outward unit normal $\boldsymbol{n}$ is directed from the −phase to the +phase. For an interface that does not move in the normal direction, $\mathbf{u}^S\cdot\boldsymbol{n}=0$ and the mass jump condition reduces to

$$\rho^+\mathbf{u}^+\cdot\boldsymbol{n}=\rho^-\mathbf{u}^-\cdot\boldsymbol{n} \tag{13.3-2}$$

To better understand the implication of this interfacial condition, let us apply it to an x_1 surface (Fig. 13.1-2) whose unit outward normal components are [1,0,0]:

$$\rho^+\left[u_1^+\;\;u_2^+\;\;u_3^+\right]\cdot\begin{bmatrix}1\\0\\0\end{bmatrix}=\rho^-\left[u_1^-\;\;u_2^-\;\;u_3^-\right]\cdot\begin{bmatrix}1\\0\\0\end{bmatrix}\Rightarrow u_1^+=\frac{\rho^-}{\rho^+}u_1^- \tag{13.3-3a,b}$$

Thus, the mass jump condition reveals a discontinuity in the normal velocity component u_1 as it crosses the x_1 surface. The situation for tangential velocity components, u_2 and u_3, is different however. According to the commonly used no-slip condition, sliding friction between the +phase and the −phase is so large that these velocities are continuous:

$$u_2^+=u_2^-,\quad u_3^+=u_3^- \tag{13.3-4a,b}$$

Momentum Balance

Excluding again the effects of surface active materials, an interface produces neither linear momentum nor contact forces of its own. However, the net rate of momentum convected across an interface from the −phase to the +phase must balanced by the normal force acting on the interface. This produces a jump condition on linear momentum:

$$\rho^+\left(\mathbf{u}^+-\mathbf{u}^S\right)\left(\mathbf{u}^+-\mathbf{u}^S\right)\cdot\boldsymbol{n}-\rho^-\left(\mathbf{u}^--\mathbf{u}^S\right)\left(\mathbf{u}^--\mathbf{u}^S\right)\cdot\boldsymbol{n}=(\mathbf{T}^+\cdot\boldsymbol{n})-(\mathbf{T}^-\cdot\boldsymbol{n}) \tag{13.3-5}$$

In the absence of mass transfer through an interface, $\rho^+(\mathbf{u}^+ - \mathbf{u}^S)\cdot \boldsymbol{n} = \rho^-(\mathbf{u}^- - \mathbf{u}^S)\cdot \boldsymbol{n} = 0$, and the momentum jump condition reduces to an equality between the normal contact forces:

$$\mathbf{T}^+ \cdot \boldsymbol{n} = \mathbf{T}^- \cdot \boldsymbol{n} \tag{13.3-6}$$

For a x_1 surface element, we can write this equation as

$$\begin{bmatrix} T_{11}^+ & T_{12}^+ & T_{13}^+ \\ T_{21}^+ & T_{22}^+ & T_{23}^+ \\ T_{31}^+ & T_{32}^+ & T_{33}^+ \end{bmatrix} \cdot \begin{bmatrix} 1 \\ 0 \\ 0 \end{bmatrix} = \begin{bmatrix} T_{11}^- & T_{12}^- & T_{13}^- \\ T_{21}^- & T_{22}^- & T_{23}^- \\ T_{31}^- & T_{32}^- & T_{33}^- \end{bmatrix} \cdot \begin{bmatrix} 1 \\ 0 \\ 0 \end{bmatrix} \Rightarrow \begin{matrix} T_{11}^+ = T_{11}^- \\ T_{21}^+ = T_{21}^- \\ T_{31}^+ = T_{31}^- \end{matrix} \tag{13.3-7a,b}$$

Therefore, in the absence of mass transfer, both the normal and shearing components of $\mathbf{T}$ are continuous across an interface.

Example 13.3-1 Multiphase Flow System

Consider an incompressible Newtonian liquid film of constant thickness flowing down a wide inclined plane (Fig. 13.3-2). Using the kinematics and stress tensor of this liquid, formulate boundary conditions at the liquid–solid and gas–liquid interfaces.

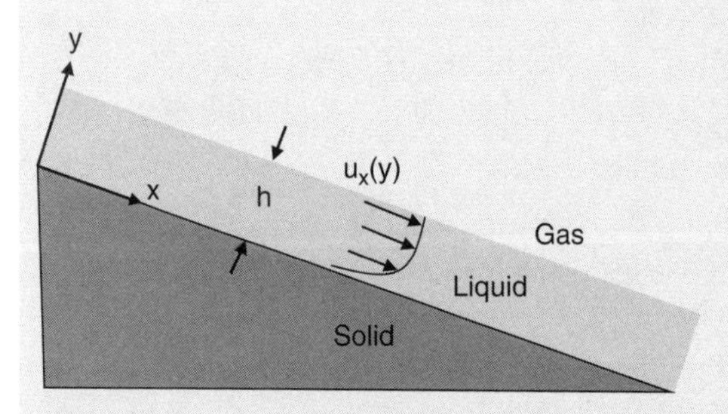

Figure 13.3-2 Flow of a liquid film down an inclined plane.

Solution

In this analysis, variables associated with the solid and gas phases outside of the film will be indicated by superscripts S and G. Variables referring to the liquid in the film will not have a superscripted designation.

Stress Tensor. For a film of constant thickness h, the fluid kinematics are similar to those in the gap between two parallel plates. In this case, however, the film is exposed to a gas phase at $y = h$ so the velocity distribution is not necessarily a linear function of y:

$$u_x = u_x(y), \quad u_y = 0, \quad u_z = 0 \tag{13.3-8a-c}$$

For this velocity field, the components of the deformation rate tensor are given by

$$\boldsymbol{\gamma} = \begin{bmatrix} 0 & \gamma_{xy} & 0 \\ \gamma_{yx} & 0 & 0 \\ 0 & 0 & 0 \end{bmatrix} = \begin{bmatrix} 0 & u_x'/2 & 0 \\ u_x'/2 & 0 & 0 \\ 0 & 0 & 0 \end{bmatrix} \tag{13.3-9}$$

where $u_x'(y) \equiv du_x/dy$. The stress tensor in the liquid film is

$$\mathbf{T} = -P\mathbf{I} + \boldsymbol{\tau} = -P\mathbf{I} + 2\mu\boldsymbol{\gamma} \Rightarrow \mathbf{T} = \begin{bmatrix} -P & \mu u'_x & 0 \\ \mu u'_x & -P & 0 \\ 0 & 0 & -P \end{bmatrix} \tag{13.3-10a,b}$$

Liquid–Solid Interface. The liquid–solid interface corresponds to an x–z plane at $y=0$. We apply the no-slip conditions in addition to the mass jump condition at this boundary:

$$\left.\begin{aligned} u_x(0) &= u_x^S(0) \\ u_z(0) &= u_z^S(0) \end{aligned}\right\} \text{ No-slip conditions}$$
$$u_y(0) = \frac{\rho^S}{\rho} u_y^S(0) \quad \text{Jump condition} \tag{13.3-11a-c}$$

For a stationary solid, $u_x^S(0) = u_y^S(0) = u_z^S(0) = 0$ and thus the velocity condition reduces to

$$u_y(0) = u_x(0) = u_z(0) = 0 \tag{13.3-12}$$

Since the kinematics assert that u_y and u_z are zero everywhere in the film, the only velocity condition that is actually needed at the liquid–solid interface is $u_x(0) = 0$. As indicated by Equation 13.3-7b, the unique stresses acting on the liquid-solid interface are T_{xy}, T_{yy} and T_{zy}. With no mass transfer across this surface, we apply the momentum jump condition (Eq. 13.3-7b) to obtain

$$T_{xy}^S(0) = T_{xy}(0), \quad T_{yy}^S(0) = T_{yy}(0), \quad T_{zy}^S(0) = T_{zy}(0) \tag{13.3-13a-c}$$

By considering the stress-deformation rate relationships in the liquid (Eq. 13.3-10), these conditions become

$$T_{xy}^S(0) = \mu u'_x(0), \quad T_{yy}^S(0) = -P(0), \quad T_{zy}^S(0) = 0 \tag{13.3-14a-c}$$

Typically, stresses in the solid are not known nor needed for solving the Navier–Stokes equation in the liquid film. Rather, Equation 13.3-14a,b provides a means of determining $T_{xy}^S(0)$ and $T_{yy}^S(0)$, once the solution for the liquid velocity and pressure fields have been found.

Gas–Liquid Interface. The gas-liquid interface consists of an x–z plane located at $y=h$. The velocity conditions at this boundary are

$$\left.\begin{aligned} u_x(h) &= u_x^G(h) \\ u_z(h) &= u_z^G(h) \end{aligned}\right\} \text{ No-slip conditions}$$
$$u_y(h) = \frac{\rho_G}{\rho} u_y^G(h) \quad \text{Jump condition} \tag{13.3-15a-c}$$

If we do not know the velocity field in the gas phase, then these conditions are of no practical use. Without mass transfer across the gas–liquid interface, the momentum jump condition indicates that

$$T_{xy}^G(h) = T_{xy}(h), \quad T_{yy}^G(h) = T_{yy}(h), \quad T_{zy}^G(h) = T_{zy}(h) \tag{13.3-16a-c}$$

Based on the stress-deformation rate relationships for the liquid (Eq. 13.3-10), we further obtain

$$\begin{aligned} T_{xy}^G(h) &= \tau_{xy}^G(h) = \mu u'_x(h) \\ T_{yy}^G(h) &= -P^G(h) + \tau_{yy}^G(h) = -P(h) \\ T_{zy}^G(h) &= \tau_{zy}^G(h) = \tau_{zy}(h) = 0 \end{aligned} \tag{13.3-17a-c}$$

Because gases have small viscosities relative to liquids, the viscous stresses in the gas phase can be neglected which leaves the following useful boundary conditions at the gas-liquid interface:

$$u'_x(h) = 0, \quad P(h) = P^G \qquad (13.3\text{-}18a,b)$$

In summary, the velocity and pressure conditions at the boundaries of the liquid film with a solid at $y = 0$ and gas at $y = h$ are

$$\begin{aligned} y = 0: &\quad u_x = 0 \\ y = h: &\quad \frac{du_x}{dy} = 0, \quad P = P^G \end{aligned} \qquad (13.3\text{-}19a,b)$$

13.3.3 Dimensionless Flow Equations

A flow model is made dimensionless by scaling the variables appearing in the equation of motion, the continuity equation, and their boundary conditions (Appendix C1). Making a model dimensionless allows us to identify terms that are of negligible importance and to minimize the number of independent parameters that ultimately must be considered.

For example, when modeling the flow of an incompressible fluid described by the continuity equation and Navier–Stokes equation, we can define the dimensionless variables:

$$t = \frac{u_c t_c}{L_c}, \quad \boldsymbol{u} = \frac{\mathbf{u}}{u_c}, \quad \nabla = r_c \nabla, \quad P = \frac{L_c \mathcal{P}}{\mu u_c}, \quad \tau = \frac{L_c \tau}{\mu u_c} \qquad (13.3\text{-}20a\text{-}e)$$

where the scale factors are viscosity μ and characteristic values of velocity u_c, pressure P_c, time t_c, and length L_c. Substituting these variables into Equations 13.1-5 and 13.2-9, we obtain a dimensionless continuity equation:

$$\nabla \bullet \boldsymbol{u} = 0 \qquad (13.3\text{-}21)$$

and a dimensionless Navier–Stokes equation:

$$\frac{1}{St}\frac{\partial \boldsymbol{u}}{\partial t} + \boldsymbol{u} \cdot \nabla \boldsymbol{u} = \frac{1}{Re}\left(-\nabla P + \nabla^2 \boldsymbol{u}\right) \qquad (13.3\text{-}22)$$

The dimensionless equation of motion contains two dimensionless parameter groups, the Strouhal number (St) and Reynolds number (Re):

$$St \equiv \frac{t_c u_c}{L_c}, \quad Re \equiv \frac{\rho u_c L_c}{\mu} \qquad (13.3\text{-}23a,b)$$

The relative importance of the terms appearing in the Navier–Stokes equation can be evaluated from the relative magnitude of these dimensionless coefficients. When $St \gg 1$, the transient term can be neglected relative to the convection term $\boldsymbol{u} \cdot \nabla \boldsymbol{u}$ that has a coefficient of unity. Thus, St can be interpreted as the ratio of convected momentum to transient momentum. When $Re \gg 1$, pressure and viscous stress terms are small compared to the convection term so that Re represents the ratio of convected momentum to contact forces.

13.4 Steady Flow Through a Tube

13.4.1 Flow of Newtonian and Power-Law Fluids

The steady flow of a homogeneous incompressible fluid through a straight cylindrical tube is a basic model used as an approximation in many biomedical applications (Fig. 13.4-1). Suppose that the axial velocity of fluid entering the tube from a quiescent reservoir has a uniform radial distribution. By retarding forward motion, friction at the tube wall creates a momentum boundary layer that thickens as fluid moves along the tube. Eventually, the boundary layer fills the tube, and the axial velocity distribution no longer changes its shape. Beyond this point, designated as the entrance length L_e, the velocity distribution is said to be fully developed.

For a Newtonian fluid of viscosity μ flowing at a volumetric rate Q through a tube of radius "a," the fully developed axial velocity profile has a parabolic shape:

$$u_z = 2u_{ave}\left[1-\left(\frac{r}{a}\right)^2\right] \tag{13.4-1}$$

where $u_{ave} = Q/\pi a^2$ is the average velocity across the tube cross section. This velocity is related to the modified pressure drop through a tube of length L by the Hagen–Poiseuille equation:

$$\Delta\mathcal{P} = \frac{8u_{ave}\mu L}{a^2} = \frac{8Q\mu L}{\pi a^4} \tag{13.4-2}$$

Over a limited range of shear rates, the non-Newtonian behavior of physiological fluids flowing through a tube can often be approximated by the power-law model. In the following illustration, we will derive the fully developed velocity distribution of such a fluid by simultaneously solving the continuity equation and the equation of motion.

Model Formulation

Fully developed laminar flow in a straight tube of constant radius is most simply described with a cylindrical coordinate system whose z coordinate is coincident with the tube centerline. Flow enters the tube at the origin of the coordinate system z = 0, and it exits the tube at z = L. The inner wall of the tube is located at r = a. Because of circumferential symmetry about the z axis, u_z does not depend on θ, and there is no velocity component in that direction. When the tube wall is not permeable, the r velocity component is also zero:

$$u_z = u_z(r,z), \quad u_r = 0, \quad u_\theta = 0 \tag{13.4-3a-c}$$

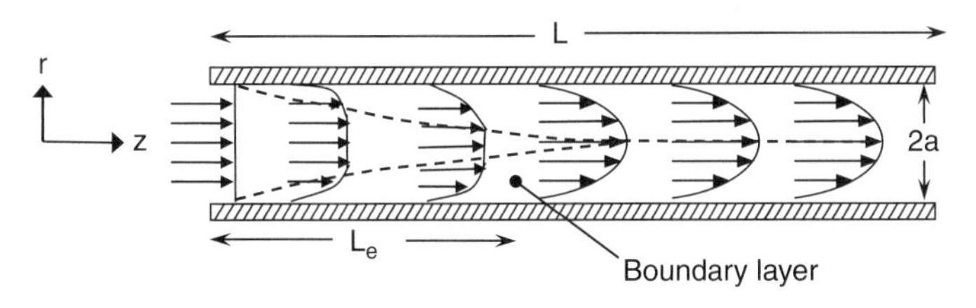

Figure 13.4-1 Velocity profile development in an entrance region ($L_e > z > 0$) of a tube.

Substituting these velocity components into the continuity equation for an incompressible fluid (Table 13.1-1), we get

$$\frac{\partial u_z}{\partial z} = 0 \Rightarrow u_z = u_z(r) \tag{13.4-4a, b}$$

From Table B4-8, we see that the only nonzero components of the deformation rate tensor corresponding to this velocity field are $\gamma(r) \equiv \gamma_{zr} = \gamma_{rz} = (1/2)du_z/dr$. Thus, this is a type of simple shearing flow. It follows from Equation 13.2-14b that the corresponding components of the viscous stress tensor, $\tau(r) \equiv \tau_{zr} = \tau_{rz}$, are given by

$$\tau(r) = 2m|2\gamma|^{n-1}\gamma = m\left|\frac{du_z}{dr}\right|^{n-1}\frac{du_z}{dr} \tag{13.4-5}$$

With τ only depending on r, the scalar components of the equation of motion (Table B4-4) reduce to

$$\text{r component}: \quad \frac{\partial \mathcal{P}}{\partial r} = 0 \Rightarrow \mathcal{P}(\theta, z) \tag{13.4-6}$$

$$\theta \text{ component}: \quad \frac{1}{r}\frac{\partial \mathcal{P}}{\partial \theta} = 0 \Rightarrow \mathcal{P}(r, z) \tag{13.4-7}$$

$$\text{z component}: \quad \frac{\partial \mathcal{P}}{\partial z} = \frac{1}{r}\frac{\partial}{\partial r}(r\tau) \tag{13.4-8}$$

Analysis

From the r and θ components of the equation of motion, $\mathcal{P}(r,z) = \mathcal{P}(\theta,z) = \mathcal{P}(z)$. Thus, $\partial\mathcal{P}/\partial z$ can only be a only function of z. Moreover, since τ is only a function of r, Equation 13.4-8 has a solution only if the pressure gradient equals a constant. We define this constant in terms of the pressure difference between $z = 0$ and $z = L$.

$$-\frac{d\mathcal{P}}{dz} = \frac{\mathcal{P}_0 - \mathcal{P}_L}{L} \equiv \frac{\Delta\mathcal{P}}{L} \tag{13.4-9}$$

Equation 13.4-8 can now be written as

$$\frac{1}{r}\frac{d}{dr}(r\tau) = -\frac{\Delta\mathcal{P}}{L} \tag{13.4-10}$$

At the tube centerline, we expect u_z to be symmetric with respect to r and to reach its maximum value. Both of these conditions are satisfied when the local r derivative of u_z vanishes:

$$r = 0: \quad \frac{\partial u_z}{\partial r} = 0 \Rightarrow \tau = 0 \tag{13.4-11}$$

Integrating Equation 13.4-10 then leads to

$$\tau = -\frac{\Delta\mathcal{P}}{2L}r \tag{13.4-12}$$

Therefore, shear stress is directly proportional to r with its maximum value occurring at the tube wall where $\tau_{wall} = -\Delta\mathcal{P}a/2L$. Notice that this result is independent of the fluid rheology. It requires only that shear stress be a function of radial position alone.

We now introduce power-law rheology by combining Equation 13.4-12 with Equation 13.4-5, to obtain

$$m\left|\frac{du_z}{dr}\right|^{n-1}\left(-\frac{du_z}{dr}\right)=\frac{\Delta\mathcal{P}}{2L}r \tag{13.4-13}$$

Because friction retards fluid motion at the tube wall, u_z should decrease as r increases so that du_z/dr is a negative quantity. This means that $|du_z/dr|^{n-1}$ is equivalent to $(-du_z/dr)^{n-1}$, and we can rearrange Equation 13.4-13 to get

$$-\frac{du_z}{dr}=\left(\frac{\Delta\mathcal{P}}{2mL}r\right)^{1/n} \tag{13.4-14}$$

According to this equation, du_z/dr will have a physically realistic value at any arbitrary value of $n>0$ provided that $\Delta\mathcal{P}>0$. That is, pressure must decrease in the direction of flow. Integrating Equation 13.4-14 with respect to r and applying a no-slip boundary condition at the tube wall

$$r=a: \quad u_z=0 \tag{13.4-15}$$

we obtain the velocity profile:

$$u_z=\left(\frac{n}{n+1}\right)\left(\frac{a^{(n+1)}\Delta\mathcal{P}}{2mL}\right)^{1/n}\left[1-\left(\frac{r}{a}\right)^{(n+1)/n}\right] \tag{13.4-16}$$

By integrating u_z over the tube cross section, we find the volumetric flow rate:

$$Q=\int_0^{2\pi}\int_0^{a}u_z r\,dr\,d\theta=\pi\left(\frac{n}{3n+1}\right)\left(\frac{a^{(3n+1)}\Delta\mathcal{P}}{2mL}\right)^{1/n} \tag{13.4-17}$$

Notice that as $n\rightarrow 1$ and $m\rightarrow\mu$, the last two equations are equivalent to the corresponding relations for a Newtonian fluid (Eqs. 13.4-1 and 13.4-2). Also from these equations, we formulate the ratio of the centerline velocity $u_x(0)$ to the average velocity $u_{ave}=Q/\pi a^2$ as

$$\frac{u_z(0)}{u_{ave}}=\frac{\left(\frac{n}{n+1}\right)\left(\frac{a^{(n+1)}\Delta\mathcal{P}}{2mL}\right)^{1/n}}{\frac{1}{a^2}\left(\frac{n}{3n+1}\right)\left(\frac{a^{(3n+1)}\Delta\mathcal{P}}{2mL}\right)^{1/n}}=\frac{3n+1}{n+1} \tag{13.4-18}$$

For a Newtonian fluid, $n=1$ and $u_z(0)/u_{ave}=2$. That is, the centerline velocity is twice the average axial velocity. For a shear-thinning fluid, $n<1$ and $u_z(0)/u_{ave}<2$, indicating that the velocity profile is less sharply pointed than for a Newtonian fluid. This makes sense since friction at the tube wall retards the local velocity less for a shear-thinning fluid than for a Newtonian fluid. The opposite is true when $n>1$ for a shear-thickening fluid.

With Equation 13.1-29, we can write that

$$\frac{d\mathcal{P}}{dz}=\frac{dP}{dz}-\rho\mathcal{G}_z \tag{13.4-19}$$

where $\mathcal{G}_z$ is the z component of the gravitational vector $\mathcal{G}$. With dP/dz constant, we can integrate Equation 13.4-19 from $z=0$ to $z=L$ to obtain

$$\Delta\mathcal{P} = \Delta P - \rho L \mathcal{G}_z \tag{13.4-20}$$

where $\Delta P \equiv (P_0 - P_L)$ is the pressure drop along the entire tube. If the flow direction z is inclined to gravity by an angle α, then $\mathcal{G}_z = \mathcal{G}\cos(\alpha)$ and

$$\Delta\mathcal{P} = \Delta P - \rho L \mathcal{G}\cos(\alpha) \tag{13.4-21}$$

where $\mathcal{G}$ is the magnitude of $\boldsymbol{\mathcal{G}}$. For a horizontal tube, $\alpha = 90^\circ$ and the gravitational body force does not contribute to $\Delta\mathcal{P}$. For downward flow through a vertical tube, $\alpha = 0^\circ$ and gravity contributes $+\rho\mathcal{G}L$. For an upward flow through a vertically inclined tube, $\alpha = 180^\circ$ and the gravitational contribution is $-\rho\mathcal{G}L$.

Example 13.4-1 Flow of Amniotic Fluid

Amniotic fluid is a serous fluid containing proteins, sulfated polysaccharides, and proteoglycans. Rheological data obtained during simple shearing flow of bovine amniotic fluid at 20°C (Dasari *et al.*, 1995) is given in Figure 13.4-2.

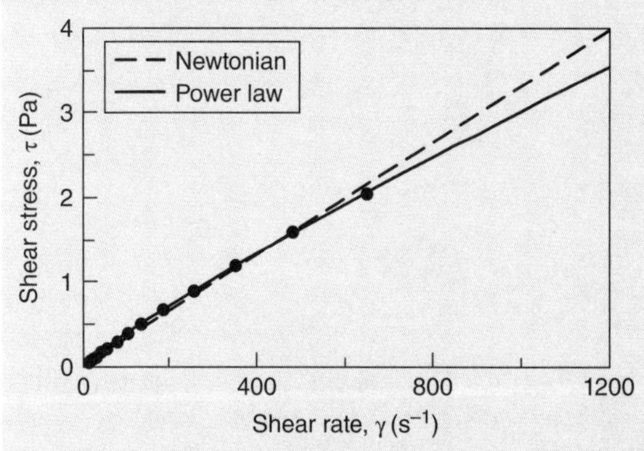

Figure 13.4-2 Rheological modeling of amniotic fluid (Data from Dasari *et al.* (1995)).

Proteins can be isolated from bovine amniotic fluid using liquid chromatography with a capillary column that has been coated with a protein-selective absorbent. Consider a vertical capillary column, 30 cm in length with a 1 mm inside diameter. We will estimate the overall pressure drop and the relative contributions of gravity and friction when amniotic fluid is pumped upward at a flow of 0.2 ml/s. Results obtained by modeling the fluid as a Newtonian liquid will be compared to those obtained for a power-law liquid model.

Solution

For a Newtonian fluid in a simple shearing flow, μ can be obtained as one half the slope of a linear regression of τ versus γ data (Eq. 13.2-6). For a power-law fluid in a simple shearing flow, n and m can be determined from a linear regression of $\log(\tau)$ versus $\log(2\gamma)$. The slope of this regression corresponds to m, while the antilog of the intercept is n (Eq. 13.2-14b).

The curves in Figure 13.4-2 show the results of these two regressions. For the Newtonian model, we estimate a $\mu = 0.00166$ Pa-s. For the power-law model, we estimate that $m = 0.00322$ Pa-s$^{0.890}$ and $n = 0.890$. The power-law model is able to capture the slight curvature of the data. Because the Newtonian model is linear, however, it tends to underestimate τ at low γ and overestimate τ at high γ. Substituting this μ value with the system parameters $a = 0.0005$ m, $L = 0.3$ m, and $Q = 2.00 \times 10^{-7}$ m^3/s into Equation 13.4-2, we determine the modified pressure drop for the Newtonian model:

$$[\Delta\mathcal{P}]_{\text{Newtonian}} = \frac{8(2\times10^{-7})(0.00166)(0.3)}{\pi(0.0005)^4} = 4050 \text{ Pa} \tag{13.4-22}$$

Similarly, we determine the modified pressure drop for the power-law model by rearranging Equation 13.4-17:

$$[\Delta\mathcal{P}]_{\text{Power-Law}} = \left\{\frac{2(0.00322)(0.3)}{(0.0005)^{[3(0.890)+1]}}\right\}\left\{\left[\frac{3(0.890)+1}{0.890}\right]\left[\frac{2\times10^{-7}}{1.29\pi}\right]\right\}^{0.890} = 3500 \text{ Pa} \tag{13.4-23}$$

As expected from the model behaviors at high shear rates in Figure 13.4-2, the power-law fluid requires a smaller pressure drop to overcome friction than the Newtonian model. Since we are considering a vertical tube in which flow is upward, we let $\alpha = 180^{\circ}$ in Equation 13.4-21 and get

$$\begin{aligned}[\Delta P]_{\text{Newtonian}} &= 4050 - 1000(9.81)(0.3)\cos(180^{\circ}) = 6990 \text{ Pa}\\ [\Delta P]_{\text{Power-Law}} &= 3500 - 1000(9.81)(0.3)\cos(180^{\circ}) = 6440 \text{ Pa}\end{aligned} \tag{13.4-24a, b}$$

where we have approximated fluid density by that of water. These pressure drops, necessary to overcome both friction and gravity, are larger by +2940 Pa than the modified pressure drops imposed by friction alone.

13.4.2 Two-Phase Annular Flow

In some problems, two or more fluids with different physical properties coexist in a tube. For example, this situation can occur during blood flow when a shear-induced rotation of red cells produces a radial force on the cells. The resulting migration of red cells can create two regions in a blood vessel: a cell suspension of high viscosity in the core of the blood vessel; and a cell-free skimming layer of relatively low viscosity adjacent to the vessel wall (Fig. 13.4-3).

In this illustration, we analyze the annular flow of two immiscible, Newtonian liquids of different viscosities in a rigid cylindrical tube. We assume steady-state, fully developed flows in the two phases and a constant thickness of the peripheral phase along the tube wall.

Model Formulation

We assign a viscosity μ_c to the core phase that occupies the center of the tube at $a_c > r \geq 0$ and a viscosity μ_p to the peripheral phase that occupies the annular space at $a > r \geq a_c$.

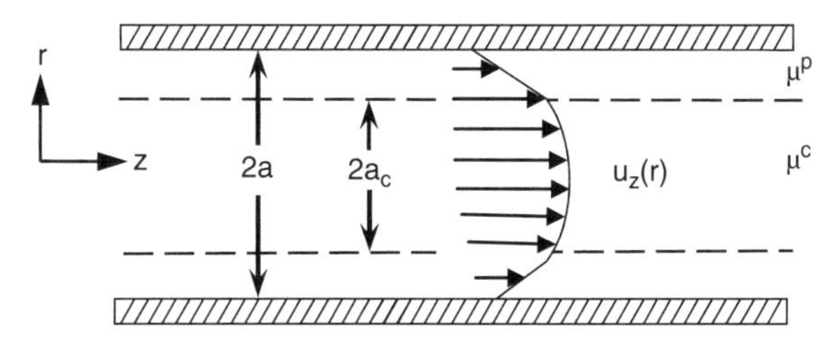

Figure 13.4-3 Fully developed annular flow in a tube.

The kinematics within the core phase as well as the peripheral phase are the same as those for fully developed, single-phase flow through a tube:

$$u_z^c = u_z^c(r), \quad u_r^c = u_\theta^c = 0$$
$$u_z^p = u_z^p(r), \quad u_r^p = u_\theta^p = 0 \qquad (13.4\text{-}25a\text{-}d)$$

As in the previous illustration, the r and θ components of the Navier–Stokes equation (Table B4-5) for the two fluids lead us to conclude that the modified pressure gradients in the core and peripheral phases are both constant. The z component of the Navier–Stokes equation for the core phase then reduces to

$$\frac{\mu_c}{r}\frac{d}{dr}\left(r\frac{du_z^c}{dr}\right) = \frac{d\mathcal{P}^c}{dz} \equiv -\frac{\Delta\mathcal{P}^c}{L} \quad (a_c > r \geq 0) \qquad (13.4\text{-}26)$$

where $\Delta\mathcal{P}^c$ is the modified pressure drop over the tube length L. The z component in the peripheral phase is similar.

$$\frac{\mu_p}{r}\frac{d}{dr}\left(r\frac{du_z^p}{dr}\right) = \frac{d\mathcal{P}^p}{dz} \equiv -\frac{\Delta\mathcal{P}^p}{L} \quad (a > r \geq a_c) \qquad (13.4\text{-}27)$$

For two second-order differential equations, we specify the following four boundary conditions related to the velocities $u_z^c(r)$ and $u_z^p(r)$. The velocity at the tube wall must satisfy the no-slip condition:

$$r = a: \quad u_z^p = 0 \qquad (13.4\text{-}28)$$

The centerline velocity must be symmetric with respect to r:

$$r = 0: \quad \frac{du_z^c}{dr} = 0 \qquad (13.4\text{-}29)$$

Velocity is continuous at the interface between the two fluids:

$$r = a_c: \quad u_z^c = u_z^p \qquad (13.4\text{-}30)$$

The z–r shear component of the total stress tensor must be continuous at the interface:

$$r = a_c: \quad T_{zr}^c = T_{zr}^p \Rightarrow \tau_{zr}^c = \tau_{zr}^p \Rightarrow \mu^c\frac{du_z^c}{dr} = \mu^p\frac{du_z^p}{dr} \qquad (13.4\text{-}31a\text{-}c)$$

We also note that the r–r normal component of the total stress tensor must be equal on the two sides of the interface. Therefore,

$$r = a_c: \quad T_{rr}^c = T_{rr}^p \Rightarrow -P^c + \tau_{rr}^c = -P^p + \tau_{rr}^p \qquad (13.4\text{-}32a,b)$$

Based on the assumed kinematics, $\gamma_{rr}^c = \gamma_{rr}^p = 0$ so that $\tau_{rr}^c = \tau_{rr}^p = 0$ and this condition becomes

$$r = a_c: \quad P^c = P^p \Rightarrow \mathcal{P}^c = \mathcal{P}^p \qquad (13.4\text{-}33a,b)$$

This condition implies that $\Delta\mathcal{P}^c = \Delta\mathcal{P}^p$ at the interface between the two fluids. Since neither $\Delta\mathcal{P}^c/L$ nor $\Delta\mathcal{P}^p/L$ depend on r, we further conclude that $\Delta\mathcal{P}^c/L = \Delta\mathcal{P}^p/L \equiv \Delta\mathcal{P}/L$ over the entire tube cross section.

Analysis

Equations 13.4-26 and 13.4-27 can be separately integrated to obtain

$$u_z^c = -\frac{\Delta\mathcal{P}}{4\mu_c L}r^2 + c_1 \ln r + c_2 \quad (13.4\text{-}34)$$

$$u_z^p = -\frac{\Delta\mathcal{P}}{4\mu_p L}r^2 + c_3 \ln r + c_4 \quad (13.4\text{-}35)$$

The constants of integration, c_1, c_2, c_3, and c_4, are evaluated using the four boundary conditions in Equations 13.4-28–13.4-31. Consequently, the velocity distributions are

$$u_z^c = \frac{a^2\Delta\mathcal{P}}{4\mu^c L}\left[\left(1-\frac{r^2}{a^2}\right) + \left(\frac{\mu^c}{\mu^p}-1\right)\left(1-\frac{a_c^2}{a^2}\right)\right] \quad (a_c > r \geq 0) \quad (13.4\text{-}36)$$

$$u_z^p = \frac{a^2\Delta\mathcal{P}}{4\mu^p L}\left(1-\frac{r^2}{a^2}\right) \quad (a > r \geq a_c) \quad (13.4\text{-}37)$$

Notice that u_z^c consists of two terms. The first term is equivalent to the Hagen–Poiseuille equation for a fluid of viscosity μ^c flowing in a tube of radius a_c. The second term is the deviation of velocity distribution in the core phase imposed by the peripheral phase. Since $a > a_c$, the deviation is positive when $\mu^c > \mu^p$, indicating that the higher viscosity core fluid is able to "slip" by the lower viscosity peripheral fluid.

An effective viscosity for two-phase annular flow can be defined for an equivalent Hagen–Poiseuille flow:

$$\mu_{eff} \equiv \left(\frac{\pi a^4 \Delta\mathcal{P}}{8L}\right)\left(\frac{1}{Q}\right) = \left(\frac{\pi a^4 \Delta\mathcal{P}}{8L}\right)\left[\int_0^{2\pi}\int_0^{a_c} u_z^c r\,dr\,d\theta + \int_0^{2\pi}\int_{a_c}^{a} u_z^p r\,dr\,d\theta\right]^{-1} \quad (13.4\text{-}38)$$

After substituting Equations 13.4-36 and 13.4-37 into Equation 13.4-38, we obtain

$$\frac{\mu_{eff}}{\mu_c} = \left[\frac{\mu^c}{\mu^p}\left(1-\frac{a_c^4}{a^4}\right) + \frac{a_c^4}{a^4}\right]^{-1} \quad (13.4\text{-}39)$$

In the limiting case $(\mu^c/\mu^p) = 1$, we find $\mu_{eff} = \mu^c$, and when $(\mu^c/\mu^p) > 1$, we find $(\mu_{eff}/\mu^c) < 1$. This indicates that a low viscosity peripheral fluid lubricates the flow of a higher viscosity core fluid during well-developed flow through rigid tube.

Example 13.4-2 Plasma Skimming Layer Thickness in Flowing Blood

Recall that the viscosity of blood generally depends on shear rate. However, even at high shear rates when whole blood is expected to behave like a Newtonian fluid, its effective viscosity during tube flow can depend on the tube radius. This is demonstrated by the data points in Figure 13.4-4 showing that effective viscosity increases as tube radius increases (Ruch and Patton, 1965). This "Fahraeus–Lindqvist" effect might be due to a cell-free plasma skimming layer of constant thickness $\delta \equiv (a - a_c)$ near the tube wall. We will determine whether this explanation is plausible by interpreting the data using our two-phase annular flow model with a constant value of δ is independent of tube radius.

Solution

Rewriting Equation 13.4-39 in terms of δ, we obtain

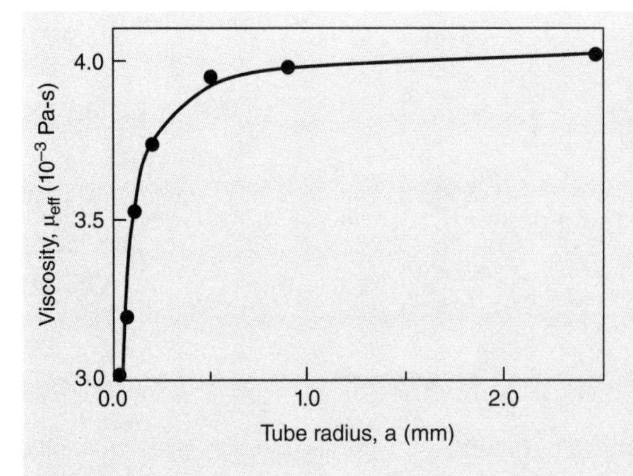

Figure 13.4-4 Effective viscosity during tube flow with a skimming layer of constant thickness (Fahraeus–Lindqvist effect) (Data from Ruch and Patton (1965)).

$$\mu_{eff} = \left[\frac{1}{\mu^p} + \left(1 - \frac{\delta}{a}\right)^4 \left(\frac{1}{\mu^c} - \frac{1}{\mu^p}\right)\right]^{-1} \quad (13.4\text{-}40)$$

A nonlinear regression of this equation to the μ_{eff}−a data set is indicated by the curve in Figure 13.4-4 for parameter estimates of $\delta = 0.00190$ mm, $\mu_c = 0.00405$ Pa-s, and $\mu_p = 0.00128$ Pa-s. It is encouraging that the core viscosity estimate is similar but somewhat larger than the viscosity of whole blood, and the peripheral viscosity estimate is almost the same as plasma viscosity (Table 13.2-1). The δ estimate is roughly the thickness of a red blood cell. Furthermore, since $\delta \ll 0.03$ mm, the smallest tube radius in which μ_{eff} was measured, the skimming layer model provides a reasonable explanation of the Fahraeus–Lindqvist effect.

References

Dasari G, Prince I, Hearn MT. Investigations into the rheological characteristics of bovine amniotic fluid. J Biochem Biophys Methods. 1995; 30:217–25.

Merrill EW, Benis AM, Gilliland ER, Sherwood TK, Salzman EW. Pressure flow relations of human blood in hollow fibers at low flow rates. J Appl Physiol. 1965; 20:954–967.

Ruch TC, Patton HD. Physiology and Biophysics. 19th ed. Philadelphia: WB Saunders Company; 1965, p 530.

Slattery, JC. Advanced Transport Phenomena. Cambridge: Cambridge University Press; 1999, p 25, 33.

Wells R, Goldstone J. Rheology of the red cell and capillary blood flow. In: Gabelnick HL, Litt M, editors. Rheology of biological systems. Springfield: Charles C. Thomas; 1973, p 7, ch. 1.

Chapter 14

Fluid Mechanics II

Complex Flows

In the previous chapter, we introduced the basic concepts of convective processes and demonstrated their application to steady, fully developed flow through tubes. In these simple shearing flows, the velocity was in the z direction and was solely a function of radial position. In this chapter, we describe more complex applications involving multidirectional and time-dependent velocity fields.

14.1 Boundary Layer Flows

14.1.1 Flow Development over a Flat Plate

In a sufficiently long, straight cylindrical tube, we can neglect the nonlinear convective processes that affect velocity and pressure fields in the vicinity of the tube entrance. In biomedical systems, blood vessels and airways are generally too short to overlook convective effects. In those cases, the kinematics are more complicated and nonlinear convective forces are important in the equation of motion. To illustrate this, we analyze the development of flow along a flat plate.

Model Formulation

Suppose that an incompressible fluid approaches the leading edge of a stationary plate at a constant, uniform velocity U in the x direction (Fig. 14.1-1). The no-slip condition specifies that the velocity of fluid at the plate surface will be zero. Consequently, viscous forces resist the forward movement of fluid in the vicinity of the surface. This results in an x-directed velocity that increases with distance y from the surface until it matches the entrance velocity U.

Biomedical Mass Transport and Chemical Reaction: Physicochemical Principles and Mathematical Modeling, First Edition. James S. Ultman, Harihara Baskaran, and Gerald M. Saidel.

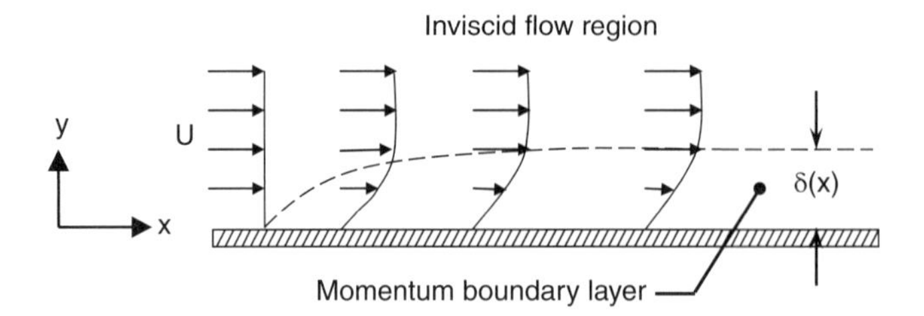

Figure 14.1-1 Development of a momentum boundary layer on a flat plate.

The region of flow in which these viscous effects occur is the momentum or velocity boundary layer. With increasing distance along the plate, viscous forces are dissipated over greater y distances so that the boundary layer thickness δ increases with x. The flow domain outside of the boundary layer, where viscous forces are unimportant, is called the inviscid flow region.

We will analyze how the boundary layer develops under steady-state conditions. The analysis will be restricted to incompressible, Newtonian flow over a wide plate, that is, one whose characteristic length in the z direction is much greater than the characteristic lengths in the x and y directions. This allows us to (i) neglect the u_z velocity component and (ii) ignore changes of u_x and u_y in the z direction due to effects at the lateral edges of the plate. Furthermore, we confine the analysis to the region $L > x > 0$ in which $\delta(x) \ll L$ (i.e., a relatively thin boundary layer). Finally, we will consider a plate that is horizontal so there is no gravitational effect.

In the inviscid flow region ($y > \delta$), the velocity vector has the components: $u_x = U$ and $u_y = u_z = 0$. Also, since the plate is horizontal, the modified pressure $\mathcal{P}$ is equal to the hydrodynamic pressure P. Inserting this information into the rectangular components of the Navier–Stokes equations for steady-state flow of an incompressible fluid (Table B4-3), we obtain

$$\frac{\partial P}{\partial x} = \frac{\partial P}{\partial y} = \frac{\partial P}{\partial z} = 0 \tag{14.1-1}$$

In other words, P remains constant throughout the inviscid region.

The velocity components within the boundary layer have the following forms: $u_x = u_x(x,y)$, $u_y = u_y(x,y)$, and $u_z = 0$. This allows us to eliminate several terms from the continuity (Table 13.1-1) and Navier–Stokes equations, which become

$$\text{Continuity:} \quad \frac{\partial u_x}{\partial x} + \frac{\partial u_y}{\partial y} = 0 \tag{14.1-2}$$

$$\text{x component:} \quad \rho\left(u_x \frac{\partial u_x}{\partial x} + u_y \frac{\partial u_x}{\partial y}\right) = \mu\left(\frac{\partial^2 u_x}{\partial x^2} + \frac{\partial^2 u_x}{\partial y^2}\right) - \frac{\partial P}{\partial x} \tag{14.1-3}$$

$$\text{y component:} \quad \rho\left(u_x \frac{\partial u_y}{\partial x} + u_y \frac{\partial u_y}{\partial y}\right) = \mu\left(\frac{\partial^2 u_y}{\partial x^2} + \frac{\partial^2 u_y}{\partial y^2}\right) - \frac{\partial P}{\partial y} \tag{14.1-4}$$

$$\text{z component:} \quad 0 = -\frac{\partial P}{\partial z} \Rightarrow P = P(x,y) \tag{14.1-5a,b}$$

Analysis

Order-of-Magnitude Approximations

The continuity equation and x and y components of the Navier–Stokes equation can be simplified by comparing the relative magnitude of terms in a dimensionless form of the model equations. For this purpose, the variables of the model are scaled using characteristic model parameters. In the boundary layer region, the domains of the independent variables are as follows: $L > x > 0$ and $\delta_L > y > 0$, where δ_L is the boundary layer thickness at $x = L$. The ranges of the velocity components are as follows: $U > u_x > 0$ and $V > u_y > 0$. The characteristic parameters δ_L and V, as of yet unknown, are defined through the following analysis.

Scaling the variables by the characteristic parameters, we define the corresponding dimensionless variables:

$$x = \frac{x}{L}, \quad y = \frac{y}{\delta_L}, \quad u_x = \frac{u_x}{U}, \quad u_y = \frac{u_y}{V} \tag{14.1-6a-d}$$

According to the Bernoulli principle, the maximum pressure rise due to fluid deceleration in the boundary layer is equal to the kinetic energy per unit volume of the undisturbed flow, $\rho U^2/2$. Thus, we define a dimensionless pressure as

$$P = \frac{P}{\rho U^2} \tag{14.1-7}$$

Introducing the dimensionless variables into the continuity equation, we have

$$\left(\frac{U\delta_L}{VL}\right)\frac{\partial u_x}{\partial x} + \frac{\partial u_y}{\partial y} = 0 \tag{14.1-8}$$

The velocity derivatives in this equation have an order-of-magnitude of one, O(1). To avoid a trivial solution, both terms in Equation 14.1-8 must have the same nonzero value so that $(U\delta_L/VL) \sim O(1)$. This leads to a relationship between the two unknown scale factors, U and V:

$$V = U\frac{\delta_L}{L} \tag{14.1-9}$$

The dimensionless x component of the Navier–Stokes equation is

$$\left(u_x\frac{\partial u_x}{\partial x} + u_y\frac{\partial u_x}{\partial y}\right) = -\frac{\partial P}{\partial x} + \left(\frac{\upsilon L}{U\delta_L^2}\right)\left[\left(\frac{\delta_L}{L}\right)^2\frac{\partial^2 u_x}{\partial x^2} + \frac{\partial^2 u_x}{\partial y^2}\right] \tag{14.1-10}$$

where $\upsilon = \mu/\rho$ is the kinematic viscosity. Both the dimensionless velocities and velocity derivatives are O(1) so that the convective terms are O(1). In addition, $(\delta_L/L)^2 \ll 1$ for a thin boundary layer, allowing us to neglect the first viscous term on the right side of this equation:

$$\left(u_x\frac{\partial u_x}{\partial x} + u_y\frac{\partial u_x}{\partial y}\right) = -\frac{\partial P}{\partial x} + \left(\frac{\upsilon L}{U\delta_L^2}\right)\frac{\partial^2 u_x}{\partial y^2} \tag{14.1-11}$$

If the remaining viscous term and the convective terms are to be of comparable importance, then $\left(\upsilon L/U\delta_L^2\right) \sim O(1)$. Therefore, a reasonable definition of the boundary layer scaling parameter is

$$\delta_L = \sqrt{\frac{\upsilon L}{U}} \tag{14.1-12}$$

This equation indicates that the boundary layer thickness increases with the square root of the kinematic viscosity and the distance from the leading edge of the plate. It also suggests that a Reynolds number based on the length of the plate provides a consistency criterion for satisfying the thin boundary layer approximation:

$$Re_L \equiv \frac{LU}{\upsilon} = \left(\frac{L}{\delta_L}\right)^2 \gg 1 \tag{14.1-13}$$

The dimensionless form of the y component of the Navier–Stokes equation leads to

$$u_x\frac{\partial u_y}{\partial x} + u_y\frac{\partial u_y}{\partial y} = -\left(\frac{L}{\delta_L}\right)^2\frac{\partial P}{\partial y} + \left[\frac{\partial^2 u_y}{\partial y^2} + \left(\frac{\delta_L}{L}\right)^2\frac{\partial^2 u_y}{\partial x^2}\right] \tag{14.1-14}$$

Following the same order-of-magnitude analysis that we used to simplify the x component, we conclude that the second viscous term is negligible and the first viscous term and the convective terms are both O(1). If the pressure term is to be of equal importance, then $(L/\delta_L)^2(\partial P/\partial y) \sim O(1)$. Since $\left(L^2/\delta_L^2\right) \gg 1$, pressure variation in the y direction is negligible:

$$\frac{\partial P}{\partial y} \approx 0 \;\Rightarrow\; P = P(x) \tag{14.1-15}$$

It follows that at any x in the boundary layer, P has the same value for all y including at the artificial boundary with the inviscid flow region. Since P is constant in the inviscid flow region and pressure is continuous between the viscous boundary layer and the inviscid region, P must be constant everywhere. This implies that $\partial P/\partial x = 0$ so that Equation 14.1-11 simplifies as

$$\left(u_x\frac{\partial u_x}{\partial x} + u_y\frac{\partial u_x}{\partial y}\right) = \frac{\partial^2 u_x}{\partial y^2} \tag{14.1-16}$$

This corresponds to the following dimensional form of the x component of the Navier–Stokes equation:

$$\left(u_x\frac{\partial u_x}{\partial x} + u_y\frac{\partial u_x}{\partial y}\right) = \upsilon\frac{\partial^2 u_x}{\partial y^2} \tag{14.1-17}$$

This equation along with the continuity equation (Eq. 14.1-2) are sufficient to describe u_x and u_y. The velocities in this problem must satisfy the following boundary conditions:

$$\begin{aligned} y = 0 &\;:\; u_x = 0 \\ y = \delta &\;:\; u_x = U \\ x = 0 &\;:\; u_x = U \end{aligned} \tag{14.1-18a-c}$$

Similarity Transformation

Integrating Equation 14.1-2 with respect to y, we obtain

$$u_y = -\int_0^y \frac{\partial u_x}{\partial x}dy \tag{14.1-19}$$

Substitution for u_y in Equation 14.1-17 reduces the model to one governing equation for $u_x(x,y)$:

$$u_x \frac{\partial u_x}{\partial x} - \left(\int_0^y \frac{\partial u_x}{\partial x} dy \right) \frac{\partial u_x}{\partial y} = \upsilon \frac{\partial^2 u_x}{\partial y^2} \tag{14.1-20}$$

This equation can be solved by the von Karman integral method which assumes the velocity distribution has the same shape at different x positions when y is scaled by $\delta(x)$. We define this similarity variable as η and also introduce dimensionless forms of the velocity, x position and boundary layer thickness:

$$\eta \equiv \frac{y}{\delta(x)}, \quad f(\eta) = \frac{u_x(x,y)}{U}, \quad x \equiv \frac{x}{L}, \quad \delta \equiv \frac{\delta(x)}{L} \tag{14.1-21a-d}$$

After transformation by the similarity variable (Appendix C6), and introduction of the other dimensionless variables, Equation 14.1-20 becomes

$$\left[f'(\eta) \int_0^\eta \xi f'(\xi) d\xi - \eta f(\eta) f'(\eta) \right] \delta \frac{d\delta}{dx} = \frac{1}{Re_L} f''(\eta) \tag{14.1-22}$$

where f' and f'' represent the first- and second-order derivatives of f with respect to η. Note that the functions of x and η are separable. In terms of the transformed variables, the first two boundary conditions are

$$\eta = 0 \; : \; f = 0, \quad \eta = 1 \; : \; f = 1 \tag{14.1-23a,b}$$

The third boundary condition, namely, that the entrance velocity U is imposed at the origin of the plate, implies that the boundary layer thickness is zero at that point:

$$x = 0 \; : \; \delta = 0 \tag{14.1-24}$$

Integrating both sides of Equation 14.1-22 over the boundary layer domain $1 \geq \eta \geq 0$ and noting that $\delta(d\delta/dx) = (1/2)d(\delta^2)/dx$, we obtain

$$\left\{ \int_0^1 \left[\int_0^\eta \xi f'(\xi) d\xi \right] f'(\eta) d\eta - \int_0^1 \eta f(\eta) f'(\eta) d\eta \right\} \frac{d\delta^2}{dx} = \frac{2}{Re_L} \int_0^1 f''(\eta) d\eta \tag{14.1-25}$$

Integrating the first term by parts and applying the boundary condition that $f(1) = 1$, we get

$$\int_0^1 \left[\int_0^\eta \xi f'(\xi) d\xi \right] f'(\eta) d\eta = \int_0^1 \eta f'(\eta) d\eta - \int_0^1 \eta f(\eta) f'(\eta) d\eta \tag{14.1-26}$$

Equation 14.1-25 now becomes

$$\frac{d(\delta)^2}{dx} = \frac{1}{Re_L} \zeta^2 \tag{14.1-27}$$

where ζ^2 is a constant given by

$$\zeta^2 = \left(2 \int_0^1 f'' d\eta \right) \left(\int_0^1 \eta f' d\eta - 2 \int_0^1 \eta f f' d\eta \right)^{-1} \tag{14.1-28}$$

Solution

Integrating Equation 14.1-27 from $x > 0$ and applying the boundary condition that $\delta(0) = 0$ yields the boundary-layer thickness in dimensionless and dimensional forms:

$$\delta(x) = \zeta\sqrt{\frac{x}{Re_L}} \Rightarrow \delta(x) = \zeta\sqrt{\left(\frac{\upsilon}{U}\right)x} \qquad (14.1\text{-}29a,b)$$

Here, the boundary layer thickness not only specifies the same square-root relationship as Equation 14.1-12, but it also provides the basis for quantitative evaluation via the coefficient ζ. To determine ζ, we approximate $f(\eta)$ as a third-order polynomial that satisfies the boundary conditions in Equation 14.1-23a,b:

$$f = a_1\eta + a_2\eta^2 + (1 - a_1 - a_2)\eta^3 \qquad (14.1\text{-}30)$$

At the artificial boundary between the viscous and inviscid regions (i.e., $y = \delta \Rightarrow \eta = 1$), the y derivatives of u_x and thus the η derivatives of f are continuous. Since $f = 1$ throughout the inviscid region, $f'(1) = f''(1) = f'''(1) = \ldots = 0$. To define the two constants in Equation 14.1-30, we apply the boundary conditions $f'(1) = f''(1) = 0$, which lead to

$$f = \frac{3}{2}\eta - \frac{1}{2}\eta^3 \qquad (14.1\text{-}31)$$

When this function is substituted in Equation 14.1-28, we find $\zeta = 4.64$. Combining Equations 14.1-21, 14.1-29, and 14.1-31, we can express u_x within the boundary layer as

$$\frac{u_x}{U} = \left(\frac{0.323}{\sqrt{\nu/U}}\right)\frac{y}{\sqrt{x}} - \left(\frac{0.00501}{\sqrt[3]{\nu/U}}\right)\frac{y^3}{\sqrt[3]{x}} \qquad (14.1\text{-}32)$$

By substituting this equation into Equation 14.1-19, we can also obtain an expression for u_y within the boundary layer:

$$\frac{u_y}{U} = \left(\frac{0.0808}{\sqrt{\nu/U}}\right)\frac{y^2}{\sqrt[3]{x}} - \left(\frac{0.00188}{\sqrt[3]{\nu/U}}\right)\frac{y^4}{\sqrt[5]{x}} \qquad (14.1\text{-}33)$$

Notice that these velocity components are independent of pressure, which is everywhere constant.

Example 14.1-1 Entrance Length in Conducting Airways

As fluid flows in the entrance region of a tube, the momentum boundary layer at the wall progressively thickens until the velocity profile reaches a fully developed shape. The axial distance required for this to occur is known as the entrance length L_e (Fig. 13.4-1). Let us determine whether the flow of air is closer to being fully developed in the trachea or in a terminal conducting airway located in the 16th generation of an adult human lung. During quiet breathing the volumetric flow of air is about $Q = 300$ ml/s. The diameter and length of the trachea are $d_t = 1.8$ cm and $L_t = 12$ cm. The diameter and length of a terminal conducting airway are $d_a = 0.06$ and $L_a = 0.17$ cm.

Solution

To the extent that we can neglect curvature of the airway wall, the developing flow along the entrance length L_e can be approximated from our boundary-layer analysis for a flat plate. We estimate L_e as that value of x when the boundary layer thickness approaches the tube radius and the velocity outside the boundary layer is equal to the average axial velocity. Substituting $x = L_e$, $\delta = d/2$, and $U = u_{ave}$ into Equation 14.1-29b, we have

$$\frac{L_e}{d} = \frac{1}{4\zeta^2}\left(\frac{u_{ave}d}{\upsilon}\right) \equiv \frac{1}{4\zeta^2}Re_d \tag{14.1-34}$$

where $Re_d = Ud/\upsilon$ is the Reynolds number based on tube diameter. According to the flat-plate analysis, $\zeta = 4.64$ so that the predicted entrance length is $L_e \approx 0.0116dRe_d$. Note that 0.0116 is lower than the value of 0.0232 obtained in more exact analyses of flow in a cylindrical tube geometry.

In computing the entrance length in an airway during quiet breathing, we can treat air as an incompressible fluid with a kinematic viscosity $\upsilon = 1.67 \times 10^{-5}$ m²/sec. Since all respired air must flow through the trachea, the average axial velocity is the volumetric flow rate divided by the tracheal cross section. Based on the tracheal diameter, we compute the average tracheal velocity and Reynolds number as

$$\begin{aligned} u_{trachea} &= \frac{(300 \times 10^{-6})}{\pi(0.018/2)^2} = 1.18 \text{ m/s} \\ [Re_d]_{trachea} &= \frac{(1.18)(0.018)}{1.67 \times 10^{-5}} = 1270 \end{aligned} \tag{14.1-35a, b}$$

If we assume a dichotomously branching lung, then flow is divided equally among the 2^{16} conducting airway branches of the 16th generation. Based on this geometry, we compute an average velocity and Reynolds number of this airway as

$$\begin{aligned} u_{terminal} &= \frac{(300 \times 10^{-6})}{\pi(0.0006/2)^2(2^{16})} = 0.0162 \text{ m/s} \\ [Re_d]_{terminal} &= \frac{(0.0162)(0.0006)}{1.67 \times 10^{-5}} = 0.582 \end{aligned} \tag{14.1-36a, b}$$

Thus, the combination of the small diameter and division of flow through the terminal conducting airways creates a marked reduction in Re_d when compared to the trachea. According to Equation 14.1-34, this smaller Re_d in addition to a smaller diameter leads to a much reduced entrance length in the terminal conducting airways as compared to the trachea:

$$\begin{aligned} [L_e]_{trachea} &= (0.0116)(1270)(0.018) = 0.265 \text{ m} = 26.5 \text{ cm} \\ [L_e]_{terminal} &= (0.0116)(0.582)(0.0006) = 4.10 \times 10^{-6} \text{ m} = 0.000410 \text{ cm} \end{aligned} \tag{14.1-37a, b}$$

Since the ratio of the actual length to the predicted entrance length of the trachea is about $L/L_e = (12/27) \approx 0.4$, the flow is not fully developed anywhere in this airway. For the terminal airway, however, $L/L_e = (0.17/4.1 \times 10^{-4}) \approx 400$, indicating that flow is developed throughout most of the airway.

14.1.2 Flow Induced by a Rotating Disk

A rotating disk is a convenient *in vitro* system for investigating the effect of shear stress on cell functions such as cell-biomaterial interactions, cell adhesion and mechano-chemical signal transduction.

Consider a circular disk that is rotating in an infinite fluid domain (Fig. 14.1-2). The spinning of the disk induces a centrifugal force that moves fluid outward, especially, in the vicinity

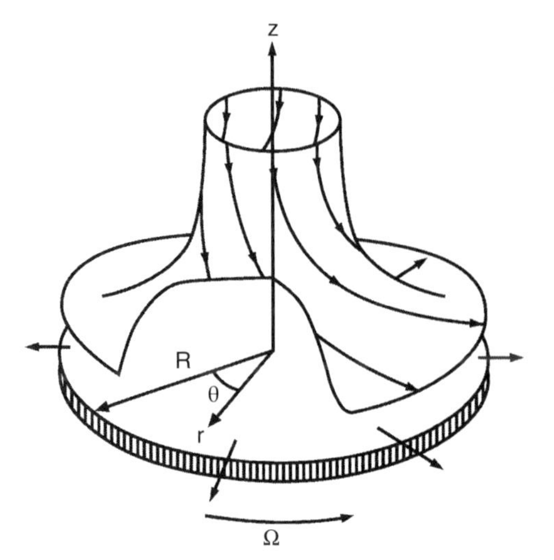

Figure 14.1-2 Flow patterns induced by a disk rotating in an unbounded liquid (Courtesy of Joanna Ramage).

of the solid surface. At sufficiently high rotational speeds, this has three consequences: (i) a thin velocity boundary layer is formed along the disk surface; (ii) the outflow along the disk surface is matched by a perpendicular inflow to the disk; and (iii) the high velocity in the boundary layer reduces pressure near the disk surface.

Model Formulation

We will analyze the steady-state flow of an incompressible, Newtonian fluid in contact with a horizontal disk of radius R. The disk is rotating at an angular speed Ω, thereby creating a relatively thin velocity boundary layer of thickness $\delta \ll R$. To take advantage of the system geometry, we describe the flow field in cylindrical coordinates whose z-axis coincides with the axis of rotation. Because of symmetry, neither the velocity components nor the pressure depends on the circumferential coordinate θ; and for a horizontal disk, there are no gravitational effects so that $\mathcal{P} = P$. For this model geometry, the continuity equation (Table 13.1-1) reduces to

$$\frac{1}{r}\frac{\partial(ru_r)}{\partial r} + \frac{\partial u_z}{\partial z} = 0 \tag{14.1-38}$$

and the scalar components of the Navier–Stokes equation (Table B4-5) reduce to

$$\text{r component:}\quad \rho\left(u_r\frac{\partial u_r}{\partial r} - \frac{u_\theta^2}{r} + u_z\frac{\partial u_r}{\partial z}\right) = -\frac{\partial P}{\partial r} + \mu\left[\frac{\partial}{\partial r}\left(\frac{1}{r}\frac{\partial(ru_r)}{\partial r}\right) + \frac{\partial^2 u_r}{\partial z^2}\right] \tag{14.1-39}$$

$$\theta\text{ component:}\quad \rho\left(u_r\frac{\partial u_\theta}{\partial r} + \frac{u_r u_\theta}{r} + u_z\frac{\partial u_\theta}{\partial z}\right) = \mu\left[\frac{\partial}{\partial r}\left(\frac{1}{r}\frac{\partial(ru_\theta)}{\partial r}\right) + \frac{\partial^2 u_\theta}{\partial z^2}\right] \tag{14.1-40}$$

$$\text{z component:}\quad \rho\left(u_r\frac{\partial u_z}{\partial r} + u_z\frac{\partial u_z}{\partial z}\right) = -\frac{\partial P}{\partial z} + \mu\left[\frac{\partial}{\partial r}\left(\frac{1}{r}\frac{\partial(ru_z)}{\partial r}\right) + \frac{\partial^2 u_z}{\partial z^2}\right] \tag{14.1-41}$$

Analysis

Order-of-Magnitude Approximations

In the boundary layer region, the domains of the independent variables are: $\delta > z > 0$ and $R > r > 0$. The ranges of the velocity components are: $\Omega R > u_\theta > 0$, $V_r > u_r > 0$ and $V_z > u_z > 0$, where V_r and V_z are scaling factors to be determined by analyzing the model in dimensionless form. Based on these ranges, we define dimensionless variables:

$$z = \frac{z}{\delta}, \quad r = \frac{r}{R}, \quad u_\theta = \frac{u_\theta}{R\Omega}, \quad u_r = \frac{u_r}{V_r}, \quad u_z = \frac{u_z}{V_z} \tag{14.1-42a-e}$$

Notice that we are treating the boundary layer thickness as a constant, which will be shown later in the analysis. We expect the maximum pressure change in the boundary layer to be limited by the kinetic energy per unit volume of fluid approaching the boundary layer, $\rho V_z^2/2$. We therefore define a dimensionless modified pressure as

$$P = \frac{P}{\rho V_z^2} \tag{14.1-43}$$

Introducing the dimensionless variables in the continuity equation, we obtain

$$\frac{1}{r}\frac{\partial(ru_r)}{\partial r} + \left(\frac{V_z R}{V_r \delta}\right)\frac{\partial u_z}{\partial z} = 0 \tag{14.1-44}$$

Because the two terms must be of equal magnitude and the dimensionless velocity and position variables are of O(1), we require that $(V_z R/V_r \delta) = O(1)$. The characteristic velocities can then be related as

$$V_z = \left(\frac{\delta}{R}\right) V_r \tag{14.1-45}$$

The dimensionless θ component of the Navier–Stokes equation is

$$u_r\frac{\partial u_\theta}{\partial r} + \frac{u_r u_\theta}{r} + u_z\frac{\partial u_\theta}{\partial z} = \frac{\upsilon R}{V_r \delta^2}\left[\frac{\partial^2 u_\theta}{\partial z^2} + \left(\frac{\delta}{R}\right)^2 \frac{\partial}{\partial r}\left(\frac{1}{r}\frac{\partial(ru_\theta)}{\partial r}\right)\right] \tag{14.1-46}$$

Since the boundary layer is thin, $(\delta/R)^2 \ll 1$, the first of the viscous terms is dominant, and Equation 14.1-46 reduces to

$$u_r\frac{\partial u_\theta}{\partial r} + \frac{u_r u_\theta}{r} + u_z\frac{\partial u_\theta}{\partial z} = \left(\frac{\upsilon R}{V_r \delta^2}\right)\frac{\partial^2 u_\theta}{\partial z^2} \tag{14.1-47}$$

The remaining viscous term should be as important as the convective terms, which implies that $\left(\upsilon R/V_r \delta^2\right) \sim O(1)$. This relationship and Equation 14.1-45 lead to the definition of the velocity scale factors:

$$V_r = \frac{\upsilon R}{\delta^2}, \quad V_z = \left(\frac{\delta}{R}\right) V_r = \frac{\upsilon}{\delta} \tag{14.1-48a,b}$$

The dimensionless form of the r component of the Navier–Stokes equation is

$$u_r\frac{\partial u_r}{\partial r} - \left(\frac{\Omega\delta^2}{\upsilon}\right)^2 \frac{u_\theta^2}{r} + u_z\frac{\partial u_r}{\partial z} = -\left(\frac{\delta}{R}\right)^2 \frac{\partial P}{\partial r} + \left[\frac{\partial^2 u_r}{\partial z^2} + \left(\frac{\delta}{R}\right)^2 \frac{\partial}{\partial r}\left(\frac{1}{r}\frac{\partial(ru_r)}{\partial r}\right)\right] \tag{14.1-49}$$

By comparison to the convective terms with coefficients of unity, the terms in this equation that are multiplied by $(\delta/R)^2 \ll 1$ can be neglected. Equation 14.1-49 then becomes

$$u_r \frac{\partial u_r}{\partial r} - \left(\frac{\Omega\delta^2}{\upsilon}\right)^2 \frac{u_\theta^2}{r} + u_z \frac{\partial u_r}{\partial z} = \frac{\partial^2 u_r}{\partial z^2} \quad (14.1\text{-}50)$$

Assuming that all three convective terms are of equal magnitude, we expect $(\Omega^2\delta^4\rho^2/\mu^2) \sim O(1)$, which provides a relationship of δ to known parameters:

$$\delta = \sqrt{\frac{\upsilon}{\Omega}} \Rightarrow V_r = R\Omega, \quad V_z = \sqrt{\upsilon\Omega} \quad (14.1\text{-}51a\text{-}c)$$

Equation 14.1-51a indicates that δ does not depend on radial position. It is only a function of the fluid properties and rotational rate. This equation also leads to a criterion for satisfying the thin boundary-layer approximation. Defining a rotational Reynolds number with a length scale R and velocity scale $R\Omega$, we obtain

$$Re_\Omega \equiv \frac{R(R\Omega)}{\upsilon} = \left(\frac{R}{\delta}\right)^2 \gg 1 \quad (14.1\text{-}52)$$

Making the z component of the Navier–Stokes equation dimensionless, we obtain

$$u_r \frac{\partial u_z}{\partial r} + u_z \frac{\partial u_z}{\partial z} = -\frac{\partial P}{\partial z} + \left[\frac{\partial^2 u_z}{\partial z^2} + \left(\frac{\delta}{R}\right)^2 \frac{1}{r}\frac{\partial}{\partial r}\left(r\frac{\partial u_z}{\partial r}\right)\right] \quad (14.1\text{-}53)$$

where the viscous term with coefficient $(\delta/R)^2$ can be neglected. Based on the definitions of the arbitrary scaling parameters (Eq. 14.1-51a-c), the dimensionless continuity and simplified components of the Navier–Stokes equation finally become

$$\begin{aligned}
&\frac{1}{r}\frac{\partial(ru_r)}{\partial r} + \frac{\partial u_z}{\partial z} = 0 \\
&\frac{\partial^2 u_r}{\partial z^2} - u_r \frac{\partial u_r}{\partial r} + \frac{u_\theta^2}{r} - u_z \frac{\partial u_r}{\partial z} = 0 \\
&\frac{\partial^2 u_\theta}{\partial z^2} - u_r \frac{\partial u_\theta}{\partial r} - \frac{u_r u_\theta}{r} - u_z \frac{\partial u_\theta}{\partial z} = 0 \\
&\frac{\partial^2 u_z}{\partial z^2} - u_r \frac{\partial u_z}{\partial r} - u_z \frac{\partial u_z}{\partial z} = \frac{\partial P}{\partial z}
\end{aligned} \quad (14.1\text{-}54a\text{-}d)$$

These equations must be simultaneously solved with the appropriate boundary conditions. At the surface of the disk, we apply the no-slip condition by asserting that the velocity components at the surface conform to the rigid body rotation of the spinning disk. We also specify a reference value for the pressure:

$$z = 0: \quad u_r = 0, \quad u_\theta = \Omega r, \quad u_z = 0, \quad P = P_0 \quad (14.1\text{-}55)$$

In the flow field far from the disk, there is an inflow that we expect to be dominated by the z-component of velocity such that

$$z \to \infty: \quad u_r = 0, \quad u_\theta = 0, \quad u_z = u_\infty \quad (14.1\text{-}56)$$

Along the axis of rotation, the radial velocity is zero since there is no source of fluid, and the rotational velocity is zero since there is no angular movement. In addition, the radial derivative of the axial velocity vanishes because of radial symmetry:

$$r = 0: \quad u_r = u_\theta = 0, \quad \frac{\partial u_z}{\partial r} = 0 \tag{14.1-57}$$

In terms of dimensionless variables, these conditions become

$$\begin{aligned} z = 0: &\quad u_\theta = r, \quad u_r = u_z = 0, \quad P = \frac{P_0}{\rho V_z^2} = \frac{P_0}{\rho \upsilon \Omega} \equiv P_0 \\ z \to \infty: &\quad u_r = u_\theta = 0, \quad u_z = \frac{u_\infty}{V_z} = \frac{u_\infty}{\sqrt{\upsilon \Omega}} \\ r = 0: &\quad u_r = u_\theta = 0, \quad \frac{\partial u_z}{\partial r} = 0 \end{aligned} \tag{14.1-58a-c}$$

Similarity Transformation

Because the governing equations do not contain any finite characteristic length, we can apply a similarity transformation:

$$u_r = rf(z), \quad u_\theta = rg(z), \quad u_z = h(z), \quad P = P(z) \tag{14.1-59a-d}$$

This allows the system to be described by functions of z alone. When Equation 14.1-59a-d are substituted into Equation 14.1-54a-d, the model is reduced to four ordinary, nonlinear differential equations:

$$\begin{aligned} &h' + 2f = 0 \\ &g'' - hg' - 2fg = 0 \\ &f'' - hf' - f^2 + g^2 = 0 \\ &hh' - h'' = (h^2)'/2 - h'' = -P' \end{aligned} \tag{14.1-60a-d}$$

where $(\ldots)' \equiv d(\ldots)/dz$. The kinematics expressed by Equation 14.1-59a-d automatically satisfy the boundary conditions at $r = 0$. Transformation of the other boundary conditions yields

$$\begin{aligned} z = 0: &\quad g = 1, \quad f = h = 0, \quad P = P_0/\rho\upsilon\Omega \\ z \to \infty: &\quad f = g = 0, \quad h = u_\infty/\sqrt{\upsilon\Omega} \end{aligned} \tag{14.1-61a,b}$$

According to the last boundary condition, the fluid inflow far from the disk surface is a uniform axial velocity field with a magnitude $u_\infty = h(z \to \infty)\sqrt{\upsilon\Omega}$.

Solution

When Equation 14.1-60d is combined with the derivative of Equation 14.1-60a to eliminate h'', it can be directly integrated with the boundary conditions at $z = 0$:

$$P = \frac{P_0}{\rho\upsilon\Omega} + \frac{1}{2}h^2 + 2f \tag{14.1-62}$$

A numerical solution to Equation 14.1-60a-c for the similarity variables $f(z)$, $g(z)$, and $h(z)$ is shown in Figure 14.1-3. This solution indicates that $-h(z \to \infty) \to 0.88$ so the axial inflow velocity far from the disk is $u_\infty = -0.88\sqrt{\upsilon\Omega}$.

As $z \to 0$, the numerical solution indicates that $f \to 0.51z$, $g \to (1 - 0.62z)$ and $h \to -0.51z^2$. The corresponding velocity components in the vicinity of the disk are

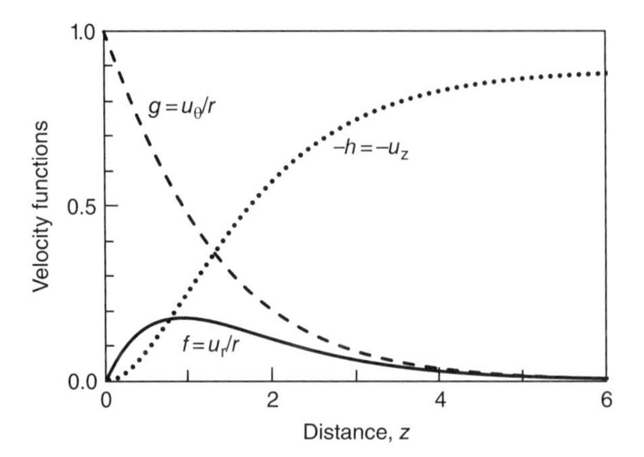

Figure 14.1-3 Velocity similarity profiles on a spinning disk.

$$u_r(z \to 0) = V_r fr = \frac{\upsilon R}{\delta_R^2}\left(0.51\frac{z}{\delta}\right)\left(\frac{r}{R}\right) = 0.51\sqrt{\frac{\Omega^3}{\upsilon}}rz$$

$$u_\theta(z \to 0) = V_\theta gr = R\Omega\left(1 - 0.62\frac{z}{\delta}\right)\left(\frac{r}{R}\right) = \Omega r\left(1 - 0.62\sqrt{\frac{\Omega}{\upsilon}}z\right) \qquad (14.1\text{-}63a\text{-}c)$$

$$u_z(z \to 0) = V_z h = \sqrt{\upsilon\Omega}\left(-0.51\frac{z^2}{\delta^2}\right) = -0.51\sqrt{\frac{\Omega^3}{\upsilon}}z^2$$

Based on these velocity relationships, the shear stresses acting on the disk surface are given by

$$\tau_{\theta z}(0) = 2\mu\gamma_{\theta z}(0) = \mu\left[\frac{\partial u_\theta}{\partial z} + \frac{1}{r}\frac{\partial u_z}{\partial \theta}\right]_{z\to 0} = -0.62r\sqrt{\rho\mu\Omega^3}$$

$$\tau_{rz}(0) = 2\mu\gamma_{rz}(0) = \mu\left[\frac{\partial u_r}{\partial z} + \frac{\partial u_z}{\partial r}\right]_{z\to 0} = 0.51r\sqrt{\rho\mu\Omega^3} \qquad (14.1\text{-}64a,b)$$

Since $\tau_{\theta z}$ is a force component in the θ direction and τ_{rz} is a force component in the r direction, the magnitude of the shear force vector acting on the surface is

$$\tau_{wall} = \sqrt{\tau_{\theta z}{}^2(0) + \tau_{rz}{}^2(0)} = 0.80r\sqrt{\rho\mu\Omega^3} \qquad (14.1\text{-}65)$$

A boundary layer thickness, more meaningful than δ, is that distance δ_{99} from the surface where u_z reaches 99% of u_∞. This is equivalent to a value of $h(z)/h(\infty) = 0.99$, which occurs at $z = 3.6$:

$$z = \frac{\delta_{99}}{\delta} = 3.6 \Rightarrow \delta_{99} = 3.6\sqrt{\frac{\upsilon}{\Omega}} = \frac{3.6R}{Re_\Omega^{1/2}} \qquad (14.1\text{-}66)$$

Example 14.1-2 Applying Shear Stress to a Cell Layer

Garcia and coworkers (1997) studied the adhesion of osteoblast-like cell monolayers to fibronectin-treated glass disks. The 15 mm diameter disks, submerged in a large volume of physiological saline solution, were rotated in order to subject the cells to shear stresses from 0 to 10 Pa. Measurements were made of the cells remaining adherent after 10 min. What rotational speed was necessary in this experiment and how would the resulting boundary layer thicknesses compare to the disk diameter?

Solution

According to Equation 14.1-65, τ_{wall} is proportional to radial position. Thus, the cells located at r = 15/2 mm are subjected to the maximum shear stress. The angular speed for these cells to reach $\tau_{wall} = 10$ Pa is

$$\Omega = \left[\frac{10}{0.80(0.015/2)\sqrt{994(6.93\times10^{-4})}}\right]^{2/3} = 34.3\,s^{-1} \tag{14.1-67}$$

where we have taken the viscosity and density for physiological saline to be same as those for water at 37°C ($\rho = 994$ kg/m^3 and $\mu = 6.93 \times 10^{-4}$ Pa-s). This corresponds to a fairly large rotational rate of $(34.3/2\pi) = 5.45$ times per second and a rotational Reynolds number of

$$Re_\Omega = \frac{(0.015/2)^2(34.3)}{(6.93\times10^{-4}/994)} = 2770 \tag{14.1-68}$$

The corresponding boundary layer thickness is

$$\delta_{99} = \frac{3.6(0.015/2)}{\sqrt{2770}} = 2.57\times10^{-4}\,m \approx 0.3\,mm \tag{14.1-69}$$

This meets the requirement that δ_{99} be much smaller than the 7.5 mm disk radius.

14.2 Creeping Flow Through a Leaky Tube

The nonlinear convective momentum term in the equation of motion limits the types of fluid flow problems that can be solved analytically. When the Reynolds number is sufficiently small, convective forces can be neglected compared to viscous forces. The governing equations are then linear and an analytical solution is possible. In this illustration, we consider creeping flow through a leaky tube as occurs, for example, in a renal tubule of a kidney nephron or in a hollow fiber of a hemodialyzer. We will assume steady-state conditions, an incompressible Newtonian fluid and a cylindrical tube with a uniform leak velocity through the tube wall. The fluid kinematics then consists of two velocity components that depend on two spatial coordinates.

Model Formulation

For this circularly symmetric problem, the model is represented in cylindrical coordinates so that the derivatives of all independent variables with respect to the angular coordinate θ are zero. In addition, the circumferential velocity component is zero, $u_\theta = 0$. Whereas axial velocity depends on both radial and axial positions, $u_z = u_z(r,z)$, radial velocity caused by the uniform filtration velocity u_{leak} at the tube wall depends only on radial position, $u_r = u_r(r)$.

For steady flow of an incompressible fluid with these kinematics, the continuity equation reduces to

$$\frac{1}{r}\frac{d(ru_r)}{dr} = -\frac{\partial u_z}{\partial z} \tag{14.2-1}$$

Neglecting all inertial terms, the three components of the steady-state Navier–Stokes equation reduce to

$$\theta \text{ component}: \quad \frac{\partial \mathcal{P}}{\partial \theta} = 0 \Rightarrow \mathcal{P} = \mathcal{P}(r,z) \tag{14.2-2}$$

$$r \text{ component}: \quad \frac{\partial \mathcal{P}}{\partial r} = \mu \frac{d}{dr}\left[\frac{1}{r}\frac{d(ru_r)}{dr}\right] \tag{14.2-3}$$

$$z \text{ component}: \quad \frac{\partial \mathcal{P}}{\partial z} = \mu \left[\frac{1}{r}\frac{\partial}{\partial r}\left(r\frac{\partial u_z}{\partial r}\right) + \frac{\partial^2 u_z}{\partial z^2}\right] \tag{14.2-4}$$

The boundary condition at the tube wall expresses continuity of the two velocity components:

$$r = a: \quad u_r = u_{leak}, \; u_z = 0 \tag{14.2-5a,b}$$

At the tube centerline, the radial velocity is zero. Also, about the centerline, the velocity derivatives with respect to r vanish because of radial symmetry:

$$r = 0: \quad u_r = 0, \; \frac{\partial u_z}{\partial r} = 0 \tag{14.2-6a,b}$$

At the entrance of the tube, we specify the volumetric flow rate and the cross-sectional average pressure:

$$z = 0: \quad Q_o = 2\pi \int_0^a u_z r dr, \; \mathcal{P}_o = \frac{2}{a^2}\int_0^a \mathcal{P} r dr \tag{14.2-7a,b}$$

Analysis

We can eliminate P from the Navier–Stokes equation by differentiating the r component with respect to z to obtain $\partial^2\mathcal{P}/\partial r\partial z$ and setting it equal to $\partial^2\mathcal{P}/\partial z\partial r$ obtained by differentiating the z component with respect to r:

$$\frac{\partial^2}{\partial z \partial r}\left[\frac{1}{r}\frac{d(ru_r)}{dr}\right] = \frac{\partial}{\partial r}\left[\frac{1}{r}\frac{\partial}{\partial r}\left(r\frac{\partial u_z}{\partial r}\right)\right] + \frac{\partial^2}{\partial r \partial z}\left(\frac{\partial u_z}{\partial z}\right) \tag{14.2-8}$$

Since u_r only depends on r, the continuity equation indicates that $\partial u_z/\partial z$ is not a function of z. Thus, the last term in Equation 14.2-8 is equal to zero. Also because u_r only depends on r, the left side of this equation is equal to zero. What remains of Equation 14.2-8 is

$$\frac{\partial}{\partial r}\left[\frac{1}{r}\frac{\partial}{\partial r}\left(r\frac{\partial u_z}{\partial r}\right)\right] = 0 \tag{14.2-9}$$

Taking the derivative of this equation with respect to z and substituting $\partial u_z/\partial z$ from the continuity equation, we get

$$\frac{d}{dr}\left\{\frac{1}{r}\frac{d}{dr}\left[r\frac{d}{dr}\left(\frac{1}{r}\frac{d(ru_r)}{dr}\right)\right]\right\} = 0 \tag{14.2-10}$$

When we successively integrate this equation with respect to r, we obtain

$$u_r = A_1 r^3 + A_2 r \ln r + A_3 r + \frac{A_4}{r} \tag{14.2-11}$$

where A_1, A_2, A_3, and A_4 are constants. Substituting this expression into the continuity equation and integrating with respect to z yields:

$$u_z = -\left[2A_1 r^2 + A_2\left(\ln r + \frac{1}{2}\right) + A_3\right]2z + U(r) \tag{14.2-12}$$

where the function of integration, $U(r) = u_z(r,0)$, represents the velocity profile at the tube entrance. By employing the conditions on u_r and u_z from Equations 14.2-5a,b and 14.2-6a, b, we obtain values for the four constants of integration as well as two boundary conditions on U(r):

$$A_1 = -\frac{u_{leak}}{a^3},\quad A_2 = 0,\quad A_3 = \frac{2u_{leak}}{a},\quad A_4 = 0$$
$$U(a) = 0,\quad \left[\frac{dU}{dr}\right]_{r=0} = 0 \tag{14.2-13a-f}$$

Combining Equations 14.2-11 and 14.2-12 with Equation 14.2-13a-d, the velocity components are

$$\frac{u_z}{u_{leak}} = \frac{U(r)}{u_{leak}} - 4\left(\frac{z}{a}\right)\left[1-\left(\frac{r}{a}\right)^2\right]$$
$$\frac{u_r}{u_{leak}} = 2\left(\frac{r}{a}\right) - \left(\frac{r}{a}\right)^3 \tag{14.2-14a,b}$$

Substituting Equation 14.2-14a into Equation 14.2-9,

$$r^3\frac{d^3U}{dr^3} + r^2\frac{d^2U}{dr^2} - r\frac{dU}{dr} = 0 \tag{14.2-15}$$

This equation has the solution:

$$U = B_1 r^2 + B_2 \ln r + B_3 \tag{14.2-16}$$

where B_1, B_2, and B_3 are constants of integration. Applying Equation 14.2-13e,f, we establish that $B_2 = 0$ and $B_3 = -B_1 a^2$. With Equation 14.2-7a,b, we obtain

$$Q_o = 2\pi\int_0^a U r\,dr = 2\pi\int_0^a B_1\left(r^2 - a^2\right) r\,dr \Rightarrow B_1 = -\frac{2Q_o}{\pi a^4} \tag{14.2-17a,b}$$

Substituting B_1, B_2, and B_3 into Equation 14.2-17b, we find the entrance velocity profile:

$$\frac{U}{u_{leak}} = \frac{2Q_o}{\pi a^2 u_{leak}}\left[1-\left(\frac{r}{a}\right)^2\right] \tag{14.2-18}$$

Finally, we combine Equation 14.2-18 with Equation 14.2-14a to obtain the axial velocity component:

$$u_z = \frac{2Q(z)}{\pi a^2}\left[1 - \left(\frac{r}{a}\right)^2\right] \quad (14.2\text{-}19)$$

where

$$Q(z) \equiv Q_o - (2\pi a u_{leak})z \quad (14.2\text{-}20)$$

is the volumetric flow at a particular axial position. Notice the similarity of Equation 14.2-19 to the velocity profile during Hagen–Poiseuille flow (Eq. 13.4-1). With a constant leak velocity, u_z retains a parabolic shape throughout the tube with a centerline velocity that is equal to twice the local value of the cross-section averaged velocity, $Q/\pi a^2$.

To determine $\mathcal{P}(r, z)$, we substitute the equations for u_r and u_z into the r and z components of the Navier–Stokes equation to obtain

$$\frac{\partial \mathcal{P}}{\partial z} = -\frac{8\mu Q(z)}{\pi a^4}, \quad \frac{\partial \mathcal{P}}{\partial r} = -\frac{8\mu u_{leak} r}{a^3} \quad (14.2\text{-}21a,b)$$

By integrating each of these equations and comparing the results, we find

$$\mathcal{P} = \frac{8\mu u_{leak}}{a^3}z^2 - \frac{8\mu Q_o}{\pi a^4}z - \frac{4\mu u_{leak} r^2}{a^3} + B_4 \quad (14.2\text{-}22)$$

We evaluate the constant of integration B_4 from the entrance condition given by Equation 14.2-7a,b:

$$\mathcal{P}_o = \frac{2}{a^2}\int_0^a [\mathcal{P}]_{z=0} r dr = \frac{2}{a^2}\int_0^a \left(B_4 - \frac{4\mu u_{leak} r^2}{a^3}\right) r dr \Rightarrow B_4 = \mathcal{P}_o + \frac{2\mu u_{leak}}{a} \quad (14.2\text{-}23a,b)$$

The pressure distribution is now expressed as

$$\mathcal{P} = \mathcal{P}_o - \frac{8\mu Q_o}{\pi a^3}\left(\frac{z}{a}\right) - \frac{8\mu u_{leak}}{a}\left[\left(\frac{z}{a}\right)^2 - \frac{1}{2}\left(\frac{r}{a}\right)^2 + \frac{1}{4}\right] \quad (14.2\text{-}24)$$

For flow through an impermeable tube, $u_{leak} = 0$ and pressure decreases linearly with z and is independent of r. In the presence of a flow leak, however, pressure decreases in a nonlinear manner with both z and r. By integrating Equation 14.2-24, we find that the difference between the average exit pressure, $\mathcal{P}_L = 2/a^2 \int \mathcal{P}(r, L) r dr$, and average entrance pressure $\mathcal{P}_o$ is given by:

$$\Delta\mathcal{P} = \left(\frac{8\mu L}{\pi a^4}\right)\left(Q_o - \frac{Q_{leak}}{2}\right), \quad Q_{leak} \equiv 2\pi a L u_{leak} \quad (14.2\text{-}25)$$

where Q_{leak} is the volumetric rate of leakage through the entire tube wall. Employing Equation 14.2-20, we find that the longitudinally average flow in a tube of length L is

$$\bar{Q} = \frac{1}{L}\int_0^z (Q_o - 2\pi a u_{leak} z) dz = Q_o - \frac{Q_{leak}}{2} \quad (14.2\text{-}26)$$

so that Equation 14.2-25 becomes

$$\Delta\mathcal{P} = \left(\frac{8\mu L \bar{Q}}{\pi a^4}\right) \quad (14.2\text{-}27)$$

Thus, pressure drop follows a Hagen–Poiseuille equation (Eq. 13.4-2) based on the average volumetric flow. Since $\bar{Q} < Q_o$ when there is a loss of fluid through the tube wall, a leaky tube requires a smaller pressure drop than an impermeable tube. In the extreme that all the incoming flow is lost through leakage, $\bar{Q} = Q_o/2$ and the pressure drop would be one half that of an impermeable tube.

Example 14.2-1 Pressure Drop in a Renal Tubule

Suppose that a total of 100 ml/min of glomerular filtrate reaches the entrance of the proximal tubules in the kidneys. Following water reabsorption into the peritubular capillaries, only 25 ml/min leaves the proximal tubules. Estimate the pressure change through a proximal tubule and compute the pressure drop if there were no water reabsorption. The typical length and diameter of a proximal tubule are 20 mm and 50 μm, respectively. Blood plasma, which is similar in composition to glomerular filtrate, has a shear viscosity of 0.0012 Pa-s and a mass density of 1020 kg/m^3 (Table 13.2-1).

Solution

Assuming there are 10^6 nephrons in the kidney, the flows associated with a single proximal tubule are

$$\begin{aligned} Q_o &= \frac{(100\times10^{-6})/60}{10^6} = 1.67\times10^{-12}\ \mathrm{m^3/s} \\ Q_{leak} &= \frac{(100-25)\times10^{-6}/60}{10^6} = 1.25\times10^{-12}\ \mathrm{m^3/s} \\ \bar{Q} &= (1.67-1.25/2)\times10^{-12} = 1.04\times10^{-12}\ \mathrm{m^3/s} \end{aligned} \tag{14.2-28a-c}$$

According to Equation 14.2-27, the pressure drops with and without reabsorption are

$$\begin{aligned} [\Delta\mathcal{P}]_{\text{with reabsorption}} &= \frac{8(0.0012)(20\times10^{-3})(1.04\times10^{-12})}{\pi(50\times10^{-6}/2)^4} = 163\ \mathrm{Pa} \\ [\Delta\mathcal{P}]_{\text{no reabsorption}} &= \frac{8(0.0012)(20\times10^{-3})(1.67\times10^{-12})}{\pi(50\times10^{-6}/2)^4} = 261\ \mathrm{Pa} \end{aligned} \tag{14.2-29a, b}$$

This modified pressure drop is the sum of contributions from hydrodynamic pressure and gravity. The maximum effect of gravity would occur if the tubules were straight and u_z had a downward or upward vertical orientation, corresponding to an inclination angle α of 0° or 180°, respectively. The gravitational contribution of $\rho GL\cos(\alpha) = \pm 1020(9.81)\,(20\times10^{-3}) = \pm 200\,\mathrm{Pa}$ would then be substantial.

14.3 Periodic Flow Along a Tube

So far, we have focused on steady flow problems. In many physiological processes, however, there are time-dependent changes that can have a dramatic effect on mass transport. Pulsatile blood flow in the cardiovascular system and cyclic gas flow in the lungs are the two important examples of such processes. In this section, we will model periodic flow in a tube, which is a composite of a time-averaged flow and an oscillatory flow driven by a periodic pressure gradient.

Model Formulation

We consider the periodic flow of an incompressible, Newtonian fluid in a rigid, cylindrical tube of radius "a" and length L. The tube is so long, $(a/L) \ll 1$, that the specific conditions at the end of the tube have little effect on the velocity field anywhere in the tube. When this flow is described in cylindrical coordinates to take advantage of symmetry, the angular velocity vanishes, $u_\theta = 0$, and the axial velocity varies only with radial position and time, $u_z = u_z(r,t)$. For a tube that is impermeable to solution, we also expect that radial velocity is zero, $u_r = 0$. The radial and circumferential components of the Navier–Stokes equation then reduce to

$$\frac{\partial \mathcal{P}}{\partial r} = \frac{1}{r}\frac{\partial \mathcal{P}}{\partial \theta} = 0 \Rightarrow \mathcal{P} = \mathcal{P}(z,t) \qquad (14.3\text{-}1a,b)$$

and the axial component reduces to

$$\rho \frac{\partial u_z}{\partial t} = \mu \frac{1}{r}\frac{\partial}{\partial r}\left(r\frac{\partial u_z}{\partial r}\right) - \frac{\partial \mathcal{P}}{\partial z} \qquad (14.3\text{-}2)$$

A periodic input function that is sufficiently smooth over a finite time interval can be represented by an infinite series of sinusoids with frequencies that are multiples of a fundamental frequency (i.e., a Fourier series). Since Equations 14.3-1 and 14.3-2 are linear, the output response to such an input can be obtained by superimposing the individual sinusoidal responses at these frequencies. Consequently, we will simplify our analysis by considering the system response to a single sinusoid of frequency ω. In particular, we specify the input as the real part of the following complex pressure gradient:

$$-\frac{\partial \mathcal{P}}{\partial z} = \frac{\Delta \mathcal{P}^o}{L} + \frac{\Delta \mathcal{P}^\omega}{L} e^{j\omega t} \qquad (14.3\text{-}3)$$

Here, the time-average $\Delta \mathcal{P}^o/L$ and the sinusoidal amplitude $\Delta \mathcal{P}^\omega/L$ of the pressure gradient as well as the oscillation frequency ω are constants. In the cardiovascular system where blood must have a net forward motion, $\Delta \mathcal{P}^o/L > 0$. In the lung airways, however, the net volumetric gas flow is very small so that $\Delta \mathcal{P}^o/L \approx 0$.

We restrict our analysis to a long enough time and a sufficiently large number of oscillation cycles that transients can be neglected. For a linear system, the output function u_z then varies sinusoidally at the same frequency as the input pressure gradient, but is shifted in phase:

$$u_z = u^o + u^\omega e^{j\omega t} \qquad (14.3\text{-}4)$$

where $u^o(r)$ is the time-average velocity component, and $u^\omega = |u^\omega| e^{j\phi_u}$ is a complex function with a modulus $|u^\omega|$ that is a function of r, and phase angle ϕ_u relative to the sinusoidal pressure gradient.

Analysis

Dimensionless Forms

Using the dimensionless variables,

$$t = \omega t, \quad r = \frac{r}{a}, \quad z = \frac{z}{L}, \quad u_z = \frac{u_z}{u_{ave}}, \quad P = \frac{\mathcal{P}}{\rho u_{ave} \omega L} \qquad (14.3\text{-}5a\text{-}e)$$

Equation 14.3-2 can be represented compactly as

$$\frac{\partial u_z}{\partial t} = \frac{1}{Wo^2 r}\frac{\partial}{\partial r}\left(r\frac{\partial u_z}{\partial r}\right) - \frac{\partial P}{\partial z} \tag{14.3-6}$$

Here, u_{ave} is the spatial average of $u^o(r)$ over the tube cross section, and Wo is a dimensionless frequency parameter called the Womersley number:

$$Wo \equiv a\sqrt{\rho\omega/\mu} = a\sqrt{\omega/\upsilon} \tag{14.3-7}$$

The quantity $\sqrt{\upsilon/\omega}$ is the characteristic thickness of a boundary or Stokes layer that develops at the tube wall because of fluid oscillation. When $Wo \gg 1$, the Stokes layer is much thinner than the tube radius and oscillatory effects occur close to the wall. When $Wo < 1$, however, the Stokes layer virtually fills the entire tube cross section. With the additional dimensionless quantities

$$u^o \equiv \frac{u^o}{u_{ave}}, \quad u^\omega \equiv \frac{u^\omega}{u_{ave}}, \quad P^o \equiv \frac{\Delta\mathcal{P}^o}{\rho u_{ave}\omega L}, \quad P^\omega \equiv \frac{\Delta\mathcal{P}^\omega}{\rho u_{ave}\omega L}, \quad \Lambda \equiv \frac{\Delta\mathcal{P}^\omega}{\Delta\mathcal{P}^o} = \frac{P^\omega}{P^o} \tag{14.3-8a-e}$$

the complex pressure driving force (Eq. 14.3-3) becomes

$$-\frac{\partial P}{\partial z} = P^o\left(1 + \Lambda e^{jt}\right) \tag{14.3-9}$$

and the complex output velocity (Eq. 14.3-4) becomes

$$u_z = u^o + u^\omega e^{jt} \tag{14.3-10}$$

Substituting Equations 14.3-9 and 14.3-10 into Equation 14.3-6 and separately grouping the time-averaged and the oscillating parts leads to

$$\left[-\frac{1}{Wo^2 r}\frac{\partial}{\partial r}\left(r\frac{\partial u^o}{\partial r}\right) - P^o\right] + \left[u^\omega j - \frac{1}{Wo^2 r}\frac{\partial}{\partial r}\left(r\frac{\partial u^\omega}{\partial r}\right) - \Lambda P^o\right]e^{jt} = 0 \tag{14.3-11}$$

To satisfy this equation, these two parts must separately be equal to zero such that

$$\frac{1}{Wo^2 r}\frac{\partial}{\partial r}\left(r\frac{\partial u^o}{\partial r}\right) = -P^o \tag{14.3-12}$$

$$\frac{1}{Wo^2 r}\frac{\partial}{\partial r}\left(r\frac{\partial u^\omega}{\partial r}\right) - ju^\omega = -\Lambda P^o \tag{14.3-13}$$

Equations 14.3-12 and 14.3-13 can be solved with boundary conditions analogous to those used for steady fully developed flow in a tube. At the tube wall, the velocity vanishes:

$$r = a: \quad u^o = u^\omega = 0 \Rightarrow r = 1: \quad u^o = u^\omega = 0 \tag{14.3-14a,b}$$

At the tube centerline, the velocity is radially symmetric:

$$r = 0: \quad \frac{\partial u^o}{\partial r} = \frac{\partial u^\omega}{\partial r} = 0 \Rightarrow r = 0: \quad \frac{\partial u^o}{\partial r} = \frac{\partial u^\omega}{\partial r} = 0 \tag{14.3-15a,b}$$

Analytical Solution

Equation 14.3-12 and its accompanying boundary conditions are equivalent to the problem statement for Hagen–Poiseuille flow. Thus, we can infer the time-average velocity in dimensionless form from Equations 13.4-1 and 13.4-2:

$$u^{o} \equiv \frac{u^{o}}{u^{ave}} = 2\left(1-r^{2}\right); \quad u_{ave} = \frac{a^{2}\Delta\mathcal{P}^{o}}{8\mu L} \tag{14.3-16a,b}$$

Combining Equations 14.3-7, 14.3-8c, and 14.3-16b, the dimensionless time-average driving pressure can be rewritten:

$$P^{o} \equiv \frac{\Delta\mathcal{P}^{o}}{\rho\omega u_{ave}L} = \frac{8\mu}{\rho\omega a^{2}} = \frac{8}{Wo^{2}} \tag{14.3-17}$$

We can express the oscillatory part of this problem as a homogeneous version of Equation 14.3-13 by defining a new dependent variable:

$$w = u^{\omega} + jAP^{o} \tag{14.3-18}$$

Recognizing that AP^{o} is a constant, this leads to a zeroth-order Bessel equation:

$$r^{2}\frac{d^{2}w}{dr^{2}} + r\frac{dw}{dr} - \left(Wo^{2}j\right)r^{2}w = 0 \tag{14.3-19}$$

with boundary conditions:

$$\begin{aligned} r = 1: &\quad w = jAP^{o} \\ r = 0: &\quad \frac{\partial w}{\partial r} = 0 \end{aligned} \tag{14.3-20a,b}$$

The general solution to Equation 14.3-19 is

$$w = A_{1}J_{0}\left(j^{3/2}rWo\right) + A_{2}Y_{0}\left(j^{3/2}rWo\right) \tag{14.3-21}$$

where J_0 is the zeroth-order Bessel function of the first kind, and Y_0 is the zeroth-order Bessel function of the second kind. Using the two boundary conditions on w with the identities that $[dJ_o(\xi)/d\xi]_{\xi=0} = -J_1(0) = 1$ and $[dY_o(\xi)/d\xi]_{\xi=0} = -Y_1(0) \rightarrow \infty$, we find

$$A_{1} = \frac{jAP^{o}}{J_{0}\left(j^{3/2}Wo\right)}, \quad A_{2} = 0 \tag{14.3-22a,b}$$

so that

$$u^{\omega} = AP^{o}j\left[\frac{J_{0}\left(j^{3/2}rWo\right)}{J_{0}\left(j^{3/2}Wo\right)} - 1\right] \tag{14.3-23}$$

By making use of Equations 14.3-10, 14.3-16a, 14.3-17, and 14.3-23 with the identity j ≡ −1/j, we obtain an expression for the complete dimensionless velocity distribution:

$$u_{z} = 2\left(1-r^{2}\right) + \frac{8A}{jWo^{2}}\left[1 - \frac{J_{0}\left(j^{3/2}rWo\right)}{J_{0}\left(j^{3/2}Wo\right)}\right]e^{jt} \tag{14.3-24}$$

In terms of dimensional variables,

$$u_z = \frac{a^2 \Delta\mathcal{P}^o}{4\mu L}\left(1 - \frac{r^2}{a^2}\right) + \frac{\Delta\mathcal{P}^\omega}{j\omega\rho L}\left[1 - \frac{J_0\left(j^{3/2} r Wo/a\right)}{J_0\left(j^{3/2} Wo\right)}\right] e^{j\omega t} \tag{14.3-25}$$

An important limiting equation for u_z occurs as $Wo \to 0$. Using the approximations that $\lim_{\xi\to 0} J_o(\xi) = 1 - \xi^2/4$ and $\lim_{\xi\to 0} J_1(\xi) = \xi/2 - \xi^3/16$, we get

$$\lim_{Wo\to 0} u_z = \frac{a^2 \Delta\mathcal{P}^o}{4\mu L}\left(1 - \frac{r^2}{a^2}\right)\left(1 + \frac{\Delta\mathcal{P}^\omega}{\Delta\mathcal{P}^o} e^{j\omega t}\right) \quad (a \geq r \geq 0) \tag{14.3-26}$$

This indicates that at very small frequencies of the imposed pressure gradient, u_z is in a pseudo-steady state with a parabolic velocity distribution at all times. In this situation, there is no phase difference between the time-average and oscillating portions of the velocity.

The volumetric flow Q is obtained by integrating u_z (Eq. 14.3-25) over the tube cross section. By employing the relation $\int r J_o(r) dr = r J_1(r)$, this leads to

$$Q = 2\pi \int_0^a u_z r dr = \left(\frac{\pi a^4}{8\mu L}\right)\Delta\mathcal{P}^o + \frac{\pi a^2 \Delta\mathcal{P}^\omega}{j\omega\rho L}\left[1 - \frac{2 J_1\left(j^{3/2} Wo\right)}{\left(j^{3/2} Wo\right) J_0\left(j^{3/2} Wo\right)}\right] e^{j\omega t} \tag{14.3-27}$$

Thus, the volumetric flow distribution consists of the time-average and oscillatory parts:

$$Q^o = \left(\frac{\pi a^4}{8\mu L}\right)\Delta\mathcal{P}^o, \quad Q^\omega = \frac{\pi a^2 \Delta\mathcal{P}^\omega}{j\omega\rho L}\left[1 - \frac{2 J_1\left(j^{3/2} Wo\right)}{\left(j^{3/2} Wo\right) J_0\left(j^{3/2} Wo\right)}\right] \tag{14.3-28a,b}$$

Complex Impedance

The resistance to flow oscillation is expressed by the complex mechanical impedance, $\Delta\mathcal{P}^\omega/Q^\omega$. We define the dimensionless complex impedance χ as the ratio of $\Delta\mathcal{P}^\omega/Q^\omega$ to the pseudo-steady impedance, $[\Delta\mathcal{P}^\omega/Q^\omega]_{Wo=0}$:

$$\chi \equiv \frac{\Delta\mathcal{P}^\omega/Q^\omega}{[\Delta\mathcal{P}^\omega/Q^\omega]_{Wo=0}} = |\chi| e^{j(\omega t + \phi)} \tag{14.3-29}$$

where $|\chi|$ and ϕ are the modulus and phase, respectively, of χ. Note that ϕ also corresponds to the phase angle of the pressure gradient oscillation relative to the flow oscillation. Taking the limit as $Wo \to 0$ of Equation 14.3-28b, we obtain the pseudo-steady impedance:

$$[\Delta\mathcal{P}^\omega/Q^\omega]_{Wo=0} = 8\mu L/\pi a^4 \tag{14.3-30}$$

Combining this with Equation 14.3-28b leads to

$$\chi = |\chi| e^{j(\omega t + \phi)} = \frac{j Wo^2}{8}\left[1 - \frac{2 J_1\left(j^{3/2} Wo\right)}{\left(j^{3/2} Wo\right) J_0\left(j^{3/2} Wo\right)}\right]^{-1} \tag{14.3-31}$$

Figure 14.3-1 shows the effect of Wo on $|\chi|$ and ϕ computed from this equation. Under pseudo-steady conditions when $Wo \to 0$, the modulus $|\chi|$ approaches one, while the phase ϕ approaches zero. At increasing values of Wo, the value of $|\chi|$ increases without bound. The value of ϕ simultaneously increases, and asymptotically approaches 90° at very large

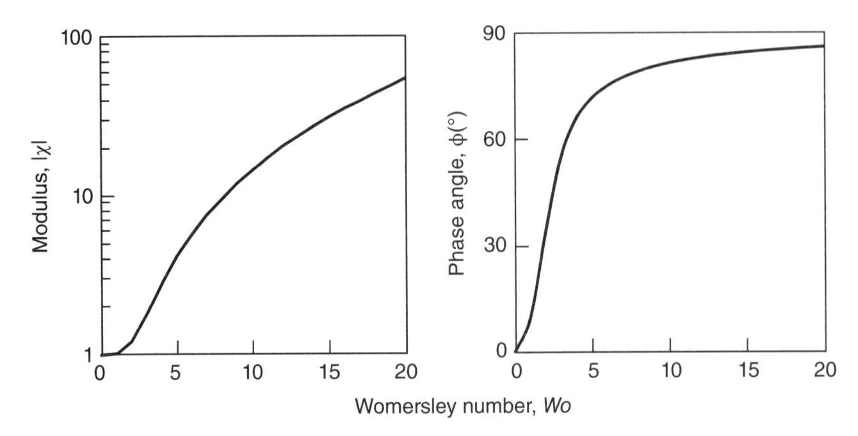

Figure 14.3-1 Complex mechanical impedance of periodic tube flow.

Wo. Since $\phi > 0$ at all $Wo > 0$, the pressure gradient oscillation always leads the flow oscillation.

Example 14.3-1 Pressure Oscillations in the Conducting Airways

During quiet respiration, a person breathes with a frequency of about 15 breaths per minute and a tidal volume of about 500 ml. Estimate the amplitude of the pressure oscillations in the trachea and a terminal conducting airway that are necessary to sustain this respiratory pattern. The transport properties for air and the airway dimensions were given in Example 14.1-1.

Solution

If we approximate the time variation of lung volume during quiet breathing as a pure sinusoid with an angular frequency ω and an amplitude V_T, then the corresponding flow amplitude of respiration will be $|Q| = (\omega/2)V_T$. In this problem, the angular frequency is $\omega = 2\pi(15)/60 = 1.57\,\text{rad/s}$, and the volume amplitude is $V_T = 500$ ml. Therefore, the amplitude of tidal flow oscillations in the trachea is

$$\left|Q^{\omega}_{\text{trachea}}\right| = 1.57\left(500\times10^{-6}\right)/2 = 3.93\times10^{-4}\ \text{m}^3/\text{s} \tag{14.3-32}$$

We assume that air density in the lungs is constant and that flow is uniformly distributed among the 2^{16} terminal conducting airways. The oscillation amplitude in a terminal airway is then given by

$$\left|Q^{\omega}_{\text{terminal}}\right| = 3.93\times10^{-4}/2^{16} = 6.00\times10^{-9}\ \text{m}^3/\text{s} \tag{14.3-33}$$

For pseudo-steady flows, Equation 14.3-30 indicates that the corresponding pressure drop amplitudes would be $|\Delta\mathcal{P}^{\omega}|_{Wo=0} = (8\mu L/\pi a^4)|Q^{\omega}|$:

$$\begin{aligned}
\left|\Delta\mathcal{P}^{\omega}_{\text{trachea}}\right|_{Wo=0} &= \frac{8\left(1.91\times10^{-5}\right)(0.12)\left(3.93\times10^{-4}\right)}{\pi(0.009)^4} = 0.350\,\text{Pa} \\
\left|\Delta\mathcal{P}^{\omega}_{\text{terminal}}\right|_{Wo=0} &= \frac{8\left(1.91\times10^{-5}\right)(0.0017)\left(6.00\times10^{-9}\right)}{\pi(0.0003)^4} = 0.0613\ \text{Pa}
\end{aligned} \tag{14.3-34a,b}$$

To adjust these values for the effect of nonzero frequencies, we first compute the Wormesley numbers in the two airways:

$$(Wo)_{\text{trachea}} \equiv 0.009\sqrt{\frac{1.57}{(1.90\times 10^{-5}/1.14)}} = 2.76$$
$$(Wo)_{\text{terminal}} \equiv 0.0003\sqrt{\frac{1.57}{(1.90\times 10^{-5}/1.14)}} = 0.0921 \qquad (14.3\text{-}35\text{a, b})$$

We then find dimensionless impedance amplitudes of $|\chi|_{\text{trachea}} = 1.64$ and $|\chi|_{\text{terminal}} = 1.00$ by inspection of Figure 14.3-1. Multiplying these values by the corresponding pseudo-steady pressure differences, we obtain the actual pressure differences:

$$\left|\Delta\mathcal{P}^{\omega}_{\text{trachea}}\right| = 1.64(0.350) = 0.583 \text{ Pa}$$
$$\left|\Delta\mathcal{P}^{\omega}_{\text{terminal}}\right| = 1.00(0.0613) = 0.0613 \text{ Pa} \qquad (14.3\text{-}36\text{a, b})$$

These results demonstrate that a much smaller modified pressure drop is required to maintain tidal flow in a terminal airway than in the trachea. Thus, pressure-flow behavior in the whole lung is dominated by the geometry of the large airways. This conclusion is valid for quiet breathing in healthy lungs. During forced vital capacity exhalation, constriction of the distensible small airway walls determines the pressure drop, especially in diseased lungs.

Reference

Garcia AJ, Ducheyne P, Boettiger D. Quantification of cell adhesion using a spinning disk device and application to surface-reactive materials. Biomaterials. 1997; 18:1091–1098.

Chapter 15

Mass Transport I

Basic Concepts and Nonreacting Systems

In Chapter 9, we analyzed unidirectional transport by employing mass conservation equations derived by species balances on rectangular, cylindrical, and spherical shells. This chapter considers an approach for more complex problems in which transport is multidimensional in nature. We begin by using vector methods to derive a general form of the convection–diffusion equation for homogeneous materials. After describing how transport processes are modeled with this equation, we present several illustrations in nonreacting systems.

15.1 Three-Dimensional Mass Balances

In Chapter 13, we derived the equation of motion with a Lagrangian control volume comprised of a fixed collection of material elements as they move through space. To derive a species mass balance, it is more convenient to use an Eulerian control volume that is fixed in space. Whereas a Lagrangian control volume is a closed system, mass can cross the stationary boundaries of an Eulerian control volume.

For the mass of a species i, to be conserved in an Eulerian control volume V enclosed by surface S (Fig. 15.1-1), its accumulation rate $d(\text{Mass}_i)/dt$ within V must be balanced by its formation rate $\dot{r}_i$ within V less its net output rate $\dot{n}_i$ across S:

$$\left[\frac{d(\text{Mass}_i)}{dt}\right]_V = \dot{r}_i - \dot{n}_i \qquad (15.1\text{-}1)$$

Biomedical Mass Transport and Chemical Reaction: Physicochemical Principles and Mathematical Modeling, First Edition. James S. Ultman, Harihara Baskaran, and Gerald M. Saidel.

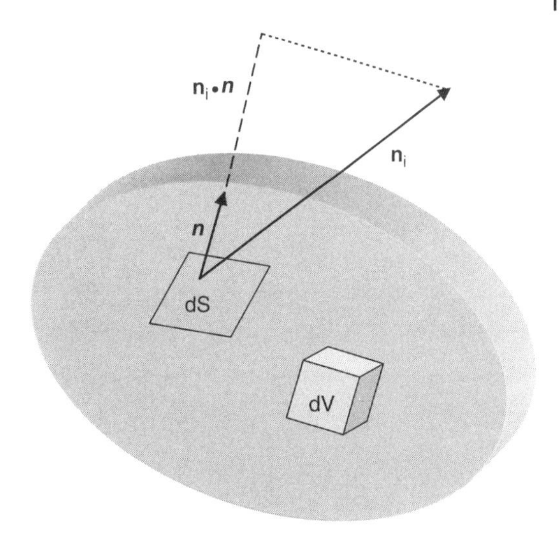

Figure 15.1-1 Fixed control volume for the species mass balance.

To evaluate each of the terms in this equation, we divide the control volume V into a collection of differential elements of volume dV. Associated with each volume element is a mass concentration ρ_i of substance i that is formed by chemical reaction at a molar rate R_i per unit volume. The mass of this substance and its mass rate of formation rate in the control volume are given by

$$\text{Mass}_i = \int_V \rho_i dV, \quad \dot{r}_i = M_i \int_V R_i dV \tag{15.1-2a,b}$$

where M_i is molecular weight. We also divide the surface S of the control volume into a collection of surface elements, each of area dS through which substance i is transported to the surroundings at a mass flux $\mathbf{n}_i$. The orientation of each surface element is identified by an outward-pointing unit vector $\boldsymbol{n}$ so that the mass output rate across the entire surface S can be expressed as the surface integral of $\mathbf{n}_i \cdot \boldsymbol{n}$. By the divergence theorem (Eq. B1-16a), we transform this surface integral into a volume integral:

$$\dot{n}_i = \int_S \mathbf{n}_i \cdot \boldsymbol{n} dS = \int_V \nabla \cdot \mathbf{n}_i dV \tag{15.1-3}$$

Interchanging the derivative and integral operations for the mass and substituting the volume integral relations into Equation 15.1-1, we find

$$\int_V \left(\frac{\partial \rho_i}{\partial t} + \nabla \cdot \mathbf{n}_i - M_i R_i \right) dV = 0 \tag{15.1-4}$$

This equation is always satisfied when the integrand vanishes:

$$\frac{\partial \rho_i}{\partial t} + \nabla \cdot \mathbf{n}_i = M_i R_i \tag{15.1-5}$$

By summing each term of this multidimensional mass concentration equation over all I species in the control volume, we obtain

Table 15.1-1 Species Molar Balances

$$\text{Rectangular: } \frac{\partial C_i}{\partial t} + \left(\frac{\partial N_{i,x}}{\partial x} + \frac{\partial N_{i,y}}{\partial y} + \frac{\partial N_{i,z}}{\partial z}\right) = R_i$$

$$\text{Cylindrical: } \frac{\partial C_i}{\partial t} + \frac{1}{r}\frac{\partial (rN_{i,r})}{\partial r} + \frac{1}{r}\frac{\partial N_{i,\theta}}{\partial \theta} + \frac{\partial N_{i,z}}{\partial z} = R_i$$

$$\text{Spherical: } \frac{\partial C_i}{\partial t} + \frac{1}{r^2}\frac{\partial (r^2 N_{i,r})}{\partial r} + \frac{1}{r}\frac{\partial N_{i,\theta}}{\partial \theta} + \frac{1}{r = \sin\theta}\frac{\partial N_{i,\phi}}{\partial \phi} = R_i$$

$$\sum_{i=1}^{I}\left(\frac{\partial \rho_i}{\partial t} + \nabla \cdot \mathbf{n}_i\right) = \sum_{i=1}^{I} M_i R_i \qquad (15.1\text{-}6)$$

Total mass can be neither created nor destroyed by conventional (i.e., nonnuclear) reactions, so that $\sum_i M_i R_i = 0$. Since summation and differentiation operations are commutative, we can use the definitions of total mass density $\sum_i \rho_i = \rho$ and mass flux $\sum_i \mathbf{n}_i = \rho\mathbf{u}$ to obtain the same continuity equation as previously derived in Chapter 13:

$$\frac{\partial \rho}{\partial t} + \nabla \cdot (\rho\mathbf{u}) = 0 \qquad (13.1\text{-}4)$$

When Equation 15.1-5 is divided by M_i, we obtain a species conservation equation in terms of the species molar concentration $C_i = \rho_i/M_i$ and molar flux $\mathbf{N}_i = \mathbf{n}_i/M_i$:

$$\frac{\partial C_i}{\partial t} + \nabla \cdot \mathbf{N}_i = R_i \qquad (15.1\text{-}7)$$

Scalar components of this equation in rectangular, cylindrical, and spherical coordinate systems (Fig. 9.1-1) are listed in Table 15.1-1. By summing each term in Equation 15.1-7 over all I and applying the definitions for total molar concentration $\sum_i C_i = c$ and molar flux $\sum_i \mathbf{N}_i = c\mathbf{u}^*$, we obtain

$$\frac{\partial c}{\partial t} + \nabla \cdot (c\mathbf{u}^*) = \sum_{i=1}^{I} R_i \qquad (15.1\text{-}8)$$

This molar equation is analogous to the mass equation given by Equation 13.1-4 except for the term $\sum_i R_i$. This difference arises because the total moles of product formed by a chemical reaction can differ from the total moles of reactions consumed, even though the total mass of products must always be equal to the total mass of reactants.

15.2 Special Cases

In the species concentration equations, we must incorporate equations for the transport fluxes. By substituting Equation 7.1-4 into Equation 15.1-5 and Equation 7.1-3 into Equation 15.1-7, we obtain

$$\frac{\partial \rho_i}{\partial t} + \nabla \cdot (\rho \omega_i \mathbf{u}) = \nabla \cdot (\mathcal{D}_i \rho \nabla \omega_i) + M_i R_i$$
$$\frac{\partial C_i}{\partial t} + \nabla \cdot (\mathbf{u}^* C_i) = \nabla \cdot (\mathcal{D}_i \nabla C_i) + R_i \qquad (15.2\text{-}1a, b)$$

In general, ρ, c, and $\mathcal{D}_i$ depend on space and time, which significantly complicates the solution of these equations. However, $\mathcal{D}_i$ and either ρ or c can usually be approximated as constants in physiological fluids (Section 2.1).

15.2.1 Constant Mass Density

Assuming that ρ is constant and applying the definition $\rho_i = \rho \omega_i$ to Equation 15.2-1a, we get

$$\frac{\partial \rho_i}{\partial t} + \nabla \cdot (\rho_i \mathbf{u}) = \nabla \cdot (\mathcal{D}_i \nabla \rho_i) + M_i R_i \qquad (15.2\text{-}2)$$

For a fluid of constant mass density, $\nabla \cdot \mathbf{u} = 0$ and the vector identity for $\nabla \cdot (\rho_i \mathbf{u})$ given in Equation B1-24 leads to

$$\nabla \cdot (\rho_i \mathbf{u}) = \mathbf{u} \cdot \nabla \rho_i \qquad (15.2\text{-}3)$$

This further simplifies the mass concentration equation for species i:

$$\frac{\partial \rho_i}{\partial t} + \mathbf{u} \cdot \nabla \rho_i = \nabla \cdot (\mathcal{D}_i \nabla \rho_i) + M_i R_i \qquad (15.2\text{-}4)$$

Dividing each term by M_i and noting that $C_i = \rho_i / M_i$, we can also express this equation as

$$\frac{\partial C_i}{\partial t} + \mathbf{u} \cdot \nabla C_i = \nabla \cdot (\mathcal{D}_i \nabla C_i) + R_i \qquad (15.2\text{-}5)$$

When the diffusion coefficient is also constant,

$$\frac{\partial C_i}{\partial t} + \mathbf{u} \cdot \nabla C_i = \mathcal{D}_i \nabla^2 C_i + R_i \qquad (15.2\text{-}6)$$

Scalar components of this equation in common systems (Fig. 9.1-1) are listed in Table 15.2-1.

Table 15.2-1 Species Concentration Equations: Constant ρ and $\mathcal{D}_A$

Rectangular coordinates:

$$\frac{\partial C_i}{\partial t} + \left(u_x \frac{\partial C_i}{\partial x} + u_y \frac{\partial C_i}{\partial y} + u_z \frac{\partial C_i}{\partial z} \right) = \mathcal{D}_i \left(\frac{\partial^2 C_i}{\partial x^2} + \frac{\partial^2 C_i}{\partial y^2} + \frac{\partial^2 C_i}{\partial z^2} \right) + R_i$$

Cylindrical coordinates:

$$\frac{\partial C_i}{\partial t} + \left(u_r \frac{\partial C_i}{\partial r} + u_\theta \frac{1}{r}\frac{\partial C_i}{\partial \theta} + u_z \frac{\partial C_i}{\partial z} \right) = \mathcal{D}_i \left[\frac{1}{r}\frac{\partial}{\partial r}\left(r \frac{\partial C_i}{\partial r} \right) + \frac{1}{r^2}\frac{\partial^2 C_i}{\partial \theta^2} + \frac{\partial^2 C_i}{\partial z^2} \right] + R_i$$

Spherical coordinates:

$$\frac{\partial C_i}{\partial t} + \left(u_r \frac{\partial C_i}{\partial r} + u_\theta \frac{1}{r}\frac{\partial C_i}{\partial \theta} + u_\phi \frac{1}{r = \sin\theta}\frac{\partial C_i}{\partial \phi} \right) = \mathcal{D}_i \left[\frac{1}{r^2}\frac{\partial}{\partial r}\left(r^2 \frac{\partial C_i}{\partial r} \right) + \frac{1}{r^2 = \sin\theta}\left(\sin\theta \frac{\partial^2 C_i}{\partial \theta} \right) + \frac{1}{r^2 \sin^2\theta}\frac{\partial^2 C_i}{\partial \phi^2} \right] + R_i$$

15.2.2 Constant Molar Density

If c is constant, the molar continuity equation (Eq. 15.1-8) becomes

$$\nabla \cdot \mathbf{u}^* = \frac{1}{c}\sum_{i=1}^{I} R_i \tag{15.2-7}$$

Combining this with Equation 15.2-1b, we get

$$\frac{\partial C_i}{\partial t} + \left(\frac{C_i}{c}\sum_{i=1}^{I} R_i + \mathbf{u}^* \cdot \nabla C_i \right) = \nabla \cdot (\mathcal{D}_i \nabla C_i) + R_i \tag{15.2-8}$$

If $\mathcal{D}_i$ is also constant, this equation becomes

$$\frac{\partial C_i}{\partial t} + \mathbf{u}^* \cdot \nabla C_i = \mathcal{D}_i \nabla^2 C_i + R_i - \frac{C_i}{c}\sum_{j=1}^{I} R_j \tag{15.2-9}$$

As discussed in Section 6.1, the value of molar average velocity $\mathbf{u}^*$ can be similar to that of mass average velocity $\mathbf{u}$, particularly in dilute solutions. Nevertheless, these two velocities are fundamentally different. Equation 15.2-6, which is based on is $\mathbf{u}$, is generally a better starting point for transport analyses than Equation 15.2-9 for a couple of reasons. First, the single reaction rate in Equation 15.2-6 is simpler to deal with than the multiple reaction rates required by Equation 15.2-9. Second, $\mathbf{u}$ is the appropriate velocity to use in the Euler equation of motion.

15.3 One-Dimensional Transport Equations

In many physiological processes, mass transport occurs while a solution flows along a tube with a permeable wall. Models of this multidimensional process can be overly complex relative to the experimental data available for their validation. Thus, it is sometimes advisable to use a simplified model with mass conservation equations that allow for a continuous axial variation of species concentration, but only permit a discrete concentration change laterally, between the fluid and the tube wall. Below we illustrate two approaches for deriving such one-dimensional models.

15.3.1 Cross-Sectional Averaging

One approach is to spatially average the microscopic three-dimensional transport equations over the coordinates for which only discrete changes in concentration are considered. Consider axisymmetric transport of species i with a constant diffusion coefficient in a constant density fluid flowing through a leaky, cylindrical tube of constant radius "a" (Fig. 15.3-1). In such a geometry, any local variable h(r,z,t) can be averaged over the circular tube cross section to obtain a variable that is not a continuous function of r:

$$\bar{h}(z,t) \equiv \frac{1}{\pi a^2}\int_0^{2\pi}\int_0^{a} h(r,z,t)\,r\,dr\,d\theta = \frac{2}{a^2}\int_0^{a} h(r,z,t)\,r\,dr \tag{15.3-1}$$

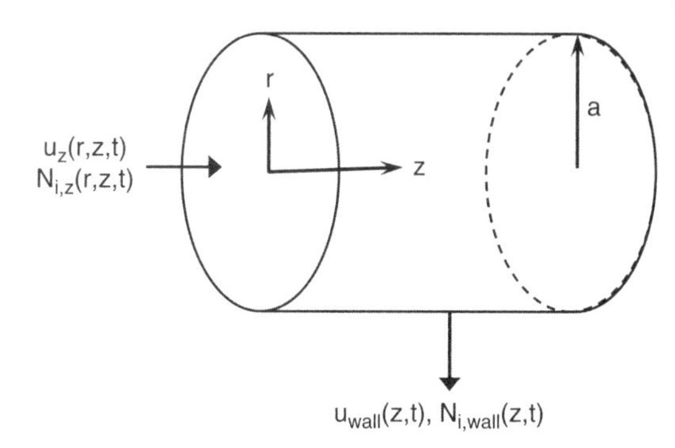

Figure 15.3-1 Axial transport through a tube with a permeable wall.

Continuity Equation

The axisymmetric velocity field for this process is specified by the cylindrical components $u_r(r,z,t)$, $u_z(r,z,t)$, and $u_\theta = 0$. From Table 13.1-1, the continuity equation for a fluid of constant density that follows these kinematics is

$$\frac{1}{r}\frac{\partial}{\partial r}(ru_r) + \frac{\partial u_z}{\partial z} = 0 \tag{15.3-2}$$

Multiplying each term by rdr and integrating over the tube radius, we obtain

$$\int_0^a \frac{\partial(ru_r)}{\partial r} dr + \int_0^a \frac{\partial u_z}{\partial z} rdr = 0 \tag{15.3-3}$$

Integrating the first term in this equation yields

$$\int_0^a \frac{\partial(ru_r)}{\partial r} dr = [ru_r]_{r=a} - [ru_r]_{r=0} = au_{wall} \tag{15.3-4}$$

Here, $u_{wall} \equiv [u_r]_{r=a}$ is the radial velocity at the tube wall. In the second term, we interchange differentiation and integration so that

$$\int_0^a \frac{\partial u_r}{\partial z} rdr = \frac{\partial}{\partial z}\int_0^a u_z(r,z,t)rdr = \frac{a^2}{2}\frac{\partial \bar{u}}{\partial z} \tag{15.3-5}$$

When these two relations are substituted into Equation 15.3-3, we obtain the one-dimensional continuity equation for a constant density fluid flowing through a rigid cylindrical tube:

$$\frac{\partial \bar{u}}{\partial z} = -\frac{2}{a}u_{wall} \tag{15.3-6}$$

Species Balance

For an axisymmetric system, the species concentration equation from Table 15.1-1 is

$$\frac{\partial C_i}{\partial t} + \frac{1}{r}\frac{\partial(rN_{i,r})}{\partial r} + \frac{\partial N_{i,z}}{\partial z} = R_i \tag{15.3-7}$$

Multiplying by rdr and integrating over the tube radius, this equation becomes

$$\int_0^a \frac{\partial C_i}{\partial t} r dr + \int_0^a \frac{\partial (rN_{i,r})}{\partial r} dr + \int_0^a \frac{\partial N_{i,z}}{\partial z} r dr = \int_0^a R_i r dr \tag{15.3-8}$$

Changing the order of differentiation and integration in the first and third terms, we find

$$\int_0^a \frac{\partial C_i}{\partial t} r dr = \frac{\partial}{\partial t} \int_0^{a(z,t)} C_i(r,z,t) r dr = \frac{a^2}{2} \frac{\partial \bar{C}_i}{\partial t}$$
$$\int_0^a \frac{\partial N_{i,z}}{\partial z} r dr = \frac{\partial}{\partial z} \int_0^a N_{i,z}(r,z,t) r dr = \frac{a^2}{2} \frac{\partial \bar{N}_i}{\partial z} \tag{15.3-9a, b}$$

Integration of the second term yields

$$\int_0^a \frac{\partial (rN_{i,r})}{\partial r} dr = [rN_{i,r}]_{r=a} - [rN_{i,r}]_{r=0} = aN_{i,wall} \tag{15.3-10}$$

where $N_{i,wall} \equiv [N_{i,r}]_{r=a}$ is the radial flux of species i at the tube wall. Also recognizing that the fourth term can be expressed as $\int R_i r dr \equiv (a^2/2)\bar{R}_i$, Equation 15.3-8 can be written in terms of cross-sectional averaged quantities:

$$\frac{\partial \bar{C}_i}{\partial t} + \frac{\partial \bar{N}_i}{\partial z} = \bar{R}_i - \frac{2}{a} N_{i,wall} \tag{15.3-11}$$

Flux Equation

To express the average flux $\bar{N}_i(zt)$ in terms of average concentration $\bar{C}_i(z,t)$, we start with the z component of the vector relation between local flux $N_{i,z}(r,z,t)$ and local concentration $C_i(r,z,t)$ when mass density is constant (Eq. 7.1-7):

$$N_{i,z} = u_z C_i - \mathcal{D}_i \frac{\partial C_i}{\partial z} \tag{15.3-12}$$

To obtain a cross-sectional average of this equation, we multiply by rdr, integrate over the tube radius, and interchange the order of differentiation and integration in the last term:

$$\bar{N}_i = \overline{u_z C_i} - \mathcal{D}_i \frac{\partial \bar{C}_i}{\partial z} \tag{15.3-13}$$

To evaluate $\overline{u_z C_i}$, we divide $u_z(r,z,t)$ and $C_i(r,z,t)$ into area-average portions, $\bar{u}$ and $\bar{C}_i$, that depend on z and t and deviating portions, u' and C_i', that depend on r as well as z and t:

$$u_z = \bar{u} + u', \quad C_i = \bar{C}_i + C_i' \tag{15.3-14a, b}$$

By these definitions, the area average of u' and C_i' are both zero so that

$$\overline{u_z C_i} = \overline{(\bar{u} + u')(\bar{C}_i + C_i')} = (\bar{u}\bar{C}_i + \overline{u' C'}_i) + (\bar{u}\bar{C'}_i + \bar{u'}\bar{C}_i) = \bar{u}\bar{C}_i + \overline{u' C'}_i \tag{15.3-15}$$

Here, $\overline{u'C'_i}$ represents the average of local deviations in convective transport resulting from the local deviations in axial velocity and concentration. A more detailed three-dimensional analysis (e.g., Taylor, 1953) indicates that in rigid cylindrical tubes, it is frequently possible to express $\overline{u'C'_i}$ as

$$\overline{u'C'_i} = -\mathcal{D}_i^* \frac{\partial \bar{C}_i}{\partial z}$$

where $\mathcal{D}_i^*$ is a dispersion coefficient that depends on the velocity profile. Substituting Equations 15.3-15 and 15.3-16 into Equation 15.3-13,

$$\bar{N}_i = \bar{u}\bar{C}_i - \left(\mathcal{D}_i + \mathcal{D}_i^*\right) \frac{\partial \bar{C}_i}{\partial z} \qquad (15.3\text{-}17)$$

Combining this with Equation 15.3-11, we obtain the one-dimensional species transport equation in a rigid cylindrical tube:

$$\frac{\partial \bar{C}_i}{\partial t} + \frac{\partial(\bar{u}\bar{C}_i)}{\partial z} = \frac{\partial}{\partial z}\left[\left(\mathcal{D}_i + \mathcal{D}_i^*\right)\frac{\partial \bar{C}_i}{\partial z}\right] - \frac{2}{a}N_{i,\text{wall}} + \bar{R}_i \qquad (15.3\text{-}18)$$

15.3.2 Generalized One-Dimensional Transport

In many problems of interest, flow and concentration fields are not axisymmetric and flow cross sections have irregular shapes. The development of one-dimensional transport equations starting from the three-dimensional transport equations, as presented above, is then not practical. As an alternative derivation, we assume at the outset that all dependent variables change with time and change continuously along the flow direction z. Orthogonal to the z direction, dependent variables can only change by a discrete jump across the tube wall. In Section 2.4, we used this method to derive one-dimensional equations for flow of a constant density fluid through a rigid, permeable tube with axial transport occurring by convection alone. The more general case presented below incorporates diffusion and dispersion in the flow direction. In addition, fluid density $\rho(z,t)$ and tube cross-sectional area $A_t(z,t)$ are allowed to vary both with time and in flow direction (Fig. 15.3-2; left).

As before, we use a control volume consisting of a short tube section between a flow inlet at z and an outlet at $z + \Delta z$ (Fig. 2.4-2; bottom). In this case, the tube section has a variable volume $\Delta V(z,t) = A_t(z,t)\Delta z$, and the balance equations for species i and overall mass (Eqs. 2.4-15 and 2.4-17) become

$$\frac{\partial(C_i A_t \Delta z)}{\partial t} = \left[\dot{N}_i\right]_z - \left[\dot{N}_i\right]_{z+\Delta z} - \Delta\dot{N}_{i,\text{wall}} + \Delta\dot{R}_i \qquad (15.3\text{-}19)$$

$$\frac{\partial(\rho A_t \Delta z)}{\partial t} = [Q\rho]_z - [Q\rho]_{z+\Delta z} - \rho\Delta Q_{\text{wall}} \qquad (15.3\text{-}20)$$

Because the length of the control volume is constant, we can divide both equations by Δz and take the limit as $\Delta z \to 0$ to obtain

$$\frac{\partial(C_i A_t)}{\partial t} + \frac{\partial \dot{N}_i}{\partial z} = \frac{\partial \dot{R}_i}{\partial z} - \frac{\partial \dot{N}_{i,\text{wall}}}{\partial z} \qquad (15.3\text{-}21)$$

$$\frac{\partial(\rho A_t)}{\partial t} + \frac{\partial(Q\rho)}{\partial z} = -\rho\frac{\partial Q_{\text{wall}}}{\partial z} \qquad (15.3\text{-}22)$$

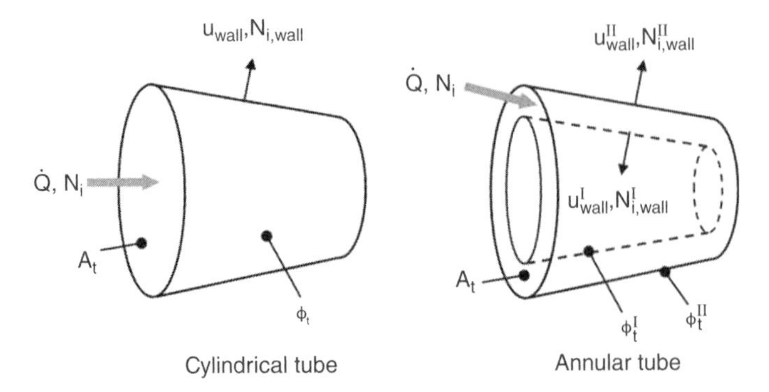

Figure 15.3-2 One-dimensional model for transport with tubular flow.

It is convenient to express some of the spatial derivatives in terms of cumulative tube volume V(z,t) and surface S(z,t) as follows:

$$\frac{\partial \dot{N}_{i,wall}}{\partial z} = \left(\frac{\partial V}{\partial z}\right)\left(\frac{\partial S}{\partial V}\right)\frac{\partial \dot{N}_{i,wall}}{\partial S} = A_t \phi_t N_{i,wall}$$
$$\frac{\partial \dot{R}_i}{\partial z} = \left(\frac{\partial V}{\partial z}\right)\frac{\partial \dot{R}_i}{\partial V} = A_t R_i \tag{15.3-23a-c}$$
$$\frac{\partial Q_{wall}}{\partial z} = \left(\frac{\partial V}{\partial z}\right)\left(\frac{\partial S}{\partial V}\right)\frac{\partial Q_{wall}}{\partial S} = A_t \phi_t u_{wall}$$

Here, we have defined the following quantities that can depend on z and t: molar flux of species i at the tube wall, $N_{i,wall} \equiv \partial \dot{N}_{i,wall}/\partial S$; molar formation rate of species i per unit tube volume, $R_i \equiv \partial \dot{R}_i/\partial V$; fluid velocity at the wall, $u_{wall} = \partial Q_{wall}/\partial S$; cross-sectional area, $A_t = dV/dz$; and surface-to-volume ratio, $\phi_t = \partial S/\partial V$. Inserting these relations into the transport equations, we find

$$\frac{1}{A_t}\frac{\partial (C_i A_t)}{\partial t} + \frac{1}{A_t}\frac{\partial (N_i A_t)}{\partial z} = R_i - \phi_t N_{i,wall} \tag{15.3-24}$$
$$\frac{1}{\rho A_t}\frac{\partial (\rho A_t)}{\partial t} + \frac{1}{\rho A_t}\frac{\partial (\rho Q)}{\partial z} = -\phi_t u_{wall} \tag{15.3-25}$$

Based on Equation 15.3-17, we assume the following form for the axial transport rate of species i:

$$\dot{N}_i = N_i A_t = u A_t C_i - \left(\mathcal{D}_i + \mathcal{D}_i^*\right) A_t \frac{\partial C_i}{\partial z} \tag{15.3-26}$$

where $u = Q/A_t$ is the average fluid velocity. The species molar balance then becomes

$$\frac{1}{A_t}\frac{\partial (C_i A_t)}{\partial t} + \frac{1}{A_t}\frac{\partial (Q C_i)}{\partial z} = \frac{1}{A_t}\frac{\partial}{\partial z}\left[\left(\mathcal{D}_i + \mathcal{D}_i^*\right) A_t \frac{\partial C_i}{\partial z}\right] + R_i - \phi_t N_{i,wall} \tag{15.3-27}$$

For the special case of an incompressible fluid flowing through a tube of constant cross section, A_t, ϕ_t, and ρ are constants. Equations 15.3-25 and 15.3-27 then become

$$\frac{\partial C_i}{\partial t} + \frac{1}{A_t}\frac{\partial (QC_i)}{\partial z} = \frac{\partial}{\partial z}\left[\left(\mathcal{D}_i + \mathcal{D}_i^*\right)\frac{\partial C_i}{\partial z}\right] + R_i - \phi_t N_{i,\text{wall}} \tag{15.3-28}$$

$$\frac{1}{A_t}\frac{\partial Q}{\partial z} = -\phi_t u_{\text{wall}} \tag{15.3-29}$$

It is sometimes useful to express these equations in terms of average axial velocity instead of the volumetric flow:

$$\frac{\partial C_i}{\partial t} + \frac{\partial (uC_i)}{\partial z} = \frac{\partial}{\partial z}\left[\left(\mathcal{D}_i + \mathcal{D}_i^*\right)\frac{\partial C_i}{\partial z}\right] - \phi_t N_{i,\text{wall}} + R_i \tag{15.3-30}$$

$$\frac{\partial u}{\partial z} = -\phi_t u_{\text{wall}} \tag{15.3-31}$$

For a cylindrical tube of radius "a," $A_t = \pi a^2$ and $\phi_t = 2/a$. These transport equations then reduce to Equations 15.3-6 and 15.3-18. Thus, under these restrictive conditions, the velocity and concentration are equivalent to cross-sectional averaged quantities.

In some applications, we are interested in modeling one-dimensional transport in a phase that is confined by two different surfaces forming an annular tube (Fig. 15.3-2; right). By distinguishing between the different geometries and transport rates associated with surface I and surface II, we extend Equations 15.3-28 and 15.3-29 and arrive at

$$\frac{\partial C_i}{\partial t} + \frac{1}{A_t}\frac{\partial (QC_i)}{\partial z} = \frac{\partial}{\partial z}\left[\left(\mathcal{D}_s + \mathcal{D}_s^*\right)\frac{\partial C_i}{\partial z}\right] - \left(\phi_t^{II} N_{i,\text{wall}}^{II} + \phi_t^{I} N_{i,\text{wall}}^{I}\right) + R_i \tag{15.3-32}$$

$$\frac{1}{A_t}\frac{\partial Q}{\partial z} = -\left(\phi_t^{II} u_{\text{wall}}^{II} + \phi_t^{I} u_{\text{wall}}^{I}\right) \tag{15.3-33}$$

Here, $\left(N_{i,\text{wall}}^{I}, N_{i,\text{wall}}^{II}\right)$ are species fluxes and $\left(u_{\text{wall}}^{I}, u_{\text{wall}}^{II}\right)$ are fluid velocities exiting through the two surfaces, and $(\phi_t^{I}, \phi_t^{II})$ are the corresponding surface-to-volume ratios of the annular space. These equations can also be written in terms of the average axial velocity:

$$\frac{\partial C_i}{\partial t} + \frac{\partial (uC_i)}{\partial z} = \frac{\partial}{\partial z}\left[\left(\mathcal{D}_s + \mathcal{D}_s^*\right)\frac{\partial C_i}{\partial z}\right] - \left(\phi_t^{II} N_{i,\text{wall}}^{II} + \phi_t^{I} N_{i,\text{wall}}^{I}\right) + R_i \tag{15.3-34}$$

$$\frac{\partial u}{\partial z} = -\left(\phi_t^{II} u_{\text{wall}}^{II} + \phi_t^{I} u_{\text{wall}}^{I}\right) \tag{15.3-35}$$

15.4 Model Formulation and Scaling of Mass Transport

15.4.1 Elements of Model Formulation

The species concentration equation with the equations of motion and continuity are the basic components for modeling the dynamics of a species concentration field in a spatially continuous gas or liquid phase. To accurately simulate the effects of a complex structure or a nonlinear function, numerical computation is required. With appropriate simplifying assumptions, however, these equations can frequently be solved analytically. The following elements are typical in modeling a mass transport process.

Velocity Field

When parameters that affect the velocity field (e.g., viscosity and mass density) depend on the composition of a fluid, the equations of motion and continuity must be solved together with the species concentration equations. When these parameters do not vary significantly in space and time, a velocity field can be determined by the continuity equation and equation of motion independent of the species concentration equation.

Coordinate System

As was the case in solving fluid mechanics problems, it is desirable to select a coordinate system with coordinate surfaces that align with the boundaries of the flow domain. This minimizes the number of spatial coordinates necessary to model a process. For example, suppose we want to analyze a drug that is delivered isotropically into a large volume of stationary fluid from the surface of a spherical pellet. When expressed in rectangular coordinates, the drug concentration at the pellet surface will be a function of x, y, and z. When spherical coordinates are used, however, drug concentration only depends on radial distance r from the center of the pellet.

Simplification

Constant transport parameters, μ, ρ, and $\mathcal{D}_S$, are frequently assumed in mass transport problems. The simplification of system geometry is also quite common. For instance, blood vessels are often treated as rigid cylindrical tubes even though they are actually elastic conduits with irregularly shaped cross sections. An order-of-magnitude analysis of the transport equations can lead to further simplification. For example, when modeling gas transport in the lungs, the transient accumulation term in the species balance can be neglected in small airways where the breathing period is long compared to the time constants for diffusion and convection. The resulting model equations are then pseudo-steady.

Boundary Conditions

A spatially distributed model of mass transport depends on the boundary conditions that involve species concentrations and concentration gradients. These conditions are based on information associated with the physical system. For example, a concentration profile might be known or approximated at the boundary where flow enters a system. At a flow outlet, it is often reasonable to neglect diffusion compared to convection such that a species concentration gradient is zero. Concentration gradients can also vanish at a point or line of geometric symmetry within the transport domain. As discussed below, concentration and flux conditions at phase interfaces are based on variations of the mass conservation principle (Slattery, 1999).

15.4.2 Interface Relationships

Consider a point lying on an arbitrarily thin interface between two immiscible fluids (+phase and −phase) moving at a velocity $\mathbf{u}^S$ (Fig. 15.4-1). According to the velocity jump condition (Eq. 13.3-1), a discontinuity in mass average velocity occurs across the interface when there is a discontinuity in mass density. Similarly, a discontinuity in a species velocity $\left(\mathbf{u}_i^+, \mathbf{u}_i^-\right)$ occurs when there is a discontinuity in its mass concentrations $\left(\rho_i^+, \rho_i^-\right)$.

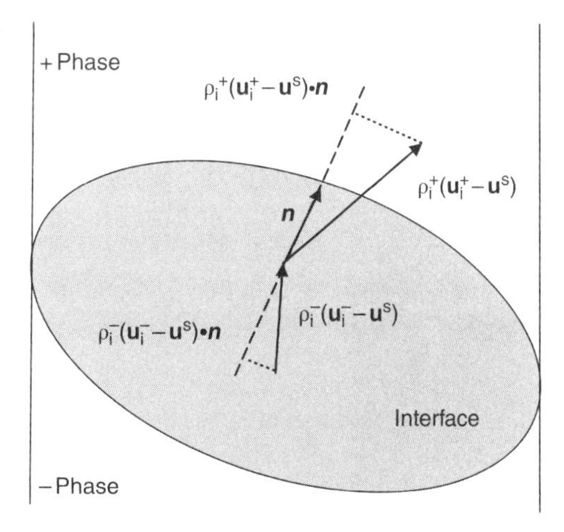

Figure 15.4-1 Mass transport through an interface moving at a velocity $\mathbf{u}^S$.

To demonstrate this, we focus on a species i that is not surface active and will not accumulate at the interface. Conservation of mass then requires the net output of species i from the interface to balance its formation rate by chemical reaction at the interface:

$$\rho_i^+ \left(\mathbf{u}_i^+ - \mathbf{u}^S\right) \cdot \boldsymbol{n} - \rho_i^- \left(\mathbf{u}_i^- - \mathbf{u}^S\right) \cdot \boldsymbol{n} = r_i^S \tag{15.4-1}$$

Here, $\boldsymbol{n}$ is the unit normal at the interface pointing outward from −phase to +phase, and r_i^S is the interfacial formation rate of a substance i in units of mass per time per surface area. Examples of interfacial reactions include receptor–ligand binding at a cell membrane surface and metabolic oxygen consumption by a thin layer of cells modeled as an interface.

By dividing Equation 15.4-1 by molecular weight M_i and noting that the species molar flux is $N_i = \rho_i u_i / M_i$ and the molar reaction rate is $R_i^S \equiv r_i^S / M_i$, we obtain

$$\left(\mathbf{N}_i^+ - C_i^+ \mathbf{u}^S\right) \cdot \boldsymbol{n} - \left(\mathbf{N}_i^- - C_i^- \mathbf{u}^S\right) \cdot \boldsymbol{n} = R_i^S \tag{15.4-2}$$

Here, $\left(\mathbf{N}_i^+, \mathbf{N}_i^-\right)$ represents a discontinuity in species molar flux crossing from the negative to the positive phases, and $\left(C_i^+, C_i^-\right)$ represents a discontinuity in species molar concentration between the two phases. If the interface does not move in the direction of $\boldsymbol{n}$, then $\mathbf{u}^S \cdot \boldsymbol{n} = 0$ and the jump condition reduces to

$$\left(\mathbf{N}_i^+ - \mathbf{N}_i^-\right) \cdot \boldsymbol{n} = R_i^S \tag{15.4-3}$$

If a z coordinate is perpendicular to the interface such that $\boldsymbol{n} = \mathbf{i}_z$, then this equation can be written in terms of the flux components in the z direction:

$$N_{i,z}^+ - N_{i,z}^- = R_i^S \tag{15.4-4}$$

Two Continuous Phases

Suppose that a species i is transported across a stationary interface between two spatially continuous liquid phases that have constant but unequal mass densities. We can express $N_{i,z}^+$ and $N_{i,z}^-$ as the sum of their convective and diffusive contributions so that Equation 15.4-4 becomes

$$\left[u_z^+ C_i^+ - \mathcal{D}_i^+ \left(\frac{\partial C_i}{\partial z}\right)^+\right] - \left[u_z^- C_i^- - \mathcal{D}_i^- \left(\frac{\partial C_i}{\partial z}\right)^-\right] = R_i^S \qquad (15.4\text{-}5)$$

In general, a discontinuity in the concentration gradient exists across the interface. Even when there is no net convection in either phase and the interface is inert (i.e., $u_z^+ = u_z^- = 0$ and $R_i^S = 0$), there will be a discontinuity in concentration gradient that depends on the relative diffusivity of species i in the two phases:

$$\left(\frac{\partial C_i}{\partial z}\right)^+ = \frac{\mathcal{D}_i^-}{\mathcal{D}_i^+}\left(\frac{\partial C_i}{\partial z}\right)^- \qquad (15.4\text{-}6)$$

When phase equilibrium exists, interfacial concentrations are also constrained by the continuity of chemical potential. For example, in the absence of electrical and nonideal concentration effects, the distribution of species i is

$$C_i^- = \lambda_i^{-,+} C_i^+ \qquad (15.4\text{-}7)$$

where $\lambda_i^{-,+}$ is the equilibrium partition coefficient between −phase and +phase.

Continuous Phase Adjacent to a Spatially Lumped Phase

Suppose that an interface separates a spatially continuous phase with $N_{i,x}^+$ expressed as the sum of convective and diffusive fluxes from a spatially lumped phase with $N_{i,x}^-$ described by the unstirred layer model of Equation 12.1-3a. The jump condition on solute s then becomes

$$\left[u_z^+ C_i^+ - \mathcal{D}_i^+ \left(\frac{\partial C_i}{\partial z}\right)^+\right] - \left[k_i^- \left(C_{ib}^- - C_i^-\right)\right] = R_i^S \qquad (15.4\text{-}8)$$

where k_i^- is a mass transfer coefficient and C_{ib}^- is the bulk (or average) species concentration in the −phase.

Continuous Phase with an Impermeable Boundary

When a +phase is in contact with a barrier impermeable to species i, we can apply Equation 15.4-8 with $k_i^- = 0$:

$$u_z^+ C_i^+ - \mathcal{D}_i^+ \frac{\partial C_i^+}{\partial z} = R_i^S \qquad (15.4\text{-}9)$$

This equation states that convection and diffusion away from the interface balances the formation of species i at the interface. If such a barrier is also chemically inert, permitting fluid flow but not the transfer of species i (i.e., $R_i^S = 0$ and $u_z^+ \neq 0$ in Eq. 15.4-9), then

$$-\mathcal{D}_i^+ \frac{\partial C_i^+}{\partial z} = -u_z^+ C_i^+ \qquad (15.4\text{-}10)$$

Thus, species diffusion $-\mathcal{D}_i^+ \left(\partial C_i^+ / \partial z\right)$ from the interface toward the +phase is balanced by convection $-u_z^+ C_i^+$ in the opposite direction. At a chemically reactive barrier that is impermeable to fluid flow as well as to species i (i.e., $R_i^S \neq 0$ and $u_x^+ = 0$ in Eq. 15.4-9),

$$-\mathcal{D}_i^+ \frac{\partial C_s^+}{\partial z} = R_i^S \qquad (15.4\text{-}11)$$

In this case, diffusion from the interface toward the +phase is balanced by the surface formation rate of species i.

15.4.3 Dimensionless Concentration Equation

As an essential method for model simplification, we express the governing equations in dimensionless form by scaling all the variables. For the spatially distributed, convection–diffusion–reaction model with constant mass density and diffusivity (Eq. 15.2-6), we scale the variables as follows:

$$\nabla = L_c \nabla, \quad \boldsymbol{v} = \frac{\mathbf{u}}{u_c}, \quad t = \frac{t}{t_c}, \quad C_i = \frac{C_i}{C_{ic}}, \quad R_i = \frac{R_i}{R_{ic}} \tag{15.4-12a-e}$$

where L_c, u_c, C_{ic}, and R_{ic} are characteristic length, velocity, species concentration, and reaction rate. Incorporating these quantities into Equation 15.2-6 results in a dimensionless equation for species i:

$$\frac{1}{St}\frac{\partial C_i}{\partial t} + \boldsymbol{u} \cdot \nabla C = \left(\frac{1}{Pe}\right)\nabla^2 C_i + \left(\frac{Da}{Pe}\right) R_i \tag{15.4-13}$$

where the dimensionless groups are defined as

$$St \equiv \frac{t_c u_c}{L_c}, \quad Pe \equiv \frac{u_c L_c}{\mathcal{D}_i}, \quad Da \equiv \frac{L_c R_{ic}}{\mathcal{D}_i C_{ic}} \tag{15.4-14a-c}$$

To interpret the dimensionless groups, we examine the ratios of the coefficients of any two terms in Equation 15.4-13. The Strouhal number (*St*) characterizes the ratio of convective transport to transient change. The Peclet number (*Pe*) represents the ratio of convective to diffusive transport. The Damkohler number (*Da*) corresponds to the ratio of chemical reaction rate to diffusive transport rate. Note that the Peclet number is the product of the Reynolds and Schmidt numbers, which are defined in Section 12.1.

To illustrate the types of dimensionless parameters that can arise from boundary conditions, we consider the jump condition for species i between a +phase in which simultaneous convection and diffusion occur and a −phase that is well mixed except for an unstirred layer at the interface. Substitution of dimensionless variables into Equation 15.4-8 yields

$$Pe\left(u^+ C_s^+\right) - \left(\frac{\partial C_s}{\partial x}\right)^+ - Bi\left(C_{sb}^- - C_s^-\right) = Da^S\left(R_s^S\right) \tag{15.4-15}$$

where two new dimensionless groups are defined as

$$Da^S \equiv \frac{L_c R_{ic}^S}{\mathcal{D}_i C_{ic}}, \quad Bi \equiv \frac{k_i^- L_c}{\mathcal{D}_i} \tag{15.4-16a,b}$$

The surface Damkohler number, Da^S, characterizes the rate of surface reaction relative to diffusion rate in the +phase. The Biot number, *Bi*, represents the ratio of mass transfer across an unstirred layer in the −phase to diffusive transport in the +phase. This is similar to Sherwood number $Sh \equiv k_i L_c / \mathcal{D}_i$ (Section 12.1) that represents mass transfer rate across an unstirred layer to the characteristic diffusion rate when both processes occur within same phase.

15.5 Diffusion and Convection in Nonreacting Systems

This section illustrates how the species concentration equations are applied to specific problems in nonreacting systems: unsteady-state diffusion in a finite domain, simultaneous convection and diffusion in a boundary layer flow, and axial dispersion by convection and diffusion. In each of these problems, we develop linear models that have analytical solutions.

15.5.1 Unsteady-State Diffusion in a Finite Domain

In this illustration, we consider unsteady-state transport by diffusion in a finite spatial domain. Drug delivery from a thin patch placed directly on the skin is one situation where this occurs. The goal of the current analysis is to predict the drug release rate to underlying tissue when transport is controlled by diffusion in the patch itself.

Model Formulation

We model the drug patch as a thin planar slab of thickness L with diffusion principally occurring perpendicular to the patch–skin interface in the z direction (Fig. 15.5-1). With neither convection nor reaction in the patch, the general species transport equation in rectangular coordinates (Table 15.2-1) can be simplified to

$$\frac{\partial C_s}{\partial t} = \mathcal{D}_s \frac{\partial^2 C_s}{\partial z^2} \quad (L > z > 0) \tag{15.5-1}$$

where $C_s(z,t)$ is local concentration of drug s and $\mathcal{D}_s$ represents its diffusion coefficient in the patch. Initially, the drug concentration is uniform:

$$t = 0 \; : \; C_s = C_{so} \tag{15.5-2}$$

At the air-patch interface ($z = 0$), there is no diffusion flux so that the concentration gradient is zero. At the patch–skin interface ($z = L$), we consider the drug concentration C_{sL} to be constant. The corresponding boundary conditions are

$$z = 0 \; : \; \frac{\partial C_s}{\partial z} = 0, \quad z = L \; : \; C_s = C_{sL} \tag{15.5-3a,b}$$

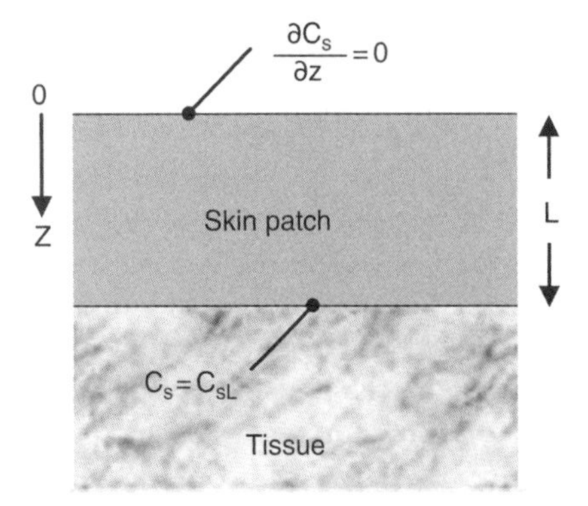

Figure 15.5-1 Drug release from a skin patch.

Analysis

Dimensionless Forms

In terms of the dimensionless variables

$$C = \frac{C_s - C_{sL}}{C_{so} - C_{sL}}, \quad z = \frac{z}{L}, \quad t = \frac{\mathcal{D}_s t}{L^2} \tag{15.5-4a-c}$$

the problem becomes

$$\frac{\partial C}{\partial t} = \frac{\partial^2 C}{\partial z^2} \quad (1 > z > 0) \tag{15.5-5}$$

$$t = 0 : C = 1, \quad z = 0 : \frac{\partial C}{\partial z} = 0, \quad z = 1 : C = 0 \tag{15.5-6a-c}$$

Solution

This linear, homogeneous boundary value problem can be solved using separation of variables. We first express C as a product of two independent functions:

$$C = T(t)Z(z) \tag{15.5-7}$$

Substituting the above expression in Equation 15.5-5 and rearranging, we obtain

$$\frac{1}{T}\frac{\mathrm{d}T}{\mathrm{d}t} = \frac{1}{Z}\frac{\mathrm{d}^2 Z}{\mathrm{d}z^2} \tag{15.5-8}$$

Since the left side of this equation cannot be a function of z and the right side cannot be a function of t, both sides must be equal to a constant. Designating this constant as $-\lambda^2$, we divide this equation into two separate ordinary differential equations:

$$\frac{\mathrm{d}^2 Z}{\mathrm{d}z^2} + \lambda^2 Z = 0 \tag{15.5-9}$$

$$\frac{\mathrm{d}T}{\mathrm{d}t} + \lambda^2 T = 0 \tag{15.5-10}$$

If $\lambda^2 = 0$, then the solution to Equation 15.5-9 would be $Z(z) = \mathrm{c}z + \mathrm{d}$ where c and d are constants of integration. Because $\mathrm{c} = \mathrm{d} = 0$ in order to satisfy the two boundary conditions, $\lambda^2 = 0$ results in a trivial solution to the problem. When $\lambda \neq 0$, the solutions to Equations 15.5-9 and 15.5-10 are

$$Z = \mathrm{a}\sin(\lambda z) + \mathrm{b}\cos(\lambda z), \quad T = \mathrm{c}\exp(-\lambda^2 t) \tag{15.5-11a,b}$$

in which a, b, and c are constants. If $\lambda^2 < 0$, then $\mathrm{T}(t)$ would be unbounded with time. Since this is not physically realistic, $\lambda^2 > 0$. Putting the solutions for Z and T together, we obtain

$$C = [\mathrm{A}\sin(\lambda z) + \mathrm{B}\cos(\lambda z)]\exp(-\lambda^2 t) \tag{15.5-12}$$

The constants ($\mathrm{A} \equiv \mathrm{ac}$, $\mathrm{B} \equiv \mathrm{bc}$, λ) must be determined from the boundary and initial conditions. When we apply the boundary condition at $z = 0$,

$$\frac{\partial C}{\partial z} = 0 = \mathrm{A}\exp(-\lambda^2 t) \Rightarrow \mathrm{A} = 0 \tag{15.5-13a,b}$$

and the solution reduces to

$$C(z,t) = \mathrm{B}\cos(\lambda z)\exp(-\lambda^2 t) \tag{15.5-14}$$

Applying the boundary condition at $z = 1$,

$$C(1,t) = 0 = B\cos(\lambda)\exp(-\lambda^2 t) \Rightarrow \cos(\lambda) = 0 \quad (15.5\text{-}15a,b)$$

Thus, λ can assume a countable infinity of eigenvalues:

$$\lambda_n = (2n-1)\frac{\pi}{2} \quad (n = 1,2,....) \quad (15.5\text{-}16)$$

To satisfy the initial condition, we must consider the complete solution given by the sum of the individual solutions at all values of n:

$$C(z,t) = \sum_{n=1}^{\infty} B_n \cos(\lambda_n z)\exp(-\lambda_n^2 t) \quad (15.5\text{-}17)$$

where $\cos(\lambda_n z)$ is the eigenfunction corresponding to eigenvalue λ_n. Applying the initial condition,

$$1 = \sum_{n=1}^{\infty} B_n \cos(\lambda_n z) \quad (15.5\text{-}18)$$

Multiplying both sides by $\cos(\lambda_m z)$ and integrating between $z = 0$ and $z = 1$,

$$\int_0^1 \cos(\lambda_m z)dz = \sum_{n=1}^{\infty} B_n \int_0^1 \cos(\lambda_m z)\cos(\lambda_n z)dz \quad (15.5\text{-}19)$$

To determine the constants B_n, we use an orthogonality relation:

$$\int_0^1 \cos(\lambda_m z)\cos(\lambda_n z)dz = \begin{cases} 0 & \text{when } m \neq n \\ \frac{1}{2} & \text{when } m = n \end{cases} \quad (15.5\text{-}20)$$

such that

$$\sum_{n=1}^{\infty} B_n \int_0^1 \cos(\lambda_m z)\cos(\lambda_n z)dz = \frac{B_m}{2} \quad (15.5\text{-}21)$$

Combining this equation with Equation 15.5-19, we find

$$B_m = 2\int_0^1 \cos(\lambda_m z)dz = \frac{4}{(2m-1)\pi}\sin\left[(2m-1)\frac{\pi}{2}\right] = \frac{4(-1)^{m+1}}{(2m-1)\pi} \quad (15.5\text{-}22)$$

The final solution for the concentration is now given by

$$C = \frac{4}{\pi}\sum_{n=1}^{\infty}\frac{(-1)^{n+1}}{(2n-1)}\cos(\lambda_n z)\exp(-\lambda_n^2 t) \quad (15.5\text{-}23)$$

The drug concentration profile is shown at various times in Figure 15.5-2. The value of C at any z position decreases as t increases and more drug is delivered from the patch to the surrounding skin.

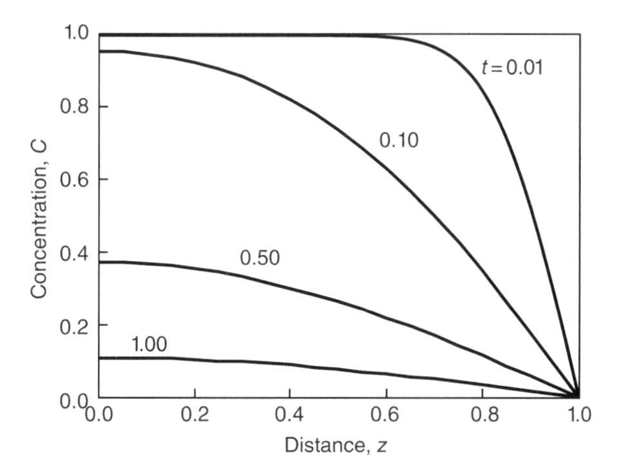

Figure 15.5-2 Concentration profiles within the skin patch.

Example 15.5-1 Albumin Release from a Hydrogel Patch

Meilander and coworkers (2004) tested a drug delivery patch consisting of lipid microtubule carriers suspended in agarose gel. After preloading the microtubules with a protein and then embedding them in the gel, the patch was submerged to a well-mixed saline solution free of drug. The amount of protein released into the solution was measured on a daily basis. The purpose of this problem is to determine the diffusion coefficients of albumin and thyroglobulin in the skin patch by using their cumulative release data (Fig. 15.5-3).

In these measurements, both surfaces of the patch were exposed to solution so that protein concentration at the center of the patch was symmetric with respect to z, and the zero gradient boundary condition (Eq. 15.5-6b) still holds. Also, the volume of the saline solution was sufficiently large that protein concentration in the solution can be neglected and $C_{sL} = 0$ at all times.

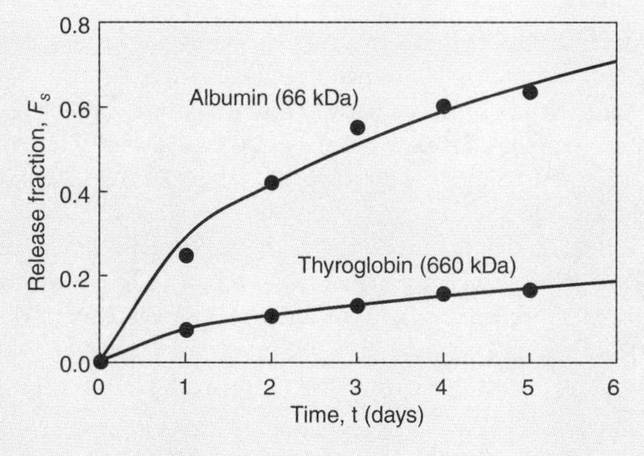

Figure 15.5-3 Fractional cumulative release of protein from the skin patch.

Solution

The drug delivery system is actually a heterogeneous material in which protein is initially associated with the microtubules. If transport is limited by protein transport through the gel rather than dissociation of protein from the tubules, then we can still use our model by interpreting $\mathcal{D}_s$ as a hindered diffusion coefficient.

To determine the flux at the skin–patch interface, we take the z derivative of the dimensionless concentration in Equation 15.5-23, evaluate it at $z = 1$, and express it in dimensional form:

$$[N_s]_{z=L} = -\frac{\mathcal{D}_s(C_{sI} - \mathcal{C}_{sL})}{L}\left[\frac{\partial C}{\partial z}\right]_{z=1} = \frac{2\mathcal{D}_s C_{sI}}{L}\sum_{n=1}^{\infty}(-1)^{n+1}\sin(\lambda_n)\exp\left(-\frac{\lambda_n^2\mathcal{D}_s}{L^2}t\right) \tag{15.5-24}$$

For a drug release patch with a surface S in contact with the saline solution, the cumulative amount of protein released up to time t is

$$S\int_0^t [N_s]_{z=L}dt = \frac{8}{\pi^2}(SLC_{sI})\sum_{n=1}^{\infty}\frac{(-1)^n\sin(\lambda_n)}{(2n-1)^2}\left[1-\exp\left(-\frac{\lambda_n^2\mathcal{D}_s}{L^2}t\right)\right] \tag{15.5-25}$$

Since SLC_{sI} is the amount of protein present at $t = 0$, the fraction of this initial amount released up to time t is

$$F_s = \frac{8}{\pi^2}\sum_{n=1}^{\infty}\frac{(-1)^{n+1}\sin(\lambda_n)}{(2n-1)^2}\left[1-\exp\left(-\frac{\lambda_n^2\mathcal{D}_s}{L^2}t\right)\right] \tag{15.5-26}$$

After performing a least-square regression of this equation to the experimental data (Fig. 15.5-3), we estimate that $\mathcal{D}_A/L^2 = 8.3\times10^{-7}\ s^{-1}$ for albumin and $\mathcal{D}_T/L^2 = 5.4\times10^{-8}\ s^{-1}$ for thyroglobulin. The ratio of these values, $\mathcal{D}_A/\mathcal{D}_T = 15$, has a magnitude that is consistent with the relative sizes of the two protein molecules.

15.5.2 Concentration Boundary Layer over a Flat Plate

Consider a solution that flows along an inert surface at $x \le 0$ and encounters an active surface at $x > 0$ where there is a source of an inert substance s (Fig. 15.5-4). This common situation occurs at the entrance of blood capillaries, membrane blood oxygenators and dialyzers, and at the surfaces of flush-mounted biosensors. Here, we model transport from a flat surface with an active region of length L and a width W that is relatively large, $W \gg L$.

Solution arrives at the active surface at $x = 0$ with a uniform solute concentration C_{so} for all $y > 0$. Solution in contact with the active surface at $y = 0$ has a constant solute concentration C_{sw} for all $L > x > 0$. As the solution travels downstream, solute diffuses away from the active surface and toward the bulk fluid where its concentration remains at its undisturbed value, C_{so}. It is the resulting concentration difference $(C_{sw} - C_{so})$ between the active surface and the bulk fluid that produces a solute flux $N_{s,wall}(x)$ into the flowing solution. We assume this concentration difference occurs within a thin concentration boundary layer of thickness $\delta^c(x) \ll L$ adjacent to the surface.

Model Formulation

In general, the velocity distribution in the solution varies in three coordinate directions. To simplify the analysis, however, we assume a steady laminar shearing flow with kinematics in

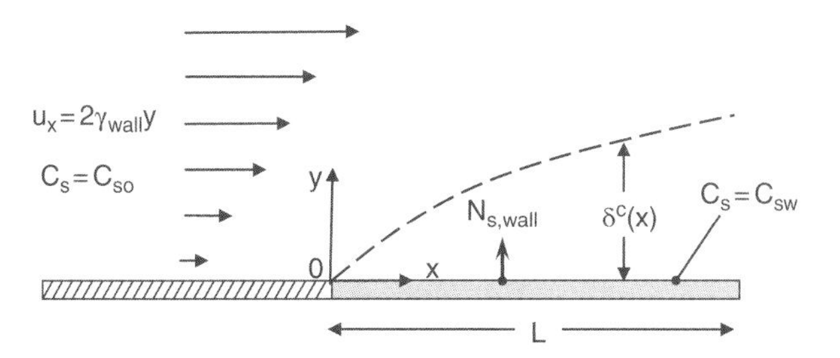

Figure 15.5-4 Development of a concentration boundary layer.

the region L > x > 0 given by $u_x = u_x(y)$ and $u_y = u_z = 0$. Because the concentration boundary layer is thin, changes in C_s are confined to a near wall region in which u_x increases linearly with distance y (Eq. 14.1-32):

$$u_x(y) \approx 2\gamma_{wall} y \qquad (15.5\text{-}27)$$

where $\gamma_{wall} = (1/2)[du_x/dy]_{y=0}$ is the x–y shear rate evaluated at the surface. For this velocity field, the steady-state species concentration equation (Table 15.2-1) in rectangular coordinates becomes

$$2\gamma_{wall} y \frac{\partial C_s}{\partial x} = \mathcal{D}_s \left(\frac{\partial^2 C_s}{\partial x^2} + \frac{\partial^2 C_s}{\partial y^2} + \frac{\partial^2 C_s}{\partial z^2} \right) \qquad (15.5\text{-}28)$$

Analysis

Order-of-Magnitude Approximations

This equation can be further simplified by an order-of-magnitude analysis in which variables are scaled by the magnitudes of their domains: L > x > 0, $\delta^c_L > y > 0$, W > z > 0, and $C_{sw} > C_s > C_{so}$. Note that δ^c_L is the concentration boundary layer thickness at x = L. Introducing the dimensionless variables,

$$C = \frac{C_s - C_{so}}{C_{so} - C_{sw}}, \quad x \equiv \frac{x}{L}, \quad y \equiv \frac{y}{\delta^c_L}, \quad z \equiv \frac{z}{W} \qquad (15.5\text{-}29a\text{-}d)$$

Equation 15.5-28 becomes

$$\left(\frac{2\gamma_{wall} L^2}{\mathcal{D}_s} \right) \left(\frac{\delta^c_L}{L} \right)^3 y \frac{\partial C}{\partial x} = \left(\frac{\delta^c_L}{L} \right)^2 \frac{\partial^2 C}{\partial x^2} + \frac{\partial^2 C}{\partial y^2} + \left(\frac{\delta^c_L}{W} \right)^2 \frac{\partial^2 C}{\partial z^2} \qquad (15.5\text{-}30)$$

Because all the dimensionless variables are scaled to have a magnitude of one, $\partial^2 C/\partial x^2$, $\partial^2 C/\partial y^2$, and $\partial^2 C/\partial z^2$ are also of magnitude one. Moreover, $W \gg L \gg \delta^c_L$ so that the first and third terms on the right side of Equation 15.5-30 can be neglected. Thus,

$$\left(\frac{2\gamma_{wall} L^2}{\mathcal{D}_s} \right) \left(\frac{\delta^c_L}{L} \right)^3 y \frac{\partial C}{\partial x} = \frac{\partial^2 C}{\partial y^2} \qquad (15.5\text{-}31)$$

Since the terms on both sides of this equation must be of the same order of magnitude to avoid a trivial result, the coefficient $(2\gamma_{wall} L^2/\mathcal{D}_s)(\delta^c_L/L)^3$ has a magnitude of one. Thus, the concentration equation reduces to

$$y\frac{\partial C}{\partial x} = \frac{\partial^2 C}{\partial y^2} \qquad (15.5\text{-}32)$$

and we can approximate the concentration boundary layer thickness as

$$\left(\frac{2\gamma_{\text{wall}}L^2}{\mathcal{D}_s}\right)\left(\frac{\delta_L^c}{L}\right)^3 = 1 \Rightarrow \delta_L^c = \sqrt[3]{\frac{\mathcal{D}_s L}{2\gamma_{\text{wall}}}} \Rightarrow \delta^c(x) = \sqrt[3]{\frac{\mathcal{D}_s x}{2\gamma_{\text{wall}}}} \qquad (15.5\text{-}33a\text{-}c)$$

Notice that the local boundary layer thickness increases with the cube root of position x along the active surface. Equation 15.5-32 must be solved with three dimensionless boundary conditions based on the three dimensional boundary conditions:

$$\begin{aligned} &x = 0 \;:\; C_s = C_{so} &&\Rightarrow\; x = 0 \;:\; C = 0 \\ &y = 0 \;:\; C_s = C_{sw} &&\Rightarrow\; y = 0 \;:\; C = 1 \\ &y \to \infty \;:\; C_s = C_{so} &&\Rightarrow\; y \to \infty \;:\; C = 0 \end{aligned} \qquad (15.5\text{-}34a\text{-}f)$$

Solution

In the absence of natural characteristic lengths in the boundary conditions, we transform the model by combining two spatial variables (x,y) into one similarity variable (Appendix C6):

$$\eta \equiv \frac{y}{\sqrt[3]{9x}} \qquad (15.5\text{-}35)$$

The governing equation then becomes a second-order ordinary differential equation:

$$\frac{d^2C}{d\eta^2} + 3\eta^2\frac{dC}{d\eta} = 0 \qquad (15.5\text{-}36)$$

and the number of boundary conditions are reduced from three to two:

$$\begin{aligned} C(\eta = 0) &= 1 \\ C(\eta \to \infty) &= 0 \end{aligned} \qquad (15.5\text{-}37a,b)$$

The solution to these equations detailed in Appendix C6 is

$$C = 1 - \frac{\int_0^{\eta} e^{-\xi^3}\, d\xi}{\int_0^{\infty} e^{-\xi^3}\, d\xi} \qquad (15.5\text{-}38)$$

With a change of variables $\zeta = \xi^3$, the denominator in this expression is a gamma function Γ with argument 4/3 (Abramowitz and Stegun, 1965):

$$\int_0^{\infty} e^{-\xi^3}\, d\xi = \frac{1}{3}\int_0^{\infty} \zeta^{-2/3} e^{-\zeta}\, d\zeta = \Gamma(4/3) = 0.89298\ldots \qquad (15.5\text{-}39)$$

Figure 15.5-5 shows the boundary layer concentration distribution from a numerical evaluation of Equation 15.5-38. This graph indicates that C decreases linearly with η for small values of η. For $\eta > 1.4$, C reaches within 1% of its ultimate value of zero.

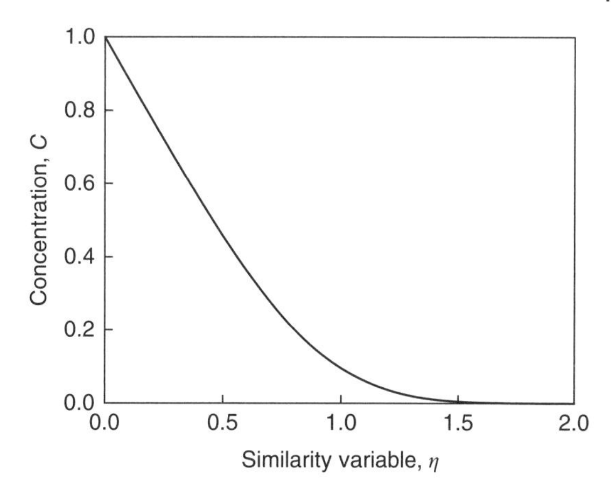

Figure 15.5-5 Concentration distribution in the boundary layer.

With $u_y = 0$, the molar flux at the active surface occurs by diffusion according to the equation:

$$N_{s,\mathrm{wall}} = \left[u_y C_s\right]_{y=0} - \mathcal{D}_s \left[\frac{\partial C_s}{\partial y}\right]_{y=0} \tag{15.5-40}$$

Converting from $C_s(y)$ to $C(y)$ and expanding the y derivative,

$$N_{s,\mathrm{wall}} = -\frac{\mathcal{D}_s(C_{sw} - C_{so})}{\delta_L^c}\left[\frac{\partial C}{\partial y}\right]_{y=0} = \sqrt[3]{\frac{2\gamma_{\mathrm{wall}}\mathcal{D}_s^2}{L}}(C_{so} - C_{sw})\left[\frac{dC}{d\eta}\frac{\partial \eta}{\partial y}\right]_{y=0} \tag{15.5-41}$$

With $dC/d\eta = -\exp(-\eta^3)/\Gamma(4/3)$ and $\partial\eta/\partial y = 1/\sqrt[3]{9x/L}$, the final expression for the flux is

$$N_{s,\mathrm{wall}} = 0.678\sqrt[3]{\frac{\gamma_{\mathrm{wall}}\mathcal{D}_s^2}{x}}(C_{sw} - C_{so}) \tag{15.5-42}$$

From this equation, we see that $N_{s,\mathrm{wall}}$ is proportional to the concentration driving force $(C_{sw} - C_{so})$ and decreases as $x^{-1/3}$ because of progressive thickening of the concentration boundary layer.

Example 15.5-2 Mass Transport of a Permeating Solute in a Tube

The solution to the concentration boundary layer problem provides us with a convenient means of deriving mass transfer coefficients in a variety of geometries. We will demonstrate this for a solute that permeates a section of a cylindrical tube through which fully developed Poiseuille flow is first established in an upstream calming section (Fig. 15.5-6).

Solution

We define a local mass transfer coefficient as the ratio of the wall flux to the concentration driving force between the wall and the undisturbed fluid:

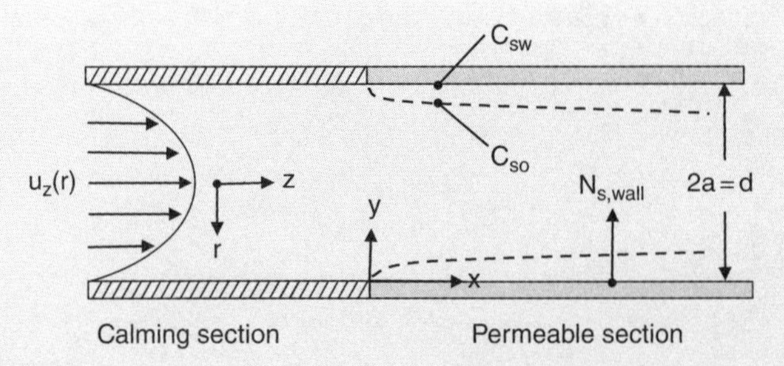

Figure 15.5-6 Solute transport in a permeable tube with fully developed flow.

$$k_{s,loc} \equiv \frac{N_{s,wall}}{(C_{sw} - C_{so})} = 0.678\sqrt[3]{\frac{\gamma_{wall}\mathcal{D}_s^2}{x}} \tag{15.5-43}$$

For fully developed, laminar flow of an incompressible, Newtonian fluid through a cylindrical tube, the velocity profile indicated by Equation 13.4-1 is

$$u_z = 2u_{ave}\left[1 - \left(\frac{2r}{d}\right)^2\right] \tag{15.5-44}$$

Here, d is the tube diameter and u_{ave} is the mean axial velocity. To determine γ_{wall}, we note that the y distance from the tube wall is related to the radial distance r from the tube centerline by $y = (d/2) - r$ such that

$$\gamma_{wall} \equiv \frac{1}{2}\left[\frac{\partial u_x}{\partial y}\right]_{y=0} = \frac{1}{2}\left[\frac{\partial u_z}{\partial r}\right]_{r=a}\left(\frac{dr}{dy}\right) = \frac{4u_{ave}}{d} \tag{15.5-45}$$

Provided that the boundary layer is thin enough to neglect curvature of the tube wall, it is reasonable to substitute Equation 15.5-45 into Equation 15.5-43, which was developed for a flat surface:

$$k_{s,loc} = 1.08\left(\frac{\mathcal{D}_s^2 u_{ave}}{xd}\right)^{1/3} \tag{15.5-46}$$

According to Equation 12.1-20, we can integrate $k_{s,loc}$ over the length of the permeable tube section to obtain an average value of the mass transfer coefficient:

$$k_s = \frac{1}{L}\int_0^L 1.08\left(\frac{\mathcal{D}_s^2 u_{ave}}{xd}\right)^{1/3} dx = 1.62\left(\frac{\mathcal{D}_s^2 u_{ave}}{Ld}\right)^{1/3} \tag{15.5-47}$$

This equation can be expressed in terms of Sherwood, Reynolds, and Schmidt numbers ($Sh_d \equiv k_s d/\mathcal{D}_s$, $Re_d \equiv u_{ave}d/\mu$, $Sc_d \equiv \mu/\mathcal{D}_s$):

$$Sh_d = 1.62 Re_d^{1/3} Sc^{1/3}\left(\frac{d}{L}\right)^{1/3} \tag{15.5-48}$$

Notice that this dimensionless solution of the model equations is identical to entry 1a of Table 12.1-1. Defining the Graetz number as $Gz \equiv (u_{ave}d/\mathcal{D}_s)(d/L)$, we can also write that

$$Sh_d = 1.62(Gz)^{1/3} \tag{15.5-49}$$

As we see by combining Equations 15.5-33b and 15.5-45, the thickness of the concentration boundary layer is inversely related to the Graetz number:

$$\frac{\delta_L^c}{d} = \sqrt[3]{\frac{1}{8Gz}} \tag{15.5-50}$$

Our analysis requires that $\delta_L^c \ll d$, which is only valid when $Gr \gg 1$. Since Gz is the ratio of the time constants for radial diffusion ($d^2/\mathcal{D}_s$) and axial convection (L/u_{ave}), radial diffusion must be a much slower process (i.e., have a much larger time constant) than axial convection for the thin boundary layer assumption to be valid.

15.5.3 Dispersion of an Inert Tracer Flowing in a Tube

Transport dynamics of physiological flow processes are often studied by introducing an easily detectable foreign substance or tracer into an inlet stream and monitoring the resulting changes in composition of an outlet stream. For example, Evans Blue, a dye that binds to plasma albumin, is convenient for studying transport dynamics in the cardiovascular system. In some cases, natural substances and their analogs can provide more information. For example, radioactively labeled sugars are used in positron emission tomography to visualize metabolic activity in the brain.

Consider the transport of an inert tracer species s flowing in a long, impermeable, cylindrical tube of diameter d (Fig. 15.5-7). Initially, there is no tracer in the tube at axial positions $z > 0$, but a uniform tracer concentration C_{so} exists at $z < 0$. In other words, at $t = 0$, a concentration front consisting of a step decrease in tracer concentration is located at $z = 0$. When $t > 0$, solution containing tracer continuously enters the tube, pushing the concentration front in the positive z direction. Because of friction, tracer near the tube wall moves more slowly than tracer near the tube centerline. This relative movement of tracer along different streamlines produces radial as well as axial concentration gradients with two opposing effects: (i) tracer spreads out along the tube more than would be predicted by axial diffusion and (ii) radial diffusion mixes tracer between streamlines thereby limiting this convection-induced enhancement of axial transport. The net result of these two coupled processes is responsible for the dispersion coefficient $\mathcal{D}_s^*$.

Model Formulation

A cross-sectional average model of axial transport of an inert tracer (i.e., $\bar{R}_s = 0$) in a rigid tube with an impermeable wall (i.e., $N_{s,wall} = u_{wall} = 0$) is obtained from Equations 15.3-6 and 15.3-18:

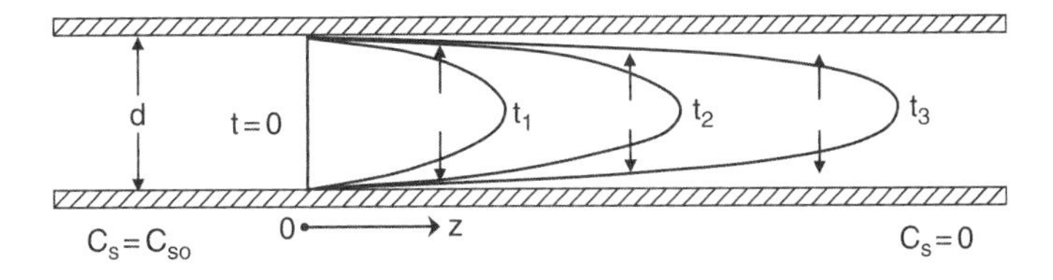

Figure 15.5-7 Convection of a tracer front with simultaneous radial diffusion in tube flow.

$$\frac{\partial \bar{C}_s}{\partial t} + \frac{\partial(\bar{u}_z \bar{C}_s)}{\partial z} = \frac{\partial^2}{\partial z}\left[\left(\mathcal{D}_s + \mathcal{D}_s^*\right)\frac{\partial \bar{C}_s}{\partial z}\right] \tag{15.5-51}$$

$$\frac{\partial \bar{u}_z}{\partial z} = 0 \tag{15.5-52}$$

Equation 15.5-52 implies that at most $\bar{u}_z$ is a function of time. Assuming that $\bar{u}_z$, $\mathcal{D}_s$, and $\mathcal{D}_s^*$ are constants, Equation 15.5-51 becomes

$$\frac{\partial \bar{C}_s}{\partial t} + \bar{u}_z \frac{\partial \bar{C}_s}{\partial z} = \left(\mathcal{D}_s + \mathcal{D}_s^*\right)\frac{\partial^2 \bar{C}_s}{\partial z^2} \quad (z > 0) \tag{15.5-53}$$

For an infinitely long tube initially free of tracer and with a tracer concentration C_{so} continually introduced at z = 0, the initial and boundary conditions are

$$t = 0 : \bar{C}_s = 0, \quad z = 0 : \bar{C}_s = \bar{C}_{so}, \quad z \to \infty : \bar{C}_s = 0 \tag{15.5-54a-c}$$

Analysis

Dimensionless Forms

In terms of the dimensionless variables,

$$t \equiv \frac{\bar{u}_z t}{d}, \quad z \equiv \frac{z}{d}, \quad C \equiv \frac{C_s}{C_o} \tag{15.5-55a-c}$$

the governing equation and its boundary and initial conditions become

$$\frac{\partial C}{\partial t} + \frac{\partial C}{\partial z} = \frac{1}{Pe_d}\frac{\partial^2 C}{\partial z^2} \quad (z > 0) \tag{15.5-56}$$

$$t = 0 : C = 0, \quad z = 0 : C = 1, \quad z \to \infty : C = 0 \tag{15.5-57a-c}$$

Here, the Peclet number Pe_d is defined as the ratio of axial convection velocity $\bar{u}_z$ to the sum of the characteristic radial diffusion and dispersion velocities, $\left(\mathcal{D}_s + \mathcal{D}_s^*\right)/d$:

$$Pe_d \equiv \frac{\bar{u}_z d}{\mathcal{D}_s + \mathcal{D}_s^*} \tag{15.5-58}$$

Solution. For this linear, dynamic model with constant coefficients and a semi-infinite spatial domain, it is convenient to obtain a solution by applying the Laplace transform (Appendix C4). Operating on both sides of Equation 15.5-56 with the Laplace operator, $L\{C(z,t)\} \equiv \widetilde{C}(z,s)$ and applying the initial condition (Eq. 15.5-57a), the governing equation changes from a partial to an ordinary differential equation:

$$\frac{d^2 \widetilde{C}}{dz^2} - Pe_d \frac{d\widetilde{C}}{dz} - (Pe_d s)\widetilde{C} = 0 \tag{15.5-59}$$

whose transformed boundary conditions are given by

$$z = 0 : \widetilde{C} = \frac{1}{s}, \quad z \to \infty : \widetilde{C} = 0 \tag{15.5-60a,b}$$

The solution in the transform domain is

$$\widetilde{C}(z,s) = A_1(s)\exp(r_1 z) + A_2(s)\exp(r_2 z) \tag{15.5-61}$$

where r_1 and r_2 are the roots of the characteristic equation associated with the differential operator of Equation 15.5-59:

$$r_1 = \frac{Pe_d}{2}\left(1-\sqrt{1+\frac{4s}{Pe_d}}\right) < 0, \quad r_2 = \frac{Pe_d}{2}\left(1+\sqrt{1+\frac{4s}{Pe_d}}\right) > 0 \tag{15.5-62a, b}$$

Since the concentration must be finite as $z \to \infty$, we set $A_2 = 0$ and the solution simplifies to

$$\tilde{C}(z,s) = A_1(s)\exp(r_1 z) \tag{15.5-63}$$

Applying the condition at $z = 0$, we find that $A_1 = 1/s$ and

$$\tilde{C}(z,s) = \frac{\exp[r_1(s)z]}{s} = \left[\exp\left(\frac{Pe_d z}{2}\right)\right]\left[\frac{1}{s}f(z,s)\right] \tag{15.5-64}$$

where

$$f(z,s) = \exp\left[-\sqrt{(Pe_d z^2)s + \left(Pe_d^2 z^2/4\right)}\right] \tag{15.5-65}$$

The inverse Laplace transform of $f(z, s)$ can be obtained from tables (Roberts and Kaufman, 1966):

$$L^{-1}\{f(z,s)\} = \sqrt{\frac{Pe_d z^2}{4\pi t^3}}\exp\left[-\frac{Pe_d t}{4}\right]\exp\left[-\frac{Pe_d z^2}{4t}\right] \tag{15.5-66}$$

Since multiplication by $1/s$ in the Laplace domain corresponds to the integration in the t domain (Table C4-1), we obtain the concentration distribution from Equations 15.5-64 and 15.5-66 as

$$C(z,t) = \sqrt{\frac{Pe_d z^2}{4\pi}}\exp\left(\frac{Pe_d z}{2}\right)\int_0^t \sqrt{\frac{1}{\xi^3}}\exp\left[-\frac{Pe_d}{4}\left(\xi+\frac{z^2}{\xi}\right)\right]d\xi \tag{15.5-67}$$

Figure 15.5-8 shows dynamic changes in the cross-sectional average tracer concentration found by numerical integration of Equation 15.5-67 when $Pe_d = 1$. As time increases, the

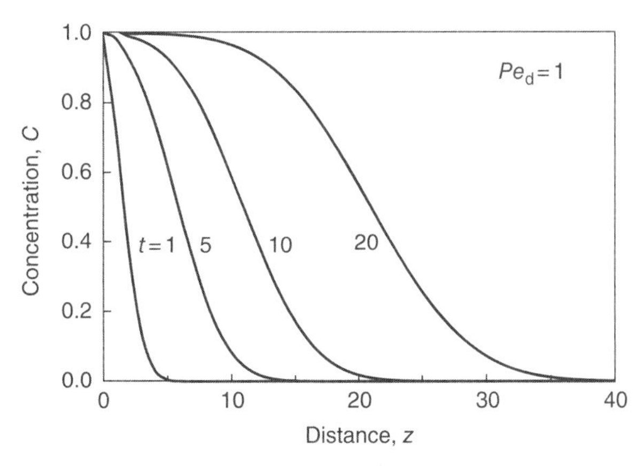

Figure 15.5-8 Tracer concentration distribution in response to a step input.

tracer concentration front moves downstream by convection and broadens by a combination axial diffusion and dispersion.

Example 15.5-3 Diffusion Coefficients from Dispersion Data

Consider an experiment with an aqueous solution of L-phenylalanine flowing through a tube of diameter 52 μm and length 50.2 cm that is initially filled with water (Sharma *et al.*, 2005). The phenylalanine solution moves at an average axial velocity of 1.32 cm/s. A detector located 39.7 cm downstream of the tube entrance records the cross-sectional average concentration leaving the tube. We can evaluate the dispersion and diffusion coefficients of phenylalanine using the measured concentration response shown in Figure 15.5-9.

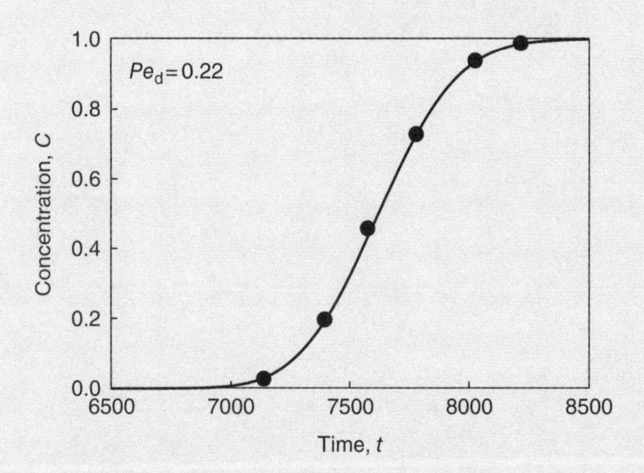

Figure 15.5-9 Comparison of experimental and simulated tracer concentrations.

Solution

The dimensionless position of the detector is $z = z/d = 39.7/0.0052 = 7640$. By substituting this value into Equation 15.5-67, we obtain

$$C(z,t) = 2154\int_0^t \sqrt{\frac{Pe_d}{\xi^3}}\exp\left[-\frac{Pe_d}{4}\left(\xi + \frac{5.83\times10^7}{\xi} - 1.53\times10^4\right)\right]d\xi \tag{15.5-68}$$

We evaluate this integral for different values of Pe_d to determine the value that best simulates the six data points. The best estimate $Pe_d = 0.22$ predicts the tracer output response shown by the curved line in Figure 15.5-9. From this value, the sum of the diffusion and dispersion coefficient is determined to be

$$\mathcal{D}_s + \mathcal{D}_s^* = \frac{\bar{u}_z d}{Pe_d} = \frac{1.32(0.0052)}{0.22} = 0.0312\ \text{cm}^2/\text{s} \tag{15.5-69}$$

If the flow is laminar and the tube is sufficiently long relative to its diameter, as is the case in this example, then dispersion coefficient reaches a constant limiting value of $\mathcal{D}_s^* = (\bar{u}_z d)^2/192\mathcal{D}_s$ (Taylor, 1953). Thus, $\mathcal{D}_s^*$ and $\mathcal{D}_s$ are also related by

$$\mathcal{D}_s^* = \frac{[1.32(0.0052)]^2}{192\mathcal{D}_s} = \frac{2.00\times10^{-7}}{\mathcal{D}_s}\ \text{cm}^2/\text{s} \tag{15.5-70}$$

We simultaneously solve Equations 15.5-69 and 15.5-70 and get

$$\mathcal{D}_s^* = 0.0312\ \text{cm}^2/\text{s}, \quad \mathcal{D}_s = 6.41 \times 10^{-6}\ \text{cm}^2/\text{s} \qquad (15.5\text{-}71\text{a,b})$$

This result indicates that dispersion produced by the interaction of nonuniform axial convection with radial diffusion is far more important than axial diffusion.

References

Abramowitz M, Stegun IA. Handbook of Mathematical Functions. Washington: US Government Printing Office; 1965, p 268.

Meilander NJ, Saidel GM, Bellamkonda RV. Lipid microtubules as sustained delivery vehicles for proteins and nucleic acids. In: Svenson S, editor. Carrier-Based Drug Delivery. ACS Symposium Series 879. Washington: American Chemical Society; 2004, ch 7, pp. 85–97.

Roberts GE, Kaufman H. Table of Laplace Transforms. Philadelphia: WB Saunders Company; 1966, p 246, (entry 14).

Sharma U, Gleason NJ, Carbeck JD. Diffusivity of solutes measured in glass capillaries using Taylor's analysis of dispersion and a commercial CE instrument. Anal Chem. 2005; 77:806–813.

Slattery JC. Advanced Transport Phenomena. Cambridge: Cambridge University Press; 1999, p 426.

Taylor GI. Dispersion of soluble matter in solvent flowing slowly through a tube. Proc Roy Soc A. 1953; 219:186–203.

Chapter 16

Mass Transport II

Chemical Reacting Systems

In this chapter, illustrations are given for transport and reaction occurring simultaneously within a single phase or within several phases, and for a reaction occurring at the interface between phases. When transport occurs in multiple phases, a species concentration equation must be separately written for each phase. Since these equations are coupled by continuity conditions at interfaces, they must be solved simultaneously. For zeroth-order and first-order reactions, analytical solutions may be obtained for linear models. For biological reaction rates that are typically nonlinear, numerical solutions of the models are usually needed. However, as shown in the following illustrations, analytical solutions can sometimes be found by using linear approximations of nonlinear reaction rate equations.

16.1 Single-Phase Processes

16.1.1 Reactive Gas Transport in the Lung Mucus Layer

When a reactive gas species such as ozone, chlorine or formaldehyde is inhaled, it can damage respiratory tract tissue leading to cell injury and abnormal lung function. As a defense mechanism against such air pollutants, the mucus layer that lines the inner walls of the conducting airways contains natural antioxidants and other substrates that detoxify these damaging gases before they can reach underlying tissue. The top of Figure 16.1-1 represents an airway mucus layer of thickness δ_m bounded by respired air at a radial position $r = a$ and by tissue at $r = a + \delta_m$. The bottom of the figure shows a section of this mucus layer with a shifted coordinate $y = r - a$ representing the penetration distance of the reactive gas species into the mucus from the air–mucus interface.

Biomedical Mass Transport and Chemical Reaction: Physicochemical Principles and Mathematical Modeling, First Edition. James S. Ultman, Harihara Baskaran, and Gerald M. Saidel.

Figure 16.1-1 Flat plate approximation of mucus layer.

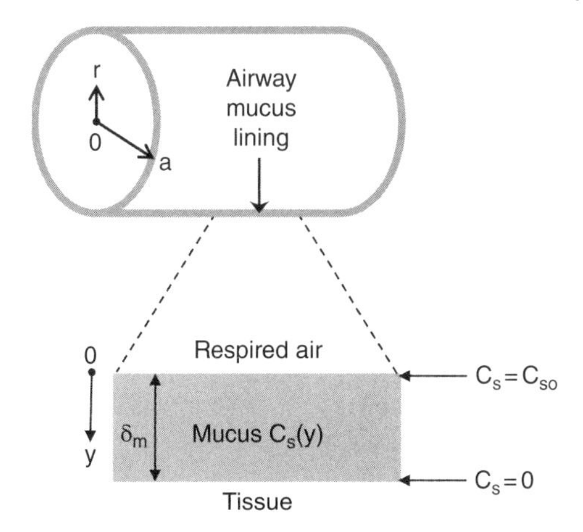

Model Formulation

In an airway section of length L, a reactive species s in the airway gas enters the mucus where it is transported by convection and diffusion. When the mucus layer is considered to be a symmetrical cylindrical film, the species concentration C_s is independent of circumferential position and varies with radial position in the domain $a + \delta_m > r > a$ and with axial position in the domain $L > z > 0$. For mucus of constant density and with a constant diffusion coefficient of the reactive gas, the continuity and species concentration equations in cylindrical coordinates (Tables 13.1-1 and 15.2-1) become

$$\frac{1}{r}\frac{\partial}{\partial r}(r u_r) + \frac{\partial u_z}{\partial z} = 0 \tag{16.1-1}$$

$$\frac{\partial C_s}{\partial t} + u_r\frac{\partial C_s}{\partial r} + u_z\frac{\partial C_s}{\partial z} = \mathcal{D}_s\left[\frac{1}{r}\frac{\partial}{\partial r}\left(r\frac{\partial C_s}{\partial r}\right) + \frac{\partial^2 C_s}{\partial z^2}\right] + R_s \tag{16.1-2}$$

where R_s is the reaction rate of gas species s with substrates it encounters in mucus.

Analysis

Order-of-Magnitude Approximations

We can express these equations in dimensionless form with the scaled variables:

$$t = \frac{t}{t_b},\quad z = \frac{z}{L},\quad y = \frac{r - a}{\delta_m},\quad u_r = \frac{u_r}{U},\quad u_z = \frac{u_z}{u_{ave}},\quad C = \frac{C_s}{C_{so}},\quad R = \frac{\delta_m^2 R_s}{\mathcal{D}_s C_{so}} \tag{16.1-3a-g}$$

where $\delta_m > y > 0$ is dimensionless penetration distance of the reactive gas. The scaling parameters in these equations are breathing period t_b, average axial velocity u_{ave} of the mucus, reactive species concentration C_{so} at the gas side of the mucus layer and a characteristic radial velocity U that is yet to be determined. In terms of these variables, Equations 16.1-1 and 16.1-2 become

$$\frac{\partial u_r}{\partial y} + \frac{u_r}{(a/\delta_m) + y} + \left(\frac{u_{ave}\delta_m}{UL}\right)\frac{\partial u_z}{\partial z} = 0 \tag{16.1-4}$$

$$\left(\frac{\delta_m^2}{\mathcal{D}_s t_b}\right)\frac{\partial C}{\partial t} + \frac{u_{ave}\delta_m^2}{\mathcal{D}_s L}\left[\left(\frac{UL}{u_{ave}\delta_m}\right)u_r\frac{\partial C}{\partial y} + u_z\frac{\partial C}{\partial z}\right] = \frac{\partial^2 C}{\partial y^2} + \frac{1}{(a/\delta_m)+y}\frac{\partial C}{\partial y} + \left(\frac{\delta_m}{L}\right)^2\frac{\partial^2 C}{\partial z^2} + R \tag{16.1-5}$$

With the radius of an airway being large compared to the thickness of its mucus layer, $(a/\delta_m) + y \gg 1$, we can approximate Equations 16.1-4 and 16.1-5 as:

$$\frac{\partial u_r}{\partial y} + \left(\frac{u_{ave}\delta_m}{UL}\right)\frac{\partial u_z}{\partial z} = 0 \tag{16.1-6}$$

$$\left(\frac{\delta_m^2}{\mathcal{D}_s t_b}\right)\frac{\partial C}{\partial t} + \frac{u_{ave}\delta_m^2}{\mathcal{D}_s L}\left[\left(\frac{UL}{u_{ave}\delta_m}\right)u_r\frac{\partial C}{\partial y} + u_z\frac{\partial C}{\partial z}\right] = \frac{\partial^2 C}{\partial y^2} + \left(\frac{\delta_m}{L}\right)^2\frac{\partial^2 C}{\partial z^2} + R \tag{16.1-7}$$

These equations, whose domain is $(1 > y > 0,\ L > z > 0)$, are equivalent to the dimensionless concentration equations in rectangular coordinates. In other words, because the curved mucus layer is relatively thin, it can be approximated as a planar slab.

The terms in Equation 16.1-6 must be of the same order of magnitude for their sum to be zero. Since the derivatives are scaled to be of the same order of magnitude, typically O(1), we expect that $(u_{ave}\delta_m/UL) \sim O(1)$. Consequently, we define the scaling factor for radial velocity as

$$U = \left(\frac{\delta_m}{L}\right)u_{ave} \tag{16.1-8}$$

Equation 16.1-7 then becomes

$$\left(\frac{\delta_m^2}{\mathcal{D}_s t_b}\right)\frac{\partial C}{\partial t} + \frac{u_{ave}\delta_m}{\mathcal{D}_s}\left(\frac{\delta_m}{L}\right)\left(u_r\frac{\partial C}{\partial y} + u_z\frac{\partial C}{\partial z}\right) = \frac{\partial^2 C}{\partial y^2} + \left(\frac{\delta_m}{L}\right)^2\frac{\partial^2 C}{\partial z^2} + R \tag{16.1-9}$$

Because each of the scaled variables and their derivatives are O(1), we can determine the relative magnitudes of the terms in this equation from the relative magnitudes of their dimensionless coefficients. Based on typical parameter values for the human lung: $(\delta_m/L) \ll 1$, $(u_{ave}\delta_m/\mathcal{D}_s)(\delta_m/L) \ll 1$, and $\left(\delta_m^2/\mathcal{D}_s t_b\right) \ll 1$. Thus, Equation 16.1-9 can be approximated as

$$\frac{d^2 C}{dy^2} = -R \tag{16.1-10}$$

Provided that there is an large excess of detoxifying agents in mucus, the reaction rate of a reactive gas will be quasi-first order with respect to the reactive gas concentration. The reaction rate equation is then

$$R = -k_r C_s \ \Rightarrow\ R = -Da C \tag{16.1-11a,b}$$

where the negative sign indicates that species s is being depleted. The Damkohler number which we have defined as

$$Da \equiv \left(\frac{k_r\delta_m^2}{\mathcal{D}_s}\right) \tag{16.1-12}$$

represents the ratio of the characteristic diffusion time in the radial direction $\left(\delta_m^2/\mathcal{D}_s\right)$ to the characteristic reaction time $(1/k_r)$. The dimensionless concentration equation now becomes

$$\frac{d^2C}{dy^2} = DaC \tag{16.1-13}$$

On the mucus side of the air–mucus interface, the reactive gas concentration is C_{so}. Assuming that reaction in tissue is much faster than in mucus, the species concentration is negligible at the mucus–tissue interface. The corresponding dimensional and dimensionless boundary conditions are

$$y = 0: \quad C_s \equiv C_{so} \Rightarrow y = 0: \quad C = 1 \tag{16.1-14a,b}$$

$$y = \delta_m: \quad C_s = 0 \Rightarrow y = 1: \quad C = 0 \tag{16.1-15a,b}$$

Solution

Since the characteristic equation associated with Equation 16.1-13 has the real roots $\pm\sqrt{Da}$, the general solution for $C(y)$ can be expressed as

$$C = A_1 \sinh\left(y\sqrt{Da}\right) + A_2 \cosh\left(y\sqrt{Da}\right) \tag{16.1-16}$$

Applying the boundary conditions, we find that $A_1 = -1/\tanh\left(\sqrt{Da_m}\right)$ and $A_2 = 1$ so that

$$C = \frac{\left[\sinh\left(\sqrt{Da}\right)\cosh\left(y\sqrt{Da}\right) - \cosh\left(\sqrt{Da}\right)\sinh\left(y\sqrt{Da}\right)\right]}{\sinh\left(\sqrt{Da}\right)} \tag{16.1-17}$$

which is equivalent to

$$C = \frac{\sinh\left[\sqrt{Da}(1-y)\right]}{\sinh\left(\sqrt{Da}\right)} \tag{16.1-18}$$

The left side of Figure 16.1-2 shows the concentration distributions predicted for a range of Da values. For $Da < 0.1$, the chemical reaction rate is slow compared to diffusion. The reactive gas concentration then decreases linearly with y, the dimensionless penetration distance into the mucus layer. For $Da > 0.1$, the reaction rate is more important, and the concentration decreases more sharply with penetration distance.

With no convective transport, the diffusion flux at any radial position in the mucus layer is

$$N_s = -\mathcal{D}_s \frac{dC_s}{dy} = -\frac{\mathcal{D}_s C_{so}}{\delta_m} \frac{dC}{dy} \tag{16.1-19}$$

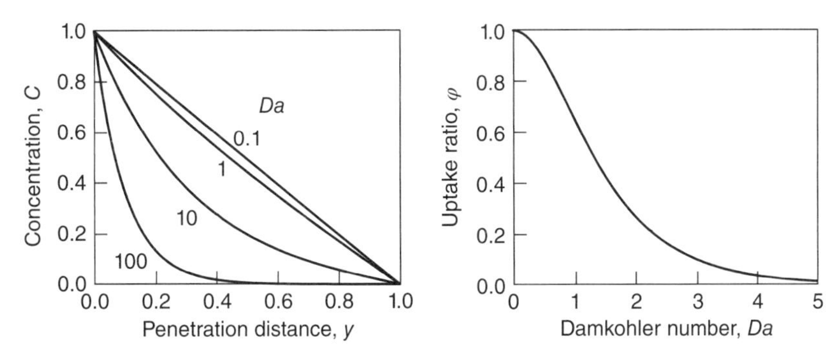

Figure 16.1-2 Reaction–diffusion process in stationary mucus.

By defining a dimensionless flux as $N \equiv (\delta_m/\mathcal{D}_s C_{so}) N_s$, Equations 16.1-18 and 16.1-19 lead to

$$N = -\frac{dC}{dy} = \frac{\sqrt{Da}\cosh\left[\sqrt{Da}(1-y)\right]}{\sinh\left(\sqrt{Da}\right)} \tag{16.1-20}$$

Relative to reactive gas transported into mucus at $y = 0$, the gas that reaches tissue at $y = 1$ is given by the flux ratio:

$$\varphi \equiv \frac{[N]_{y=1}}{[N]_{y=0}} = \frac{1}{\cosh\left(\sqrt{Da}\right)} \tag{16.1-21}$$

This function, shown on the right side of Figure 16.1-2, reveals that only a small fraction of species s absorbed at the gas–mucus interface is able to reach tissue when reaction in mucus is so rapid that $Da > 4$.

Example 16.1-1 Mucus Layer Protection in Lung Airways

The thickness of mucus δ_m in a human trachea is about 10 μm, but is only 1 μm in a terminal conducting airway. How protective against tissue damage by ozone (O_3) is tracheal mucus compared to mucus in a terminal airway? The O_3 diffusion coefficient in mucus is $\mathcal{D}_{O_3} = 2.7 \times 10^{-5}\,\text{cm}^2/\text{s}$, and the quasi-first-order rate constant of O_3 with substrates in mucus has been estimated as $k_r = 1200\,\text{s}^{-1}$ (Miller *et al.*, 1985).

Solution

We compute $Da \equiv k_r\delta^2/\mathcal{D}_{O_3}$ for the mucus layer in the two types of airways as:

$$Da_{\text{trachea}} = \frac{1500(10)^2}{2.7\times10^{-5}\left(10^4\right)^2} = 55.6, \quad Da_{\text{terminal}} = \frac{1500(1)^2}{2.7\times10^{-5}\left(10^4\right)^2} = 0.556 \tag{16.1-22a,b}$$

We can characterize the relative protection the mucus layer provides in these airways in two ways. First, we compute the dimensionless penetration required for C to reach some arbitrarily low value, say 0.01. Solving Equation 16.1-18 for that penetration distance, we obtain

$$y_{0.01} = 1 - \frac{1}{\sqrt{Da}}\sinh^{-1}\left[0.01\sinh\left(\sqrt{Da}\right)\right] = \begin{cases} 0.617 & \text{trachea} \\ 0.989 & \text{terminal} \end{cases} \tag{16.1-23}$$

We see from this result that O_3 penetrates a smaller fraction of the mucus thickness in the trachea than in the terminal conducting airways. Second, we compute the fraction of absorbed O_3 that reaches underlying tissue. From Equation 16.1-21, we determine that

$$\varphi = \frac{1}{\cosh\left(\sqrt{Da}\right)} = \begin{cases} 0.00116 & \text{trachea} \\ 0.775 & \text{terminal} \end{cases} \tag{16.1-24}$$

Thus, the mucus layer in the trachea eliminates over 99% of the O_3 that crosses the gas–mucus interface, whereas the mucus layer in the terminal airways eliminates only about 22%. Judging from either the penetration distance or the flux ratio, tissue is much better protected by the thicker mucus layer in the trachea.

16.1.2 Urea Uptake by an Encapsulated Enzyme

An experimental method for the extracorporeal removal of urea from blood utilizes small capsules containing a solution of urease enclosed by a membrane that is highly permeable to urea but impermeable to this enzyme. When the capsules are suspended in blood, urea diffuses into the capsules where it is hydrolyzed according to the reaction:

$$NH_2CONH_2 + H_2O \xrightarrow{\text{urease}} 2NH_3 + CO_2 \tag{16.1-25}$$

To prevent exposure to blood, the ammonia product is immobilized inside the capsules by a binding agent. The first step in designing artificial kidney devices that utilize such capsules is to model simultaneous reaction and diffusion within a single capsule exposed to a constant external urea concentration.

Model Formulation

We consider a spherical capsule of radius "a" in which the urea–urease reaction occurs by Michaelis–Menten kinetics. Because of spherical symmetry, urea concentration in the capsule C_u varies only with radial position r from the center of the capsule. We assume that convection is negligible and fluid density as well as the urea diffusion coefficient are constants. In addition, we focus on long times when C_u reaches a pseudo-steady-state distribution. The governing equations for this situation are identical to those in Section 9.4 for diffusion of a metabolic substrate into a cell:

$$\frac{1}{r^2}\frac{d}{dr}\left(r^2\frac{dC}{dr}\right) = Da\left(\frac{C}{K+C}\right) \tag{9.4-6}$$

$$r = 1: \quad C = 1, \qquad r = 0: \quad C = \text{finite} \tag{9.4-7a,b}$$

where the dimensionless quantities are defined as:

$$C = \frac{C_u}{C_{uo}}, \quad r \equiv \frac{r}{a}, \quad Da = \frac{V_m a^2}{\mathcal{D}_u C_{uo}}, \quad K = \frac{K_m}{C_{uo}} \tag{16.1-26a-d}$$

Here, $C_u(r)$ is urea concentration within the capsule, C_{uo} is urea concentration at the capsule surface, V_m and K_m are Michaelis–Menten parameters, and $\mathcal{D}_u$ is the urea diffusion coefficient. Because this nonlinear, boundary-value problem does not have an exact analytical solution, we consider two special cases that do have analytical solutions.

Analysis

Zeroth-Order Kinetics

We first find a solution when $C \gg K$ everywhere in the capsule such that reaction occurs at its maximum rate. For this zeroth-order reaction case,

$$\frac{1}{r^2}\frac{d}{dr}\left(r^2\frac{dC}{dr}\right) = Da \tag{9.4-8}$$

As given in Section 9.4, the solution to equation and its boundary conditions is

$$C = 1 - \frac{Da}{6}\left(1 - r^2\right) \tag{16.1-27}$$

This equation is subject to two constraints. Because negative concentrations are predicted at $r = 0$ when Da exceeds a value of six, the first constraint is that $6 \geq Da > 0$. Once this constraint is satisfied, we must still ensure that $C \gg K$ at all r. Equation 16.1-27 indicates that $C(\mathrm{r}) \geq 1 - Da/6$, and thus, the second constraint is that $(1 - Da/6) \gg K \geq 0$.

First-Order Kinetics

We now find a solution when $C << K$ such that the reaction rate is proportional to urea concentration. Since the largest urea concentration occurs at the capsule surface where $C\,(r = 1) = 1$, the condition for first-order kinetics requires $K \gg 1$. For this first-order reaction case,

$$\frac{1}{r^2}\frac{\mathrm{d}}{\mathrm{d}r}\left(r^2\frac{\mathrm{d}C}{\mathrm{d}r}\right) = DaC \tag{16.1-28}$$

Introducing a change of variables $C = g(r)/r$ into Equation 16.1-28, we transform the governing equation so that it has constant coefficients:

$$\frac{K}{Da}\frac{\mathrm{d}^2 g}{\mathrm{d}r^2} - g = 0 \tag{16.1-29}$$

Because the associated characteristic equation has the real roots $\pm\sqrt{Da/K}$, the solution to Equation 16.1-29 can be written as

$$g = \mathrm{k}_1 \cosh\left(r\sqrt{Da/K}\right) + \mathrm{k}_2 \sinh\left(r\sqrt{Da/K}\right) \tag{16.1-30}$$

Applying the transformed boundary conditions

$$r = 1: \quad g = 1, \quad r = 0: \quad g = 0 \tag{16.1-31a,b}$$

we get the following solution when the reaction kinetics are first order:

$$C = \frac{1}{r}\frac{\sinh\left(r\sqrt{Da/K}\right)}{\sinh\left(\sqrt{Da/K}\right)} \quad \text{when } K \gg 1 \tag{16.1-32}$$

Concentration Distributions

As shown in Figure 16.1-3, approximations of the urea–urease reaction by first-order kinetics predicts that urea concentration continuously decreases from the capsule surface at $r = 1$ to the capsule center at $r = 0$. This concentration drop becomes more pronounced as Da becomes larger or K becomes smaller. In either case, C reaches zero at $r = 0$ as $Da/K \to \infty$. With zeroth-order reaction, urea concentration also decreases from the capsule surface to the capsule center. Unlike the first-order solution, $C(0)$ only reaches zero when $Da = 6$. The dashed line drawn when $Da = 6$ indicates that this is the largest Damkohler number possible to avoid negative C values in the capsule.

Urea Influx

In terms of dimensional quantities, the molar diffusion rate of urea out of the capsule surface is

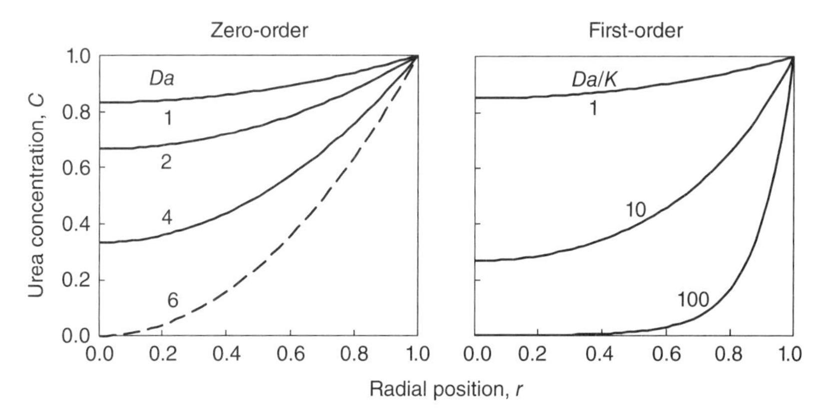

Figure 16.1-3 Urea distributions for zeroth- and first-order reactions.

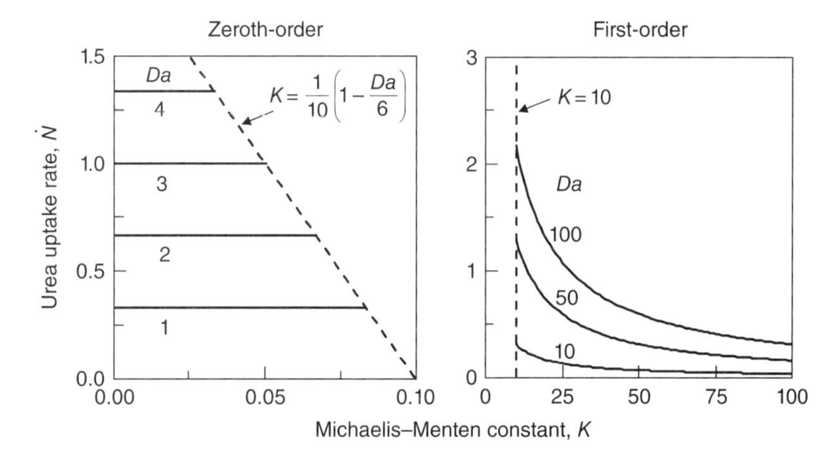

Figure 16.1-4 Urea uptake for zeroth-order and first-order reactions.

$$\dot{N}_u = -4\pi a^2 \mathcal{D}_u \left(\frac{dC_u}{dr}\right)_{r=a} \tag{16.1-33}$$

Defining a dimensionless urea influx as $\dot{N} = -\dot{N}_u / 4\pi a \mathcal{D}_u C_{uo}$, we arrive at

$$\dot{N} = \left(\frac{dC}{dr}\right)_{r=1} \tag{16.1-34}$$

Substituting the derivatives of Equations 16.1-27 and 16.1-32, we get

$$\dot{N} = \begin{cases} \dfrac{Da}{3} & (\text{zeroth order: } Da \le 6,\ K \ll 1 - Da/6) \\ \sqrt{Da/K}\coth\left(\sqrt{Da/K}\right) - 1 & (\text{first order: } K \gg 1) \end{cases} \tag{16.1-35}$$

The flux behavior is illustrated by the solid curves Figure 16.1-4. The slanted dashed line on the left graph represents an upper bound of $K = (1 - Da/6)/10$ that satisfies the constraint for the zeroth-order reaction that $K \ll (1 - Da/6)$. Similarly, the vertical dashed line on the right graph corresponds to a lower bound of $K = 10$ that satisfies the constraint that $K \gg 1$ for the first-order reaction. As reaction rate increases so that Da becomes larger, $\dot{N}$ increases for the zeroth-order as well as the first-order reactions. At a particular Da, the effect of K on $\dot{N}$ is irrelevant for the zeroth-order reaction, and decreases with increasing K for the first-order reaction.

Example 16.1-2 Urease Bioreactor

A flow-through reactor is designed to remove urea from blood of patients with kidney failure (Fig. 16.1-5). Blood from a patient continuously flows into the reactor containing many small capsules filled with urease solution. The blood leaving the reactor with a reduced urea concentration is returned to the patient.

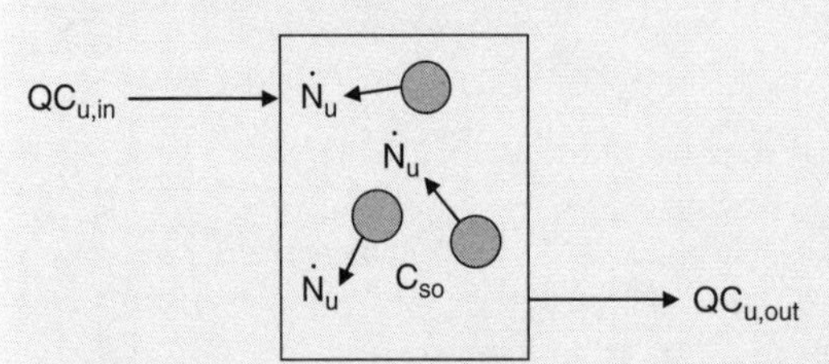

Figure 16.1-5 Well-mixed urea reactor with urease capsules suspended in blood.

This artificial kidney is intended to operate under conditions similar to those used for an extracorporeal hemodialyzer: a urea removal efficiency (Eq. 12.3-44) of $\zeta_u = 0.3$, a blood flow of 300 ml/min, and a urea concentration of 50 mM entering the device. Our objective is to determine the capsule size, total capsule volume V_{cap}, and urease concentration C_{urease} that will satisfy these conditions.

Rate constants for the urea–urease reaction available at 25°C and pH = 7 are $K_m =$ 2.47 mM and an enzyme turnover number of $k_{cat} \equiv V_m/C_{urease} = 0.0928\ s^{-1}$ (Qin and Cabral, 1994). The diffusion coefficient of urea in a dilute aqueous solution at 25°C is $\mathcal{D}_u = 1.4 \times 10^{-5}\ cm^2/s$ (Table A4-2).

Solution

In the reactor shown in Figure 16.1-5, Q is the volumetric blood flow, $C_{u,in}$ and $C_{u,out}$ are urea concentrations in blood entering and exiting the reactor, and $-\dot{N}_u$ is the urea removal rate by each of the n capsules in the device. Removal efficiency ζ_u is defined as the ratio of total removal rate $-n\dot{N}_u$ to the input rate $QC_{u,in}$ (12.3-44):

$$\zeta_u \equiv \frac{-n\dot{N}_u}{QC_{u,in}} \quad \Rightarrow \quad n = \frac{Q\zeta_u C_{u,in}}{-\dot{N}_u} \tag{16.1-36a,b}$$

During most of the dialysis treatment lasting several hours, the device operates at pseudo-steady state. Consequently, urea in the blood inflow is balanced by urea in the blood outflow and urea loss to all the capsules:

$$QC_{u,in} = QC_{u,out} - n\dot{N}_u \tag{16.1-37}$$

Urea concentration in the internal fluid at the capsule surface, C_{uo}, is approximately equal to $C_{u,out}$. This assumes complete mixing of urea in the blood phase and equal solubility of urea in blood and capsule fluid. Combining Equations 16.1-36a and 16.1-37, we get

$$C_{uo} = C_{u,out} = (\zeta_u - 1)C_{u,in} \tag{16.1-38}$$

For the operating conditions of this example, $C_{uo} = (1-0.3)50 = 35$ mM. Since $K_m = 2.47$ mM is much smaller than this, $K \ll 1$ and the first-order reaction approximation is not appropriate. Instead, we apply the zeroth-order reaction approximation. This has an added advantage: the urea reaction rate and thus the urea uptake is maximized when $C \gg K$. From Equation 16.1-35, the uptake rate per capsule is

$$\dot{N} = \frac{Da}{3} \quad \Rightarrow \quad -\dot{N}_u = \frac{4}{3}\pi a^3 V_m \tag{16.1-39a, b}$$

Combining this with Equation 16.1-36b, we get the total volume of all n capsules:

$$V_{cap} \equiv n\left(\frac{4}{3}\pi a^3\right) = \frac{QC_{u,in}\zeta_u}{V_m} \tag{16.1-40}$$

By selecting a factor of 1/10 to satisfy the constraint that $K \ll 1 - Da/6$, we define the maximum allowable radius a_{max} at which zeroth-order reaction kinetics occurs everywhere within a capsule:

$$\frac{K_m}{C_{uo}} = \frac{1}{10}\left(1 - \frac{V_m a_{max}^2}{6\mathcal{D}_u C_{uo}}\right) \quad \Rightarrow \quad a_{max} = \sqrt{\frac{6\mathcal{D}_u}{V_m}(C_{uo} - 10K_m)} \tag{16.1-41a, b}$$

Substituting $k_{cat}\, C_{urease}$ for V_m, in Equations 16.1-40 and 16.1-41b, we get the final design equations for V_{cap} and a_{max}:

$$V_{cap} = \frac{QC_{u,in}\zeta_u}{k_{cat}C_{urease}} \tag{16.1-42}$$

$$a_{max} = \sqrt{\frac{6\mathcal{D}_u}{k_{cat}C_{urease}}(C_{uo} - 10K_m)} \tag{16.1-43}$$

With values of K_m, k_{cat}, $\mathcal{D}_u$, and C_{uo} specified, we are at still at liberty to choose V_{cap}, a_{max}, and C_{urease}. The relationships between these parameters, dictated by Equations 16.1-42 and 16.1-43, are shown in Figure 16.1-6. The left graph indicates that as enzyme concentration C_{urease} increases, causing a proportional increase in the reaction rate, less total capsule volume

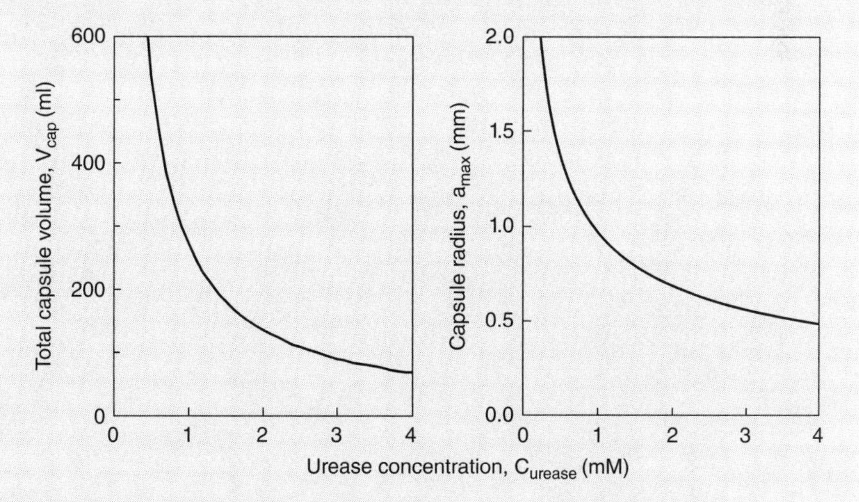

Figure 16.1-6 Relationship between capsule size and urease concentration in design of an extracorporeal urea reactor.

is required. The right graph shows that a larger C_{urease} also requires that smaller capsules be used to maintain zeroth-order reaction conditions.

In principle, a urea removal fraction of 0.3 when blood urea level is 50 mM can be achieved with various combinations of C_{urease}, V_{cap}, and a_{max} values. However, there are additional considerations in making a specific choice. For example, to restrict the volume of a patient's blood retained in the device, the maximum capsule volume should be limited to a few hundred milliliters. In addition, the radius of the capsule must be sufficiently large that it is practical to manufacture. A reasonable combination of design parameters that meet the performance objectives of this device are $C_{urease} \approx 1$ mM and $a_{max} \approx 1$ mm.

16.2 Multiphase Processes

16.2.1 Reactive Gas Transport in a Lung Airway Wall

In a previous section, we developed a single-phase model for the diffusion and reaction of an inhaled reactive gas in the mucus layer that coats the inner wall of a respiratory airway. More generally, a reactive gas diffuses and reacts in the mucus layer and then enters the tissue layer where it also diffuses and reacts before reaching the capillaries (Fig. 16.2-1). Assuming that the gas is much more reactive in blood than in tissue, its concentration in at the tissue–blood interface will be very small.

Model Formulation

We can treat the two curved layers as a planar slabs since they are thin compared to the radius of an airway. Also, the time scale of diffusion–reaction processes are short compared to the breathing period, so that transport will be pseudo-steady. Applying these conditions to both mucus (j = m) and tissue (j = t), we can write Equation 16.1-13 for the two layers as:

$$\frac{d^2C^j}{dy^2} - Da_j C^j = 0 \quad \begin{cases} j = m & \text{when } 1 > y \geq 0 \\ j = t & \text{when } \delta \geq y \geq 1 \end{cases} \tag{16.2-1}$$

where the dimensionless quantities are defined as:

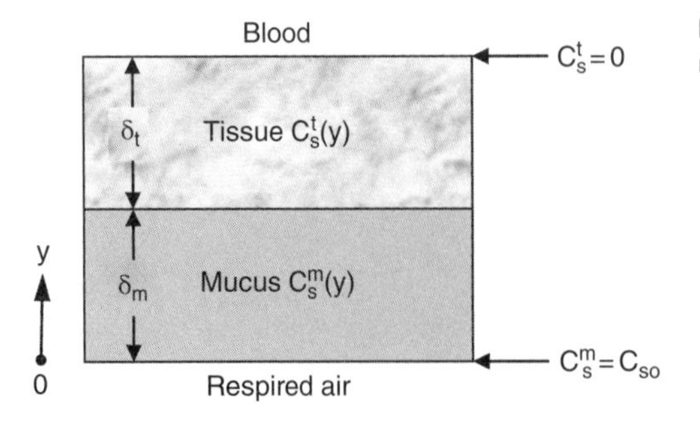

Figure 16.2-1 Reactive gas transport in airway mucus and tissue.

$$C^j \equiv \frac{C_s^j}{C_{so}}, \quad y \equiv \frac{y}{\delta_m}, \quad Da_j \equiv \frac{k_r^j \delta_m^2}{\mathcal{D}_s^j}, \quad \delta = \frac{\delta_m + \delta_t}{\delta_m} \tag{16.2-2a-d}$$

Here, δ_j, $\mathcal{D}_s^j$, and k_r^j are the thicknesses, diffusion coefficient, and reaction rate coefficient associated with layer j. Taking C_{so} to represent the reactive gas concentration at the mucus side of the air–mucus interface leads to the boundary condition:

$$y = 0: \quad C^m = 1 \tag{16.2-3}$$

At the mucus–tissue interface, the dimensionless concentrations in the two phases are related by local equilibrium and the diffusion flux is continuous:

$$y = 1: \quad C^m = \lambda_s^{m,t} C^t, \quad \frac{dC^m}{dy} = D \frac{dC^t}{dy} \tag{16.2-4a, b}$$

where $D \equiv \mathcal{D}_s^t / \mathcal{D}_s^m$. With the assumption of rapid reaction in blood, we consider the reactive gas concentration at the tissue–blood interface to be negligible:

$$y = \delta: \quad C^t = 0 \tag{16.2-5}$$

Analysis

The general solution to Equation 16.2-1 whose roots are $\pm\sqrt{Da_j}$ can be written as

$$C^j = A_j \sinh(\sqrt{Da_j} y) + B_j \cosh(\sqrt{Da_j} y) \quad (j = m, t) \tag{16.2-6}$$

The four unknown coefficients A_j and B_j can be determined by applying the four boundary conditions:

$$\begin{aligned}
&y = 0: \quad B_m = 1 \\
&y = 1: \quad \lambda_s^{m,t} \left[\frac{A_t \sinh\sqrt{Da_t} + B_t \cosh\sqrt{Da_t}}{A_m \sinh\sqrt{Da_m} + B_m \cosh\sqrt{Da_m}} \right] = 1 \\
&y = 1: \quad D\sqrt{\frac{Da_t}{Da_m}} \left[\frac{A_t \cosh\sqrt{Da_t} + B_t \sinh\sqrt{Da_t}}{A_m \cosh\sqrt{Da_m} + B_m \sinh\sqrt{Da_m}} \right] = 1 \\
&y = \delta: \quad A_t \sinh\left(\sqrt{Da_t}\delta\right) + B_t \cosh\left(\sqrt{Da_t}\delta\right) = 0
\end{aligned} \tag{16.2-7a-d}$$

After substituting $B_m = 1$ into Equation 16.2-7b–d, we are left with three equations in the three unknowns A_m, A_t, and B_t:

$$\begin{aligned}
&s_t A_t + c_t B_t - \frac{s_m}{\lambda_s^{m,t}} A_m = \frac{c_m}{\lambda_s^{m,t}} \\
&c_t A_t + s_t B_t - \frac{c_m}{D}\sqrt{\frac{Da_m}{Da_t}} A_m = \frac{s_m}{D}\sqrt{\frac{Da_m}{Da_t}} \\
&s_\delta A_t + c_\delta B_t = 0
\end{aligned} \tag{16.2-8a-c}$$

where we have defined

$$\begin{aligned}
&s_t \equiv \sinh\sqrt{Da_t}, \quad s_m \equiv \sinh\sqrt{Da_m}, \quad s_\delta \equiv \sinh\left(\sqrt{Da_t}\delta\right) \\
&c_t \equiv \cosh\sqrt{Da_t}, \quad c_m \equiv \cosh\sqrt{Da_m}, \quad c_\delta \equiv \cosh\left(\sqrt{Da_t}\delta\right)
\end{aligned} \tag{16.2-9a-f}$$

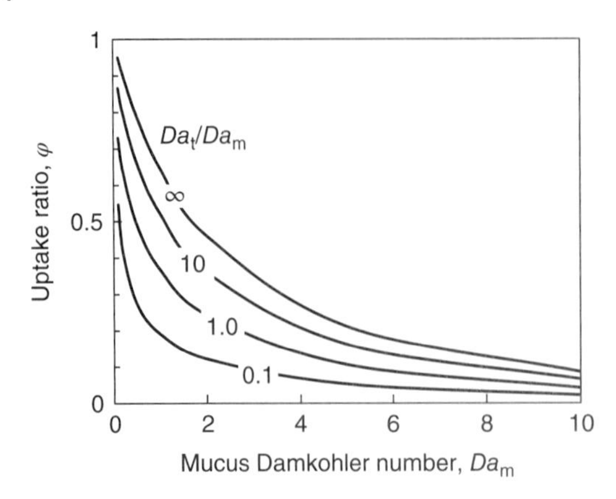

Figure 16.2-2 Uptake of reactive gas into tissue relative to total uptake.

These equations can be expressed in a compact matrix form:

$$[\mathrm{W}][\mathrm{x}] = [\mathrm{b}] \tag{16.2-10}$$

where the matrices have the components

$$[\mathrm{W}] \equiv \begin{bmatrix} s_t & c_t & -s_m/\lambda_s^{m,t} \\ c_t & s_t & -\dfrac{c_m}{D}\sqrt{\dfrac{Da_m}{Da_t}} \\ s_\delta & c_\delta & 0 \end{bmatrix}, \quad [\mathrm{x}] \equiv \begin{bmatrix} A_t \\ B_t \\ A_m \end{bmatrix}, \quad [\mathrm{b}] \equiv \begin{bmatrix} c_m/\lambda_s^{m,t} \\ \dfrac{s_m}{D}\sqrt{\dfrac{Da_m}{Da_t}} \\ 0 \end{bmatrix} \tag{16.2-11a-c}$$

It is convenient to solve these equations numerically to evaluate the constants of integration represented by the vector x. The dimensionless flux of the reactive species at any y can then be determined from the derivative of Equation 16.2-6 as:

$$N^j \equiv -\frac{dC^j}{dy} = -\sqrt{Da_j}\left[A_j \cosh\left(\sqrt{Da_j}y\right) + B_j \sinh\left(\sqrt{Da_j}y\right)\right] \tag{16.2-12}$$

We can also compute the molar rate at which reactive gas reaches the mucus–tissue interface at $y = 1$ relative to its absorption rate at the gas–mucus interface at $y = 0$:

$$\varphi = \frac{[N^m]_{y=1}}{[N^m]_{y=0}} = \cosh\sqrt{Da_m} + \frac{B_m}{A_m}\sinh\sqrt{Da_m} \tag{16.2-13}$$

Figure 16.2-2 shows model predictions for φ at approximate, but realistic, parameter values: $\delta_t/\delta_m = 100$, $\lambda_s^{m,t} = 1$, $D = 1$ and $Da_t/Da_m = 0.1$, 1.0, 10. For comparison, predictions are shown for the corresponding single-layer model from Equation 16.1-21 (labeled as $Da_t/Da_m = \infty$). Similar to the single-layer model, the fraction of reactive gas that reaches the tissue layer increases as Da_m decreases. The fraction reaching the tissue layer also increases as Da_t increases relative to Da_m.

16.2.2 Nutrient Transport and Reaction in Perfused Tissue: the Krogh Model

Many tissues are perfused by a large number of parallel capillaries, each surrounded by extravascular fluid and cells. The Krogh model (Fig. 16.2-3) treats each element of this system as a

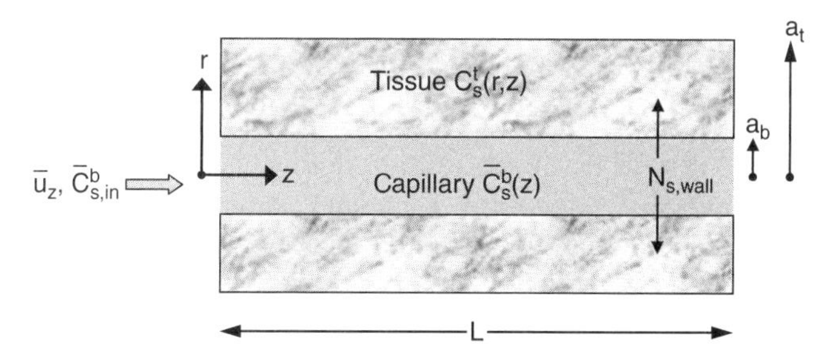

Figure 16.2-3 Krogh model of solute transport in perfused tissue.

single cylindrical capillary surrounded by an annular tissue space. The capillary wall is modeled as a membrane between the capillary lumen and the extravascular cells. As nutrients are transported by blood flow through the capillary, they diffuse into the adjoining tissue space where they are metabolized. Convection within the tissue is negligible and mass density, which is primarily due to water, is constant and uniform everywhere. For cells to remain viable, the rate at which nutrients diffuse to them in the tissue space must match their metabolic requirements. This condition cannot be met if extravascular cells are too distant from capillaries or concentration gradients are too small to provide adequate diffusion rates. An important application of the Krogh model is to examine how system parameters affect cell viability with increasing distance from capillary blood.

Model Formulation

In one form of the Krogh model, one-dimensional convection–diffusion occurs in blood flowing through a capillary of radius a_b, while a two-dimensional diffusion–reaction process occurs in a concentric tissue region of inner radius a_b and outer radius a_t. In both phases, transport occurs under steady-state conditions.

Capillary Processes

For blood of constant mass density flowing in the axial direction with negligible fluid loss across the cylindrical capillary membrane, a cross-sectional average solution balance (Eq. 15.3-6) reduces to

$$\frac{d\bar{u}_z}{dz} = 0 \tag{16.2-14}$$

This implies that $\bar{u}_z$, the axial velocity averaged over the circular capillary cross section, is constant. Now consider the cross-sectional average concentration $\bar{C}_s^b$ of a nutrient s that is nonreactive in blood. The nutrient is transported axially by convection, diffusion, and dispersion, as well as by radial diffusion across the capillary membrane. Incorporating these processes in a one-dimensional species balance for the nutrient (Eq. 15.3-18) leads to

$$\bar{u}_z \frac{d\bar{C}_s^b}{dz} - \frac{d}{dz}\left[\left(\mathcal{D}_s^c + \mathcal{D}_s^*\right)\frac{d\bar{C}_s^b}{dz}\right] = -\frac{2}{a_b} N_{s,wall} \quad (a_b > r > 0, \;\; L > z > 0) \tag{16.2-15}$$

Here, $\mathcal{D}_s^b$ and $\mathcal{D}_s^*$ are the nutrient diffusion and dispersion coefficients in capillary blood; $N_{s,wall}$ is the passive nutrient flux across the capillary membrane and its adjacent blood boundary layer. In terms of its bounding nutrient concentrations, $N_{s,wall}$ is

$$N_{s,wall}(z) = K_s\left(\bar{C}_s^b - \lambda_s^{b,t}\left[C_s^t\right]_{r=a_b}\right) \tag{16.2-16}$$

where K_s is an overall mass transfer coefficient, $\lambda_s^{b,t}$ is the equilibrium distribution coefficient of nutrient between blood and tissue, and $\left[C_s^t\right]_{r=a_b}$ is the nutrient concentration at the capillary–tissue interface. Substituting this rate equation into the species balance, we obtain

$$\bar{u}_z \frac{d\bar{C}_s^b}{dz} - \frac{d}{dz}\left[\left(\mathcal{D}_s^b + \mathcal{D}_s^*\right)\frac{d\bar{C}_s^b}{dz}\right] = -\frac{2K_s}{a_b}\left(\bar{C}_s^b - \lambda_s^{b,t}\left[C_s^t\right]_{r=a_b}\right) \tag{16.2-17}$$

Tissue Processes

In the extravascular tissue, nutrient diffuses with a diffusion coefficient $\mathcal{D}_s$ and is consumed by metabolic reactions at an extensive rate $-R_s$. For an axisymmetric distribution of nutrient, the equation for concentration $C_s^t(r,z)$ in cylindrical coordinates (Table 15.2-1) reduces to

$$0 = \mathcal{D}_s^t\left[\frac{1}{r}\frac{\partial}{\partial r}\left(r\frac{\partial C_s^t}{\partial r}\right) + \frac{\partial^2 C_s^t}{\partial z^2}\right] + R_s \quad (a_t > r > a_b,\ \ L > z > 0) \tag{16.2-18}$$

When nutrient depletion occurs in this tissue space by an enzymatic reaction that follows Michaelis–Menten kinetics, then

$$-R_s(r,z) = \frac{V_m C_s^t}{K_m + C_s^t} \tag{16.2-19}$$

and

$$\mathcal{D}_s^t\left[\frac{1}{r}\frac{\partial}{\partial r}\left(r\frac{\partial C_s^t}{\partial r}\right) + \frac{\partial^2 C_s^t}{\partial z^2}\right] = \frac{V_m C_s^t}{K_m + C_s^t} \tag{16.2-20}$$

Analysis

Dimensionless Forms

The species balances can be expressed in dimensionless form using the dimensionless variables:

$$C^t = \frac{C_s^t}{\bar{C}_{in}^b},\quad C^b = \frac{\bar{C}_s^b}{\bar{C}_{s,in}^b},\quad z = \frac{z}{L},\quad r = \frac{r}{a_b} \tag{16.2-21a-d}$$

where $\bar{C}_{s,in}^b$ is the nutrient concentration entering the capillary and L is the length of the model. Consequently, the dimensionless governing equations are

$$\frac{dC^b}{dz} - \frac{d}{dz}\left(\frac{\mathcal{D}_s^b + \mathcal{D}_s^*}{\bar{u}_z L}\frac{dC^b}{dz}\right) = -K_s\left(C^b - \lambda_s^{b,t}[C^t]_{r=1}\right) \quad (1 > r > 0,\ \ 1 > z > 0) \tag{16.2-22}$$

$$\frac{1}{r}\frac{\partial}{\partial r}\left(r\frac{\partial C^t}{\partial r}\right) + \left(\frac{a_b}{L}\right)^2\frac{\partial^2 C^t}{\partial z^2} = \frac{V_m C^t}{K_m + C^t} \quad (1 + a > r > 1,\ \ 1 > z > 0) \tag{16.2-23}$$

where dimensionless parameters are defined as:

$$\mathcal{K}_s \equiv \frac{2L}{\bar{u}_z a_b} K_s, \quad \mathcal{K}_m \equiv \frac{K_m}{\bar{C}^b_{in}}, \quad \mathcal{V}_m \equiv \frac{a_b^2}{\mathcal{D}_s^t \bar{C}^b_{s,in}} V_m, \quad a \equiv \frac{a_t}{a_b} \tag{16.2-24a – d}$$

Typically in a capillary bed, the dimensionless groupings $(\mathcal{D}_s^b + \mathcal{D}_s^*)/\bar{u}_z L \ll 1$ and $a_b \ll L$ so Equations 16.2-22 and 16.2-23 can be approximated as:

$$\frac{dC^b}{dz} = -\mathcal{K}_s \left(C^b - \lambda_s^{b,t} [C^t]_{r=1} \right) \tag{16.2-25}$$

$$\frac{1}{r}\frac{\partial}{\partial r}\left(r \frac{\partial C^t}{\partial r} \right) = \frac{\mathcal{V}_m C^t}{\mathcal{K}_m + C^t} \tag{16.2-26}$$

To complete the model, we express the boundary conditions in dimensional as well as in dimensionless forms. At the inlet to the capillary, we specify the nutrient concentration:

$$z = 0: \quad \bar{C}^b_s = \bar{C}^b_{s,in} \quad \Rightarrow \quad z = 0: \quad C^b = 1 \tag{16.2-27a, b}$$

We visualize an entire capillary bed as a parallel arrangement of Krogh models with their adjacent surfaces touching each other. For a nutrient that is metabolized in the tissue space, we expect C^t to reach its minimum value between the models at $r = a_t$:

$$r = a_t: \quad \frac{\partial C^t}{\partial r} = 0 \quad \Rightarrow \quad r = a: \quad \frac{\partial C^t}{\partial r} = 0 \tag{16.2-28a, b}$$

At the capillary–tissue interface, the flux leaving the capillary must equal the flux entering the tissue

$$r = a_b: \quad K_s\left(\bar{C}^b - \lambda_s^{b,t} C^t \right) = -\mathcal{D}_s^t \frac{\partial C^t}{\partial r} \quad \Rightarrow$$
$$r = 1: \quad \mathcal{K}_s \left(C^b - \lambda_s^{b,t} C^t \right) = -2\beta \frac{\partial C^t}{\partial r} \tag{16.2-29a, b}$$

where $\beta \equiv \mathcal{D}_s^t L / \bar{u}_z a_b^2$ represents the radial diffusion rate in tissue relative to the axial convection rate through the capillary.

Solution

For many metabolites of interest, $\mathcal{K}_m \ll C_s^t$ such that nutrient concentration in tissue is depleted by a zeroth-order reaction rate, and Equation 16.2-26 becomes

$$\frac{\partial}{\partial r}\left(r \frac{\partial C^t}{\partial r} \right) = \mathcal{V}_m r \tag{16.2-30}$$

Integration of this equation yields

$$\frac{\partial C^t}{\partial r} = \frac{\mathcal{V}_m}{2} r + \frac{A_1(z)}{r} \tag{16.2-31}$$

Applying the condition at the outer tissue boundary where $r = a$, $A_1(z) = -\mathcal{V}_m a^2/2$ so that

$$\frac{\partial C^t}{\partial r} = \frac{\mathcal{V}_m}{2}\left(r - \frac{a^2}{r} \right) \quad \Rightarrow \quad \left[\frac{\partial C^t}{\partial r} \right]_{r=1} = \frac{\mathcal{V}_m}{2}\left(1 - a^2 \right) \tag{16.2-32a, b}$$

When this is inserted into the inner tissue boundary condition at $r = 1$, the result is

$$C^{\mathrm{b}}(z) - \lambda_{\mathrm{s}}^{\mathrm{b,t}} C^{\mathrm{t}}(1,z) = \frac{\beta V_{\mathrm{m}}}{K_{\mathrm{s}}}(a^2 - 1) \tag{16.2-33}$$

To determine $C^{\mathrm{t}}(1, z)$, we integrate Equation 16.2-32a and find

$$C^{\mathrm{t}} = \frac{V_{\mathrm{m}}}{2}\left(\frac{r^2}{2} - a^2 \ln r\right) + A_2(z) \tag{16.2-34}$$

which at $r = 1$ becomes

$$C^{\mathrm{t}}(1,z) = \frac{V_{\mathrm{m}}}{4} + A_2(z) \tag{16.2-35}$$

Substituting this equation into Equation 16.2-33 leads to

$$A_2(z) = \frac{C^{\mathrm{b}}(z)}{\lambda_{\mathrm{s}}^{\mathrm{b,t}}} - \frac{V_{\mathrm{m}}}{2}\left[\frac{1}{2} + \frac{2\beta}{\lambda_{\mathrm{s}}^{\mathrm{b,t}} K_{\mathrm{s}}}(a^2 - 1)\right] \tag{16.2-36}$$

To determine $C^{\mathrm{b}}(z)$, we combine Equation 16.2-25 with Equations 16.2-29b and 16.2-32b:

$$\frac{\mathrm{d}C^{\mathrm{b}}}{\mathrm{d}z} = -K_{\mathrm{s}}\left(C^{\mathrm{b}} - \lambda_{\mathrm{s}}^{\mathrm{b,t}} C^{\mathrm{t}}\big|_{r=1}\right) = 2\beta\left[\frac{\partial C^{\mathrm{t}}}{\partial r}\right]_{r=1} = -\beta V_{\mathrm{m}}(a^2 - 1) \tag{16.2-37}$$

Integrating this equation and applying the entrance condition at $z = 0$, we find that nutrient concentration decreases in a linear fashion along the capillary:

$$C^{\mathrm{b}}(z) = 1 - \beta V_{\mathrm{m}}(a^2 - 1)z \tag{16.2-38}$$

By inserting this relationship into a combination of Equations 16.2-34 and 16.2-36, we obtain the nutrient distribution in the surrounding tissue:

$$C^{\mathrm{t}} = \frac{1}{\lambda_{\mathrm{s}}^{\mathrm{b,t}}} + \frac{V_{\mathrm{m}}}{2}\left[\frac{r^2 - 1}{2} - a^2 \ln r - \frac{2\beta}{\lambda_{\mathrm{s}}^{\mathrm{b,t}}}(a^2 - 1)\left(\frac{1}{K_{\mathrm{s}}} + z\right)\right] \tag{16.2-39}$$

These equations indicate that at any radial distance, the concentration in blood and tissue will be the smallest at $z = 1$.

Figure 16.2-4 shows dimensionless concentration distributions when $K_{\mathrm{s}} \to \infty$, that is, in the absence of a diffusion resistance between blood and extravascular tissue. In that case, the boundary condition at the inside surface of the tissue (Eq. 16.2-29b) reduces to an equilibrium expression, $C^{\mathrm{b}}(z) \approx \lambda_{\mathrm{s}}^{\mathrm{b,t}} C^{\mathrm{t}}(1,z)$. Other parameters are fixed at $\beta = 1$, $\lambda_{\mathrm{s}}^{\mathrm{b,t}} = 1$ and $V_{\mathrm{m}} = 0.02$. The left graph shows the change in nutrient concentration with position at a fixed tissue-to-capillary radius ratio of $a = 5.6$. The nutrient concentration continuously decreases with axial position along the capillary and decreases further with radial position in the tissue space. The graph on the right illustrates the effect of a on radial concentration distribution at the capillary outlet, $z = 1$. When a is increased, such that the tissue space is thicker relative to the capillary radius, more nutrient is depleted in both the capillary and the tissue. An increase in a also magnifies the decrease in C^{t} with r.

In both graphs, the nutrient concentration in the extravascular space is reduced to zero at the most remote point from the blood inlet ($r = 5.6$, $z = 1$). Beyond this point, cells would not be viable. When $a > 5.6$ in the right graph or $r > 5.6$ in the left graph, nutrient concentration becomes negative. This breakdown in the zeroth-order reaction model occurs because the

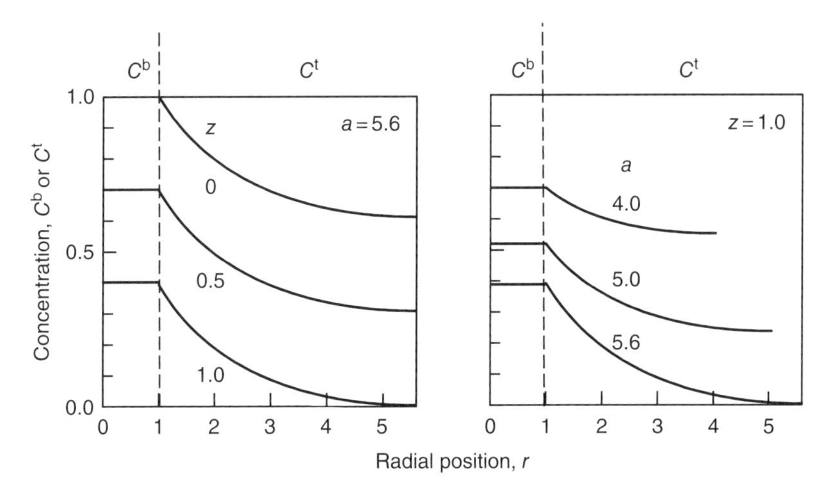

Figure 16.2-4 Solute concentration distribution in capillary and tissue.

constraint that $K_m \ll C_g^t$ cannot be met as nutrient concentration approaches smaller and smaller values.

16.2.3 Oxygenation of Pulmonary Capillary Blood

In the Krogh model, blood is viewed as a single-phase material. Here, we analyze pulmonary O_2 transport with a more exact two-phase model of blood consisting of red blood cells (RBC) suspended in plasma. As blood flows through a pulmonary capillary, O_2 diffuses across the capillary wall from alveolar gas into plasma. Simultaneously, O_2 is transported from plasma into the RBC where it can bind to hemoglobin (Fig. 16.2-5). Both heme groups Hb and oxyheme groups HbO_2 formed by O_2 binding are confined to the RBC.

We idealize the pulmonary capillary bed as a representative tube of length L enclosing a RBC core surrounded by an annular layer of plasma. The RBC core has a volume V^r whose surface area $S^{p,r}$ consists of the total surface of all RBCs in contact with plasma. The plasma layer between the RBC core and the alveolar gas has a volume V^p bounded by an inner surface $S^{p,r}$ in contact with the RBC core and an outer surface $S^{a,p}$ in contact with the alveolar space. The total blood volume is $V^b \equiv V^r + V^p$. The individual phase volumes can be expressed in terms of the fractional RBC hematocrit $\varepsilon_H \equiv V^r/V^b$ so that $V^p = (1-\varepsilon_H)V^b$ and $V^r = \varepsilon_H V^b$.

Model Formulation

Two-Phase Blood Model

The RBC and plasma phases flow at average velocities u^r and u^p, respectively. With negligible solution transfer across the plasma–RBC and alveolar gas–plasma interfaces, these velocities are constant.

Assuming a well-mixed alveolar gas, a complete model must also account for O_2 transport in the plasma phase and simultaneous transport and reaction in the RBC core. We represent these processes with one-dimensional species balances.

In the annular plasma layer, we assume that axial dispersion and diffusion of O_2 are negligible compared to convective transport. Applying a one-dimensional species transport equation (Eq. 15.3-34) to dissolved O_2 in this layer, we get

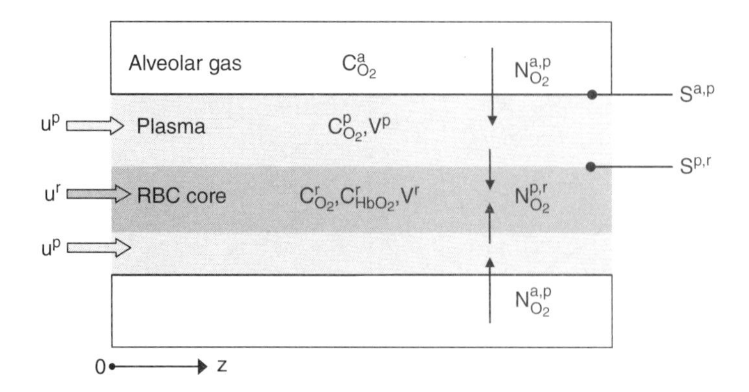

Figure 16.2-5 Two-phase model of pulmonary capillary blood oxygenation.

$$\frac{\partial C^p_{O_2}}{\partial t} + u^p \frac{\partial C^p_{O_2}}{\partial z} = \frac{1}{V^b(1-\varepsilon_H)}\left(S^{a,p}N^{a,p}_{O_2} - S^{p,r}N^{p,r}_{O_2}\right) \tag{16.2-40}$$

where $N^{a,p}_{O_2}$ and $N^{p,r}_{O_2}$ are molar fluxes at the alveolar gas–plasma and plasma–RBC interfaces, respectively. The corresponding surface areas per unit volume of plasma are $S^{a,p}/V^b(1-\varepsilon_H)$ and $S^{p,r}/V^b(1-\varepsilon_H)$.

In the RBC core, we also assume that axial diffusion and dispersion are negligible relative to convection. The simultaneous transport and reaction processes of O_2 and HbO_2 can then be represented by a simplified form of Equation 15.3-30:

$$\frac{\partial C^r_{O_2}}{\partial t} + u^r \frac{\partial C^r_{O_2}}{\partial z} = \frac{S^{p,r}}{\varepsilon_H V^b} N^{p,r}_{O_2} + R^r_{O_2} \tag{16.2-41}$$

$$\frac{\partial C^r_{HbO_2}}{\partial t} + u^r \frac{\partial C^r_{HbO_2}}{\partial z} = R^r_{HbO_2} \tag{16.2-42}$$

Here, $S^{p,r}/\varepsilon_H V^b$ is the surface of the plasma–RBC interface per unit RBC volume; $R^r_{O_2}$ and $R^r_{HbO_2}$ are the O_2 and HbO_2 formation rates, respectively, per unit RBC volume. Because O_2 binding with Hb to form HbO_2 is a one-to-one process, $R^r_{HbO_2} = -R^r_{O_2}$.

Representing HbO_2 by the O_2 saturation fraction, $S_{O_2} = C^r_{HbO_2}/T^r_{Hb}$, in which $T^r_{Hb} \equiv C^r_{HbO_2} + C^r_{Hb}$ is constant, the HbO_2 concentration equation (Eq. 16.2-42) becomes

$$\frac{\partial S_{O_2}}{\partial t} + u^r \frac{\partial S_{O_2}}{\partial z} = -\frac{R^r_{O_2}}{T^r_{Hb}} \tag{16.2-43}$$

The interfacial O_2 fluxes can be expressed in terms of overall mass transfer coefficients, $K^{a,p}_{O_2}$ and $K^{p,r}_{O_2}$, and equilibrium distribution coefficients, $\lambda^{a,p}_{O_2}$ and $\lambda^{p,r}_{O_2}$:

$$N^{p,r}_{O_2,wall} = K^{p,r}_{O_2}\left(C^p_{O_2} - \lambda^{p,r}_{O_2} C^r_{O_2}\right), \quad N^{a,p}_{O_2,wall} = K^{a,p}_{O_2}\left(C^a_{O_2} - \lambda^{a,p}_{O_2} C^p_{O_2}\right) \tag{16.2-44a, b}$$

With these relations, the concentration equations for dissolved O_2 in plasma (Eq. 16.2-40) and in RBC (Eq. 16.2-41) are written as

$$\frac{\partial C^p_{O_2}}{\partial t} + u^p \frac{\partial C^p_{O_2}}{\partial z} = \frac{S^{a,p} K^{a,p}_{O_2}}{V^b(1-\varepsilon_H)}\left(C^a_{O_2} - \lambda^{a,p}_{O_2} C^p_{O_2}\right) - \frac{S^{p,r} K^{p,r}_{O_2}}{V^b(1-\varepsilon_H)}\left(C^p_{O_2} - \lambda^{p,r}_{O_2} C^r_{O_2}\right) \quad (16.2\text{-}45)$$

$$\frac{\partial C^r_{O_2}}{\partial t} + u^r \frac{\partial C^r_{O_2}}{\partial z} = \frac{S^{p,r} K^{p,r}_{O_2}}{\varepsilon_H V^b}\left(C^p_{O_2} - \lambda^{p,r}_{O_2} C^r_{O_2}\right) + R^r_{O_2} \quad (16.2\text{-}46)$$

Reaction Rate

To complete the model, we must specify the rate of oxyhemoglobin binding in terms of the concentration of the reactant species present in the RBC core. As an illustration, we represent the binding kinetics by a Hill-type model. Assuming that "n" oxygen molecules simultaneously bind to a heme cluster Hb_n, the binding reaction is given by

$$Hb_n + nO_2 \rightleftharpoons (HbO_2)_n \quad (5.5\text{-}7)$$

By analogy to Equation 8.2-5, the rate of O_2 formation by this reaction is obtained as:

$$R^r_{O_2} = -nk_r\left[\left(C^r_{O_2}\right)^n C^r_{Hbn} - \frac{1}{\kappa_c} C^r_{(HbO_2)n}\right] \quad (16.2\text{-}47)$$

where κ_c is the reaction equilibrium constant and k_r is the rate constant for the forward reaction. Since $C^r_{Hbn} = C^r_{Hb}/n$, $C^r_{(HbO_2)n} = C^r_{HbO_2}/n$, $C^r_{Hb} = T^r_{Hb} - C^r_{HbO_2}$, and $C^r_{HbO_2} = T^r_{Hb} S_{O_2}$, this rate can also be written as:

$$R^r_{O_2} = T^r_{Hb} k_r \left[\frac{S_{O_2}}{\kappa_c} - (1 - S_{O_2})\left(C^r_{O_2}\right)^n\right] \quad (16.2\text{-}48)$$

Single-Phase Blood Model

We now consider the conditions necessary for the two-phase plasma–RBC model to reduce to a homogeneous, single-phase blood model, which we will denote with a superscript b. The dissolved O_2 concentration in the single-phase model should be the volume average over the RBC and plasma phases.

$$C^b_{O_2} \equiv \varepsilon_H C^r_{O_2} + (1-\varepsilon_H) C^p_{O_2} \quad (16.2\text{-}49)$$

Normally, the RBC and the plasma velocities are comparable, $u^r = u^p \equiv u^b$. With this condition, we eliminate $\left(S^{p,r} K^{p,r}_{O_2}/V^b\right)\left(C^p_{O_2} - \lambda^{p,r}_{O_2} C^r_{O_2}\right)$ between Equations 16.2-45 and 16.2-46 to get

$$\frac{\partial C^b_{O_2}}{\partial t} + u^b \frac{\partial C^b_{O_2}}{\partial z} = \frac{S^{a,p}}{V^b} K^{a,p}_{O_2}\left(C^a_{O_2} - \lambda^{a,p}_{O_2} C^p_{O_2}\right) + \varepsilon_H R^r_{O_2} \quad (16.2\text{-}50)$$

To arrive at a single-phase blood model, the plasma O_2 concentration $C^p_{O_2}$ must be related to the combined O_2 concentration $C^b_{O_2}$. This can be done provided that O_2 transport between plasma and RBC phases is so rapid that interfacial equilibrium is approached, $C^p_{O_2} \to \lambda^{p,r}_{O_2} C^r_{O_2}$. When substituted into Equation 16.2-49 with the elimination of $C^r_{O_2}$, this yields the relations

$$C^p_{O_2} = \lambda^{p,b}_{O_2} C^b_{O_2}; \quad \lambda^{p,b}_{O_2} \equiv \frac{\lambda^{p,r}_{O_2}}{\varepsilon_H + \lambda^{p,r}_{O_2}(1-\varepsilon_H)} \quad (16.2\text{-}51a,b)$$

Here, $\lambda_{O_2}^{p,b}$ is the effective partition coefficient between the plasma phase and the combined blood phase. Equation 16.2-50 now becomes

$$\frac{\partial C_{O_2}^{b}}{\partial t} + u^{b}\frac{\partial C_{O_2}^{b}}{\partial z} = \frac{S^{a,b}}{V^{b}} K_{O_2}^{a,p}\left(C_{O_2}^{a} - \lambda_{O_2}^{a,p}\lambda_{O_2}^{p,b}C_{O_2}^{b}\right) + \varepsilon_H R_{O_2}^{r} \tag{16.2-52}$$

For a single blood phase, the alveolar gas–plasma surface is the same as the alveolar gas–blood surface, $S^{a,p} = S^{a,b}$; the partition coefficients are related as $\lambda_{O_2}^{a,b} \equiv \lambda_{O_2}^{a,p}\lambda_{O_2}^{p,b}$; and $R_{O_2}^{b} \equiv \varepsilon_H R_{O_2}^{r}$ is the O_2 formation rate per unit volume of the blood phase. Incorporating these relations, we obtain

$$\frac{\partial C_{O_2}^{b}}{\partial t} + u^{b}\frac{\partial C_{O_2}^{b}}{\partial z} = \frac{S^{a,b}}{V^{b}} K_{O_2}^{a,p}\left(C_{O_2}^{a} - \lambda_{O_2}^{a,b}C_{O_2}^{b}\right) + R_{O_2}^{b} \tag{16.2-53}$$

Similarly, $T_{O_2}^{b} \equiv \varepsilon_H T_{Hb}^{r}$ is the total Hb concentration per unit volume of the blood phase. Thus, $R_{O_2}^{r}/T_{Hb}^{r} = R_{O_2}^{b}/T_{Hb}^{b}$ and Equation 16.2-43 can be written as:

$$\frac{\partial S_{O_2}}{\partial t} + u^{b}\frac{\partial S_{O_2}}{\partial z} = -\frac{R_{O_2}^{b}}{T_{Hb}^{b}} \tag{16.2-54}$$

Also for the single-phase model, $C_{O_2}^{r} = \lambda_{O_2}^{r,p}C_{O_2}^{p} = \lambda_{O_2}^{r,p}\left(\lambda_{O_2}^{p,b}C_{O_2}^{b}\right)$ so that Equation 16.2-48 for the O_2 formation rate per unit blood volume becomes

$$R_{O_2}^{b} = T_{Hb}^{b}k_r\left[\frac{S_{O_2}}{\kappa_c} - (1 - S_{O_2})\left(\lambda_{O_2}^{r,p}\lambda_{O_2}^{p,b}C_{O_2}^{b}\right)^{n}\right] \tag{16.2-55}$$

The model now consists of O_2 transport (Eq. 16.2-53), hemoglobin saturation (Eq. 16.2-54 and reaction rate (Eq. 16.2-55) equations containing variables and parameters that are almost all associated with the combined blood phase. One exception to this is the mass transfer coefficient $K_{O_2}^{a,p}$ that, in theory, should express O_2 transport from alveolar gas to plasma alone rather than from alveolar gas to the combined blood phase.

Analysis

Dimensionless Forms: Single-Phase Model

To express the transport equations for the single-phase blood model in dimensionless form, we define the variables:

$$\mathcal{C}^{b} \equiv \frac{\lambda_{O_2}^{a,b}C_{O_2}^{b}}{C_{O_2}^{a}},\quad \mathcal{R}^{b} \equiv \frac{LR_{O_2}^{b}}{u^{b}C_{O_2}^{a}},\quad \mathcal{z} \equiv \frac{z}{L},\quad \mathcal{t} \equiv \frac{u^{b}}{L}t \tag{16.2-56a-d}$$

The dimensionless concentration equation for dissolved O_2 (Eq. 16.2-53) and the dimensionless equation for the occupied hemoglobin site fraction (Eq. 16.2-54) are

$$\frac{\partial \mathcal{C}^{b}}{\partial \mathcal{t}} + \frac{\partial \mathcal{C}^{b}}{\partial \mathcal{z}} = \lambda_{O_2}^{a,b}\mathcal{K}^{a,p}\left(1 - \mathcal{C}^{b}\right) + \lambda_{O_2}^{a,b}\mathcal{R}^{b} \tag{16.2-57}$$

$$\frac{\partial S_{O_2}}{\partial \mathcal{t}} + \frac{\partial S_{O_2}}{\partial \mathcal{z}} = -\frac{\mathcal{R}^{b}}{\mathcal{T}^{b}} \tag{16.2-58}$$

and the dimensionless reaction rate is

$$R^{b} = T^{b}k\left[\frac{S_{O_2}}{\kappa} - (1 - S_{O_2})(C^{b})^{n}\right] \tag{16.2-59}$$

The dimensionless groups appearing in these equations are

$$K^{a,p} \equiv \frac{LS^{a,p}}{V^{b}u^{b}}K_{O_2}^{a,p}, \quad T^{b} \equiv \frac{T_{Hb}^{b}}{C_{O_2}^{a}}$$
$$k \equiv \frac{k_r L}{u^{b}}\left(C_{O_2}^{a}/\lambda_{O_2}^{a,r}\right)^{n}, \quad \kappa \equiv \kappa_c\left(C_{O_2}^{a}/\lambda_{O_2}^{a,r}\right)^{n} \tag{16.2-60a-d}$$

where $\lambda_{O_2}^{a,r} \equiv \lambda_{O_2}^{a,b}/\lambda_{O_2}^{r,p}\lambda_{O_2}^{p,b}$ is the effective partition coefficient between alveolar air and the RBC core in the two-phase model.

Simulation: Single-Phase Blood Model

Breathing period is typically much longer than the residence time of O_2 molecules in pulmonary capillary blood as well as the characteristic time for O_2 diffusion across the capillary wall. Under these conditions, variations over time of O_2 alveolar concentration $C_{O_2}^{a}$ are relatively small, and pseudo-state conditions exist for O_2 concentrations and O_2 saturations within a capillary.

Setting time derivatives to zero in the single-phase model equations and substituting the dimensionless reaction rate, we obtain the steady-state equations.

$$\frac{dC^{b}}{dz} = \lambda_{O_2}^{a,b}K^{a,p}(1 - C^{b}) + \lambda_{O_2}^{a,b}kT^{b}\left[\frac{S_{O_2}}{\kappa} - (1 - S_{O_2})(C^{b})^{n}\right] \tag{16.2-61}$$

$$\frac{dS_{O_2}}{dz} = k\left[(C^{b})^{n}(1 - S_{O_2}) - \frac{S_{O_2}}{\kappa}\right] \tag{16.2-62}$$

with entrance conditions:

$$z = 0: \quad S_{O_2} = S_{O_2}(0), \quad C^{b} = C^{b}(0) \tag{16.2-63}$$

Under normal resting conditions, the dimensionless parameters in this model have approximate values: $\lambda_{O_2}^{a,b}K^{a,p} = 100$ and $\lambda_{O_2}^{a,b}T^{b} = 70$. Equilibrium S_{O2} measurements on hemoglobin solutions (Example 5.5-1) provide Hill constants of $n = 2.8$ and $\kappa \approx 100$. At the entrance of the pulmonary capillaries, $C^{b}(0) = 0.35$ and $S_{O_2}(0) = 0.73$. With these parameter values, we obtain the numerical simulations shown in Figure 16.2-6 for various k values, corresponding to the residence time of blood in a pulmonary capillary relative to the oxyhemoglobin reaction time.

The left graph shows a continuous increase of dissolved O_2 concentration C^{b} along a capillary. The steep rise in C^{b} at very small distances from the capillary inlet is due to rapid O_2 diffusion from the alveolar compartment to blood. At $k = 10$ and 100, C^{b} tends toward an asymptotic value of one (equivalent to $C_{O_2}^{b} = C_{O_2}^{a}/\lambda_{O_2}^{a,b}$) as z increases. This agrees with the common understanding that, under resting conditions, dissolved O_2 in end-pulmonary capillary blood is in interfacial equilibrium with alveolar gas. However, at the lowest value of $k = 1$, C^{b} levels out below one. This is consistent with the notion that interfacial equilibrium

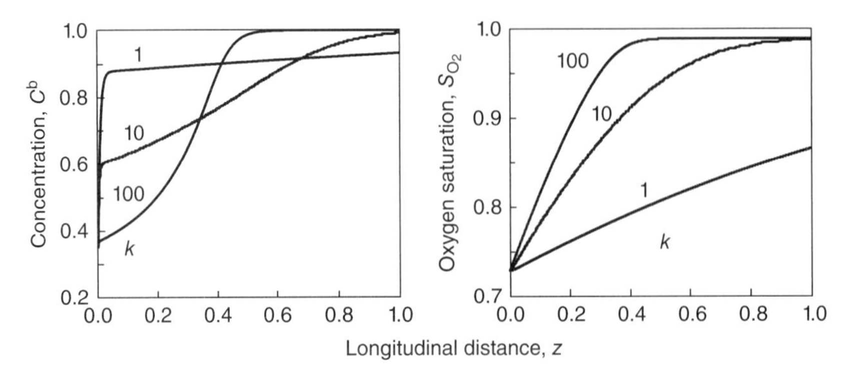

Figure 16.2-6 Oxygen distribution along a pulmonary capillary.

is not reached during heavy exercise because of a diminished residence time L/u^b of blood in pulmonary capillaries.

The right graph indicates that S_{O2} approaches an asymptotic value close to one at both $k = 10$ and $k = 100$. That is, at rapid reaction rates, oxyhemoglobin binding is close to equilibrium at the end of a pulmonary capillary. This is clearly not the case when $k = 1$.

16.3 Processes with Interfacial Reaction

16.3.1 Solute Transport to a Rapidly Rotating Disk with Surface Reaction

As shown in Section 14.1, a rapidly rotating disk submerged in a large volume of liquid is a convenient apparatus for applying a known shear stress to the fluid. According to that analysis based on the continuity and Navier–Stokes equations, a velocity boundary layer of uniform thickness $\delta = \sqrt{\upsilon/\Omega}$ is formed on the surface of a disk that spins at an angular speed Ω in a fluid of kinematic viscosity υ. We now analyze transport to such a disk when a species is depleted by a first-order reaction on its surface. In that case, a concentration boundary layer is formed in addition to the velocity boundary layer.

Model Formulation

For the cylindrical coordinate system shown in Figure 16.3-1, the concentration $C_s(r,t)$ of solute in a solution surrounding a disk of radius R is symmetric about the z axis of rotation. Although the solute reacts on the disk surface, it is inert in the solution itself. Assuming that the solution density and the solute diffusion coefficient $\mathcal{D}_s$ are constants, the steady-state convection–diffusion equation in cylindrical coordinates (Table 15.2-1) becomes

$$u_r \frac{\partial C_s}{\partial r} + u_z \frac{\partial C_s}{\partial z} = \mathcal{D}_s \left[\frac{\partial^2 C_s}{\partial r^2} + \frac{1}{r}\frac{\partial C_s}{\partial r} + \frac{\partial^2 C_s}{\partial z^2} \right] \quad (R > r > 0,\ z > 0) \tag{16.3-1}$$

This equation can be expressed in terms of the dimensionless variables

$$z = \frac{z}{\delta^c}, \quad r = \frac{r}{R}, \quad u_r = \left(\frac{\delta^2}{\upsilon R}\right) u_r, \quad u_z = \left(\frac{\delta}{\upsilon}\right) u_z, \quad C = \frac{C_s}{C_{s\infty}} \tag{16.3-2a-e}$$

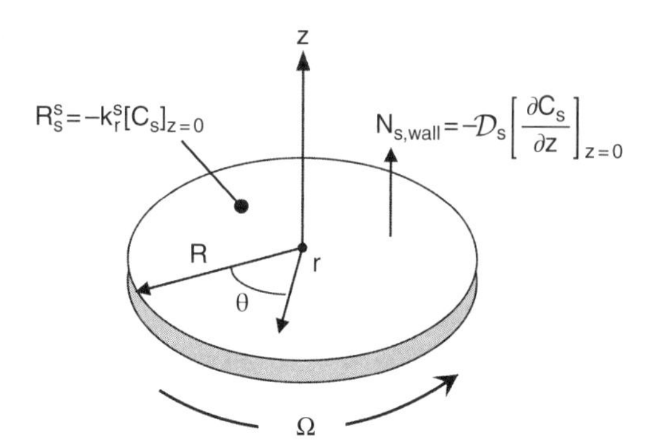

Figure 16.3-1 First-order reaction of solute on the surface of a rotating disk.

Here, the scaling factors for u_r and u_z are those used for the velocity boundary layer analysis (Eq. 14.1-48a,b), and $C_{s\infty}$ is the solute concentration far from the disk surface. The concentration boundary thickness δ^c, a scaling factor for distance from the disk surface, is determined by substituting the dimensionless variables into Equation 16.3-1:

$$\left(\frac{\delta^c}{\delta}\right)u_r\frac{\partial C}{\partial r} + u_z\frac{\partial C}{\partial z} = \frac{1}{Sc}\frac{\delta}{\delta^c}\left[\left(\frac{\delta^c}{R}\right)^2\left(\frac{\partial^2 C}{\partial r^2} + \frac{1}{r}\frac{\partial C}{\partial r}\right) + \frac{\partial^2 C}{\partial z^2}\right] \quad (1 > r > 0,\ z > 0) \qquad (16.3\text{-}3)$$

where the Schmidt number is defined as $Sc \equiv \upsilon/\mathcal{D}_s$. If we assume a thin concentration boundary layer, then $(\delta^c/R) \ll 1$ and the radial diffusion term is negligible compared to the axial diffusion term. Since axial diffusion is as important as axial convection, their coefficients should be equal. In that case, $(\delta/\delta^c)(1/Sc) = 1$ and

$$\frac{\delta_c}{\delta} = \frac{1}{Sc} \quad \Rightarrow \quad \delta^c = \frac{\sqrt{\upsilon/\Omega}}{\upsilon/\mathcal{D}_s} = \frac{\mathcal{D}_s}{\sqrt{\upsilon\Omega}} \qquad (16.3\text{-}4a,b)$$

For diffusion in aqueous solution, $Sc > 100$ so that $\delta^c/\delta \ll 1$. That is, the concentration boundary layer is much thinner than the velocity boundary layer. This indicates that on the left side of Equation 16.3-3, the radial convection term can be neglected compared to the axial convection term. Consequently, the convection–diffusion equation reduces to

$$u_z\frac{\partial C}{\partial z} = \frac{\partial^2 C}{\partial z^2} \qquad (16.3\text{-}5)$$

As obtained in Section 14.1, the axial component of the velocity as $z \rightarrow 0$ is

$$u_z \approx -0.51\sqrt{\upsilon\Omega}\left(\frac{z}{\delta}\right)^2 \qquad (14.1\text{-}63)$$

Since $z < \delta^c \ll \delta$ within the concentration boundary layer, it is reasonable to apply this velocity approximation to the convection–diffusion equation. Expressing Equation 14.1-63 in dimensionless form,

$$u_z \approx -0.51\sqrt{\upsilon\Omega}\left(\frac{\delta}{\upsilon}\right)\left(\frac{\delta^c z}{\delta}\right)^2 = -\left(\frac{0.51}{Sc^2}\right)z^2 \qquad (16.3\text{-}6)$$

and Equation 16.3-5 can be approximated as:

$$\frac{d^2C}{dz^2} + \left(\frac{0.51}{Sc^2}\right) z^2 \frac{dC}{dz} = 0 \qquad (16.3\text{-}7)$$

We require two boundary conditions for a specific solution. Far from the disk, the solute concentration is constant:

$$z \to \infty: \; C_s \to C_{s\infty} \; \Rightarrow \; z \to \infty: \; C \to 1 \qquad (16.3\text{-}8a,b)$$

Because no solute accumulates at the surface of an impermeable disk, the solute diffusion flux in the positive z direction, $N_{s,wall} = -\mathcal{D}_s(\partial C_s/\partial z)_{z=0}$, is balanced by its production rate per unit surface area, R_s^S. For a first-order surface reaction in which solute is being depleted, $R_s^S = -k_r^S C_s(0)$ where k_r^S [m/s] is a rate constant that also accounts for solute distribution between the solution and the surface. Equating $N_{s,wall}$ to R_s^S, we obtain the surface condition on the disk:

$$z = 0: \; -\mathcal{D}_s \frac{dC_s}{dz} = -k_r^S C_s \; \Rightarrow \; z = 0: \; \frac{dC}{dz} = Da^S C \qquad (16.3\text{-}9a,b)$$

Here, Da^S is a surface Damkohler number defined as:

$$Da^S \equiv \frac{k_r^S \delta^c}{\mathcal{D}_s} = \frac{k_r^S}{\sqrt{\upsilon\Omega}} \qquad (16.3\text{-}10)$$

Analysis

Concentration Distribution

By successive integrations of Equation 16.3-7 and applying the boundary conditions, we obtain the complete solution:

$$C(z) = C(0)\left[1 + Da^S \int_0^z \exp(-\alpha\xi^3)\, d\xi\right] \qquad (16.3\text{-}11)$$

where $\alpha \equiv 0.17/Sc^2$ and

$$C(0) = \left[1 + Da^S \int_0^\infty \exp(-\alpha\xi^3)\, d\xi\right]^{-1} \qquad (16.3\text{-}12)$$

With a change of variables $\zeta = \alpha\xi^3$, we evaluate the integral as

$$\int_0^\infty \exp(-\alpha\xi^3)\, d\xi = \left(\frac{1}{\alpha}\right)^{1/3}\left[\frac{1}{3}\int_0^\infty \zeta^{-2/3} e^{-\zeta} d\zeta\right] = \alpha^{-1/3}\Gamma(4/3) \qquad (16.3\text{-}13)$$

Here, $\Gamma(4/3) = 0.893$ is a Gamma function. Thus, the dimensionless concentration at the surface is

$$C(0) = \left(1 + 1.61 Sc^{2/3} Da^S\right)^{-1} \qquad (16.3\text{-}14)$$

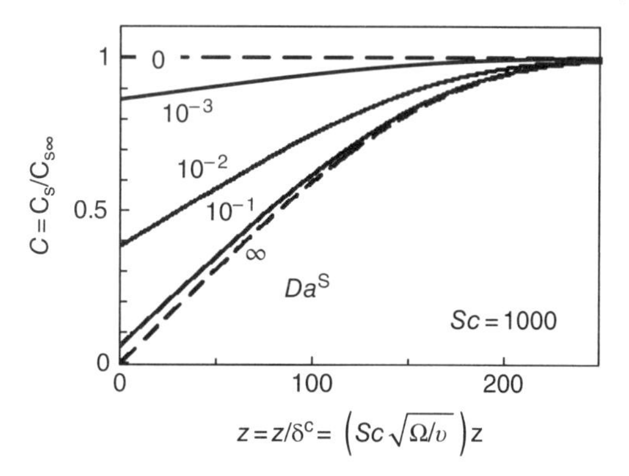

Figure 16.3-2 Solute concentration distribution above a rotating disk.

and the dimensionless concentration distribution within the fluid is

$$C(z) = \frac{1}{(1 + 1.61Sc^{2/3}Da^S)}\left[1 + Da^S\int_0^z \exp\left(-\frac{0.17}{Sc^2}\xi^3\right)d\xi\right] \tag{16.3-15}$$

We determine the effect of Da on the solute distribution by numerical evaluation of Equation 16.3-15 when $Sc = 1000$. In Figure 16.3-2, the solid lines illustrate $C(z)$ behavior for $Da^S>0$. The dashed lines show concentration distributions at the limiting values of $Da^S \rightarrow 0$ and $Da^S \rightarrow \infty$. With increasing Da^S, solute depletion at the surface is more rapid. This causes $C(0)$ to decrease, thereby increasing the concentration gradient in the adjacent solution. When $Da^S \rightarrow 0$, there is no reaction, so that $C(0) = 1$ and the concentration gradient is zero. When $Da^S \rightarrow \infty$, the reaction is so fast that $C(0) = 0$ and the concentration gradient is at its maximum.

Solute Processes at the Disk Surface

Since the solute flux in the positive z direction is equal to its formation rate by chemical reaction,

$$N_{s,wall} = R_s^S = -k_r^S C_s(0) = -k_r^S C_{s\infty} C(0) \tag{16.3-16}$$

The negative flux indicates that solute actually diffuses toward the surface where it is depleted. By substituting for k_r^S and $C(0)$ using Equations 16.3-10 and 16.3-14, respectively, we find that

$$N_{s,wall} = -\left(\frac{Da^S}{1 + 1.61Sc^{2/3}Da^S}\right)C_{s\infty}\sqrt{\upsilon\Omega} \tag{16.3-17}$$

To characterize this transport process further, we define a mass transfer coefficient k_s as the ratio of the wall flux to its driving force:

$$k_s \equiv \frac{N_{s,wall}}{C_s(0) - C_{s\infty}} = -k_r^S\left[\frac{C(0)}{C(0) - 1}\right] = \frac{\sqrt{\upsilon\Omega}}{1.61Sc^{2/3}} \tag{16.3-18}$$

We make this equation dimensionless with a Sherwood number Sh and a rotational Reynolds number Re_Ω, both based on the disk diameter $d = 2R$:

$$Sh_d \equiv 1.24Re_\Omega^{1/2}Sc^{1/3}; \quad Sh_d \equiv \frac{k_s d}{\mathcal{D}_s}, \quad Re_\Omega \equiv \frac{d^2\Omega}{4\upsilon} \tag{16.3-19a-c}$$

Notice that Equation 16.3-19, identical to entry 6 of Table 12.1-1, does not depend on Da^S because the mass transfer coefficient has been scaled by the concentration driving force.

Example 16.3-1 Rotating Disk as a Polarographic Electrode

The model of transport to a spinning disk requires that the rotational speed be fast enough to maintain a thin boundary layer but slow enough to avoid turbulent flow effects in the surrounding liquid. Garcia and coworkers (1997) examined whether this was the case for a 15 mm diameter platinum disk spinning at a rotational speed in the range 400–2500 rpm by configuring it as a polarographic cathode. A large platinum anode was placed far from the disk, and both electrodes were submerged in a large volume of electrolyte solution containing potassium ferrocyanide and potassium ferricyanide, each at a concentration of $C_{s\infty} = 0.01$ M.

With a negative bias voltage applied to the spinning cathode, electrons were consumed by the reduction of ferrocyanide to ferricyanide ion. The reverse reaction released electrons at the anode:

$$Fe(CN_6)^{3-} + e^- \rightleftharpoons Fe(CN_6)^{4-} \tag{16.3-20}$$

Completion of the electrical circuit within the solution occurred by the equimolar counterdiffusion of $Fe(CN_6)^{3-}$ toward the cathode and $Fe(CN_6)^{4-}$ toward the anode. At a sufficiently large bias voltage, the cathode reaction became so rapid that the electrical current i between the electrodes reached a limiting value. Given a diffusion coefficient of $\mathcal{D}_s = 7.5 \times 10^{-6}$ cm^2/s for the ferrocyanide ion (Konopka and McDuffle, 1970), we can compare the current flux data of Garcia *et al.* (Fig. 16.3-3) to the predictions of the transport model.

Solution

The current per unit surface S is proportional to molar flux of ferrocyanide from the solution to the disk surface, that is, in the negative z direction:

$$\frac{i}{S} = -\mathcal{F}N_{s,\text{wall}} \tag{16.3-21}$$

With a fast reaction at the disk surface, $N_{s,\text{wall}}$ is found from Equation 16.3-17 by taking the limit as $Da^S \to \infty$. The corresponding limiting current flux is

$$\frac{i}{S} = \frac{\mathcal{F}C_{s\infty}}{1.61 Sc^{2/3}}\sqrt{\upsilon\Omega} \tag{16.3-22}$$

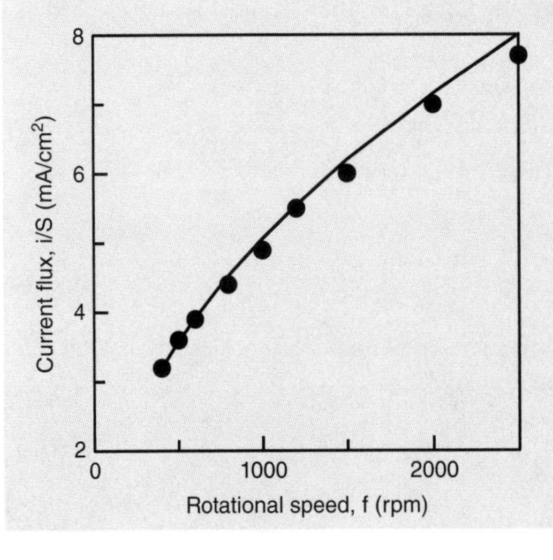

Figure 16.3-3 Limiting current from a rotating disk polarographic electrode (Data from Garcia *et al.* (1997)).

Taking the kinematic viscosity for the electrolyte solution to be $\upsilon = 10^{-2}\,cm^2/s$ (the same as that of water at 20°C), we express the current flux as

$$\frac{i}{S}(mA/cm^2) = \frac{9.65 \times 10^7}{1.61}\left(\frac{10^{-2}}{7.5 \times 10^{-6}}\right)^{-2/3}\left(\frac{0.01}{1000}\right)\sqrt{10^{-2}\left(\frac{2\pi}{60}f[rpm]\right)} = 0.160\sqrt{f[rpm]} \tag{16.3-23}$$

where $f[rpm] = (60/2\pi)\Omega[s^{-1}]$ is the rotational speed in revolutions per minute. This predicted relationship between i/S and f, which is shown by the curve in Figure 16.3-3, is in good agreement with the data points.

16.3.2 Solute Transport with Surface Reaction in a Blood Vessel

The transport of solutes that react on the walls of blood vessels is an important problem in physiology and medicine. For example, the activation of factor X to factor Xa in the arterial wall is a key step in blood coagulation. Another example is the slow elution of a therapeutic drug from a vascular stent implanted on the inner surface of an artery. Although the convection–diffusion equations that describe such processes are essentially the same, the kinetics associated with the surface reactions can be quite different.

Here, we will model steady-state transport of factor X in blood and its activation by a membrane-bound enzyme in the smooth muscle cells of the arterial wall. We assume that an artery is a circular cylinder, blood flows according to the Hagen–Poiseuille equation, and factor X activation follows Michaelis–Menten kinetics (Fig. 16.3-4).

Model Formulation

For an axisymmetric, steady-state concentration distribution $C_x(z,r)$ in a fluid of constant density and diffusion coefficient, the steady-state convection–diffusion equation in cylindrical coordinates (Table 15.2-1) reduces to

$$u_r\frac{\partial C_x}{\partial r} + u_z\frac{\partial C_x}{\partial z} = \mathcal{D}_x\left[\frac{1}{r}\frac{\partial}{\partial r}\left(r\frac{\partial C_x}{\partial r}\right) + \frac{\partial^2 C_x}{\partial z^2}\right] \quad (a > r > 0,\ z > 0) \tag{16.3-24}$$

where "a" is the vessel radius. In Hagen–Poiseuille flow at an average velocity u_{ave} through a tube, $u_z = 2u_{ave}[1 - (r/a)^2]$ and $u_r = 0$. Also, when the radius of the blood vessel is small

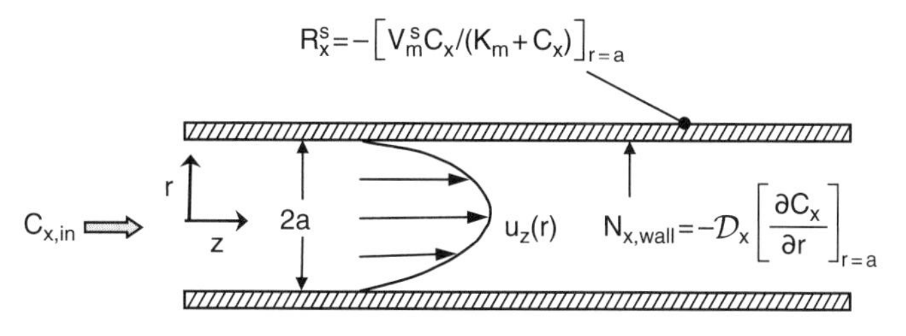

Figure 16.3-4 Transport of factor X in a blood vessel.

compared to its length, diffusion in the radial direction dominates axial diffusion. Equation 16.3-24 then becomes

$$u_{ave}\left[1-\left(\frac{r}{a}\right)^2\right]\frac{\partial C_x}{\partial z}=\mathcal{D}_x\left[\frac{1}{r}\frac{\partial}{\partial r}\left(r\frac{\partial C_x}{\partial r}\right)\right] \quad (a>r>0,\ z>0) \tag{16.3-25}$$

The solution to this equation requires the specification of one boundary condition on z and two more on r. For factor X supplied at a constant concentration in blood, we have the entrance condition

$$z=0:\quad C_x=C_{x,in} \tag{16.3-26}$$

Because the concentration field is axisymmetric, the condition at the center of the vessel is

$$r=0:\quad \frac{\partial C_x}{\partial r}=0 \tag{16.3-27}$$

Provided that factor X does not accumulate at the artery wall, its influx by diffusion, $N_{x,wall}=-\mathcal{D}_s[\partial C_x/\partial r]_{r=a}$, must equal its depletion rate per unit area, $-R_x^S=V_m^S[C_x/(K_m+C_x)]_{r=a}$. This leads to the following condition at the wall:

$$r=a:\quad \mathcal{D}_s\frac{\partial C_x}{\partial r}=\frac{V_m^S C_x}{K_m+C_x} \tag{16.3-28}$$

Analysis

Dimensionless Forms

Introducing the dimensionless variables

$$C=\frac{C_x}{C_{x,in}},\quad z=\frac{\mathcal{D}_x}{a^2u_{ave}}z,\quad r=\frac{r}{a} \tag{16.3-29a-c}$$

and the dimensionless parameters

$$Da^S\equiv\frac{V_m^S a}{\mathcal{D}_s C_{x,in}},\quad K\equiv\frac{K_m}{C_{x,in}} \tag{16.3-30a,b}$$

the dimensionless governing equation is

$$\left(1-r^2\right)\frac{\partial C}{\partial z}=\left[\frac{1}{r}\frac{\partial}{\partial r}\left(r\frac{\partial C}{\partial r}\right)\right] \quad (1>r>0,\ z>0) \tag{16.3-31}$$

and the dimensionless boundary conditions are

$$z=0:\ C=1,\quad r=0:\ \frac{\partial C}{\partial r}=0,\quad r=1:\ \frac{\partial C}{\partial r}=\frac{Da^S C}{K+C} \tag{16.3-32a-c}$$

Simulation

Figure 16.3-5 shows the dimensionless concentration distributions of factor X obtained by numerical solution of these equations with a fixed value of $K=0.01$. The left graph indicates that at any axial position z, C decreases between the tube centerline at $r=0$ and the tube wall at $r=1$. As z increases, this radial concentration distribution flattens. This occurs because an increasing amount of factor X is depleted from the bulk fluid as blood moves along the tube. The right graph reinforces this conclusion by showing that there is a decrease in centerline concentration as z increases, a trend that is amplified when Da^S increases.

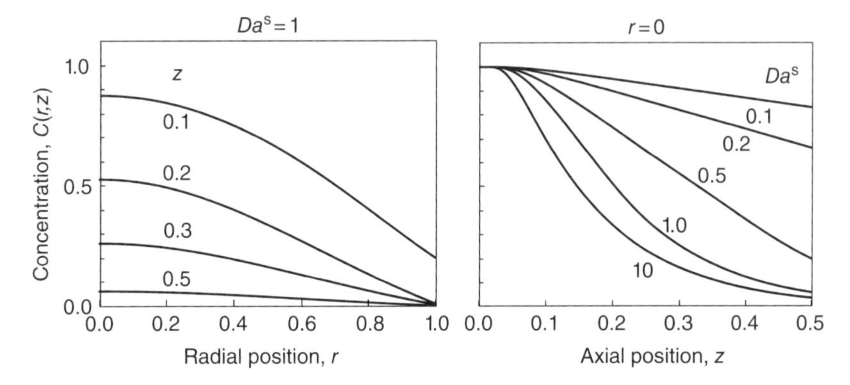

Figure 16.3-5 Distribution of factor X activated by reaction at a blood vessel surface.

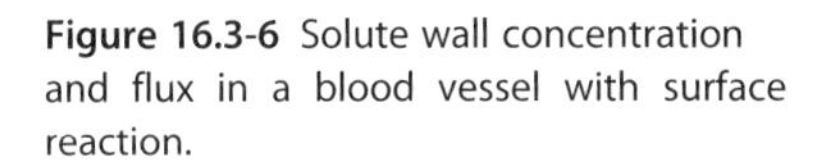

Figure 16.3-6 Solute wall concentration and flux in a blood vessel with surface reaction.

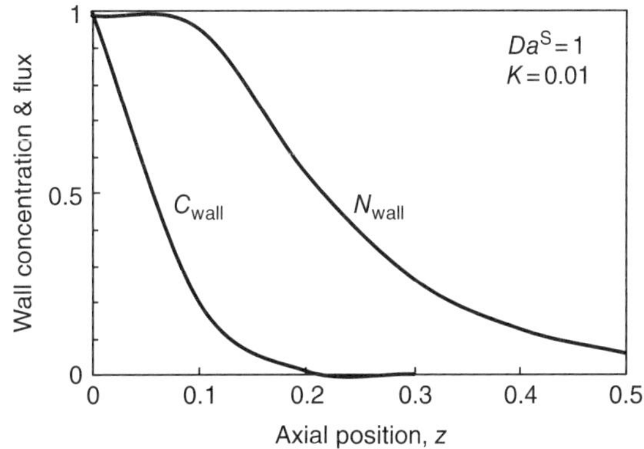

A dimensionless flux toward the blood vessel wall can be obtained from the dimensionless wall concentration $C_{\text{wall}} \equiv [C]_{r=1}$ as follows:

$$N_{\text{wall}} \equiv \left(\frac{\mathrm{a}}{\mathcal{D}_{\mathrm{x}} \mathrm{C}_{\mathrm{x,in}}}\right) \mathrm{N}_{\mathrm{x,wall}} = -\left(\frac{\mathrm{a}}{\mathcal{D}_{\mathrm{x}} \mathrm{C}_{\mathrm{x,in}}}\right) \mathrm{R}_{\mathrm{x}}^{\mathrm{S}} = \frac{Da\, C_{\text{wall}}}{K + C_{\text{wall}}} \tag{16.3-33}$$

Figure 16.3-6 shows the dimensionless wall concentration and diffusion flux of factor X toward the artery wall obtained by numerical simulation. As z increases, C_{wall} and therefore N_{wall} decreases from a scaled value of one at $z = 0$ toward a lower limit of zero as $z \to \infty$.

References

Garcia AJ, Ducheyne P, Boettiger D. Quantification of cell adhesion using a spinning disk device and application to surface-reactive materials. Biomaterials. 1997; 18:1091–1098.

Konopka SJ, McDuffle B. Diffusion coefficients of ferri and ferrocyanide ions in aqueous media, using twin-electrode thin-layer electrochemistry. Anal Chem. 1970; 42:1741–1746.

Miller FJ, Overton, JH, Jaskot RH, Menzel DB. A model of the regional uptake of gaseous pollutants in the lung. Toxicol Appl Pharmacol. 1985; 79:11–27.

Qin Y, Cabral JMS. Kinetic studies of the urease-catalyzed hydrolysis of urea in a buffer-free system. Appl Biochem Biotechnol. 1994; 49:217–240.

Chapter 17

Cell Population Dynamics

The dynamics of cell populations depend on metabolic activity and can change in space as a consequence of fate and transport processes. Fate processes include proliferation, death, and differentiation. Common transport processes involve random motility and taxis. These processes depend on local concentrations of soluble substances such as growth factors as well as insoluble substances such as extracellular matrix proteins. In addition, the transport of nutrients and waste products is coupled to the metabolic activity of cells. Cell type is a major determinant of metabolic activity. For example, a hepatocyte in the liver consumes oxygen at a much higher rate than a chondrocyte in cartilage.

In this chapter, we describe how number balances of single and multiple cell phenotypes are used in conjunction with mass balances on metabolites and chemical agents to analyze the dynamics of cell populations. These balances are based on continuum approximations of the transport and fate processes. The models presented here describe cell populations that vary in space as well as time. They do not, however, account for forced convection or for different metabolic states and sizes within a particular cell type.

17.1 Cell Number Balances

As a continuum approximation, cell populations are considered to be a spatial ensemble of material elements, each consisting of a cell group and its surrounding extracellular fluid. The number of any cell type i per unit volume is represented as a cell number density Ω_i, which is a continuous function of time and position. Just as a chemical species balance leads to a molar concentration equation (Eq. 15.1-7), a cell number balance leads to a number density equation:

$$\frac{\partial \Omega_i}{\partial t} + \nabla \cdot \Theta_i = \Xi_i \qquad (17.1\text{-}1)$$

Biomedical Mass Transport and Chemical Reaction: Physicochemical Principles and Mathematical Modeling, First Edition. James S. Ultman, Harihara Baskaran, and Gerald M. Saidel.

Here, Θ_i is the cell transport flux vector and Ξ_i is the cell formation rate due to fate processes. Whereas Ω_i is analogous to the molar concentration C_i of a chemical species, Θ_i and Ξ_i are analogous to the transport flux $\mathbf{N}_i$ and production rate R_i, respectively. If changes in cell density occur along a single Cartesian coordinate z, then the cell density equation reduces to

$$\frac{\partial \Omega_i}{\partial t} + \frac{\partial \Theta_i}{\partial z} = \Xi_i \qquad (17.1\text{-}2)$$

where Θ_i is the transport flux in the z direction. For a cell population in a closed, well-mixed system of constant volume, the spatial derivatives are zero and the cell number density equation reduces even more to

$$\frac{d\Omega_i}{dt} = \Xi_i \qquad (17.1\text{-}3)$$

17.2 Cell Transport and Fate Processes

17.2.1 Cell Movement

Cell migration is important in several physiological and pathological processes. During embryogenesis, for example, unspecialized cells migrate and differentiate into specialized cells to form distinct features of an organism. In metastasis, cancer cells migrate from a primary tumor to remote locations where secondary tumors develop. *In vitro*, some cells migrate along a solid surface by the protrusion and attachment of actin-rich lamellipodia that provide a traction force.

Cells move by random motility when their number density varies in space. Cells can also move by directed migration or taxis along a concentration gradient of either an insoluble factor such as an adhesion site (i.e., haptotaxis) or a soluble chemoattractant such as a cytokine (i.e., chemotaxis). For example, axons exhibit haptotaxis in the presence of a gradient of laminin, an extracellular matrix protein. As part of the inflammatory response to tissue injury, chemotaxis of neutrophils occurs against a gradient of interleukin-8.

Analogous to molecular diffusion, the cell flux due to random motility is proportional to the gradient of the cell number density:

$$\Theta_i^{rand} = -\mu_i^{rand} \nabla \Omega_i \qquad (17.2\text{-}1)$$

where μ_i^{rand}, the random motility coefficient, is a function of temperature and may depend on the concentration of biological agents such as chemokines. As is true for diffusion, the negative sign indicates that Θ_i^{rand} is positive in the direction of decreasing cell number density.

Directed cell migration by either chemotaxis or haptotaxis is represented by the following flux equation:

$$\Theta_i^{taxis} = \Omega_i \left(\mu_i^{taxis} - \frac{1}{2}\frac{d\mu_i^{rand}}{dC_A} \right) \nabla C_A \qquad (17.2\text{-}2)$$

Here, C_A is the molar concentration of attractant A, and μ_i^{taxis} is the taxis coefficient that can depend on C_A. The first term on the right side of this equation is the primary source of taxis, while the second term, known as secondary taxis or kinesis, only occurs when μ_i^{rand} depends on C_A. While primary taxis always occurs in the direction of increasing attractant concentration, kinesis would occur in the opposite direction if $d\mu_i/dC_A$ were positive. Because the adhesion sites responsible for haptotaxis are immobile, C_A is time invariant as are taxis and kinesis.

17.2.2 Cell Division and Proliferation

A distinct feature of cells is their ability to divide and form genetically identical offspring by mitosis. Mitotic division is an important contributor to a cell population balance when the time scale of interest is large compared to the time scale for mitosis. Whereas cell division by mitosis takes 12 h or more for animal (eukaryotic) cells, it only requires 30 min for bacterial (procaryotic) cells when they proliferate in culture. A simple model of cell proliferation assumes first-order kinetics:

$$\Xi_i^{prolif} = k_i^{prolif}\Omega_i \tag{17.2-3}$$

The specific growth rate k_i^{prolif} primarily depends on nutrient concentration, but can also be affected by growth and inhibitory factors. Frequently, cell proliferation exhibits saturation kinetics with respect to a rate-limiting nutrient identified as substrate S. This may be modeled by a Monod equation, which is analogous to Michaelis–Menten kinetics:

$$k_i^{prolif} = \frac{V_i^{prolif}C_S}{K_i^{prolif} + C_S} \tag{17.2-4}$$

where V_i^{prolif} is the maximum value of k_i^{prolif} and K_i^{prolif} is the substrate concentration at which $k_i^{prolif} = V_i^{prolif}/2$. Even when nutrient is readily available, the proliferation rate of mammalian organisms exhibits an upper bound because of cell crowding. For such contact inhibition, the specific proliferation rate can be expressed as

$$k_i^{prolif} = \varsigma_i^{prolif}\left(1 - \frac{\Omega_i}{\Omega_i^{cc}}\right) \tag{17.2-5}$$

In this model, it is presumed that the carrying capacity Ω_i^{cc} is greater than Ω_i so that cells proliferate until Ω_i is equal to Ω_i^{cc}. For example, if cells are initially seeded with $\Omega_i << \Omega_i^{cc}$ on an adherent surface, then $k_i^{prolif} \approx \varsigma_i^{prolif}$ and a monolayer will begin to develop by cell proliferation. As Ω_i increases, k_i^{prolif} decreases and cell propagation will continue at a decreasing rate until $\Omega_i = \Omega_i^{cc}$ at all points on the surface. The monolayer is then said to be confluent.

Example 17.2-1 Development of a Cell Monolayer with Contact Inhibition

Human endothelial cells (EC) are grown in a monolayer on the bottom surface of a Petri dish containing a nutrient medium. The surface is initially seeded with a uniform cell distribution at a number density Ω_o. Thereafter, sufficient time is allowed for the cells to proliferate to confluency. Because the cells tightly adhere to the surface, cell transport from the dish surface to the adjacent nutrient medium or along the surface is negligible. Using the contact-inhibition model for EC, we can determine the cell density dynamics.

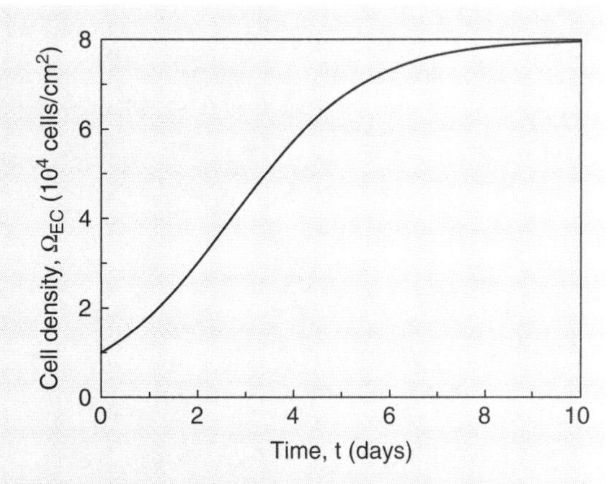

Figure 17.2-1 Cell monolayer dynamics with contact inhibition.

Solution

In the absence of cell transport, we describe the dynamics of Ω_{EC}, the number of endothelium per unit surface area, by combining Equations 17.1-3, 17.2-3, and 17.2-5:

$$\frac{d\Omega_{EC}}{dt} = \varsigma_{EC}^{prolif}\Omega_{EC}\left(1 - \frac{\Omega_{EC}}{\Omega_{EC}^{cc}}\right) \tag{17.2-6}$$

After separation of variables, this equation can be integrated to get the logistic equation:

$$\Omega_{EC} = \frac{\Omega_o \exp\left(\varsigma_{EC}^{prolif} t\right)}{1 - \left(\Omega_o/\Omega_{EC}^{cc}\right)\left[1 - \exp\left(\varsigma_{EC}^{prolif} t\right)\right]} \tag{17.2-7}$$

where Ω_o is the initial cell density. The cell number increases when $\Omega_o < \Omega_{EC}^{cc}$ or decreases when $\Omega_o > \Omega_{EC}^{cc}$. Assuming $\Omega_{EC}^{cc} = 8 \times 10^4$ cells/cm^2, $\Omega_o = 10^4$ cells/cm^2, and $\varsigma_{EC}^{prolif} = 0.03\,h^{-1} = 0.72$ day, Equation 17.2-7 generates the proliferation dynamics shown in Figure 17.2-1.

Example 17.2-2 Nutrient-Limited Cell Proliferation

Commonly, the rate-limiting step in the anaerobic fermentation of sucrose by yeast cells (Y) is the enzymatic hydrolysis of sucrose (S) to form glucose and fructose. This is followed by the partial oxidation of the glucose and fructose to ethanol together with the proliferation of yeast cells. The stoichiometric ratio of the rates of cell proliferation Ξ_Y^{prolif} and sucrose utilization per unit volume $-R_S$ is the yield coefficient which is often a constant:

$$\eta_Y \equiv -\frac{\Xi_Y^{prolif}}{R_S} \tag{17.2-8}$$

Consider a well-mixed batch reactor containing a constant suspension volume with an initial number concentration Ω_o of yeast cells and an initial molar concentration C_o of sucrose. With dynamic models for cell proliferation and substrate utilization, we can characterize the system behavior.

Solution

With sucrose acting as a rate-limiting nutrient that is hydrolyzed by Michaelis–Menten kinetics, a population balance on yeast cells Y and a species balance on sucrose S lead to

$$\frac{d\Omega_Y}{dt} = \Xi_Y^{prolif} = \left(\frac{V_Y^{prolif} C_S}{K_Y^{prolif} + C_S}\right)\Omega_Y \tag{17.2-9}$$

$$\frac{dC_S}{dt} = R_S = -\frac{\Xi_Y^{prolif}}{\eta_Y} = -\frac{1}{\eta_Y}\left(\frac{V_Y^{prolif} C_S}{K_Y^{prolif} + C_S}\right)\Omega_Y \tag{17.2-10}$$

The initial conditions are

$$t = 0: \quad \Omega_Y = \Omega_o, \quad C_S = C_o \tag{17.2-11}$$

After substituting the dimensionless variables:

$$\mathit{\Omega} \equiv \frac{\Omega_Y}{\Omega_o}, \quad C \equiv \frac{C_S}{C_o}, \quad t \equiv V_Y^{prolif} t \tag{17.2-12a-c}$$

the governing equations and their initial conditions have the forms:

$$\frac{d\mathit{\Omega}}{dt} = \left(\frac{C}{K + C}\right)\mathit{\Omega}$$

$$\frac{dC}{dt} = -\frac{1}{\eta}\left(\frac{C}{K + C}\right)\mathit{\Omega} \tag{17.2-13a-c}$$

$$t = 0: \quad \mathit{\Omega} = 1, \quad C = 1$$

The dimensionless parameters are defined as:

$$\eta \equiv \frac{C_o}{\Omega_o}\eta_Y, \quad K \equiv \frac{K_Y^{prolif}}{C_o} \tag{17.2-14a,b}$$

Taking the ratio of the balance equations and integrating subject to the condition that $\mathit{\Omega} = 1$ when $C = 1$, we obtain

$$\frac{d\mathit{\Omega}}{dC} = -\eta \;\Rightarrow\; \mathit{\Omega} = (1 + \eta) - \eta C \tag{17.2-15a,b}$$

Substituting this relation into the sucrose concentration equation,

$$\frac{dC}{dt} = \frac{C[C - (1 + 1/\eta)]}{K + C} \tag{17.2-16}$$

After separation of variables, we integrate this equation with the initial condition that $C(0) = 1$:

$$t = \frac{1}{1 + \eta}[(1 + \eta + K\eta)\ln(1 + \eta - \eta C) - K\eta \ln C] \tag{17.2-17}$$

As an alternative to evaluating this implicit equation for $C(t)$, we could solve the differential equations for $\mathit{\Omega}(t)$ and $C(t)$ numerically. In either case, model simulations for several values of K and η are shown in Figure 17.2-2. In the left graph, the curves for cell number density are characteristic of experimental data (not shown), which display a lag phase over a short time, an exponential phase over an intermediate period, and a stationary phase after a sufficiently long time. An increase in K at fixed η prolongs the exponential phase, whereas an increase in η at fixed K increases the ultimate cell density in the stationary phase. In the right graph, each

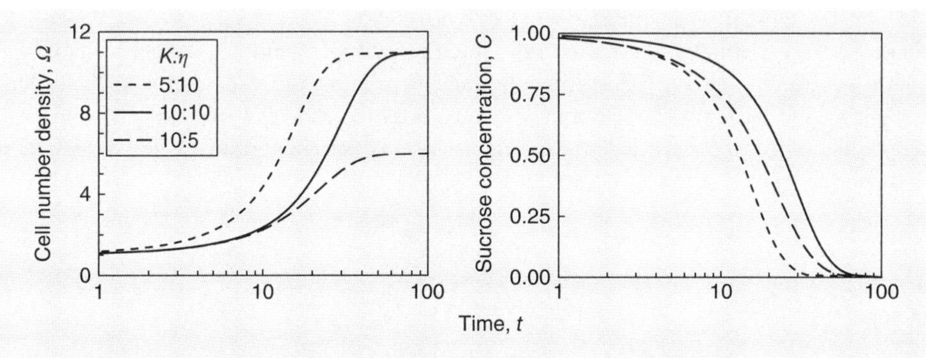

Figure 17.2-2 Dynamics of nutrient-limited cell proliferation.

substrate utilization curve reveals a progressive depletion of sucrose, whose concentration reaches zero when cell number density reaches its stationary phase. Either an increase in K at fixed η or an increase in η at fixed K slows the sucrose depletion rate.

17.2.3 Cell Death

Under homeostatic conditions, a stable cell density is reached by a balance between cell proliferation and cell death. The programmed death of targeted cells without damage to nearby cells and tissues is called apoptosis. For example, most nucleated animal cells secrete biochemical factors, allowing them as well as other nearby cells to proliferate. A controlled reduction of these survival factors below a critical level triggers the production of proteolytic enzymes that stimulate the breakdown of DNA and structural cell proteins.

Cell death can also occur as a result of injury, infection, lack of nutrients, or presence of intolerable levels of metabolic waste products. Such abnormal cell death is known as necrosis. Unlike apoptosis, necrosis is manifest by cell swelling and subsequent bursting. Necrosis is typically followed by inflammatory processes that can damage nearby cells and tissues.

Whether it occurs by apoptosis or necrosis, a simple relationship to describe cell death is

$$\Xi_i^{death} = -k_i^{death}\Omega_i \tag{17.2-18}$$

The specific death rate k_i^{death} is a parameter that depends on the levels of nutrients and metabolic waste products. This parameter can also depend on densities of interacting cells of other types.

17.2.4 Cell Differentiation

Differentiation is a process in which an unspecialized cell becomes a specialized cell. Cell differentiation is important in embryogenesis, growth, and maintenance of an organism. Differentiation is usually denoted by morphological or phenotypical changes in the cell. The unspecialized cell is sometimes referred to as the progenitor cell. Whereas most differentiation processes are irreversible, a few specialized cells can dedifferentiate into their progenitor cells.

Differentiation involves a minimum of two types of cells. To model this process, we need to make a population balance for each cell type. For example, consider the simple situation where a progenitor type 1 cell irreversibly transforms into a differentiated type 2 cell. This is analogous to a chemical reaction in which there is one reactant 1 that isomerizes to form one product 2. A simple expression that can be used to describe the rate of cell differentiation is then

$$\Xi_1^{diff} = -\Xi_2^{diff} = -k_{12}^{diff}\Omega_1 \tag{17.2-19}$$

The specific rate of differentiation where k_{12}^{diff} can depend on levels of nutrients and growth factors.

17.3 Single Cell Population Dynamics

In the previous section, we modeled cell dynamics in spatially uniform systems. In this section, we illustrate the dynamics of single cell populations in which cell transport and fate processes give rise to spatial changes of cell number density.

17.3.1 Axon Growth by Haptotaxis

Enhancing the growth of axons is a potential treatment for neuronal wound healing as needed to treat spinal cord injuries. This enhancement can be achieved by cell signaling with a substance such as laminin, an extracellular matrix protein that attracts and binds axonal receptors. When immobilized on a solid substrate, a gradient of laminin can induce haptotaxis, which directs the extension of axons from a neural cell body.

Model Formulation

We model the axial migration of axons (ax) through a tube of length L and cross-sectional area A. The tube contains a linear distribution of laminin (lam) concentration immobilized by an agarose gel (Fig. 17.3-1). At the proximal end of the tube where the laminin concentration is lowest, a uniform distribution of neuronal cell bodies is anchored in an arbitrarily small axial region from $z = 0$ to $z_o \ll L$. Axon tips migrate along the tube from the cell bodies

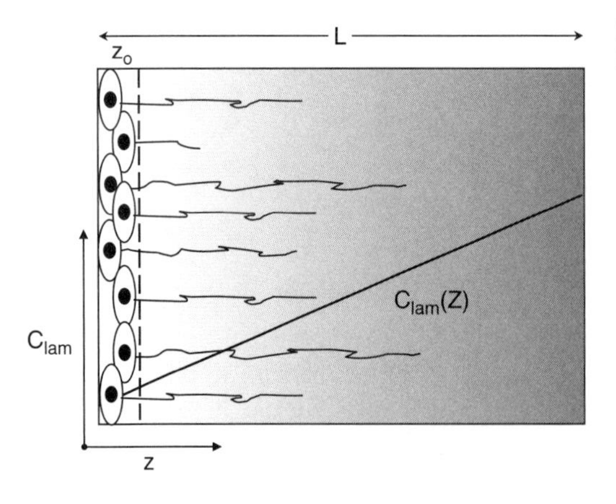

Figure 17.3-1 Axon tip migration with haptotaxis.

toward the highest laminin concentration at the distal end of the tube. At that point, the axon tips become anchored to a targeted tissue or a synthetic substrate and they cease to be active.

Neuronal cells typically do not undergo mitosis, and for experiments over a sufficiently short time, we can ignore cell death. Therefore, we assume that the number of axons does not change. Nevertheless, the number density of active axon tips Ω_{ax} that appear at a particular axial position z changes because of random motility Θ_{ax}^{rand} and haptotaxis Θ_{ax}^{taxis} along the tube. These dynamic changes occurring when $\Xi_i = 0$ can be found by applying Equation 17.1-2:

$$\frac{\partial \Omega_{ax}}{\partial t} + \frac{\partial\left(\Theta_{ax}^{rand} + \Theta_{ax}^{taxis}\right)}{\partial z} = 0 \quad (0 < z < L) \tag{17.3-1}$$

The migration flux is related to the axon number density according to

$$\Theta_{ax}^{rand} + \Theta_{ax}^{taxis} = -\mu_{ax}^{rand}\frac{\partial \Omega_{ax}}{\partial z} + \Omega_{ax}\left(\mu_{ax}^{taxis} - \frac{1}{2}\frac{d\mu_{ax}^{rand}}{dC_A}\right)\frac{\partial C_{lam}}{\partial z} \tag{17.3-2}$$

With a linear laminin concentration distribution, dC_{lam}/dz is constant. Assuming that μ_{ax}^{rand} and μ_{ax}^{taxis} are also constant, we combine these equations to obtain

$$\frac{\partial \Omega_{ax}}{\partial t} + \left(\mu_{ax}^{taxis}\frac{dC_{lam}}{dz}\right)\frac{\partial \Omega_{ax}}{\partial z} - \mu_{ax}^{rand}\frac{\partial^2 \Omega_{ax}}{\partial z^2} = 0 \quad (0 < z < L) \tag{17.3-3}$$

The initial distribution of axon tips corresponds to the distribution of the anchored cell bodies:

$$t = 0: \quad \Omega_{ax} = \begin{cases} \Omega_o & \text{when } z_o > z \geq 0 \\ 0 & \text{when } L > z \geq z_o \end{cases} \tag{17.3-4}$$

At the proximal boundary, where they are attached to the cell bodies, the axon tips cannot leave the tube so that their flux is zero:

$$z = 0: \quad \mu_i^{rand}\frac{\partial \Omega_{ax}}{\partial z} - \left(\mu_{ax}^{taxis}\frac{dC_{lam}}{dz}\right)\Omega_{ax} = 0 \tag{17.3-5}$$

At the distal boundary z = L, where the axon tips become attached and are no longer active, we set their concentration equal to zero.

$$z = L: \quad \Omega_{ax} = 0 \tag{17.3-6}$$

The effectiveness of axon migration can be characterized by the cumulative number of axon tips that have migrated across the entire length of the tube at any time. This is the difference between the initial number of axon tips and the number remaining in the tube:

$$\Gamma_{ax}^{cumul}(t) = z_o A \Omega_o - \int_0^L \Omega_{ax}(z,t) A\,dz \tag{17.3-7}$$

Analysis

Dimensionless Forms

When the dimensionless quantities

$$\mathit{\Omega} = \frac{\Omega_{ax}}{\Omega_o}, \quad \mathit{z} = \frac{z}{L}, \quad \mathit{t} = \frac{\mu_{ax}^{rand}}{L^2}t \tag{17.3-8a-c}$$

are substituted into the model equations, we obtain

$$\frac{\partial \Omega}{\partial t} + Pe_{ax}\frac{\partial \Omega}{\partial z} = \frac{\partial^2 \Omega}{\partial z^2} \tag{17.3-9}$$

The Peclet number Pe_{ax} indicates the relative importance of haptotaxis to random motility:

$$Pe_{ax} \equiv \left(\mu_{ax}^{taxis}\frac{dC_{lam}}{dz}\right) \Big/ \left(\frac{\mu_{ax}^{rand}}{L}\right) \tag{17.3-10}$$

The dimensionless initial and boundary conditions are

$$\begin{aligned} t=0:\ & \Omega = \begin{cases} 1 & \text{when } z_o/L > z \geq 0 \\ 0 & \text{when } 1 > z \geq z_o/L \end{cases} \\ z=0:\ & \frac{\partial \Omega}{\partial z} - Pe_{ax}\Omega = 0 \\ z=1:\ & \Omega = 0 \end{aligned} \tag{17.3-11a-c}$$

The fraction of the initial axon tips that have migrated across the tube is

$$\Gamma^{cumul}(t) \equiv \frac{\Gamma_{ax}^{cumul}}{z_o A \Omega_o} = 1 - \frac{L}{z_o}\int_0^1 \Omega(z,t)dz \tag{17.3-12}$$

Simulation

By numerical solution of the dimensionless model equations with $z_o/L = 0.01$ and different values of Pe_{ax}, we simulate the effect of haptotaxis on the distribution of axon tips (Fig. 17.3-2). In the left graph, the number density of axon tips remaining in the tube exhibit a progressively diminishing peak concentration and increased spatial broadening as time progresses. The right graph indicates that axon tips reach the distal boundary sooner as Pe_{ax} increases, that is, when there is greater haptotaxis relative to random migration. Eventually, all the axon tips adhere at the boundary so that $\Gamma^{cumul} = 1$ as $t \rightarrow \infty$.

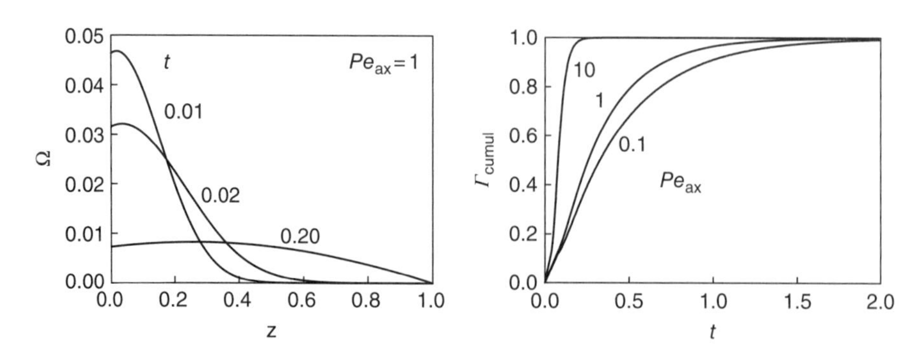

Figure 17.3-2 Axon tip distribution dynamics.

17.3.2 Endothelial Cell Migration

Endothelial cell migration is important in wound healing and tissue engineering of vascular grafts. A fence assay can be used for studying the effect of different material surfaces and chemical factors on cell motility and two-dimensional growth on a surface (Fig. 17.3-3). Endothelial cells (EC) are initially seeded within a circular fence of radius "a," typically made of a nonadherent polymer such as Teflon. The fence is subsequently removed and cells are allowed to migrate over the entire surface.

Model Formulation

In this model, the dominant processes are random motility and proliferation. Under these conditions, changes in cell number density are described by

$$\frac{\partial \Omega_{EC}}{\partial t} + \nabla \cdot \Theta_{EC}^{rand} = \Xi_{EC}^{prolif} \tag{17.3-13}$$

Substituting the random motility flux and assuming proliferation is limited by contact inhibition, we obtain

$$\frac{\partial \Omega_{EC}}{\partial t} - \mu_{EC}^{rand} \nabla^2 \Omega_{EC} = \varsigma_{EC}^{prolif} \Omega_{EC} \left(1 - \frac{\Omega_{EC}}{\Omega_{EC}^{cc}}\right) \tag{17.3-14}$$

where μ_{EC}^{rand} is constant. Further assuming isotropic migration from its initial circular boundary, cell density remains circularly symmetric and varies only with radial position and time. In cylindrical coordinates, Equation 17.3-14 becomes

$$\frac{\partial \Omega_{EC}}{\partial t} = \mu_{EC}^{rand} \frac{1}{r} \frac{\partial}{\partial r}\left(r \frac{\partial \Omega_{EC}}{\partial r}\right) + \varsigma_{EC}^{prolif} \Omega_{EC} \left(1 - \frac{\Omega_{EC}}{\Omega_{EC}^{cc}}\right) \tag{17.3-15}$$

The initial and boundary conditions are

$$\begin{aligned}
t = 0: \quad & \Omega_{EC} = \begin{cases} \Omega_o & \text{when } a \geq r \geq 0 \\ 0 & \text{when } r > a \end{cases} \\
r = 0: \quad & \frac{\partial \Omega_{EC}}{\partial r} = 0 \\
r \to \infty: \quad & \Omega_{EC} = 0
\end{aligned} \tag{17.3-16a-c}$$

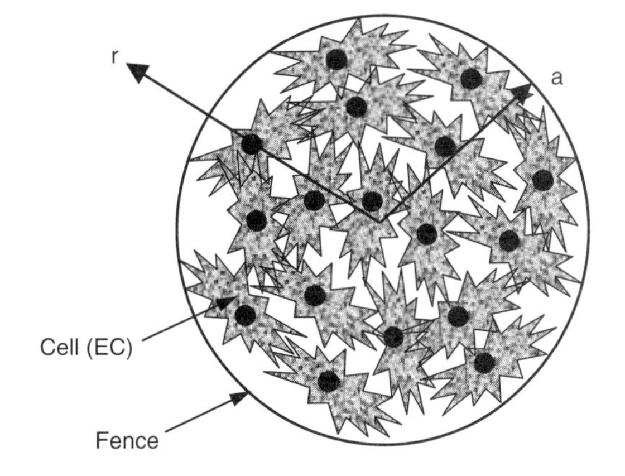

Figure 17.3-3 Endothelial cells on a surface within a circular fence.

Initially, the cell number density is a positive constant Ω_o within the fence and is zero outside the fence. By radial symmetry, the gradient at the center must be zero. If the time of an experiment is short enough, cells never reach the end of the surface so that their number density remains zero far from the center.

Analysis

Dimensionless Forms

In terms of the dimensionless variables

$$\mathit{\Omega} = \frac{\Omega_{EC}}{\Omega_o}, \quad r = \frac{\mathrm{r}}{\mathrm{a}}, \quad t = \left(\frac{\mu_{EC}^{rand}}{\mathrm{a}^2}\right)\mathrm{t} \tag{17.3-17a-c}$$

Equation 17.3-15 becomes

$$\frac{\partial \mathit{\Omega}}{\partial t} - \frac{1}{r}\frac{\partial}{\partial r}\left(r\frac{\partial \mathit{\Omega}}{\partial r}\right) = Da_{EC}\mathit{\Omega}\left(1 - \frac{\mathit{\Omega}}{\mathit{\Omega}^{cc}}\right) \tag{17.3-18}$$

where

$$Da_{EC} \equiv \frac{\mathrm{a}^2 \varsigma_{EC}^{prolif}}{\mu_{EC}^{rand}}, \quad \mathit{\Omega}^{cc} = \frac{\Omega_{EC}^{cc}}{\Omega_o} \tag{17.3-19a,b}$$

Here, Da_{EC} is a Damkohler number, the ratio of the characteristic proliferation rate to the transport rate by random motility. The parameter $\mathit{\Omega}^{cc}$ is the carrying capacity as a multiple of the density of seeded cells. If the seeded cells are initially grown to confluence within the fence, then the Ω_o will correspond to Ω_{EC}^{cc} so that $\mathit{\Omega}^{cc} = 1$ and Equation 17.3-18 reduces to

$$\frac{\partial \mathit{\Omega}}{\partial t} - \frac{1}{r}\frac{\partial}{\partial r}\left(r\frac{\partial \mathit{\Omega}}{\partial r}\right) = Da_{EC}\left(\mathit{\Omega} - \mathit{\Omega}^2\right) \tag{17.3-20}$$

The dimensionless initial and boundary conditions are

$$t = 0: \quad \mathit{\Omega} = \begin{cases} 1 & \text{when } 1 \geq r \geq 0 \\ 0 & \text{when } r > 1 \end{cases}$$

$$r = 0: \quad \frac{\partial \mathit{\Omega}}{\partial r} = 0 \tag{17.3-21a-c}$$

$$r = \infty: \quad \mathit{\Omega} = 0$$

Simulation

We solve this problem (Eqs. 17.3-20 and 17.3-21a–c) numerically after specifying that Da_{EC} is either 0.1 or 10 (Fig. 17.3-4). At both Da_{EC} values, cell surface density at any time continuously decreases from a maximum value at the center of the cell layer at $r = 0$ to zero as $r \rightarrow \infty$. When $Da_{EC} = 0.1$, cell density decreases with time at all $r < 1.5$. This is due to spreading of the initially seeded cells by random motility. Because of the opposite effect of cell proliferation, cell density slightly increases with time at $r > 1.5$. At the much larger $Da_{EC} = 10$, cell surface density increases substantially with dimensionless time throughout the cell layer. Thus, for most wound healing applications, a high proliferation rate relative to the random motility rate is desirable.

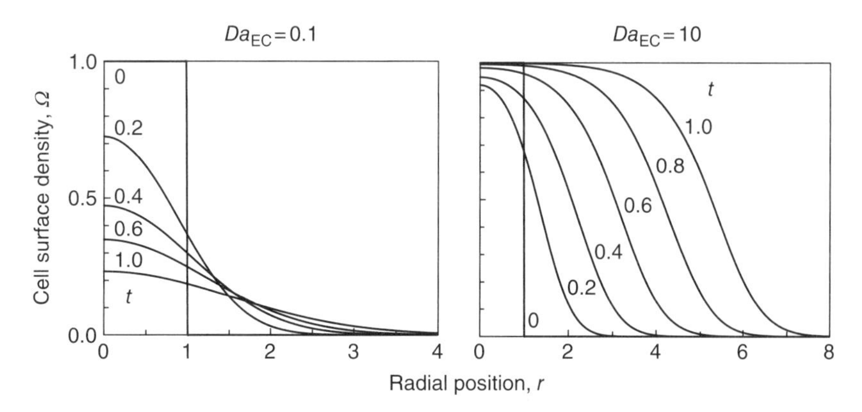

Figure 17.3-4 Endothelial cell migration in the fence assay.

17.4 Multiple Cell Population Dynamics

More general models of cell population dynamics include multiple cell types whose number densities vary spatially as determined by various transport and fate processes as well as by concentrations of chemical species.

17.4.1 Tumor Vascularization and Growth

In this illustration, based on the work of Liotta *et al.* (1977), we model dynamic changes of tumor cells implanted in normal tissue. Tumor cell proliferation depends on the local availability of nutrients supplied by surrounding capillaries. Another key contributor to growth is tumor vascularization that involves the proliferation and migration of blood vessels toward the region of highest tumor density. This process is induced by tumor angiogenic factor (TAF) that is supplied by tumor cells as they proliferate.

Model Formulation

The model consists of a spherically symmetric system of tumor cells and capillary cells (Fig. 17.4-1). The number densities of the tumor and capillary cells change with time and location in and around the tumor. This type of model is intended to aid in the clinical prediction of solid tumor growth and in the design of chemotherapy regimens. Tumor cells are considered as a single cell type (i = T) and capillary cells as a second cell type (i = C). Both cell types undergo proliferation, death, and migration by random motility. Following Equation 17.1-1, the number density dynamics of the two cell types are represented as:

$$\frac{\partial \Omega_i}{\partial t} + \nabla \cdot \Theta_i^{rand} = \Xi_i^{prolif} + \Xi_i^{death} \quad (i = T, C) \tag{17.4-1}$$

With the rate relationships in Equations 17.2-1, 17.2-3, and 17.2-18, this equation becomes

$$\frac{\partial \Omega_i}{\partial t} - \mu_i^{rand} \nabla^2 \Omega_i = k_i^{prolif} \Omega_i - k_i^{death} \Omega_i \quad (i = T, C) \tag{17.4-2}$$

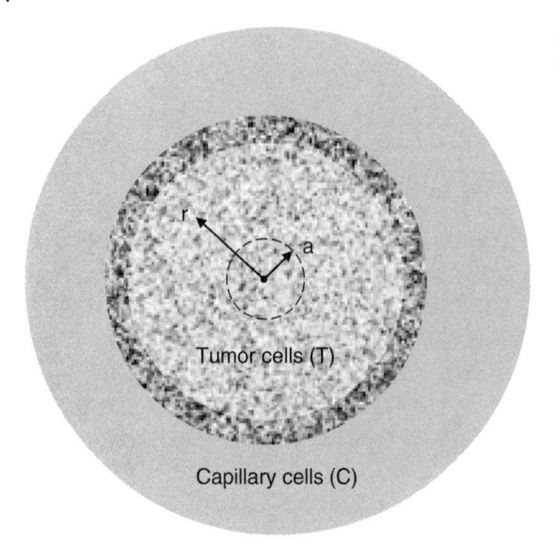

Figure 17.4-1 Invasion of a spherical tumor into normal tissue.

where μ_i^{rand} is a constant. For a spherically symmetric system about the center of the tumor cells, spatial changes in cell number densities vary only with radial position so that Equation 17.4-2 becomes

$$\frac{\partial \Omega_i}{\partial t} - \frac{\mu_i^{rand}}{r^2}\frac{\partial}{\partial r}\left(r^2 \frac{\partial \Omega_i}{\partial r}\right) = k_i^{prolif}\Omega_i - k_i^{death}\Omega_i \quad (i = T, C) \tag{17.4-3}$$

Initially, the tumor cells are assumed to be located in a radius "a" with a constant number density, Ω_{To}. Surrounding these tumor cells is a relatively large mass of normal vascularized tissue with a constant capillary cell density Ω_{Co}:

$$t = 0: \quad \Omega_T = \begin{cases} \Omega_{To} & \text{when } r \le a \\ 0 & \text{when } r > a \end{cases}, \quad \Omega_C = \begin{cases} 0 & \text{when } r \le a \\ \Omega_{Co} & \text{when } r > a \end{cases} \tag{17.4-4}$$

Because of spherical symmetry, the boundary condition at the center of the tumor is

$$r = 0: \quad \frac{\partial \Omega_T}{\partial r} = \frac{\partial \Omega_C}{\partial r} = 0 \tag{17.4-5}$$

Far from the center, no tumor cells are present and the capillary cell density remains at its initial level:

$$r \to \infty: \quad \Omega_T = 0, \; \Omega_C = \Omega_{Co} \tag{17.4-6}$$

Analysis

Dimensionless Forms

This model can be recast in terms of the dimensionless variables

$$\mathit{t} \equiv \frac{\mu_T^{rand}}{a^2} t, \quad \mathit{r} \equiv \frac{r}{a}, \quad \mathit{\Omega}_i \equiv \frac{\Omega_i}{\Omega_{io}} \quad (i = T, C) \tag{17.4-7a-c}$$

so that the dimensionless governing equations are

$$\frac{\partial \Omega_T}{\partial t} - \frac{1}{r^2}\frac{\partial^2}{\partial r}\left(r^2 \frac{\partial \Omega_T}{\partial r}\right) = k_T^{prolif}\Omega_T - k_T^{death}\Omega_T$$
$$\frac{\partial \Omega_C}{\partial t} - \frac{\mu}{r^2}\frac{\partial^2}{\partial r}\left(r^2 \frac{\partial \Omega_C}{\partial r}\right) = k_C^{prolif}\Omega_C - k_C^{death}\Omega_C \quad (17.4\text{-}8a,b)$$

The dimensionless initial and boundary conditions are

$$t=0: \quad \Omega_T = \begin{cases} 1 & \text{when } r \le 1 \\ 0 & \text{when } r > 1 \end{cases}, \quad \Omega_C = \begin{cases} 0 & \text{when } r \le 1 \\ 1 & \text{when } r > 1 \end{cases} \quad (17.4\text{-}9)$$

$$r=0: \quad \frac{\partial \Omega_T}{\partial r} = \frac{\partial \Omega_C}{\partial r} = 0 \quad (17.4\text{-}10)$$

$$r \to \infty: \quad \Omega_T = 0, \quad \Omega_C = 1 \quad (17.4\text{-}11)$$

The dimensionless parameters defined by

$$\mu \equiv \frac{\mu_C^{rand}}{\mu_T^{rand}}, \quad k_i^{prolif} \equiv \frac{a^2}{\mu_T^{rand}} k_i^{prolif}, \quad k_i^{death} \equiv \frac{a^2}{\mu_T^{rand}} k_i^{death} \quad (i = T, C) \quad (17.4\text{-}12a\text{-}c)$$

are the characteristic dimensionless motility and cell fate coefficients.

Fate Processes

Phenomenological equations for k_i^{prolif} and k_i^{death} can account for the interaction of the tumor cells and endothelial surfaces. Tumor cell proliferation is directly related to the availability of nutrients, which depends on capillary surface area in contact with tumor cells. Availability of nutrients, however, is reduced by a higher density of tumor cells. Therefore, we assume that k_T^{prolif} is a function of the capillary cell density per tumor cell number density, Ω_C/Ω_T. Since proliferation is limited by inherent cellular factors, we express k_T^{prolif} as a saturation function:

$$k_T^{prolif} = \frac{\alpha_T^{prolif}(\Omega_C/\Omega_T)}{\beta_T^{prolif} + (\Omega_C/\Omega_T)} \quad (17.4\text{-}13)$$

The death rate of tumor cells increases with their metabolic waste production relative to waste elimination by blood flow. Thus, k_T^{death} is directly related to tumor cell density but inversely related to capillary density:

$$k_T^{death} = \alpha_T^{death}\left(\frac{\Omega_T}{\beta_T^{death} + \Omega_C}\right)^2 \quad (17.4\text{-}14)$$

When the capillary cell density is negligibly small $(\beta_T^{death} \gg \Omega_C)$, we see that $k_T^{death} \approx \alpha_T^{death}(\Omega_T/\beta_T^{death})^2$. This behavior corresponds to the rapid onset of tumor cell necrosis occurring in highly dense, poorly perfused tumor tissue.

Capillary cell proliferation is linked to secretion rate of TAF by tumor cells as they proliferate at a rate $k_T^{prolif}\Omega_T$. As with tumor cell proliferation, inherent factors limit capillary proliferation that we also express as saturation function:

$$k_C^{prolif} = \alpha_C^{prolif}\left(\frac{k_T^{prolif}\Omega_T}{\beta_C^{prolif} + k_T^{prolif}\Omega_T}\right) \quad (17.4\text{-}15)$$

Loss of capillary surface results in large part from toxic species released during tumor cell death. Thus, we assume that the specific rate coefficient for capillary cell death is proportional to the rate of tumor cell death:

$$k_{\mathrm{C}}^{\mathrm{death}} = \alpha_{\mathrm{C}}^{\mathrm{death}} \left(k_{\mathrm{T}}^{\mathrm{death}} \Omega_{\mathrm{T}} \right) \tag{17.4-16}$$

Simulation

Numerical solution of this model was obtained with parameter values based on experimental data with mice (Liotta *et al.* 1977): $\alpha_{\mathrm{T}}^{\mathrm{prolif}} = 200$, $\beta_{\mathrm{T}}^{\mathrm{prolif}} = 0.1$; $\alpha_{\mathrm{T}}^{\mathrm{death}} = 0.5$, $\beta_{\mathrm{T}}^{\mathrm{death}} = 0.1$; $\alpha_{\mathrm{C}}^{\mathrm{prolif}} = 100$, $\beta_{\mathrm{C}}^{\mathrm{prolif}} = 1000$; and $\alpha_{\mathrm{C}}^{\mathrm{death}} = 0.01$. Simulations show the effect of the motility μ of capillary cells relative to tumor cells on cellular distributions at several times (Fig. 17.4-2). As time increases, tumor cell growth (solid curves) is directed toward the vasculature, which provides nutrients and remove metabolic waste products. Simultaneously, the capillaries (dashed curves) grow toward the tumor cells, which are the source of TAF. Because of insufficient vascularization, the center of the tumor, located approximately at $0.5 > z \geq 0$, becomes progressively necrotic. Maximum tumor and capillary cell densities occur near $r = 1$, the initial boundary between tumor and normal tissue. With slower capillary cell motility (right graph), the tumor density tends to be smaller at the center of the tumor because of more necrosis. This supports treatment strategies that restrict capillary motility to reduce tumor mass.

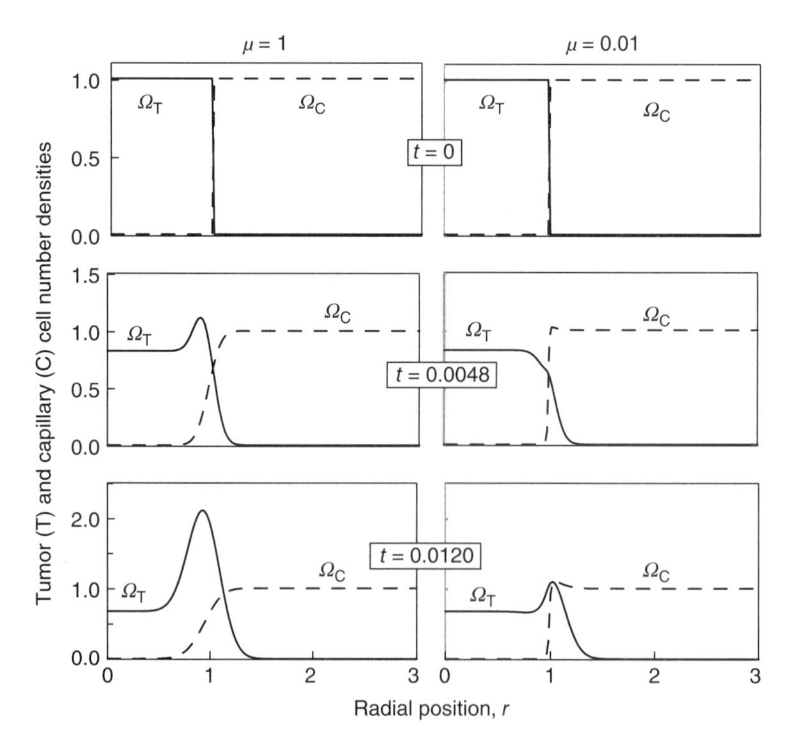

Figure 17.4-2 Interaction of tumor cells and capillaries.

17.4.2 Chemotaxis with an Inflammatory Response

A model of cell number dynamics can be the basis of optimizing drug treatments to eliminate bacterial infection. As part of a tissue's inflammatory response to infection, neutrophils originating from blood migrate along a chemoattractant gradient produced directly by the bacteria or indirectly by bacterial tissue destruction. We can analyze the fate of bacteria (B) initially at an exposed skin surface as they migrate and interact with neutrophils (N) in the skin.

Skin is composed of the epidermis, an unperfused stratum of keratinized cells, and the dermis, a perfused stratum of connective tissue, hair follicles, and various glands. For this analysis, however, we consider skin to be a single tissue layer of thickness h that is exposed to the environment at one surface and to a sheath of capillary blood at the other surface (Fig. 17.4-3).

Model Formulation

Mature neutrophils are terminally differentiated and do not proliferate, but they are transported and undergo cell death. Their number density Ω_N changes according to random migration Θ_N^{rand}, chemotaxis associated with the bacterial chemoattractant Θ_N^{taxis}, and death Ξ_N^{death}:

$$\frac{\partial \Omega_N}{\partial t} + \nabla \cdot \left(\Theta_N^{rand} + \Theta_N^{taxis}\right) = \Xi_N^{death} \tag{17.4-17}$$

The number density of bacteria Ω_B changes by random migration Θ_B^{rand}, proliferation Ξ_B^{prolif}, and death Ξ_B^{death} according to

$$\frac{\partial \Omega_B}{\partial t} + \nabla \cdot \Theta_B^{rand} = \Xi_B^{prolif} + \Xi_B^{death} \tag{17.4-18}$$

We model fate processes as:

$$\begin{aligned} \Xi_i^{death} &= -k_i^{death}\Omega_i \quad (i = N, B) \\ \Xi_B^{prolif} &= k_B^{prolif}\Omega_B \end{aligned} \tag{17.4-19a,b}$$

where the bacterial proliferation rate coefficient k_B^{prolif} is constant, but death rate coefficients depend on number density of the opposing cell types:

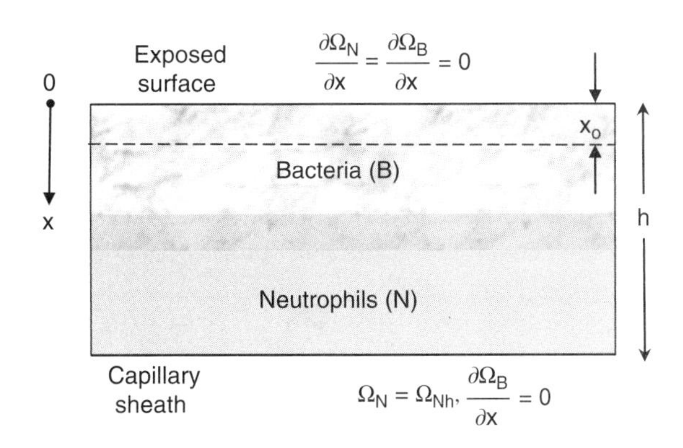

Figure 17.4-3 Bacteria and neutrophil transport and interaction in skin.

$$k_N^{death} = \alpha_N^{death} + \beta_N^{death}\Omega_B, \quad k_B^{death} = \beta_B^{death}\Omega_N \tag{17.4-20a,b}$$

Here, α_N^{death} determines neutrophil apoptosis that occurs even in the absence of bacteria, β_N^{death} determines the death of neutrophils after they phagocytize bacteria, and β_B^{death} determines loss of bacteria by interaction with neutrophils.

The transport fluxes, which have constant coefficients, are dependent on number densities and chemoattractant concentration, C_A:

$$\begin{aligned} \Theta_j^{rand} &= -\mu_j^{rand}\nabla\Omega_j \quad (j = N, B) \\ \Theta_N^{taxis} &= \mu_N^{taxis}\Omega_N\nabla C_A \end{aligned} \tag{17.4-21a,b}$$

With these fate and transport processes, the cell density equations become

$$\frac{\partial\Omega_N}{\partial t} - \mu_N^{rand}\nabla^2\Omega_N + \mu_N^{taxis}\nabla\cdot(\Omega_N\nabla C_A) = -\left(\alpha_N^{death} + \beta_N^{death}\Omega_B\right)\Omega_N \tag{17.4-22}$$

$$\frac{\partial\Omega_B}{\partial t} - \mu_B^{rand}\nabla^2\Omega_B = k_B^{prolif}\Omega_B - \left(\beta_B^{death}\Omega_N\right)\Omega_B \tag{17.4-23}$$

Because skin is typically thin compared to its surface curvature, the cell number densities and chemoattractant concentration primarily change along one coordinate direction x that is perpendicular to the skin surface. Thus, we restrict these cell number density equations to the one-dimensional domain $h > x > 0$:

$$\frac{\partial\Omega_N}{\partial t} + \mu_N^{taxis}\frac{\partial}{\partial x}\left(\Omega_N\frac{\partial C_A}{\partial x}\right) = \mu_N^{rand}\frac{\partial^2\Omega_N}{\partial x^2} - \left(\alpha_N^{death} + \beta_N^{death}\Omega_B\right)\Omega_N \tag{17.4-24}$$

$$\frac{\partial\Omega_B}{\partial t} = \mu_B^{rand}\frac{\partial^2\Omega_B}{\partial x^2} + \left(k_B^{prolif} - \beta_B^{death}\Omega_N\right)\Omega_B \tag{17.4-25}$$

If the production rate of chemoattractant is proportional to the bacteria number density, then we write the chemoattractant concentration equation in the same domain as:

$$\frac{\partial C_A}{\partial t} = \mathcal{D}_A\frac{\partial^2 C_A}{\partial x^2} + k_A\Omega_B \tag{17.4-26}$$

where $\mathcal{D}_A$ is the diffusion coefficient of the chemoattractant.

We assume that bacteria initially invade a small distance x_o into the skin from the exposed surface at $x = 0$. We also assume that neutrophils and chemoattractant are initially absent from the skin:

$$t = 0: \quad \Omega_N = 0, \quad C_A = 0, \quad \Omega_B = \begin{cases} \Omega_{Bo} & \text{when } x_o \geq x \geq 0 \\ 0 & \text{when } h \geq x > x_o \end{cases} \tag{17.4-27}$$

where Ω_{Bo} is a constant. After their initial invasion, bacteria no longer cross the skin surface. In addition, attractant and neutrophils do not cross this skin boundary. Therefore, we impose zero-flux conditions at the skin surface which lead to

$$x = 0: \quad \begin{aligned} &\frac{\partial\Omega_B}{\partial x} = 0, \quad \frac{\partial C_A}{\partial x} = 0 \\ &\mu_N^{rand}\frac{\partial\Omega_N}{\partial x} - \mu_N^{taxis}\Omega_N\frac{\partial C_A}{\partial x} = 0 \Rightarrow \frac{\partial\Omega_N}{\partial x} = 0 \end{aligned} \tag{17.4-28}$$

At x = h, where skin is in contact with capillary walls, we assume that neutrophil number density is constant and is in equilibrium with neutrophils in blood, Ω_{nh}. We also assume that capillary walls provide an effective barrier against bacteria so that the spatial gradient that drives bacterial migration is negligible. Since any chemoattractant entering blood flows away quickly, the diffusion flux of attractant across the capillary walls is simply the product of an overall mass transfer coefficient, K_A, and the local chemoattractant concentration in the skin:

$$x = h: \quad \Omega_N = \Omega_{Nh}, \quad \frac{\partial \Omega_B}{\partial x} = 0, \quad -\mathcal{D}_A \frac{\partial C_A}{\partial x} = K_A C_A \tag{17.4-29}$$

Analysis

Dimensionless Forms

We introduce the following dimensionless variables:

$$\mathit{\Omega}_N = \frac{\Omega_N}{\Omega_{Nh}}, \quad \mathit{\Omega}_B = \frac{\Omega_B}{\Omega_{Bo}}, \quad \mathit{C}_A = \frac{C_A}{C_o}, \quad \mathit{x} = \frac{x}{h}, \quad t = \left(\frac{h^2}{\mathcal{D}_A}\right)\mathit{t} \tag{17.4-30a-e}$$

Since a characteristic concentration C_o is not designated in the model equations, we define it on the basis of a combination of other parameters as $C_o = k_A \Omega_{Bo} h^2/\mathcal{D}_A$. The dimensionless governing equations are

$$\frac{\partial \mathit{\Omega}_N}{\partial \mathit{t}} + \mu_N Pe_N \frac{\partial}{\partial \mathit{x}}\left(\mathit{\Omega}_N \frac{\partial \mathit{C}_A}{\partial \mathit{x}}\right) = \mu_N \frac{\partial^2 \mathit{\Omega}_N}{\partial \mathit{x}^2} - (\alpha_N + \beta_N \mathit{\Omega}_B)\mathit{\Omega}_N \tag{17.4-31}$$

$$\frac{\partial \mathit{\Omega}_B}{\partial \mathit{t}} = \mu_B \frac{\partial^2 \mathit{\Omega}_B}{\partial \mathit{x}^2} + (\mu_B Da_B - \beta_B \mathit{\Omega}_N)\mathit{\Omega}_B \tag{17.4-32}$$

$$\frac{\partial \mathit{C}_A}{\partial \mathit{t}} = \frac{\partial^2 \mathit{C}_A}{\partial \mathit{x}^2} + \mathit{\Omega}_B \tag{17.4-33}$$

and the dimensionless conditions are

$$t = 0: \quad \mathit{\Omega}_N = 0, \quad \mathit{C}_A = 0, \quad \mathit{\Omega}_B = \begin{cases} 1 & \text{when } x_o/h \ge \mathit{x} \ge 0 \\ 0 & \text{when } 1 \ge \mathit{x} > x_o/h \end{cases} \tag{17.4-34}$$

$$x = 0: \quad \frac{\partial \mathit{\Omega}_N}{\partial \mathit{x}} = 0, \quad \frac{\partial \mathit{\Omega}_B}{\partial \mathit{x}} = 0, \quad \frac{\partial \mathit{C}_A}{\partial \mathit{x}} = 0 \tag{17.4-35}$$

$$\mathit{x} = 1: \quad \mathit{\Omega}_N = 1, \quad \frac{\partial \mathit{\Omega}_B}{\partial \mathit{x}} = 0, \quad -\frac{\partial \mathit{C}_A}{\partial \mathit{x}} = Sh_A \mathit{C}_A \tag{17.4-36}$$

Here, we have defined the dimensionless parameters

$$\alpha_N = \frac{\alpha_N^{death} h^2}{\mathcal{D}_A}, \quad \beta_N \equiv \frac{\beta_N^{death}\Omega_{Bo} h^2}{\mathcal{D}_A}, \quad \beta_B \equiv \frac{\beta_B^{death}\Omega_{Nh} h^2}{\mathcal{D}_A}$$
$$\mu_N \equiv \frac{\mu_N^{rand}}{\mathcal{D}_A}, \quad \mu_B \equiv \frac{\mu_B^{rand}}{\mathcal{D}_A} \tag{17.4-37a-h}$$
$$Pe_N \equiv \frac{\mu_N^{taxis} C_o}{\mu_N^{rand}}, \quad Da_B \equiv \frac{k_B^{prolif} h^2}{\mu_B^{rand}}, \quad Sh_A \equiv \frac{K_A h}{\mathcal{D}_A}$$

Simulation

The dynamics of cell number density and the chemoattractant concentration distributions (Fig. 17.4-4) were obtained from numerical solutions of the model equations with the following parameter values: $x_o/h = 0.01$, $\alpha_N = 0.1$, $\beta_N = 0.1$, $\beta_B = 0.5$, $\mu_N = 0.01$, $\mu_B = 0.05$, $Da_B = 10$,

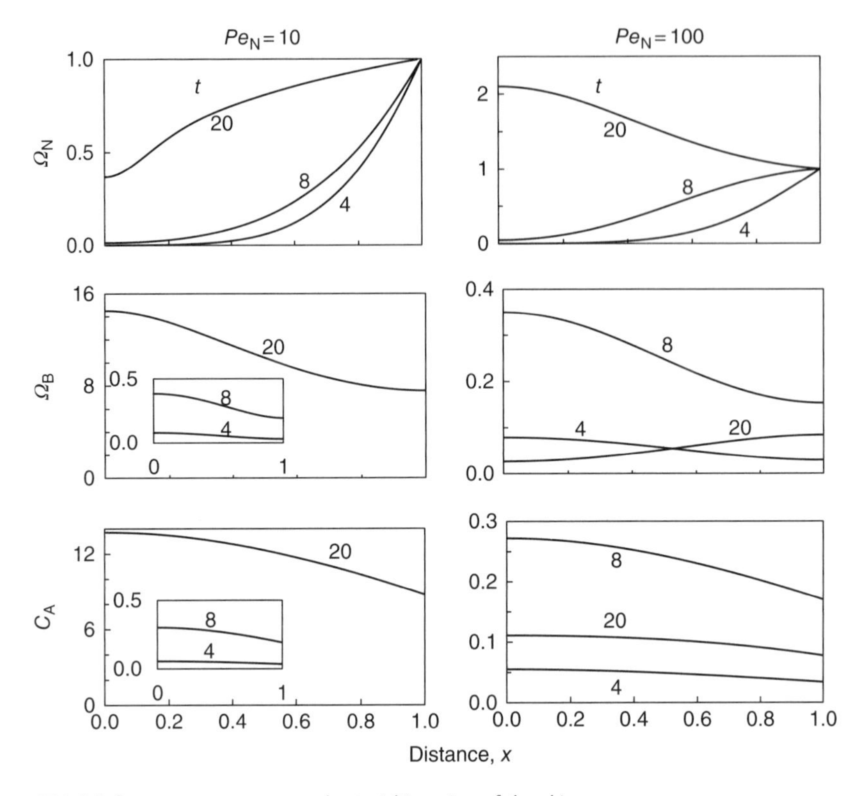

Figure 17.4-4 Inflammatory response to bacterial invasion of the skin.

and $Sh_A = 1$. To explore the effect of neutrophil chemotaxis relative to random motility, simulations were performed with Pe_N values of 10 and 100.

When $Pe_N = 10$, Ω_N decreases with increasing distance from the capillary sheath at $x = 1$, the source of neutrophils. Simultaneously, Ω_B and C_A decrease with distance from the skin surface at $x = 0$, the source of bacteria. With increasing time, Ω_B increases dramatically by bacterial proliferation. The corresponding increase in bacterial production of chemoattractant substantially increases neutrophil chemotaxis, causing a concomitant increase in Ω_N.

When $Pe_N = 100$, the behavior of the Ω_N, Ω_B and C_A distributions at shorter times ($t =$ 4,8) is similar to their behavior when $Pe_N = 10$. At a long time ($t = 20$), however, the increased haptotaxis associated with the higher Pe_N causes neutrophils to aggregate where C_A is highest, near the skin surface. Consequently, Ω_B is lower closer to the skin surface and higher closer to the capillary sheath. Most importantly, we see that increased neutrophil haptotaxis leads to higher neutrophil levels and much lower bacterial levels throughout the skin.

17.4.3 Stem Cells for Cartilage Tissue Engineering

The differentiation of stem cells derived from bone marrow is a potential method for fabricating replacements for damaged or dysfunctional tissue. For example, stem cell differentiation into chondrocytes is a possible treatment for osteoarthritis. To form a cartilage construct, stem cells (SC) are placed in a polymeric scaffold and cultured in a medium containing chondrogenic factors (F) such as TGF-β that can trigger SC differentiation into chondrocytes (C).

Model Formulation

Consider a scaffold disk of thickness 2 h that is initially loaded with SCs at their carrying capacity Ω_{SC}^{cc} so that cell migration is negligible. This SC-loaded scaffold is submerged in a medium containing a factor F that is transported through the scaffold by molecular diffusion. If the lateral dimensions of the scaffold are substantially greater than its thickness, then differentiation of SC and diffusion of F occur in the one-dimensional domain $h > x > 0 > -h$ with $x = 0$ located at the center of the disk (Fig. 17.4-5).

When the SC differentiation rate Ξ_{SC}^{diff} is much greater than its proliferation rate and the SCs do not migrate, the number density of stem cells $\Omega_{SC}(x, t)$ changes with location and time according to the cell density equation

$$\frac{\partial \Omega_{SC}}{\partial t} = \Xi_{SC}^{diff}(x,t) \tag{17.4-38}$$

Assuming that SC differentiation into C does not involve other cell types, then $\Xi_{C}^{diff} = -\Xi_{SC}^{diff}$, and the number density of chondrocytes $\Omega_C(z, t)$ changes as:

$$\frac{\partial \Omega_C}{\partial t} = \Xi_{C}^{diff}(x,t) = -\Xi_{SC}^{diff}(x,t) \tag{17.4-39}$$

The initial conditions for these two equations are

$$t = 0: \quad \Omega_{SC} = \Omega_{SC}^{cc}, \quad \Omega_C = 0 \tag{17.4-40}$$

Adding Equations 17.4-38 and 17.4-39 reveals that $\partial(\Omega_{SC} + \Omega_C)/\partial t = 0$. Thus, total cell number density does not depend on time and must always be equal to its initial, spatially uniform value:

$$\Omega_{SC} + \Omega_C = \Omega_{SC}^{cc} \tag{17.4-41}$$

Consequently, the cell densities can be predicted with only one differential equation. Assuming that SC differentiation is a first-order process with respect to Ω_{SC} once the concentration of chondrogenic factor C_F is greater than some critical value C_{Fc},

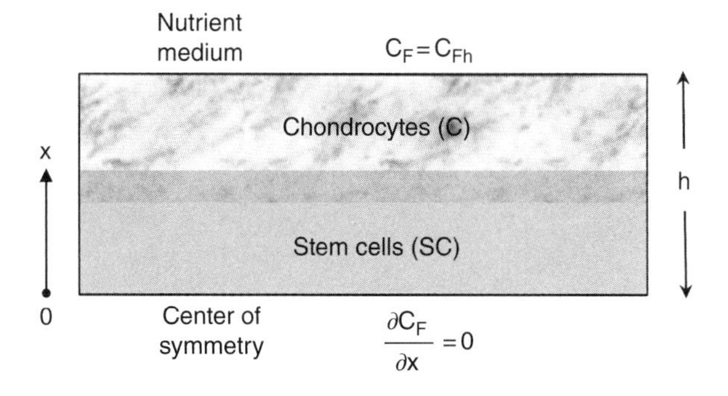

Figure 17.4-5 Cartilage formation from stem cells.

$$\Xi_{SC}^{diff} = \begin{cases} 0 & \text{when } C_F < C_{Fc} \\ -k_{SC,C}^{diff}\Omega_{SC} & \text{when } C_F \geq C_{Fc} \end{cases} \tag{17.4-42}$$

where $k_{SC,C}^{diff}$ is a constant. Incorporating this rate with Equation 17.4-41 into the governing equation for the chondrocyte cell density, we find

$$\frac{\partial \Omega_C}{\partial t} = \begin{cases} 0 & \text{when } C_F < C_{Fc} \\ k_{SC,C}^{diff}\left(\Omega_{SC}^{cc} - \Omega_C\right) & \text{when } C_F \geq C_{Fc} \end{cases} \tag{17.4-43}$$

If F is independent of the cell processes and is not taken up by the cells, then its distribution by diffusion into the scaffold is determined by

$$\frac{\partial C_F}{\partial t} = \mathcal{D}_F \frac{\partial^2 C_F}{\partial x^2} \tag{17.4-44}$$

Here, $\mathcal{D}_F$ represents the effective diffusion coefficient of F in the cell-loaded scaffold. The initial and boundary conditions for this equation are

$$\begin{aligned} t = 0: &\quad C_F = 0 \\ x = 0: &\quad \frac{\partial C_F}{\partial x} = 0 \\ x = h: &\quad C_F = C_{Fh} > C_{Fc} \end{aligned} \tag{17.4-45a-c}$$

At $x = 0$, the concentration derivative is zero since the C_F distribution in the $0 > x > -h$ half-domain is a mirror image of that in the $h > x > 0$ half-domain. At $x = h$, C_F is a constant in equilibrium with the concentration in the medium, C_{Fh}, which must be greater than the critical value.

Analysis

The analytical solution of the diffusion model for chondrogenic factor F is equivalent to one previously found by an eigenfunction solution (Eq. 15.5-23):

$$C_F = C_{Fh}\left[1 - \frac{4}{\pi}\sum_{n=1}^{\infty}\frac{(-1)^{n+1}}{(2n-1)}\cos\left(\frac{\lambda_n x}{h}\right)\exp\left(-\frac{\lambda_n^2 \mathcal{D}_F t}{h^2}\right)\right] \tag{17.4-46}$$

where the eigenvalues are $\lambda_n = (2n-1)\pi/2$. To determine the time when factor F reaches its critical concentration at any location x, we set $C_F = C_{Fc}$ and compute the root $t = t_c(x)$ by a numerical solution of

$$\frac{4}{\pi}\sum_{n=1}^{\infty}\frac{(-1)^{n+1}}{(2n-1)}\cos\left(\frac{\lambda_n x}{h}\right)\exp\left(-\frac{\lambda_n^2 \mathcal{D}_{Fc} t_c}{h^2}\right) = 1 - \frac{C_{Fc}}{C_{Fh}} \tag{17.4-47}$$

Then, the rate of increase of C at any x can be determined from Equation 17.4-43 as:

$$\frac{\partial \Omega_C}{\partial t} = \begin{cases} 0 & \text{when } t_c(x) > t \geq 0 \\ k_{SC,C}^{diff}\left(\Omega_{SC}^{cc} - \Omega_C\right) & \text{when } t \geq t_c(x) \end{cases} \tag{17.4-48}$$

whose solution yields

$$\frac{\Omega_C}{\Omega_{SC}^{cc}} = \begin{cases} 0 & \text{when } t_c(x) > t \geq 0 \\ 1 - \exp\left[-k_{SC,C}^{diff}(t - t_c)\right] & \text{when } t \geq t_c(x) \end{cases} \tag{17.4-49}$$

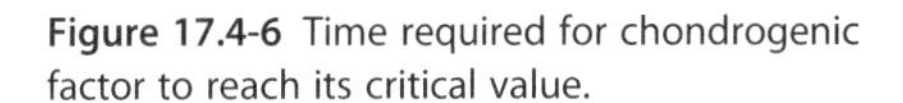
Figure 17.4-6 Time required for chondrogenic factor to reach its critical value.

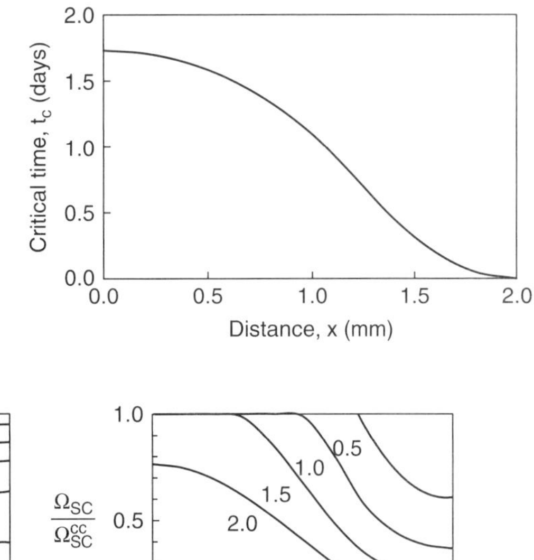

Figure 17.4-7 Number density distributions of stem cells and chondrocytes.

From numerical simulation of the model with the parameter values $C_{Fh} = 2C_{Fc}$, h = 2 mm, $\mathcal{D}_F = 10^{-6}\ cm^2/s$ and $k^{diff}_{SC,C} = 1\ day^{-1}$, we obtain the critical times at any x when $C_F = C_{Fc}$ (Fig. 17.4-6). For this calculation, we truncated the series in Equation 17.4-47 when an additional term was an order of magnitude less than the preceding term. The intercept of this graph indicates that 1.7 days must elapse before there is sufficient diffusion of F for some stem cell differentiation to occur throughout the scaffold up to its center of symmetry at x = 0.

Using these calculated $t_c(x)$ values, we computed $\Omega_C(x,t)/\Omega^{cc}_{SC}$ and $\Omega_{SC}(x,t)/\Omega^{cc}_{SC}$ for the $h > x > 0$ half-domain in Figure 17.4-7. As time proceeds, cell differentiation occurs further from the surface at x = 2 mm exposed to the nutrient. Almost complete SC differentiation occurs throughout the scaffold by day 3.

Reference

Liotta LA, Saidel GM, Kleinerman J. Diffusion model of tumor vascularization and growth. Bull Math Biol. 1977; 39:117–128.

Part VI
Compartmental Modeling

Chapter 18

Compartment Models I

Basic Concepts and Tracer Analysis

In this chapter, we describe dynamic changes in species concentration as a consequence of transport and reaction processes associated with one or more interconnected compartments. Typically, a compartment is assumed to be perfectly mixed so that chemical composition inside the compartment is uniform and is the same as that leaving the compartment. Such models are particularly convenient for examining the concentration outputs (responses) of biomedical systems to imposed concentration or mass inputs (disturbances). For example, we can investigate the effect of a constant input on a system's output over a long time so that the process variables reach their steady-state values. More information about a system is available, however, by imposing a prescribed time-dependent input. For dynamic analysis, step or pulse inputs are commonly used.

We consider two types of compartmental models, compartmental pool models and physiologically based compartmental models. As a basis for analyzing these models, we develop dynamic mass balances that account for changes within each compartment as well as in compartment inputs and outputs. This results in a system of ordinary differential equations and their associated initial conditions. In this chapter, we focus on systems associated with low concentrations of chemical species (e.g., tracers) that are governed by linear models with time-invariant parameters. The behavior of such systems can be conveniently analyzed by Laplace transform methods (Appendix C4).

18.1 Compartmental Modeling Concepts

18.1.1 Pool Models and Physiologically Based Models

In compartmental pool modeling, each compartment represents a hypothetical volume (or pool) of a particular chemical species that can actually contain multiple tissues and organs.

Biomedical Mass Transport and Chemical Reaction: Physicochemical Principles and Mathematical Modeling,
First Edition. James S. Ultman, Harihara Baskaran, and Gerald M. Saidel.

A material stream between two compartments containing the same substance represents transport of this substance between the pools. A stream between two compartments containing different substances corresponds to a chemical transformation from one substance to another. For example, a simple model of glucose distribution in the human body consists of a pool of glucose and a pool of CO_2. A stream between these pools represents metabolic oxidation of glucose to form CO_2. An input stream to the glucose pool represents glucose supplied by the environment. An output stream from the CO_2 pool corresponds to CO_2 elimination to the environment.

In a physiologically based compartmental model, each compartment corresponds to an identifiable anatomical domain such as a cell, tissue, or organ. Within a compartment, there can be many substances corresponding to those found in the real system. Chemical transformation between substances occurs by specific biochemical reactions within the compartment. The interconnecting streams between compartments are associated with a particular physiological transport process such as blood flow or membrane permeation. For example, in modeling the metabolism of ethanol, the body can be represented by interconnected gastrointestinal, liver, muscle/fat, and central blood regions. Digestion of ethanol occurs in the gastrointestinal compartment, enzymatic oxidation occurs in the liver compartment, and storage exists primarily in the muscle/fat and central blood compartments. These compartments are connected by convective streams that are assigned physiologically realistic blood flows.

Figure 18.1-1 illustrates a generic compartment j of volume V_j containing a molar concentration C_j of a chemical species. This species can accumulate in the compartment because of a molar inflow from another compartment n at rate $\dot{I}_{nj}$ or from an external source at rate $\dot{S}_j$. It can also accumulate by chemical reaction at a rate $\dot{R}_j$. Conversely, a substance can be depleted by a molar outflow to another compartment m at a rate $\dot{O}_{jm}$ or an external discharge to the environment at a rate $\dot{D}_j$. Corresponding to these processes is the molar balance equation for a solute in compartment j:

$$\frac{d(C_j V_j)}{dt} = \left(\sum_n \dot{I}_{nj} - \sum_m \dot{O}_{jm} \right) + \left(\dot{S}_j - \dot{D}_j \right) + \dot{R}_j \tag{18.1-1}$$

In physiologically based models, $\dot{R}_j$ depends on the concentrations of all substances that participate in a chemical reaction and are located in compartment j. It is also proportional to the compartment volume V_j. In pool models, $\dot{R}_j$ is incorporated as input–output streams that do not depend on compartment volume.

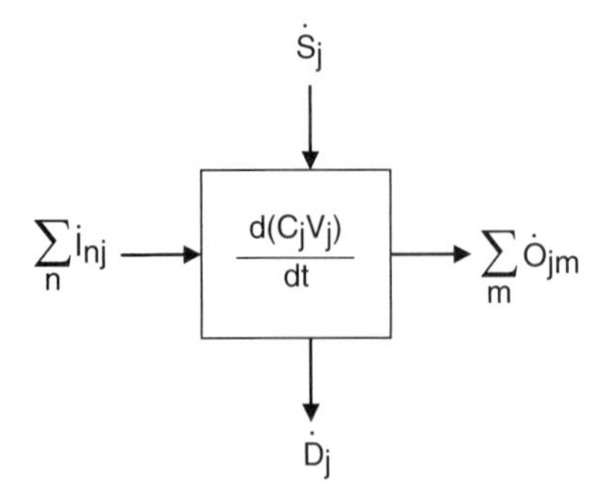

Figure 18.1-1 Compartment model with inputs and outputs.

Example 18.1-1 Pool and Physiologically Based Models of an Isolated Liver

An isolated, perfused rat liver is being used to investigate the metabolism of a new drug. To prevent flow through the portal vein and the major lymphatics, these vessels are ligated. A constant volumetric flow rate Q of physiological saline is perfused into the hepatic artery, and the effluent perfusate is collected from the hepatic vein at the same flow rate. A typical experiment is initiated by continually infusing drug into the hepatic artery at a low concentration C_o. Subsequently, the response of the drug concentration $C_1(t)$ is monitored in the hepatic vein. To analyze this experiment, assume that the drug is metabolized in extravascular tissue only. Our goal is to represent experimental data by a two-compartment pool model and compare it to a corresponding physiologically based model of the experiment.

Solution

An illustration of the two models is shown in Figure 18.1-2. In the pool model, drug is distributed between an intravascular pool 1 and an extravascular pool 2. The respective volumes and drug concentrations in the two pools are $V_{i,pool}$ and C_i ($i = 1,2$). Pool 1 has source and discharge rates of drug by convective transport of QC_0 and QC_1, respectively. As is usually the case for pool models, drug transport and reaction streams between the pools are represented as first-order processes. One-way transport of drug occurs from pool 1 to pool 2 at a rate $k_{12}C_1$ and from pool 2 to pool 1 at a rate $k_{21}C_2$. Metabolism of drug to products P is represented by the outflow from pool 2 at a first-order rate $k_{2P}C_2$. Molar balances of drug and associated initial conditions for pools 1 and 2 are

$$\frac{d(C_1V_{1,pool})}{dt} = Q(C_0 - C_1) + k_{21}C_2 - k_{12}C_1; \quad C_1(0) = 0 \tag{18.1-2a,b}$$

$$\frac{d(C_2V_{2,pool})}{dt} = k_{12}C_1 - (k_{21} + k_{2P})C_2; \quad C_2(0) = 0 \tag{18.1-3a,b}$$

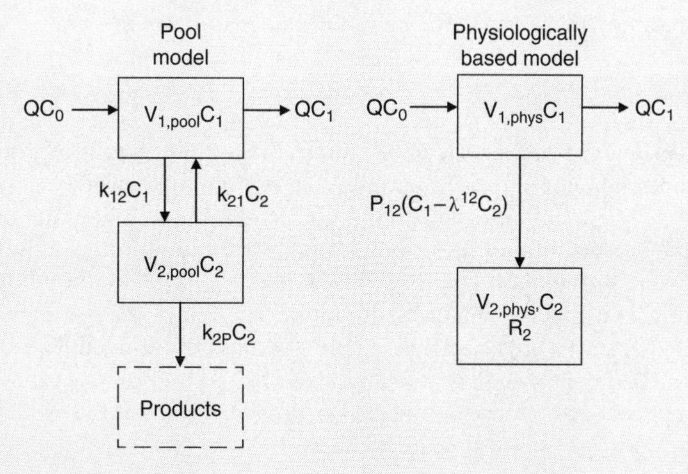

Figure 18.1-2 Alternative compartment models for drug metabolism.

In a physiologically based model, we assume that the drug is distributed between an intravascular compartment 1 of volume $V_{1,phys}$ representing the blood sinusoids in a liver lobule and an extravascular compartment 2 of volume $V_{2,phys}$ corresponding to the cell plates. Diffusion of drug between the two compartments is represented by a single transport stream with a flux equation that accounts for intracompartmental permeability P_{12} and includes an equilibrium partition coefficient λ^{12}. Drug metabolism occurs within compartment 2 by an enzymatic reaction at a rate $\dot{R}_2(C_2)$ that most likely follows nonlinear Michaelis–Menten kinetics. Molar balances in the two compartments and their initial conditions are

$$\frac{d(V_{1,phys}C_1)}{dt} = Q(C_0 - C_1) - P_{12}(C_1 - \lambda^{12}C_2); \;\; C_1(0) = 0 \tag{18.1-4a, b}$$

$$\frac{d(V_{2,phys}C_2)}{dt} = P_{12}(C_1 - \lambda^{12}C_2) - \frac{k_r V_{2,phys} C_2}{C_2 + K_m}; \;\; C_2(0) = 0 \tag{18.1-5a, b}$$

These two models are mathematically equivalent when $V_{1,pool} = V_{1,phys}$, $V_{2,pool} = V_{2,phys}$, $k_{12} = P_{12}$, and $k_{21} = \lambda^{12}P_{12}$. In addition, drug metabolism must reduce to a first-order process (i.e., $C_2 \ll K_m$) with $k_{2P} = k_r V_{2,phys}$. Even when these special circumstances exist, the physiologically based model has a distinct advantage: Its parameters are tied to specific anatomical regions and molecular mechanisms. For this problem, $V_{1,phys}$ could be estimated by measuring the blood volume in the liver; λ^{12} could be evaluated by *in vitro* equilibration experiments using tissue and blood samples; k_r could be found by reaction kinetics measurements in tissue specimens. To the extent that these parameters can be evaluated from such independent sources, fewer unknown parameters are associated with the physiologically based model than with the corresponding pool model.

18.1.2 Tracer Inputs to a Flow-Through Model

As described previously (Section 11.3), radiolabeled compounds are useful for quantifying membrane transport since they can be detected at small concentrations that do not alter inherent system dynamics. More generally, any compound with physical properties (e.g., opacity, fluorescence, radioactivity) that can be monitored at such small concentrations can be used as a tracer. A tracer can be an analog to a natural component such as a fluorescein isothiocyanate conjugate or it can be a foreign substance such Evan's blue dye. In flow-through systems such as lung airways or blood vessels, the input–output behavior of an inert tracer that does not undergo chemical reaction allows us to quantify internal mixing processes.

Consider the injection of a tracer solution into a system inlet stream that already contains some naturally occurring tracer (Fig. 18.1-3). The injected solution is introduced at a mass rate $\dot{S}_m$, and tracer within this solution enters at a molar rate $\dot{S}$. We represent this process using a mixing node with negligible volume that separates the solution flow Q_0 and tracer concentration C_0 in the system inlet from the flow Q_{in} and tracer concentration C_{in} in the mixed stream that directly enters the system. For a mixing node that is treated as a perfectly mixed compartment, the solution and a tracer balances are

$$\frac{d(\rho_{node}V_{node})}{dt} = Q_0\rho_0 + \dot{S}_m - Q_{in}\rho_{in} \tag{18.1-6}$$

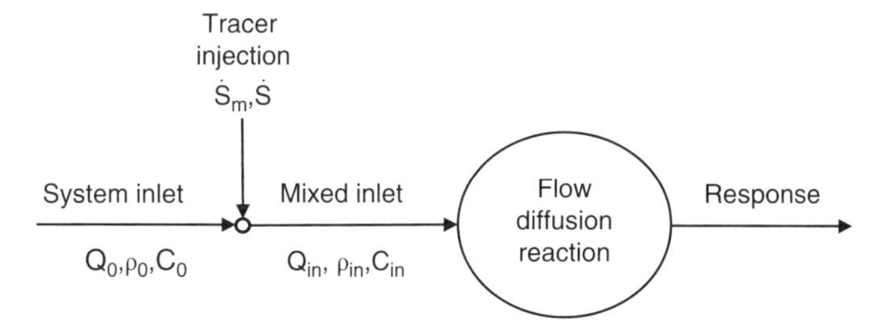

Figure 18.1-3 Tracer injection into a system inlets.

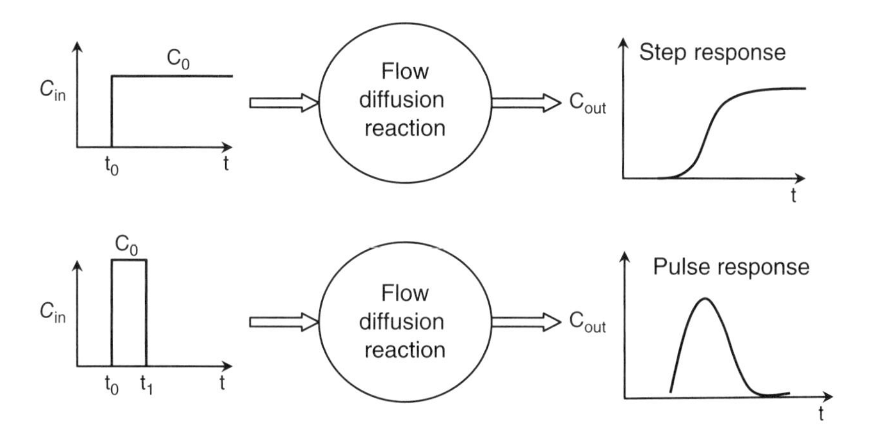

Figure 18.1-4 System responses to step and pulse concentration inputs.

$$\frac{d(C_{node} V_{node})}{dt} = Q_0 C_0 + \dot{S} - Q_{in} C_{in} \tag{18.1-7}$$

Since the mixing node has negligible volume, these equations reduce to

$$Q_0 \rho_0 + \dot{S}_m - Q_{in} \rho_{in} = 0 \tag{18.1-8}$$

$$Q_0 C_0 + \dot{S} - Q_{in} C_{in} = 0 \tag{18.1-9}$$

Throughout this chapter, we will make the following assumptions: (i) Fluid density is constant (in time) and uniform (in space), $\rho_0 = \rho_{in}$; (ii) the mass rate at which tracer solution is injected is much smaller than the mass rate of the system inlet flow, $\dot{S}_m \ll Q_0 \rho_0$; and (iii) the system inlet stream does not contain tracer, $C_0 = 0$. In that case, the solution and tracer balances indicate that

$$Q_{in} = Q_0, \quad C_{in} = \frac{\dot{S}}{Q_{in}} \tag{18.1-10a, b}$$

Two tracer injection schemes commonly used to interrogate the input response of a system are shown in Figure 18.1-4: (i) a step input corresponding to a sudden and sustained increase in the rate at which a substance is introduced into a system and (ii) a pulse input corresponding to a rapid injection (relative to the time period of change in the system output) of a known amount of substance into a system. In mathematical terms, a unit step input of tracer initiated at time t_0 is indicated by

$$H_0(t) = H(t - t_0) \equiv \begin{cases} 0 & \text{when } t < t_0 \\ 1 & \text{when } t > t_0 \end{cases} \tag{18.1-11}$$

This Heaviside function is a dimensionless quantity. A pulse input of tracer applied at time t_0 but discontinued after time t_1 can be constructed from two Heaviside functions:

$$\Pi_{0,1}(t) \equiv \frac{1}{t_1 - t_0}[H(t - t_0) - H(t - t_1)] = \begin{cases} 0 & \text{when } t < t_0 \\ 1/(t_1 - t_0) & \text{when } t_1 > t > t_0 \\ 0 & \text{when } t > t_1 \end{cases} \tag{18.1-12}$$

With a height of $1/(t_1 - t_0)$ and a width of $(t_1 - t_0)$, this "box-car" function has units of reciprocal time and an integral of unity. When a pulse input occurs very rapidly such that $(t_1 - t_0) \to 0$, it can be approximated by the Dirac delta or unit impulse function:

$$\delta(t - t_0) \equiv \lim_{t_1 \to t_o} \Pi_{0,1}(t) \equiv \begin{cases} 0, & t < t_0 \\ \uparrow \infty, & t = t_0 \\ 0, & t > t_0 \end{cases} \tag{18.1-13}$$

As is the case for a unit pulse, the unit impulse function has units of reciprocal time and an integral of unity.

For a step input of tracer introduced at time t_o at a constant molar rate $\dot{m}$, the source term $\dot{S}$ is

$$\dot{S}(t) = \dot{m}H(t - t_0) \tag{18.1-14}$$

The source corresponding to an impulse input of m moles of tracer at time t_0 is given by

$$\dot{S}(t) = m\delta(t - t_0) \tag{18.1-15}$$

These ideal sources can also be expressed in the Laplace transform domain (Table C4-2):

$$\tilde{S}(s) \equiv \boldsymbol{L}\{\dot{S}(t)\} \equiv \int_0^{\infty} e^{-st}\dot{S}(t)dt = \begin{cases} \dfrac{\dot{m}}{s}e^{-st_0} & \text{step input} \\ me^{-st_0} & \text{impulse input} \end{cases} \tag{18.1-16}$$

Substituting these source terms into the Laplace transform of Equation 18.1-10b ($\tilde{C}_{in} = \tilde{S}/Q_{in}$), we obtain the corresponding tracer concentrations at the mixed inlet of a system:

$$\tilde{C}_{in}(s) \equiv \boldsymbol{L}[\dot{S}(t)/Q] = \begin{cases} \dfrac{\dot{m}}{Q_{in}s}e^{-st_0} & \text{step input} \\ \dfrac{me^{-st_0}}{Q_{in}} & \text{impulse input} \end{cases} \tag{18.1-17}$$

If these inputs are introduced at t = 0, then these their equations reduce to a simpler form:

$$\tilde{C}_{in}(s) = \begin{cases} \dfrac{\dot{m}}{Q_{in}s} & \text{step input} \\ \dfrac{m}{Q_{in}} & \text{impulse input} \end{cases} \tag{18.1-18}$$

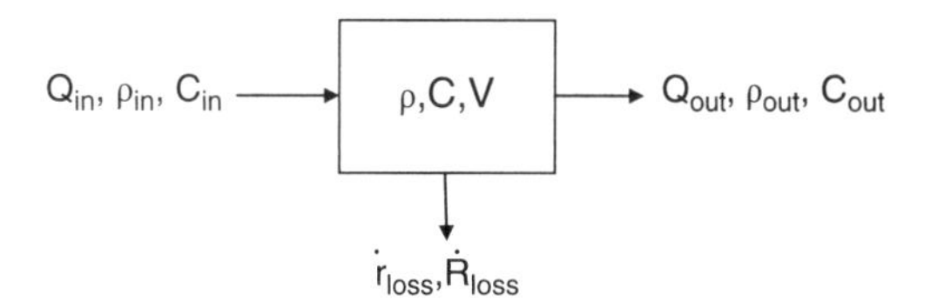

Figure 18.1-5 Single flow-through compartment.

18.1.3 Dynamic Responses of a Single-Compartment Model

Consider a compartment of volume V that contains fluid with a mass density ρ and tracer concentration C (Fig. 18.1-5). The single inlet stream has a volumetric flow Q_{in}, mass density $\rho_{in,}$ and tracer concentration C_{in}; a convective outlet stream has a flow Q_{out}, a mass density ρ_{out}, and a tracer concentration C_{out}. There is also a mass rate of solution loss $\dot{r}_{loss}$ and molar rate of tracer loss $\dot{R}_{loss}$ from the compartment. For this system, the solution mass balance and tracer molar balance are

$$\frac{d(\rho V)}{dt} = (\rho_{in} Q_{in} - \rho_{out} Q_{out}) - \dot{r}_{loss} \tag{18.1-19}$$

$$\frac{d(CV)}{dt} = (Q_{in} C_{in} - Q_{out} C_{out}) - \dot{R}_{loss}(C) \tag{18.1-20}$$

For a constant and uniform fluid mass density in a perfectly mixed compartment, we set $\rho = \rho_{out} = \rho_{in}$ and $C_{out} = C$ so that the solution mass balance becomes

$$\rho \frac{dV}{dt} = \rho(Q_{in} - Q_{out}) - \dot{r}_{loss} \tag{18.1-21}$$

For a constant compartment volume V and relatively small mass loss rate, $dV/dt = 0$ and $\dot{r}_{loss} \ll \rho Q_{in}$. The solution balance then reduces to $Q_{out} = Q_{in} = Q$, and the tracer mass balance leads to

$$V \frac{dC}{dt} = Q(C_{in} - C) - \dot{R}_{loss}(C) \tag{18.1-22}$$

Alternatively, we can express this equation as

$$\theta \frac{dC}{dt} = (C_{in} - C) - \frac{\dot{R}_{loss}(C)}{Q}; \quad \theta \equiv \frac{V}{Q} \tag{18.1-23a,b}$$

where the time constant θ corresponds to the average residence time of a fluid element (or tracer molecules) in the compartment. In applications with an inert tracer and an impermeable compartment wall, $\dot{R}_{loss} = 0$, whereas for applications with a first-order chemical reaction or diffusion across the compartment wall, $\dot{R}_{loss} \propto C$.

No Tracer Loss

When $\dot{R}_{loss} = 0$, Equation 18.1-23a becomes

$$\theta \frac{dC}{dt} + C = C_{in} \tag{18.1-24}$$

Without any significant loss of generality, we assume that the compartment is initially free of tracer so that $C(0) = 0$. Taking the Laplace transform (Table C4-1) of the tracer balance equation yields

$$\left[\theta s\tilde{C}(s) + \cancel{C(0)}\right] + \tilde{C}(s) = \tilde{C}_{in}(s) \tag{18.1-25}$$

The ratio of the transformed output concentration, $\tilde{C}_{out}(s) = \tilde{C}(s)$, to the transformed input concentration, $\tilde{C}_{in}(s)$, is the dimensionless transfer function:

$$\tilde{g}(s) \equiv \frac{\tilde{C}_{out}(s)}{\tilde{C}_{in}(s)} = \frac{\tilde{C}(s)}{\tilde{C}_{in}(s)} = \frac{1}{1+\theta s} \tag{18.1-26}$$

Given $\tilde{g}(s)$ and a specific input concentration $\tilde{C}_{in}(s)$, we can determine the output concentration as $\tilde{C}_{out}(s) = \tilde{g}(s)\tilde{C}_{in}(s)$. For example, a tracer input at $t = 0$ of either a step (Eq. 18.1-18a) or an impulse (Eq. 18.1-18b) leads to

$$\tilde{C}_{out}(s) = \begin{cases} \dfrac{\dot{m}}{Q}\left(\dfrac{1}{s}\right)\left(\dfrac{1}{1+\theta s}\right) & \text{step response} \\[2ex] \dfrac{m}{Q}\left(\dfrac{1}{1+\theta s}\right) & \text{impulse response} \end{cases} \tag{18.1-27}$$

Taking inverse Laplace transforms, we obtain the corresponding responses in the time domain:

$$\frac{C_{out}(t)}{C_S} = \begin{cases} [1-\exp(-t/\theta)]; \; C_S \equiv \dot{m}/Q & \text{step response} \\ \exp(-t/\theta) \quad ; \; C_S \equiv m/V & \text{impulse response} \end{cases} \tag{18.1-28}$$

For the step response, C_S is the steady-state tracer concentration in the limit as $t \to \infty$. Since there is no loss of tracer, the steady-state output concentration is equal to the entering tracer concentration. For the impulse response, the initial concentration $C_S = m/V$ is the mathematical equivalent of mixing the m moles of injected tracer throughout the compartment volume V.

Figure 18.1-6 provides a comparison between these two idealized responses. The dimensionless step response rises exponentially from zero and reaches within 5% of its steady-state value of one when $t = 3\theta$. The dimensionless impulse response decays exponentially from one to within 5% of its asymptotic value of zero at $t = 3\theta$.

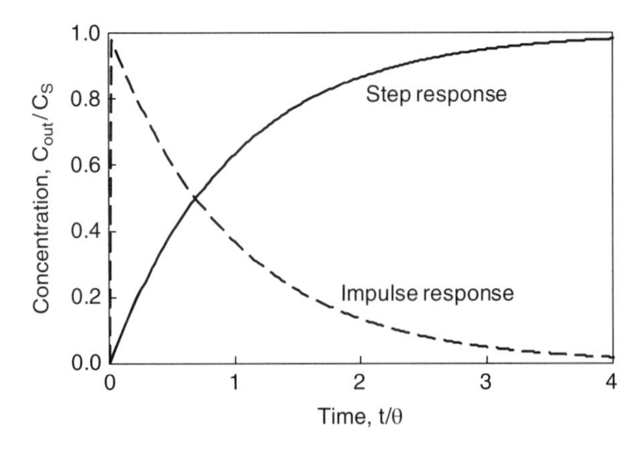

Figure 18.1-6 Responses of a perfectly mixed compartment to an inert tracer.

Notice that the time derivative of the step response in Equation 18.1-28a is equal to the impulse response given in Equation 18.1-28b. This is a general relationship for any time-invariant, linear system that is appropriately scaled. For such a system, a unit step input $\widetilde{C}_{in}(s) = 1/s$ results in an output $\widetilde{C}_{out}(s) = \widetilde{g}(s)/s$, whereas a unit impulse input $\widetilde{C}_{in}(s) = 1$ generates an output $\widetilde{C}_{out}(s) = \widetilde{g}(s)$. Thus, in the Laplace domain, the unit impulse response is "s" times the unit-step response. In the time domain, this is equivalent to taking the derivative of the unit-step response (Table C4-1).

Tracer Loss Present

We now consider a first-order rate of tracer loss expressed as $\dot{R}_{loss} = k_{loss}C$ so that Equation 18.1-23a becomes

$$\theta\frac{dC}{dt} + \left(1 + \frac{k_{loss}}{Q}\right)C = C_{in} \tag{18.1-29}$$

With tracer loss by chemical reaction, k_{loss} is proportional to the compartment volume, but with tracer loss by diffusion, k_{loss} is proportional to the compartment surface. In either case, we assume that both compartment volume and surface are constants so that k_{loss} is independent of time. Transforming to the Laplace domain with $C(0) = 0$, we obtain the transfer function

$$\widetilde{g}(s) = \frac{1/(1 + k_{loss}/Q)}{1 + \theta_{loss}s}; \quad \theta_{loss} = \frac{\theta}{1 + k_{loss}/Q} \tag{18.1-30a,b}$$

For tracer input prescribed as an ideal step or impulse applied at $t = 0$, this transfer function leads to the following outputs in the time domain:

$$\frac{C_{out}(t)}{C_S} = \begin{cases} [1 - \exp(-t/\theta_{loss})]; \; C_S \equiv \dot{m}/(Q + k_{loss}) & \text{step response} \\ \exp(-t/\theta_{loss}) \quad ; \; C_S \equiv m/V & \text{impulse response} \end{cases} \tag{18.1-31}$$

Thus, compared to inert tracer dynamics (Eq. 18.1-28), tracer loss causes a decrease in the steady-state concentration $\dot{m}/(Q + k_{loss})$ of the step response, but does not affect the initial concentration m/V of the impulse response. Because of a reduction in the exponential time constant (i.e., $\theta_{loss} < \theta$), tracer loss also speeds up the system dynamics.

18.2 Multiple-Compartment Models

Models of biomedical systems commonly require more than one compartment in order to simulate dynamic outputs and quantitatively characterize the effects of mixing processes. In this section, we consider transport of an inert tracer through systems that have a single input and a single output (SISO).

18.2.1 Two Compartments in Series

A series of two perfectly mixed compartments of volumes V_1 and V_2 (Fig. 18.2-1) is an extension of the single-compartment model that accounts for imperfect mixing in the principal flow direction. When this model has the same basic assumptions as the single-compartment

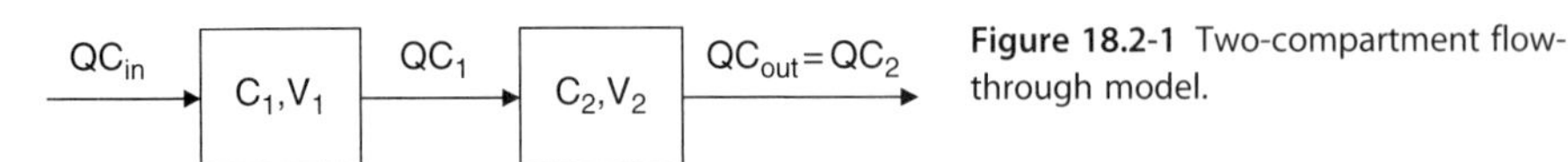

Figure 18.2-1 Two-compartment flow-through model.

model, solution mass balances on each of the compartments indicate that the volumetric flows of all streams are equal to Q. A molar balance for an inert tracer flow through each compartment then leads to

$$\theta_1 \frac{dC_1}{dt} + C_1 = C_{in} \tag{18.2-1}$$

$$\theta_2 \frac{dC_2}{dt} + C_2 = C_1 \tag{18.2-2}$$

The compartment residence times, θ_1 and θ_2, and the overall residence time θ of tracer in the system are defined as

$$\theta_j \equiv \frac{V_j}{Q} \; (j = 1,2), \quad \theta \equiv \frac{V}{Q} = \frac{V_1 + V_2}{Q} = \theta_1 + \theta_2 \tag{18.2-3a,b}$$

By taking the Laplace transforms of Equations 18.2-1 and 18.2-2 with the initial conditions $C_1(0) = C_2(0) = 0$, we obtain the individual transfer function of each compartment.

$$\tilde{g}_1(s) = \frac{\tilde{C}_1(s)}{\tilde{C}_{in}(s)} = \frac{1}{1 + \theta_1 s}, \quad \tilde{g}_2(s) = \frac{\tilde{C}_2(s)}{\tilde{C}_1(s)} = \frac{1}{1 + \theta_2 s} \tag{18.2-4a,b}$$

Notice that each of these transfer functions has the same form as that of the single-compartment model (Eq. 18.1-26). Since the output concentration of the entire system equals the output concentration of the second compartment, $\tilde{C}_{out}(s) = \tilde{C}_2(s)$, the overall transfer function is the product of the individual transfer functions:

$$\frac{\tilde{C}_{out}(s)}{\tilde{C}_{in}(s)} = \frac{\tilde{C}_1(s)\,\tilde{C}_2(s)}{\tilde{C}_{in}(s)\,\tilde{C}_1(s)} \Rightarrow \tilde{g}(s) = \tilde{g}_1(s)\tilde{g}_2(s) = \left(\frac{1}{1 + \theta_1 s}\right)\left(\frac{1}{1 + \theta_2 s}\right) \tag{18.2-5a,b}$$

With injection of tracer into the first compartment at $t = 0$ as either as a step or an impulse, the overall transfer function provides the following outputs in the Laplace domain:

$$\tilde{C}_{out}(t) = \begin{cases} \dfrac{\dot{m}}{Q}\left(\dfrac{1}{s}\right)\left(\dfrac{1}{1 + \theta_1 s}\right)\left(\dfrac{1}{1 + \theta_2 s}\right) & \text{step response} \\ \dfrac{m}{Q}\left(\dfrac{1}{1 + \theta_1 s}\right)\left(\dfrac{1}{1 + \theta_2 s}\right) & \text{impulse response} \end{cases} \tag{18.2-6}$$

The corresponding responses in the time domain are

$$\frac{C_{out}(t)}{C_S} = \begin{cases} \left[1 - \left(\dfrac{1}{\theta_1 - \theta_2}\right)\left(\theta_1 e^{-t/\theta_1} - \theta_2 e^{-t/\theta_2}\right)\right]; \; C_S \equiv \dfrac{\dot{m}}{Q} & \text{step response} \\ \left(\dfrac{\theta}{\theta_1 - \theta_2}\right)\left(e^{-t/\theta_1} - e^{-t/\theta_2}\right); \; C_S \equiv \dfrac{m}{V} & \text{impulse response} \end{cases} \tag{18.2-7}$$

Figure 18.2-2 illustrates the step and impulse responses when compartment 2 has twice the residence time as compartment 1. In that case, $\theta = 3\theta_1 = 3\theta_2/2$, so the responses are

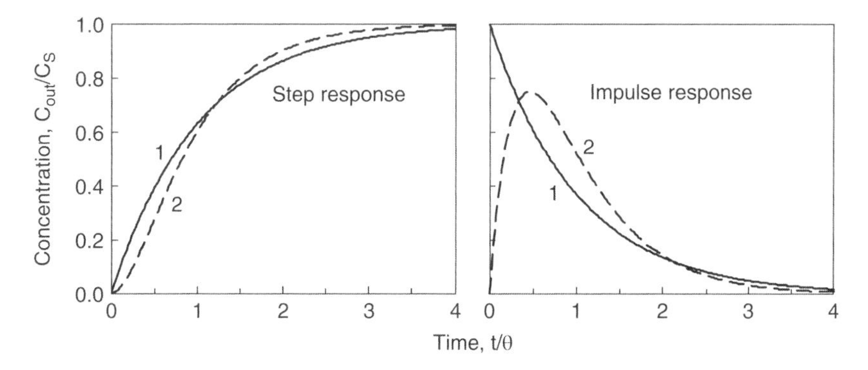

Figure 18.2-2 Responses of single-compartment (1) and two-compartment (2) models.

$$\frac{C_{out}(t)}{C_S} = \begin{cases} 1 - 2e^{-3t/2\theta} + e^{-3t/\theta} & \text{step response} \\ 3\left(e^{-3t/2\theta} - e^{-3t/\theta}\right) & \text{impulse response} \end{cases} \tag{18.2-8}$$

This figure also compares the responses of the two-compartment model to that of a single-compartment. The step response of the two-compartment model shows an inflection point, which is not present in the single-compartment model. For the impulse responses, the time to reach a maximum concentration is longer in the two-compartment than is the single-compartment model.

The moles of an inert substance injected into a SISO system must eventually leave the system whether it is represented by a single-compartment or two-compartment model. This means that the injected moles m should be equal to the integral of the molar output rate QC_{out} over all times after tracer injection:

$$m = \int_0^\infty QC_{out}dt = Q\int_0^\infty C_{out}dt \tag{18.2-9}$$

For an impulse response, in particular, the dimensionless output concentration, $C_{out}/C_S = (V/m)C_{out}$, integrated over the dimensionless time, $t/\theta = (Q/V)t$ is given by

$$\int_0^\infty \frac{C_{out}}{C_S} d\left(\frac{t}{\theta}\right) = \int_0^\infty \left(\frac{V}{m}C_{out}\right) d\left(\frac{Qt}{V}\right) = \frac{Q}{m}\left(\int_0^\infty C_{out}dt\right) = 1 \tag{18.2-10}$$

Thus, the areas under either of the impulse-response curves on the right side of Figure 18.2-2 must be equal to one.

18.2.2 Multiple Compartments in Series

We can extend our two-compartment model to describe J compartments arranged in series (Fig. 18.2-3) by finding the overall transfer function $\tilde{g}(s)$ from the product of the individual compartment transfer functions. By induction from Equation 18.2-5b,

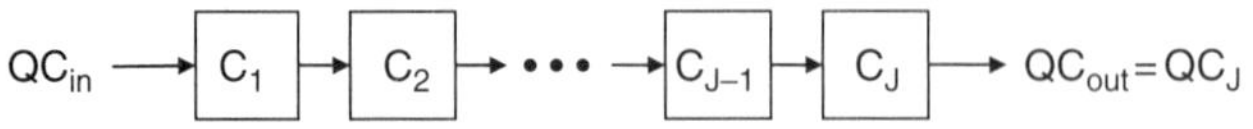

$QC_{in} \rightarrow C_1 \rightarrow C_2 \rightarrow \cdots \rightarrow C_{J-1} \rightarrow C_J \rightarrow QC_{out} = QC_J$

Figure 18.2-3 Series of J flow-through compartments.

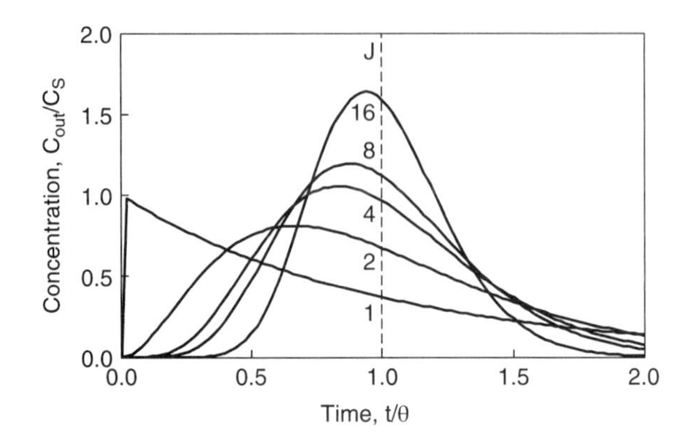

Figure 18.2-4 Impulse response of J identical compartments in series.

$$\tilde{g}(s) = \prod_{j=1}^{J} \left(\frac{1}{1 + \theta_j s} \right)^j \tag{18.2-11}$$

where $\theta_j \equiv V_j/Q$ is the residence time in compartment j. For the special case of J equal-sized compartments of total volume V and overall residence time $\theta = V/Q$, the compartment volumes are V/J and compartmental residence times are $\theta_j = (V/J)/Q = \theta/J$. The resulting transfer function is

$$\tilde{g}(s) = \left[\frac{1}{1 + (\theta/J)s} \right]^J \tag{18.2-12}$$

The response of this system to an impulse input into compartment 1 for which $\tilde{C}_{in} = m/Q$ is

$$\tilde{C}_{out}(s) = \frac{m}{Q} \left[\frac{1}{1 + (\theta/J)s} \right]^J \tag{18.2-13}$$

This corresponds to a dimensionless time response of

$$\frac{C_{out}(t)}{C_S} = \frac{J(Jt/\theta)^{J-1} \exp(-Jt/\theta)}{(J-1)!}; \quad C_S \equiv \frac{m}{V} \tag{18.2-14a, b}$$

Figure 18.2-4 illustrates the effect on the impulse response of increasing the number of compartments J, while keeping the same value of the overall residence time of the system. Perfect mixing occurs when $J = 1$. As more compartments are added, mixing is reduced because the fluid is segregated into additional, smaller volumes. This causes the impulse-response curve to become more symmetric and narrower with a peak that is closer to $t = \theta$ and a tail that decays more rapidly toward $C_{out} = 0$. As is the case for single- or two-compartment models, the area under each of the dimensionless response curves in the figure must be equal to one since Equation 18.2-10 still applies.

In the limiting case of an infinite number of compartments of equal volume, the transfer function becomes

$$\tilde{g}(s) = \lim_{J\to\infty}\left[\frac{1}{1+(\theta/J)s}\right]^J = \lim_{J\to\infty}\left[1+\left(\frac{\theta s}{J}\right)\right]^{-J} = \lim_{J\to\infty}\left[\exp\left(\frac{\theta s}{J}\right)\right]^{-J} = e^{-\theta s} \qquad (18.2\text{-}15)$$

Thus, for any arbitrary input,

$$\tilde{C}_{out}(s) = e^{-\theta s}\tilde{C}_{in}(s) \qquad (18.2\text{-}16)$$

The inverse transform yields

$$C_{out}(t) = C_{in}(t-\theta) \qquad (18.2\text{-}17)$$

In other words, an infinite series of identical compartments does not produce any mixing of tracer. It only acts to delay the output concentration by the residence time θ. This is mathematically equivalent to tracer transport through a tube of length L at a uniform velocity L/θ without any axial mixing.

18.2.3 Parallel Compartments without Interaction

Parallel arrangements of flow-through compartments can represent some physiological systems. For example, the circulatory system is often modeled by the parallel flow of blood through the vital organs. We consider a SISO system in which inert tracer flows into a splitting node that distributes the flow between two parallel compartments without any change in the inlet tracer concentration C_{in} (Fig. 18.2-5). The compartments discharge their solutions to a common mixing node, but they do not directly exchange tracer. As before, we assume constant solution density and compartment volumes so that the inlet flow to each compartment is equal to its outlet flow. A solution balance around the entire system indicates that

$$Q_{out} = Q_{in} \equiv Q \qquad (18.2\text{-}18)$$

A tracer balance around the downstream mixing node that has no volume yields

$$Q_{out}C_{out} = Q_1C_1 + Q_2C_2 \;\Rightarrow\; C_{out} = \frac{Q_1C_1 + Q_2C_2}{Q} \qquad (18.2\text{-}19a,b)$$

The Laplace transform of this equation is as follows:

$$\tilde{C}_{out} = \mathcal{Q}_1\tilde{C}_1 + \mathcal{Q}_2\tilde{C}_2; \quad \mathcal{Q}_j \equiv \frac{Q_j}{Q} \quad (j = 1,2) \qquad (18.2\text{-}20a,b)$$

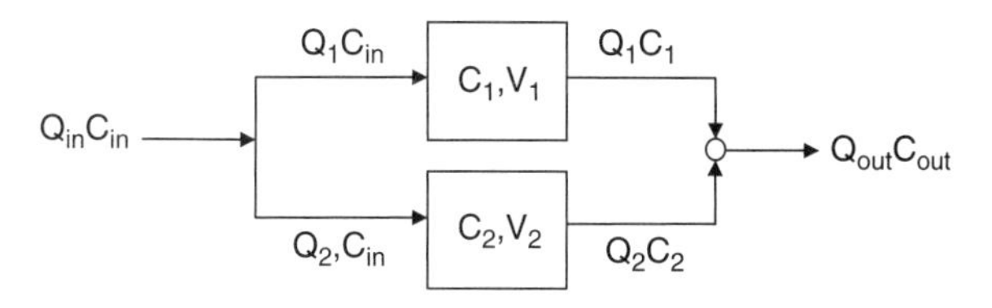

Figure 18.2-5 Parallel-compartment model without interaction.

where Q_j is the fraction of the total flow distributed to compartment j. The transfer function of each compartment subject to a zero-concentration initial condition is the same as the transfer function of the single-compartment model:

$$\tilde{g}_j(s) = \frac{\tilde{C}_j}{\tilde{C}_{in}} = \frac{1}{1+\theta_j s}; \quad \theta_j \equiv \frac{V_j}{Q_j} \quad (j = 1,2) \tag{18.2-21a,b}$$

where θ_j represents the residence time of tracer in compartment j. By combining Equations 18.2-20a and 18.2-21a, we arrive at the overall transfer function:

$$\tilde{g}(s) = \frac{\tilde{C}_{out}}{\tilde{C}_{in}} = \frac{Q_1}{1+\theta_1 s} + \frac{Q_2}{1+\theta_2 s} \tag{18.2-22}$$

For a step input of tracer, $\tilde{C}_{in} = \dot{m}/Q$ and the output in the Laplace domain is

$$\frac{\tilde{C}_{out}(s)}{C_S} = \frac{1}{s}\left(\frac{Q_1}{1+\theta_1 s} + \frac{Q_2}{1+\theta_2 s}\right); \quad C_S \equiv \frac{\dot{m}}{Q} \tag{18.2-23a,b}$$

From the inverse Laplace transform, we get the dimensionless response in the time domain:

$$\frac{C_{out}(t)}{C_S} = \left(1 - Q_1 e^{-t/\theta_1} - Q_2 e^{-t/\theta_2}\right) \tag{18.2-24}$$

We can analyze this response for any specific parameter set. For example, consider a system in which the volumetric flows through the compartments are equal such that $Q_1 = Q_2 = 1/2$ which implies that $\theta_2/\theta_1 = V_2/V_1$. The step response then becomes

$$\frac{C_{out}(t)}{C_S} = \left\{1 - \frac{1}{2}\exp\left[-\frac{t}{\theta_1}\right] - \frac{1}{2}\exp\left[-\left(\frac{\theta_1}{\theta_2}\right)\frac{t}{\theta_1}\right]\right\} \tag{18.2-25}$$

When $\theta_1 = \theta_2$, this step response is the same as that of a single-compartment model (Eq. 18.1-28a) whose residence time is equal to θ_1. For a sufficiently short time, we can approximate the exponentials as $\exp(\alpha t) \approx 1 - \alpha t$ and the response simplifies to

$$\frac{C_{out}(t)}{C_S} = \left(1 + \frac{\theta_1}{\theta_2}\right)\frac{t}{\theta_1} \tag{18.2-26}$$

According to Equation 18.2-25, a plot of dimensionless concentration C/C_s versus dimensionless time t/θ_1 depends only on the residence time ratio θ_1/θ_2. Figure 18.2-6 illustrates these dynamics. Consistent with Equation 18.2-26, the initial slopes of the curves, $(1 + \theta_1/\theta_2)$, increase with the decreasing residence time ratio θ_2/θ_1.

When $\theta_2/\theta_1 = 0.02$, tracer washout from compartment 2 is so rapid relative to compartment 1 that tracer appears almost instantaneously at its flow outlet. In contrast, transport through compartment 1 does not influence the response until later times. This situation results in a biphasic response curve, that is, a rapid rise in tracer concentration over a short time and a relatively long-time approach to steady state. As θ_2/θ_1 becomes greater than 0.02, tracer washout from compartment 2 takes longer. Thus, the concentration response changes more slowly and its biphasic behavior disappears.

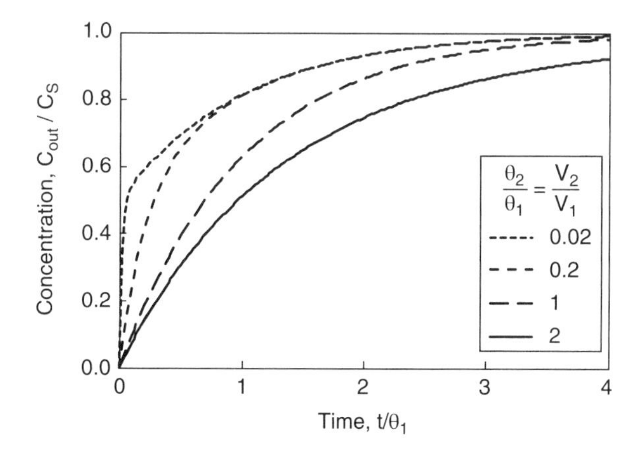

Figure 18.2-6 Step response of noninteractive model with equal compartment flows.

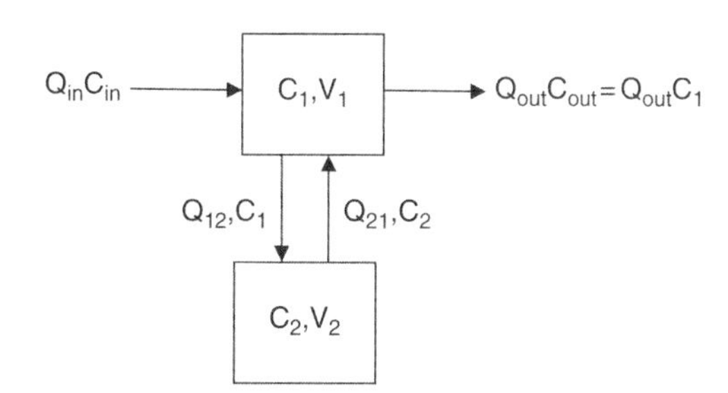

Figure 18.2-7 Parallel-compartment model with flow interaction.

18.2.4 Parallel Compartments with Flow Interaction

We now consider a parallel-compartment model in which solutes can be transported directly between the compartments by convection at volumetric flow rates Q_{12} and Q_{21} (Fig. 18.2-7). Such a model could represent the convective movement of blood between a well-perfused and a poorly perfused region of a particular tissue. With constant mass densities and compartment volumes, solution balances around compartment 2 and around the entire system indicate that

$$Q_{12} = Q_{21}, \quad Q_{out} = Q_{in} \equiv Q \tag{18.2-27a,b}$$

The corresponding solute balances are

$$V_1 \frac{dC_1}{dt} = (Q_{in}C_{in} + Q_{21}C_2) - (Q_{out}C_1 + Q_{12}C_1) \tag{18.2-28}$$

$$V_2 \frac{dC_2}{dt} = Q_{12}C_1 - Q_{21}C_2 \tag{18.2-29}$$

Combining these four equations, we obtain the final model equations:

$$\theta_1 \frac{dC_1}{dt} + (1 + \mathcal{Q}_{12})C_1 = C_{in} + \mathcal{Q}_{12}C_2 \tag{18.2-30}$$

$$\theta_2 \frac{dC_2}{dt} + C_2 = C_1 \tag{18.2-31}$$

where the inter-compartmental flow ratio and the compartment residence times are defined as

$$\mathcal{Q}_{12} = \frac{Q_{12}}{Q}, \quad \theta_1 \equiv \frac{V_1}{Q}, \quad \theta_2 \equiv \frac{V_2}{Q_{12}} \qquad (18.2\text{-}32a\text{–}c)$$

The interaction between compartments is evident by a coupling between the differential model equations, which are linear and have constant coefficients. By taking the Laplace transforms of the two model equations with the initial condition $C_1(0) = C_2(0) = 0$, and then solving for the overall transfer function, we obtain

$$\tilde{g}(s) \equiv \frac{\tilde{C}_{out}}{\tilde{C}_{in}} = \frac{\tilde{C}_1}{\tilde{C}_{in}} = \frac{(1/\theta_1)(s + 1/\theta_2)}{s^2 + (1/\theta_1 + 1/\theta_2 + \mathcal{Q}_{12}/\theta_1)s + 1/\theta_1\theta_2} \qquad (18.2\text{-}33)$$

The roots of the quadratic equation in the denominator are as follows:

$$r_1, r_2 = -\frac{1}{2}\left[(1/\theta_1 + 1/\theta_2 + \mathcal{Q}_{12}/\theta_1) \pm \sqrt{(1/\theta_1 + 1/\theta_2 + \mathcal{Q}_{12}/\theta_1)^2 - 4/\theta_1\theta_2}\right] \qquad (18.2\text{-}34)$$

so that the transfer function can be rewritten as

$$\tilde{g}(s) = \frac{(s + 1/\theta_2)}{\theta_1(s - r_1)(s - r_2)} \qquad (18.2\text{-}35)$$

For a step input introduced at $t = 0$, the dimensionless system response in the Laplace domain is

$$\frac{\tilde{C}_{out}(s)}{C_S} = \frac{(s + 1/\theta_2)}{\theta_1 s(s - r_1)(s - r_2)}; \quad C_S \equiv \frac{\dot{m}}{Q} \qquad (18.2\text{-}36a,b)$$

According to the initial value theorem, the dynamic behavior of $C_{out}(t)/C_S$ at short times is equivalent to the inverse transform of $s\tilde{C}_{out}(s)/C_S$ when s is very large (Table C4-1):

$$\lim_{s\to\infty}\left(\frac{s\tilde{C}_{out}}{C_S}\right) = \frac{1}{\theta_1 s} \Rightarrow \lim_{t\to 0}\left(\frac{C_{out}}{C_S}\right) = \boldsymbol{L}^{-1}\left[\frac{1}{\theta_1 s}\right] = \frac{t}{\theta_1} \qquad (18.2\text{-}37a,b)$$

To find the dynamic behavior at all times, we take the inverse Laplace transform of Equation 18.2-36a:

$$C_{out}(t) = C_S\left\{1 + \frac{1 + (\theta_2/\theta_1)\mathcal{r}_1}{(\theta_2/\theta_1)\mathcal{r}_1^2 - 1}e^{\mathcal{r}_1(t/\theta_1)} + \frac{1 + \mathcal{r}_2}{(\theta_1/\theta_2)\mathcal{r}_2^2 - 1}e^{(\theta_1/\theta_2)\mathcal{r}_2(t/\theta_1)}\right\} \qquad (18.2\text{-}38)$$

Here, we have defined dimensionless roots $\mathcal{r}_1 = \theta_1 r_1$ and $\mathcal{r}_2 = \theta_2 r_2$ whose product $\mathcal{r}_1\mathcal{r}_2$ is equal to one. To demonstrate the behavior of this response, consider what happens when inter-compartmental flow Q_{12} is the same as the overall flow Q so that $\mathcal{Q}_{12} = 1$ and $\theta_2/\theta_1 = V_2/V_1$. In that case,

$$\mathcal{r}_1 = -\frac{1}{2}\left[(2 + \theta_1/\theta_2) + \sqrt{(2 + \theta_1/\theta_2)^2 - 4\theta_1/\theta_2}\right]$$

$$\mathcal{r}_2 = -\frac{1}{2}\left[(1 + 2\theta_2/\theta_1) - \sqrt{(1 + 2\theta_2/\theta_1)^2 - 4\theta_2/\theta_1}\right] \qquad (18.2\text{-}39a,b)$$

Consequently, C_{out}/C_S only depends on t/θ_1 and the θ_2/θ_1 ratio as illustrated in Figure 18.2-8. These conditions are analogous to those for a two-compartment noninteractive model in

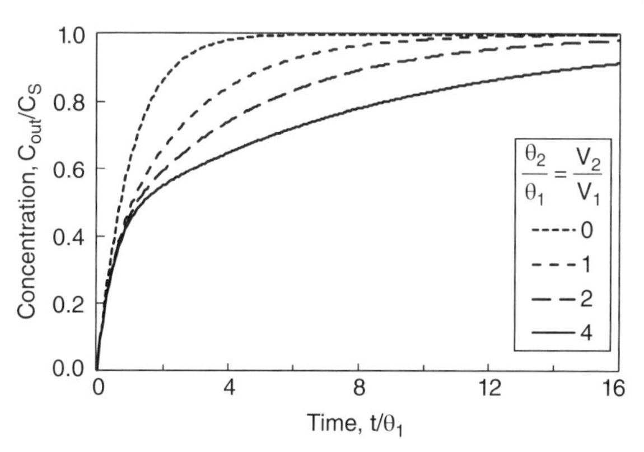

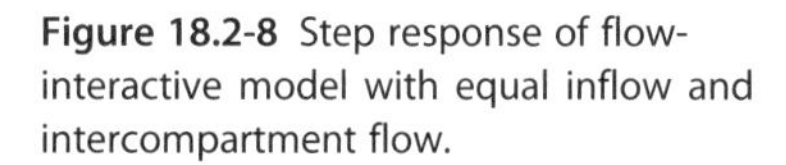
Figure 18.2-8 Step response of flow-interactive model with equal inflow and intercompartment flow.

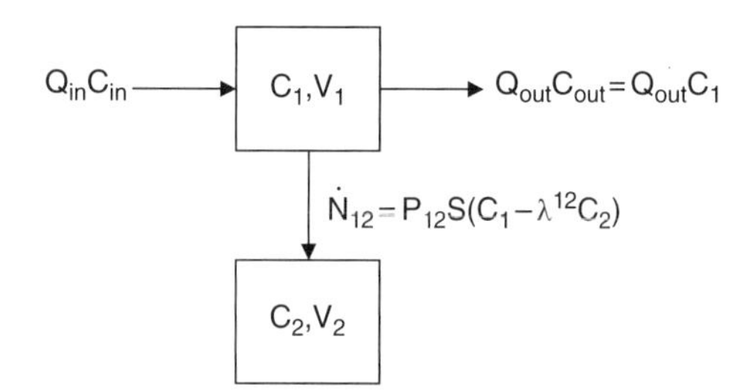

Figure 18.2-9 Parallel-compartment model with diffusion interaction.

Figure 18.2-6. Similar to the behavior of the noninteractive model, the response of the interactive model becomes slower as θ_2/θ_1 increases. Unlike the noninteractive model, however, the response of the interactive model has an initial slope of one at all θ_2/θ_1 values (Eq. 18.2-37b). The interactive model also tends to exhibit a biphasic response at the largest rather than the smallest θ_2/θ_1. This occurs because the relative washout of compartment 2 at a large θ_2/θ_1 is so slow that tracer appears much earlier from compartment 1.

18.2.5 Parallel Compartments with Diffusion Interaction

Mass transport through capillaries and surrounding tissue can be modeled with two interacting compartments (Fig. 18.2-9). While tracer moves through capillary compartment 1 by convective transport, it can be transported into tissue compartment 2 at a molar rate $\dot{N}_{12}$ by diffusion and facilitated transport. We assume that compartment volumes and fluid density are constant, and the mass transfer rate of solution between the two compartments is negligible compared to the mass input rate. A solution balance on the entire system then indicates that input and output flows are equal, $Q_{out} = Q_{in} \equiv Q$. Tracer balances on the individual compartments are

$$V_1 \frac{dC_1}{dt} + Q_{out}C_1 = Q_{in}C_{in} - \dot{N}_{12} \tag{18.2-40}$$

$$\frac{dC_2}{dt} = \frac{\dot{N}_{12}}{V_2} \tag{18.2-41}$$

For an inert tracer that is transported by diffusion only

$$\dot{N}_{12} = P_{12}S\left(C_1 - \lambda^{12}C_2\right) \tag{18.2-42}$$

where $P_{12}S$ represents a permeability-surface area product, and λ^{12} is the equilibrium partition coefficient between the two compartments. The tracer concentration dynamics can now be represented by

$$\theta_1 \frac{dC_1}{dt} + (1 + P_{12})C_1 = C_{in} + P_{12}\lambda^{12}C_2 \tag{18.2-43}$$

$$\theta_2 \frac{dC_2}{dt} + \lambda^{12}C_2 = C_1 \tag{18.2-44}$$

where

$$P_{12} \equiv \frac{P_{12}S}{Q}, \quad \theta_1 \equiv \frac{V_1}{Q}, \quad \theta_2 \equiv \frac{V_2}{P_{12}S} \tag{18.2-45a–c}$$

Taking the Laplace transform of the tracer balances subject to the initial condition $C_1(0) = C_2(0) = 0$, we find the overall transfer function:

$$\tilde{g}(s) = \frac{\tilde{C}_{out}}{\tilde{C}_{in}} = \frac{\tilde{C}_1}{\tilde{C}_{in}} = \frac{(1/\theta_1)\left(s + \lambda^{12}/\theta_2\right)}{s^2 + \left(1/\theta_1 + \lambda^{12}/\theta_2 + P_{12}/\theta_1\right)s + \lambda^{12}/\theta_2\theta_1} \tag{18.2-46}$$

Responses corresponding to this transfer function can be found by the same approach applied to the flow-interactive two-compartment model. When $\lambda^{12} = 1$, the diffusion interaction reduces to $P_{12}S(C_1 - C_2)$, which has the same form as a flow interaction $Q_{12}(C_1 - C_2)$. Under this condition, the input–output behaviors of the interactive diffusion and interactive flow models are the same.

18.3 Nonideal Inputs and Moment Analysis

Evaluating parameters such as inter-compartmental flows and permeabilities is an important objective of modeling biomedical processes. One way of achieving this is to minimize the error (e.g., in a least-squares sense) between simulated model outputs and corresponding experiment measurements. This method may not be successful with limited, noisy data. An alternative approach that characterizes key aspects of the system dynamics is based on a moment analysis of unimodal tracer outputs.

18.3.1 Moments of Dynamic Inputs and Outputs

When measured input and output functions of a tracer, $C_{in}(t)$ and $C_{out}(t)$, have only one maximum (Fig. 18.3-1), it is possible to characterize them based on their moments defined as

$$\mu_{i,in} \equiv \int_0^\infty t^i C_{in} dt, \quad \mu_{i,out} \equiv \int_0^\infty t^i C_{out} dt \quad (i = 0, 1, 2, 3, \ldots) \tag{18.3-1a,b}$$

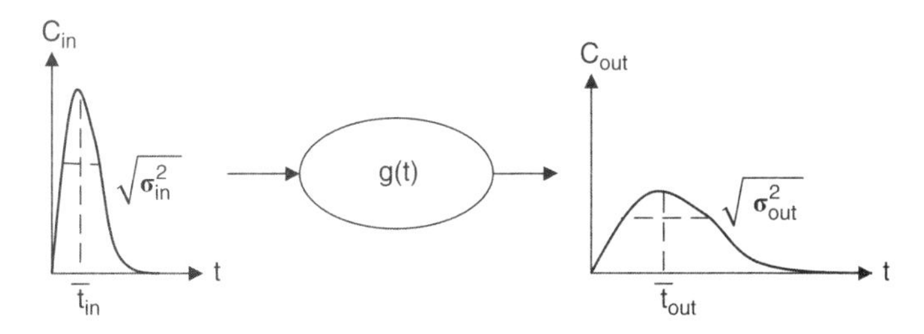

Figure 18.3-1 Input–output behavior of a (single-input)–(single-output) flow-through system.

The moments exist only when the integrands go to zero as time goes to infinity. In concept $\mu_{i,in}$ and $\mu_{i,out}$ of any order, i can be computed from experiments in which $C_{in}(t)$ and $C_{out}(t)$ are simultaneously recorded. Typically, however, data are too noisy to compute accurate moments above the second order. From the three lower-order moments (i = 0,1,2), we can determine two normalized moments at the inlet and at the outlet tracer sampling sites:

$$\bar{t}_j \equiv \frac{\mu_{1,j}}{\mu_{0,j}}, \quad \sigma_j^2 \equiv \frac{\mu_{2,j}}{\mu_{0,j}} - \bar{t}_j^2 \quad (j = \text{in, out}) \tag{18.3-2a, b}$$

The means, $\bar{t}_{in}$ and $\bar{t}_{out}$, represent the average times at which tracer molecules appear at inlet and outlet sampling sites, respectively. The square root of the time variances, $\sqrt{\sigma_{in}^2}$ and $\sqrt{\sigma_{out}^2}$, are the characteristic widths of the tracer distributions about the mean times at the two sites. A dimensionless measure of the time variance is the coefficient of variation:

$$CV_j = \frac{\sqrt{\sigma_j^2}}{\bar{t}_j} \quad (j = \text{in, out}) \tag{18.3-3}$$

To characterize changes within a system, we can evaluate changes in mean and variance between output and input sites:

$$\Delta\bar{t} \equiv \bar{t}_{out} - \bar{t}_{in} = \frac{\mu_{out,1}}{\mu_{out,0}} - \frac{\mu_{in,1}}{\mu_{in,0}} \tag{18.3-4}$$

$$\Delta\sigma^2 \equiv \sigma_{out}^2 - \sigma_{in}^2 = \left(\frac{\mu_{out,2}}{\mu_{out,0}} - \bar{t}_{out}^2\right) - \left(\frac{\mu_{in,2}}{\mu_{in,0}} - \bar{t}_{in}^2\right) \tag{18.3-5}$$

Whereas $\Delta\bar{t}$ is the average transit time of tracer molecules between its inlet and outlet sampling sites, $\sqrt{\Delta\sigma^2}$ represents the increased deviation of transit times from their mean during transport between the sites.

18.3.2 Relationship of Transfer Function to Impulse-Response Function

Since tracer concentrations are usually very low, the input–output relation of a system as indicated by tracer concentration is linear. Furthermore, if any changes in the characteristic time constants of the system are small within the duration of an experiment, then the input–output relation is also time invariant. As we have illustrated previously, any time-invariant, linear SISO system with a zero-initial tracer concentration has a concentration response that can be found by using the input–output relation in the Laplace transform domain:

$$\tilde{C}_{out}(s) = \tilde{g}(s)\tilde{C}_{in}(s) \tag{18.3-6}$$

In the time domain, this is equivalent to the convolution integral (Table C4-1):

$$C_{out}(t) = \int_0^t g(t-t_o)C_{in}(t_o)dt_o \tag{18.3-7}$$

where g(t) is the inverse Laplace transform of $\tilde{g}(s)$. If an impulse of m moles $m\delta(t)$ is introduced into a system with a flow Q, then $\tilde{C}_{in}(t) = m/Q$ and

$$\tilde{g}(s) = \frac{Q}{m}\tilde{C}_{out}(s) \Leftrightarrow g(t) = \frac{Q}{m}C_{out}(t) \tag{18.3-8a, b}$$

Therefore, the transfer function $\tilde{g}(s)$ is proportional to the Laplace transform of an impulse-response; its inverse, g(t), is proportional to an impulse-response function in the time domain.

18.3.3 Moment Relationships for a Nonideal Input Response

The Laplace transform of a function f(t) is defined by Equation C4-1:

$$\tilde{f}(s) \equiv \int_0^\infty e^{-st}f(t)dt \tag{18.3-9}$$

By taking the i-th derivative with respect to s, we obtain

$$\frac{d^i\tilde{f}(s)}{ds^i} = \frac{d^i}{ds^i}\left(\int_0^\infty e^{-st}f(t)dt\right) = (-1)^i\int_0^\infty e^{-st}t^if(t)dt \tag{18.3-10}$$

and in the limit of $s \to 0$,

$$\lim_{s\to 0}\left(\frac{d^i\tilde{f}}{ds^i}\right) = (-1)^i\int_0^\infty t^if(t)dt = (-1)^i\mu_{i,f} \tag{18.3-11}$$

This result provides a framework for determining the moments of any function of time from the Laplace transform of the function. If f(t) corresponds to a time-varying tracer concentration $C_j(t)$ at sampling point j, then

$$\lim_{s\to 0}\left(\frac{d^i\tilde{C}_j}{ds^i}\right) = (-1)^i\int_0^\infty t^iC_j(t)dt = (-1)^i\mu_{i,j} \quad (i = 0, 1, 2;\ j = \text{in, out}) \tag{18.3-12}$$

If f(t) corresponds to the time-domain transfer function g(t), then

$$\lim_{s\to 0}\left(\frac{d^i\tilde{g}}{ds^i}\right) = (-1)^i\int_0^\infty t^ig(t)dt \equiv (-1)^i\mu_{i,g} \tag{18.3-13}$$

Example 18.3-1 Moments of a Single-Compartment Model

In this problem, we find the moments of the time-domain transfer function, $\bar{t}_g$ and σ_g^2, for an inert tracer flowing at a volumetric rate Q through a single well-mixed compartment of volume V for which there is a first-order loss of tracer.

Solution

From Equation 18.1-30, the transfer function for this compartment is

$$\tilde{g}(s) = \frac{Q/V}{s + (Q + k_{loss})/V} \tag{18.3-14}$$

According to Equation 18.3-13, the leading moments of $\tilde{g}(s)$ are

$$\begin{aligned} \mu_{0,g} &= \tilde{g}(0) = \frac{1}{(1 + k_{loss}/Q)} \\ \mu_{1,g} &= -\left[\frac{d\tilde{g}}{ds}\right]_{s\to 0} = \frac{V/Q}{(1 + k_{loss}/Q)^2} \\ \mu_{2,g} &= \left[\frac{d^2\tilde{g}}{ds^2}\right]_{s\to 0} = \frac{2(V/Q)^2}{(1 + k_{loss}/Q)^3} \end{aligned} \tag{18.3-15a–c}$$

Equation 18.3-15a indicates that the inert tracer is completely recovered at the outlet of the system when $k_{loss} = 0$. Otherwise, there is a fractional loss by the factor $\mu_{0,h} = 1/(1 + k_{loss}/Q)$. Substituting Equation 18.3-15b,c into Equations 18.3-2a,b and 18.3-3, we obtain the dimensionless mean, variance, and coefficient of variation of g(t):

$$\bar{t}_g = \frac{V/Q}{1 + k_{loss}/Q}, \quad \sigma_g^2 = \left(\frac{V/Q}{1 + k_{loss}/Q}\right)^2, \quad CV_g = 1 \tag{18.3-16a–c}$$

By applying these equations to experimentally determined values of $\bar{t}_g$ and σ_g^2, we can quantify the ratios V/Q and k_{loss}/Q. If Q is known independently, then V and k_{loss} can be evaluated.

To evaluate a transport model, we need to relate the output moments $\mu_{i,out}$ and input moments $\mu_{i,in}$ of experimentally observed concentration dynamics to the theoretical moments $\mu_{i,g}$ of the transfer function $\tilde{g}$. Taking the zeroth, first, and second derivatives with respect to s of Equation 18.3-6,

$$\begin{aligned} \tilde{C}_{out} &= \tilde{g}\tilde{C}_{in} \\ \frac{d\tilde{C}_{out}}{ds} &= \tilde{g}\frac{d\tilde{C}_{in}}{ds} + \frac{d\tilde{g}}{ds}\tilde{C}_{in} \\ \frac{d^2\tilde{C}_{out}}{ds^2} &= \tilde{g}\frac{d^2\tilde{C}_{in}}{ds^2} + 2\frac{d\tilde{g}}{ds}\frac{d\tilde{C}_{in}}{ds} + \frac{d^2\tilde{g}}{ds^2}\tilde{C}_{in} \end{aligned} \tag{18.3-17a–c}$$

In the limit as $s \to 0$, these equations in consideration of Equations 18.3-12 and 18.3-13 lead to

$$\begin{aligned} \mu_{0,out} &= \mu_{0,g}\mu_{0,in} \\ \mu_{1,out} &= \mu_{0,g}\mu_{1,in} + \mu_{1,g}\mu_{0,in} \\ \mu_{2,out} &= \mu_{0,g}\mu_{2,in} + 2\mu_{1,g}\mu_{1,in} + \mu_{2,g}\mu_{0,in} \end{aligned} \tag{18.3-18a–c}$$

The moles of tracer transported into and out of a SISO system by a volumetric flow Q are $m_{in} = \int QC_{in}dt = Q\mu_{0,in}$ and $m_{out} = \int QC_{out}dt = Q\mu_{0,out}$. According to Equation 18.3-18a,

$\mu_{0,g} = \mu_{0,out}/\mu_{0,in} = m_{out}/m_{in}$, indicating that the zeroth moment of g(t) corresponds to the fractional recovery of tracer from the system. For an inert tracer that is completely recovered, $\mu_{0,g} = 1$. For a tracer that is lost from a system, we expect $\mu_{0,g} < 1$. Combining Equation 18.3-18a–c with Equations 18.3-4 and 18.3-5, we find

$$\bar{t}_{out} = \bar{t}_{in} + \bar{t}_g, \quad \sigma^2_{out} = \sigma^2_{in} + \sigma^2_g \tag{18.3-19a,b}$$

That is, the mean of a concentration response obtained from a SISO system is the sum of the mean of the concentration input pattern with the mean of g(t). This summation rule also applies to the variance of the concentration response. Consider an impulse input appearing at $t = 0$. Since its concentration distribution has zero width, both the mean and variance of this input are zero. Consequently, $\bar{t}_{out} = \bar{t}_g$ and $\sigma^2_{out} = \sigma^2_g$. This important observation indicates that the mean and variance of g(t) are equivalent to the mean and variance of a system's ideal impulse response.

These results can be generalized to a system composed of SISO subsystems of one or more compartments arranged in series. Suppose $\bar{t}_j$ and σ^2_j represent the mean and variance at the output of subsystem j in response to an arbitrary input, whereas $\left(\bar{t}_j\right)_g$ and $\left(\sigma^2_j\right)_g$ are the mean and variance of subsystem j in response to an impulse introduced at $t = 0$. We can then relate the output of subsystem j to its input with Equation 18.3-19a,b:

$$\bar{t}_j = \bar{t}_{j-1} + \left(\bar{t}_j\right)_g, \quad \sigma^2_j = \sigma^2_{j-1} + \left(\sigma^2_j\right)_g \tag{18.3-20a,b}$$

For a system consisting of J such subsystems, we find the mean time and variance of the system as a whole by induction:

$$\begin{aligned} \bar{t}_g &= \bar{t}_{out} - \bar{t}_{in} = \bar{t}_J - \bar{t}_0 \equiv \sum_{j=1}^{J}\left(\bar{t}_j - \bar{t}_{j-1}\right) = \sum_{j=1}^{J}\left(\bar{t}_j\right)_g \\ \sigma^2_g &= \sigma^2_{out} - \sigma^2_{in} = \sigma^2_J - \sigma^2_0 \equiv \sum_{j=1}^{J}\left(\sigma^2_j - \sigma^2_{j-1}\right) = \sum_{j=1}^{J}\left(\sigma^2_j\right)_g \end{aligned} \tag{18.3-21a,b}$$

This is a powerful result indicating that the residence time and variance of a system in response to an impulse input can be determined by summing the residence times and variances of the impulse responses associated with underlying in-series subsystems.

Formulas for $\bar{t}_g$ and σ^2_g in various SISO models with a volumetric flow Q and a total volume V are provided in Table 18.3-1. These models are restricted to zero tracer loss so that $\mu_{0,g} = 1$. As required by Equation 18.3-21a,b, entries 3 and 4 for compartments in series have $\bar{t}_g$ and σ^2_g that are the sum of the $\left(\bar{t}_j\right)_g$ and the $\left(\sigma^2_j\right)_g$ associated with their individual compartments. For compartments in parallel, the relationship of the overall $\bar{t}_g$ and σ^2_g to the individual $\left(\bar{t}_j\right)_g$ and $\left(\sigma^2_j\right)_g$ is more complicated. For example, with diffusion interaction, we must account for a possible difference in tracer solubility between compartments. This is reflected by the effective system volume, $V \equiv V_1 + V_2/\lambda^{12}$, that contributes to the mean appearance time in entry 7 of the table.

Table 18.3-1 Moments of a Unit Impulse Response with Zero Loss in SISO Models

Model	Transfer Function, $\tilde{g} = \tilde{C}_{out}/\tilde{C}_{in}$	Mean Appearance Time, $\bar{t}_g$	Variance, σ_g^2
1. Transport delay	Equation 18.2-17	V/Q	0
2. Single compartment (Fig. 13.1-5)	Equation 18.1-26	V/Q	$(V/Q)^2$
3. J Series compartments of unequal volumes V_j (Fig. 13.2-1; J = 2)	Equation 18.2-11	$\frac{V}{Q} = \frac{\sum_{j=1}^{J} V_j}{Q}$	$\sum_{j=1}^{J}\left(\frac{V_j}{Q}\right)^2$
4. J Series compartments of equal volumes V_J (Fig. 13.2-3)	Equation 18.2-12	$\frac{V}{Q} = \frac{JV_J}{Q}$	$J\left(\frac{V_J}{Q}\right)^2$
5. 2 Parallel compartments of volumes V_1 and V_2 without interaction (Fig. 13.2-5)	Equation 18.2-22	$\frac{V}{Q} = \frac{V_1 + V_2}{Q_1 + Q_2}$	$\left(\frac{V}{Q}\right)^2 \times \left[\frac{2Q}{Q_1}\left(\frac{V_1}{V}\right)^2 + \frac{2Q}{Q_2}\left(\frac{V_2}{V}\right)^2 - 1\right]$
6. 2 Parallel compartments of volumes V_1 and V_2 with flow interaction (Fig. 13.2-7)	Equation 18.2-35	$\frac{V}{Q} = \frac{V_1 + V_2}{Q}$	$\left(\frac{V}{Q}\right)^2\left[1 + 2\frac{Q}{Q_{12}}\left(\frac{V_2}{V}\right)^2\right]$
7. 2 Parallel compartments of volumes V_1 and V_2 with diffusion interaction (Fig. 13.2-9)	Equation 18.2-46	$\frac{V_1 + V_2/\lambda^{12}}{Q}$	$\left(\frac{V_1 + V_2/\lambda^{12}}{Q}\right)^2 \times \left[1 + \frac{2Q}{P_{12}S}\left(\frac{V_2}{\lambda^{12}V_1 + V_2}\right)^2\right]$

Example 18.3-2 Tracer Distribution Through an Isolated Tumor

To simultaneously study transport through the intravascular and extravascular spaces of an isolated tumor, Eskey and coworkers (1994) used two inert tracers: (i) fluorescein isothiocyanate (FITC)-labeled albumin that was largely confined to the intravascular space and (ii) deuterium-labeled water (D_2O) that diffuses into the extravascular space. The isolated tumor had a single inlet and a single outlet for perfusate flow. After a mixture of FITC-albumin and D_2O was injected into the entering perfusate, samples were collected from the entering and exiting perfusate at multiple time points.

In Figure 18.3-2, the resulting concentration data for each tracer have been normalized by their zeroth moments. On this basis, the input dynamics of the two tracers were virtually the same so that only the FTIC input data is shown (left graph). The output dynamics for the two tracers were quite distinctive (right graph).

A model of the isolated tumor (Fig. 18.3-3) consists of J identical subsystems arranged in series. Each subsystem is a parallel, two-compartment, diffusion interactive model (Fig. 18.2-9) that has an intravascular compartment volume of V_J and an extravascular compartment of volume V_J'. In this example, we estimate model parameter values by using the method of moments. We then use the convolution integral with the albumin input data to see how well the measured albumin response is simulated by the model.

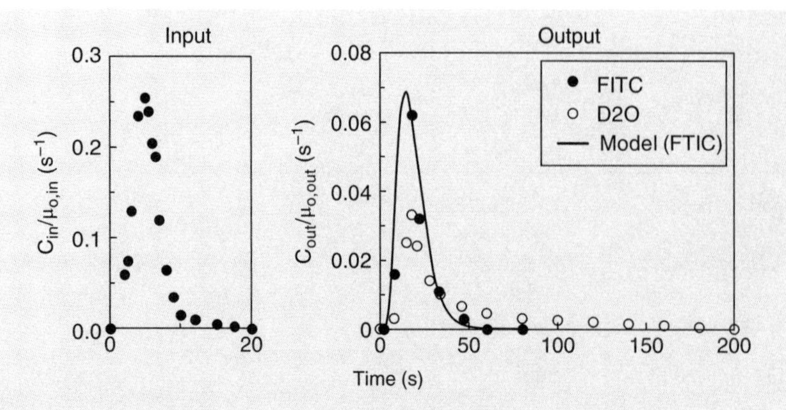

Figure 18.3-2 Concentration input and outputs measured for an isolated tumor.

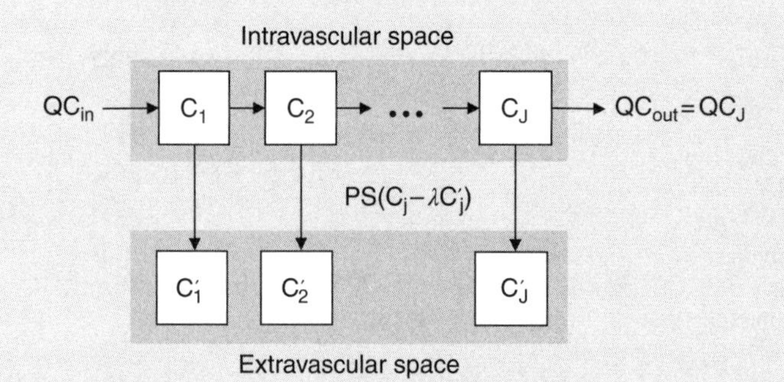

Figure 18.3-3 Model of dual-tracer transport in an isolated tumor.

Solution

Measured Moments. We obtain the first three moments of the input and the two output data sets by numerical integration according to Equation 18.3-1. After converting these values to means and variances by employing Equation 18.3-2a,b, we find

$$\begin{aligned}
\bar{t}_{in}(\text{FITC}) &= \bar{t}_{in}(D_2O) = 5.47 \text{ s} \\
\bar{t}_{out}(\text{FITC}) &= 21.0 \text{ s}, \ t_{out}(D_2O) = 48.3 \text{ s} \\
\sigma_{in}^2(\text{FITC}) &= \sigma_{in}^2(D_2O) = 5.99 \text{ s}^2 \\
\sigma_{out}^2(\text{FITC}) &= 74.4 \text{ s}^2, \ \sigma_{out}^2(D_2O) = 1630 \text{ s}^2
\end{aligned} \tag{18.3-22a–d}$$

Using these values in Equation 18.3-19a,b, we predict the means and variances for an ideal impulse response of the entire tumor:

$$\bar{t}_g(\text{FITC}) = 21.0 - 5.47 = 15.5 \text{ s}, \quad \sigma_g^2(\text{FITC}) = 74.4 - 5.99 = 68.5 \text{ s}^2 \tag{18.3-23a, b}$$

$$\bar{t}_g(D_2O) = 48.3 - 5.47 = 42.8 \text{ s}, \quad \sigma_g^2(D_2O) = 1630 - 5.99 = 1620 \text{ s}^2 \tag{18.3-24a, b}$$

Theoretical Moments. For D_2O that diffuses between the intravascular and extravascular compartments, the mean and variance of the unit impulse response of each subsystem are given by entry 7 of Table 18.3-1. After identifying the parameters in Figure 18.2-9 with the corresponding subsystem parameters in the tumor model (i.e., $V_1 \to V_J$, $V_2 \to V'_J$, $\lambda^{12} \to \lambda$, $P_{12}S \to PS$), we can express the moments of g for each subsystem as

$$\bar{t}_j(D_2O)_g = \frac{V_J + V'_J/\lambda}{Q}, \quad \sigma_j^2(D_2O)_g = \left(\frac{V_J + V'_J/\lambda}{Q}\right)^2 + \frac{2Q}{PS}\left(\frac{V'_J}{\lambda Q}\right)^2 \qquad (18.3\text{-}25a,b)$$

With J identical subsystems, Equation 18.3-21a,b predict that the mean and variance of g for the entire system is J times the mean and variance of one subsystem.

$$\bar{t}(D_2O)_g = J\left(\frac{V_J + V'_J/\lambda}{Q}\right), \quad \sigma(D_2O)_g^2 = J\left[\left(\frac{V_J + V'_J/\lambda}{Q}\right)^2 + \frac{2Q}{PS}\left(\frac{V'_J}{\lambda Q}\right)^2\right] \qquad (18.3\text{-}26a,b)$$

For FITC that does not permeate into the extravascular space, we obtain the corresponding moments of g by setting V'_J equal to zero:

$$\bar{t}(FITC)_g = \frac{JV_J}{Q}, \quad \sigma^2(FITC)_g = J\left(\frac{V_J}{Q}\right)^2 \qquad (18.3\text{-}27a,b)$$

Parameter Estimates. Substituting the experimentally derived values for FITC mean and variance (Eq. 18.3-23) into Equation 18.3-27a,b, we get

$$J\left(\frac{V_J}{Q}\right) = 15.5\ s, \quad J\left(\frac{V_J}{Q}\right)^2 = 68.5\ s^2 \qquad (18.3\text{-}28a,b)$$

This leads to the parameter estimates

$$\frac{V_J}{Q} = \frac{68.5}{15.5} = 4.42\ s, \quad J = \frac{15.5}{4.42} = 2.85\ \text{compartments} \qquad (18.3\text{-}29a,b)$$

Substituting these results as well as the D_2O mean and variance values (Eq. 18.3-24a,b) into Equation 18.3-26, we further obtain

$$\frac{V'_J}{\lambda Q} = 10.6s \quad \frac{PS}{Q} = 0.395 \qquad (18.3\text{-}30a,b)$$

These results indicate that (i) the model should contain $J \approx 3$ subsystems to explain the albumin input response; (ii) since D_2O has essentially the same solubility in the intravascular and extravascular spaces, $\lambda = 1$ and the extravascular volume V'_J is about twice as large as the intravascular volume V_J; and (iii) the permeability-area parameter PS for diffusive interaction is about 40% of the volumetric flow Q through the system.

Goodness of Fit. We will examine the fit of the model to the FTIC albumin input-response data by directly comparing the predicted $C_{out}(t)$ to its measured values. For a series of J identical compartments of total volume $V = JV_J$, overall residence time $\theta = JV_J/Q$ and with no loss of FTIC, Equation 18.2-14 indicates that the unit impulse response is

$$C_{out}(t) = \left(\frac{1}{V}\right)\frac{J(Jt/\theta)^{J-1}\exp(-Jt/\theta)}{(J-1)!} = \frac{t^{J-1}\exp(-Qt/V_J)}{V_J(V_J/Q)^{J-1}(J-1)!} \qquad (18.3\text{-}31)$$

It follows from Equation 18.3-8b that for a unit impulse input,

$$g(t) = QC_{out}(t) = \frac{t^{J-1}\exp(-Qt/V_J)}{(V_J/Q)^J(J-1)!} \qquad (18.3\text{-}32)$$

With this relationship for g(t), we can obtain the response of the model to an arbitrary input concentration by convolution (Eq. 18.3-7):

$$C_{out}(t) = \int_0^t \frac{(t-t_o)^{J-1}\exp[-Q(t-t_o)/V_J]}{(V_J/Q)^J(J-1)!} C_{in}(t_o)dt_o \qquad (18.3\text{-}33)$$

Since the experimental data have been normalized by the area under the curve, which is equivalent to a zero moment, we need to modify this equation. With no loss of FITC from the intravascular space, $\mu_{0,out} = \mu_{0,in}$ and Equation 18.3-33 can be written in terms of the scaled input and output concentration functions as

$$\left\{\frac{C_{out}(t)}{\mu_{0,out}}\right\} = \int_0^t \frac{(t-t_o)^{J-1}\exp[-Q(t-t_o)/V_J]}{(V_J/Q)^J(J-1)!} \left\{\frac{C_{in}(t_o)}{\mu_{0,in}}\right\} dt_o \qquad (18.3\text{-}34)$$

Using parameter estimates of $J = 3$ and $V_J/Q = 4.42$, we numerically integrate this equation at the times when $C_{in}(t)/\mu_{0,in}$ was measured. A smoothed curve through the resulting $C_{out}(t)/\mu_{0,out}$ values is shown in the figure. This curve is a reasonable match to the FITC output data points.

Reference

Eskey CJ, Wolmark N, McDowell C L, Domach MM, Jain RK. Residence time distributions of various tracers in tumors: implications for drug delivery and blood flow measurement. J Natl Cancer Inst. 1994; 86:293–299.

Chapter 19

Compartment Models II

Analysis of Physiological Systems

In this chapter, a variety of pool and physiologically based compartmental models are applied to physiological systems. Some of these models describe inert tracer dynamics in a single organ. Others model the distribution and metabolism of reactive substances among different tissues and organs. The first section of this chapter deals with open-loop systems in which feedback is not considered. Models of the second section incorporate recirculation and feedback, which are essential in the regulation of physiological systems.

19.1 Open-Loop Models

19.1.1 Multipool Model of Glucose Metabolism

Monitoring the dynamics of glucose radioactively tagged with C^{14} is a useful method for studying metabolism and diagnosing metabolic disorders. Experimental data consists of time-varying changes of C^{14} in blood plasma and in expired gas subsequent to the injection of C^{14}-glucose into venous blood. During aerobic metabolism when an excess of O_2 is available for glucose oxidation, these data can be analyzed by the whole-body, multipool model shown in Figure 19.1-1. In this model, glucose and its reaction products are assigned to four well-mixed stores: (1) unreacted glucose in blood; (2) substances reversibly formed from glucose, primarily glycogen; (3) dissolved CO_2 and bound CO_2, mostly HCO_3^-, irreversibly formed by complete oxidation of glucose; (4) other substances such as lactic acid produced irreversibly by non-oxidative metabolism. The model also accounts for the production of CO_2 by the oxidation of untagged lipids and proteins that are present in pool 5. The input stream from the environment is the source term F_S representing the intestinal absorption rate of untagged glucose plus injected tagged glucose. The only output to the environment F_E

Biomedical Mass Transport and Chemical Reaction: Physicochemical Principles and Mathematical Modeling,
First Edition. James S. Ultman, Harihara Baskaran, and Gerald M. Saidel.

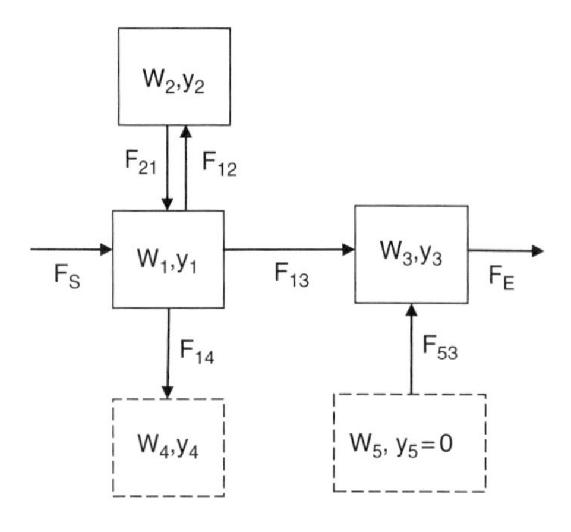

Figure 19.1-1 Multipool model of glucose metabolism.

represents the molar expiration rate of CO_2 from the lungs. The clearance of HCO_3^- by the kidneys from pool 3 is negligible over the time course of the experiment. In addition to F_E, the observable system responses are the radioactivity of the C^{14} fraction $y_1(t)$ measured in blood and the radioactivity of the C^{14} fraction $y_3(t)$ measured in expired breath.

Model Formulation

We define the size of pool j as W_j, the total gram-atoms of tagged and untagged carbon that are associated with substances in the pool. We designate the mole fraction of nuclide y_j as the gram-atoms of C^{14} divided by the total gram-atoms of carbon within pool j. Thus, the amount of C^{14} in pool j is $W_j y_j$. The molar flow rate of all carbon atoms from pool j to pool m is F_{jm} and the corresponding molar flow rate of C^{14} alone is $F_{jm} y_j$. Similarly, the molar flow rates of all carbon atoms and of C^{14} from pool n to pool j are F_{nj} and $F_{nj} y_n$, respectively. The balance equations for total carbon and for C^{14} in pool j are:

$$\frac{dW_j}{dt} = \sum_n F_{nj} - \sum_m F_{jm}, \quad \frac{d\left(W_j y_j\right)}{dt} = \sum_n F_{nj} y_n - \sum_m F_{jm} y_j \tag{19.1-1a,b}$$

For an experimental time sufficiently short relative to the characteristic time of glucose metabolism, the sizes W_j of pools 1, 2, and 3 are approximately constant so their balance equations become

$$\sum_n F_{nj} = \sum_m F_{jm}, \quad W_j \frac{dy_j}{dt} = \sum_n F_{nj} y_n - \sum_m F_{jm} y_j \tag{19.1-2a,b}$$

Applying a total carbon balance (Eq. 19.1-2a) to pools 1, 2, and 3, we obtain

$$F_{12} = F_{21}, \quad F_{13} + F_{14} = F_S, \quad F_E = F_{13} + F_{53} \tag{19.1-3a-c}$$

With a bolus injection of m moles (or gram-atoms) of C^{14} considered as an impulse function, $F_S = m\delta(t)$ and the C^{14} carbon balances (Eq. 19.1-2b) for pools 1, 2, and 3 are

$$W_1 \frac{dy_1}{dt} = -(F_{12} + F_{13} + F_{14})y_1 + F_{12} y_2 + m\delta(t) \tag{19.1-4}$$

$$W_2 \frac{dy_2}{dt} = F_{12}(y_1 - y_2) \quad (19.1\text{-}5)$$

$$W_3 \frac{dy_3}{dt} = F_{13}y_1 - F_E y_3 \quad (19.1\text{-}6)$$

In the absence of nuclide prior to injection of the tagged glucose, the initial conditions are

$$t = 0: \quad y_1 = y_2 = y_3 = 0 \quad (19.1\text{-}7)$$

By transferring the effect of the impulse input to the initial condition (Appendix C5), we can write Equation 19.1-4 in the form

$$W_1 \frac{dy_1}{dt} = -(F_{12} + F_{13} + F_{14})y_1 + F_{12}y_2 \quad (19.1\text{-}8)$$

with the modified initial condition

$$t = 0: \quad y_1 = \frac{m}{W_1}, \quad y_2 = y_3 = 0 \quad (19.1\text{-}9)$$

Equations 19.1-5 and 19.1-6 along with either Equation 19.1-4 or 19.1-8 are sufficient to solve for the measureable system outputs, y_1 and y_3.

Analysis

Specific radioactivity y_i^* in pool i, expressed in units of nuclear disintegrations per total carbon gram-atoms [Bq/g-atom], is proportional to the mole fraction $y_i(t)$ of C^{14} in the pool multiplied by an exponential correction for its nuclear decay over a time interval {0,t}:

$$y_i^* = (\mathcal{A}\ln(2)/T_{1/2})\exp\left[-\frac{\ln(2)}{T^{1/2}}t\right]y_i(t) \quad (19.1\text{-}10)$$

where $\mathcal{A}$ is Avogadro's number and $T_{1/2}$ is the half-life in seconds of the nuclide. Because the half-life of C^{14} is so long relative to the experiments we are modeling, the exponential correction term approaches unity, and the specific radioactivity reduces to $y_j^* \equiv [\mathcal{A}\ln(2)/T_{1/2}]y_j$. With this relation, we can write the model equations (Eqs. 19.1-5, 19.1-6, and 19.1-8) in terms of the y_i^*. In matrix form,

$$\frac{d[y^*]}{dt} = [A][y^*] \quad (19.1\text{-}11)$$

with initial conditions (Eq. 19.1-9) represented by a matrix $[y^*(0)]$. In terms of their elements, these matrices are

$$[y^*] = \begin{bmatrix} y_1^* \\ y_2^* \\ y_3^* \end{bmatrix}, \quad [A] = \begin{bmatrix} A_{11} & A_{12} & 0 \\ A_{21} & A_{22} & 0 \\ A_{31} & 0 & A_{33} \end{bmatrix}, \quad [y^*(0)] = \begin{bmatrix} y_1^*(0) \\ 0 \\ 0 \end{bmatrix} \quad (19.1\text{-}12a\text{-}c)$$

where $A_{11} \equiv -(F_{12} + F_{13} + F_{14})/W_1$, $A_{12} \equiv F_{12}/W_1$, $A_{21} \equiv F_{12}/W_2$, $A_{22} \equiv -F_{12}/W_2$, $A_{31} \equiv F_{13}/W_3$, $A_{33} \equiv -F_E/W_3$, and $y_1^*(0) \equiv [\mathcal{A}\ln(2)/T_{1/2}]m/W_1$. The analytical solution to this set of linear ordinary differential equations is as follows:

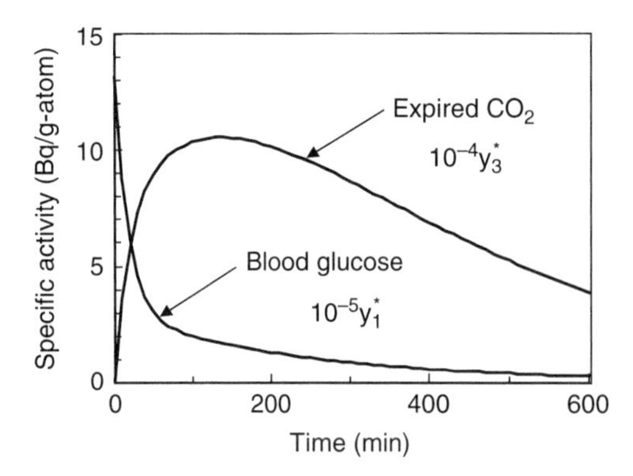

Figure 19.1-2 Response to C^{14}-labeled glucose impulse.

$$y_1^* = y_1^*(0)\left[\frac{(r_1 + A_{21})}{(r_1 - r_2)}e^{r_1 t} + \frac{(r_2 + A_{21})}{(r_2 - r_1)}e^{r_2 t}\right] \tag{19.1-13}$$

$$y_3^* = y_1^*(0)\left[\frac{A_{31}(A_{21} + A_{33})}{(r_1 + A_{33})(r_2 + A_{33})}e^{A_{33}t} - \frac{A_{31}(r_1 - A_{21})}{(r_1 - r_2)(A_{33} + r_1)}e^{r_1 t} - \frac{A_{31}(r_2 - A_{21})}{(r_2 - r_1)(A_{33} + r_2)}e^{r_2 t}\right] \tag{19.1-14}$$

where r_1 and r_2, the roots of a characteristic quadratic equation, are given by

$$r_1, r_2 = \frac{1}{2}\left[(A_{11} - A_{21}) \pm \sqrt{(A_{21} - A_{11})^2 + 4A_{21}(A_{11} + A_{12})}\right] \tag{19.1-15}$$

From an experiment on one human subject in which the measured CO_2 flow was $F_E = 0.00857$ g-atom carbon/min and the radioactivity of the injected bolus was $W_1 y_1^*(0) = 7.23 \times 10^5$ Bq, intercompartmental carbon flows of $F_{12} = 0.0158$, $F_{13} = 0.00538$, and $F_{14} = 0.00254$ g-atom carbon/min were estimated (Spencer *et al.*, 1971). The estimated pool sizes were $W_1 = 0.550$, $W_2 = 1.04$ and $W_3 = 1.57$ g-atom carbon. Figure 19.1-2 shows the specific radioactivity in blood and in expired breath predicted by the model with these experimentally determined parameters. The monotonic decay of y_1^* is the impulse response expected from the perfectly mixed CO_2 stores in the blood pool. The peaked response of y_3^* occurs because the impulse input must pass between both the blood glucose and the CO_2 stores before being expired from the body.

19.1.2 Multibreath Lung Washout

The washout (or washin) dynamics of a slightly soluble, inert gas tracer from the lungs depends on airway geometry and ventilation distribution. A multibreath washout of nitrogen, in particular, is a non-invasive clinical test used for identifying abnormalities in gas distribution between the environment and the alveolar spaces. While a person breathes from a source of pure oxygen, N_2 concentration and flow at the mouth are continuously measured. With every breath, N_2 in the lungs becomes more dilute until the expired N_2 concentration falls to a small fraction of its initial value, at which point the test is terminated.

The Bohr model of gas transport in healthy lungs consists of two compartments in series (Fig. 19.1-3): a dead space (i.e., non-gas-exchanging airways) and an alveolar region (i.e., gas-exchanging spaces). The dead space is assumed to have a constant volume through which transport occurs by convection such that no gas mixing occurs. Consequently, this perfectly unmixed compartment acts as a time delay t_D. The alveolar region, whose volume changes during breathing, is assumed to be well mixed, primarily by diffusion.

Normally, the dead-space volume is a third or less of the tidal volume. During inhalation, the dead space becomes completely filled with gas inhaled from the environment, but the alveolar region contains a mixture of the inhaled gas and residual gas from the previous breath. During exhalation, the dead space becomes filled with gas from the alveolar spaces. The N_2 washout dynamics for this model of healthy lungs provides a reference for N_2 washout from lungs with an abnormal ventilation distribution.

Model Formulation

Volume Relationships

The total lung volume is the sum of the constant dead-space volume V_D and the variable alveolar volume $V_A(t)$:

$$V_L(t) = V_D + V_A(t) \quad \Rightarrow \quad \frac{dV_L}{dt} = \frac{dV_A}{dt} \qquad (19.1\text{-}16a,b)$$

Thus, the alveolar volumes at $t = 0$, the beginning of inhalation, and at $t = t_I$, the end of inhalation, are given by

$$V_A(0) = V_L(0) - V_D, \quad V_A(t_I) = V_L(t_I) - V_D \qquad (19.1\text{-}17a,b)$$

Defining $\dot{V}(t)$ as the volumetric flow of inhaled gas at the airway openings, the nose and/or mouth, the instantaneous volume of inhaled gas is

$$V(t) \equiv \int_0^t \dot{V}(t)dt \qquad (19.1\text{-}18)$$

For calculations with a common basis, $\dot{V}$ and V should be adjusted to a standard pressure and temperature. Inhaled tidal volume is defined as the change in lung volume over the duration of an inspiration, $t_I > t > 0$.

$$V_T = \int_0^{t_I} \dot{V}(t)dt \qquad (19.1\text{-}19)$$

Dead-space volume is defined as the volume of inhaled gas needed to displace residual gas from the dead space. This occurs over the initial interval of inhalation $t_D > t > 0$:

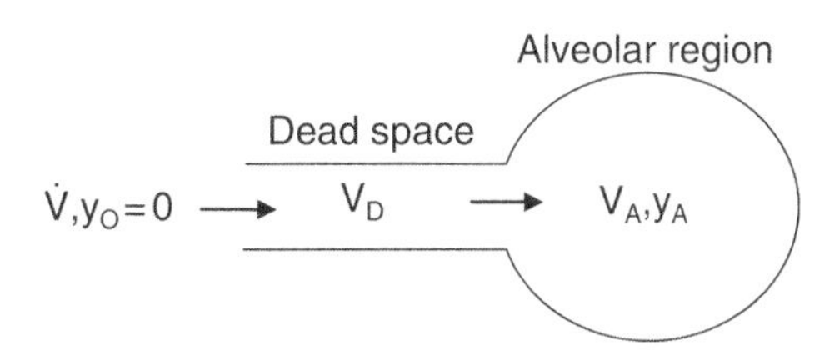

Figure 19.1-3 Pulmonary model with inhalation of nitrogen-free gas.

$$V_D = \int_0^{t_D} \dot{V} dt \tag{19.1-20}$$

During breathing without excessive force, pressure, and temperature changes in the lungs are relatively small so that the molar density of the gas mixture is everywhere uniform and constant. Also, the net molar transport of gases and water vapor between airspaces and surrounding tissue is negligible. Consequently, the molar balance for the gas mixture in the lungs leads to

$$\frac{dV_L}{dt} = \dot{V} \Rightarrow \frac{dV_A}{dt} = \dot{V} \tag{19.1-21a,b}$$

The lung volume at the end of inhalation can be obtained by integration:

$$\int_{V_L(0)}^{V_L(t_I)} dV_L = \int_0^{t_I} \dot{V} dt \Rightarrow V_L(t_I) = V_L(0) + V_T \tag{19.1-22a,b}$$

When combined with Equation 19.1-17b, we obtain the volume relationship:

$$V_A(t_I) = V_L(0) - V_D + V_T \tag{19.1-23}$$

Nitrogen Balances

Because N_2 is only slightly soluble and non-reactive in blood, we can neglect transport between alveolar gas and blood. A N_2 molar balance in the alveolar compartment during a single inhalation then leads to

$$\frac{d(y_A V_A)}{dt} = \dot{V} y_D \quad (t_I > t \geq 0) \tag{19.1-24}$$

where $y_A(t)$ is the N_2 mole fraction within the alveolar region and $y_D(t)$ is the N_2 mole fraction that enters the alveolar region from the dead space.

At the beginning of any inhalation, all the gas in the dead space has a N_2 mole fraction $y_A(0)$ that existed at the end of the previous exhalation. Therefore, the N_2 mole fraction entering the alveolar region during inhalation is $y_D = y_A(0)$ until time reaches t_D. For times greater than t_D, the dead space is filled with inhaled O_2 so that $y_D = 0$. Equation 19.1-24 now becomes

$$\frac{d(y_A V_A)}{dt} = \begin{cases} \dot{V} y_A(0) & \text{when } t_D \geq t > 0 \\ 0 & \text{when } t_I \geq t > t_D \end{cases} \tag{19.1-25}$$

During the following exhalation, a molar N_2 balance in the alveolar compartment leads to

$$\frac{d(y_A V_A)}{dt} = \dot{V} y_A \Rightarrow V_A \frac{dy_A}{dt} + \frac{dV_A}{dt} y_A = \dot{V} y_A \Rightarrow \frac{dy_A}{dt} = 0 \tag{19.1-26a-c}$$

Therefore, alveolar N_2 fraction does not change during exhalation and equals the alveolar N_2 fraction $y_A(t_I)$ at the end of previous inhalation. Provided that exhalation time is long enough to allow alveolar gas to completely fill the dead space, the N_2 fraction at the end of the expired breath will be $y_E = y_A(t_I)$. Using all of the above information, we can evaluate the change of y_E at the end of successive breaths.

Analysis

To determine the change of alveolar N_2 fraction over a single inhalation, we integrate Equation 19.1-25 from 0 to t_I:

$$y_A(t_I)V_A(t_I) - y_A(0)V_A(0) = y_A(0)\int_0^{t_D} \dot{V}dt \quad (19.1\text{-}27)$$

or equivalently

$$y_E V_A(t_I) - y_A(0)V_A(0) = y_A(0)V_D \quad (19.1\text{-}28)$$

Substituting the volume relations from Equations 19.1-17a and 19.1-23, we obtain

$$y_A(t_I) = \frac{y_A(0)}{1+\chi(0)}; \quad \chi_0 \equiv \frac{V_T - V_D}{V_L(0)} \quad (19.1\text{-}29a,b)$$

where χ_0 is a dilution factor. The corresponding ratio of end-expired N_2 fractions for any two successive breaths, k and k + 1, is

$$y_E(k+1) = \frac{y_E(k)}{1+\chi_k}; \quad \chi_k \equiv \frac{V_T(k) - V_D}{V_L(k)} \quad (k = 0, 1, 2, \ldots) \quad (19.1\text{-}30a,b)$$

where χ_k is the dilution fraction that accounts for the inhaled tidal volume and initial lung volume associated with breath k. Over k breaths of a N_2 washout, we can relate the end-expired N_2 fraction at any breath k to the initial end-expired N_2 fraction $y_E(0)$:

$$\frac{y_E(k)}{y_E(0)} = \left[\frac{y_E(k)}{y_E(k-1)}\right]\left[\frac{y_E(k-1)}{y_E(k-2)}\right]\cdots\left[\frac{y_E(1)}{y_E(0)}\right] = \prod_{i=0}^{k}\left(\frac{1}{1+\chi_i}\right)^i \quad (19.1\text{-}31)$$

When inhaled tidal volume and initial lung volume have the same values for each breath, the dilution fraction $\chi_k = \chi$ is a constant so that

$$\frac{y_E(k)}{y_E(0)} = \left(\frac{1}{1+\chi}\right)^k; \quad \chi \equiv \frac{(V_T - V_D)}{V_L} \quad (19.1\text{-}32a,b)$$

or

$$\ln\left[\frac{y_E(k)}{y_E(0)}\right] = -k\ln(1+\chi) \quad (19.1\text{-}33)$$

According to this model, a multibreath N_2 washout presented as a semilogarithmic plot of the normalized mole fraction $y_E(k)/y_E(0)$ versus k is expected to be a straight line with a negative slope of $\ln(1+\chi)$. The steepness of the slope indicates the magnitude of alveolar ventilation, that is, the extent of dilution of the residual gas in the lungs.

Figure 19.1-4 shows washout data from a subject with healthy lungs breathing a N_2-free gas (Saidel *et al.*, 1975). For this subject, $V_L = 2400$ ml, $V_T = 880$ ml, and the breathing frequency was 14 min^{-1}. From the slope determined by linear regression, the dilution factor is $\chi = 0.270$ corresponding to a dead space of $V_D = 230$ ml. For a subject with inhomogeneous ventilation, the washout data plotted this way could deviate considerately from the straight line. Such lung dysfunction can be simulated using more complex compartmental models consisting of parallel transport paths with asynchronous ventilation among different paths.

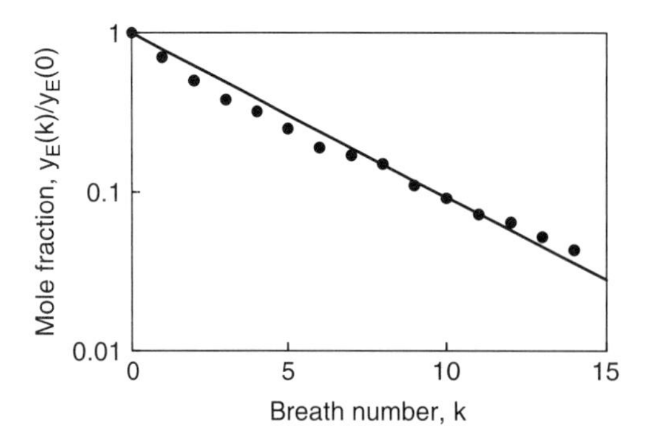

Figure 19.1-4 Multibreath washout of a human subject with healthy lungs (Data from Saidel *et al.* (1975)).

19.1.3 Pulmonary Ventilation, Diffusion, and Perfusion

Another way to quantify and distinguish abnormal from normal pulmonary function is to establish input–output relationships for several inert gases with different solubilities in blood and in tissues. Data of this type comes from experiments with constant venous infusion of inert tracer gases, which lead to steady-state relationships (Saidel, 1982). As a means of interpreting the data, we model the pulmonary system with compartments representing a dead space with constant volume V_D, an alveolar region with alveolar ventilation $\dot{V}_A$, and pulmonary capillaries with blood flow Q (Fig. 19.1-5). To account for abnormal gas exchange, the model incorporates a shunt flow Q_S of blood that bypasses the pulmonary circulation.

Model Formulation

Under steady-state conditions, inert gas concentrations vary spatially within the system, but not from breath to breath. When the input-output relations of the lung are averaged over a breath, the cyclic inflow–outflow can be represented as if it were a continuous input-output flow. To show this, we start with a molar balance on the dead space over a single breath. We define C_I, C_E, and C_A as inspired, expired, and alveolar inert gas concentrations, respectively. We assume that tidal volume V_T is equal during inhalation and exhalation. During inhalation, V_TC_I moles enter the dead space from the airway opening, and $V_DC_A + (V_T - V_D)C_I$ moles exit to the alveolar region. During the following exhalation, when the flow direction is reversed, V_TC_A moles enter the dead space from the alveolar region, and V_TC_E moles exit the dead space through the airway opening. Since we are assuming that a steady state exists from breath to breath, the total input must be equal to the total output:

$$V_TC_I + V_TC_A = V_DC_A + (V_T - V_D)C_I + V_TC_E \tag{19.1-34}$$

Upon rearranging, this equation becomes

$$V_T(C_I - C_E) = (V_T - V_D)(C_I - C_A) \tag{19.1-35}$$

At a respiration frequency f, we define the total ventilation rate as $\dot{V}_T \equiv fV_T$ and alveolar ventilation rate as

$$\dot{V}_A \equiv f(V_T - V_D) \tag{19.1-36}$$

so that the cyclic balance in Equation 19.1-35 is transformed to an effective continuous input-output relationship:

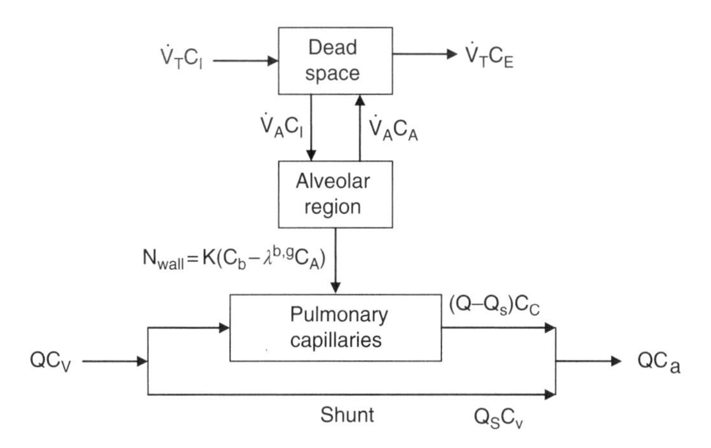

Figure 19.1-5 Compartment model of the pulmonary system.

$$\dot{V}_T(C_I - C_E) = \dot{V}_A(C_I - C_A) \tag{19.1-37}$$

Similarly, from a species molar balance on a subsystem consisting of the alveolar compartment and the pulmonary capillaries, we obtain the steady-state relationship:

$$(Q - Q_s)(C_c - C_v) = \dot{V}_A(C_I - C_A) \tag{19.1-38}$$

where $(Q - Q_s)$ is the net pulmonary capillary blood flow, and $(C_c - C_v)$ is the difference between the inert gas concentrations of end-capillary and mixed-venous blood. At the node where the shunt and end-capillary flows merge to produce the total blood flow Q and mixed-arterial concentration C_a, the species molar balance is

$$(Q - Q_s)C_c + Q_s C_v = QC_a \tag{19.1-39}$$

The spatially distributed species concentration that occurs in pulmonary capillaries because of their large length-to-diameter ratio could be simulated by several perfectly mixed compartments in series. As an alternate approach, we use a one-dimensional transport model in a straight conduit of uniform cross section A_c, surface area S_c, and length L_c. Since there is no solution transport across the capillary walls, Equation 15.3-29 indicates that the volumetric flow of blood though the capillaries is constant. Also neglecting axial dispersion, Equation 15.3-28 leads to the steady-state equation of inert species concentration distribution in capillary blood, $C_b(z)$:

$$(Q - Q_s)\frac{dC_b}{dz} = \phi_c A_c N_{wall} \quad (L_c > z > 0) \tag{19.1-40}$$

where z is distance from the flow inlet, $\phi_c = S_c / A_c L_c$ is the surface-to-volume ratio of the capillaries and N_{wall} is the local species flux from the alveolar compartment into the blood. We can express N_{wall} as the product of an overall mass transfer coefficient and a local concentration driving force $(C_b - \lambda^{b,g} C_A)$, where $\lambda^{b,g}$ is an equilibrium partition coefficient. With the mass transfer resistance of the gas phase being very small, it is reasonable to assume that the overall mass transfer coefficient is essentially equal to the individual mass transfer coefficient k_b in blood. Therefore, we can write

$$\phi_c A_c N_{wall} = -\frac{k_b S_c}{L_c}\left(C_b - \lambda^{b,g} C_A\right) \tag{19.1-41}$$

Substituting this equation into Equation 19.1-40, the species concentration in capillary blood becomes

$$\left(\frac{Q - Q_s}{k_b S_c / L_c}\right)\frac{dC_b}{dz} + C_b = \lambda^{b,g} C_A \quad (L_c > z > 0) \tag{19.1-42}$$

When integrated between the capillary inlet where $C_b(0) = C_v$, and at the capillary outlet where $C_b(L_c) = C_c$, we get the input-output relationship for the capillary bed:

$$C_c = \beta C_v + (1 - \beta)\lambda^{b,g} C_A; \;\; \beta = \exp[-k_b S_c/(Q - Q_s)] \tag{19.1-43a,b}$$

Analysis

We write the independent set of linear model equations (Eqs. 19.1-37–19.1-39, and 19.1-43a) in terms of dimensionless concentrations by dividing them by C_v and Q, as needed. In matrix form, these four equations in four unknowns become

$$[a][C] = [b] \tag{19.1-44}$$

where

$$[a] \equiv \begin{bmatrix} 1 & -\frac{\dot{V}_A}{\dot{V}_T} & 0 & 0 \\ 0 & \frac{\dot{V}_A}{Q} & 1 - \frac{Q_s}{Q} & 0 \\ 0 & 0 & 1 - \frac{Q_s}{Q} & -1 \\ 0 & -(1-\beta)\lambda^{b,g} & 1 & 0 \end{bmatrix}, \; [b] = \begin{bmatrix} \left(1 - \frac{\dot{V}_A}{\dot{V}_T}\right)\frac{C_I}{C_v} \\ \left(1 - \frac{Q_s}{Q}\right) + \frac{\dot{V}_A}{Q}\frac{C_I}{C_v} \\ -\frac{Q_s}{Q} \\ \beta \end{bmatrix}, \; [C] \equiv \begin{bmatrix} \frac{C_E}{C_v} \\ \frac{C_A}{C_v} \\ \frac{C_c}{C_v} \\ \frac{C_a}{C_v} \end{bmatrix} \tag{19.1-45a-c}$$

In applications where tracer species are infused intravenously and not inhaled, $C_I = 0$ and we find

$$\begin{aligned} \frac{C_A}{C_v} &= \frac{1}{\gamma + \lambda^{b,g}}; \;\; \gamma \equiv \frac{1}{1-\beta}\left(\frac{\dot{V}_A}{Q - Q_s}\right) \\ \frac{C_E}{C_v} &= \left(\frac{\dot{V}_A}{\dot{V}_T}\right)\frac{C_A}{C_v} \\ \frac{C_c}{C_v} &= 1 - \left(\frac{\dot{V}_A}{Q - Q_s}\right)\frac{C_A}{C_v} \\ \frac{C_a}{C_v} &= 1 - \left(\frac{\dot{V}_A}{Q}\right)\frac{C_A}{C_v} \end{aligned} \tag{19.1-46a-e}$$

Experimental evaluation of the dimensionless parameter groups in this model can be used to diagnose functional abnormalities: $\dot{V}_A/\dot{V}_T = (1 - V_D/V_T) < 0.6$ indicates ventilation inhomogeneity in the pulmonary gas spaces; $(\dot{V}_A/Q) < 0.7$ indicates ventilation-perfusion mismatching between the alveolar region and the capillaries; $\beta > 0.1$ indicates a perfusion-diffusion limitation; and $(Q_S/Q) > 0.05$ indicates apparent blood shunting as a consequence of inadequate gas exchange.

In the multiple inert gas elimination method, a subject inhales tracer-free air during the continuous venous infusion of a series of six tracer gases with $\lambda^{b,g}$ ranging from 0.004 for sulfur hexafluoride to 300 for acetone. After allowing sufficient time for steady state to be reached, mixed expired gas and mixed-venous blood are analyzed to obtain the excretion ratio C_E/C_v of each inert tracer. If pulmonary arterial blood is sampled, requiring a rather invasive catheterization, then retention ratio C_a/C_v can also be evaluated. From Equation 19.1-46a,b,e, we obtain these observable quantities as

$$\frac{C_E}{C_v} = \left(\frac{\dot{V}_A}{\dot{V}_T}\right)\frac{1}{\gamma + \lambda^{b,g}}$$
$$\frac{C_a}{C_v} = 1 - \left(\frac{\dot{V}_A}{Q}\right)\frac{1}{\gamma + \lambda^{b,g}} \qquad (19.1\text{-}47a,b)$$

When $\lambda^{b,g} \to \infty$, these equations reach the limits $C_a/C_v \to 1$ and $C_E/C_v \to 0$. This means that a highly soluble gas is completely retained in pulmonary blood and cannot be excreted through the airway opening. For a poorly soluble gas, $\lambda^{b,g} \to 0$ so that the excretion and retention ratios are given by

$$\lim_{\lambda^{b,g} \to 0} \frac{C_E}{C_v} = \left(\frac{\dot{V}_A}{\dot{V}_T}\right)\frac{1}{\gamma} = (1-\beta)\left(\frac{Q-Q_s}{\dot{V}_T}\right)$$
$$\lim_{\lambda^{b,g} \to 0} \frac{C_a}{C_v} = 1 - \left(\frac{\dot{V}_A}{Q}\right)\frac{1}{\gamma} = \frac{Q_s}{Q} - (1-\beta)\left(1 - \frac{Q_s}{Q}\right) \qquad (19.1\text{-}48a,b)$$

Figure 19.1-6 presents the excretion and retention ratios representative of adults with $V_T = 600$ ml, $V_D = 150$ ml, $Q = 6000$ ml/min and $f = 15\ \text{min}^{-1}$. For a healthy adult indicated by the solid curves, we assume negligible blood shunting $(Q_S/Q = 0)$ and no diffusion limitation $(\beta = 0)$. Results are also shown when respiratory disease causes either a finite diffusion limitation of $\beta = 0.4$ (dotted curves), or a blood flow shunt of $Q_S/Q = 0.2$ (dashed curves).

In all three cases, the excretion ratio is largest for poorly soluble gases that are thermodynamically driven to leave the blood. The excretion ratio is smallest for highly soluble gases

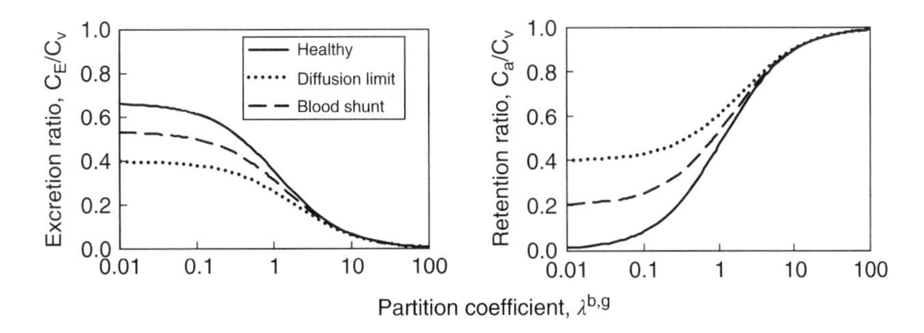

Figure 19.1-6 Predicted excretion and retention curves.

that are tightly held by blood. The opposite is true of the retention ratio. Both diffusion limitation and blood shunting lower the excretion ratio while increasing the retention ratio, indicating a loss in the efficiency of gas transfer.

19.1.4 Urea Dynamics with Hemodialysis

Modeling the dynamics of urea transport in a subject who periodically undergoes hemodialysis provides a quantitative understanding the underlying physiological responses and provides the basis for improved treatment efficiency and effectiveness. An important aspect of this modeling is to distinguish between urea in intracellular fluid (ICF) and in extracellular fluid (ECF) compartments as presented by Grandi *et al.* (1995). These compartments have time-varying volumes due to the accumulation of ingested water between hemodialysis treatments and the removal of this excess water during the treatment.

We will consider a model structure consisting of perfectly mixed ICF and ECF compartments of volumes V_I and V_E containing urea concentrations C_I and C_E (Fig. 19.1-7). During a complete treatment cycle, the patient undergoes a period $T_D \geq t \geq 0$ with dialysis followed by a period $T \geq t \geq T_D$ without dialysis. Throughout this cycle, passive urea transport occurs between the ICF and ECF compartments at a molar rate $(KS)_{IE}(C_I - C_E)$, where K_{IE} is an overall mass transfer coefficient and S_{IE} is the intercompartment contact area. When there is no dialysis, water is ingested into ECF at a constant volumetric rate Q_w, and urea is generated by protein metabolism in ICF at a constant molar rate $\dot{R}_I$.

Regarding dialyzer performance, we assume that volumetric rate of fluid filtration Q_F is constant, and transport occurs under pseudo-steady conditions such that urea clearance Q_C is constant. With ECF as the source of blood circulation to the dialyzer, there is a net solution flow Q_F and a net urea transport rate $Q_C C_E$ from ECF to the dialyzer.

Model Formulation

Total body volume is the sum of the volumes of the ECF and ICF compartments, $V(t) \equiv V_I(t) + V_E(t)$. With osmotic water shifts between ICF and ECF reaching osmotic equilibrium very rapidly, the osmolarity of the two compartments are nearly equal. Thus, the volume ratio of the two compartments $V_E(t)/V_I(t)$ is equal to the molar ratio of their osmotically active material. Since the number of osmoles remains essentially constant in the ICF and ECF, $V_E(t)/V_I(t)$ is also constant. Noting that $V/V_I = 1 + V_E/V_I$, we define a constant compartment intracellular volume fraction as $\varphi = V_I/V = 1 - V_E/V$, such that

$$\frac{dV_I}{dt} = \varphi \frac{dV}{dt}, \quad \frac{dV_E}{dt} = \frac{d(V - V_I)}{dt} = (1 - \varphi)\frac{dV}{dt} \tag{19.1-49a,b}$$

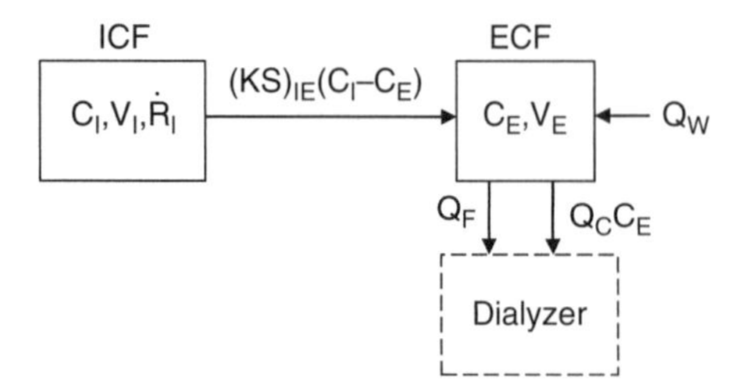

Figure 19.1-7 Model of urea dynamics with dialysis.

A mass balance of the constant density fluid over the entire body yields

$$\frac{dV}{dt} = \begin{cases} -Q_F & \text{when } T_D > t \geq 0 \quad \text{dialysis} \\ +Q_w & \text{when } T > t \geq T_D \quad \text{no dialysis} \end{cases} \tag{19.1-50}$$

By integrating these equations with the initial condition $V(0) = V_o$, we get

$$V(t) = \begin{cases} V_o - Q_F t, & T_D > t \geq 0 \\ (V_o - Q_F T_D) + Q_w t, & T > t \geq T_D \end{cases} \tag{19.1-51}$$

In the absence of kidney function, it is desirable for hemodialysis to eliminate a fluid volume, $Q_F T_D$, that matches the fluid volume retained in the body between dialysis treatments $Q_w(T - T_D)$ so that

$$Q_F = \left(\frac{T - T_D}{T_D}\right) Q_w \tag{19.1-52}$$

Since urea is always being produced, its molar balance in ICF is

$$\frac{d(C_I V_I)}{dt} + (KS)_{IE}(C_I - C_E) = \dot{R}_I, \quad T > t \geq 0 \tag{19.1-53}$$

The urea molar balance in ECF with and without dialysis is

$$\frac{d(C_E V_E)}{dt} - (KS)_{IE}(C_I - C_E) = \begin{cases} -Q_C C_E, & T_D > t \geq 0 \\ 0, & T > t \geq T_D \end{cases} \tag{19.1-54}$$

We can expand the time derivatives in these equations and substitute the volumes and volume derivatives from Equations 19.1-50 and 19.1-51. During dialysis when $T_D > t \geq 0$, we obtain

$$\begin{aligned} \varphi(V_o - Q_F t)\frac{dC_I}{dt} + \left[(KS)_{IE} - \varphi Q_F\right] C_I &= (KS)_{IE} C_E + \dot{R}_I \\ (1-\varphi)(V_o - Q_F t)\frac{dC_E}{dt} + \left[(KS)_{IE} - (1-\varphi)Q_F + Q_C\right] C_E &= (KS)_{IE} C_I \end{aligned} \tag{19.1-55a, b}$$

In the absence of dialysis when $T > t \geq T_D$, we get

$$\begin{aligned} \varphi[V_o - (T - T_D - t)Q_w]\frac{dC_I}{dt} + \left[(KS)_{IE} + \varphi Q_w\right] C_I &= (KS)_{IE} C_E + \dot{R}_I \\ (1-\varphi)[V_o - (T - T_D - t)Q_w]\frac{dC_E}{dt} + \left[(KS)_{IE} + (1-\varphi)Q_w\right] C_E &= (KS)_{IE} C_I \end{aligned} \tag{19.1-56a, b}$$

The initial conditions are $C_I(0) = C_{Io}$ and $C_E(0) = C_{Eo}$.

Analysis

By numerical solution of this model, we simulate body fluid volume and urea concentration changes for an adult human during a cycle consisting of 0.2 days of dialysis treatment followed by 1.8 days with no dialysis. Tissue volume parameters are set at typical values of $V_o = 40$ L and $\varphi = 0.6$. For uremic patients on a low-protein diet, urea production is about $\dot{R}_I = 200\,\text{mmol/day} = 0.14\,\text{mmol/min}$ and a peak urea concentration is about $C_{1o} = C_{2o} =$ 20 mM when dialysis begins. A normal kidney output is roughly 1.5 L/day. To eliminate

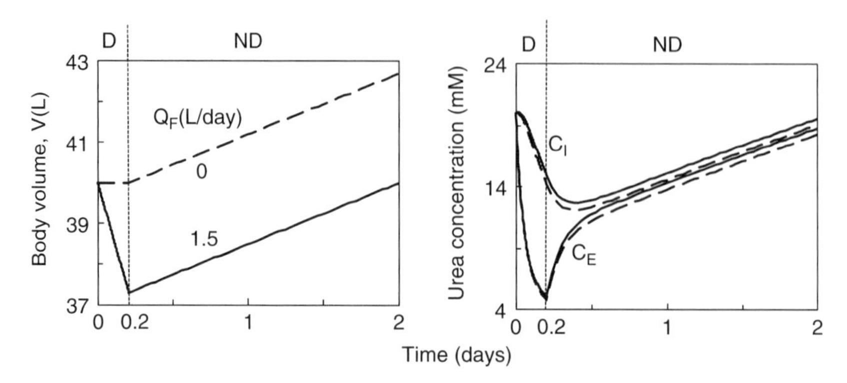

Figure 19.1-8 Urea ICF and ECF concentrations during (D) and after dialysis (ND).

the fluid retained during kidney failure, the required dialyzer filtration is $Q_F = 1.5(1.8)/0.2 = 14$ L/day = 9.4 ml/min.

Achieving this Q_F in a typical hollow fiber hemodialyzer would require a blood flow of 360 ml/min and dialysate flow of 200 ml/min (Table 12.3-2). This would result in a clearance of 170 ml/min. Although the value of $(KS)_{IE}$ is unknown, we require that the urea concentrations $C_I(2)$ and $C_E(2)$ at the end of the no-dialysis period be close to their concentrations $C_I(0) = C_E(0) = 20$ mM at the beginning of the previous dialysis period. Simulations using the estimated dialyzer operating parameters indicated that this would be possible if we set $(KS)_{IE} = 100$ ml/min.

The final simulations shown in Figure 19.1-8 compare the body volume and urea concentration dynamics at $Q_F = 0$ (dashed curves) to those obtained at $Q_F = 1.5$ L/day (solid curves). Without dialyzer filtration, fluid accumulation over a complete treatment cycle is large and can produce edema. Fluid accumulation also causes a small decrease in C_I and C_E, due to a dilution of urea in both IC and EC compartments.

During dialysis, C_E falls rapidly but C_I falls at a slower rate, whether or not there is filtration by the hemodialyzer. This occurs because intercompartmental urea transport from IC to EC is slower than transport from EC to the dialyzer. Even after dialysis, C_I continues to lag behind C_E because intercompartmental transport is insufficient to equilibrate newly generated urea in the IC compartment with the surrounding EC fluid. Therefore, an evaluation of patient status by sampling blood (an extracellular fluid) underestimates urea concentration in intracellular fluid.

19.2 Models with Feedback and Recirculation

19.2.1 Cardiovascular Recirculation of a Tracer

Tracer-response experiments in the intact cardiovascular system are complicated by the recirculation of tracer past its sampling sites. Figure 19.2-1 presents data in which a pulse of an indicator dye was infused into the superior vena cava of a dog (Nicholes *et al.*, 1964). Concentration responses were simultaneously monitored just downstream of the vena cava in the pulmonary artery (PA) and further downstream in the aortic root (AR). These two

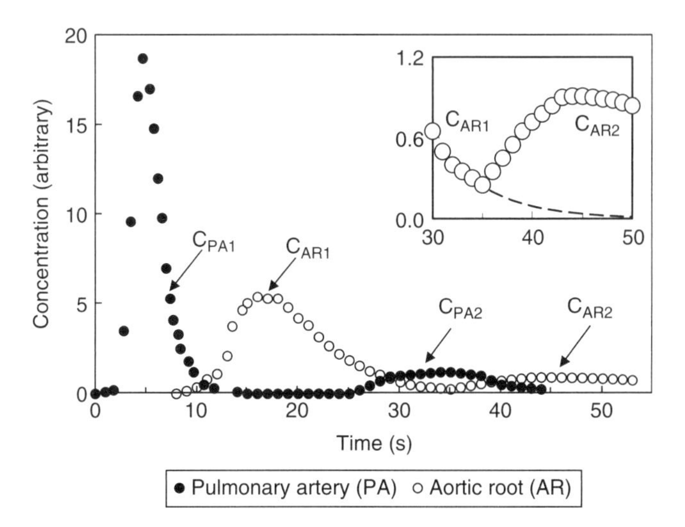

Figure 19.2-1 Pulse response of tracer in the circulatory system of a dog (Data from Nicholes *et al.* (1964)).

sample sites provide a means of separately analyzing the pulmonary circulation as blood flows from PA to AR, and analyzing the systemic circulation as blood flows from AR to PA.

At both the PA and AR sampling sites, the data exhibit a pair of concentration peaks representing first and second passes of tracer past the sites. At the PA site, the first-pass peak C_{PA1} is caused almost exclusively by tracer infusion; the second-pass peak C_{PA2} is due to tracer circulating around the entire cardiovascular loop. At the AR site, the first-pass peak C_{AR1} is due to transport of injected tracer through the pulmonary circulation; the second-pass peak C_{AR2} is due to tracer circulating twice through the pulmonary circulation and once through the systemic circulation.

We analyze these data with a physiologically based model in which the spatially distributed tracer in the pulmonary circulation is represented by J identical compartments in series (Fig. 19.2-2). The systemic circulation is represented as a single well-mixed compartment. Transport delay elements are present in both of these circulations to account for the time shifts between the first-pass and second-pass peaks seen in the data. The model assumes that the tracer is conserved in the pulmonary circulation, but is lost from the systemic circulation in proportion to tracer concentration.

Model Formulation

In this model, the blood density, compartment volumes, and volumetric blood flow Q through the system are constant. The tracer molar balance for the j-th pulmonary compartment of volume V_p describes the tracer concentration $C_{p,j}(t)$:

$$\frac{V_p}{Q}\frac{dC_{p,j}}{dt} + C_{p,j} = C_{p,j-1} \quad (j = 1,2,\ldots,J) \tag{19.2-1}$$

where $C_{p,0} = C_{PA}$ is the tracer concentration observed at the PA sampling site. The tracer molar balance on the systemic compartment of volume V_s containing a tracer concentration C_s is

$$\frac{V_s}{Q}\frac{dC_s}{dt} + \left(1 + \frac{k_{loss}}{Q}\right)C_s = C_{AR} \tag{19.2-2}$$

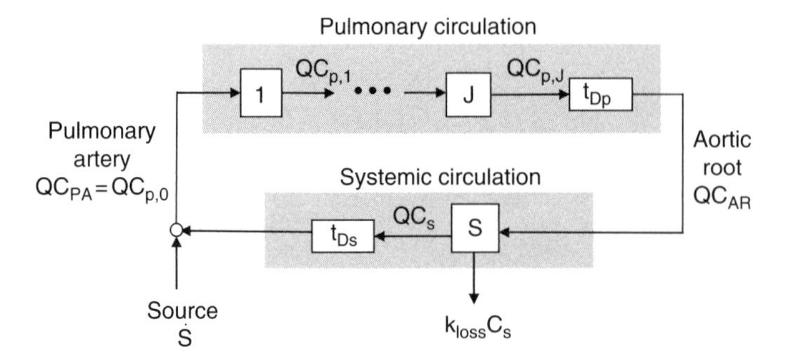

Figure 19.2-2 Closed-loop model of tracer transport in the circulatory system.

where C_{AR} is tracer concentration recorded at the AR sampling site, and k_{loss} is a rate coefficient of tracer loss. The pulmonary tracer concentration $C_{p,J}$ passes through a pulmonary transport delay t_{Dp} before reaching the AR sampling site. Similarly, the systemic tracer concentration C_s passes through a systemic delay t_{Ds} before reaching the PA site. We can formally express these relationships as

$$C_{AR}(t) = C_{p,J}(t - t_{Dp})$$
$$C_{PA}(t) = C_s(t - t_{Ds}) \tag{19.2-3a,b}$$

Initially, all tracer concentrations in the circulation are zero.

Analysis

The parameters associated with this mathematical model are the number of pulmonary compartments, J, and the characteristic times: V_p/Q, V_s/Q, t_{Dp}, t_{Ds}, and k_{loss}/Q. Some of these parameters can be estimated by computing moments of those data that have a sufficiently high signal-to-noise ratio (S/N). Estimates of other parameters can be obtained by direct fitting of model simulations to the data.

Before trying to compute mathematical moments, we must consider the nature of the experimental data. The tracer concentration $C_{PA1}(t)$ of the first pulmonary peak decays to zero before the second PA peak occurs. Consequently, $C_{PA1}(t)$ is not affected by the reappearance of tracer that has circulated through the vascular system. This is not the case for the first aortic peak, whose concentration $C_{AR1}(t)$ does not completely decay to zero in its trailing portion (Fig. 19.2-1; inset). However, after performing an exponential extrapolation of the tail (dashed curve), we consider the modified $C_{AR1}(t)$ pattern to adequately represent the first pass of tracer through the pulmonary circulation alone.

Based on S/N considerations, we judged that reliable values for the zeroth, first, and second moments of the C_{PA1} and modified C_{AR1} peaks could be computed. However, only the zeroth moment of the $C_{PA2}(t)$ peak was determined, and no moments were computed from the $C_{AR2}(t)$ data. Table 19.2-1 contains the computed moments of the response data.

For a single pass through the pulmonary and systemic circulations, which is largely unaffected by recirculation, the output/input ratio of zeroth moments is an approximation of the zeroth moment of a unit impulse-response function. In addition, the input–output differences of the mean and the variance approximate the mean and variance of an impulse-response function:

Table 19.2-1 Moments of Concentration Peaks

Moment	C_{PA1} (15 > t > 0 s)	C_{PA2} (54 > t > 25 s)	C_{AR1} (t > 8 s)
μ_0 (arb)	67.1	16.1	60.9
Mean, $\bar{t}$ (s)	5.4	—	19.7
Variance, σ^2 (s^2)	3.2	—	32.6

$$\begin{aligned}
\left(\mu_{0,g}\right)_{pul} &= \mu_{0,AR1}/\mu_{0,PA1} = 60.9/67.1 = 0.91 \\
\left(\mu_{0,g}\right)_{sys} &= \mu_{0,PA2}/\mu_{0,AR1} = 16.1/60.9 = 0.26 \\
\left(\bar{t}_g\right)_{pul} &= \bar{t}_{AR1} - \bar{t}_{PA1} = 19.7 - 5.4 = 14.3\,s \\
\left(\sigma_g^2\right)_{pul} &= \sigma_{AR1}^2 - \sigma_{PA1}^2 = 32.6 - 3.2 = 29.4\,s^2
\end{aligned} \qquad (19.2\text{-}4a\text{-}d)$$

The zeroth moment ratio of 0.91 through the pulmonary circulation suggests that there is either a small loss of tracer or that exponential extrapolation of the AR1 peak prematurely forced the concentration to zero. The zeroth moment ratio of 0.26 through the systemic circulation can be used in Equation 18.3-15a to estimate a loss coefficient:

$$0.26 = \frac{1}{(1 + k_{loss}/Q)} \quad \Rightarrow \quad \frac{k_{loss}}{Q} = 2.92 \qquad (19.2\text{-}5a,b)$$

This substantial tracer loss could be due to tracer binding to blood vessel walls or tracer permeation through the vessel walls.

For the pulmonary subsystem, $\left(\bar{t}_{pul}\right)_g$ and $\left(\sigma_{pul}^2\right)_g$ are equal to sums of the means and the variances of J equal-sized compartments in series with a transport delay (Table 18.3-1; entries 1 and 4):

$$\left(\bar{t}_{pul}\right)_g = J\frac{V_P}{Q} + t_{Dp}, \quad \left(\sigma_{pul}^2\right)_g = J\left(\frac{V_P}{Q}\right)^2 + \left(\sigma_g^2\right)_{Dp} \qquad (19.2\text{-}6a,b)$$

Combining these two equations, we obtain an expression for the number of pulmonary compartments.

$$\frac{\left[\left(\bar{t}_{pul}\right)_g - t_{Dp}\right]^2}{\left(\sigma_{pul}^2\right)_g - \left(\sigma_{Dp}^2\right)_g} = \frac{\left(J\,V_P/Q\right)^2}{J\left(V_P/Q\right)^2} = J \qquad (19.2\text{-}7)$$

Figure 19.2-1 indicates that the C_{PA1} data rises above zero at t = 1.5 s, whereas nonzero C_{AR1} values do not appear until t = 9 s so that $t_{Dp} \approx 9.0 - 1.5 = 7.5\,s$. Recognizing further that the variance $\left(\sigma_{Dp}^2\right)_g$ of a transport delay is zero and using the moment values in Equation 19.2-4c,d, we get

$$J = \frac{(14.3 - 7.5)^2}{29.4 - 0} = 1.57 \qquad (19.2\text{-}8)$$

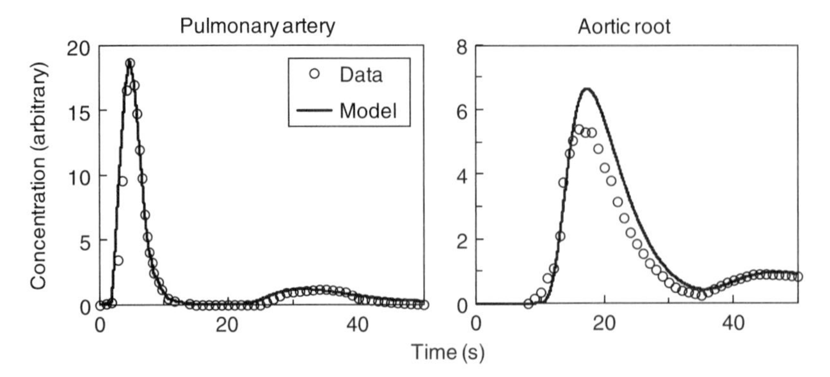

Figure 19.2-3 Model simulation of tracer circulation data.

Because the model requires an integer number of compartments, we round this number to $J = 2$ compartments. Using this value, we can determine V_p/Q on the basis of $\left(\bar{t}_{pul}\right)_g$ (Eq. 19.2-6a):

$$\frac{V_p}{Q} = \frac{\left(\bar{t}_{pul}\right)_g - t_{pD}}{J} = \frac{14.3 - 7.5}{2} = 3.4\,s \tag{19.2-9}$$

An alternative computation of V_p/Q on the basis of $\left(\sigma^2_{pul}\right)_g$ (Eq. 19.2-6b) yields $V_p/Q = 3.8\,s$. These values are somewhat different because J has been forced to be an integer. To accommodate this inconsistency, we specify an intermediate value of $V_p/Q = 3.6\,s$.

The values of the remaining parameters, V_s/Q and t_{Ds} cannot be obtained from a moment analysis alone. Instead, we carried out numerical simulations of the model (Eqs. 19.2-1–19.2-3) by fixing $J = 2$, $V_p/Q = 3.6\,s$, $k_{loss}/Q = 2.92$, and $t_{Dp} = 7.5\,s$ while using various trial values of V_s/Q and t_{Ds}. The best fit to the data occurred when $V_s/Q = 19.6$ s and $t_{Ds} = 10$ s (Fig. 19.2-3).

19.2.2 Control of Ventilation by Carbon Dioxide

Carbon dioxide is dissolved throughout body tissues and blood, where it is also present in bound forms, mainly as bicarbonate ion. Under steady-state conditions, the metabolic production of CO_2 is balanced by its excretion from the lungs so that CO_2 is maintained at its physiologically appropriate level. When metabolic rate changes, a corresponding change of CO_2 partial pressure in the arterial blood of the brain is transported to cerebral-spinal fluid where it stimulates pH-sensitive chemoreceptors. From these neurons, a signal is transmitted to other neurons that stimulate nerves of the respiratory muscles to alter ventilation rate. If CO_2 partial pressure in blood increases above its steady-state level, then ventilation increases. This, in turn, increases CO_2 excretion, which diminishes CO_2 partial pressure toward its initial steady-state level. A decrease in CO_2 partial pressure below its steady-state level is compensated by a neurological signal that decreases ventilation rate.

Such ventilatory regulation of CO_2 can be considered to be a feedback control system in which there is an optimal set point for CO_2 partial pressure. A model of this process can be used to examine the sensitivity of the control system as indicated by the change in ventilation due to a metabolic change in CO_2 partial pressure. Such information provides a quantitative

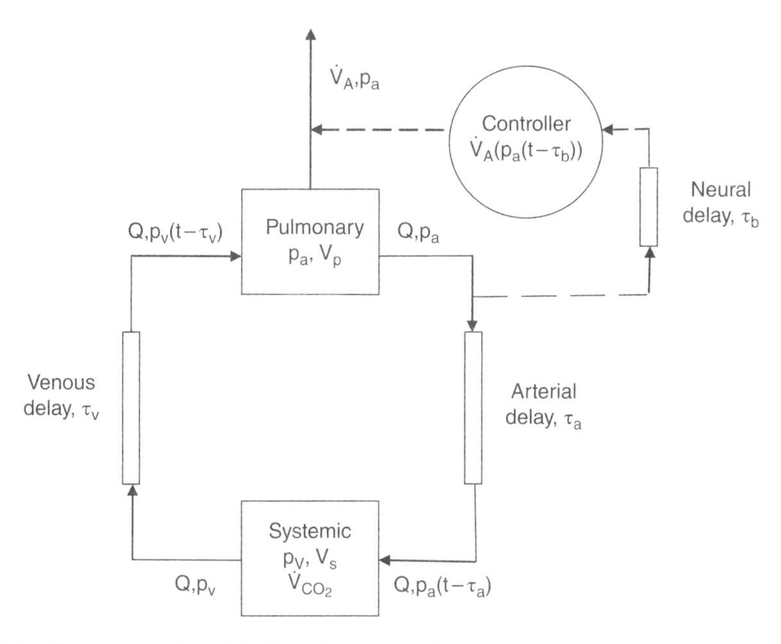

Figure 19.2-4 Compartmental model of ventilatory control.

understanding of system behavior, for example, the short-term periodicities in ventilation observed under some pathological conditions.

To simulate CO_2 respiratory control, we use a physiologically based compartmental model adapted from elHefnawy *et al.* (1988). The model consists of a closed circulatory loop between a pulmonary compartment and a systemic tissue compartment that are separated by arterial and venous transport time delays (Figure 19.2-4). Expiration of CO_2 occurs from the pulmonary compartment, while metabolic CO_2 production occurs in the systemic compartment. The characteristic time scale of this dynamic model is on the order of several breathing periods. With respect to this time scale, we assume that the average alveolar gas volume is constant and that the breath-averaged respiratory gas inflow and outflow are both continuous rather than alternating.

Model Formulation

The pulmonary compartment consists of two well-mixed subcompartments: an alveolar gas subcompartment of constant breath-averaged volume V_{alv} and a well-mixed tissue subcompartment of volume V_{tis} that contains both capillary blood and extravascular tissue. In alveolar gas containing a CO_2 concentration C_{alv}, a molar species balance yields

$$\frac{d(V_{alv}C_{alv})}{dt} = -\dot{V}_A C_{alv} + \dot{N}_{pul} \tag{19.2-10}$$

Here, $\dot{V}_A$ is the pseudo-continuous ventilation rate (Eq. 19.1-36) at which CO_2 is eliminated to the environment, and $\dot{N}_{pul}$ is the molar rate of CO_2 transport from pulmonary capillary blood. As exists under typical conditions, this equation assumes no CO_2 in inhaled air.

In the pulmonary capillary-tissue subcompartment, we assume that the transport and reversible CO_2 binding reactions are so fast that we can represent the total CO_2 in dissolved and bound forms by a single uniform equilibrium concentration C_{tis}. Under most physiological conditions, the rate of metabolic CO_2 production in the pulmonary tissues is negligible compared to $\dot{N}_{pul}$. A molar balance of capillary-tissue CO_2 then leads to:

$$\frac{d(V_{tis}C_{tis})}{dt} = Q(C_v(t-\tau_v) - C_a) - \dot{N}_{pul} \tag{19.2-11}$$

where $C_v(t-\tau_v)$ is the input concentration of mixed-venous CO_2 that is delayed by a time τ_v from the systemic compartment; $C_a(t)$ is the output concentration of arterial CO_2; and Q is the pulmonary blood flow that is essentially the cardiac output. The CO_2 concentration in the well-mixed pulmonary tissue is essentially equal to the CO_2 concentration exiting in arterial blood, $C_{tis} = C_a$. Adding Equations 19.2-10 and 19.2-11 to obtain the total CO_2 molar balance in the pulmonary compartment, we get

$$\frac{d}{dt}(V_{alv}C_{alv} + V_{tis}C_a) = Q(C_v(t-\tau_v) - C_a) - \dot{V}_A C_{alv} \tag{19.2-12}$$

Rather than using molar concentrations, this equation can be expressed more simply in terms of the CO_2 partial pressure in alveolar gas and CO_2 volume content in blood. The partial pressure in alveolar gas is $p_{alv} = (P^o/c_G^o)C_{alv}$ and the blood content is $\hat{C}_i(p_i) = C_i(p_i)/c_G^o$. Dividing Equation 19.2-12 by c_G^o, we obtain

$$\frac{d}{dt}\left(\frac{V_{alv}p_{alv}}{P^o} + V_{tis}\hat{C}_a\right) = Q(\hat{C}_v(t-\tau_v) - \hat{C}_a) - \frac{\dot{V}_A p_{alv}}{P^o} \tag{19.2-13}$$

We assume that a single equilibrium function describes the relation of dissolved and bound CO_2 content to CO_2 partial pressure: $\hat{C}_i = \hat{C}(p_i)$ for i = a,v. We also assume that phase equilibrium exists between CO_2 in alveolar gas and in arterial blood so that $p_{alv} = p_a$. The pulmonary CO_2 balance then becomes

$$\left(\frac{V_{alv}}{P^o} + V_{tis}\frac{d\hat{C}(p_a)}{dp_a}\right)\frac{dp_a}{dt} = Q[\hat{C}(p_v(t-\tau_v)) - \hat{C}(p_a)] - \left(\frac{1}{P^o}\right)\dot{V}_A p_a \tag{19.2-14}$$

The well-mixed systemic compartment consists of both capillary blood and extravascular tissue in which CO_2 is produced at a molar rate $\dot{M}_{CO_2}$. The CO_2 molar balance in this compartment is as follows:

$$V_s\frac{dC_v}{dt} = Q(C_a(t-\tau_a) - C_v) + \dot{M}_{CO_2} \tag{19.2-15}$$

where C_v is the CO_2 output concentration at time t, $C_a(t-\tau_a)$ is the systemic input concentration from the pulmonary compartment delayed by τ_a, and V_s is the total volume of systemic blood and extravascular tissue. For dissolved and bound CO_2 in the systemic compartment following the same equilibrium function $\hat{C}(p_i)$ as in the pulmonary tissue–blood subcompartment, the CO_2 balance can be expressed in terms of CO_2 content and partial pressures as

$$V_s\frac{d\hat{C}(p_v)}{dp_v}\frac{dp_v}{dt} = Q[\hat{C}(p_a(t-\tau_a)) - \hat{C}(p_v)] + \dot{V}_{CO_2} \tag{19.2-16}$$

where $\dot{V}_{CO_2} = \dot{M}_{CO_2}/c_G^o$ is the volumetric rate of metabolic CO_2 production in the system.

To complete the model, a relationship must be specified for the neurological control of ventilation by an arterial CO_2 signal received by the brain. This relationship will be a function of $p_a(t-\tau_b)$, where τ_b is the transit time of the arterial signal to the brain and a brain signal to the respiratory muscles. We expect τ_b to be shorter than the CO_2 transport delay τ_a.

Analysis

Perturbation Variables

The behavior of this nonlinear system can be analyzed by considering deviations of metabolic CO_2 production, $\dot{V}'_{CO_2}$, from a desired steady-state operating (or set) point, $\dot{V}^s_{CO_2}$:

$$\dot{V}'_{CO_2} \equiv \dot{V}_{CO_2} - \dot{V}^s_{CO_2} \tag{19.2-17}$$

This will result in a perturbation of the other state variables from their steady-state levels:

$$\begin{aligned} p'_i &= p_i - p^s_i, && (i = a, v) \\ \hat{C}'(p_i) &\equiv \hat{C}(p_i) - \hat{C}^s_i && (i = a, v) \\ \dot{V}'_A(p_a(t-\tau_b)) &\equiv \dot{V}_A(p_a(t-\tau_b)) - \dot{V}^s_A \end{aligned} \tag{19.2-18a-c}$$

Substitution of these perturbation variables into the dynamic compartment balances (Eqs. 19.2-14 and 19.2-16) yields

$$\begin{aligned} \left(\frac{V_{alv}}{p^o} + V_{tis}\frac{d\hat{C}'(p'_a)}{dp'_a}\right)\frac{dp'_a}{dt} &= Q\left[\hat{C}'(p'_v(t-\tau_v)) - \hat{C}'(p'_a)\right] + Q\left(\hat{C}^s_v - \hat{C}^s_a\right) \\ &\quad - \frac{1}{p^o}\left(\dot{V}'_A + \dot{V}^s_A\right)(p'_a + p^s_a) \end{aligned} \tag{19.2-19}$$

$$\begin{aligned} \left(V_s\frac{dC'(p'_v)}{dp'_v}\right)\frac{dp'_v}{dt} &= Q\left[\hat{C}'_a(p'_a(t-\tau_a)) - \hat{C}'_v(p'_v)\right] \\ &\quad + Q\left(\hat{C}^s_a - \hat{C}^s_v\right) + \dot{V}'_{CO_2} + \dot{V}^s_{CO_2} \end{aligned} \tag{19.2-20}$$

To obtain the steady-state relationships between state variables, we set the time derivatives in these equations to zero. Noting also that $p^s_i(t) = p^s_i(t-\tau_i)$ at steady state, we get

$$0 = Q\left(\hat{C}^s_v - \hat{C}^s_a\right) - \left(\frac{\dot{V}^s_A}{p^o}\right)p^s_a \tag{19.2-21}$$

$$0 = Q\left(\hat{C}^s_a - \hat{C}^s_v\right) + \dot{V}^s_{CO_2} \tag{19.2-22}$$

By adding Equations 19.2-21 and 19.2-22, we also find that at steady state

$$\frac{\dot{V}^s_A}{p^o} = \frac{\dot{V}^s_{CO_2}}{p^s_a} \tag{19.2-23}$$

When these relations are substituted into the balance equations for the pulmonary and systemic compartments, we get

$$\left(\frac{V_{alv}}{P^o} + V_{tis}\frac{d\hat{C}'(p'_a)}{dp'_a}\right)\frac{dp'_a}{dt}$$
$$= Q\left[\hat{C}'(p'_v(t-\tau_v)) - \hat{C}'(p'_a)\right] - \frac{\dot{V}'_A p'_a}{P^o} - \frac{\dot{V}^s_{CO_2} p'_a}{p^s_a} - \frac{\dot{V}'_A p^s_a}{P^o} \tag{19.2-24}$$

$$\left(V_s\frac{dC'(p'_v)}{dp'_v}\right)\frac{dp'_v}{dt} = Q\left[\hat{C}'_a(p'_a(t-\tau_a)) - \hat{C}'_v(p'_v)\right] + \dot{V}'_{CO_2} \tag{19.2-25}$$

Linearization

We can simplify the analysis by linearizing the model equations for small deviations (perturbations) of CO_2 partial pressure about a steady-state set point s of the controller. We approximate the CO_2 content in the blood-tissue compartments by a Taylor series expansion:

$$\hat{C}_i(p_i) = \hat{C}^s_i + \left[\frac{d\hat{C}_i}{dp_i}\right]_{p=p^a_i}(p_i - p^s_i) \;\Rightarrow\; \hat{C}'_i(p'_i) = \beta p'_i \tag{19.2-26a, b}$$

where $\beta \equiv [d\hat{C}_i/dp_i]_{p=p^a_i}$ represents an equilibrium coefficient about the steady-state operating point, which has the same value for arterial (i=a) and venous (i=v) blood. A similar linearization of $\dot{V}_A$ imposed by the respiratory controller yields

$$\dot{V}_A(p_a(t-\tau_b) - p^s_a) = \dot{V}^s_A + \left(\frac{d\dot{V}_A}{dp_a}\right)_{p_a = p^s_a}(p_a(t-\tau_b) - p^s_a) \;\Rightarrow\; \dot{V}'_A = Gp'_a(t-\tau_b) \tag{19.2-27a, b}$$

where $G \equiv (d\dot{V}_A/dp_a)$ is the proportional gain. Substituting these expressions into the model equations and neglecting the product of perturbation variables compared to terms that are linear in these variables, we obtain

$$\frac{dp'_a}{dt} = \frac{Q}{V_p}(p'_v(t-\tau) - p'_a) - \left(\frac{\dot{V}^s_{CO_2}}{\beta V_p p^s_a}\right)p'_a - \left(\frac{Gp^s_a}{\beta V_p P^o}\right)p'_a(t-\tau_b) \tag{19.2-28}$$

$$\frac{dp'_v}{dt} = \frac{Q}{V_s}(p'_a(t-\tau_a) - p'_v v) + \frac{\dot{V}'_{CO_2}}{\beta V_s} \tag{19.2-29}$$

If the system is at steady state to begin with, then $p'_a(0) = p'_v(0) = 0$ are the initial conditions necessary to solve these equations.

Consider an impulse disturbance of metabolic CO_2 production in the systemic compartment such that $\dot{V}'_{CO_2} = \Delta V_{CO_2}\delta(t)$, where ΔV_{CO_2} is the volume change of CO_2. In that case, the governing equation for the systemic compartment (Eq. 19.2-29) and its initial condition can be written in an equivalent form (Appendix C5) as

$$\frac{dp'_v}{dt} = \frac{Q}{V_s}p'_a(t-\tau_a) - \frac{Q}{V_s}p'_v$$
$$t = 0:\; p'_v = \frac{\Delta V_{CO_2}}{\beta V_s} \tag{19.2-30a, b}$$

Dimensionless Forms

In terms of the dimensionless variables,

$$t \equiv \frac{Q}{V_p} t, \quad p_i \equiv \frac{p_i'}{p_v'(0)} = \frac{\beta V_s}{\Delta V_{CO_2}} p_i' \quad (i = a, v) \tag{19.2-31a,b}$$

we can express the linearized model as

$$\frac{dp_a}{dt} = p_v(t - \tau_v) - \left(1 + \dot{V}_{CO_2}\right) p_a - G p_a(t - \tau_b) \tag{19.2-32}$$

$$\frac{dp_v}{dt} = V\left(p_a(t - \tau_a) - p_v\right) \tag{19.2-33}$$

and the initial conditions as

$$t = 0: \quad p_a = 0, \quad p_v = 1 \tag{19.2-34}$$

The dimensionless parameter groups in these equations are defined as

$$G \equiv \frac{G p_a^s}{P^o \beta Q} \quad V = \frac{V_P}{V_s} = \frac{V_{tis} + V_{alv}/P^o\beta}{V_s}, \quad \dot{V}_{CO_2} \equiv \frac{\dot{V}_{CO_2}^s}{Q\beta p_a^s}$$
$$\tau_i = \frac{Q}{V_p} \tau_i = \frac{Q}{V_{tis} + V_{alv}/P^o\beta} \tau_i \quad (i = a, v, b) \tag{19.2-35a-d}$$

Simulation

Based on physiological data for an adult male, typical values of the parameters can be estimated: $Q = 5\ \text{L/min}$, $V_s = 17\ \text{L}$, $V_{tis} = 2\ \text{L}$, $V_{alv} = 3\ \text{L}$, $p_a^s = 5.1\ \text{kPa}$, $\dot{V}_A^s = 4.9\ \text{L/min}$, $\dot{V}_{CO_2}^s = 0.25\ \text{L/min}$, $\tau_a = 0.13$ min, $\tau_v = 0.28$ min, $\tau_b = 0.065$, $\beta = 0.032\text{kPa}^{-1}$ (estimated from Eq. 5.5-39), and $P^o = 101\text{kPa}$. The corresponding dimensionless parameter groups have the following values: $V = 0.172$, $\dot{V}_{CO_2} = 0.306$ $\tau_a = 0.222$, $\tau_v = 0.478$, $\tau_b = 0.111$, and $G = 0.315G$ [L/(min-Pa)]. Since controller gain plays an important role in the stability of the system, we simulated various dimensionless values in the range $16 > G \geq 0$.

With these parameter values, we numerically solve the model equations for $p_a = p_a'/p_v'(0)$. This variable represents the deviation of arterial CO_2 relative to the initial disturbance in venous CO_2 caused by a small change in systemic CO_2 production. Depending on the controller gain G, simulated outputs exhibit stable responses (over-damped or under-damped) and unstable responses.

Figure 19.2-5 shows simulations for G of ten or less. In the absence of feedback control when $G = 0$, an over-damped response to the impulse disturbance in CO_2 production causes a rapidly increasing deviation in arterial CO_2 to a peak value. This is followed by a relatively slow decay toward the initial steady state of $p_a = 0$. For a controlled system at $10 \geq G > 0$, the peak CO_2 deviation is lower and the return to the initial steady state is quicker. Also, the behavior of p_a changes from strictly over-damped when G is small to somewhat under-damped when $G = 10$.

Figure 19.2-6 shows that at a higher gain of $G = 15$, the under-damped oscillations increase markedly in both frequency and amplitude. Finally, at $G = 16$, the impulse response of p_a becomes unstable, that is, it increases without limit. Since the function of feedback control is to return a system to a steady state as quickly as possible without producing markedly under-damped behavior, $G = 10$ is the maximum gain we expect for counteracting small CO_2 disturbances. Simulations with a nonlinear model, however, would possibly show stable responses for larger values of G.

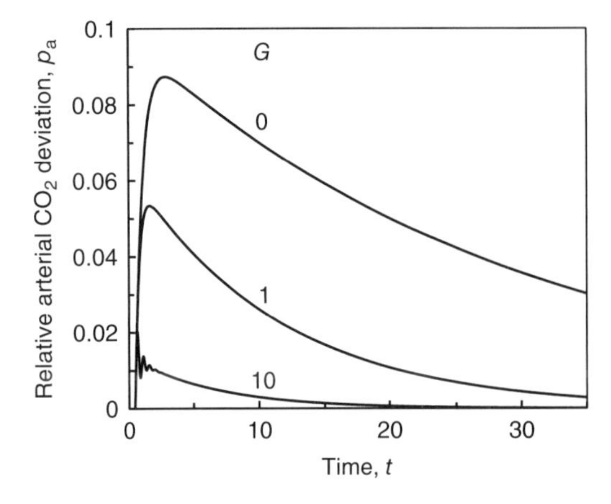

Figure 19.2-5 Effect of CO_2 ventilatory controller gain on arterial partial pressure.

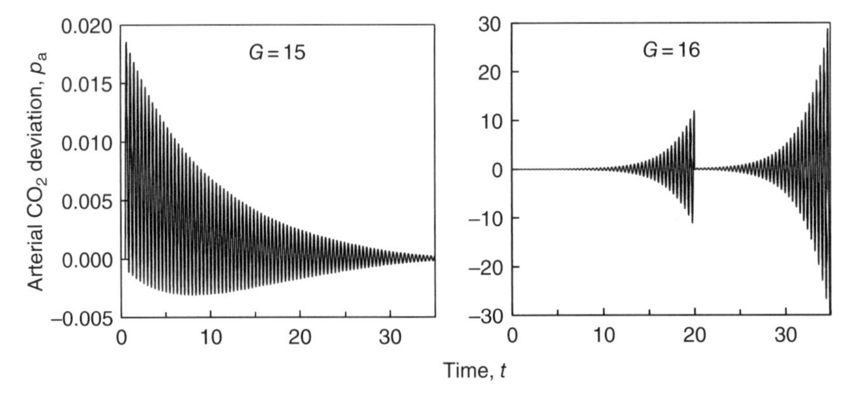

Figure 19.2-6 Under-damped (left) and unstable (right) behavior of CO_2 ventilatory control model.

19.2.3 Perfusion-Controlled Ethanol Metabolism

In developing treatments for the toxic effects of excessive alcohol consumption, it is useful to apply transport models that describe the distribution and metabolism of ethanol in major tissues and organs of the body. Following ingestion of ethanol, the stomach contents empty into the intestines from which essentially all the ethanol is absorbed into the bloodstream and reaches the liver through the portal and hepatic arteries. Within the liver, the enzymatic oxidation of ethanol creates a reaction intermediate, acetaldehyde, that accounts for much of ethanol's toxicity.

Here, we develop a physiologically based model consisting of a stomach (S) compartment and four perfused systemic compartments: gut (G), liver (L), muscle (M), and central (C) circulation (Fig. 19.2-7). The gut compartment contains the entire intestinal tract, and the muscle compartment includes striated muscle and fat. The central circulation compartment lumps all other tissues and organs. This model is a modification of the model developed by Umulis *et al.* (2005).

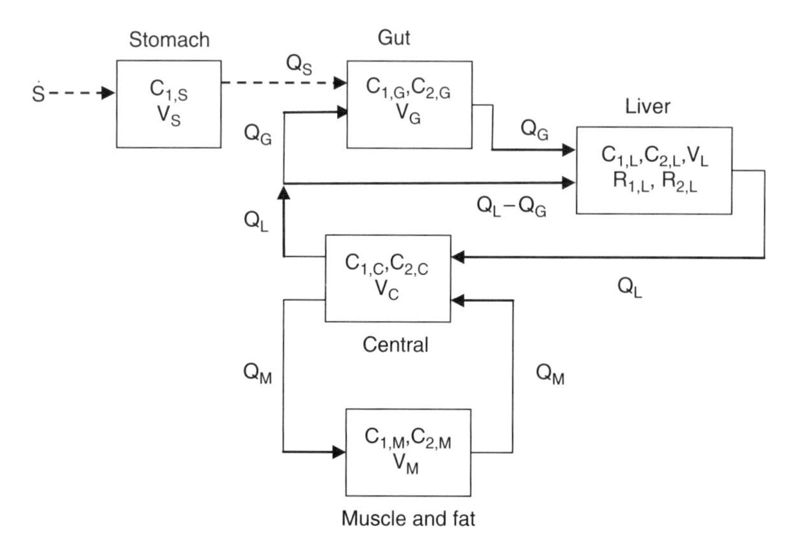

Figure 19.2-7 Physiologically based model of ethanol distribution and metabolism.

Model Formulation

The variables Q_j, V_j, $C_{i,j}$ and $R_{i,j}$ associated with ethanol ($i = 1$) or acetaldehyde ($i = 2$) and with compartments $j = S$, G, L, C, or M represent volumetric flow rate, compartment volume, molar concentration, and molar production rate per unit volume, respectively. Except for the stomach, each compartment volume is constant. Whereas blood flow rates between compartments (solid lines) are constant, ethanol ingestion and stomach emptying (dashed lines) vary with time. With a uniform mass density throughout the non-stomach compartments, overall mass conservation dictates that volumetric inflows and outflows of blood are matched in each of these four perfused compartments. Consequently, the model has three independent blood flows: Q_G, Q_L, and Q_M.

Stomach Compartment

We model the stomach as temporary storage for ingested ethanol. Acetaldehyde formed in the liver does not reach the stomach. We consider what happens following ethanol ingestion, when the stomach empties its contents at a volumetric rate Q_S. A total mass balance then leads to

$$\frac{dV_S}{dt} = -Q_S \tag{19.2-36}$$

with the initial condition that $V_S(0) = V_0$. Emptying the contents of the stomach is dependent on its expansion relative to a fasting state. We approximate the emptying rate as the linear function:

$$Q_S = k_S[V_S(t) - V_{Sf}] \tag{19.2-37}$$

where k_S is a rate constant, and V_{Sf} is the fasting-state volume. Now, Equation 19.2-36 can be written as

$$\frac{dV_S}{dt} + k_S V_S = k_S V_{Sf} \tag{19.2-38}$$

Integrating this equation with the initial condition $V_S(0)=V_{So}$, we get

$$V_S(t) = (V_{So} - V_{Sf})\exp(-k_S t) + V_{Sf} \;\Rightarrow\; Q_S = k_S(V_{So} - V_{Sf})\exp(-k_S t) \qquad (19.2\text{-}39a, b)$$

The molar balance for ethanol in the stomach is as follows:

$$\frac{d(V_S C_{1,S})}{dt} = \dot{S} - Q_S C_{1,S} \qquad (19.2\text{-}40)$$

where $\dot{S}$ is the molar rate of ethanol ingestion. Expanding the derivative and substituting Equation 19.2-36 into this equation yields

$$\frac{dC_{1,S}}{dt} = \frac{\dot{S}}{V_S} \qquad (19.2\text{-}41)$$

Systemic Compartments

Within each of the four well-mixed systemic compartments, we assume that the diffusion of ethanol and acetaldehyde within and between the blood and extravascular tissue is so rapid that they are both in phase equilibrium. Consequently, the transport of these chemical species is limited by perfusion through the compartments. From molar balances on ethanol ($i = 1$) and acetaldehyde ($i = 2$), we obtain concentration dynamics for the gut compartment:

$$\begin{aligned} \frac{dC_{1,G}}{dt} &= \frac{Q_G}{V_G}(C_{1,C} - C_{1,G}) + \frac{Q_S}{V_G}C_{1,S} \\ \frac{dC_{2,G}}{dt} &= \frac{Q_G}{V_G}(C_{2,C} - C_{2,G}) \end{aligned} \qquad (19.2\text{-}42a, b)$$

liver compartment:

$$\frac{dC_{i,L}}{dt} = \frac{Q_L}{V_L}(C_{i,C} - C_{i,L}) + \frac{Q_G}{V_L}(C_{i,G} - C_{i,C}) + R_{i,L} \qquad (19.2\text{-}43)$$

central compartment:

$$\frac{dC_{i,C}}{dt} = \frac{Q_L}{V_C}(C_{i,L} - C_{i,C}) + \frac{Q_M}{V_C}(C_{i,M} - C_{i,C}) \qquad (19.2\text{-}44)$$

and muscle compartment:

$$\frac{dC_{i,M}}{dt} = \frac{Q_M}{V_M}(C_{i,C} - C_{i,M}) \qquad (19.2\text{-}45)$$

Here, $R_{i,L}$ represents the molar formation rates of ethanol ($i = 1$) and acetaldehyde ($i = 2$) per unit volume of liver.

Once in the liver, ethanol (CH_3CH_2OH) is enzymatically oxidized in a reversible fashion by alcohol dehydrogenase (ADH) to form acetaldehyde (CH_3CHO) which is irreversibly oxidized by acetaldehyde dehydrogenase (ACDH) to form methyl acetate (CH_3COOH). The overall stoichiometry of the ethanol and acetaldehyde reactions is

$$\begin{aligned} &CH_3CH_2OH + \frac{1}{2}O_2 \xrightarrow{ADH} CH_3CHO + H_2O \\ &CH_3CHO + \frac{3}{2}O_2 \xrightarrow{ACDH} \frac{1}{2}CH_3COOH + CO_2 + H_2O \end{aligned} \qquad (19.2\text{-}46a, b)$$

The oxidation kinetics of ethanol in the liver are complicated by the dependence of ADH and ACDH on a cosubstrate, nicotinamide adenine dinucleotide (NAD). Under reasonable conditions, the inherent rates of these non-elementary reactions can be expressed as:

$$\bar{r}_{ADH} = -\frac{V_{f,ADH}C_{1,L} - V_{r,ADH}C_{2,L}}{K_{f,ADH} + C_{1,L} + K_{r,ADH}C_{2,L}}, \quad \bar{r}_{ACDH} = -\frac{V_{f,ACDH}C_{2,L}}{K_{f,ACDH} + C_{2,L}} \qquad (19.2\text{-}47a,b)$$

where V_f and V_r represent maximum reaction rates for forward and reverse reactions, and K_f and K_r are the corresponding Michaelis constants. These parameters depend on concentrations of the NAD, both in oxidized and reduced forms. The overall production rates of ethanol and acetaldehyde per unit liver volume are

$$R_{1,L} = -\bar{r}_{ADH}, \quad R_{2,L} = \bar{r}_{ADH} - \bar{r}_{ACDH} \qquad (19.2\text{-}48a,b)$$

To complete the model, we specify that ethanol and acetaldehyde are initially absent from the perfused compartments:

$$t = 0: \quad C_{i,G} = C_{i,L} = C_{i,C} = C_{i,M} = 0 \quad (i = 1,2) \qquad (19.2\text{-}49)$$

Analysis

Before solving for the system dynamics, we must provide the rate $\dot{S}$ at which ethanol is ingested. Suppose a person rapidly drinks m moles of ethanol at $t = 0$. Modeling $\dot{S}$ by the ideal impulse $m\delta(t)$, Equation 19.2-41 becomes

$$\frac{dC_{1,S}}{dt} = \frac{m\delta(t)}{V_S} \qquad (19.2\text{-}50)$$

with the prior condition that $C_{1,S} = 0$ for $t < 0$. Upon integration, we see that $C_{1,S}(t) = m/V_{So}$ for $t \geq 0$. Combining this result with Equation 19.2-39b, we obtain the molar rate of ethanol output from the stomach that is required in the molar balance for the gut (Eq. 19.2-42a):

$$Q_S C_{1,S} = \frac{m}{V_{So}}[k_S(V_{So} - V_{Sf})\exp(-k_S t)] \qquad (19.2\text{-}51)$$

Figure 19.2-8 shows a numerical solution of Equations 19.2-42–19.2-45 for ethanol and acetaldehyde distribution dynamics following the consumption of 1 oz of ethanol, equivalent to m = 450 mol. Experimental measurements indicate that this amount of ingestion is associated with a rate constant for stomach emptying of $k_S = 0.04\ \text{min}^{-1}$ (Umulis *et al.*, 2005; table 3). Provided that the person drinks the ethanol shot on a full stomach, we can assume that $V_{So} >> V_{Sf}$. The remaining parameter values used to simulate the responses of a 70 kg person are

$$\begin{aligned} &V_G = 2.4,\ V_L = 1.08,\ V_C = 11.6,\ V_M = 25.8\,\text{L} \\ &Q_G = 0.900,\ Q_L = 1.35,\ Q_M = 0.95\,\text{L/min} \\ &V_{f,ADH} = 3.19,\ V_{r,ADH} = 47.3,\ V_{f,ACDH} = 3.92\,\text{mmol/min} \\ &K_{f,ADH} = 0.581,\ K_{r,ADH} = 1.45,\ K_{f,ACDH} = 0.00170\,\text{mM} \end{aligned} \qquad (19.2\text{-}52)$$

These simulations predict that the effects of sudden ingestion last for more than an hour, both with respect to the primary substance, ethanol, and its toxic oxidation product, acetaldehyde. Because of its closer proximity to the source, the gut exhibits a higher ethanol level than the liver. By contrast, acetaldehyde is formed in the liver so its concentration is higher and its peak concentration appears earlier than in the gut. To indicate the significance of the ethanol level,

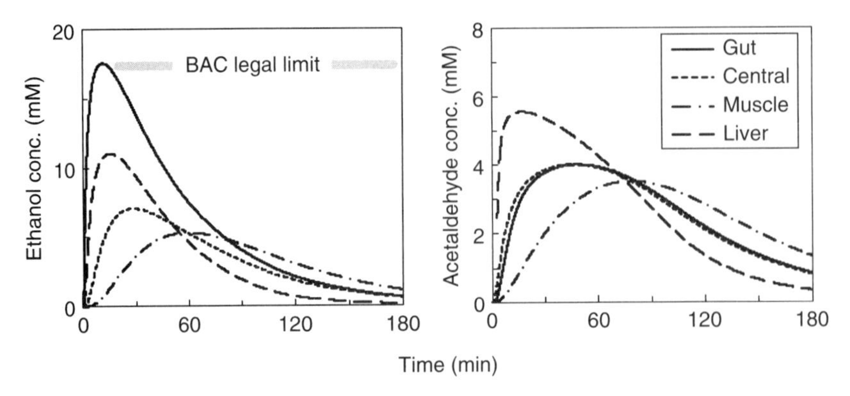

Figure 19.2-8 Distribution dynamics of ethanol and acetaldehyde after sudden ethanol ingestion.

the figure also displays the blood alcohol content (BAC) of 0.08% ≈ 17 mM that is a common legal threshold for "driving under the influence." For the ingestion of 1 oz of alcohol typical of a single drink, the peak ethanol level that is predicted in the central compartment (the source of blood samples for BAC determination) is about one half of this legal limit.

References

elHefnawy A, Saidel GM, Bruce EN. CO_2 control of the respiratory system: plant dynamics and stability analysis. Ann Biomed Eng. 1988; 16:445–461.

Grandi, F, Avanzolini G, Cappello A. Analytical solution of the variable-volume double pool urea kinetics model applied to parameter estimation in hemodialysis. Comput Biol Med. 1995; 25:505–518.

Nicholes, KRK, Warner HR, Wood EH. Study of dispersion of an indicator in the circulation. Ann N Y Acad Sci. 1964; 115:721–737.

Saidel, GM. Alveolar-capillary diffusion and ventilation-perfusion inhomogeneity: a mathematical model. Med Biol Eng Comput. 1982; 20:269–273.

Saidel GM, Salmon RB, Chester EH. Moment analysis of multibreath lung washout. J Appl Physiol. 1975; 38:328–334.

Spencer JL, Long CL, Kinney JM. A model of glucose metabolism in man. Ind Eng Chem Fundam. 1971; 10:2–12.

Umulis DM, Gurmen NM, Singh P, Fogler HS. A physiologically based model for ethanol and acetaldehyde metabolism in human beings. Alcohol. 2005; 35:3–12.

Part VII
Advanced Biomedical Applications

Chapter 20

Therapies for Tissue and Organ Dysfunction

An important area of biomedical engineering is the development of man-made devices or tissue constructs to support or replace physiological function compromised by disease or trauma. This chapter presents therapeutic analyses that emphasize the impact of design decisions on a patient's health and on the effectiveness of a device. The development of the underlying models incorporates convective and diffusive mass transport of chemical species at tissue and cellular levels including cellular population dynamics.

20.1 Dynamics of Urea Clearance in a Patient During Hemodialysis

In Section 19.1, a dynamic model of urea concentration and tissue volume was presented for a patient undergoing dialysis treatment of end-stage kidney disease. The body fluids were represented by an intracellular fluid (ICF) compartment in which urea is formed by protein catabolism and an extracellular fluid (ECF) compartment from which urea and water are removed by extracorporeal hemodialysis. In that model, we assumed that the hemodialyzer has a constant urea clearance Q_C. In fact, Q_C can only be constant when urea transport processes are so rapid within the dialyzer compared to those in ICF and ECF that the device operates in a pseudo-steady state, and the flows of blood, dialysate, and filtrate do not vary with time. Here, a more general closed-loop model will account for urea dynamics within the dialyzer as well as within the body fluids. The dialyzer will be described by the model developed in Section 12.3 for countercurrent blood and dialysate flow in a device with a uniform filtration velocity across the membrane.

As shown in Figure 20.1-1, blood from the body consisting of well-mixed ICF and ECF compartments circulates through the hemodialyzer that contains an axial distribution of urea concentration. Urea produced in ICF at a constant rate $\dot{R}_I$ diffuses across an

Biomedical Mass Transport and Chemical Reaction: Physicochemical Principles and Mathematical Modeling,
First Edition. James S. Ultman, Harihara Baskaran, and Gerald M. Saidel.

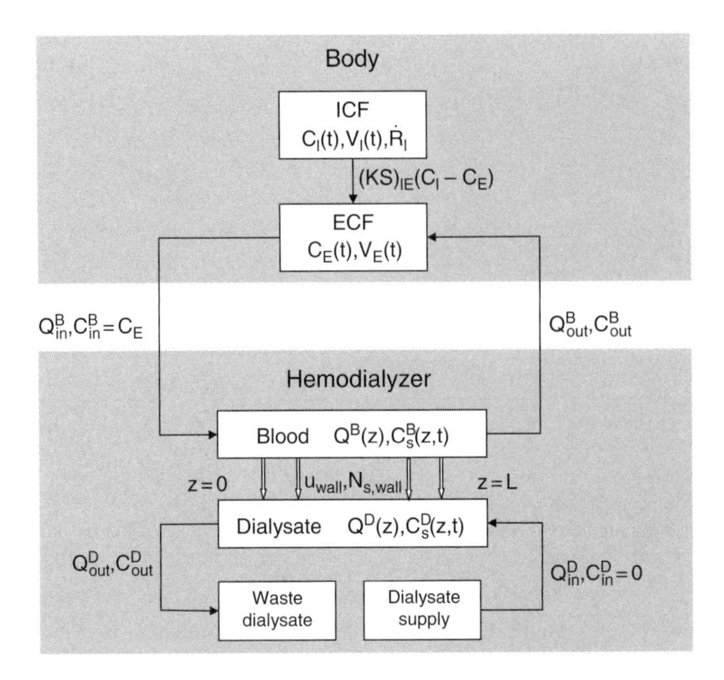

Figure 20.1-1 Hemodialysis model with body fluids and a countercurrent exchanger.

intercompartment membrane into the ECF. Blood leaving ECF with a urea concentration $C_E(t)$ enters the dialyzer at a constant volumetric flow rate Q^B_{in} and a urea concentration $C^B_{in}(t) = C_E(t)$. Blood enters the ECF from the dialyzer exit at a constant volumetric flow rate Q^B_{out} and a urea concentration $C^B_{out}(t)$.

Within the dialyzer, blood flow $Q^B(z)$ and dialysate flow $Q^D(z)$ occurring in the z and – z directions, respectively, are separated by a membrane with a surface area S and a length L. By diffusion of urea through the membrane, its concentration in exiting blood is reduced to $C^B_{out}(t) < C^B_{in}(t)$. Because of filtration of water through the membrane, blood from the dialyzer is returned to ECF at a reduced flow rate $Q^B_{out} < Q^B_{in}$. Dialysate from a supply reservoir enters the dialyzer at a constant flow rate Q^D_{in} and urea concentration $C^D_{in} = 0$. The dialysate leaves the dialyzer as waste product at a constant flow rate $Q^D_{out} > Q^D_{in}$ and urea concentration $C^D_{out}(t) > C^D_{in}(t)$.

Model Formulation

Volume and Flow Equations

The total fluid volume of the body V(t) is the sum of the ICF volume V_I and the ECF volume V_E. The volume ratio of the ICF and ECF compartments is constant, which implies a constant ICF volume fraction, $\varphi = V_I/V$. The fluids everywhere in this model have the same constant mass density. As in Section 19.1, we assume that the total filtration flow Q_F by the hemodialyzer does not vary with time. Consequently, the mass balance of the body fluids leads to

$$\frac{dV}{dt} = -Q_F \Rightarrow V = V_o - Q_F t \Rightarrow \begin{cases} V_I = \varphi(V_o - Q_F t) \\ V_E = (1-\varphi)(V_o - Q_F t) \end{cases} \quad (20.1\text{-}1a\text{-}c)$$

where V_o is the initial fluid volume in the body.

For a membrane with a uniformly distributed filtration velocity, the volumetric blood flow Q^B and dialysate flow Q^D change with position along the membrane according to

$$Q^B(z) = Q^B_{in} - Q_F(z/L), \quad Q^D(z) = Q^D_{out} - Q_F(z/L) \quad (12.3\text{-}58a,b)$$

where z is longitudinal distance relative to the blood entry point. The flows in the dialyzer at z = L are

$$Q^B_{out} = Q^B_{in} - Q_F, \quad Q^D_{in} = Q^D_{out} - Q_F \quad (12.3\text{-}59a,b)$$

Urea Concentration Dynamics

The molar balances for urea in the ICF and ECF compartments are

$$\frac{d(C_I V_I)}{dt} = -(KS)_{IE}(C_I - C_E) + \dot{R}_I \quad (20.1\text{-}2)$$

$$\frac{d(C_E V_E)}{dt} = (Q^B_{out} C^B_{out} - Q^B_{in} C_E) + (KS)_{IE}(C_I - C_E) \quad (20.1\text{-}3)$$

where $(KS)_{IE}$ is an overall mass transfer coefficient K_{IE} multiplied by the contact surface S_{IE} between the two body compartments. By substituting the compartment volume equations into these material balances, we obtain

$$\varphi(V_o - Q_F t)\frac{dC_I}{dt} + \left[(KS)_{IE} - \varphi Q_F\right]C_I = (KS)_{IE} C_E + \dot{R}_I \quad (20.1\text{-}4)$$

$$(1-\varphi)(V_o - Q_F t)\frac{dC_E}{dt} + \left[(KS)_{IE} + Q^B_{in} - (1-\varphi)Q_F\right]C_E = (KS)_{IE} C_I + \left(Q^B_{in} - Q_F\right)C^B_{out} \quad (20.1\text{-}5)$$

Assuming axial transport is dominated by convection, the one-dimensional urea balances in blood and dialysate are

$$A^B \frac{\partial C^B_s}{\partial t} + \frac{\partial (Q^B C^B_s)}{\partial z} = -\frac{S}{L} N_{s,wall}$$

$$A^D \frac{\partial C^D_s}{\partial t} - \frac{\partial (Q^D C^D_s)}{\partial z} = \frac{S}{L} N_{s,wall} \quad (12.3\text{-}32a,b)$$

Here, $C^B_s(z,t)$ and $C^D_s(z,t)$ are the urea concentration distributions along the dialyzer; $N_{s,wall}(z,t)$ is the local urea flux; A^B and A^D are the total cross-sectional areas available for flow on the blood and dialysate sides of the device.

The rate equation for $N_{s,wall}$ must account for transport across a membrane by parallel convection and diffusion. A small molecule like urea has a reflection coefficient close to zero, and $N_{s,wall}$ is approximated from Eq. 10.2-6 as:

$$N_{s,wall} = \frac{u_{wall}}{1 - \exp(-u_{wall}/K_s)}\left[C^B_s - C^D_s \exp(-u_{wall}/K_s)\right] \quad (12.3\text{-}42)$$

where u_{wall} is the constant solution velocity across the membrane; K_s is an overall mass transfer coefficient that accounts for diffusion resistances across the membrane and its adjacent blood and dialysate boundary layers. By combining this rate equation for flux with the urea concentration equations (Eq. 12.3-32a,b) and flow equations (Eq. 12.3-58a,b), and noting that $Q_F = u_{wall}S$ for a uniformly distributed transmembrane velocity, we get consolidated concentration equations for the dialyzer:

$$\frac{V^B}{L}\frac{\partial C_s^B}{\partial t} + \left(Q_{in}^B - Q_F\frac{z}{L}\right)\frac{\partial C_s^B}{\partial z} + \frac{e/L}{1-e}Q_F C^B = \frac{e/L}{1-e}Q_F C_s^D \tag{20.1-6}$$

$$\frac{V^D}{L}\frac{\partial C_s^D}{\partial t} - \left[Q_{in}^D + Q_F\left(1-\frac{z}{L}\right)\right]\frac{\partial C_s^D}{\partial z} + \frac{1/L}{1-e}Q_F C_s^D = \frac{1/L}{1-e}Q_F C_s^B \tag{20.1-7}$$

Here, $V^B = A^B L$ and $V^D = A^D L$ are the blood and dialysate volumes in the dialyzer, and

$$e \equiv \exp(-u_{wall}/K_s) = \exp(-Q_F/K_s S) \tag{20.1-8}$$

Initially, urea concentration is C_o in ICF, ECF, and in the primed blood within the dialyzer. No urea is initially present in the dialysate:

$$t = 0: \quad C_E = C_I = C_o, \quad C_s^B(z,0) = C_o, \quad C_s^D(z,0) = 0 \tag{20.1-9}$$

The urea concentration in blood entering the dialyzer is equal to its concentration in ECF, and dialysate entering the dialyzer contains no urea:

$$z = 0: \quad C_B^{in} = C_E, \qquad z = L: \quad C_D^{in} = 0 \tag{20.1-10}$$

The effectiveness of dialysis treatment during a period {0,t} can be judged from the total moles of urea removed from the two body compartments:

$$U(t) = C_o V_o - (V_o - Q_F t)[\varphi C_I(t) + (1-\varphi)C_E(t)] \tag{20.1-11}$$

where $C_o V_o$ are the moles of urea initially in the body. Note that particular names have been assigned to the variables at the boundaries of the hemodialyzer: $Q_{in}^B \equiv Q^B(0,t)$, $Q_{out}^B \equiv Q^B(L,t)$, $Q_{in}^D \equiv Q^D(L,t)$, $Q_{out}^D \equiv Q^D(0,t)$, $C_{in}^B \equiv C_s^B(0,t)$, $C_{out}^B \equiv C_s^B(L,t)$, $C_{in}^D \equiv C_s^D(L,t)$, and $C_{out}^D \equiv C_s^D(0,t)$.

Analysis

Dimensionless Forms

The model equations can be expressed in terms of dimensionless variables:

$$t = \frac{Q_{in}^B}{V_B}t, \quad z = \frac{z}{L}, \quad C_I(t) = \frac{C_I(t)}{C_o}, \quad C_E(t) = \frac{C_E(t)}{C_o}$$
$$C^B(z,t) = \frac{C^B(z,t)}{C_o}, \quad C^D(z,t) = \frac{C^D(z,t)}{C_o}, \quad C_{out}^B(t) = \frac{C^B(L,t)}{C_o} \tag{20.1-12a-g}$$

The dimensionless equations for urea concentration in the ICF and ECF are

$$\varphi(V_o - Q_F t)\frac{dC_I}{dt} + [(KS)_{IE} - Q_F\varphi]C_I = (KS)_{IE}C_E + \dot{R}_I \tag{20.1-13}$$

$$(1-\varphi)(V_o - Q_F t)\frac{dC_E}{dt} + [(KS)_{IE} + 1 - (1-\varphi)Q_F]C_E = (KS)_{IE}C_I + (1-Q_F)C_{out}^B \tag{20.1-14}$$

For the dialyzer, the dimensionless urea equations for the blood and dialysate are

$$\frac{\partial C^{B}}{\partial t} + (1 - Q_{F}z)\frac{\partial C^{B}}{\partial z} + \left(\frac{Q_{F}e}{1-e}\right)C^{B} = \left(\frac{Q_{F}e}{1-e}\right)C^{D} \tag{20.1-15}$$

$$V^{D/B}\frac{\partial C^{D}}{\partial t} - \left[Q_{in}^{D/B} + Q_{F}(1-z)\right]\frac{\partial C^{D}}{\partial z} + \left(\frac{Q_{F}}{1-e}\right)C^{D} = \left(\frac{Q_{F}}{1-e}\right)C^{B} \tag{20.1-16}$$

Here, we have defined the dimensionless parameters:

$$Q_{F} \equiv \frac{Q_{F}}{Q_{in}^{B}}, \quad Q_{in}^{D/B} \equiv \frac{Q_{in}^{D}}{Q_{in}^{B}}, \quad (KS)_{IE} \equiv \frac{(KS)_{IE}}{Q_{in}^{B}}, \quad (KS) \equiv \frac{K_{s}S}{Q_{in}^{B}}$$
$$V_{o} \equiv \frac{V_{o}}{V^{B}}, \quad V^{D/B} \equiv \frac{V^{D}}{V^{B}}, \quad \dot{R}_{I} \equiv \frac{\dot{R}_{I}}{C_{o}Q_{in}^{B}}, \quad e = \exp[-Q_{F}/(KS)] \tag{20.1-17a-h}$$

The dimensionless boundary conditions are

$$z = 0: \; C_{in}^{B} = C_{E}, \quad z = 1: \; C_{in}^{D} = 0 \tag{20.1-18}$$

and the dimensionless initial conditions are

$$t = 0: \; C_{E} = C_{I} = 1, \; C^{D} = 0, \; C^{B} = 1 \tag{20.1-19}$$

When expressed as a fraction of initial amount of urea, the dimensionless dialyzer effectiveness is

$$U(t) = \frac{U(t)}{C_{o}V_{o}} = 1 - \left(1 - \frac{Q_{F}}{V_{o}}t\right)[\varphi C_{I}(t) + (1-\varphi)C_{E}(t)] \tag{20.1-20}$$

Simulation

This model was solved numerically using parameter values typical of an adult patient (V_I = 24 L, V_E = 16 L, $(KS)_{IE}$ = 70 ml/min, $\dot{R}_I = 0.014$ mmol/min, C_o = 20 mM) and a commercial hollow fiber hemodialyzer (V^B = 100 ml, V^D = 100 ml, K_sS = 500 ml/min). While blood inlet and filtration flows were fixed at $Q_B^{in} = 360$ ml/min and Q_F = 9.4 ml/min, three alternative dialysate input flows shown in Table 20.1-1 were considered. Also shown in the table are the clearances that would occur if a hemodialyzer with the same K_sS, Q_B^{in}, Q_D^{in}, and Q_F was operated at steady state.

The solid curves in Figure 20.1-2 show the dynamics of urea loss from the body predicted with the current model during a 4 h dialysis treatment. These simulations indicate that the effectiveness of urea removal from a particular device operating for a given time can be improved by using a higher dialysate flow. Alternatively, the treatment time necessary to achieve a targeted amount of urea removal can be reduced by increasing dialysate flow. The dashed lines in this figure are simulations of the constant clearance model (Section 19.1) obtained with the Q_C values in Table 20.1-1. While the dashed curves follow the same trends as the solid curves, the constant clearance model consistently overpredicts

Table 20.1-1 Dialyzer Clearance for Steady-State Operation

Q_D^{in} (ml/min)	50	100	500
Q_C (ml/min)	59	107	230

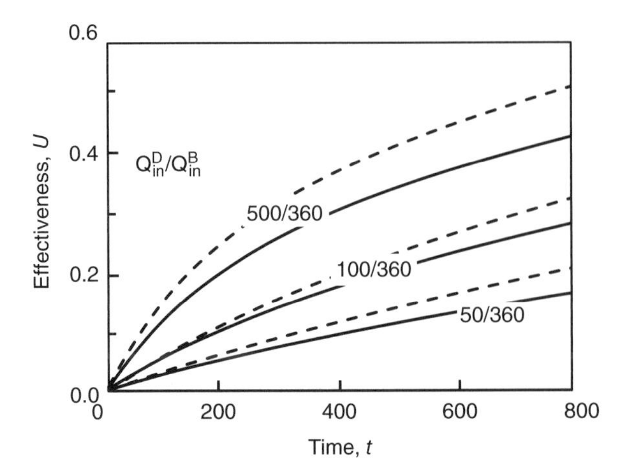

Figure 20.1-2 Effect of dialysate flow on the dynamics of urea removal: variable clearance (solid curves) compared to constant clearance (dashed curves).

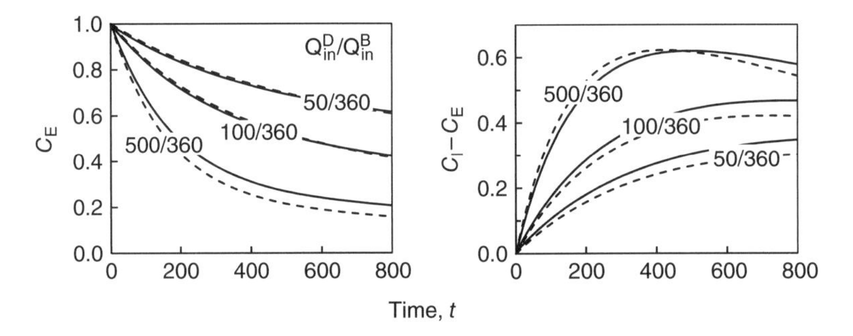

Figure 20.1-3 Effect of dialyzer/blood flow ratio on urea ICF and ECF concentrations: variable clearance (solid curves) compared to constant clearance (dashed curves).

the dialyzer effectiveness compared to the more comprehensive model. Thus, assuming a constant clearance when predicting the time course of urea removal during *in vivo* dialysis can be misleading.

As shown in Figure 20.1-3, the concentration dynamics in the body compartments also follow the same trends for the two models. The left graph demonstrates how much faster the reduction of C_E by dialysis can be made by increasing dialysate–blood flow ratio. The right graph reveals a persistent disequilibrium in transport between the two body compartments such that $C_I > C_E$. As concluded in Section 19.1, this indicates that urea concentration monitored in (extracellular) blood is not necessarily a good indicator of urea concentration in cells where the effects of uremia originate.

20.2 Hemodialyzer Performance with Varying Filtration

The filtration rate of water across a membrane separating blood from dialysate can play an important role in the function of a hemodialyzer as well as in the hemodynamic stability of a patient. In preceding models, we assumed that filtration rate is zero or is uniformly distributed along the membrane. These approximations rule out the possible contribution of spatial variations in transmembrane hydrodynamic and osmotic pressure differences. We now model the steady-state performance of a countercurrent hemodialyzer (Fig. 12.3-4) including these effects.

Model Formulation

Filtration and Pressure–Flow Relations

In addition to fluid pressure and osmolarity differences between bulk solutions, water filtration from blood to dialysate can be affected by concentration polarization (Section 12.2). This phenomenon elevates protein concentration in the vicinity of the membrane–blood interface, resulting in a local increase in oncotic pressure and encouraging protein deposition. Neglecting these effects, the Starling equation for thermodynamically ideal solutions (Eq. 6.5-32) is sufficient to represent the transmembrane filtration velocity $u^* \to u_{wall}$:

$$u_{wall}(z) = L_P\left[\left(P^B - P^D\right) - \sum_i \sigma_i \mathcal{R}T\left(C_i^B - C_i^D\right)\right] \tag{20.2-1}$$

where L_P is the hydraulic conductivity of the membrane; $P^B(z)$ and $P^D(z)$ are the pressures in blood and dialysate; σ_i is the reflection coefficient of solute i; and $C_i^B(z)$ and $C_i^D(z)$ are the local concentrations of solute i in blood and dialysate. Large molecules, primarily proteins, do not permeate the membrane, while electrolytes and other small molecules freely pass through the walls. Consequently, the reflection coefficients are approximately one for proteins ($\sigma_P = 1$) and zero for small solutes ($\sigma_i = 0$; $i \neq P$), and the osmotic driving force reduces to

$$\sum_i \sigma_i \mathcal{R}T\left(C_i^B - C_i^D\right) \approx \mathcal{R}T\left(C_P^B - C_P^D\right) \tag{20.2-2}$$

where $C_P^B(z)$ and $C_P^D(z)$ are the protein concentrations in blood and dialysate. Normally, the dialysate has no proteins. With this additional simplification, Equation 20.2-1 becomes

$$u_{wall}(z) = L_P\left[\left(P^B - P^D\right) - \mathcal{R}TC_P^B\right] \tag{20.2-3}$$

Evaluating the local pressure difference across a dialyzer membrane requires separate pressure–flow relationships for blood and dialysate. We previously derived the pressure gradient established by low Reynolds number flow of a Newtonian flow through a cylindrical tube with a uniform leak (Section 14.2). To generalize this result to a dialyzer in which the channels for blood flow Q^B and dialysate flow Q^D may not be cylindrical, we replace the circular radius "a" in Equation 14.2-21a with one half of the hydraulic diameter d_H as defined in Equation 12.1-23. Noting that the pressures only depend on z, and neglecting gravitational effects, we then have

$$\frac{dP^B}{dz} = -\xi^B Q^B(z), \quad \frac{dP^D}{dz} = \xi^D Q^D(z) \tag{20.2-4a,b}$$

where

$$\xi^B \equiv \frac{128\mu^B}{\pi\left(d_H^B\right)^4}, \quad \xi^D \equiv \frac{128\mu^D}{\pi\left(d_H^D\right)^4} \tag{20.2-5a,b}$$

Here, the sign differences in the pressure drops occur because Q^B is positive in the +z direction, whereas the countercurrent flow Q^D is taken to be positive in the −z direction. Combining these equations with the z derivative of u_{wall} from Equation 20.2-3 yields

$$\frac{du_{wall}}{dz} = -L_P\left[\xi^B Q^B + \xi^D Q^D + \mathcal{R}T\frac{dC_p^B}{dz}\right] \tag{20.2-6}$$

Dialysate Flow and Concentration Distributions

This model assumes steady-state transport in blood and dialysate. For a membrane of surface area S and length L in the direction, the steady-state flow equations with constant fluid density are

$$\frac{dQ^D}{dz} = \frac{dQ^B}{dz}, \quad \frac{dQ^B}{dz} = -\frac{S}{L}u_{wall} \tag{12.3-31a,b}$$

Since protein cannot penetrate the membrane, it remains in the blood and its steady-state molar balance (Eq. 12.3-33b) yields

$$\frac{d\left(Q^B C_p^B\right)}{dz} = 0 \Rightarrow Q^B C_p^B = Q_{in}^B C_{p,in}^B \tag{20.2-7a,b}$$

By expansion of the derivative and substitution of the flow gradient from Equation 12.3-31b, we obtain

$$Q^B\frac{dC_p^B}{dz} + C_p^B\frac{dQ^B}{dz} = 0 \Rightarrow \frac{dC_p^B}{dz} = \left(\frac{S}{L}\right)\frac{u_{wall}C_p^B}{Q^B} = \left(\frac{Q_{in}^B C_{p,in}^B S}{L}\right)\frac{u_{wall}}{\left(Q^B\right)^2} \tag{20.2-8a,b}$$

Incorporating this protein concentration gradient into Equation 20.2-6, the governing equation for filtration velocity becomes

$$\frac{du_{wall}}{dz} = -L_P\left[\xi Q^B + \xi^D Q^D + \mathcal{R}T\left(\frac{Q_{in}^B C_{p,in}^B S}{L}\right)\frac{u_{wall}}{\left(Q^B\right)^2}\right] \tag{20.2-9}$$

From Equation 12.3-33a,b, we write the steady-state urea balances in blood and in dialysate as

$$\frac{d(Q^B C_s^B)}{dz} = -\left(\frac{S}{L}\right)N_{s,wall}, \quad \frac{d(Q^D C_s^D)}{dz} = -\left(\frac{S}{L}\right)N_{s,wall} \tag{20.2-10a,b}$$

where $N_{s,wall}(z)$ is the local urea flux from blood to dialysate. Upon expansion of the derivatives and substitution of the flow gradients from Equations 12.3-31b and 12.3-33b, we obtain the governing equations for the urea concentration distributions:

$$\begin{aligned}\frac{dC_s^B}{dz} - \left(\frac{S}{L}\right)\frac{u_{wall}}{Q^B}C_s^B &= -\left(\frac{S}{L}\right)\frac{N_{s,wall}}{Q^B}\\ \frac{dC_s^D}{dz} - \left(\frac{S}{L}\right)\frac{u_{wall}}{Q^D}C_s^D &= -\left(\frac{S}{L}\right)\frac{N_{s,wall}}{Q^D}\end{aligned} \tag{20.2-11a,b}$$

To complete this model, we must relate $N_{s,wall}$ to local blood and dialysate concentrations as well as to local filtration rate. Assuming that the reflection coefficient for urea is essentially zero, we can use Equation 12.3-42, which we rewrite as

$$N_{s,wall} = u_{wall}\left(\frac{C_s^B - eC_s^D}{1-e}\right); \quad e \equiv \exp\left(-\frac{u_{wall}}{K_s}\right) \qquad (20.2\text{-}12a,b)$$

By affecting the local blood and dialysate velocities, transmembrane filtration can influence the individual mass transfer coefficients that contribute to the overall mass transfer coefficient K_s (Eq. 12.3-40b). When this effect is relatively small, K_s is constant. Incorporating the urea transport rate into the urea concentration equations, we find

$$\frac{dC_s^B}{dz} = \left(\frac{S}{L}\right)\frac{eu_{wall}}{(1-e)Q^B}\left(C_s^D - C_s^B\right) \qquad (20.2\text{-}13)$$

$$\frac{dC_s^D}{dz} = \left(\frac{S}{L}\right)\frac{u_{wall}}{(1-e)Q^D}\left(C_s^D - C_s^B\right) \qquad (20.2\text{-}14)$$

This model consists of ordinary differential equations for Q^B, Q^D, u_{wall}, C_s^B, and C_s^D and algebraic relations for ξ^B, ξ^D and e. The boundary conditions are as follows:

$$z = 0: \quad Q^B = Q_{in}^B, \quad C_s^B = C_{s,in}^B, \quad u_{wall} \equiv u_o$$
$$z = L: \quad Q^D = Q_{in}^D, \quad C_s^D = 0 \qquad (20.2\text{-}15a,b)$$

where the value of u_o is determined from Equation 20.2-3 by the pressures P_{in}^B of entering blood and P_{out}^D of exiting dialysate:

$$u_o = L_P\left[\left(P_{in}^B - P_{out}^D\right) - \mathcal{R}TC_p^B(0)\right] \qquad (20.2\text{-}16)$$

Two useful performance parameters that characterize dialyzer performance are the total filtration rate Q_F, and the urea clearance Q_C (Eq. 12.3-46b):

$$Q_F = Q_{in}^B - Q_{out}^B, \quad Q_C = Q_{in}^B - Q_{out}^B C_{s,out}^B / C_{s,in}^B \qquad (20.2\text{-}17a,b)$$

Here, Q_{out}^B and $C_{s,out}^B$ correspond to the outlet blood flow and blood urea concentration at z = L.

Analysis

Dimensionless Forms

We can express the model equations using the dimensionless variables:

$$z = \frac{z}{L}, \quad u = \frac{u_{wall}}{K_s}, \quad C^j = \frac{C_s^j}{C_{s,in}^B}, \quad Q^j = \frac{Q^j}{Q_{in}^j} \quad (j = B, D) \qquad (20.2\text{-}18a\text{-}d)$$

The dimensionless blood and dialysate flows change according to

$$\frac{dQ^B}{dz} = -\frac{u}{Q_{in}^B}, \quad \frac{dQ^D}{dz} = -\frac{u}{Q_{in}^D} \qquad (20.2\text{-}19a,b)$$

where

$$Q_{in}^j \equiv \frac{Q_{in}^j}{K_s S} \quad (j = B, D) \qquad (20.2\text{-}20)$$

The dimensionless filtration flow varies with location according to

$$\frac{du}{dz} = -\left(\frac{\beta}{Q_{in}^B}\right)\frac{u}{(Q^B)^2} - \left(\beta\alpha^B Q_{in}^B\right)Q^B - \left(\beta\alpha^D Q_{in}^D\right)Q^D \tag{20.2-21}$$

where

$$\beta \equiv \frac{\mathcal{R}TL_P C_{p,in}^B}{K_s}, \quad \alpha^j \equiv \frac{K_s SL}{\mathcal{R}TC_{p,in}^B}\xi^j \quad (j = B, D) \tag{20.2-22a,b}$$

The dimensionless urea concentrations in blood and dialysate change as

$$\frac{dC^B}{dz} = \left(\frac{e/Q_{in}^B}{1-e}\right)\frac{u}{Q^B}\left(C^D - C^B\right) \tag{20.2-23}$$

$$\frac{dC^D}{dz} = \left(\frac{1/Q_{in}^D}{1-e}\right)\frac{u}{Q^D}\left(C^D - C^B\right) \tag{20.2-24}$$

where

$$e \equiv \exp(-u) \tag{20.2-25}$$

The dimensionless boundary conditions are

$$\begin{aligned} z = 0: &\quad Q^B = 1, \quad C^B = 1, \quad u = u_o \\ z = 1: &\quad Q^D = 1, \quad C^D = 0 \end{aligned} \tag{20.2-26a,b}$$

The dimensionless performance parameters are

$$Q_F \equiv \frac{Q_F}{Q_{in}^B} = 1 - \left[Q^B\right]_{z=1}, \quad Q_C \equiv \frac{Q_C}{Q_{in}^B} = 1 - \left[Q^B C^B\right]_{z=1}, \quad u_o = \frac{u_o}{K_s} \tag{20.2-27a-c}$$

Simulation

For numerical simulations of the model equations, we selected typical parameter values for a hollow fiber hemodialyzer (Fig. 12.3-4): T = 310°C, $Q_{in}^B = 360\,\text{ml/min}$, $Q_{in}^D = 100\,\text{ml/min}$, $P_{in}^B - P_{out}^D = 12\,\text{kPa}$, $\mu^B = 3.3$ kPa-s, $\mu^D = 0.8$ kPa-s, $C_{p,in}^B = 0.8\,\text{mM}$, n_f(fiber number) = 17,000, L = 25 cm, d_f(fiber diameter) = 0.25 mm, d(shell diameter) = 4.7 cm, and $K_s = 0.012$ cm/min. Given these parameters, $S = 3.3\,\text{m}^2$, $d_H^B = 0.25\,\text{mm}$, and $d_H^D = 0.27\,\text{mm}$ (Example 12.1-1). To examine the effects of a change in filtration flow, we set $L_p = 4.4 \times 10^{-4}$ or 4.4×10^{-3} mm/(kPa-min), corresponding to $\beta = 0.007$ or 0.07 (Eq. 20.2-22a).

Figure 20.2-1 shows the dimensionless blood, dialysate, and filtration flow distributions along the dialyzer (or fiber) from z = 0 where blood enters to $z = 1$ where dialysate enters. From the left and middle graphs, we see that blood and dialysate flows both decrease with increasing z. Also, the difference between local blood and dialysate flows increase as the dimensionless hydraulic permeability β increases. The right graph shows that filtration velocity decreases from $z = 0$ to $z = 1$. This occurs because the frictional loss in blood progressively reduces the pressure difference between blood and dialysate. This effect is amplified by an increased hydraulic permeability.

Figure 20.2-2 shows the decreases of urea concentration in blood and dialysate with distance from the blood inlet. Although the difference between these concentrations becomes larger with increasing z, it is virtually unaffected by a tenfold change in the hydraulic permeability of the fiber wall.

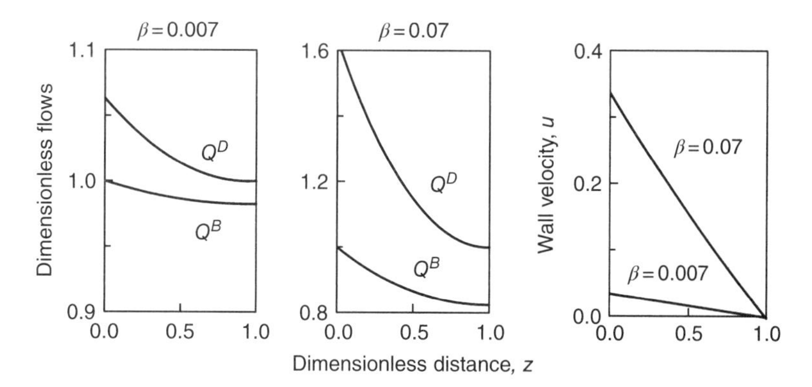

Figure 20.2-1 Effect of hydraulic permeability on dialyzer flow distributions.

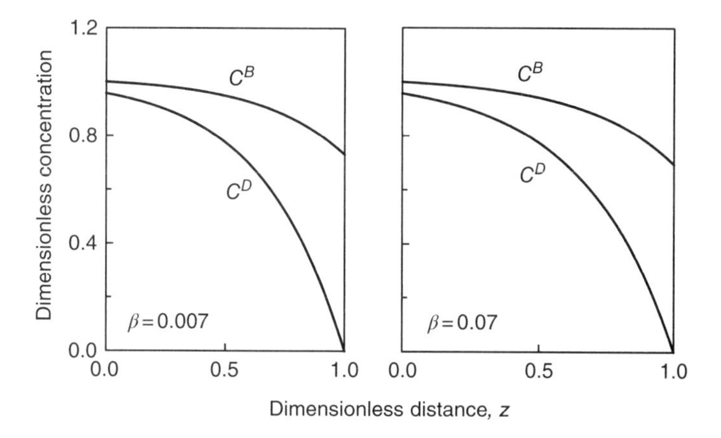

Figure 20.2-2 Effect of hydraulic permeability on blood and dialysate urea concentrations.

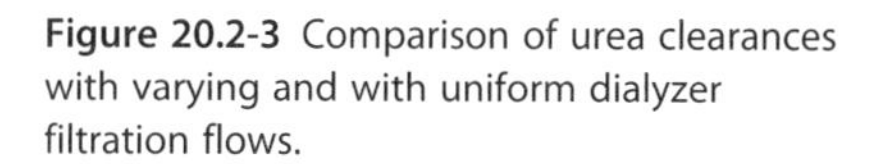

Figure 20.2-3 Comparison of urea clearances with varying and with uniform dialyzer filtration flows.

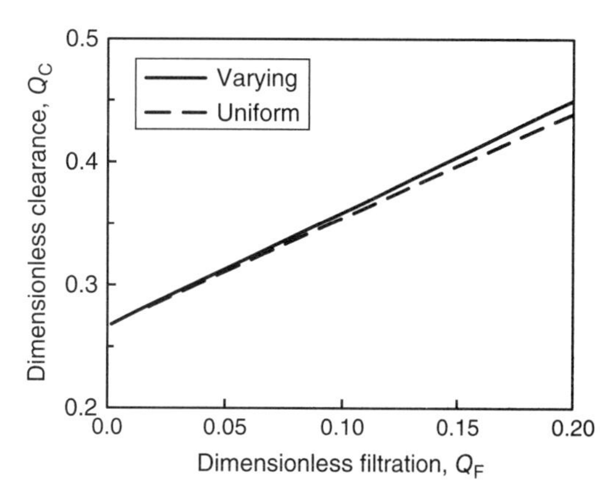

Figure 20.2-3 is a cross-plot of the device clearance and overall filtration flow predicted by the current model (solid line). For comparison, predictions of the previous uniform filtration rate model in Section 12.2 are also shown (dashed line). The dimensionless clearances Q_C predicted by both models are essentially a linear function of the dimensionless overall

filtration Q_F. The varying filtration model predicts slightly higher Q_C values than the uniform filtration model. For a typical clinical values $Q_F \leq 0.03$, the Q_C values predicted by the two models are virtually the same.

20.3 Gas Exchange in an Intravascular Lung Device

For subjects whose lungs cannot provide sufficient gas exchange, an extracorporeal membrane oxygenator (ECMO) is temporarily useful. However, the long-term use of ECMO can lead to serious infection and blood damage. A desirable alternative would be an implantable O_2–CO_2 exchanger. In this illustration, we model steady-state O_2 uptake and CO_2 removal by such a device implanted in a large blood vessel (Snider *et al.*, 1994). At the core of the device is a U-tube catheter that is inserted into the femoral artery and guided to the inferior vena cava (Fig. 20.3-1). Oxygen enriched gas is delivered to one lumen at the proximal end of this double-lumen catheter. The gas then flows to the distal end where it makes a U-turn to enter the other lumen. When this reversed gas flow reaches the proximal end of the catheter, it exhausts to the atmosphere.

Along the catheter are several manifolds, each outfitted with many hollow fibers that have gas-permeable walls (Fig. 20.3-2; top). Each fiber has an open end attached to a manifold and a sealed end submerged in the blood stream. Perforations in the wall of the gas supply lumen allow gaseous O_2 to flow into a manifold. Oxygen subsequently diffuses along a fiber lumen (Fig. 20.3-2; bottom) from which it permeates the fiber wall to reach the blood. Carbon dioxide is transported in the opposite direction, from blood through the fibers and manifolds, and then into the catheter supply lumen.

The time scale for gas transport along a fiber is much longer than for transport in the catheter, manifolds, and blood. Thus, we model transport in each catheter segment containing a manifold at two length scales. On a macroscale, we treat each manifold j with its portion of the gas supply lumen as a perfectly mixed compartment. The surrounding blood flow is treated as another perfectly mixed compartment. Between these compartments, exchange of gas i ($i = O_2$ or CO_2) occurs through all interconnecting fibers. On the microscale of a single fiber, transport of O_2 and CO_2 occurs along the fiber lumen in the z direction at a molar flux $N_{i,j}(z)$ and across the fiber wall at a flux $N_{i,j,wall}(z)$.

Model Formulation

Manifold Transport and Reaction Processes

On the macroscale, there are j = 1, 2, ..., J segments, each containing a well-mixed gas compartment G and a well-mixed blood compartment B. With respect to compartment B, the

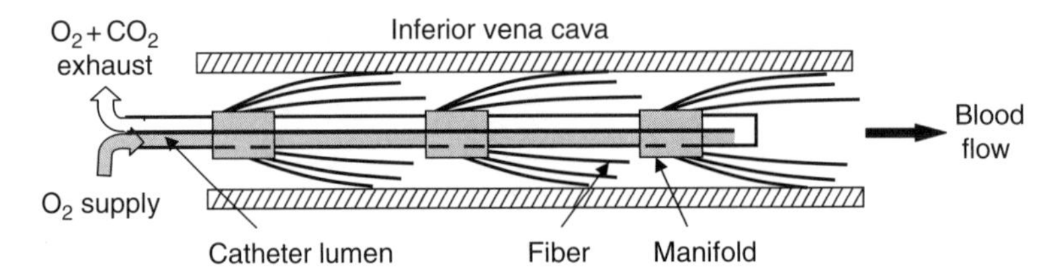

Figure 20.3-1 Catheter-mounted intravascular O_2 and CO_2 exchange device.

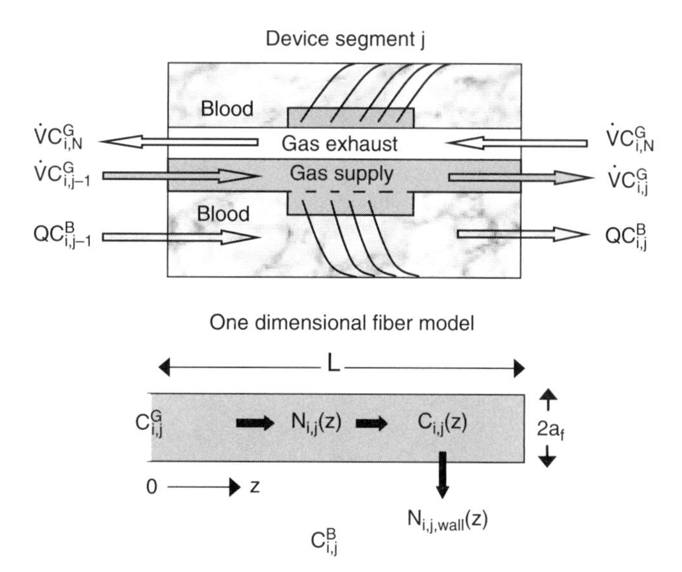

Figure 20.3-2 Macroscale manifold model and microscale fiber model.

mass density of blood is essentially constant. Consequently, a steady-state mass balance on blood leads to the volumetric flow rate relationship $Q_j \approx Q_{j-1} \equiv Q$. With respect to compartments G, the molar density of gas is essentially constant and the effects of O_2 and CO_2 exchange on gas flow are negligible. Consequently, a steady-state molar balance on gas leads to the volumetric flow rate relationship $\dot{V}_j \approx \dot{V}_{j-1} \equiv \dot{V}$.

For compartment G of segment j, the steady-state molar balances on O_2 and CO_2 are as follows:

$$\dot{V}\left(C^G_{i,j-1} - C^G_{i,j}\right) - \dot{N}_{i,j} = 0 \quad (i = O_2, CO_2;\ j = 1, 2, \ldots, J) \tag{20.3-1}$$

where $C^G_{i,j}$ is molar concentration of gas i exiting the compartment. For compartment B of segment j, steady-state molar balances for O_2 and HbO_2 are

$$\begin{aligned} Q\left(C^B_{O_2,j-1} - C^B_{O_2j}\right) + \dot{N}_{O_2,j} + \dot{R}_{O_2,j} &= 0 \\ Q\left(C^B_{HbO_2,j-1} - C^B_{HbO_2,j}\right) + \dot{R}_{HbO_2,j} &= 0 \end{aligned} \qquad (j = 1, 2, \ldots, J) \tag{20.3-2a, b}$$

where $C^B_{O_2,j}$ and $C^B_{HbO_2,j}$ are molar concentrations exiting the compartment, and $\dot{R}_{O_2,j}$ and $\dot{R}_{HbO_2,j}$ are molar rates of formation within compartment B. Because of the 1 : 1 stoichiometric reaction of oxyhemoglobin formation, $\dot{R}_{HbO_2,j} = -\dot{R}_{O_2,j}$ and we combine these equations to eliminate the reaction rates:

$$Q\left[\left(C^B_{O_2,j-1} + C^B_{HbO_2,j-1}\right) - \left(C^B_{O_2,j} + C^B_{HbO_2,j}\right)\right] + \dot{N}_{O_2,j} = 0 \tag{20.3-3}$$

Also for compartment B of segment j, the steady-state molar balances for CO_2 and its reaction products (dominated by HCO_3^-) are

$$Q\left(C^{B}_{CO_2,j-1}-C^{B}_{CO_2,j}\right)+\dot{N}_{CO_2,j}+\dot{R}_{CO_2,j}=0$$
$$Q\left(C^{B}_{HCO_3,j-1}-C^{B}_{HCO_3,j}\right)+\dot{R}_{HCO_3,j}=0 \qquad (20.3\text{-}4a,b)$$

Given the 1 : 1 stoichiometric reaction associated with the hydration of CO_2 to form HCO_3^-, we add these equations to obtain

$$Q\left[\left(C^{B}_{CO_2,j-1}+C^{B}_{HCO_3,j-1}\right)-\left(C^{B}_{CO_2,j}+C^{B}_{HCO_3,j}\right)\right]+\dot{N}_{CO_2,j}=0 \qquad (20.3\text{-}5)$$

Introducing the total blood content of O_2 as $\frown{C}^{B}_{O_2,j}=\left(C^{B}_{O_2,j}+C^{B}_{HbO_2,j}\right)/c^{o}_{G}$ and of CO_2 as $\frown{C}^{B}_{CO_2,j}=\left(C^{B}_{CO_2,j}+C^{B}_{HCO_3,j}\right)/c^{o}_{G}$, the molar balances for segment j (Eqs. 20.3-3 and 20.3-5) become

$$Qc^{o}_{G}\left(\frown{C}^{B}_{i,j-1}-\frown{C}^{B}_{i,j}\right)+\dot{N}_{i,j}=0 \quad (i=O_2,CO_2;\ j=1,2,\ldots,J) \qquad (20.3\text{-}6)$$

where c^{o}_{G} is the gas-phase mole density under standard T and P conditions.

Hollow Fiber Transport Processes

At a microscale, transport of O_2 and CO_2 between blood and gas compartments of a particular segment occurs by axial diffusion along the interior of the interconnecting thin-walled fibers and radial diffusion through the fiber walls. Applying a one-dimensional, steady-state molar balance (Eq. 15.3-30) on species i in a single fiber lumen of segment j leads to

$$\frac{\partial C_{i,j}}{\partial t}+\frac{\partial(uC_{i,j})}{\partial z}=\frac{\partial}{\partial z}\left[\left(\mathcal{D}_s+\mathcal{D}^*_s\right)\frac{\partial C_{i,j}}{\partial z}\right]-\frac{2}{a}N_{i,j,wall} \quad (i=O_2,CO_2;\ j=1,2,\ldots,J) \qquad (20.3\text{-}7)$$

where $C_{i,j}(z)$ and u are cross-sectional average concentration and axial velocity; $N_{i,j,wall}(z)$ is the local flux across a fiber wall; $\mathcal{D}_i$ and $\mathcal{D}^*_i$ are gas-phase diffusion and dispersion coefficients; and 2/a is the surface-to-volume ratio of the cylindrical fiber wall. With axial transport of O_2 and CO_2 occurring at essentially equal molar rates but in opposite directions, u and $\mathcal{D}^*_i$ can be neglected. Assuming further that $\mathcal{D}_i$ is constant, Equation 20.3-7 becomes

$$\frac{\partial C_{i,j}}{\partial t}=\mathcal{D}_i\frac{\partial^2 C_{i,j}}{\partial z^2}-\frac{2}{a}N_{i,j,wall} \qquad (20.3\text{-}8)$$

The radial flux of species i can be expressed in terms of an overall mass transfer coefficient K_i that accounts for transport resistances of the fiber wall and the adjacent blood boundary layer:

$$N_{i,j,wall}=K_i\left(C_{i,j}-\lambda^{G,B}_i C^{B}_{i,j}\right) \qquad (20.3\text{-}9)$$

where $\lambda^{G,B}_i$ is an equilibrium partition coefficient between gas and blood. Since we consider blood surrounding a fiber to be perfectly mixed, $C^{B}_{i,j}$ is independent of z. Under steady-state conditions, the combination of these last two equations yields

$$L_i^2\frac{d^2C_{i,j}}{dz^2}-C_{i,j}=-\lambda^{G,B}_i C^{B}_{i,j} \qquad (20.3\text{-}10)$$

where $L_i\equiv\sqrt{\mathcal{D}_i a/2K_i}$ is a characteristic diffusion length along the fiber. At the mouth of the fiber, the gas concentration is continuous with that in the associated manifold compartment:

$$z = 0: \quad C_{i,j} = C_{i,j}^{G} \tag{20.3-11}$$

Since there is no diffusion across the closed end of a fiber of length L, the axial concentration gradient must be zero:

$$z = L: \quad \frac{dC_{i,j}}{dz} = 0 \tag{20.3-12}$$

From the analytical solution to this linear model, we obtain the concentration of species i in any fiber of segment j:

$$C_{i,j}(z) = \lambda_i^{G,B} C_{i,j}^{B} + \left[\cosh\left(\frac{z}{L_i}\right) - \tanh\left(\frac{L}{L_i}\right)\sinh\left(\frac{z}{L_i}\right)\right]\left(C_{i,j}^{G} - \lambda_i^{G,B} C_{i,j}^{B}\right) \tag{20.3-13}$$

By evaluating the derivative of this equation at z = 0, we get the flux of gas i at the mouth of a fiber:

$$N_{i,j}(0) = -\mathcal{D}_i \left[\frac{dC_{i,j}}{dz}\right]_{z=0} = \frac{\mathcal{D}_i}{L_i}\tanh\left(\frac{L}{L_i}\right)\left(C_{i,j}^{G} - \lambda_i^{G,B} C_{i,j}^{B}\right) \tag{20.3-14}$$

Manifold Gas and Blood Equations

The molar diffusion rate of species i into the n_f thin-walled fibers of radius a_f attached to manifold j can be found from Equation 20.3-14:

$$\dot{N}_{i,j} = \left(\pi a_f^2 n_f\right) N_{i,j}(0) = \frac{\pi a^2 n_f \mathcal{D}_i}{L_i}\tanh\left(\frac{L}{L_i}\right)\left(C_{i,j}^{G} - \lambda_i^{G,B} C_{i,j}^{B}\right) \tag{20.3-15}$$

Thus, we can relate the molar convection rates of species i through the gas compartments (Eq. 20.3-1) and blood compartments (Eq. 20.3-6) of segment j to the molar transport rates along the fibers (Eq. 20.3-15):

$$\dot{V}\left(C_{i,j-1}^{G} - C_{i,j}^{G}\right) = \frac{\pi a_f^2 n_f \mathcal{D}_i}{L_i}\tanh\left(\frac{L}{L_i}\right)\left(C_{i,j}^{G} - \lambda_i^{G,B} C_{i,j}^{B}\right)$$

$$Q c_G^{o}\left(\widehat{C}_{i,j-1}^{B} - \widehat{C}_{i,j}^{B}\right) = -\frac{\pi a_f^2 n_f \mathcal{D}_i}{L_i}\tanh\left(\frac{L}{L_i}\right)\left(C_{i,j}^{G} - \lambda_i^{G,B} C_{i,j}^{B}\right) \tag{20.3-16a, b}$$

For comparison with experimental blood–gas data, it is often more useful to express a model in terms of partial pressures rather than molar concentrations. Using the ideal gas and phase equilibrium relationships, $p_{i,j}^{G} = \mathcal{R}TC_{i,j}^{G}$ and $p_{i,j}^{B} = \mathcal{R}T\lambda_i^{G,B}C_{i,j}^{B}$, we write Equation 20.3-16a,b as

$$\dot{V}\left(p_{i,j-1}^{G} - p_{i,j}^{G}\right) = \frac{n_f \pi a^2 \mathcal{D}_i}{L_i}\tanh\left(\frac{L}{L_i}\right)\left(p_{i,j}^{G} - p_{i,j}^{B}\right)$$

$$Q\left(\widehat{C}_{i,j-1}^{B} - \widehat{C}_{i,j}^{B}\right) = \frac{n_f \pi a^2 \mathcal{D}_i}{c_G \mathcal{R}TL_i}\tanh\left(\frac{L}{L_i}\right)\left(p_{i,j}^{B} - p_{i,j}^{G}\right) \quad (i = O_2, CO_2;\ j = 1, 2, \ldots, J) \tag{20.3-17a, b}$$

To complete the model, O_2 and CO_2 contents $\widehat{C}_{i,j}^{B}$ in the blood compartments must be related to their partial pressures $p_{i,j}^{B}$. Assuming that the binding of O_2 to Hb and the hydration of CO_2 to HCO_3^- are rapid reactions, we employ the quasi-equilibrium relations developed in Section 5.5 for capillary gas exchange.

Analysis

Parameter Values

Consider a device functioning in the inferior vena cava where Q = 2 L/min. At this blood flow, we estimate $K_{O_2} \approx 0.002$ mm/s and $K_{CO_2} \approx 0.03$ mm/s based on experiments with 0.3 mm diameter, microporous, polypropylene fibers mounted on a 5 mm diameter manifold (Baskaran, 1997). At body temperature, $\mathcal{D}_{O_2} \approx \mathcal{D}_{CO_2} = 16\ mm^2/s$ and the characteristic fiber diffusion lengths are $L_{O_2} \approx 25$ mm and $L_{CO_2} \approx 6$ mm. A reasonable gas supply is $\dot{V} = 1$ L/min.

Equation 20.3-15 indicates that the transport rate across a fiber wall is proportional to $\tanh(L/L_i)$, a quantity that monotonically approaches a value of 1.0 for very large L/L_i. If $L = L_i$ and $L = 2L_i$, then $\tanh(L/L_i) = 0.76$ and 0.96, respectively. Thus, choosing a fiber longer than $L = 2L_i$ does not significantly increase the transport rate of substance i. For O_2 transport, this means that there is little benefit in making fibers longer than 50 mm. For CO_2, a much shorter length of 12 mm is sufficient. As a compromise between these two values, we select a fiber length L = 25 mm.

About 16,000 fibers with a 25 mm length and 0.3 mm diameter can be close-packed in an adult's inferior vena cava, typically 20 mm in diameter. To provide additional space for blood flow around the fibers, we consider a device with a total of 10,000 fibers. A manifold of length 5 mm and diameter 5 mm (one-fourth the diameter of the vena cava) would provide sufficient surface for attachment of 1000 fibers. Ten manifolds of this length would still allow a 5 mm spacing between manifolds in a 100 mm long vena cava. Therefore, we specify that $n_f = 1000$ and J = 10.

Quantitative Relationships for Composition Variables

With these parameter values, the species balance in the gas compartment of segment j (Eq. 20.3-17a) becomes

$$\begin{aligned} p^B_{O_2,j} - 49.37 p^G_{O_2,j} &= -48.37 p^G_{O_2,j-1} \\ p^B_{CO_2,j} - 9.846 p^G_{CO_2,j} &= -8.846 p^G_{CO_2,j-1} \end{aligned} \qquad (20.3\text{-}18a,b)$$

and the species balance in the corresponding blood compartment (Eq. 20.3-17b) leads to

$$\begin{aligned} \frown{C}_{O_2,j} + 0.0102 p^B_{O_2,j} - 0.0102 p^G_{O_2,j} &= \frown{C}_{O_2,j-1} \\ \frown{C}_{CO_2,j} + 0.0558 p^B_{CO_2,j} - 0.0558 p^G_{CO_2,j} &= \frown{C}_{CO_2,j-1} \end{aligned} \qquad (20.3\text{-}19a,b)$$

By also stipulating a hemoglobin mass concentration in blood of $\rho_{Hb_4} = 15$ g/dL, O_2 and CO_2 reaction equilibria (Eqs. 5.5-44–5.5-47) are described by

$$\begin{aligned} \frown{C}_{CO_2,j} &= 33.7 + 3.32 p^B_{CO_2,j} - 5.36 S_j \\ \frown{C}_{CO_2,j} &= \left[0.503 - \frac{0.226}{\left(2.244 - 0.422 S^B_j\right)(8.740 - pH_j)}\right]\left(1 + 10^{pH_j - 6.1}\right) p^B_{CO_2,j} \\ \frown{C}_{O_2,j} &= 0.0219 p^B_{O_2,j} + 20.9 S_j \\ S_j &= \frac{\left[0.313 \times 10^{0.4(pH_j - 7.4)} p^B_{O_2,j} / \left(p^B_{CO_2,j}\right)^{0.06}\right]^{2.8}}{1 + \left[0.313 \times 10^{0.4(pH_j - 7.4)} p^B_{O_2,j} / \left(p^B_{CO_2,j}\right)^{0.06}\right]^{2.8}} \end{aligned} \qquad (20.3\text{-}20a\text{-}d)$$

where pH_j is the pH and S_j is the O_2 saturation fraction in the blood compartment of device segment j. The numerical values in the final model equations (Eqs. 20.3-18–20.3-20) all require specific units of partial pressure [kPa] and O_2 and CO_2 contents [ml/dL].

Simulation

Beginning with inlet gas partial pressures corresponding to a pure O_2 supply gas ($p^G_{O_2,0} = 101\,\text{kPa}$, $p^G_{CO_2,0} = 0$) and inlet blood partial pressures typical of venous blood conditions ($p^B_{O_2,0} = 5.3$ kPa, $p^B_{CO_2,0} = 6.2\,\text{kPa}$), we sequentially solve the model equations, segment by segment. Cumulative volumetric O_2 uptake and CO_2 excretion along the catheter is then computed from steady-state molar balances between the supply gas inlet and the outlet of manifold j:

$$\dot{V}_{O_2,j} = \frac{\dot{V}}{P}\left(p^G_{O_2,0} - p^G_{O_2,j}\right),\quad \dot{V}_{CO_2,j} = \frac{\dot{V}}{P}\left(p^G_{O_2,j} - p^G_{O_2,0}\right) \qquad (20.3\text{-}21a,b)$$

The results of these computations are shown in Figure 20.3-3. The left graph demonstrates that the partial pressure driving force for O_2 transport declines from manifold to manifold because of O_2 loss from the gas phase and O_2 gain in blood. The CO_2 driving force also decreases because CO_2 depleted from blood is gained in the gas phase. The right graph shows a continual increase in cumulative gas exchange with additional device manifolds. As a result of the diminishing driving forces, the incremental increase in gas exchange for each additional manifold decreases as more device segments are added.

These results point to two shortcomings of this intravascular lung. First, the cumulative O_2 transport rate of 108 ml/min may be insufficient to sustain the metabolism of a patient with a totally dysfunctional lung. The reasons for this are as follows: (i) the limited blood flow that passes through the vena cava, (ii) a restriction on the maximum fiber surface that can be accommodated in the vena cava (less than one half of the surface area that is typically used in extracorporeal blood oxygenators), and (iii) the substantial resistance to transport through a fiber. The second shortcoming is the small respiratory quotient (i.e., ratio of CO_2 to O_2 transport) of 0.35 for this device compared to 0.8–1.0 for healthy lungs. This occurs because the characteristic diffusion length in a fiber is much smaller for CO_2 than for O_2.

The most direct way of improving overall O_2 transport rate would be to lengthen the device so that it extends into the superior vena cava or the right atrium. This would provide room for

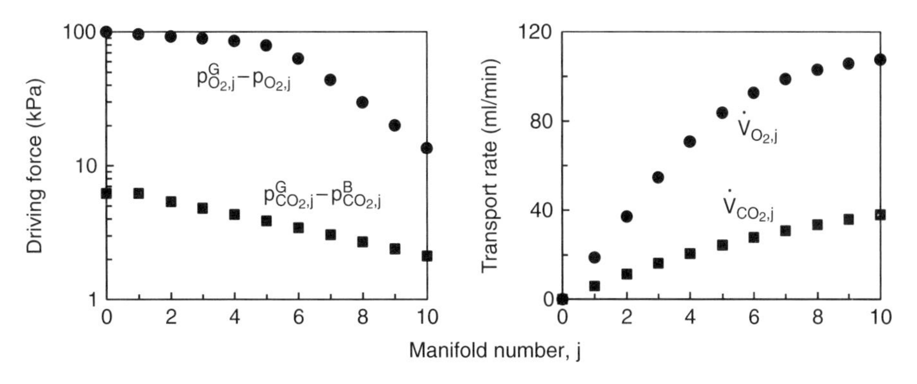

Figure 20.3-3 Performance of a 10,000-fiber intravascular lung in an inferior vena cava.

an increased number of manifolds as well as a greater blood flow over the fibers. Improvement of the respiratory quotient could be achieved by shortening the fiber, thereby decreasing O_2 transport with little effect on CO_2 transport. This strategy would, of course, exacerbate the problem of meeting a patient's O_2 demand. Another way to increase the respiratory quotient, often used in extracorporeal blood oxygenation, is to allow the CO_2 partial pressure in the patient's blood to become greater than normal, thereby increasing the driving force for CO_2 transport within the device. Modification of the fiber material is yet another way of increasing the respiratory quotient. For example, immobilization of carbonic anhydrase in a membrane can augment CO_2 diffusion by 75% (Kaar *et al.*, 2007).

20.4 Separation of Blood Components by Apheresis

Apheresis is a process for removing undesirable blood components in blood plasma that occur in diseases such as rheumatoid arthritis, cardiac myopathy and familial hypercholesterolemia. A typical extracorporeal apheresis process consists of the following simultaneous steps: Blood is removed from a patient and separated into plasma and formed elements (i.e., red cells and platelets); the plasma is passed through a packed column containing adsorbent particles that take up undesirable components; and the treated plasma is remixed with the formed elements and infused back into the patient.

The adsorbent particles in an apheresis column usually have a microporous structure with a large surface-to-volume ratio. Often they are composed of an inert material such as silica gel or agarose with pore surfaces that act as a support for a receptor, an agent that specifically binds with the ligand targeted for removal from plasma. Examples of receptors include bacterial protein A used to remove antibodies associated with autoimmune diseases and dextran sulfate to remove low density lipoproteins to reduce hypercholesterolemia. A less effective alternative is a generic adsorbent such as activated charcoal that binds a broad spectrum of bioactive molecules.

Model Formulation

A column of length L and constant cross-sectional area A is filled with a mixture of stationary porous particles and plasma (Fig. 20.4-1). In the column, the plasma volume fraction is ε, the

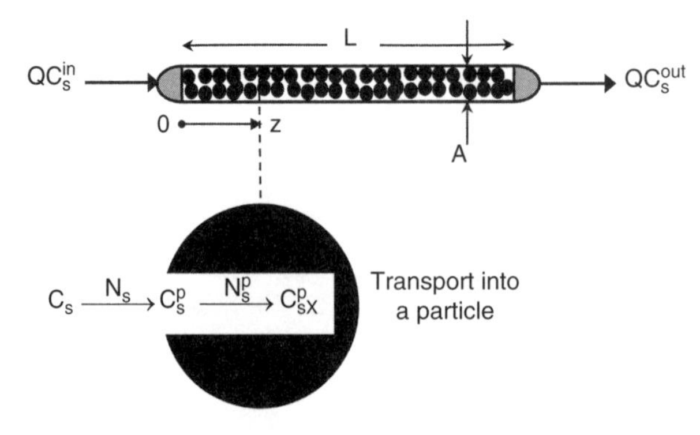

Figure 20.4-1 Packed adsorption column for apheresis.

particle volume fraction is $(1 - \varepsilon)$, the particle-to-plasma volume ratio is $(1 - \varepsilon)/\varepsilon$, and the cross-sectional area through which plasma flows is $A\varepsilon$.

In modeling transport in this device, we consider plasma and particles to be separate continuous phases. The plasma flow around the stationary particles can be described by a one-dimensional solution balance (Eq. 15.3-29). For a plasma phase with a constant mass density and no net flow of plasma into the particles, we find for a steady plasma input flow Q that

$$\frac{\partial Q}{\partial z} = 0 \Rightarrow Q = \text{constant} \tag{20.4-1}$$

From a one-dimensional species balance (Eq. 15.3-28) on ligand s in the plasma phase,

$$\frac{\partial C_s}{\partial t} + \frac{Q}{A\varepsilon}\frac{\partial C_s}{\partial z} = \frac{\partial}{\partial z}\left[\left(\mathcal{D}_s + \mathcal{D}_s^*\right)\frac{\partial C_s}{\partial z}\right] - \left(\frac{1-\varepsilon}{\varepsilon}\right)\phi N_s \tag{20.4-2}$$

where C_s is ligand concentration averaged over the cross section of the plasma phase; $\mathcal{D}_s$ and $\mathcal{D}_s^*$ are constant axial diffusion and dispersion coefficients, respectively; N_s is ligand flux from plasma into the openings of the particle pores; and ϕ is the cross-sectional area of the pore openings in contact with plasma relative to the particle volume.

We define C_s^p as the moles of unbound ligand contained in plasma-filled pores per unit particle volume. The value of C_s^p increases because of ligand transport from external plasma into particle pores and decreases by ligand binding to the pore walls. A molar balance on the dissolved ligand leads to

$$\frac{\partial C_s^p}{\partial t} = \phi N_s - \phi^p N_s^p \tag{20.4-3}$$

Here, N_s^p is the rate of ligand binding per unit internal surface of a pore, and ϕ^p is the ratio of the internal pore surface to the particle volume. The concentration of ligand bound to surface sites X in the pores increases by adsorption from the pore space:

$$\frac{\partial C_{sX}^p}{\partial t} = \phi^p N_s^p \tag{20.4-4}$$

Here, the bound ligand concentration C_{sX}^p is expressed as moles per unit particle volume. Adding the last two equations eliminates N_s^p and yields the change in total ligand concentration in bound and unbound forms:

$$\frac{\partial\left(C_s^p + C_{sX}^p\right)}{\partial t} = \phi N_s \tag{20.4-5}$$

We consider the simple case of ligand binding on the pore surface by a reversible monovalent process that is very rapid compared to diffusion through the pores. Then, C_{sX}^p and C_s^p are close to the binding equilibrium expressed by Equation 5.4-3:

$$C_{sX}^p = \frac{T^p \kappa C_s^p}{1 + \kappa C_s^p} \tag{20.4-6}$$

where κ is a binding constant, and T^p represents the total moles of surface adsorption sites (both free and occupied by ligand) per unit particle volume. With this equation, the time derivative in Equation 20.4-5 can be written as

$$\frac{\partial\left(C_s^p + C_{sX}^p\right)}{\partial t} = \left(1 + \frac{dC_{sX}^p}{dC_s^p}\right)\frac{\partial C_s^p}{\partial t} = \left[1 + \frac{T^p \kappa}{\left(1 + \kappa C_s^p\right)^2}\right]\frac{\partial C_s^p}{\partial t} \tag{20.4-7}$$

The ligand flux N_s from plasma into the openings can be given in terms of an overall mass transfer coefficient K_s that accounts for diffusion within the particle pores as well as through the unstirred plasma layer on the external particle surface:

$$N_s = K_s\left(C_s - C_s^p/\varepsilon^p\right) \tag{20.4-8}$$

Here, ε^p is the pore volume per unit particle volume so that C_s^p/ε^p represents the molar concentration of unbound ligand within a pore. Substituting this equation into Equation 20.4-2, we obtain the governing equation for ligand concentration in the flowing plasma:

$$\frac{\partial C_s}{\partial t} + \frac{Q}{A\varepsilon}\frac{\partial C_s}{\partial z} = \left(\mathcal{D}_s + \mathcal{D}_s^*\right)\frac{\partial^2 C_s}{\partial z^2} - K_s\phi\left(\frac{1-\varepsilon}{\varepsilon}\right)\left(C_s - \frac{C_s^p}{\varepsilon^p}\right) \tag{20.4-9}$$

Inserting Equations 20.4-7 and 20.4-8 into Equation 20.4-5 yields the governing equation for ligand concentration in the stationary particle pores:

$$\frac{\partial C_s^p}{\partial t} = \phi K_s\left[1 + \frac{T^p\kappa}{\left(1+\kappa C_s^p\right)^2}\right]^{-1}\left(C_s - \frac{C_s^p}{\varepsilon^p}\right) \tag{20.4-10}$$

We examine the dynamics of this process when the device is initially free of ligand:

$$t=0: \quad C_s = 0, \quad C_s^p = 0 \tag{20.4-11}$$

and the inlet ligand concentration in plasma is always the same:

$$z=0: \quad C_s = C_s^{in} \tag{20.4-12}$$

At the outlet of the column, the axial gradient of ligand concentration is expected to be negligible:

$$z=L: \quad \frac{\partial C_s}{\partial z} = 0 \tag{20.4-13}$$

Analysis

Dimensionless Forms

Using the following dimensionless variables:

$$C \equiv \frac{C_s}{C_s^{in}}, \quad C^p \equiv \frac{C_s^p}{C_s^{in}}, \quad z \equiv \frac{z}{L}, \quad t \equiv \frac{Qt}{AL} \tag{20.4-14a-d}$$

the dimensionless governing equations for ligand concentration in plasma and in particles are

$$\frac{\partial C}{\partial t} + \frac{1}{\varepsilon}\frac{\partial C}{\partial z} = \frac{1}{Pe}\frac{\partial^2 C}{\partial z^2} - \alpha\left(\frac{1-\varepsilon}{\varepsilon}\right)\left(C - \frac{C^p}{\varepsilon^p}\right) \tag{20.4-15}$$

$$\frac{\partial C^p}{\partial t} = \alpha\left[1 + \frac{\beta_1\beta_2 C^p}{(1+\beta_2 C^p)^2}\right]^{-1}\left(C - \frac{C^p}{\varepsilon^p}\right) \tag{20.4-16}$$

The dimensionless initial and boundary conditions are

$$t=0: \quad C=0, \quad C^p = 0 \tag{20.4-17}$$

$$z=0: \quad C=1, \qquad z=1: \quad \frac{\partial C}{\partial z} = 0 \tag{20.4-18a,b}$$

The dimensionless parameters in these equations are defined as

$$Pe \equiv \frac{QL}{(\mathcal{D}_s + \mathcal{D}_s^*)A}, \quad \alpha \equiv \frac{\phi K_s AL}{Q}, \quad \beta_1 \equiv \frac{T^P}{C_s^{in}}, \quad \beta_2 \equiv \kappa C_s^{in} \tag{20.4-19a-d}$$

Here, Pe reflects the importance of axial convection relative to axial dispersion in the flowing plasma; α is the relative importance of the rate of adsorption to convection; β_1 indicates the relative number of adsorption sites; and β_2 indicates the binding potential of adsorption sites.

In the design of this ligand absorption system, it is helpful to consider a dimensionless measure of surface occupancy. Based on Equation 20.4-6, we define a local occupancy fraction of particle binding sites, $Y = C_{sX}^P/T^P$, that can be integrated over the column length to obtain an overall occupancy fraction $Y_{overall}$:

$$Y(z,t) = \frac{\beta_2 C^P(z,t)}{1+\beta_2 C^P(z,t)}, \quad Y_{overall}(t) \equiv \int_0^1 \left[\frac{\beta_2 C^P(z,t)}{1+\beta_2 C^P(z,t)}\right] dz \tag{20.4-20a, b}$$

The long-time limits of the governing equations (Eqs. 20.4-15 and 20.4-16) lead to the steady-state relation $C = C^P/\varepsilon^P$ such that ligand is no longer transported into particle pores. Thus, ligand concentration everywhere in the plasma eventually becomes equal to the inlet concentration, and $C = C^P/\varepsilon^P = 1$. Utilizing this information in Equation 20.4-20a,b, we see that the steady-state limit of the occupancy fraction is less than one:

$$Y(z, t \to \infty) = Y_{overall}(t \to \infty) = \frac{\beta_2}{1+\beta_2} \tag{20.4-21}$$

Simulation

Figure 20.4-2 presents the numerical solutions of the model equations for $\alpha = \beta_1 = \beta_2 = 1$, $Pe = 100$, $\varepsilon = 0.36$ for randomly packed spherical particles, and $\varepsilon^P \to 1$ for highly porous particles. The left graph indicates that the axial distribution of occupied adsorption sites, $Y(z,t)$, decreases in essentially a linear fashion at any given time and increases with time. The right graph shows the axial distribution of ligand concentration $C(z,t)$ in the plasma. At a short time

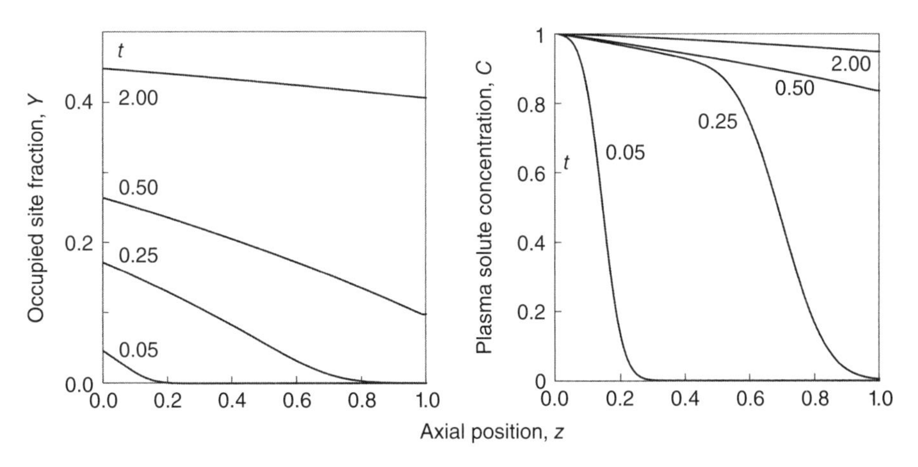

Figure 20.4-2 Spatial distributions of solute with time in an apheresis column ($\alpha = \beta_1 = \beta_2 = 1$, $Pe = 100$, $\varepsilon = 0.36$).

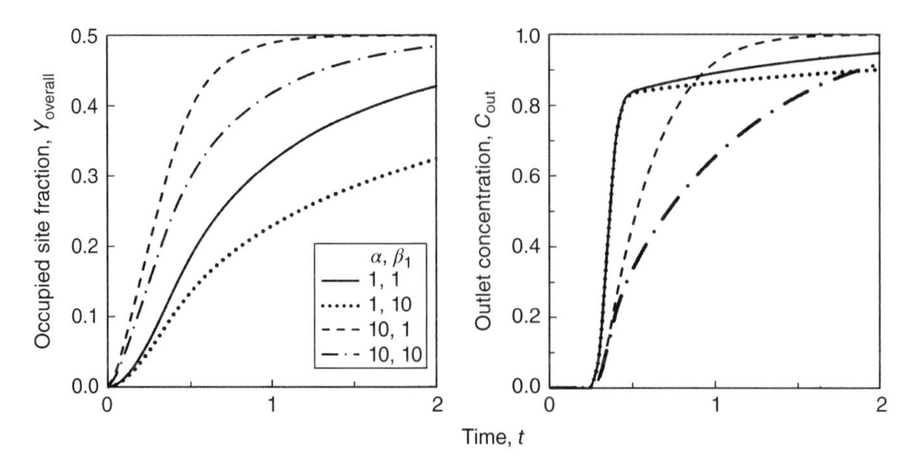

Figure 20.4-3 Dynamic performance of an apheresis device ($\beta_2 = 1$, $Pe = 100$, $\varepsilon = 0.36$).

of $t = 0.05$, when the fraction of occupied sites is small everywhere in the column, there is a sharp dividing front in plasma between an upstream ligand concentration of $C = 1$ and a downstream concentration of $C = 0$. At an intermediate time of $t = 0.25$, when upstream adsorption sites have accumulated more ligand, this front moves downstream and becomes more dispersed. At $t \geq 0.5$, the C distribution becomes linear and shifts upward toward its steady-state value of one.

Figure 20.4-3 illustrates the simulated dynamics for $2 \geq t \geq 0$ with different combinations of α and β_1 when $\beta_2 = 1$, $Pe = 100$ and $\varepsilon = 0.36$. The left graph shows that at short times when the ligand concentration in the particles is low, overall binding site occupancy $Y_{overall}(z,t)$ continuously increases with time. At very long times, however, all four curves must approach a constant steady state of $Y_{overall} = 1/(1 + 1) = 1/2$. This asymptote is reached more rapidly when the dimensionless mass transfer coefficient α is increased at a fixed β_1, but more slowly when the dimensionless binding site density β_1 is increased at a constant α.

The right graph shows the dynamic behavior of the ligand concentration in plasma exiting the device, $C^{out}(t) = C(1,t)$, the so-called breakthrough curve. Each breakthrough curve continuously increases from an initial value of zero toward its steady-state value of one, corresponding to complete occupancy of all particle binding sites. A striking feature of the breakthrough curves is their sensitivities to β_1 at the different α. When $\alpha = 10$, the breakthrough curve at $\beta_1 = 10$ is considerably lower than the curve at $\beta_1 = 1$. When $\alpha = 1$, however, the breakthrough curves reach a limiting behavior that is quite insensitive to β_1.

20.5 Epidermal Regeneration in Tissue-Engineered Skin

Human skin, which has an outer epidermal layer primarily of keratinocytes, acts as a protective barrier to damaging environmental agents such as microbes and chemicals. The inner dermal layer, which contains blood vessels and several cell types, supplies water and nutrients to the epidermis and provides mechanical strength. Skin injuries can occur from various causes such as burns, diabetic ulcers, and pressure sores. Whereas grafting of natural skin is a viable option for skin injuries, tissue-engineered skin constructs are indispensable in treating extensive injuries.

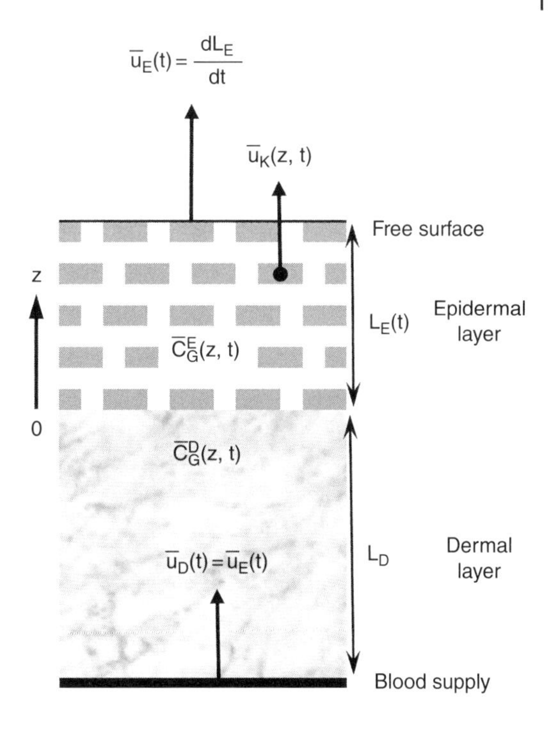

Figure 20.5-1 Tissue-engineered composite skin model.

To illustrate the development of tissue-engineered skin, we consider the one-dimensional growth of a skin construct with two adjacent regions: a porous dermal region of constant thickness L_D and an epidermal region containing a stack of keratinocyte monolayers that are each surrounded by extracellular fluid. In an incompletely formed monolayer, keratinocytes proliferate laterally until confluence is reached (Fig. 20.5-1). Further proliferation produces a population pressure that displaces keratinocytes out of the layer, thereby seeding a new cell layer. A blood supply at the basal surface of the dermis is a source of water and glucose. Transport of these materials into the epidermal layer is essential for keratinocytes to proliferate and increase the epidermal thickness $L_E(t)$.

Model Formulation

Epidermal Processes

We treat the epidermal layer as a heterogeneous mixture of keratinocytes and extracellular fluid of constant density. The z component of a volume-averaged solution balance on this mixture specifies its volume-averaged velocity $\bar{u}_E$:

$$\frac{\partial \bar{u}_E}{\partial z} = 0 \Rightarrow \bar{u}_E = \bar{u}_E(t) \tag{20.5-1}$$

Thus, the volume-averaged velocity of the epithelial layer as a whole is independent of z. In contrast, the volume-averaged velocity $\bar{u}_K$ of keratinocytes alone varies with z and t. Thus, $\bar{u}_E$ and $\bar{u}_K$ are not directly related except at the free epithelial surface where they are both equal to the rate of increase of the epidermal thickness:

$$z = L_E(t): \quad \frac{dL_E}{dt} = \bar{u}_E(t) = \bar{u}_K(z,t) \qquad (20.5\text{-}2)$$

The keratinocyte number density Ω_K changes because of a cell transport flux Θ_K in the z direction and a proliferation rate Ξ_K^{prolif} per unit volume. This is reflected in the following keratinocyte number balance obtained from Equation 17.1-2:

$$\frac{\partial \Omega_K}{\partial t} + \frac{\partial \Theta_K}{\partial z} = \Xi_K^{prolif} \qquad (20.5\text{-}3)$$

With cell tranport in the z direction restricted to convection at velocity $\bar{u}_K$, $\Theta_K = \bar{u}_K \Omega_K$ so that

$$\frac{\partial \Omega_K}{\partial t} + \bar{u}_K \frac{\partial \Omega_K}{\partial z} + \Omega_K \frac{\partial \bar{u}_K}{\partial z} = \Xi_K^{prolif} \qquad (20.5\text{-}4)$$

The change in keratinocyte velocity $\bar{u}_K$ with respect to z depends on $\Omega_K(z,t)$ relative to the maximum cell density Ω_K^{cc} (i.e., the cell carrying capacity). When Ω_K is less than Ω_K^{cc}, keratinocytes proliferate laterally within a cell layer. In that case, the increasing number of cells does not contribute to $\bar{u}_K$ so that $\partial \bar{u}_z / \partial z = 0$. When $\Omega_K \rightarrow \Omega_K^{cc} = \text{constant}$, however, Equation 20.5-4 predicts that cell crowding caused by proliferation gives rise to a nonzero longitudinal velocity gradient, $\partial \bar{u}_K / \partial z = \Xi_K^{prolif} / \Omega_K^{cc}$. Thus, the velocity gradient of the keratinocyte layer is described as

$$\frac{\partial \bar{u}_K}{\partial z} = \begin{cases} 0 & \text{when } \Omega_K(z,t) < \Omega_K^{cc} \\ \Xi_K^{prolif} / \Omega_K^{cc} & \text{when } \Omega_K(z,t) \geq \Omega_K^{cc} \end{cases} = \frac{\Xi_K^{prolif}}{\Omega_K^{cc}} H\left(\Omega_K - \Omega_K^{cc}\right) \qquad (20.5\text{-}5)$$

where $H\left(\Omega_K - \Omega_K^{cc}\right)$ is the unit step function (Eq. 18.1-11). Proliferation rate depends on keratinocyte number density and is limited by local glucose concentration $\bar{C}_G^E$:

$$\Xi_K^{prolif} = \left(\frac{V_K^{prolif} \bar{C}_G^E}{K^{prolif} + \bar{C}_G^E} \right) \Omega_K \qquad (20.5\text{-}6)$$

Here, V_K^{prolif} is the maximum value of the specific proliferation rate, and K^{prolif} is the value of $\bar{C}_G^E$ at which half the maximum rate is reached. Together with this rate relation, the explicit effect of keratinocyte number density on its velocity gradient is as follows:

$$\frac{\partial \bar{u}_K}{\partial z} = \left(\frac{V_K^{prolif} \bar{C}_G^E}{K_K^{prolif} + \bar{C}_G^E} \right) \frac{\Omega_K}{\Omega_K^{cc}} H\left(\Omega_K - \Omega_K^{cc}\right) \quad (L_E > z > 0) \qquad (20.5\text{-}7)$$

and the keratinocyte number balance becomes

$$\frac{\partial \Omega_K}{\partial t} + \bar{u}_K \frac{\partial \Omega_K}{\partial z} = \left[1 - \frac{\Omega_K}{\Omega_K^{cc}} H\left(\Omega_K - \Omega_K^{cc}\right) \right] \left(\frac{V_K^{prolif} \bar{C}_G^E}{K^{prolif} + \bar{C}_G^E} \right) \Omega_K \qquad (20.5\text{-}8)$$

Glucose Concentration Distribution

We treat the dermal layer as a heterogeneous medium with an immobile phase consisting of insoluble structural elements and a mobile phase comprised of an aqueous glucose solution.

The mobile phase flows through the interstitial space at a volume average filtration velocity $\bar{u}_D$ prescribed by the following solution balance:

$$\frac{\partial \bar{u}_D}{\partial z} = 0 \Rightarrow \bar{u}_D = \bar{u}_D(t) \qquad (20.5\text{-}9a, b)$$

The glucose concentration changes in the dermal layer and epidermal layer are described by the z component of the volume-averaged convection–diffusion equation (Eq. B3-35):

$$\frac{\partial \bar{C}_G^D}{\partial t} + \bar{u}_D \frac{\partial \bar{C}_G^D}{\partial z} = \bar{\mathcal{D}}_G^D \frac{\partial^2 \bar{C}_G^D}{\partial z^2} \quad (0 > z > -L_D) \qquad (20.5\text{-}10)$$

$$\frac{\partial \bar{C}_G^E}{\partial t} + \bar{u}_E \frac{\partial \bar{C}_G^E}{\partial z} = \bar{\mathcal{D}}_G^E \frac{\partial^2 \bar{C}_G^E}{\partial z^2} + \bar{R}_G^E \quad (L_E > z > 0) \qquad (20.5\text{-}11)$$

In these two equations, $\bar{C}_G^i(z,t)$ represents volume-averaged glucose concentrations in the dermal (i = D) or epithelial (i = E) phase, $\bar{\mathcal{D}}_G^i$ is the corresponding glucose diffusion coefficient, and $-\bar{R}_G^E(z,t)$ is the consumption rate of glucose per unit volume of the epithelial phase.

Glucose consumption by keratinocyte is necessary for their proliferation as well as homeostatic metabolic functions. We assume that the glucose consumption required for proliferation is in proportion to the proliferation rate:

$$-\bar{R}_G^{prolif} = \frac{\Xi_K^{prolif}}{\eta_K} = \left(\frac{V_K^{prolif} \bar{C}_G^E}{K^{prolif} + \bar{C}_G^E} \right) \frac{\Omega_K}{\eta_K} \qquad (20.5\text{-}12)$$

where η_K is a constant yield coefficient. If the metabolic component of glucose consumption is directly proportional to glucose-limited, Michaelis–Menten kinetics, then

$$-\bar{R}_G^{metab} = \left(\frac{V_K^{metab} \bar{C}_G^E}{K^{metab} + \bar{C}_G^E} \right) \Omega_K \qquad (20.5\text{-}13)$$

With $\bar{R}_G^E(z,t) = \bar{R}_G^{prolif} + \bar{R}_G^{metab}$, the governing equation for glucose becomes

$$\frac{\partial \bar{C}_G^E}{\partial t} + \bar{u}^E \frac{\partial \bar{C}_G^E}{\partial z} = \bar{\mathcal{D}}_G^E \frac{\partial^2 \bar{C}_G^E}{\partial z^2} - \left[\left(\frac{V_K^{prolif} \bar{C}_G^E}{K^{prolif} + \bar{C}_G^E} \right) \frac{1}{\eta_K} + \left(\frac{V_K^{metab} \bar{C}_G^E}{K^{metab} + \bar{C}_G^E} \right) \right] \Omega_K \qquad (20.5\text{-}14)$$

Initial and Boundary Conditions

For an initially thin epithelial layer of thickness L_{Eo} with uniform keratinocyte cell density $\Omega_{Ko} < \Omega_K^{cc}$ seeded on the dermal layer, the initial conditions are

$$t = 0: \quad L_E = L_{Eo}, \quad \Omega_K = \Omega_{Ko} H(z) \qquad (20.5\text{-}15)$$

Initially, glucose concentrations in the dermal and epidermal layers are assumed equal:

$$t = 0: \quad \bar{C}_G^E = \bar{C}_G^D = C_{Go} \qquad (20.5\text{-}16)$$

At the basal surface of the dermal layer, we assume that blood perfusion in capillaries provides sufficient glucose to maintain the initial glucose concentration:

$$z = -L_D: \quad \bar{C}_G^D = C_{Go} \qquad (20.5\text{-}17)$$

At the dermal–epidermal interface, keratinocyte proliferation is impeded by the structural elements in the dermal layer. As a result, the following condition applies:

$$\mathrm{z}=0: \quad \bar{\mathrm{u}}_{\mathrm{K}}=0, \quad \Omega_{\mathrm{K}}=\Omega_{\mathrm{Ko}} \tag{20.5-18}$$

Also at this interface, filtration flow from the dermis provides the liquid volume for producing new cells in the epidermal layer. Therefore, the volume average filtration velocity through the dermis must match the epithelial layer velocity. In addition, the glucose concentration, and glucose transport flux are continuous, and this leads to a continuous diffusion flux:

$$\mathrm{z}=0: \quad \bar{\mathrm{u}}_{\mathrm{D}}=\bar{\mathrm{u}}_{\mathrm{E}}, \quad \bar{\mathrm{C}}_{\mathrm{G}}^{\mathrm{D}}=\bar{\mathrm{C}}_{\mathrm{G}}^{\mathrm{E}}, \quad \bar{\mathcal{D}}_{\mathrm{G}}^{\mathrm{D}}\frac{\partial \bar{\mathrm{C}}_{\mathrm{G}}^{\mathrm{D}}}{\partial \mathrm{z}}=\bar{\mathcal{D}}_{\mathrm{G}}^{\mathrm{E}}\frac{\partial \bar{\mathrm{C}}_{\mathrm{G}}^{\mathrm{E}}}{\partial \mathrm{z}} \tag{20.5-19}$$

At the free boundary, glucose cannot diffuse across the epithelial–air interface:

$$\mathrm{z}=\mathrm{L}_{\mathrm{E}}: \quad \frac{\partial \bar{\mathrm{C}}_{\mathrm{G}}^{\mathrm{E}}}{\partial \mathrm{z}}=0 \tag{20.5-20}$$

Analysis

Dimensionless Forms

To make the model dimensionless, we use the variables

$$t=\frac{\bar{\mathcal{D}}_{\mathrm{G}}^{\mathrm{E}}\mathrm{t}}{\mathrm{L}_{\mathrm{Eo}}^{2}}, \quad z=\frac{\mathrm{z}}{\mathrm{L}_{\mathrm{Eo}}}, \quad u_i=\frac{\mathrm{L}_{\mathrm{Eo}}\bar{\mathrm{u}}_{\mathrm{i}}}{\bar{\mathcal{D}}_{\mathrm{G}}^{\mathrm{E}}} \quad (\mathrm{i}=\mathrm{E},\mathrm{K})$$
$$C^{\mathrm{i}}=\frac{\bar{\mathrm{C}}_{\mathrm{G}}^{\mathrm{i}}}{\mathrm{C}_{\mathrm{Go}}} \quad (\mathrm{i}=\mathrm{E},\mathrm{D}), \quad \Omega=\frac{\Omega_{\mathrm{K}}}{\Omega_{\mathrm{K}}^{\mathrm{cc}}}, \quad L_{\mathrm{E}}=\frac{\mathrm{L}_{\mathrm{E}}}{\mathrm{L}_{\mathrm{Eo}}} \tag{20.5-21a-f}$$

In the epidermal layer ($L_{\mathrm{E}}(t)>z>0$), the dimensionless equations for the velocities are

$$u_{\mathrm{E}}=\frac{dL_{\mathrm{E}}}{dt}=u_{\mathrm{K}}(L_{\mathrm{E}},t) \tag{20.5-22}$$

$$\frac{\partial u_{\mathrm{K}}}{\partial z}=\left(\frac{Da^{\mathrm{prolif}}C^{\mathrm{E}}}{K^{\mathrm{prolif}}+C^{\mathrm{E}}}\right)\Omega H(\Omega-1) \tag{20.5-23}$$

The dimensionless equation for the keratinocyte number density is

$$\frac{\partial \Omega}{\partial t}+u_{\mathrm{K}}\frac{\partial \Omega}{\partial z}=\left(\frac{Da^{\mathrm{prolif}}C^{\mathrm{E}}}{K^{\mathrm{prolif}}+C^{\mathrm{E}}}\right)\Omega[1-\Omega H(\Omega-1)] \tag{20.5-24}$$

The dimensionless glucose concentration changes according to

$$\frac{\partial C^{\mathrm{E}}}{\partial t}+u_{\mathrm{E}}\frac{\partial C^{\mathrm{E}}}{\partial z}-\frac{\partial^2 C^{\mathrm{E}}}{\partial z^2}=-\left[\frac{Da^{\mathrm{metab}}C^{\mathrm{E}}}{K^{\mathrm{metab}}+C^{\mathrm{E}}}+\frac{\beta_1}{\beta_2}\frac{Da^{\mathrm{prolif}}\Omega C^{\mathrm{E}}}{K^{\mathrm{prolif}}+C^{\mathrm{E}}}\right]\Omega \tag{20.5-25}$$

In the dermal layer ($0>z>-L_{\mathrm{D}}$) where $u_{\mathrm{D}}(t)=u_{\mathrm{E}}(t)$, the dimensionless glucose concentration changes according to

$$\frac{\partial C^{\mathrm{D}}}{\partial t}+u_{\mathrm{E}}\frac{\partial C^{\mathrm{D}}}{\partial z}=\beta_3\frac{\partial^2 C^{\mathrm{D}}}{\partial z^2} \tag{20.5-26}$$

The dimensionless conditions to solve simultaneously for the epidermal and dermal equations are

$$
\begin{aligned}
&t = 0: \quad L_E = 1, \quad \Omega = \beta_2 H(z), \quad C^E = C^D = 1 \\
&z = -L_D: \quad C^D = 1 \\
&z = 0: \quad u_K = 0, \quad C^D = C^E, \quad \Omega = \beta_2, \quad \frac{\partial C^E}{\partial z} = \beta_3 \frac{\partial C^D}{\partial z} \\
&z = L_E: \quad \frac{\partial C^E}{\partial z} = 0
\end{aligned}
\tag{20.5-27a-d}
$$

The dimensionless parameters are

$$
\begin{aligned}
&Da^{prolif} \equiv \frac{V_K^{prolif} L_{Eo}^2}{\bar{\mathcal{D}}_G^E}, \quad Da^{metab} \equiv \frac{V_K^{metab} L_{Eo}^2 \Omega_K^{cc}}{\bar{\mathcal{D}}_G^E C_{Go}}, \quad L_D \equiv \frac{L_D}{L_{Eo}} \\
&K^{prolif} \equiv \frac{K_K^{prolif}}{C_{Go}}, \quad K^{metab} \equiv \frac{K_K^{metab}}{C_{Go}}, \quad \beta_1 \equiv \frac{\Omega_{Ko}}{\eta_K C_{Go}}, \quad \beta_2 \equiv \frac{\Omega_{Ko}}{\Omega_K^{cc}}, \quad \beta_3 \equiv \frac{\bar{\mathcal{D}}_G^D}{\bar{\mathcal{D}}_G^E}
\end{aligned}
\tag{20.5-28a-h}
$$

Simulation

This nonlinear system with a moving boundary at $z = L_E(t)$ requires numerical solution. The values of the dimensionless parameters to simulate the response are based on realistic estimates: $Da^{prolif} = 6 \times 10^{-6}$, $K^{prolif} = 0.5$, $Da^{metab} = 6 \times 10^{-4}$, $K^{metab} = 10^{-6}$, $\beta_1 = 50$ and $L_D = 200$.

Figure 20.5-2 shows the spatial distributions of keratinocyte number density Ω at various times between $t = 50{,}000$ and $t = 1{,}000{,}000$. In these simulations, the initial seeding density is 50% of the maximum cell number density ($\beta_2 = 0.5$), and the diffusion coefficient in the dermis is 20% less than that in the epidermis ($\beta_3 = 0.8$). The endpoints of each curve correspond to the free interface of the epidermal layer positioned at $z = L_E$. Cell number density generally increases with z at any t. At a t of 50,000, 100,000, and 150,000, the curves terminate at $z = 1$ where $\Omega < 1$. Thus, at these relatively short times, cells do not reach confluence at the edge of the seeding layer, and the epithelial layer is unable to grow beyond its seeding thickness. At t of 400,000 and 1,000,000, however, the curves terminate at $z > 1$ where $\Omega = 1$. At these later

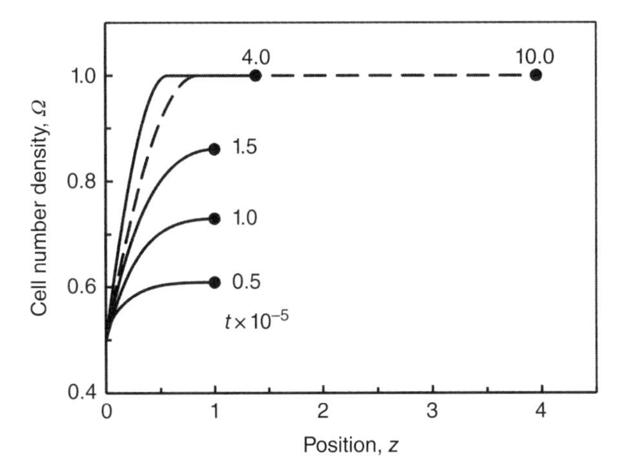

Figure 20.5-2 Cell number density distribution in a growing epidermal layer.

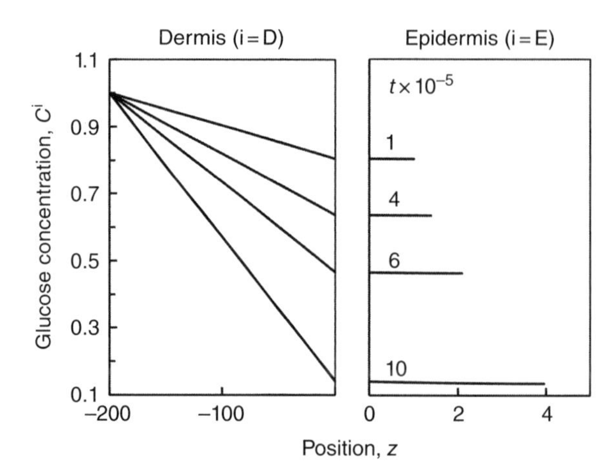

Figure 20.5-3 Glucose concentration distributions in the skin model.

times, cell number density does reach its maximum in the vicinity of the free interface, and the resulting population pressure provides impetus for the epithelial layer to thicken.

Figure 20.5-3 shows simulations of dermal glucose concentration C^D and epidermal glucose concentration C^E at various times between $t = 100{,}000$ and $t = 1{,}000{,}000$ using the same parameter values as those in Figure 20.5-2. Each of the glucose concentration curves in the dermis has a constant slope dC^D/dz, which is negative. This occurs because glucose diffusion dominates convective transport by proliferation in this layer (i.e., $\left(Da^{prolif}/\beta_3\right) = \left(V_K^{prolif} L_{Eo}^2/\bar{\mathcal{D}}_G^D\right) \ll 1$). With increasing time, dC_G/dz in the dermal layer becomes more negative to compensate for an increased transport resistance in the epidermal layer as it thickens. Although not apparent in the right graph, the $C^E(z)$ curves are slightly concave. This is due to glucose utilization during keratinocyte propagation. On the average, $-dC^E/dz$ is about one-third the value of $-dC^D/dz$. This indicates that glucose transport through the skin construct is limited more by diffusion in the dermis than by diffusion-reaction processes in the epidermis.

Figure 20.5-4 shows the dynamics of epidermal growth for different values of the glucose diffusion ratio $\beta_3 = \bar{\mathcal{D}}_G^D/\bar{\mathcal{D}}_G^E$ when initial seeding is at half of its maximum ($\beta_2 = 0.5$; left graph) or at its maximum ($\beta_2 = 1$; right graph). When $\beta_2 = 0.5$, epithelial layer growth occurs in three stages: An initial induction phase when the edge of the seeding layer is not yet confluent so that epithelial thickening cannot occur, an intermediate exponential phase in which keratinocyte propagation leads to rapid epithelial layer thickening, and a final quiescent phase in which epithelial thickening occurs very slowly. When $\beta_2 = 1$, cells are always confluent the edge of the seeding layer so there is no induction phase. Also, epithelial growth requires less time to reach its quiescent phase as compared to the $\beta_2 = 0.5$ simulation. At both of the initial seeding densities, more rapid diffusion in the dermis associated with the larger β_3 increases the L_E levels achieved during the quiescent growth phase. This suggests that during this phase, keratinocyte propagation is limited by glucose diffusion from the dermal layer.

Since there is no restriction on the amount of glucose that can be supplied by the subdermal blood supply, the epidermal layer does not reach a constant thickness. Instead, it

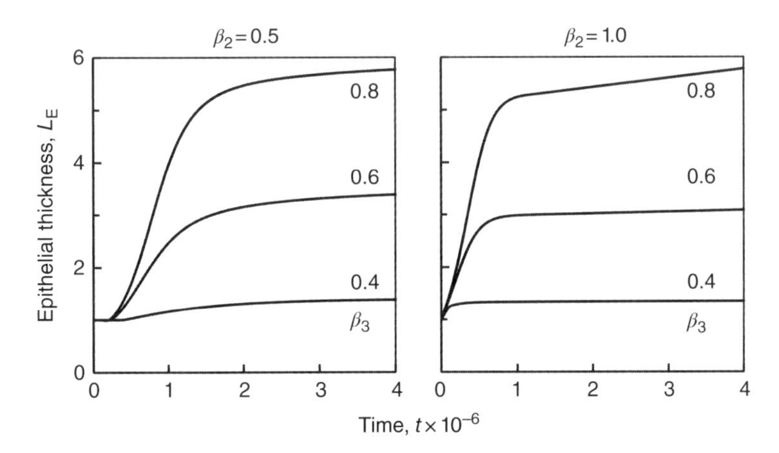

Figure 20.5-4 Effect of glucose diffusivity ratio on epithelial growth dynamics.

has a small positive slope throughout the quiescent phase. For a constant thickness to be attained, the model would need to include a limitation on glucose supply or the incorporation of cell death. Cell death could be due to a critically low glucose concentration. It could also result from metabolic waste products that accumulate at the skin–air boundary.

References

Baskaran H. Mass transfer in blind-ended hollow fiber prototypes of the Penn State Intravascular Lung [dissertation]. University Park: Pennsylvania State University; 1997, p 64. Available from: University Microfilms, Ann Arbor, MI; AAT 9802583.

Kaar JL, Oh H-I, Russell AJ, Federspiel WJ. Towards improved artificial lungs through biocatalysis. Biomaterials. 2007; 28:3131–3139.

Snider M, High KM, Richard RB, Panol G, Campbell EA, Service CV, Stene JK, Ultman JS. Small intrapulmonary artery lung prototypes: design, construction and in vitro water testing. ASAIO J. 1994; 40:M533–M539.

Chapter 21

Drug Release, Delivery, and Distribution

The controlled delivery of a drug and its active metabolite to those sites in the body where treatment is required is a challenging problem. Success in achieving this goal depends on the design of the drug delivery device, the entry route of the drug (i.e., ingestion, injection, infusion, inhalation, implantation, or dermal application), and its subsequent transport and chemical reactions in tissues and organs.

Delivery devices can be active or passive. Mechanical infusion pumps, for example, are active systems that continuously force drug directly into a catheterized vein. In the case of personal insulin infusion pumps used to treat diabetes, the device is small enough to be conveniently worn under clothing. By periodic self-monitoring and adjustment of the drug delivery rate, patients are able to maintain their insulin blood levels within a therapeutic range. In passive delivery methods, the driving force for introducing a drug is an electrochemical potential gradient formed between the device and the body. A transdermal patch, for example, induces a large drug concentration gradient within the skin, resulting in transport to underlying tissue. It is sometimes possible to enhance transdermal transport by introducing either an electric field to increase the electrochemical potential gradient or acoustic energy to weaken barrier properties across the skin.

The illustrations in this chapter will address mechanisms of drug delivery where mass transport plays a critical role in the design of a delivery device and in the distribution of a drug within the body.

21.1 Drug Release From an Agglomerated Tablet

Drug release can be sustained from an ingested tablet made of crystals of a pure drug imbedded in a matrix of polymer fibers such as micronized cellulose. We consider a tablet

Biomedical Mass Transport and Chemical Reaction: Physicochemical Principles and Mathematical Modeling,
First Edition. James S. Ultman, Harihara Baskaran, and Gerald M. Saidel.

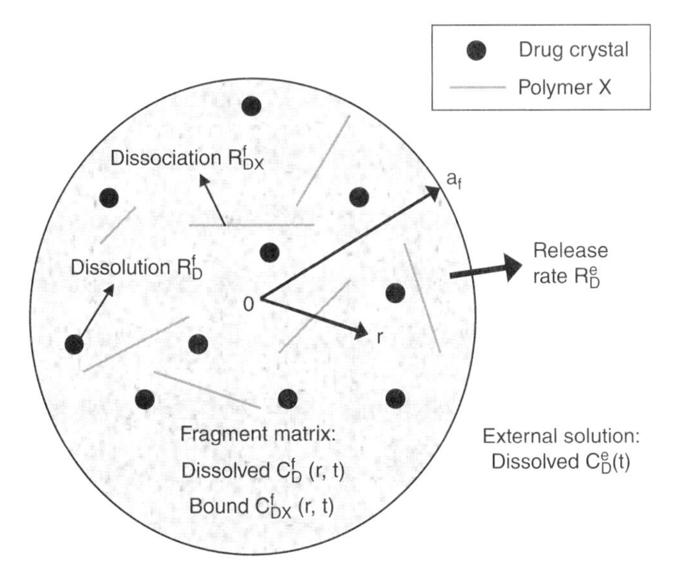

Figure 21.1-1 Components of a tablet fragment for controlled drug release.

fabricated by compressing a mixture of drug crystals and polymeric powder. When the tablet comes in contact with gastric fluids, it imbibes water, swells to a greater volume, and disintegrates into small fragments. The fragmentation and swelling processes can be very rapid. Within a swollen fragment, drug crystals continuously dissolve into the hydrated fiber matrix (Fig. 21.1-1). The dissolved drug molecules then reversibly bind to the entrapped polymer fibers or diffuse to the fragment surface where they are released into surrounding gastrointestinal fluid. The time interval over which the drug is released depends on its interactions with the fragment components.

Model Formulation

Here, we model the rate of drug release from tablet fragments suspended in a constant volume of well-mixed solution. The average of all fragments resulting from the disintegration of a tablet can be idealized as a spherical fragment that has a constant radius a_f. Contained within the fragment are concentrations $C^f_D(r,t)$ of dissolved drug and $C^f_{DX}(r,t)$ of drug attached to binding sites X on the polymer fibers.

Dissolved Drug Dynamics

If drug crystals and polymer fibers are sufficiently small compared to fragment volume and are uniformly distributed, then we can treat the fragment matrix as a continuous fluid phase with crystal dissolution and drug–polymer dissociation represented as spatially constant sources of drug. With one-dimensional radial diffusion occurring in a spherical fragment, a dynamic molar balance on dissolved drug leads to

$$\frac{\partial C^f_D}{\partial t} = \frac{\mathcal{D}^f_D}{r^2}\frac{\partial}{\partial r}\left(r^2\frac{\partial C^f_D}{\partial r}\right) + R^f_D + R^f_{DX} \tag{21.1-1}$$

where R_D^f and R_{DX}^f are the rates of crystal dissolution and drug-polymer dissociation per unit fragment volume; and $\mathcal{D}_D^f$ is a constant diffusion coefficient. Based upon a molar balance on the drug–polymer complex, we also obtain

$$\frac{\partial C_{DX}^f}{\partial t} = -R_{DX}^f \tag{21.1-2}$$

By adding these two equations, we eliminate R_{DX}^f.

$$\frac{\partial\left(C_D^f + C_{DX}^f\right)}{\partial t} = \frac{\mathcal{D}_D^f}{r^2}\frac{\partial}{\partial r}\left(r^2\frac{\partial C_D^f}{\partial r}\right) + R_D^f \tag{21.1-3}$$

If reversible binding of drug to polymer is rapid compared to drug dissolution from solid crystals and drug release into surrounding fluid, then $C_{DX}^f(r,t)$ is directly related to $C_D^f(r,t)$ by a quasi-equilibrium relationship. It follows from the chain rule that $\left(\partial C_D^f/\partial t\right) = \left(dC_{DX}^f/dC_D^f\right)\left(\partial C_D^f/\partial t\right)$, and we can write Equation 21.1-3 as

$$\left(1 + \frac{dC_{DX}^f}{dC_D^f}\right)\frac{\partial C_D^f}{\partial t} = \frac{\mathcal{D}_D^f}{r^2}\frac{\partial}{\partial r}\left(r^2\frac{\partial C_D^f}{\partial r}\right) + R_D^f \tag{21.1-4}$$

Initially, no dissolved drug exists in a fragment.

$$t = 0: \quad C_D^f = 0 \tag{21.1-5}$$

At the center of a symmetric fragment, the concentration gradient is zero.

$$r = 0: \quad \frac{\partial C_D^f}{\partial r} = 0 \tag{21.1-6}$$

At the interface between a fragment and external solution, the drug is in phase equilibrium.

$$r = a_f: \quad C_D^f = \lambda_D^{f,e} C_D^e \tag{21.1-7}$$

Here, $\lambda_D^{f,e}$ is the concentration partition coefficient of dissolved drug between a matrix fragment and the external solution.

The drug concentration $C_D^e(t)$ in a well-mixed external solution of constant volume V_e that contains a total volume V_f of tablet fragments varies according to

$$\frac{d\left(C_D^e V_e\right)}{dt} = V_f R_D^e \Rightarrow \frac{dC_D^e}{dt} = \frac{\varepsilon}{1-\varepsilon} R_D^e \tag{21.1-8a,b}$$

Here, R_D^e is the rate of drug release from a fragment to the external solution per unit fragment volume and ε is the fragment volume fraction so that $V_f/V_e = \varepsilon/(1-\varepsilon)$. For an external solution that is initially free of drug, $C_D^e(0) = 0$.

Drug Crystal Dissolution Dynamics

We assume that the number density $\Omega(t)$ of drug crystals is uniform in the tablet fragment and each drug crystal can be idealized as an average sphere of pure drug with a radius a(r,t). Initially, drug crystals have a radius a_o and are present in a fragment with number density $\Omega(0) = \Omega_o$. As dissolution continues, however, the crystals may all vanish as indicated by a crystal radius of zero.

$$\Omega(r,t) = \begin{cases} \Omega_o & \text{when } a_o \geq a(r,t) > 0 \\ 0 & \text{when } a(t) = 0 \end{cases} \tag{21.1-9}$$

The moles of solid drug in a crystal with constant molar concentration c_c and volume $4\pi a^3/3$ is $4\pi c_c \Omega a^3/3$. A molar balance on drug present in all the crystals yields

$$\frac{4\pi}{3}\frac{\partial(c_c\Omega a^3)}{\partial t} = -R_D^f \Rightarrow \frac{\partial a}{\partial t} = \begin{cases} -R_D^f/4\pi\Omega_o c_c a^2 & \text{when } a_o \geq a > 0 \\ 0 & \text{when } a = 0 \end{cases} \tag{21.1-10a,b}$$

Equilibrium and Rate Relationships

To complete this model, we need to specify an equilibrium relationship for the dissociation of drug from a polymer binding site. We also require rate equations for the dissolution of a drug crystal into a fragment matrix and the release of drug from a fragment into the surrounding solution. If a group of n drug molecules simultaneously bind to a polymer fiber, then the equilibrium relationship between the concentrations C_{DX}^f of bound drug and C_D^f of dissolved drug is analogous to the Hill model of oxyhemoglobin saturation (Eq. 5.5-10):

$$C_{DX}^f(r,t) = \frac{nT_X\left(\kappa C_D^f\right)^n}{1+\left(\kappa C_D^f\right)^n} \Rightarrow \frac{dC_{DX}^f}{dC_D^f} = \frac{T_X\kappa n^2\left(\kappa C_D^f\right)^{n-1}}{\left[1+\left(\kappa C_D^f\right)^n\right]^2} \tag{21.1-11a,b}$$

where T_X represents the moles of polymer per unit volume of a fragment, and κ is a binding constant.

The dissolution rate of drug crystals per unit volume of tablet fragment is proportional to the difference between a saturation concentration $C_{D^*}^f$ and the dissolved drug concentration C_D^f in the fiber matrix:

$$R_D^f = 4\pi a^2\Omega_o k_c\left(C_{D^*}^f - C_D^f\right) \tag{21.1-12}$$

where k_c is a mass transfer coefficient at a crystal surface, and $4\pi\Omega_o a^2$ is the total surface area of drug crystals per fragment volume. Since the concentration driving force $\left(C_{D^*}^f - C_D^f\right)$ cannot be negative, $R_D^f > 0$. With the radius of a drug crystal varying with time and position, we also expect k_c to vary with r and t. We can relate these two variables by means of a Sherwood number (*Sh*).

$$k_c(r,t) = \frac{\mathcal{D}_d}{2a(r,t)} Sh \tag{21.1-13}$$

Although *Sh* generally depends on local velocity as well as geometry, it is constant for diffusion in a stationary medium (e.g., entry 5a of Table 12.1-1), which approximates the situation here.

The drug release rate from a fragment to the well-mixed external solution equals the internal diffusion rate to the fragment surface at $r = a_f$.

$$V_f R_D^e = -S_f \mathcal{D}_d \left[\frac{\partial C_D^f}{\partial r}\right]_{r=a_f} \Rightarrow R_D^e = -\mathcal{D}_D^f \frac{3}{a_f}\left[\frac{\partial C_D^f}{\partial r}\right]_{r=a_f} \tag{21.1-14a,b}$$

where S_f/V_f is the total fragment surface per total fragment volume, which is the same as the surface-to-volume ratio $3/a_f$ of a single spherical fragment.

Consolidated Equations

Upon substitution of the rate and equilibrium relations, we can consolidate the governing equations. The dissolved drug concentration changes with time and location in the fragment as

$$\left\{1+\frac{T_X\kappa n^2\left(\kappa C_D^f\right)^{n-1}}{\left[1+\left(\kappa C_D^f\right)^n\right]^2}\right\}\frac{\partial C_D^f}{\partial t}=\frac{\mathcal{D}_D^f}{r^2}\frac{\partial}{\partial r}\left(r^2\frac{\partial C_D^f}{\partial r}\right)+2\pi a\Omega_o\mathcal{D}_d Sh\left(C_D^*-C_D^f\right)\quad(a_f>r>0)\tag{21.1-15}$$

The average drug crystal radius changes with time as

$$\frac{\partial a}{\partial t}=-\frac{\mathcal{D}_d Sh}{2c}\left(\frac{C_D^*-C_D^f}{a}\right)\quad(a_o\geq a>0)\tag{21.1-16}$$

The drug concentration in the solution external to the fragment changes as

$$\frac{dC_D^e}{dt}=-\frac{3\varepsilon_f\mathcal{D}_D^f}{a_f(1-\varepsilon_f)}\left[\frac{\partial C_D^f}{\partial r}\right]_{r=a_f}\tag{21.1-17}$$

Analysis

Dimensionless Forms

If we define the following dimensionless variables:

$$r=\frac{r}{a_f},\quad t=\frac{t\mathcal{D}_D^f}{a_f^2},\quad a=\frac{a}{a_o},\quad C=\frac{C_D^f}{C_{D*}^f}\tag{21.1-18a-b}$$

then the governing equations, Equations 21.1-15–21.1-17, become

$$\left[1+\frac{\alpha_1\alpha_2C^{n-1}}{(1+\alpha_2C^n)^2}\right]\frac{\partial C}{\partial t}=\frac{1}{r^2}\frac{\partial}{\partial r}\left(r^2\frac{\partial C}{\partial r}\right)+3\beta_1\beta_2a(1-C)\tag{21.1-19}$$

$$\frac{\partial a}{\partial t}=\begin{cases}-\beta_2\dfrac{(1-C)}{a} & \text{when } 1\geq a>0\\ 0 & \text{when } a=0\end{cases}\tag{21.1-20}$$

$$\frac{dC_e}{dt}=-\beta_3\left[\frac{\partial C}{\partial r}\right]_{r=1}\tag{21.1-21}$$

The initial and boundary conditions are

$$\begin{aligned}t=0:&\quad C=0,\quad C_e=0,\quad a=1\\ r=0:&\quad \frac{\partial C}{\partial r}=0\\ r=1:&\quad C=C_e\end{aligned}\tag{21.1-22a-c}$$

The dimensionless parameters are

$$\begin{aligned}&\alpha_1=T_X\kappa n^2,\quad \alpha_2=\kappa C_{D^*}^f,\quad C_e=\frac{\lambda_D^{f,e}C_D^e}{C_{D^*}^f}\\ &\beta_1=\frac{4\pi a_o^3c_c\Omega_o}{3C_{D^*}^f},\quad \beta_2=Sh\frac{C_{D^*}^f}{2c}\left(\frac{a_f}{a_o}\right)^2,\quad \beta_3=\frac{3\varepsilon_f\lambda_D^{f,e}}{(1-\varepsilon_f)}\end{aligned}\tag{21.1-23a-f}$$

After solving the three governing equations for C, C^e, and a, we can compute additional outputs that describe the dimensionless concentration of bound drug:

$$C_x \equiv \frac{C^f_{DX}}{C^f_{D^*}} = \frac{\alpha_1 \alpha_2^{n-1} C^n}{n(1+\alpha_2^n C^n)} \tag{21.1-24}$$

the dimensionless concentration of the drug in crystalline form:

$$C_c \equiv \frac{(4\pi a^3/3) c_c \Omega_o}{C^f_{D^*}} = \beta_1 a^3 \tag{21.1-25}$$

and the dimensionless drug release rate into the external solution.

$$R_e \equiv \frac{a_f^2}{3\mathcal{D}^f_D C^f_{D^*}} \left[R^e_D\right]_{r=a_f} = -\left[\frac{\partial C}{\partial r}\right]_{r=1} \tag{21.1-26}$$

Simulation

From a numerical solution of the model equations, we can simulate various aspects of the drug delivery process. The simulations of this model were obtained with $\alpha_1 = \alpha_2 = 1$, $\beta_1 = 3$, $10 > \beta_2 > 0.1$, $\beta_3 = 0.001$, and $n = 2$, which are reasonable for this system.

Figure 21.1-2 shows the spatial distributions of concentrations C_c, C_x, and C within a matrix fragment at three dimensionless times when $\beta_2 = 1$. Because of drug loss to surrounding fluid at $r = 1$, all three distributions decrease with distance from the fragment center at $r = 0$.

Figure 21.1-3 displays the simulated dynamics of the drug release process. At a fixed position of $r = 0.5$ and when $\beta_2 = 1$, the left graph shows that C_c continuously decreases with time as drug crystals dissolve. Under the same conditions, C and C_x initially increase as crystal dissolution causes the accumulation of both dissolved and bound drug in the fragment matrix. Eventually, C and C_x decrease as drug diffuses out of the fragment. The right graph illustrates how drug release rate to external fluid is affected by the dimensionless dissolution rate coefficient β_2. Decreasing β_2 when all other dimensionless parameters are unchanged reduces the magnitude of R_e but sustains drug release over a longer time. Thus, the release time of a drug can be manipulated by modifying fragment radius, the only dimensional parameter that contributes solely to β_2.

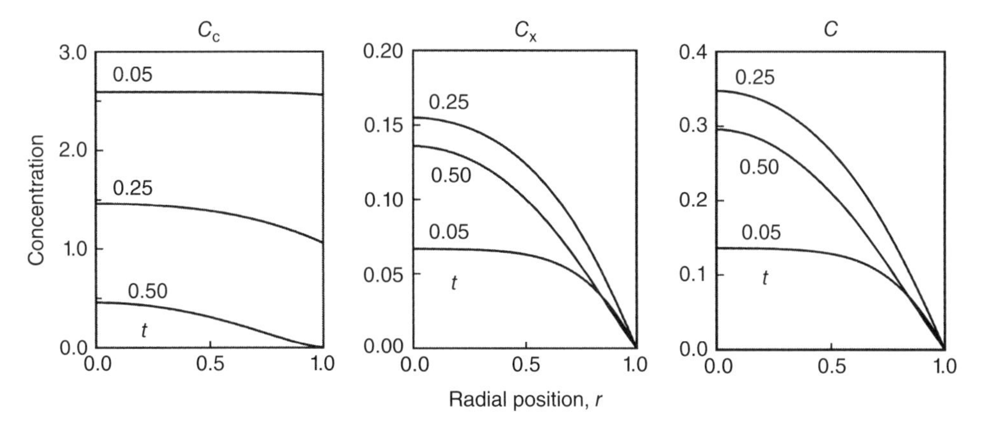

Figure 21.1-2 Spatial distributions of drug concentrations in a matrix fragment.

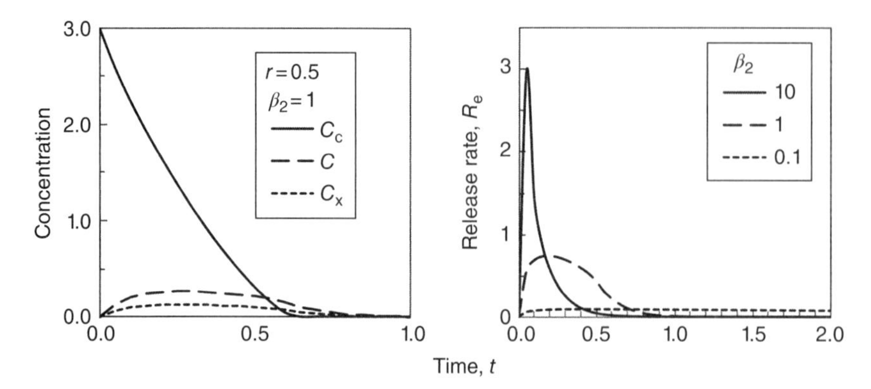

Figure 21.1-3 Drug concentration and drug release dynamics.

21.2 Drug Release From an Osmotic Pump Device

An osmotic pressure driving force between a drug delivery device and its surroundings can be used for sustained drug release. Such osmotic pumping systems have an advantage over devices that depend on drug diffusion limited by small concentration gradients.

Figure 21.2-1 is a prototype of an osmotic pump that can be miniaturized for oral ingestion or implantation in a target tissue (Theeuwes and Yum, 1976). The device has an outer osmotic compartment O and an inner drug compartment D. A rigid, semipermeable, outer membrane separates compartment O from the surroundings, and a collapsible, impermeable, inner membrane separates compartment O from compartment D. Initially, compartment O contains a slurry of soluble sodium chloride crystals in a saturated aqueous solution. Compartment D contains drug in an aqueous solution.

The high osmolarity of compartment O relative to its surroundings causes water to flow into this compartment through the rigid outer membrane. The resulting increase in hydrostatic pressure causes the inner membrane to collapse and drug to be ejected from compartment D through a small diameter dip tube. Since no liquid enters compartment D, its solution contains a constant drug concentration. Consequently, drug delivery to the surroundings is determined by volumetric outflow from the dip tube.

Model Formulation

To quantify drug delivery dynamics, we analyze an *in vitro* experiment in which the device is placed within a large reservoir R of isotonic salt solution with a uniform and constant salt concentration and pressure. Our principal assumptions are that (i) the solutions in compartments O and D are well mixed and have equal and constant mass densities, (ii) salt crystal volume in compartment O is small relative to its solution volume, (iii) hydrostatic pressure difference across the inner collapsible membrane is negligible, (iv) drug solution moves through the dip tube by a flow that is proportional to fluid pressure drop, and (v) salt transport across the outer rigid membrane is dominated by convection.

Volume and Flow Relationships

Because the outer membrane is rigid, the volume of the device is constant. With constant and equal solution mass densities and negligible crystal volume, the summed volume V of

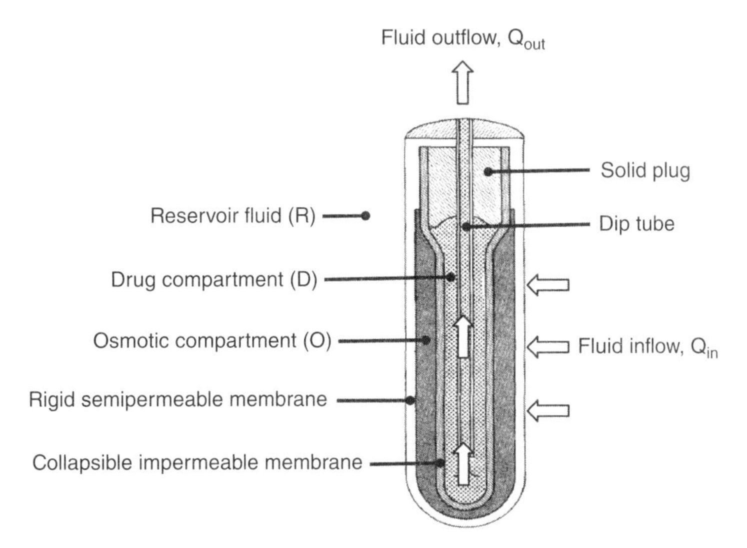

Figure 21.2-1 Osmotic pump device.

solution in the compartments is also constant. The solution volume V_O in the O compartment and V_D in the D compartment do, however, vary with time so that

$$V = V_O(t) + V_D(t) \tag{21.2-1}$$

A total mass balance on the device with constant solution density and volume reduces to

$$\frac{dV}{dt} = 0 = Q_{in} - Q_{out} \Rightarrow Q_{in} = Q_{out} \equiv Q(t) \tag{21.2-2a,b}$$

where Q_{in} is the inflow across the outer membrane and Q_{out} is the outflow through the dip tube. From a solution balance on compartment O, we get the inflow–volume relation:

$$\frac{dV_O}{dt} = Q_{in}(t) = Q(t) \tag{21.2-3}$$

The volumetric outflow of the drug through the dip tube from compartment D to the surrounding reservoir is

$$Q_{out}(t) = L_t S_t \left[P^D(t) - P^R \right] = Q(t) \tag{21.2-4}$$

where P^D and P^R are the fluid pressures in compartment D and reservoir R, S_t is the inner cross-sectional area of the dip tube, and L_t is a hydraulic conductivity of flow through the tube.

Volume Flow into the Osmotic Compartment

The inflow of water from the surrounding fluid reservoir R to compartment O is related to the hydrostatic and osmotic pressure differences across the outer membrane with the Starling equation for transmembrane flow (Eq. 6.5-31). Using the Van't Hoff equation (Eq. 4.3-16) to relate osmotic pressure difference to the concentration difference of salt, we obtain

$$Q_{in}(t) = L_m S_m \left[\left(P^R - P^O \right) + 2\mathcal{R}T\sigma_s \left(C_s^O(t) - C_s^R \right) \right] = Q(t) \tag{21.2-5}$$

Here, P^i and C_s^i (i = O,R) are the compartment pressures and salt concentrations, respectively; L_m and S_m are hydraulic permeability and surface area of the membrane; σ_s is the salt reflection coefficient; and the factor of two accounts for the dissociation of sodium chloride into two osmotically active ions. Since the inner collapsible membrane does not support a pressure difference, we note that $P^D \approx P^O$. By combining Equations 21.2-4 and 21.2-5, we eliminate the pressure difference to obtain

$$Q = 2\mathcal{R}T\sigma_s\widetilde{LS}\left(C_s^O - C_s^R\right); \quad \widetilde{LS} \equiv \left(\frac{1}{L_tS_t} + \frac{1}{L_mS_m}\right)^{-1} \qquad (21.2\text{-}6a,b)$$

Here, we have defined a mean product of hydraulic permeability and surface area, $\widetilde{LS}$. If L_tS_t and L_mS_m have substantially different values, it is the smaller of the two that limits the flow of solution through the dip tube. The maximum osmotic flow Q_{max} across the outer membrane occurs when $C_s^O = C_{s^*}$, the saturation concentration of salt in water.

$$Q_{max} = 2\mathcal{R}T\sigma_s\widetilde{LS}\left(C_{s^*} - C_s^R\right) \qquad (21.2\text{-}7)$$

The volume flow relative to its maximum is

$$\frac{Q}{Q_{max}} = \frac{\left(C_s^O - C_s^R\right)}{\left(C_{s^*} - C_s^R\right)} \Rightarrow C_s^O - C_s^R = \frac{Q}{Q_{max}}\left(C_{s^*} - C_s^R\right) \qquad (21.2\text{-}8a,b)$$

By taking the time derivative of this equation, we can relate the salt concentration dynamics to the volume flow dynamics.

$$\frac{dC_s^O}{dt} = \frac{\left(C_{s^*} - C_s^R\right)}{Q_{max}}\frac{dQ}{dt} \qquad (21.2\text{-}9)$$

Salt Crystal Dynamics

In compartment O, the molar dissolution rate $\dot{R}_s$ of salt crystals of constant molar density c_c results in a decrease in crystal volume $V_c(t)$ according to the molar balance:

$$c_c\frac{dV_c}{dt} = -\dot{R}_s \qquad (21.2\text{-}10)$$

With dissolution limited by convection and diffusion processes near the crystal surface, $\dot{R}_s$ is directly proportional to the difference between saturation and dissolved salt concentrations.

$$\dot{R}_s = V_c\phi_c k_c\left(C_{s^*} - C_s^O\right) \qquad (21.2\text{-}11)$$

where k_c is an individual mass transfer coefficient, and ϕ_c is the surface-to-volume ratio of the crystalline drug. Unlike our analysis of drug release from a tablet fraction (Section 21.1), we assume that k_c does not depend on the size of the drug crystals.

We eliminate C_s^O from Equation 21.2-11 by substitution from Equation 21.2-8a,b:

$$\dot{R}_s = V_c\phi_c k_c\left(C_{s^*} - C_s^R\right)\left(1 - \frac{Q}{Q_{max}}\right) \qquad (21.2\text{-}12)$$

When this is substituted into the crystal molar balance (Eq. 21.2-10) and combined with the relationship of the salt concentration to flow (Eq. 21.2-8a,b), we get

$$\frac{dV_c}{dt} = -\frac{V_c\phi_c k_c\left(C_{s^*} - C_s^R\right)}{c_c}\left(1 - \frac{Q}{Q_{max}}\right) \qquad (21.2\text{-}13)$$

Salt Concentration in the Osmotic Compartment

In compartment O, dissolved salt can accumulate by dissolution of salt crystals or influx of salt from reservoir R. From a molar salt balance, we obtain

$$\frac{d(C_s^O V_O)}{dt} = \dot{R}_s + \dot{N}_s \tag{21.2-14}$$

Expansion of the time derivative and substitution of Equation 21.2-3 yields the dynamics of the salt concentration:

$$V_O \frac{dC_s^O}{dt} = \dot{R}_s + \dot{N}_s - QC_s^O \tag{21.2-15}$$

When the molar rate of salt transport $\dot{N}_s$ through the semipermeable membrane depends more on convection than diffusion, it can be expressed by the combination of Equations 10.2-11 and 10.2-13b as

$$\dot{N}_s = Q\left[C_s^R(1-\sigma_s)\right] \tag{21.2-16}$$

Drug Output Flow

We combine Equations 21.2-8a,b and 21.2-9 with Equation 21.2-15 to obtain

$$V_O\left(\frac{C_{s^*} - C_s^R}{Q_{max}}\right)\frac{dQ}{dt} = (\dot{R}_s + \dot{N}_s) - Q\left[C_s^R + \frac{Q}{Q_{max}}(C_{s^*} - C_s^R)\right] \tag{21.2-17}$$

Now eliminating C_s^O by substituting the rate equations for dissolution (Eq. 21.2-12) and membrane transport of salt (Eq. 21.2-16), we get a governing equation for the flow dynamics.

$$\frac{dQ}{dt} = \frac{\phi_c k_c Q_{max}}{V_O}\left(1 - \frac{Q}{Q_{max}}\right)V_c - \frac{Q_{max}}{V_O}\left(\frac{Q}{Q_{max}} + \frac{\sigma_s C_s^R}{C_{s^*} - C_s^R}\right)Q \tag{21.2-18}$$

At this point, the model has been reduced to the minimum number of state variables: $V_O(t)$, $V_c(t)$, and $Q(t)$. The corresponding system of equations, Equations 21.2-3, 21.2-13, and 21.2-18, are solved simultaneously with the initial conditions:

$$t = 0: \quad V_O = V_O(0), \quad Q = Q_{max}, \quad V_c = n_c/c_c \tag{21.2-19}$$

Here, n_c is the initial number of moles of salt in crystalline form. The time-varying rate of drug delivery can be computed by multiplication of $Q(t)$ with the constant drug concentration in compartment D.

Analysis

Dimensionless Forms

Defining the dimensionless variables:

$$\mathit{t} \equiv \frac{Q_{max}}{V_O(0)}t, \quad \mathrm{Q} \equiv \frac{Q}{Q_{max}}, \quad \mathrm{V}_O \equiv \frac{V_O}{V_O(0)}, \quad \mathrm{V}_c \equiv \frac{V_c}{V_O(0)} \tag{21.2-20a-b}$$

we express the model equations in dimensionless form:

$$\frac{dV_O}{dt} = Q \tag{21.2-21}$$

$$\frac{dQ}{dt} + \frac{Q}{V_O}(Q+\beta_1) = \beta_2(1-Q)\frac{V_c}{V_O} \tag{21.2-22}$$

$$\frac{dV_c}{dt} = -\beta_2\beta_4(1-Q)V_c \tag{21.2-23}$$

with dimensionless initial conditions:

$$t = 0: \quad V_O = 1, \quad Q = 1, \quad V_c = \beta_3 \tag{21.2-24}$$

The dimensionless parameters, all of which have positive values, are

$$\beta_1 \equiv \frac{\sigma_s C_s^R}{\left(C_{s^*} - C_s^R\right)}, \quad \beta_2 \equiv \frac{\phi_c k_c V_O(0)}{Q_{max}}, \quad \beta_3 \equiv \frac{n_c/c_c}{V_O(0)}, \quad \beta_4 = \frac{C_{s^*} - C_s^R}{c_c} \tag{21.2-25a-d}$$

Simulation

From a numerical solution of this nonlinear system, we simulate the effects of model parameters on drug delivery as determined by the outflow $Q(t)$. As a reference case, we assume an ideal semipermeable membrane for which $\sigma_s = 1$; a salt solution in the reservoir with isotonic concentration $C_s^R = 0.15$M; a salt saturation concentration $C_{s^*} = 6.5$M; and a molar density $c_c = 36.9$M of salt crystals. With these values, $\beta_1 = 0.024$ and $\beta_4 = 0.172$, which are fixed in the simulations (Fig. 21.2-2). The parameters β_2 and β_3 are varied because these can be readily changed by the device design and initial conditions.

The left graph shows the effect of varying β_3, the mole fraction of salt crystals that are initially present in the osmotic compartment. At all three β_3 values, the dimensionless outflow is initially at its maximum value of $Q = 1$. When $t > 0$, external water enters compartment O, diluting the osmotic solution and causing salt crystals to dissolve. In the simulations at $\beta_3 = 10$, the salt crystals never completely dissolve during the time interval shown. Thus, the osmotic solution remains close to its saturation concentration C_{s^*}, and Q is reduced only slightly from its maximum value. In the $\beta_3 \ll 10$ simulations, much less crystalline salt is initially present. In those cases, enough water enters compartment O to cause complete

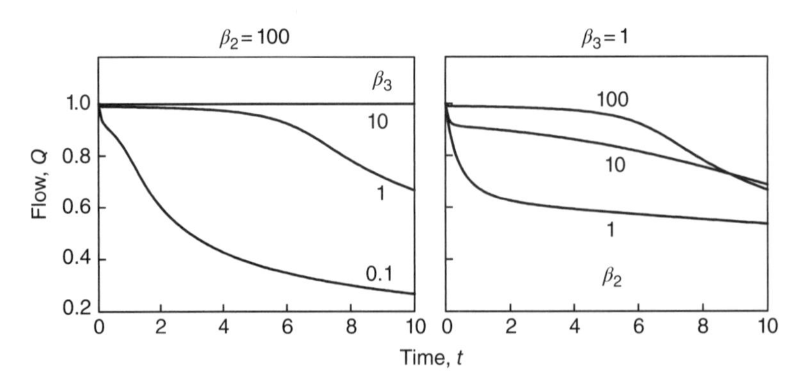

Figure 21.2-2 Dynamics of an osmotic drug delivery system (Theeuwes and Yum (1976). Reproduced with permission from Springer).

dissolution of the crystals. As external water continues to enter, there is a substantial decrease in solution osmolarity that progressively reduces Q.

The right graph shows the effect of varying β_2, the characteristic rate of crystal dissolution relative to the maximum rate of transmembrane flow. Since β_3 is the same for all three curves, the initial amount of crystalline salt is fixed. The curve with the largest $\beta_2 = 100$ exhibits a longest period of a near-maximum Q because, early in the simulation, the rapid crystal dissolution counteracts dilution of the osmotic solution by transmembrane water flow. At later times, however, the salt crystals disappear when $\beta_2 = 100$ before they do when $\beta_2 = 10$. This explains why Q predicted at $\beta_2 = 100$ eventually drops below Q at $\beta_2 = 10$.

This model provides a guide for operating an osmotic pump device to sustain the release of a drug solution at a nearly constant flow rate over a limited time.

21.3 Intestinal Drug Transport

For drugs taken orally, quantifying the rate of absorption by the intestine is essential for predicting drug delivery to the systemic circulation, as well as local distribution along the intestinal wall. In this illustration, we consider the ingestion of a drug tablet or capsule that disintegrates into a multitude of small, insoluble fragments in the stomach before entering the intestine in suspension with chyme. As chyme travels along the intestine at a constant velocity u, drug released from each fragment diffuses at a flux $N_{D,wall}$ into the intestinal wall where it enters the systemic circulation. The intestinal lumen is represented as a rigid tube of length L and constant flow cross section A (Fig. 21.3-1).

Model Formulation

Drug Release

In Section 21.1, we developed a detailed model of drug release from a tablet fragment consisting of a matrix of solid crystals and polymeric adsorption sites in an aqueous suspending medium. In that model, drug reached an external solution by a combination of diffusion within the fragment matrix and interfacial transport across the fragment surface. Here, we use a simplified model in which there are no drug crystals in the fragment, and diffusion processes are captured by an overall mass transfer coefficient K_f. In particular, we express the drug release rate per unit volume of a fragment as

$$R_D = K_f \phi_f \left(C_D^f - C_D/\lambda_D \right) \tag{21.3-1}$$

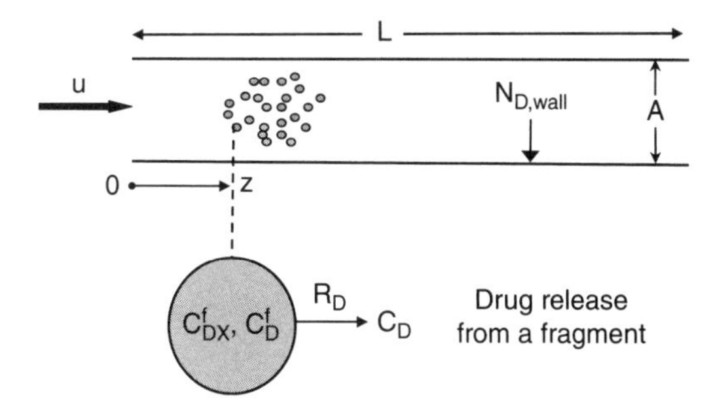

Figure 21.3-1 Drug release and uptake in an intestinal model.

where $C_D^f(z,t)$ is the concentration of dissolved drug in a fragment; $C_D(z,t)$ is the concentration dissolved in chyme; λ_D is the equilibrium concentration ratio of dissolved drug between the chyme and the fragment; and ϕ_f is the surface-to-volume ratio of a fragment.

As we have done previously, we will assume that binding of drug to a polymer site X is rapid compared to diffusion so that adsorption equilibrium is reached. For simplicity, we also assume that drug binding is a monovalent reaction that is far from saturation such that it is a first-order process so that

$$C_{DX}^f = \kappa C_D^f \tag{21.3-2}$$

where C_{DX}^f are the moles of bound drug per fragment volume, and κ is an equilibrium binding coefficient. The total concentration C_{DT}^f of drug in a fragment is given by the sum of its dissolved and bound forms.

$$C_{DT}^f = C_D^f + C_{DX}^f \tag{21.3-3}$$

Combining these last two equations allows us to write

$$C_{DT}^f = (1+\kappa)C_D^f \tag{21.3-4}$$

so that the release rate becomes

$$R_D = \frac{\phi_f K_f}{\lambda_D}\left[\left(\frac{\lambda_D}{1+\kappa}\right)C_{DT}^f - C_D\right] \tag{21.3-5}$$

Drug Dissolved in Chyme

Axial changes in $C_D(z, t)$ can be described by a one-dimensional solute balance in a rigid tube (Eq. 15.3-30). In addition to axial convection, longitudinal dispersion due to peristalsis will dominate axial diffusion, and we can write

$$\frac{\partial C_D}{\partial t} + u\frac{\partial C_D}{\partial z} = \mathcal{D}_t^* \frac{\partial^2 C_D}{\partial z^2} + (V_f \Omega)R_D - \phi_{wall} N_{D,wall} \tag{21.3-6}$$

In this equation, Ω is the fragment number density and V_f is volume of each fragment so that ΩV_f is the total fragment volume per unit volume of chyme and $(V_f\Omega)R_D$ is the release rate from all fragments per unit volume of chyme. Also, $\mathcal{D}_t^*$ is a constant dispersion coefficient of drug dissolved in chyme; and ϕ_{wall} is the intestinal surface-to-volume ratio that accounts for surface augmentation by villa protruding from the inner wall.

Because drug concentration in the richly perfused intestinal wall is negligibly small, the driving force for $N_{D,wall}$ is dominated by drug concentration in chyme. Therefore, we assume that

$$N_{D,wall} = K_{wall} C_D \tag{21.3-7}$$

where K_{wall} is an overall mass transfer coefficient between the intestinal lumen and the capillaries. After substitution of the rate expressions for $N_{D,wall}$ and R_D into Equation 21.3-6, we get

$$\frac{\partial C_D}{\partial t} + u\frac{\partial C_D}{\partial z} - \mathcal{D}_t^* \frac{\partial^2 C_D}{\partial z^2} + (\phi_{wall} K_{wall})C_D = \frac{\phi_f K_f V_f}{\lambda_D}\left[\left(\frac{\lambda_D}{1+\kappa}\right)C_{DT}^f - C_D\right]\Omega \tag{21.3-8}$$

Initially, there is no drug in the chyme so that

$$t = 0: \quad C_D = 0 \tag{21.3-9}$$

At $t > 0$, there is no drug in chyme at the proximal end of the intestine. At the distal end, the axial gradient of drug dissolved in chyme is negligible.

$$z = 0: \quad C_D = 0$$
$$z = L_t: \quad \frac{\partial C_D}{\partial z} = 0 \tag{21.3-10a, b}$$

The surface-to-length ratio of the intestines is given by the product between its surface-to-volume ratio ϕ_{wall} and cross-sectional area A. Thus, at an axial position z and time t, $(\phi_{wall}A)N_{D,wall}(z, t)$ is the drug uptake rate per unit length of the intestinal wall. During any time interval {0,t}, the accumulated moles of drug absorbed per unit intestinal length at position z are determined by

$$\hat{U}_D(z,t) = A\phi_{wall}\int_0^t N_{D,wall}(z,\xi)d\xi = A\phi_{wall}K_{wall}\int_0^t C_D(z,\xi)d\xi \tag{21.3-11}$$

Another design quantity of interest is the accumulated moles of drug uptake into the entire intestine occurring after a long enough time that all tablet fragments initially released by the stomach have exited and no more drug remains in chyme.

$$U_D = \int_0^L \hat{U}_D(z, t \to \infty)dz \tag{21.3-12}$$

Dynamics of Fragment Number Density

From a population balance (Eq. 17.1-2), we obtain dynamic changes of fragment number density due to axial transport in the absence of source and sink terms.

$$\frac{\partial \Omega}{\partial t} + \frac{\partial \Theta}{\partial z} = 0 \tag{21.3-13}$$

Because fragment transport occurs by a combination of convection and dispersion, its axial flux is

$$\Theta = u\Omega - \mathcal{D}_f^* \frac{d\Omega}{dz} \tag{21.3-14}$$

Assuming the fragment dispersion coefficient $\mathcal{D}_f^*$ is constant, the population balance becomes

$$\frac{\partial \Omega}{\partial t} + u\frac{\partial \Omega}{\partial z} = \mathcal{D}_f^* \frac{d^2\Omega}{dz^2} \tag{21.3-15}$$

With no fragments initially present

$$t = 0: \quad \Omega = 0 \tag{21.3-16}$$

At the proximal end of the intestine, the stomach empties fluid containing a fragment number density Ω_o over a time interval Δt_o. This leads to the condition

$$z = 0: \quad \Omega = \begin{Bmatrix} \Omega_o & \text{when} & t_o \geq t > 0 \\ 0 & \text{when} & t > t_o \end{Bmatrix} = \Omega_o[H(t-0) - H(t-t_o)] \tag{21.3-17}$$

where H is the unit step function (Eq. 18.1-11). At the distal end, axial diffusion of fragments is assumed to be negligible compared to convection:

$$z = L: \quad \frac{\partial \Omega}{\partial z} = 0 \tag{21.3-18}$$

Dynamics of Fragment Drug Concentration

Drug bound to fragments is transported with fragments and is released to the surrounding chyme. A molar balance on drug in the fragments contained in chyme leads to

$$\frac{\partial\left(\Omega V_f C^f_{DT}\right)}{\partial t} + \frac{\partial N^f_{DT}}{\partial z} = -(V_f\Omega)R_D \tag{21.3-19}$$

where $\Omega V_f C^f_{DT}$ represents the moles of dissolved and bound drug in fragments per unit volume of chyme; and N^f_{DT} is the longitudinal flux of fragment-associated drug per unit cross-sectional area of the intestine. This flux can be expressed as the fragment flux Θ (Eq. 21.3-14) multiplied by the total moles of drug contained in each fragment, $V_f C^f_{DT}$.

$$N^f_{DT} = \left(V_f C^f_{DT}\right)\Theta = uV_f C^f_{DT}\Omega - \mathcal{D}^*_f V_f C^f_{DT}\frac{\partial \Omega}{\partial z} \tag{21.3-20}$$

After dividing Equation 21.3-19 by a constant V_f, we expand the time derivative and substitute $\partial\Omega/\partial t$ from Equation 21.3-15 and N^f_D from Equation 21.3-20:

$$\Omega\frac{\partial C^f_{DT}}{\partial t} + C^f_{DT}\left(\mathcal{D}^*_f\frac{\partial^2\Omega}{\partial z^2} - u\frac{\partial\Omega}{\partial z}\right) + \frac{\partial}{\partial z}\left(uC^f_{DT}\Omega - \mathcal{D}^*_f C^f_{DT}\frac{\partial\Omega}{\partial z}\right) = -\Omega R_D \tag{21.3-21}$$

After expanding $\partial\left(\Omega C^f_{DT}\right)/\partial z$ and $\partial\left(C^f_{DT}\partial\Omega/\partial z\right)\partial z$, we substitute R_D to get the final governing equation for fragment-associated drug.

$$\frac{\partial C^f_{DT}}{\partial t} + u\frac{\partial C^f_{DT}}{\partial z} = \mathcal{D}^*_f\frac{\partial \ln\Omega}{\partial z}\frac{\partial C^f_{DT}}{\partial z} - \frac{K_f\phi_f}{\lambda_D}\left[\left(\frac{\lambda_D}{1+\kappa}\right)C^f_{DT} - C_D\right] \tag{21.3-22}$$

This equation is subject to initial and boundary conditions that parallel those for the fragment number density.

$$t = 0: \quad C^f_{DT} = 0$$

$$z = 0: \quad C^f_{DT} = \begin{Bmatrix} C^f_o & \text{when } t_o \geq t > 0 \\ 0 & \text{when } \quad t > t_o \end{Bmatrix} = C^f_o\left[H(t-0) - H(t-t_o)\right] \tag{21.3-23a-c}$$

$$z = L: \quad \frac{\partial C^f_{DT}}{\partial z} = 0$$

Analysis

Dimensionless Forms

We define the following dimensionless variables

$$\mathit{t} = \frac{u}{L}t, \quad \mathit{z} = \frac{z}{L}, \quad \mathit{C} = \frac{C_D}{C^f_o}, \quad \mathit{C}^f = \frac{C^f_{DT}}{C^f_o},$$
$$\mathit{\Omega} = \frac{\Omega}{\Omega_o}, \quad \hat{\mathit{U}} \equiv \frac{\hat{U}_D}{V_f\Omega_o C^f_o u A t_o/L} \tag{21.3-24a-f}$$

The quantity $V_f\Omega_o C_o^f$ represents the moles of fragment-associated drug per unit volume of the entering pulse of chyme; uAt_o is the volume of the entering pulse; and $V_f\Omega_o C_o^f uAt_o/L$ represents the moles of drug inputted at z = 0 per unit length of intestine. Thus, $\hat{U}(z, t)$ represents the local intestinal uptake during time interval {0,*t*} relative to the entering moles of drug. Similarly, we define $U \equiv U_D/V_f\Omega_o C_o^f uAt_o$ as the total intestinal uptake relative to the entering moles of drug.

The dimensionless model equations are then given by

$$\frac{\partial C}{\partial t}+\frac{\partial C}{\partial z}-\frac{1}{Pe_t}\frac{\partial^2 C}{\partial z^2}+K_w C=K_f\varepsilon_f\left(\lambda' C^f-C\right)\Omega \qquad (21.3\text{-}25)$$

$$\frac{\partial C^f}{\partial t}+\frac{\partial C^f}{\partial z}=\frac{1}{Pe_f}\frac{\partial \ln\Omega}{\partial z}\frac{\partial C^f}{\partial z}-K_f\left(\lambda' C^f-C\right) \qquad (21.3\text{-}26)$$

$$\frac{\partial \Omega}{\partial t}+\frac{\partial \Omega}{\partial z}-\frac{1}{Pe_p}\frac{\partial^2 \Omega}{\partial z^2}=0 \qquad (21.3\text{-}27)$$

$$\hat{U}(z,t)=\frac{K_w}{\varepsilon_f t_o}\int_0^t C(z,\xi)\mathrm{d}\xi,\quad U=\int_0^1 \hat{U}(z,t\to\infty)\mathrm{d}z \qquad (21.3\text{-}28\text{a,b})$$

The dimensionless initial and boundary conditions are

$$t=0:\; C=C^f=\Omega=0$$

$$z=0:\; C=0,\;\; C^f=\Omega=H(\mathrm{t}-0)-H(\mathrm{t}-\mathrm{t_o}) \qquad (21.3\text{-}29\text{a-c})$$

$$z=1:\; \frac{\partial C}{\partial z}=\frac{\partial \Omega}{\partial z}=0$$

The dimensionless parameters in these model equations are defined as:

$$Pe_t\equiv\frac{uL}{\mathcal{D}_t^*},\quad Pe_f\equiv\frac{uL}{\mathcal{D}_f^*},\quad \varepsilon_f\equiv V_f\Omega_o,\quad \lambda'\equiv\frac{\lambda_D}{1+\kappa}$$

$$K_w\equiv\frac{L}{u}\phi_{wall}K_{wall},\quad K_f\equiv\frac{L\phi_f K_f}{u\;\lambda_D},\quad t_o=\frac{ut_o}{L} \qquad (21.3\text{-}30\text{a-g})$$

where ε_f is the void fraction of fragments in chyme introduced at z = 0, and λ' is an dimensionless equilibrium distribution coefficient that accounts for both interfacial and binding equilibria.

Simulation

We carried out numerical solutions of the model equations with a fragment volume fraction of $\varepsilon_f = 0.001$ and a dimensionless partition coefficient of $\lambda' = 1$. Péclet numbers for fragments in chyme of $Pe_p = 80$ and for drugs in chyme of $Pe_t = 50$ were estimated from observed residence time distributions (Malagelada *et al.*, 1984, fig. 5). To provide a smooth transition from 1 to 0 in the vicinity of $t = t_o$, the discrete pulse input at $z = 0$ was approximated as a continuous time function

$$z=0:\; C^f=\Omega=\frac{1}{1+\exp[-10(1-t)]} \qquad (21.3\text{-}31)$$

This function, shown on the left of Figure 21.3-2, has a pulse width of $t_o = 0.01$ corresponding to a dimensional time t = 0.01 L/u. In other words, the stomach emptying time that generates

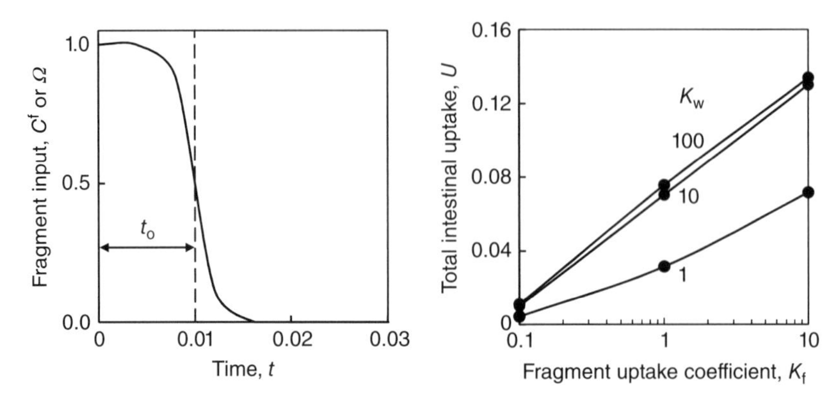

Figure 21.3-2 Particle input and total intestinal uptake.

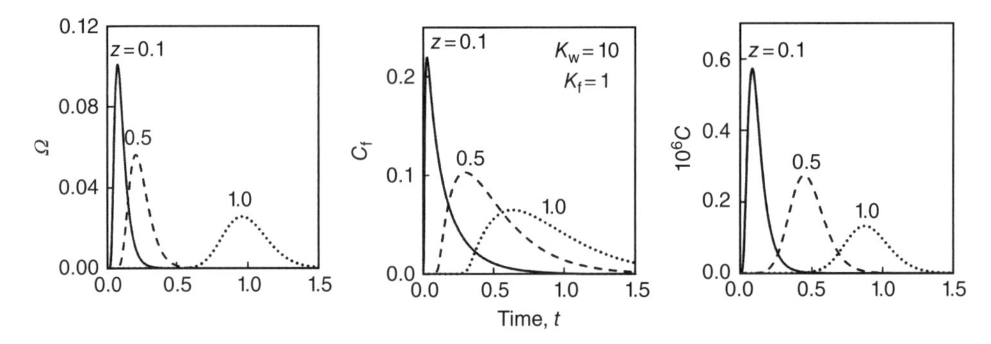

Figure 21.3-3 Breakthrough curves of Ω, C^f, and C at three axial positions.

the pulse input is 1% of the residence time L/u of chyme in the intestine. Shown on right of this figure is the effect of different values of the fragment release rate coefficient K_f and the intestinal uptake rate coefficient K_w on the total intestinal uptake per unit drug input, U. As we expect, an increase in either of these coefficients leads to a larger U. For values of K_w beyond 10, however, there is little improvement in uptake. This indicates that, for large K_w, the transport resistance of the intestinal wall is so small, that the limiting factor in uptake is drug release from the tablet fragments.

Figure 21.3-3 shows the simulated dynamics of dimensionless number density Ω, fragment drug concentration C^f, and chyme drug concentration C when $K_w = 10$ and $K_f = 1$ such that drug release is the controlling factor. The unit pulse in which tablet fragments enter at $z = 0$ is progressively attenuated and dispersed as the pulse moves through three successive positions $z = 0.1$, 0.5, and 1.0 (left graph). Since the fragments are insoluble, the area under the three curves is nearly equal. The drug concentration within the fragments is also introduced as a unit pulse. In this case, the areas under the curves decrease from position to position because of drug transport from fragments to chyme (middle graph). These curves have discernable tails due to dispersion processes that cause a greater effect on fragment-associated drug than on the fragments themselves. The drug concentration curves for chyme (right graph) are highly symmetric compared to the fragment-associated drug concentration curves. The lack of tailing in these curves is due to transport across the intestinal wall. Notice that at

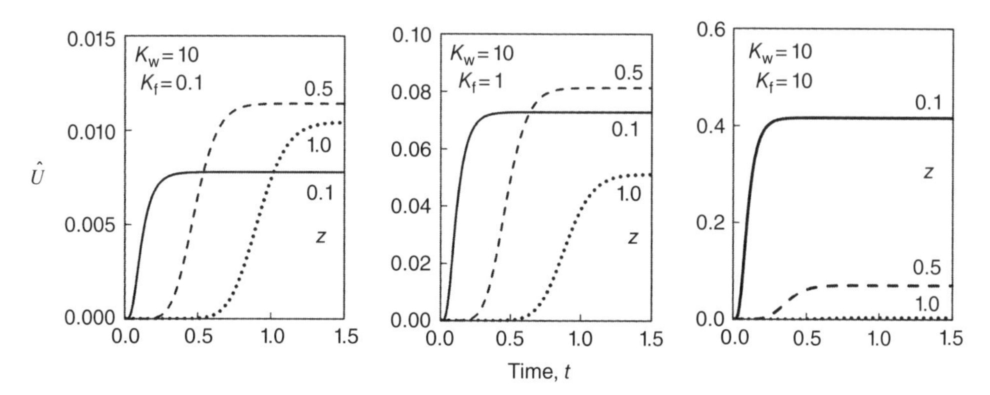

Figure 21.3-4 Accumulated uptake per unit length at three axial positions.

dimensionless times $t > 1.5$, virtually all fragments have exited the intestines at $z = 1$ and no drug remains in the chyme. However, a small concentration of drug is still contained in the exiting fragments.

The variable $\hat{U}$ is a measure of the accumulated uptake at a position z relative to the amount of drug that was introduced at $z = 0$. Figure 21.3-4 shows the dynamic distributions of $\hat{U}$ at a fixed value of $K_w = 10$ when K_f is equal to either 0.1, 1.0 or 10.0. At any $z > 0$, $\hat{U}$ is equal to zero before any fragments arrive, and progressively rises to a plateau when there is sufficient time for all fragments and drug-bearing chyme to pass by. The transition in $\hat{U}$ between zero and its plateau has a duration that depends on the extent of fragment dispersion and therefore increases at successive z positions.

Figure 21.3-4 also shows that the spatial distribution pattern of uptake depends on K_f. This is an important consideration for drugs that are intended to treat intestinal diseases by acting in a specific region. When $K_f = 0.1$, local uptake $\hat{U}$ is somewhat greater at the center of the intestine ($z = 0.5$) than at either the input end ($z = 0.1$) or the output end ($z = 1$). At the larger release coefficient of $K_f = 1$, uptake at the output end falls below $\hat{U}$ at both the center and input ends. At the largest $K_f = 10$, drug release from the fragments is so rapid that most of the uptake occurs near the input end. Therefore, all else being equal, an increase in K_f shifts drug uptake toward the input end of the intestine. From a design perspective, this suggests that drug targeting is strongly influenced by the fragment surface-to-volume ratio, which has a direct effect on K_f.

21.4 Drug Distribution in Ablated Tissues

A solid malignant tumor can be destroyed by inserting a probe into the tumor that emits energy (e.g., radio frequency or laser light) to increase the tumor temperature. After thermal ablation, some malignant cells can remain viable at the tumor margins. These cells can be killed by delivering a chemotherapy drug directly into the tumor. To be effective, the drug must reach the tumor margins at a sufficiently high drug concentration over a sustained period. One delivery device for achieving this is a drug-containing polymer millirod that is implanted in the center of the thermally ablated tissue region (Qian *et al.*, 2002). Drug diffuses outward from the millirod, first through the ablated tissue and then through non-ablated,

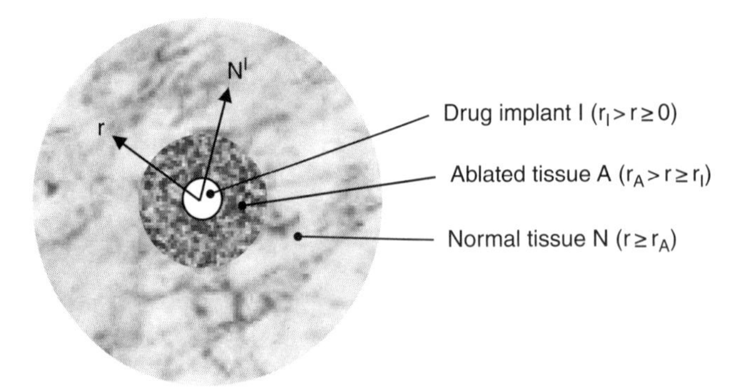

Figure 21.4-1 Cross-sectional view of drug delivery through tissue.

mostly normal, tissue (Fig. 21.4-1). The design of this device requires a mathematical model that describes drug transport and cellular reaction processes in the ablated and normal tissue regions.

Model Formulation

To develop a model that is tractable, we make several key simplifications. In both the ablated and the normal tissue, drug is assumed to readily pass between extracellular and intracellular fluids and to bind with cellular and noncellular fragments. Thus, we treat the ablated tissue (A) and the non-ablated normal tissue (N) as homogeneous, incompressible phases in which drug dissolution, binding, intracellular metabolism, and elimination by perfusion are spatially continuous processes. The drug is present in dissolved and bound forms in both tissues. While the binding sites are fixed in space, the dissolved drug is free to diffuse.

Drug Distribution Dynamics

Because the millirod implant (I) has a length-to-diameter ratio that is much greater than one, we expect the spatial distribution of drug concentration to be cylindrically symmetric about a centerline at $r = 0$. We assume that osmotic and hydrodynamic pressure gradients in all tissue are absent so there is no convective transport.

Within the ablated region ($r_A > r > r_I$), viable cells and blood circulation do not exist so that drug concentrations change only by diffusion and binding. The concentrations of dissolved drug, $C_D^A(r,t)$, and drug bound to polymer X, $C_{DX}^A(r,t)$, are governed by solute molar balances. For an axisymmetric spatial distribution, we obtain the concentration dynamics in cylindrical coordinates (Table 15.2-1) as

$$\frac{\partial C_D^A}{\partial t} = \frac{\mathcal{D}_D^A}{r}\frac{\partial}{\partial r}\left(r\frac{\partial C_D^A}{\partial r}\right) + R_D^A \tag{21.4-1}$$

$$\frac{\partial C_{DX}^A}{\partial t} = R_{DX}^A \tag{21.4-2}$$

Here, $\mathcal{D}_D^A$ is a constant diffusion coefficient of dissolved drug, and R_i^A is the formation rates of dissolved drug ($i = D$) or bound drug ($i = DX$) due to the drug–polymer binding reaction.

When drug diffuses into the surrounding non-ablated tissue ($r > r_A$), it not only binds with cells but also is eliminated by perfusion and intracellular metabolism. The governing equation for the dissolved drug concentration $C_D^N(r,t)$ is

$$\frac{\partial C_D^N}{\partial t} = \frac{\mathcal{D}_D^N}{r}\frac{\partial}{\partial r}\left(r\frac{\partial C_D^N}{\partial r}\right) + R_D^N - \gamma C_D^N \tag{21.4-3}$$

where R_D^N is the drug appearance rate by drug–polymer dissociation and $-\gamma C_D^N$ is the combined elimination rate expressed as a first-order process with a rate coefficient γ.

The governing equation for the bound drug concentration $C_{DX}^N(r,t)$ is

$$\frac{\partial C_{DX}^N}{\partial t} = R_{DX}^N \tag{21.4-4}$$

where R_{DX}^N is the appearance rate of bound drug. Typically, only the total concentration of dissolved and bound drug can be measured in experiments.

$$C_{DT}^j(r,t) = C_D^j(r,t) + C_{DX}^j(r,t) \quad (j = A, N) \tag{21.4-5}$$

Since the formation rate of bound drug is equivalent to the depletion rate of dissolved drug due to binding, $R_{DX}^j = -R_D^j$ $(j = A, N)$, we sum the dissolved and bound concentration equations to remove the reaction terms.

$$\frac{\partial C_{DT}^A}{\partial t} = \frac{\mathcal{D}_D^A}{r}\frac{\partial}{\partial r}\left(r\frac{\partial C_D^A}{\partial r}\right) \tag{21.4-6}$$

$$\frac{\partial C_{DT}^N}{\partial t} = \frac{\mathcal{D}_D^N}{r}\frac{\partial}{\partial r}\left(r\frac{\partial C_D^N}{\partial r}\right) - \gamma C_D^N \tag{21.4-7}$$

To uniquely relate the dissolved and total concentrations, we follow the same approach as in the intestinal drug delivery illustration. That is, we assume that drug binding is a first-order monovalent process that is everywhere in a quasi-equilibrium state so that

$$C_{DX}^j = \kappa^j C_D^j \quad (j = A, N) \tag{21.4-8a,b}$$

Combining this relation with Equation 21.4-5, we obtain the total concentration in terms of the dissolved concentration:

$$C_{DT}^j = \left(1 + \kappa^j\right) C_D^j \quad (j = A, N) \tag{21.4-9}$$

Consolidated Equations

Substituting Equation 21.4-9 into Equations 21.4-6 and 21.4-7, we find

$$\frac{\partial C_{DT}^A}{\partial t} = \frac{\mathcal{D}'^A_D}{r}\frac{\partial}{\partial r}\left(r\frac{\partial C_{DT}^A}{\partial r}\right) \tag{21.4-10}$$

$$\frac{\partial C_{DT}^N}{\partial t} = \frac{\mathcal{D}'^N_D}{r}\frac{\partial}{\partial r}\left(r\frac{\partial C_{DT}^N}{\partial r}\right) - \gamma' C_{DT}^N \tag{21.4-11}$$

Here, $\mathcal{D}'^{j}_{D} = \mathcal{D}^{j}_{D}/(1+\kappa^{j})$ are apparent diffusivities in ablated (j = A) and non-ablated (j = N) tissues, and $\gamma' = \gamma/(1+\kappa^{N})$ is an apparent elimination rate coefficient in the non-ablated region. To solve these equations, we must specify initial and boundary conditions. Initially, no drug exists in the tissue regions.

$$t = 0: \quad C^{A}_{DT} = C^{N}_{DT} = 0 \tag{21.4-12}$$

At the surface of the implant, dissolved drug diffuses through the ablated tissue at a flux that is equal to the flux N^{I} at which drug is released.

$$r = r_{I}: \quad N^{I}(t) = -\mathcal{D}^{A}_{D}\frac{\partial C^{A}_{D}}{\partial r} \Rightarrow N^{I}(t) = -\mathcal{D}'^{A}_{D}\frac{\partial C^{A}_{DT}}{\partial r} \tag{21.4-13a,b}$$

At the boundary between the ablated and normal tissue regions, there is phase equilibrium and flux continuity of dissolved drug:

$$r = r_{A}: \begin{cases} C^{A}_{D} = \lambda^{A,N}_{D} C^{N}_{D} & \Rightarrow \; C^{A}_{DT} = \lambda^{A,N}_{D}\dfrac{(1+\kappa^{A})}{(1+\kappa^{N})} C^{N}_{DT} \\ \mathcal{D}^{A}_{D}\dfrac{\partial C^{A}_{D}}{\partial r} = \mathcal{D}^{N}_{D}\dfrac{\partial C^{N}_{D}}{\partial r} & \Rightarrow \; \mathcal{D}'^{A}_{D}\dfrac{\partial C^{A}_{DT}}{\partial r} = \mathcal{D}'^{N}_{D}\dfrac{\partial C^{N}_{DT}}{\partial r} \end{cases} \tag{21.4-14}$$

where $\lambda^{A,N}_{D}$ is the equilibrium partition coefficient of dissolved drug between tissues A and B. Sufficiently far from the polymer millirod, the total drug concentration in normal tissue will be negligible because of elimination by the blood flow and intracellular metabolism.

$$r \to \infty: \quad C^{N}_{DT} = 0 \tag{21.4-15}$$

Analysis

Pseudo-Steady Solution

We expect the characteristic time scale for sustained release to be much greater than that of diffusion (i.e., drug diffusion in the ablated and non-ablated regions is substantially faster than concentration or flux changes at the millirod boundary). Consequently, for a sufficiently long period, N^{I} is approximately constant and the pseudo-steady form of the model equations apply.

$$r\frac{d^2 C^{A}_{DT}}{dr^2} + \frac{dC^{A}_{DT}}{dr} = 0 \quad (r_{A} > r > r_{I}) \tag{21.4-16}$$

$$\frac{d^2 C^{N}_{DT}}{dr^2} + \frac{1}{r}\frac{dC^{N}_{DT}}{dr} - \frac{\gamma'}{\mathcal{D}'^{N}_{D}} C^{N}_{DT} = 0 \quad (r > r_{A}) \tag{21.4-17}$$

The general solutions for these equations are

$$C^{A}_{DT} = b_1 \ln r + b_2 \tag{21.4-18}$$

$$C^{N}_{DT} = b_3 K_0(\beta r) + b_4 I_0(\beta r) \tag{21.4-19}$$

where b_j (j = 1,2,3,4) are integration constants, $K_0(\beta r)$ and $I_{-0}(\beta r)$ are modified Bessel functions of zero order, and $\beta = \sqrt{\gamma'/\mathcal{D}'^N_D}$. In the ablated region, we apply the flux condition at $r = r_I$ and the concentration condition at $r = r_A$ to obtain

$$C^A_{DT} = \frac{r_I}{\mathcal{D}'^A_D} N^I \ln\left(\frac{r_A}{r}\right) + \lambda_D^{A,N}\left(\frac{1+\kappa^A}{1+\kappa^N}\right) C^N_{DT}(r_A) \tag{21.4-20}$$

In the non-ablated region, we apply the concentration condition at $r \to \infty$. Because $\lim_{r\to\infty}[I(\beta r)] \to \infty$, we require $b_4 = 0$ so that

$$C^N_{DT}(r) = b_3 K_0(\beta r) \Rightarrow C^N_{DT}(r_A) = b_3 K_0(\beta r_A) \tag{21.4-21a,b}$$

Applying the remaining flux continuity condition at $r = r_A$ to evaluate b_3, we find

$$b_3 = \frac{r_I}{r_A} \frac{N^I}{\mathcal{D}'^N_D \beta K_1(\beta r_A)} \tag{21.4-22}$$

where $K_1(\beta r_A)$ is the modified Bessel function of order one. The final equation for the overall drug concentration distribution C(r) can be written in piecewise form:

$$C_{DT}(r) = \begin{cases} r_I N^I \left[\dfrac{1}{\mathcal{D}'^A_D}\ln\left(\dfrac{r_A}{r}\right) + \left(\dfrac{1+\kappa^A}{1+\kappa^N}\right)\dfrac{\lambda_d^{A,N} K_0(\beta r_A)}{\beta r_A \mathcal{D}^N K_1(\beta r_A)}\right] & \text{when } r_A > r > r_I \\ r_I N^I \left(\dfrac{K_0(\beta r)}{\beta r_A \mathcal{D}'^N_D K_1(\beta r_A)}\right) & \text{when } r > r_A \end{cases} \tag{21.4-23}$$

where $C_{DT}(r_A > r > r_I) \equiv C^A_{DT}(r)$ and $C_{DT}(r > r_A) \equiv C^N_{DT}(r)$.

Dimensionless Forms

If we define the dimensionless variables

$$C = \frac{\mathcal{D}'^A_D}{r_I N^I} C_{DT}, \quad r = \frac{r}{r_A} \tag{21.4-24a,b}$$

then the dimensionless form of the pseudo-steady concentration distribution is

$$C = \begin{cases} \ln\left(\dfrac{1}{r}\right) + \dfrac{\alpha_1 K_0(\sqrt{Da})}{\sqrt{Da}K_1(\sqrt{Da})} & 1 > r > \alpha_2 \\ \dfrac{\alpha_1 K_0(r\sqrt{Da})}{\sqrt{Da}K_1(\sqrt{Da})} & r > 1 \end{cases} \tag{21.4-25}$$

The dimensionless parameters are

$$\sqrt{Da} \equiv \beta r_A = r_A\sqrt{\frac{\gamma'}{\mathcal{D}'^N_D}}, \quad \alpha_1 \equiv \frac{\mathcal{D}'^A_D}{\mathcal{D}'^N_D}, \quad \alpha_2 \equiv \frac{r_I}{r_A}, \quad \alpha_3 \equiv \lambda_D^{A,N}\left(\frac{1+\kappa^A}{1+\kappa^N}\right) \tag{21.4-26a-d}$$

Notice that the dimensionless millirod radius α_2 does not explicitly appear in Equation 21.4-25. It only sets the position of the millirod-ablated tissue boundary.

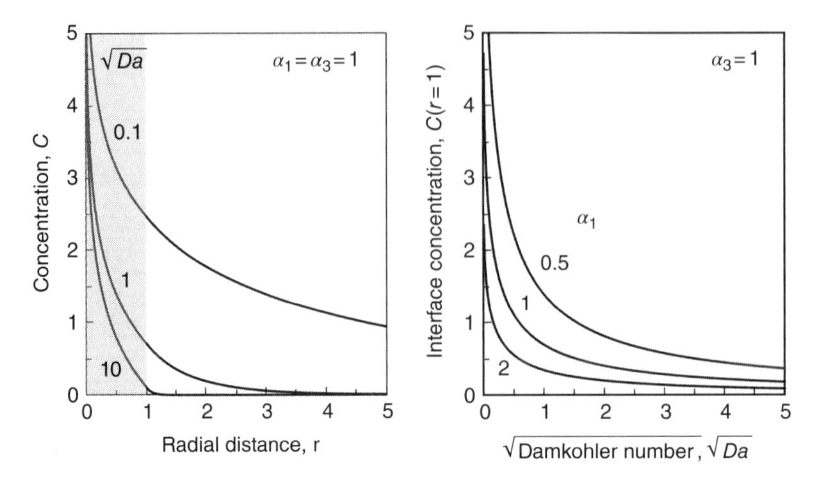

Figure 21.4-2 Effects of model parameters on the pseudo-state drug concentration.

Simulation

The left graph in Figure 21.4-2 shows how the dimensionless drug elimination rate $\sqrt{Da}$ affects the concentration distribution $C(r)$ when the apparent diffusion coefficients are equal so that $\alpha_1 = 1$. We also assume equal drug solubilities and binding coefficients of drug in the two tissues so that $\alpha_3 = 1$. For convenience, we arbitrarily set α_2 very close to zero so that ablated tissue occupies essentially the entire shaded region in the graph.

In order to support diffusion, $C(r)$ generally decreases with r. This decrease becomes greater as $\sqrt{Da}$ increases, until at $\sqrt{Da} = 10$, drug elimination from the non-ablated tissue ($r > 1$) is essentially complete. Further increases in $\sqrt{Da}$ do not have a significant effect on $C(r)$.

The right graph focuses on drug concentration at the interface ($r = 1$) between ablated and non-ablated tissue when there are changes in $\sqrt{Da}$ and α_1 while $\alpha_3 = 1$ is constant. Consistent with the left graph, increases in $\sqrt{Da}$ cause a reduction in the interface concentration that approaches zero for large $\sqrt{Da}$. As α_1 increases, the greater apparent diffusion coefficient in ablated tissue causes $C(1)$ to decrease more sharply with increasing $\sqrt{Da}$.

For this adjuvant chemotherapy treatment to be effective, the drug concentration in the region of this boundary between ablated and non-ablated tissue must be above a therapeutic threshold. Controlling the drug release rate N^I from the polymer implant is the most direct way of reaching a desired concentration (Eq. 21.4-24a,b). To achieve a more realistic design of a delivery device, one must obtain a more general, dynamic solution to the model equations that incorporates time-varying drug release rates.

21.5 Intracranial Drug Delivery and Distribution

Brain tissue is composed of a vasculature with capillaries and other blood vessels interspersed with extravascular cells and their interstitial fluid. Due to the blood–brain barrier, efficient and controlled delivery of chemotherapy drugs into brain requires special techniques and mechanistic models to describe the underlying processes (e.g., Kalyanasundaram *et al.*, 1997). Here, we consider a drug-laden material that, after surgical implantation, releases

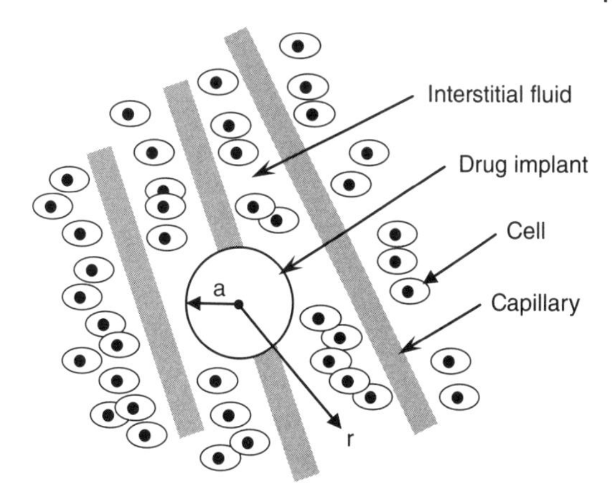

Figure 21.5-1 Spherical drug implant in brain tissue.

an osmotically active (but otherwise inert) species A in addition to the drug D into surrounding tissue (Fig. 21.5-1).

We model the brain as a porous medium consisting of three phases: an extravascular cellular matrix of constant volume (cell); the interstitial fluid that bathes the cells (int); and the blood capillaries (cap). The spherical implant (I) of constant radius "a" is a well-mixed homogeneous phase containing species A and D in dissolved form. As drug D is transported from the implant through the interstitial fluid, it enters the surrounding tumor cells where it metabolizes so rapidly that its intracellular concentration is negligible. Drug D does not enter capillaries.

Species A penetrates neither cells nor capillaries. After release from the implant, species A remains in the interstitial fluid where it elevates osmotic pressure, thereby inducing flow into the interstitial space from capillary blood. The resulting convective transport enhances drug delivery from the implant to distant tumor cells compared to diffusion alone.

We assume that fluid movement between the capillary and interstitial fluid is a steady-state process. That is, the osmotic water loss from the capillaries is balanced by a net inflow of blood, and fluid gain by the interstitial space is balanced by drainage into peripheral lymphatics. Thus, in addition the cellular matrix, the capillaries and intracellular spaces each occupy a constant volume.

Model Formulation

Macroscopic Transport Equations

To describe this multiphase system, we apply volume-average mass balances and flux equations (Appendix B3). In the interstitial fluid, the volume-average solution velocity $\bar{\mathbf{u}}^{\text{int}}$ changes according to

$$\nabla \cdot \bar{\mathbf{u}}^{\text{int}} = \frac{\phi_{\text{cap}} u_{\text{wall}}^{\text{cap}}}{\varepsilon_{\text{int}}} \tag{B3-14a}$$

where ϕ_{cap} is the surface area of the capillaries per tissue volume, $u_{\text{wall}}^{\text{cap}}$ is the local velocity across the interface from capillaries into interstitial fluid, and ε_{int} is the volume fraction of the interstitial space.

In general, the average interstitial concentration $\bar{C}_i^{int}$ of a solute i = D,A can change with convection, diffusion, reaction, and transport between phases:

$$\frac{\partial \bar{C}_i^{int}}{\partial t} + \beta_i^{int} \nabla \cdot \left(\bar{\mathbf{u}}^{int} \bar{C}_i^{int} \right) = \bar{\mathcal{D}}_i^{int} \nabla^2 \bar{C}_i^{int} + \bar{R}_i^{int} + \frac{1}{\varepsilon_{int}} \left(\phi_{cell} N_{i,wall} + \phi_{cap} N_{i,wall} \right) \quad (i = D, A) \tag{B3-30}$$

where β_i^{int} is a velocity factor and $\bar{\mathcal{D}}_i^{int}$ is a hindered diffusion coefficient that correct for wall interactions; ϕ_{cell} is the cell surface area per tissue volume; and $N_{i,wall}^{cell}$ and $N_{i,wall}^{cap}$ are molar solute fluxes from cell and capillary phases into the interstitial space.

To specify this concentration equation separately for species A and D, we note that $N_{D,wall}^{cap} = 0$ since D is not transported into capillaries; $N_{A,wall}^{cell} = N_{A,wall}^{cap} = 0$ since species A is not transported into cells or capillaries; $\bar{R}_D^{int} = \bar{R}_A^{int} = 0$ since neither A nor D react in the interstitial fluid; and the solutes are small compared to the interstitial space so that $\beta_s^{int} \approx 1$. Consequently,

$$\frac{\partial \bar{C}_D^{int}}{\partial t} + \nabla \bullet \left(\bar{\mathbf{u}}^{int} \bar{C}_D^{int} \right) = \bar{\mathcal{D}}_D^{int} \nabla^2 \bar{C}_D^{int} + \frac{\phi_{cell}}{\varepsilon_{int}} N_{D,wall}^{cell} \tag{21.5-1}$$

$$\frac{\partial \bar{C}_A^{int}}{\partial t} + \nabla \bullet \left(\bar{\mathbf{u}}^{int} \bar{C}_A^{int} \right) = \bar{\mathcal{D}}_A^{int} \nabla^2 \bar{C}_A^{int} \tag{21.5-2}$$

Assuming spherically symmetric transport surrounding the drug implant, spatial changes can be described in the radial direction only. The magnitude $\bar{u}^{int}$ of the radially directed $\bar{\mathbf{u}}^{int}$ velocity vector in Equation B3-14a then varies according to

$$\frac{1}{r^2} \frac{\partial (r^2 \bar{u}^{int})}{\partial r} = \frac{\phi_{cap}}{\varepsilon_{int}} u_{wall}^{cap} \tag{21.5-3}$$

and the interstitial concentrations of D and A in Equations 21.5-1 and 21.5-2 vary as

$$\frac{\partial \bar{C}_D^{int}}{\partial t} + \frac{1}{r^2} \frac{\partial}{\partial r} \left(r^2 \bar{u}^{int} \bar{C}_D^{int} \right) = \frac{\bar{\mathcal{D}}_D}{r^2} \frac{\partial}{\partial r} \left(r^2 \frac{\partial \bar{C}_D^{int}}{\partial r} \right) + \frac{\phi_{cell}}{\varepsilon_{int}} N_{D,wall}^{cell} \tag{21.5-4}$$

$$\frac{\partial \bar{C}_A^{int}}{\partial t} + \frac{1}{r^2} \frac{\partial}{\partial r} \left(r^2 \bar{u}^{int} \bar{C}_A^{int} \right) = \frac{\bar{\mathcal{D}}_A}{r^2} \frac{\partial}{\partial r} \left(r^2 \frac{\partial \bar{C}_A^{int}}{\partial r} \right) \tag{21.5-5}$$

A molar balance on species i = D,A within the well-mixed drug implant leads to the change in concentration $C_i^I(t)$ with time.

$$V^I \frac{dC_i^I}{dt} = -S^I N_i^I(a,t) \Rightarrow \frac{dC_i^I}{dt} = -\frac{3}{a} N_i^I(a,t) \quad (i = D, A) \tag{21.5-6a, b}$$

Here, $N_i^I(a,t)$ is the molar flux of i into the interstitial fluid at the implant surface (r = a) and $(S^I/V^I) = 3/a$ is the surface-to-volume ratio of the spherical implant.

Rate Equations

To complete the governing equations, we require rate equations for transport between phases: $u_{wall}^{cap}(r,t)$, $N_D^I(a,t)$, $N_A^I(a,t)$, and $N_{D,wall}^{cell}(r,t)$. At the implant surface, solute diffusion into the interstitial fluid can be limited by a membrane or some other transport resistance around the implant. Generally, the resulting transmembrane flux depends on solute concentrations in the implant and interstitial fluid:

$$N_i^I(a,t) = P_i\left[C_i^I(t) - \lambda_i \bar{C}_i^{int}(a,t)\right] \quad (i = D, A) \tag{21.5-7}$$

where P_i is constant permeability, and λ_i is an equilibrium partition coefficient of solute i between the implant and interstitial phases. Also, the molar flux leaving the implant must match the diffusion flux entering the interstitial space.

$$N_i^I(a,t) = -\bar{\mathcal{D}}_i\left[\frac{\partial \bar{C}_i^{int}}{\partial r}\right]_{r=a} \quad (i = D, A) \tag{21.5-8}$$

Solution flow across the capillary wall occurs because of imbalances in hydrodynamic pressure as well as osmotic pressures between blood and interstitial fluid. Neglecting the effect of concentration polarization on osmotic pressure, the transcapillary fluid velocity is described by the Starling equation (Eq. 6.5-31). In this application, species A creates the dominant osmotic pressure in the interstitial space, which we relate to $\bar{C}_A^{int}$ by the Van't Hoff equation (Eq. 4.3-16). The Starling equation can then be written as

$$u_{wall}^{cap} = L_p^{cap}\left[\left(P^{cap} - P^{int}\right) + \sigma_A^{cap}\mathcal{R}T\bar{C}_A^{int}\right] \tag{21.5-9}$$

Here, L_p^{cap} and σ_A^{cap} are the hydraulic conductivity and reflection coefficient of A imposed by the capillary wall; and P^{cap} and P^{int} are hydrodynamic pressures in the capillary and interstitial spaces. Since species A does not cross the capillary wall, its reflection coefficient is 1.

We formulate the radial pressure gradient in the interstitial space with a Darcy-type flow model. From Equation 7.4-36, we see that pressure drop per unit distance in the pores of a heterogeneous medium is proportional to fluid viscosity μ and intrinsic velocity $\bar{u}^{int}$. Therefore,

$$\frac{\partial P^{int}}{\partial r} = -\left(\frac{\mu}{K_m}\right)\bar{u}^{int} \tag{21.5-10}$$

where K_m is a lumped parameter that depends only on geometry.

The molar transport rate of D between the interstitial and intracellular fluids is dependent only on the interstitial drug concentration because the intracellular drug concentration is negligible.

$$N_{D,wall}^{cell} = -K_D\bar{C}_D^{int} \tag{21.5-11}$$

where K_D is a mass-transfer coefficient. The negative sign is necessary because $N_{D,wall}^{cell}$ has been defined as drug efflux from the cells.

Consolidated Equations

Substitution of the transcapillary velocity Equation 21.5-9 into Equation 21.5-3 for interstitial flow yields

$$\frac{1}{r^2}\frac{\partial(r^2\bar{u}^{int})}{\partial r} = \frac{\phi_{cap}}{\varepsilon_{int}}L_p^{cap}\left[\left(P^{cap} - P^{int}\right) + \mathcal{R}T\bar{C}_A^{int}\right] \tag{21.5-12}$$

To eliminate pressure from the model equations, we differentiate this equation with respect to r and substitute the pressure gradient of Equation 21.5-10 and obtain

$$\frac{\partial}{\partial r}\left[\frac{1}{r^2}\frac{\partial(r^2\bar{u}^{int})}{\partial r}\right]-\left(\frac{\mu\phi^{cap}L_p^{cap}}{\varepsilon_{int}K_m}\right)\bar{u}^{int}=\frac{\mathcal{R}T\phi_{cap}L_p^{cap}}{\varepsilon_{int}}\frac{\partial\bar{C}_A^{int}}{\partial r} \tag{21.5-13}$$

To obtain the final transport equation for interstitial drug, we combine Equation 21.5-11 with Equation 21.5-4:

$$\frac{\partial\bar{C}_D^{int}}{\partial t}+\frac{1}{r^2}\frac{\partial}{\partial r}\left(r^2\bar{u}^{int}\bar{C}_D^{int}\right)-\frac{\bar{\mathcal{D}}_D}{r^2}\frac{\partial}{\partial r}\left(r^2\frac{\partial\bar{C}_D^{int}}{\partial r}\right)=-\frac{\phi^{cell}}{\varepsilon_{int}}K_D\bar{C}_D^{int} \tag{21.5-14}$$

To describe solute concentrations changes in the implant, we combine Equations 21.5-6a,b and 21.5-8:

$$\frac{dC_i^I}{dt}=-\frac{3\bar{\mathcal{D}}_i}{a}\left[\frac{\partial\bar{C}_i^{int}}{\partial r}\right]_{r=a}\quad(i=D,A) \tag{21.5-15}$$

Equations 21.5-13–21.5-15 for $\bar{u}^{int}$, $\bar{C}_D^{int}$, C_D^I, and C_A^I must be solved simultaneously with Equation 21.5-5 for $\bar{C}_A^{int}$.

Initial and Boundary Conditions

Initial conditions for concentrations in the implant and interstitial fluid are

$$t=0:\quad \bar{C}_i^{int}=0,\quad C_i^I=C_{io}^I\quad(i=D,A) \tag{21.5-16}$$

Far from the spherical implant, the fluid flow as well as the concentrations of D and A are negligible.

$$r\to\infty:\quad \bar{u}^{int}=0,\quad \bar{C}_i^{int}=0\quad(i=D,A) \tag{21.5-17}$$

Across the implant surface, the velocity is zero.

$$r=a:\quad \bar{u}^{int}=0 \tag{21.5-18}$$

Also across the implant surface, we will consider the special case when the resistance to mass transport is negligible such that $P_i\to\infty$. If the flux given by Equation 21.5-7 is to remain finite, then

$$r=a:\quad \bar{C}_i^{int}(a,t)\to C_i^I(t)/\lambda_i\quad(i=D,A) \tag{21.5-19}$$

Analysis

Dimensionless Forms

By defining the dimension variables

$$r=\frac{r}{a},\quad t=\frac{\bar{\mathcal{D}}_A}{a^2}t,\quad u=\frac{\bar{u}^{int}}{\mathcal{R}TL_p^{cap}C_{Ao}^I}$$
$$C_i^I=\frac{C_i^I}{C_{io}^I},\quad C_i=\frac{\bar{C}_i^{int}}{C_{io}^I}\quad(i=D,A) \tag{21.5-20a-e}$$

the governing equations are transformed to

$$\frac{\partial}{\partial r}\left[\frac{1}{r^2}\frac{\partial(r^2u)}{\partial r}\right]-\beta_1 u=\beta_2\frac{\partial C_A}{\partial r} \tag{21.5-21}$$

$$\frac{\partial C_D}{\partial t}+\frac{\beta_3}{r^2}\frac{\partial}{\partial r}(r^2uC_D)-\frac{D}{r^2}\frac{\partial}{\partial r}\left(r^2\frac{\partial C_D}{\partial r}\right)+(ShD)C_D=0 \tag{21.5-22}$$

$$\frac{\partial C_A}{\partial t}+\frac{\beta_3}{r^2}\frac{\partial}{\partial r}(r^2uC_A)-\frac{1}{r^2}\frac{\partial}{\partial r}\left(r^2\frac{\partial C_A}{\partial r}\right)=0 \tag{21.5-23}$$

$$\frac{dC_i^I}{dt}=-3\left[\frac{\partial C_i}{\partial r}\right]_{r=1}\quad (i=D,A) \tag{21.5-24}$$

The dimensionless initial conditions for interstitial solute concentrations $(i=D,A)$ are

$$t=0:\ \ C_i=0,\ \ C_i^I=1\ \ (i=D,A) \tag{21.5-25}$$

Far from the implant, the dimensionless conditions are

$$r\rightarrow\infty:\ \ u=0,\ \ C_i=0\ \ (i=D,A) \tag{21.5-26}$$

At the implant surface, the dimensionless conditions applicable for $t>0$ are

$$r=1:\ \ u=0,\ \ C_i=C_i^I/\lambda_i\ \ (i=D,A) \tag{21.5-27}$$

In this model, the dimensionless parameters are

$$Sh=\frac{\phi^{cell}K_Da^2}{\varepsilon_{int}\bar{\mathcal{D}}_D},\quad D=\frac{\bar{\mathcal{D}}_D}{\bar{\mathcal{D}}_A}$$
$$\beta_1=\frac{\mu\phi^{cap}L_p^{cap}a^2}{\varepsilon_{int}K_m},\quad \beta_2=\frac{\phi^{cap}a}{\varepsilon_{int}},\quad \beta_3=\frac{RTL_p^{cap}C_{Ao}^Ia}{\bar{\mathcal{D}}_A} \tag{21.5-28a-e}$$

Simulation

A numerical solution of the dimensionless model was obtained for specific parameter values $(Sh=10, D=1, \beta_1=100, \beta_2=100, \beta_3=1, \lambda_A=\lambda_D=1)$. Figure 21.5-2 shows radial distributions of interstitial fluid velocity and interstitial concentrations of D and A originating from the implant surface $(r=1)$ at several times. The left graph indicates that the interstitial fluid velocity distribution steeply rises to a maximum and more gradually falls to zero. At longer times,

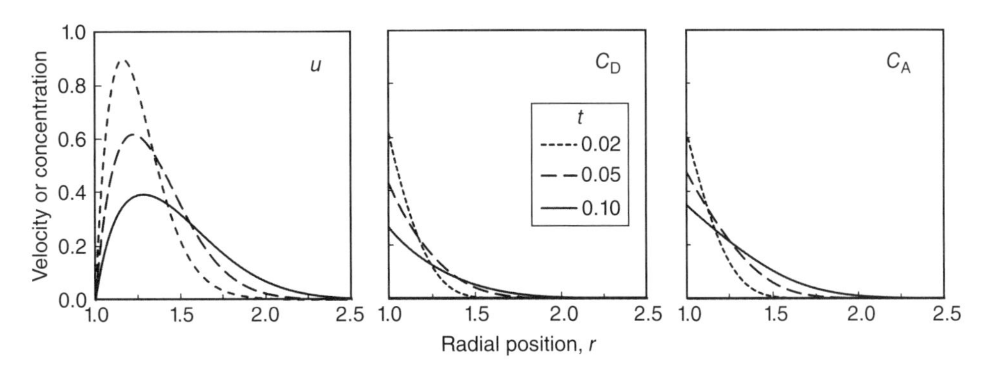

Figure 21.5-2 Spatial variations of velocity, drug, and osmotically active species in brain interstitial fluid.

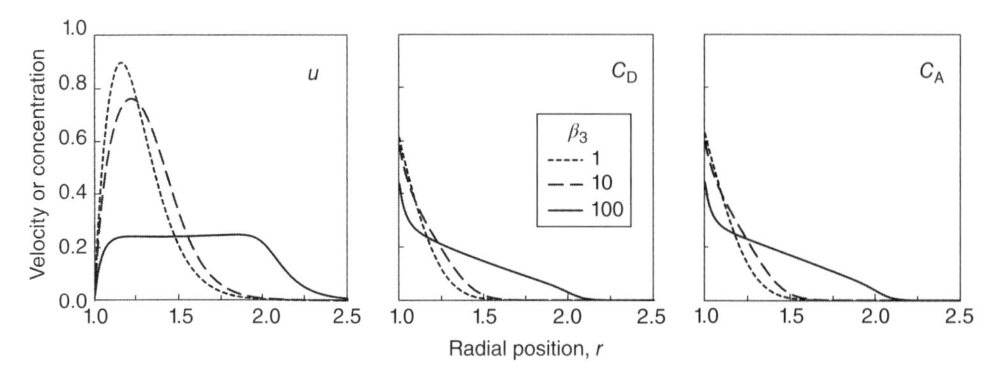

Figure 21.5-3 Effect of osmotically generated interstitial flow on velocity and drug distributions at $t = 0.02$.

the velocity peak diminishes and broadens. In the middle and right graphs, C_D and C_A decrease with radial position at any time as the implant becomes depleted. With progressing time, D and A penetrate further into the brain tissue.

Figure 21.5-3 shows the effect of increased concentrations of osmotically active species in the implant (indicated by β_3) on interstitial velocity and concentration distributions at $t = 0.02$. In the left graph, an increase of β_3 leads to greater convection of interstitial fluid further from the implant into the tissue. Consequently, as evident from the center and right graphs, D and A concentrations reach higher levels away from the implant. Thus, when more osmotically active substance A is loaded in the implant, drug delivery and penetration in brain tissue can be significantly increased.

21.6 Whole-Body Methotrexate Distribution

Modeling the distribution of drugs and their active metabolites in the whole body can play an important role in drug testing and treatment protocols. In addition, transport analysis of environmental pollutants in the whole body can improve quantitative evaluation of their toxicity to provide a better foundation for regulatory policies. Modeling and simulation can address several important questions about the distribution of a drug or toxicant. How is the distribution affected by the portal of entry (e.g., oral, respiratory, venous infusion, skin contact)? Which organ or tissue targets receive disproportionately large doses? How can dose distributions from tests on laboratory animals be extrapolated to humans?

As discussed in Chapter 18, physiologically based pharmacokinetic (PBPK) models consist of compartments corresponding to specific organs and tissues. Typically, these models use well-defined anatomical, physiological, and physical–chemical parameters to simulate measurable and nonmeasurable responses. Model parameters can include organ and tissue volumes, perfusion and ventilation rates, chemical binding coefficients, metabolic rate constants, and urinary and fecal excretion rates. The PBPK model presented here, adapted from Bischoff *et al.* (1971), describes the physiological transport and distribution of methotrexate (MTX), a drug used in treatment of cancer, autoimmune diseases, and ectopic pregnancy. The MTX content in blood and organs is governed mainly by nonspecific binding to proteins and has a very slow loss by metabolic degradation.

In the model (Fig. 21.6-1), MTX is injected into the venous return entering a compartment representing blood and tissue of the heart and lungs combined (HL). Systemic circulation occurs through liver (L), kidney (K), gut wall (G), gut lumen (GL), muscle (M), and fat (F) compartments. In addition to transport by blood flow, the K compartment excretes urine and the GL compartment discharges feces to the environment. A dramatic increase in MTX concentration that has been observed in the gut lumen indicates that biliary secretion of MTX by the liver into chyme should be accounted for in the model. As chyme flows through the GL compartment, MTX is reabsorbed into the G compartment. Compartment M is composed of well-perfused tissues, primarily muscle, that have a significant capacity for MTX. In contrast, compartment F contains adipose tissue that is poorly perfused and has a low MTX capacity.

The HL, L, K, M, and F compartments have constant volumes. They are each subdivided into two well-mixed subcompartments, one representing blood and the other extravascular tissue. Because of the complex mesenteric blood supply to the intestinal wall, we also represent the G compartment by a pair of well-mixed subcompartments of blood and extravascular tissue. In contrast, the gut lumen is a long thin-walled tube that transports MTX by a directed flow of chyme. Therefore, we model the GL compartment as a one-dimensional, spatially distributed conduit.

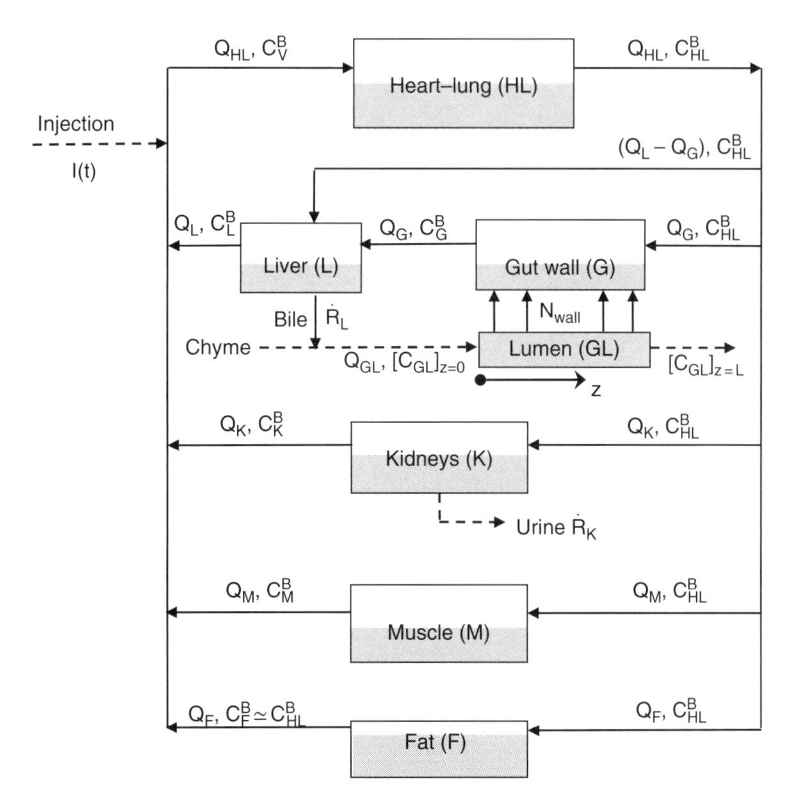

Figure 21.6-1 PBPK model for methotrexate distribution dynamics.

Model Formulation

Transport and Binding Processes

We assume that MTX transport between subcompartments and MTX binding within these subcompartments are very rapid processes compared to transport by blood flow through the compartment. In such a flow-limited situation, physically dissolved MTX will be in equilibrium between in the subcompartments. In addition, bound MTX will be in equilibrium with dissolved MTX within each subcompartment.

Volumetric Flow Relationships

The well-mixed, flow-limited compartments have constant volumes and the same constant mass density. Solution mass balances around any of these compartments require that the sum of volumetric inflows be equal to the sum of the outflows. Thus, the blood inflows to compartments M and F are identical to their blood outflows. Since the flows associated with MTX injection, urine excretion, and reabsorbed intestinal water are negligibly small relative to blood flow, compartments HL, K, and G also have blood inflows that are essentially equal to their blood outflows. As is evident from Figure 21.6-1, the sum of venous outflows from the systemic compartments equals the HL blood flow:

$$Q_{HL} = Q_M + Q_F + Q_L + Q_K \tag{21.6-1}$$

Since bile flow is also negligible compared to blood flow, a mass balance on the L compartment with a blood outflow Q_L and a blood inflow Q_G from the gut indicates that the blood inflow from the hepatic artery is $(Q_L - Q_G)$.

MTX Concentration Balances in the Well-Mixed Compartments

The total concentrations of dissolved plus bound MTX in blood (j = B) and in extravascular tissue (j = T) subcompartments associated with compartment i are denoted as $C_i^j(t)$. Since each blood subcompartment is well-mixed, C_i^B also represents the total MTX concentration within the bloodstream exiting subcompartment i. The volume average of the subcompartment concentrations in a compartment containing a volume fraction ε_i of blood is given by

$$C_i = \varepsilon_i C_i^B + (1 - \varepsilon_i) C_i^T \quad (i = HL, M, F, L, K, G) \tag{21.6-2}$$

The molar rate of MTX transport in the blood return to the HL compartment is the sum of the molar injection rate I(t) and the venous output rates from the systemic compartments:

$$Q_{HL} C_V^B = Q_M C_M^B + Q_F C_F^B + Q_L C_L^B + Q_K C_K^B + I(t) \tag{21.6-3}$$

To analyze the effect of inputs over different times, we consider n_I moles of MTX that are introduced at a constant rate $n_t/\Delta t$ during an interval Δt. We can represent such a pulse in terms of a unit up-step $H(t - 0)$ at t = 0 and a unit down-step $H(t - \Delta t)$ at Δt (Eq. 18.1-12).

$$I(t) = \frac{n_I}{\Delta t}[H(t-0) - H(t-\Delta t)] \tag{21.6-4}$$

The MTX molar balance on a flow-limited compartment i of volume V_i relates the volume-average MTX concentration C_i within the compartment to the MTX concentrations in exiting blood, C_i^B, and in entering blood, C_j^B ($j \neq i$). A MTX molar balance on the HL compartment can be expressed as

$$
\begin{aligned}
V_{HL}\frac{dC_{HL}}{dt} &= Q_{HL}C_V^B - Q_{HL}C_{HL}^B \\
&= -Q_{HL}C_{HL}^B + Q_M C_M^B + Q_F C_F^B + Q_L C_L^B + Q_K C_K^B + I(t)
\end{aligned} \tag{21.6-5}
$$

For the MTX molar balance on the fat compartment, the accumulation rate is so small that C_F changes negligibly, and consequently,

$$
V_F\frac{dC_F}{dt} = Q_F C_{HL}^B - Q_F C_F^B \simeq 0 \Rightarrow C_{HL}^B \simeq C_F^B \tag{21.6-6}
$$

With this relationship, we can simplify the HL equation.

$$
V_{HL}\frac{dC_{HL}}{dt} = -Q'_{HL}C_{HL}^B + Q_M C_M^B + Q_F C_F^B + Q_L C_L^B + Q_K C_K^B + I(t) \tag{21.6-7}
$$

where $Q'_{HL} = Q_{HL} - Q_F = Q_M + Q_L + Q_K$. For compartments M and L, the MTX molar balances are as follows:

$$
V_M\frac{dC_M}{dt} = Q_M C_{HL}^B - Q_M C_M^B \tag{21.6-8}
$$

$$
V_L\frac{dC_L}{dt} = (Q_L - Q_G)C_{HL}^B + Q_G C_G^B - Q_L C_L^B - \dot{R}_L \tag{21.6-9}
$$

where $\dot{R}_L$ is the molar rate of MTX excretion in liver bile. We model $\dot{R}_L$ as a first-order process with respect to C_L^B and introduce a delay τ, representing the time required for bile to exit the liver through the bile ducts:

$$
\dot{R}_L = k_L C_L^B(t-\tau) \tag{21.6-10}
$$

For compartment K,

$$
V_K\frac{dC_K}{dt} = Q_K C_{HL}^B - Q_K C_K^B - \dot{R}_K \tag{21.6-11}
$$

where the rate of MTX excretion from the kidneys, $\dot{R}_K$, is first order with respect to blood concentration:

$$
\dot{R}_K = k_K C_K^B \tag{21.6-12}
$$

For the gut wall compartment G,

$$
V_G\frac{dC_G}{dt} = Q_G C_{HL}^B - Q_G C_G^B + \dot{N}_{wall} \tag{21.6-13}
$$

where $\dot{N}_{wall}$ is the rate of MTX transport from the gut lumen to gut wall.

Equilibrium Relationships in the Well-Mixed Compartments

Within any blood or tissue subcompartment, the MTX concentration is the sum of its dissolved (d) and bound (b) concentrations.

$$
C_i^j = C_{i,d}^j + C_{i,b}^j \quad (i = HL, M, F, L, K, G;\ j = B, T) \tag{21.6-14}
$$

With flow-limited behavior, phase equilibrium exists between the dissolved MTX in blood and tissue with a distribution coefficient, $\lambda_i^{T,B} \equiv C_{i,d}^T / C_{i,d}^B$. Similarly, we introduce a binding equilibrium coefficient, $\beta_i^j = C_{i,b}^j / C_{i,d}^j$, that relates dissolved and bound MTX within a blood (j = B) or an extravascular (j = T) subcompartment. We can now express the combined MTX concentrations in individual subcompartments in terms of the MTX dissolved in blood.

$$C_i^B = \left(1+\beta_i^B\right)C_{i,d}^B, \quad C_i^T = \left(1+\beta_i^T\right)\lambda_i^{T,B}C_{i,d}^B \tag{21.6-15a,b}$$

With these relationships, the volume-average concentration defined by Equation 21.6-2 can be written in terms of the blood concentration:

$$C_i = \left[\varepsilon_i + \lambda_i^{T,B}(1-\varepsilon_i)\frac{\left(1+\beta_i^T\right)}{\left(1+\beta_i^B\right)}\right]C_i^B \tag{21.6-16}$$

In this analysis, ε_i, $\lambda_i^{T,B}$, β_i^B, and β_i^T are assumed to be constants. This relationship can be expressed in a more compact form by introducing a lumped constant γ_I.

$$C_i = \gamma_i C_i^B; \quad \gamma_i \equiv \varepsilon_i + \lambda_i^{T,B}(1-\varepsilon_i)\frac{\left(1+\beta_i^T\right)}{\left(1+\beta_i^B\right)} \tag{21.6-17a,b}$$

Consolidated Compartment Equations

Employing Equation 21.6-17a,b to eliminate C_i^B from the MTX molar balances on the well-mixed compartments and inserting the rate expressions for I(t), $\dot{R}_L$ and $\dot{R}_K$, we obtain

$$V_{HL}\frac{dC_{HL}}{dt} = -\frac{Q'_{HL}}{\gamma_{HL}}C_{HL} + \frac{Q_M}{\gamma_M}C_M + \frac{Q_L}{\gamma_L}C_L + \frac{Q_K}{\gamma_K}C_K + \frac{n_I}{\Delta t}[H(t-0)-H(t-\Delta t)] \tag{21.6-18}$$

$$V_M\frac{dC_M}{dt} = -\frac{Q_M}{\gamma_M}C_M + \frac{Q_M}{\gamma_{HL}}C_{HL} \tag{21.6-19}$$

$$V_L\frac{dC_L}{dt} = -\frac{Q_L}{\gamma_L}\left[C_L + \frac{k_L}{Q_L}C_L(t-\tau)\right] + \frac{Q_L-Q_G}{\gamma_{HL}}C_{HL} + \frac{Q_G C_G}{\gamma_G} \tag{21.6-20}$$

$$V_K\frac{dC_K}{dt} = -\frac{Q_K}{\gamma_K}\left(1+\frac{k_K}{Q_K}\right)C_K + \frac{Q_K}{\gamma_{HL}}C_{HL} \tag{21.6-21}$$

$$V_G\frac{dC_G}{dt} = -\frac{Q_G}{\gamma_G}C_G + \frac{Q_G}{\gamma_{HL}}C_{HL} + \dot{N}_{wall} \tag{21.6-22}$$

If no MTX is initially present in the body, then

$$t=0: \quad C_i = 0 \quad (i = HL, M, L, K, G) \tag{21.6-23}$$

Gut Lumen

We idealize the GI lumen as a rigid tube of volume V_{GL}, length L_{GL} and cross-sectional area $A_{GL} = V_{GL}/L_{GL}$. The tube has a surface-to-volume ratio ϕ_{GL} that is augmented by the surface area of intestinal villi. Within the lumen, we treat chyme as a homogeneous phase with a constant mass density. Neglecting flow through the intestinal wall, the one-dimensional GL mass balance on chyme (Eq. 15.3-29) reduces to:

$$\frac{\partial Q_{GL}}{\partial z} = 0 \tag{21.6-24}$$

where Q_{GL} is the volumetric flow of chyme through the lumen. This equation indicates that Q_{GL} is constant if variations of inlet flow of chyme with time are not significant.

In formulating a MTX molar balance, we assume that there is no binding or chemical reaction of MTX in chyme. We also assume that axial dispersion of MTX is much greater than axial diffusion. A one-dimensional molar balance (Eq. 15.3-28) then leads to the governing equation for dissolved MTX concentration $C_{GL}(z, t)$:

$$\frac{\partial C_{GL}}{\partial t} + \frac{Q_{GL}}{A_{GL}}\frac{\partial C_{GL}}{\partial z} = \mathcal{D}^*_{GL}\frac{\partial^2 C_{GL}}{\partial z^2} - \phi_{GL} N_{wall} \tag{21.6-25}$$

where $\mathcal{D}^*_{GL}$ is a constant dispersion coefficient and $N_{wall}(z,t)$ is the local MTX flux from the GL to the G compartment. Assuming that N_{wall} occurs by active transport originating at the luminal side of the lumen-blood barrier, its rate equation will follow saturation kinetics with respect to the MTX concentration. The molar rate of MTX transport per unit luminal volume is

$$N_{wall} = \frac{k_{GL} C_{GL}}{K_{GL} + C_{GL}} \tag{21.6-26}$$

Over a differential length dz, the wall surface available for transport is $\phi_{GL} A_{GL} dz$, and the molar transfer rate from the GL to the G compartment is $\phi_{GL} A_{GL} N_{wall} dz$. The molar transport rate over the entire intestinal lumen (required by Eq. 21.6-22) is given by

$$\dot{N}_{wall}(t) = \int_0^{L_{GL}} \phi_{GL} A_{GL} N_{wall} dz = \frac{V_{GL} k_{GL} \phi_{GL}}{L_{GL}} \int_0^{L_{GL}} \frac{C_{GL}}{K_{GL} + C_{GL}} dz \tag{21.6-27}$$

With no MTX initially present in the body,

$$t = 0: \quad C_{GL} = 0 \tag{21.6-28}$$

Since the chyme originating in the stomach contains no MTX, the convective transport of MTX into the entrance of the GL compartment must equal MTX excretion in bile.

$$z = 0: \quad Q_{GL} C_{GL}(t) = \dot{R}_L(t) = k_L C_L^B(t - \tau) \tag{21.6-29}$$

At the GL exit, the MTX gradient is taken to be negligible.

$$z = L_{GL}: \quad \frac{\partial C_{GL}}{\partial z} = 0 \tag{21.6-30}$$

Analysis

Dimensionless Forms

The model can be made dimensionless by using the following dimensionless variables:

$$t = \frac{Q'_{HL}}{V_{HL}} t, \quad z = \frac{z}{L_{GL}}, \quad C_i = \frac{V_{HL}}{n_I} C_i \quad (i = HL, M, L, K, G, GL) \tag{21.6-31a-c}$$

The dimensionless model equations are

$$Q_M + Q_L + Q_K = 1 \tag{21.6-32}$$

$$\frac{dC_{HL}}{dt} = -\frac{1}{\gamma_{HL}} C_{HL} + \frac{Q_L}{\gamma_L} C_L + \frac{Q_K}{\gamma_K} C_K + \frac{Q_M}{\gamma_M} C_M + \frac{1}{\Delta t}[H(t-0) - H(t-\Delta t)] \tag{21.6-33}$$

$$\frac{dC_M}{dt} + \frac{Q_M}{\gamma_M V_M} C_M = \frac{Q_M}{\gamma_{HL} V_M} C_{HL} \tag{21.6-34}$$

$$\frac{dC_L}{dt} + \frac{Q_L}{\gamma_L V_L}(1 + k_L) C_L = \frac{Q_L - Q_G}{\gamma_{HL} V_L} C_{HL} + \frac{Q_G}{\gamma_G V_L} C_G \tag{21.6-35}$$

$$\frac{dC_K}{dt} + \frac{Q_K}{\gamma_K V_K}(1 + k_K)C_K = \frac{Q_K}{\gamma_{HL} V_K} C_{HL} \tag{21.6-36}$$

$$\frac{dC_G}{dt} + \frac{Q_G}{\gamma_G V_G} C_G = \frac{Q_G}{\gamma_{HL} V_G} C_{HL} + \frac{V_{GL} k_{GL}}{V_G} \int_0^1 \frac{C_{GL}}{K_{GL} + C_{GL}} dz \tag{21.6-37}$$

$$\frac{\partial C_{GL}}{\partial t} - D\frac{\partial^2 C_{GL}}{\partial z^2} + \frac{Q_{GL}}{V_{GL}}\frac{\partial C_{GL}}{\partial z} + \frac{k_{GL} C_{GL}}{K_{GL} + C_{GL}} = 0 \tag{21.6-38}$$

with the following initial and boundary conditions:

$$t = 0: \quad C_i = 0 \quad (\mathrm{i = HL, M, L, K, G, GL}) \tag{21.6-39}$$

$$z = 0: \quad C_{GL} = \frac{k_L Q_L}{Q_{GL}} C_L(t - \tau) \tag{21.6-40}$$

$$z = 1: \quad \frac{\partial C_{GL}}{\partial z} = 0 \tag{21.6-41}$$

The dimensionless parameters are

$$\begin{aligned}
&Q_i = \frac{\mathrm{Q_i}}{\mathrm{Q'_{HL}}}, \quad V_i = \frac{\mathrm{V_i}}{\mathrm{V_{HL}}} \quad (\mathrm{i = M, L, K, G, GL}) \\
&k_{GL} = \left(\frac{\mathrm{V_{HL}^2}}{\mathrm{n_I Q'_{HL}}}\right)\phi_{GL}\mathrm{k_{GL}}, \quad k_i = \frac{\mathrm{k_i}}{\mathrm{Q_i}} \quad (\mathrm{i = K, L}) \\
&D = \frac{\mathcal{D}^*_{GL}\mathrm{V_{HL}}}{\mathrm{Q'_{HL} L_{GL}^2}}, \quad K_{GL} = \left(\frac{\mathrm{V_{HL}}}{\mathrm{n_I}}\right)\mathrm{K_{GL}} \\
&\Delta t = \frac{\mathrm{Q'_{HL}}}{\mathrm{V_{HL}}}\Delta \mathrm{t}, \quad \tau \equiv \frac{\mathrm{Q'_{HL}}}{\mathrm{V_{HL}}}\mathrm{\tau}
\end{aligned} \tag{21.6-42a-h}$$

Table 21.6-1 Parameter Values for a 70 kg Adult[a]

Geometric Parameters		*Distribution Ratios*	
V_{HL} (L)	3	γ_{HL}	1
V_M (L)	35	γ_M	0.15
V_K (L)	0.28	γ_K	3
V_L (L)	1.35	γ_L	3
V_G (L)	0.1	γ_G	1
V_{GL} (L)	2.1		
L_{GL} (m)	6.0		
Blood Flows (L/min)		*Rate Parameters*	
$Q_{HL} - Q_F$	3.22	k_K (L/min)	0.19
Q_M	0.72	k_L (L/min)	0.20
Q_K	1.17	$\phi_{GL}k_{GL}$ (μmol/L/min)	2.0
Q_L	1.33	K_{GL} (μmol/L)	440
Q_G	1.17	τ (min)	10
Q_{GL}	0.035	$\mathcal{D}^*_{GL}$ (cm^2/min)	0.01

[a] Modified from Bischoff *et al.* (1971) to be consistent with the current model.

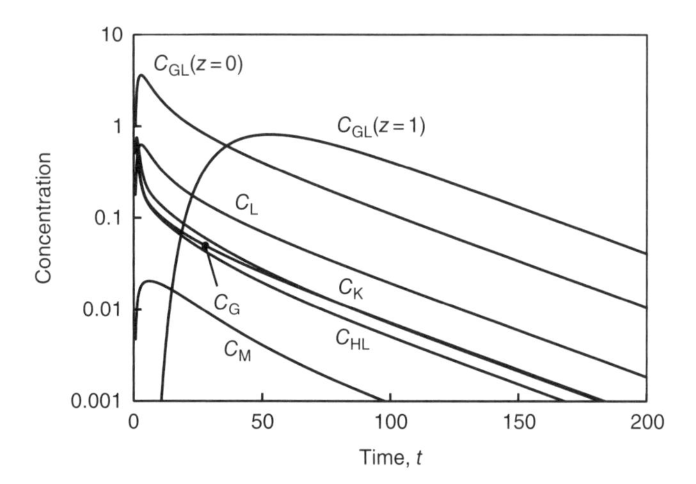

Figure 21.6-2 Methotrexate dynamics in various tissues for a moderate delivery rate ($\Delta t = 10$).

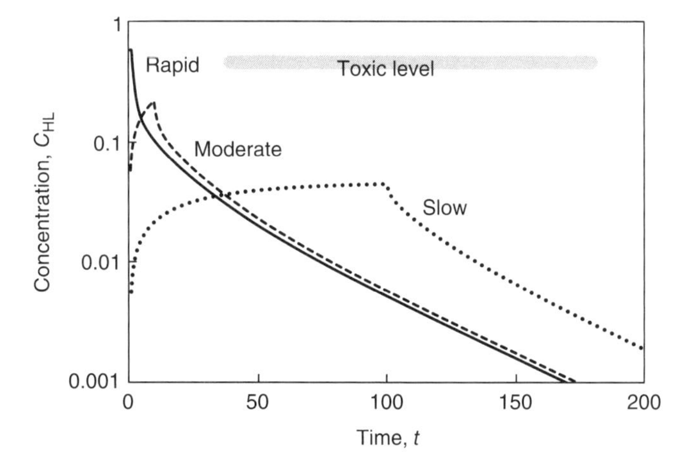

Figure 21.6-3 Methotrexate dynamics in the HL compartment for rapid ($\Delta t = 1$), moderate ($\Delta t = 10$), and slow ($\Delta t = 100$) delivery rates.

Simulation

The dynamics of MTX distributions in the different compartments were simulated by numerical solution of the model equations. Each simulation used the same set of dimensionless parameters computed from the physiological values in Table 21.6-1 and a typical MTX dose of $n_I = 30$ mg = 66 μmol.

Figure 21.6-2 shows model simulations with a dimensionless pulse duration of $\Delta t = 10$. Methotrexate levels in heart–lung (C_{HL}), kidneys (C_K), liver (C_L), muscle (C_M), gut tissue (C_G), and gut lumen (C_{GL} at $z = 0$ and $z = 1$) are shown as a function of time t. After pulse injection is initiated, MTX levels increase with time in all compartments. After peaking at some later time, which is different for each compartment, MTX levels fall off.

Figure 21.6-3 shows simulated MTX dynamics in the HL compartment for three durations of pulse delivery: $\Delta t = 1$ (rapid delivery), 10 (moderate delivery), or 100 (slow delivery). In terms of the dimensionless variables, the pulse amplitude is $1/\Delta t$ so that the dimensionless amount of injected drug is $(\Delta t)(1/\Delta t) = 1$ at all three Δt values. Clearly, slower delivery at a fixed amount of injected MTX, results in a lower peak concentration and a prolonged release time. Thus, the selection of the delivery time is an important consideration in ensuring that MTX reaches its therapeutic level while preventing it from exceeding a toxic level (about 10 μM for methotrexate).

References

Bischoff KB, Dedrick RL, Zaharko DS, Longstreth JA. Methotrexate pharmacokinetics. J Pharm Sci. 1971; 60:1128–1133.

Kalyanasundaram S, Calhoun VD, Leong KW, A finite element model for predicting the distribution of drugs delivered intracranially to the brain. Am J Physiol. 1997; 273(5 Pt 2):R1810–R1821.

Malagelada J-R, Robertson JS, Brown ML, Remington M, Duenes JA, Thomforde GM, Carryer PW. Intestinal transit of solid and liquid components of a meal in health. Gastroenterology. 1984; 87:1255–1263.

Qian F, Saidel GM, Sutton DM, Exner A, Gao J. Combined modeling and experimental approach for the development of dual-release polymer millirods. J Control Release. 2002; 83: 427–435.

Theeuwes F, Yum SI. Principles of the design and operation of generic osmotic pumps for the delivery of semisolid or liquid drug formulations. Ann Biomed Eng. 1976; 4:343–353.

Chapter 22

Diagnostics and Sensing

The study and diagnosis of disease processes require *in vivo* measurements as well as mathematical models with which to interpret and improve the measurements. In this chapter, we model transport and reaction processes in two illustrations of direct chemical sensing: analyte concentration in tissue by microdialysis and velocity and O_2 concentration in blood with a catheter-mounted electrode. We also develop models in three illustrations of indirect sensing methods: blood ethanol concentration estimated from expired gas, myocyte O_2 metabolism in an intact subject inferred from pulmonary uptake, and regional glucose metabolism in tissue by positron emission tomography (PET) imaging. Finally, we model a laboratory assay for cancer cell migration that indicates the potential for metastasis.

22.1 Chemical Monitoring of Tissue by Microdialysis

The purpose of microdialysis is to monitor specific analytes in tissue by implanting a very small probe that minimizes disturbance to its surroundings. In early studies, microdialysis was used to monitor glucose levels in adipose tissues. Later, endogenous metabolite and neurotransmitter concentrations were measured in the human brain. Currently, microdialysis is used to monitor chemical concentrations in other tissues including the myocardium, lung, and tumors. Microdialysis is particularly important in drug development because it can continuously measure concentration changes in accessible tissues after drug enters the circulation.

Figure 22.1-1 illustrates a typical probe design consisting of a dialysate supply tube surrounded by a concentric recovery tube of radius a_D. Analyte-free solution is pumped into the dialysate inlet, makes a U-turn at the probe tip, and flows through the recovery

Biomedical Mass Transport and Chemical Reaction: Physicochemical Principles and Mathematical Modeling,
First Edition. James S. Ultman, Harihara Baskaran, and Gerald M. Saidel.

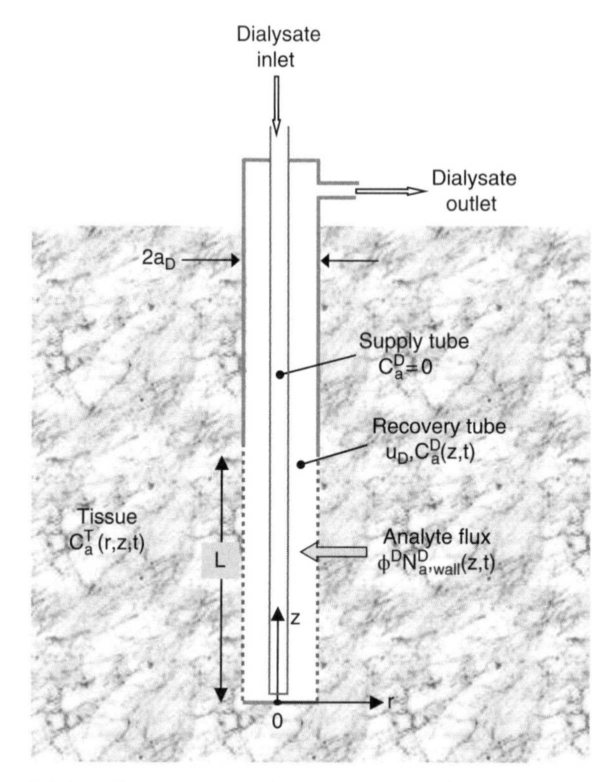

Figure 22.1-1 Microdialysis probe extracting analyte from extravascular tissue.

tube to the dialysate outlet. There is no loss of solution through the walls of either the supply or recovery tubes. The supply tube is also impermeable to analyte along its entire length. The wall of the recovery tube is permeable to analyte at its active section where $L > z > 0$ but is impermeable everywhere else. Analyte from surrounding tissue diffuses into the recovery tube through the active section and is carried by convection to the dialysate outlet. The analyte can undergo intracellular reaction in the tissue but does not react in the dialysate.

Here, we develop a mathematical model to relate analyte concentration at the dialysate outlet to analyte concentration in the tissue.

Model Formulation

Dynamic Concentration Distributions

A one-dimensional solution balance (Eq. 15.3-35) applied along the recovery tube indicates that the axial velocity u_D of a constant density dialysate D is independent of axial position z. A one-dimensional analyte balance (Eq. 15.3-34) in the active section of the recovery tube then leads to

$$\frac{\partial C_a^D}{\partial t} + u_D \frac{\partial C_a^D}{\partial z} = \left(\mathcal{D}_a^D + \mathcal{D}_a^{*D}\right) \frac{\partial^2 C_a^D}{\partial z^2} + \phi^D N_{a,\mathrm{wall}}^D \quad (L > z > 0) \tag{22.1-1}$$

where $C_a^D(z,t)$ is the concentration of analyte "a" in the dialysate, $\mathcal{D}_a^D$ and $\mathcal{D}_a^{*D}$ are constant diffusion and axial dispersion coefficients, ϕ^D is the surface-to-volume ratio of the active section, and $N_{a,wall}^D$ is the analyte flux from tissue into the active section.

Tissue consists of capillaries that perfuse an extravascular space containing cells surrounded by interstitial fluid. From a macroscopic perspective, tissue can be regarded as a homogeneous phase with a continuously distributed analyte source emanating from capillary blood and a continuously distributed analyte sink produced by cellular metabolism. For a cylindrically symmetric concentration field $C_a^T(r,z,t)$ in stationary tissue T surrounding the cylindrical probe, an analyte balance (Table 15.2-1) yields

$$\frac{\partial C_a^T}{\partial t} = \mathcal{D}_a^T\left[\frac{1}{r}\frac{\partial}{\partial r}\left(r\frac{\partial C_a^T}{\partial r}\right) + \frac{\partial^2 C_a^T}{\partial z^2}\right] + R_a^{cap}(r,z,t) - R_a^{cell}(r,z,t) \quad (r > a_D,\ L > z > 0) \qquad (22.1\text{-}2)$$

where $\mathcal{D}_a^T$ is a constant diffusion coefficient through the tissue, $R_a^{cap}(r,z,t)$ is the analyte release rate from capillaries, and $R_a^{cell}(r,z,t)$ is uptake rate by cellular metabolism. Both R_a^{cap} and R_a^{cell} represent molar rates per unit tissue volume.

Rate Processes

The diffusion flux of analyte across the active section of the recovery tube can be expressed as:

$$N_{a,wall}^D(z,t) = K_a^D\left[C_a^T(a_D,z,t) - C_a^D(z,t)\right] \qquad (22.1\text{-}3)$$

where K_a^D is an overall mass transfer coefficient. If cellular metabolism of analyte follows Michaelis–Menten kinetics, then

$$R_a^{cell} = \frac{\varepsilon^{cell} V_m C_a^T}{K_m + C_a^T} \qquad (22.1\text{-}4)$$

where ε^{cell} is the volume fraction of the tissue occupied by cells, V_m is the maximum metabolic rate of analyte per unit cell volume, and K_m is the Michaelis constant. The transport rate of analyte from capillaries to the extravascular space is proportional to the ratio of capillary surface to tissue volume ϕ^{cap}, an overall mass transfer coefficient K_i^{cap}, and the analyte concentration difference between capillaries and tissue:

$$R_a^{cap}(r,z,t) = \phi^{cap} K_a^{cap}\left[C_a^{cap}(t) - C_a^T(r,z,t)\right] \qquad (22.1\text{-}5)$$

Here, we have taken the equilibrium partition coefficient of analyte between capillary blood and extravascular space to be one. If the analyte is a drug that is injected into a major blood vessel as a bolus, then its capillary concentration will decrease with time:

$$C_a^{cap}(t) = C_{ao}^{cap}\exp(-t/\tau_a) \qquad (22.1\text{-}6)$$

Consolidated Equations

After substituting the rate equations into the governing equations, the analyte concentration equation in the active section of the recovery tube becomes

$$\frac{\partial C_a^D}{\partial t} + u_D\frac{\partial C_a^D}{\partial z} = \left(\mathcal{D}_a^D + \mathcal{D}_a^{*D}\right)\frac{\partial^2 C_a^D}{\partial z^2} + \phi^D K_a^D\left[C_a^T(z,t) - C_a^D(a_D,z,t)\right] \qquad (22.1\text{-}7)$$

and in the tissue becomes

$$\frac{\partial C_a^T}{\partial t} = \mathcal{D}_a^T \left[\frac{1}{r}\frac{\partial}{\partial r}\left(r\frac{\partial C_a^T}{\partial r} \right) + \frac{\partial^2 C_a^T}{\partial z^2} \right] + \phi^{cap} K_a^{cap} \left[C_{ao}^{cap} \exp(-t/\tau_a) - C_a^T(r,z,t) \right] - \frac{\varepsilon^{cell} V_m C_a^T}{K_m + C_a^T} \tag{22.1-8}$$

Initial and Boundary Conditions

If the analyte is a drug that is initially absent from dialysate and tissue, then

$$t = 0: \quad C_a^D = 0, \quad C_a^T = 0 \tag{22.1-9}$$

To simplify the model with respect to boundary conditions, we restrict our analysis to the tissue domain $\{r_\infty > r > a_D,\ L > z > 0\}$ where $r_\infty \gg a_D$. At $z = 0$ where dialysate enters the recovery tube, the analyte concentration is zero. At $z = 0$ in the tissue, we specify that spatial changes in analyte concentration are negligible:

$$z = 0: \quad C_a^D = 0, \quad \frac{\partial C_a^T}{\partial z} = 0 \tag{22.1-10}$$

At $z = L$ where dialysate exits the recovery tube, axial diffusion and dispersion are negligible compared to convection. At $z = L$ in the tissue, we again specify that there are no spatial changes in analyte concentration:

$$z = L: \quad \frac{\partial C_a^D}{\partial z} = 0, \quad \frac{\partial C_a^T}{\partial z} = 0 \tag{22.1-11}$$

We assume that transport in the recovery tube at $z > L$ continues to be dominated by convection. Therefore, the analyte concentration at the outlet of the device is the same as $C_a^D(r,L)$.

At the interface between the recovery tube and surrounding tissue, the radial flux of analyte across the permeable wall must match its diffusion flux in tissue:

$$r = a_D: \quad K_a^D \left(C_a^D - C_a^T \right) = -\mathcal{D}_a^T \frac{\partial C_a^T}{\partial r} \tag{22.1-12}$$

Far from the microdialysis probe, the concentration field in the tissue is not disturbed by the device so that

$$r = r_\infty: \quad \frac{\partial C_a^T}{\partial r} = 0 \tag{22.1-13}$$

Analysis

Dimensionless Forms

Substituting the dimensionless variables

$$C^D(z,t) \equiv \frac{C_a^D}{C_{ao}^{cap}}, \quad C^T(r,z,t) \equiv \frac{C_a^T}{C_{ao}^{cap}}, \quad t \equiv \frac{\mathcal{D}_a^D + \mathcal{D}_a^{*D}}{L^2} t, \quad r \equiv \frac{r}{a_D}, \quad z \equiv \frac{z}{L} \tag{22.1-14a-e}$$

into the consolidated governing equations, we obtain the dimensionless equation for the analyte concentrations in dialysate:

$$\frac{\partial C^{\mathrm{D}}}{\partial t}+Pe\frac{\partial C^{\mathrm{D}}}{\partial z}=\frac{\partial^2 C^{\mathrm{D}}}{\partial z^2}-\phi^{\mathrm{D}}Sh^{\mathrm{D}}\left[C^{\mathrm{D}}-C^{\mathrm{T}}(1,z)\right]\quad(1>z>0) \tag{22.1-15}$$

and in tissue:

$$\begin{aligned}\frac{\partial C^{\mathrm{T}}}{\partial t}&=D\left[\frac{L^2}{r}\frac{\partial}{\partial r}\left(r\frac{\partial C^{\mathrm{T}}}{\partial r}\right)+\frac{\partial^2 C^{\mathrm{T}}}{\partial z^2}\right]\\&\quad+D(\phi Sh)^{\mathrm{cap}}\left[\exp(-t/\tau)-C^{\mathrm{T}}\right]-\frac{V_{\mathrm{m}}C^{\mathrm{T}}}{K_{\mathrm{m}}+C^{\mathrm{T}}}\quad(r>1,\ \ 1>z>0)\end{aligned} \tag{22.1-16}$$

The dimensionless initial and boundary conditions are

$$\begin{aligned}t=0:&\quad C^{\mathrm{D}}=0,\ \ C^{\mathrm{T}}=0\\ r=1:&\quad \frac{Sh^{\mathrm{D}}}{L}\left(C^{\mathrm{D}}-C^{\mathrm{T}}\right)=-D\frac{\partial C^{\mathrm{T}}}{\partial r}\\ r=r_\infty:&\quad \frac{\partial C^{\mathrm{T}}}{\partial r}=0\\ z=0:&\quad C^{\mathrm{D}}=0,\ \ \frac{\partial C^{\mathrm{T}}}{\partial z}=0\\ z=1:&\quad \frac{\partial C^{\mathrm{D}}}{\partial z}=0,\ \ \frac{\partial C^{\mathrm{T}}}{\partial z}=0\end{aligned} \tag{22.1-17a-e}$$

In these equations, the dimensionless parameters are

$$\begin{aligned}&L\equiv\frac{\mathrm{L}}{\mathrm{a_D}},\quad r_\infty\equiv\frac{\mathrm{r}_\infty}{\mathrm{a_D}},\quad \tau\equiv\frac{\left(\mathcal{D}_{\mathrm{a}}^{\mathrm{D}}+\mathcal{D}_{\mathrm{a}}^{*\mathrm{D}}\right)\tau_{\mathrm{a}}}{\mathrm{L}^2},\quad D\equiv\frac{\mathcal{D}_{\mathrm{a}}^{\mathrm{T}}}{\mathcal{D}_{\mathrm{a}}^{\mathrm{D}}+\mathcal{D}_{\mathrm{a}}^{*\mathrm{D}}}\\ &\phi^{\mathrm{D}}\equiv\phi^{D}L,\quad Sh^{\mathrm{D}}\equiv\frac{\mathrm{K_a^D L}}{\mathcal{D}_{\mathrm{a}}^{\mathrm{D}}+\mathcal{D}_{\mathrm{a}}^{*\mathrm{D}}},\quad (\phi Sh)^{\mathrm{cap}}\equiv(\phi^{\mathrm{cap}}\mathrm{L})\frac{\mathrm{K_a^{cap}L}}{\mathcal{D}_{\mathrm{a}}^{\mathrm{T}}}\\ &Pe\equiv\frac{\mathrm{u_D L}}{\mathcal{D}_{\mathrm{a}}^{\mathrm{D}}+\mathcal{D}_{\mathrm{a}}^{*\mathrm{D}}},\quad V_{\mathrm{m}}\equiv\frac{\mathrm{L^2}\varepsilon^{\mathrm{cell}}\mathrm{V_{ma}}}{\left(\mathcal{D}_{\mathrm{a}}^{\mathrm{D}}+\mathcal{D}_{\mathrm{a}}^{*\mathrm{D}}\right)\mathrm{C_{ao}^{cap}}},\quad K_{\mathrm{m}}\equiv\frac{\mathrm{K_{ma}}}{\mathrm{C_{a\infty}^{cap}}}\end{aligned} \tag{22.1-18a-j}$$

Simulation

For model simulations in Figure 22.1-2, the base values of the parameters are $L = 10$, $r_\infty = 11$, $\tau = 0.5$, $D = 0.1$, $Pe = 10$, $\phi^{\mathrm{D}} = 10$, $Sh^{\mathrm{D}} = 1$, $(\phi Sh)^{\mathrm{cap}} = 1000$, $V_{\mathrm{m}} = 100$, and $K_{\mathrm{m}} = 1$. The left graph shows the radial distribution of a tissue concentration ratio, $TCR(r,t) = C^{\mathrm{T}}(r,0.5,t)/C^{\mathrm{T}}(11,0.5,t)$. This variable, evaluated at the axial midpoint of the tissue space, represents the local analyte concentration relative to its undisturbed concentration. Near the probe at $2 > r > 1$, we see that $TCR < 1$, indicating that tissue concentration is reduced from its undisturbed value. Further from the probe at $r > 2$, $TCR = 1$ so the device no longer influences transport and reaction processes in the tissue. The right graph compares dynamic changes in $C^{\mathrm{D}}(1,t)$, $C^{\mathrm{T}}(11,0.5,t)$, and $C^{\mathrm{cap}}(t)$. Because of cellular metabolism and analyte diffusion processes in tissue, the undisturbed tissue concentration $C^{\mathrm{T}}(11,0.5,t)$ is less than the capillary concentration $C^{\mathrm{cap}}(t)$. As a consequence of continuous removal of analyte by the probe, $C^{\mathrm{T}}(11,0.5,t)$ is always greater than the exiting dialysate concentration $C^{\mathrm{D}}(1, t)$.

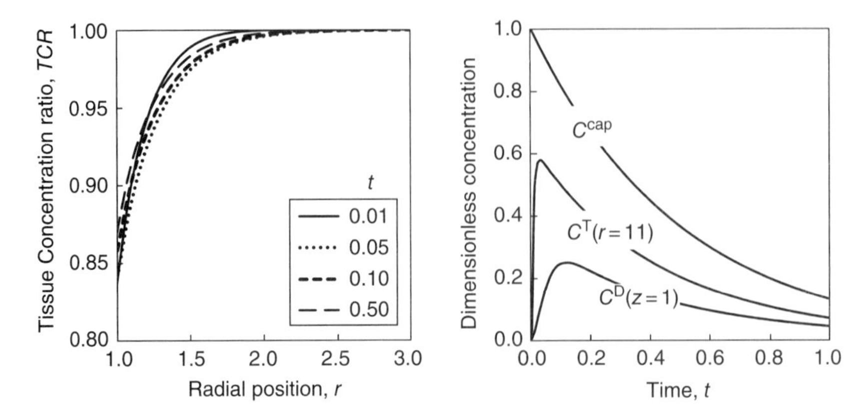

Figure 22.1-2 Analyte concentration variations in position and time.

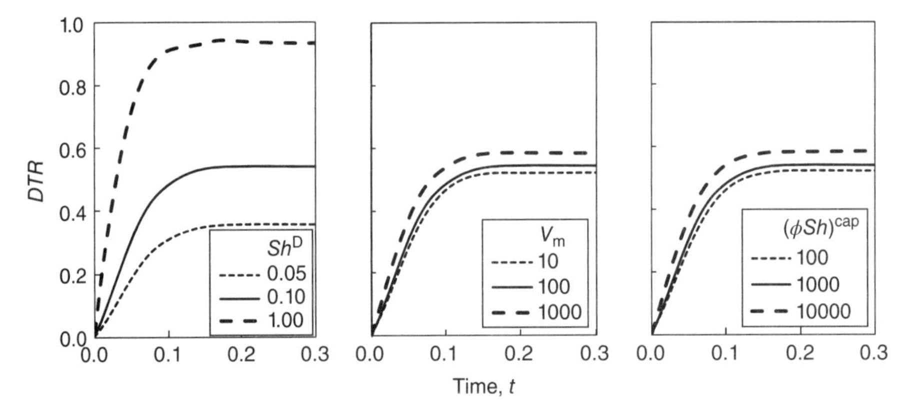

Figure 22.1-3 Effect of parameter changes on dialyzer-to-tissue concentration ratio (*DTR*).

The simulations in Figure 22.1-3 show the effects of key model parameters on the dynamics of a dialyzer–tissue concentration ratio, $DTR(t) = C^D(1,t)/C^T(10,0.5,t)$. This variable indicates how well the microdialyzer measurement tracks the undisturbed tissue concentration. The identical solid curves in the three panels of this graph result from the simulation with the base parameter values, which produce an asymptotic value of $DTR = 0.5$ when $t > 0.2$. The dashed and dotted curves show the effects of changing one of these parameters. According to the left graph, a larger dimensionless solute permeability Sh^D, corresponding to an increased transport rate across the interface between tissue and the recovery tube, causes DTR to rise to a higher value. The middle and right graphs show that different values of tissue parameters $(\phi Sh)^{\text{cap}}$ and V_m have little effect on DTR. Therefore, for the particular parameter values we have chosen, the accuracy of a microdialyzer measurement is limited by convection–diffusion processes associated with the probe rather than by diffusion–reaction in the tissue space.

We can view $1/DTR$ as a calibration factor that, when multiplied by the dynamic response of the dialysate output $C^D(1,t)$, provides the actual analyte concentration dynamics $C^T(11,0.5,t)$ in tissue. Use of a microdialyzer is practical when $1/DTR$ is a constant, which occurs when $t \geq 0.2$. Since $\mathrm{t} = \tau_a t$, the actual time required for the calibration factor to become constant is 0.2 times the characteristic time for analyte delivery from its injection point in a major vein to its measurement point in tissue.

22.2 Dual-Electrode Measurement of Blood Flow and Oxygen

Polarographic electrode systems have been used to monitor O_2 levels in a variety of biomedical tissues and fluids. As previously described (Example 9.2-1), a cathode biased at a negative voltage relative to a reference anode produces an electrical current by the reaction of O_2 with electrons and water to form hydroxyl ions. By appropriate selection of the bias voltage, this reaction rapidly eliminates O_2 at the cathode surface, making the current output proportional to O_2 partial pressure in the surrounding medium. In a flowing medium, the current is also affected by convection. To reduce this effect, O_2 electrodes are usually coated with a membrane that has a relatively low O_2 permeable (Example 12.2-1). The presence of a membrane can also reduce fouling of the electrodes by biofilm formation.

Here, we model a monitoring device consisting of two polarographic cathodes mounted on a cylindrical catheter designed for retrograde insertion to the flow in a major blood vessel. The simulated device has the same geometry as a commercial neonatal oxygen probe previously tested as a dual-cathode system (Kim *et al.*, 1981). A small frontal electrode (FE) with a thick membrane coating is centered on the catheter tip (Fig. 22.2-1). A much larger peripheral electrode (PE) with a thinner membrane coating surrounds the tip. Because of the difference in cathode sizes, membrane thicknesses, and surface orientations, we expect O_2 transport to the FE to be less sensitive to convection than transport to the PE. Thus, it should be possible to infer blood velocity as well as O_2 partial pressure from the two electrode output currents. To demonstrate this concept, we simulate steady-state O_2 transport to the electrode surfaces when the catheter is centered in a rigid tube with its tip facing a constant blood flow.

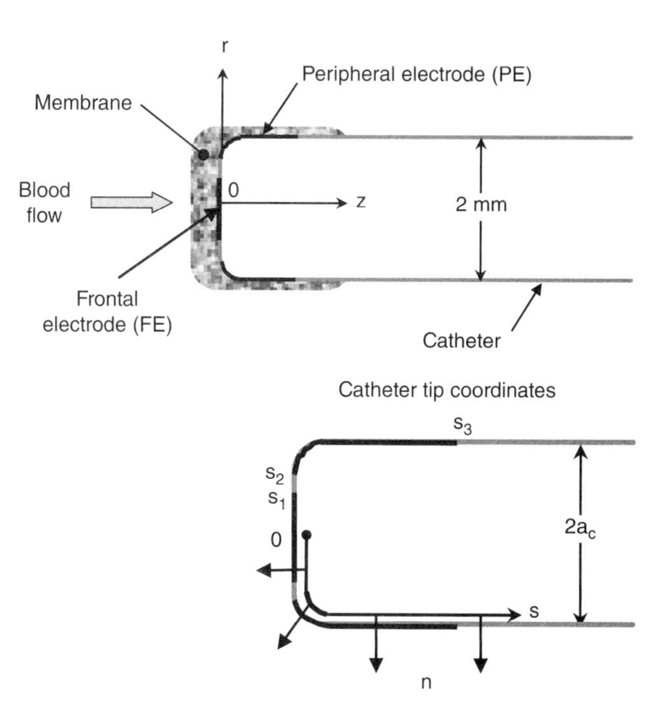

Figure 22.2-1 Catheter-mounted dual-electrode system.

Model Formulation

Equations of Fluid Motion

We consider a dual-electrode catheter of radius a_c and length $L_c \gg a_c$ mounted in a tube of radius a_t and length $L_t \gg a_t$. The downstream ends of the tube and catheter are aligned. However, the tube length is greater than the catheter length, so that the catheter tip is recessed by a distance $(L_t - L_c)$ from the tube entrance. With an axial coordinate z originating at the center of the catheter tip, the tube entrance is located at $z = -(L_t - L_c)$, the catheter tip is located at $z = 0$, and both the catheter and tube terminate at $z = L_c$.

We assume that the spatial distributions of velocity and O_2 concentration between the catheter and the tube wall are axisymmetric and that blood acts as a homogeneous, incompressible Newtonian fluid. Under steady-state conditions, the continuity (Table 13.1-1) and Navier–Stokes (Table B4-5) equations in cylindrical coordinates reduce to

$$\frac{1}{r}\frac{\partial}{\partial r}(ru_r) + \frac{\partial u_z}{\partial z} = 0 \tag{22.2-1}$$

$$\rho\left(u_r\frac{\partial u_r}{\partial r} + u_z\frac{\partial u_r}{\partial r}\right) = -\frac{\partial P}{\partial r} + \mu\left[\frac{\partial}{\partial r}\left(\frac{1}{r}\frac{\partial(ru_r)}{\partial r}\right) + \frac{\partial^2 u_r}{\partial z^2}\right] \tag{22.2-2}$$

$$\rho\left(u_r\frac{\partial u_z}{\partial r} + u_z\frac{\partial u_z}{\partial r}\right) = -\frac{\partial P}{\partial z} + \mu\left[\frac{\partial}{\partial r}\left(\frac{1}{r}\frac{\partial(ru_z)}{\partial r}\right) + \frac{\partial^2 u_z}{\partial z^2}\right] \tag{22.2-3}$$

O_2 Concentration Distribution in Blood

To account for all the O_2 in blood, we model the transport of both dissolved O_2 and hemoglobin-bound O_2 (HbO_2). Approximating blood as a homogeneous fluid, we express the volume average concentrations of dissolved O_2 as C_{O_2} and bound O_2 as C_{HbO_2}. We represent the rate per unit blood volume at which O_2 is formed by the dissociation of HbO_2 as R_{O_2}.

In cylindrical coordinates, the axisymmetric steady-state equation for dissolved O_2 in a fluid of constant density and constant diffusion coefficient (Table 15.2-1) is

$$u_r\frac{\partial C_{O_2}}{\partial r} + u_z\frac{\partial C_{O_2}}{\partial z} = \mathcal{D}_{O_2}\left[\frac{1}{r}\frac{\partial}{\partial r}\left(r\frac{\partial C_{O_2}}{\partial r}\right) + \frac{\partial^2 C_{O_2}}{\partial z^2}\right] + R_{O_2} \tag{22.2-4}$$

Since HbO_2 is confined to red blood cells, its transport occurs primarily by convection. Also, the binding of one O_2 molecules results in the formation of one hemoglobin molecule, $R_{O_2} = -R_{HbO_2}$. Consequently, the HbO_2 concentration distribution is governed by

$$u_r\frac{\partial C_{HbO_2}}{\partial r} + u_z\frac{\partial C_{HbO_2}}{\partial z} = R_{HbO_2} = -R_{O_2} \tag{22.2-5}$$

Adding these concentration equations, we obtain

$$u_r\frac{\partial}{\partial r}(C_{O_2} + C_{HbO_2}) + u_z\frac{\partial}{\partial z}(C_{O_2} + C_{HbO_2}) = \mathcal{D}_{O_2}\left[\frac{1}{r}\frac{\partial}{\partial r}\left(r\frac{\partial C_{O_2}}{\partial r}\right) + \frac{\partial^2 C_{O_2}}{\partial z^2}\right] \tag{22.2-6}$$

The dissolved O_2 concentration can be expressed as in terms of its phase equilibrium partial pressure p_{O_2} using the Bunsen solubility α_{O_2}:

$$C_{O_2} = c_G\alpha_{O_2}p_{O_2} \tag{22.2-7}$$

Since the binding reaction to form HbO_2 in red blood cells is very fast and reversible, the HbO_2 concentration C_{HbO_2} can be related to C_{O_2} by the Hill model for O_2 equilibrium saturation S_{O_2} (5.5-11):

$$C_{HbO_2} \equiv C_T S_{O_2} = \frac{C_T (\kappa_p p_{O_2})^n}{1 + (\kappa_p p_{O_2})^n} \tag{22.2-8}$$

where c_G is the molar gas density, C_T is the total concentration of heme groups in bound and unbound forms per unit blood volume, κ_p is an equilibrium constant, and n is the number of O_2 molecules that simultaneously bind to a hemoglobin molecule. Employing these relations, we express Equation 22.2-6 as

$$u_r \frac{\partial p_{O_2}}{\partial r} + u_z \frac{\partial p_{O_2}}{\partial z} = \frac{\mathcal{D}_{O_2}}{f(\kappa_p p_{O_2})} \left[\frac{1}{r} \frac{\partial}{\partial r} \left(r \frac{\partial p_{O_2}}{\partial r} \right) + \frac{\partial^2 p_{O_2}}{\partial z^2} \right] \tag{22.2-9}$$

Here, the dimensionless function

$$f(\kappa_p p_{O_2}) \equiv 1 + \frac{C_T \kappa_p}{c_G \alpha_{O_2}} \frac{(\kappa_p p_{O_2})^{n-1}}{\left[1 + (\kappa_p p_{O_2})^n\right]^2} \tag{22.2-10}$$

accounts for the retardation of dissolved O_2 diffusion by very rapid and reversible oxyhemoglobin binding.

Boundary Conditions

To complete the model formulation, we need to specify conditions for u_r, u_z, and p_{O_2} around the entire boundary of the transport domain. This boundary occurs at the flow entrance, $z = -(L_t - L_c)$, the flow exit, $z = L_c$, the inner surface of the tube, $r = a_t$, and the surface of the catheter including its electrodes. To describe locations on the catheter surface, we introduce a curvilinear coordinate s(r,z) that follows the contours of the surface and another coordinate n (not to be confused with n of the Hill model in Eq. 22.2-10) that is normal to s (Fig. 22.2-1; bottom).

At the tube entrance, we assume fully developed Poiseuille flow at an average axial velocity $\bar{u}$, no radial velocity, and a constant O_2 partial pressure p_o:

$$z = -(L_t - L_c): \quad u_z = 2\bar{u}\left(1 - \frac{r^2}{a_t^2}\right), \quad u_r = 0, \quad p_{O_2} = p_o \tag{22.2-11}$$

At the tube outlet, we assume that radial velocity as well as axial gradients of axial velocity and O_2 partial pressure are negligible. This is equivalent to stating that velocity and partial pressure distributions are fully developed in the exiting blood:

$$z = L_c, \quad a_t > r > a_c: \quad u_r = 0, \quad \frac{\partial u_z}{\partial z} = 0, \quad \frac{\partial p_{O_2}}{\partial z} = 0 \tag{22.2-12}$$

Everywhere along the tube surface, the velocity components and O_2 flux are zero:

$$r = r_t: \quad u_z = 0, \quad u_r = 0, \quad \frac{\partial p_{O_2}}{\partial r} = 0 \tag{22.2-13}$$

Along the entire catheter surface, the blood velocities are zero:

$$L_c > s \geq 0: \quad u_z = u_r = 0 \tag{22.2-14}$$

To establish the boundary conditions for O_2 partial pressure on the catheter surface, $[p_{O_2}]_{n=0}$, we must distinguish between regions with electrodes and regions without electrodes. Where FE is present at $s_1 > s \geq 0$ or PE is present at $s_3 > s \geq s_2$, O_2 influx through the membrane covering are

$$N^j_{O_2}(s) = P^j_{O_2} \Delta p_{O_2}(s) \quad (j = FE, PE) \tag{22.2-15}$$

where $P^j_{O_2}$ is membrane permeability to O_2 and Δp_{O_2} is the O_2 partial pressure across the membrane at a position s. We take the origin of the normal coordinate n to be the membrane surface exposed to blood where the O_2 partial pressure has a nonzero value of $[p_{O_2}]_{n=0}$. At the opposite membrane surface that contacts an electrode, O_2 partial pressure is forced to a negligible level by the electrode bias voltage. Thus, $\Delta p_{O_2}(s)$ is approximately equal to $[p_{O_2}]_{n=0}$, and the O_2 flux toward the electrode is $N^j_{O_2}(s) = P^j_{O_2}[p_{O_2}]_{n=0}$. By continuity, this flux must equal the O_2 diffusion flux from the blood:

$$n = 0: \begin{cases} -\alpha_{O_2} \mathcal{D}_{O_2} \left(\dfrac{\partial p_{O_2}}{\partial n}\right)_s = N^{FE}_{O_2} = P^{FE}_{O_2} p_{O_2} & (s_1 > s \geq 0) \\ -\alpha_{O_2} \mathcal{D}_{O_2} \left(\dfrac{\partial p_{O_2}}{\partial n}\right)_s = N^{PE}_{O_2} = P^{PE}_{O_2} p_{O_2} & (s_3 > s \geq s_2) \end{cases} \tag{22.2-16}$$

In regions without electrodes, the O_2 flux to the surface is zero:

$$n = 0: \quad \left(\frac{\partial p_{O_2}}{\partial n}\right)_s = 0 \quad (s_2 > s \geq s_1 \text{ or } L_c > s \geq s_3) \tag{22.2-17}$$

Once the model equations are solved for the O_2 distribution, the current output from an electrode of surface S_j is found by using Equation 9.2-12:

$$i_j = -4\mathcal{F}\dot{N}^j_{O_2} = -4\mathcal{F}\int_{S_j} N^j_{O_2} dS = -4\mathcal{F}P^j_{O_2}\int_{S_j} [p_{O_2}]_{n=0} dS \quad (j = FE, PE) \tag{22.2-18}$$

Analysis

We performed numerical simulations of this two-dimensional steady-state model for a 50 mm long catheter segment positioned along the centerline of a 55 mm long tube such that the catheter tip was recessed by 5 mm from the flow inlet. The 1 mm diameter catheter had electrodes with surface boundaries of $s_1 = 0.5$ mm, $s_2 = 0.7$ mm, and $s_3 = 2.8$ mm (Fig. 22.1-1; bottom). Assuming that the FE membrane coating is 100 times thicker than the PE coating, we assigned permeabilities of $P^G_{O_2,FE} = 1\,\mu m/s$ and $P^G_{O_2,PE} = 100\,\mu m/s$. Thermodynamic and transport parameter values were $c_G = 0.0446$ M, $\alpha_{O_2} = 2.8 \times 10^{-4}$ kPa^{-1}, $C_T = 0.00217$ M, $\kappa_p = 0.283\,kPa^{-1}$, $n = 2.8$, $\rho = 1.1$ kg/L, and $\mu = 0.0035$ Pa-s.

Figure 22.2-2 shows the results along the first 5 mm of a catheter placed in a tube with a 10 mm inner diameter. The distributions of O_2 partial pressure on the membrane surface in contact with blood, $[p_{O_2}]_{n=0}$, are shown in the left graph for inlet O_2 partial pressures of $p_o = 6$ and $p_o = 12$ kPa and average inlet velocities $\bar{u} = 8$ and $\bar{u} = 16$ cm/s. The corresponding isocontours of velocity magnitude $|\mathbf{u}|$ at an entrance velocity of $\bar{u} = 16$ cm/s are shown in the right graph. As expected, the $[p_{O_2}]_{n=0}$ distributions along FE are virtually independent of $\bar{u}$,

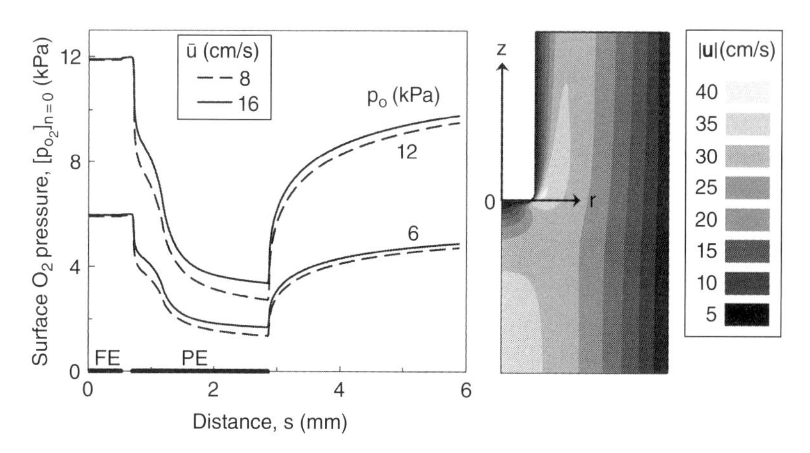

Figure 22.2-2 Membrane surface distributions of O_2 (left) and isocontours of velocity magnitude (right).

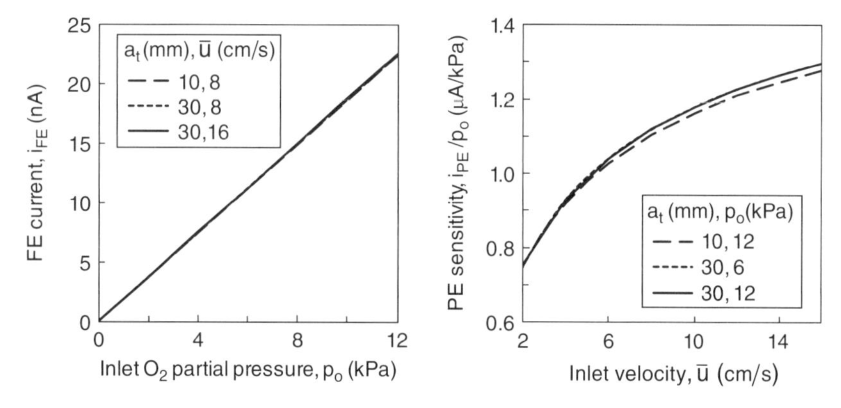

Figure 22.2-3 Current outputs of frontal (i_{FE}) and peripheral (i_{PE}) electrodes.

whereas $\left[p_{O_2}\right]_{n=0}$ along PE increases somewhat with $\bar{u}$. We also see that the stagnation flow over FE (apparent from the thick dark isocontour region adjacent to the electrode surface at $z = 0$) creates a $\left[p_{O_2}\right]_{n=0}$ that is virtually independent of s position. In contrast, $\left[p_{O_2}\right]_{n=0}$ values along PE decrease with s. This is due to a progressively increasing boundary layer resistance to O_2 transport (apparent from the progressive thickening of the dark isocontour region along the electrode surface at $z > 0$). Downstream of PE where O_2 is no longer reacts, $\left[p_{O_2}\right]_{n=0}$ is able to continuously increase. Although not apparent from this figure truncated at $z = 5$ mm, $\left[p_{O_2}\right]_{n=0}$ eventually returns to its input value p_o at the downstream end of the tube where $z = 50$ mm.

Electrical current outputs from the two electrodes are shown in Figure 22.2-3 over a range of physiologically relevant p_o and $\bar{u}$ values. These simulations were repeated for tube diameters of 30 and 10 mm, corresponding to catheter-to-tube radius ratios a_c/a_t of 0.067 and 0.200, respectively. The left graph illustrates that FE current is quite independent of $\bar{u}$ and a_c/a_t. It also indicates that the O_2 sensitivity of this electrode, i_{FE}/p_o, is constant. Thus, given an appropriate membrane coating, one can infer p_o from a FE current measured at a $\bar{u}$ typical for large blood vessels. The right graph indicates that PE exhibits an O_2 sensitivity that is virtually independent of p_o and a_c/a_t but depends substantially on $\bar{u}$. Thus, given p_o measurement with FE, it is possible to estimate blood velocity from the PE current.

22.3 Detection of Ethanol in Blood from Exhaled Gas

Driving under the influence of alcohol (ethanol) is a serious problem that can result in property damage, bodily injury, and even death. As a preventative measure or a means of legal judgment, a breathalyzer is often used for rapid, noninvasive measure of blood alcohol content (BAC). Ideally, this device determines the alcohol level in exhaled gas that comes from the gas-exchanging (alveolar) region where ethanol in the gas is presumed to be in equilibrium with ethanol in pulmonary blood. In practice, a person breathes through a mouthpiece, inhales a deep breath for about 2 s, and then blows out at a high flow rate for about 5 s. The ethanol partial pressure measured near the end of the breath originates in large part from the alveolar region.

To examine how the estimated BAC is affected by alveolar blood–gas disequilibrium and nonuniform distributions of alveolar volume and alveolar ventilation, we apply a transport model of expiration from a lung consisting of two parallel alveolar gas compartments with different properties connected to a representative airway tube that conducts gas to the mouth (Fig. 22.3-1).

Model Formulation

Volume and Flow Equations

The total gas volume in the lung is the sum of the time-varying alveolar volumes V_i ($i = 1,2$) and a constant volume V_3 of the conducting airway system. When breathing occurs without significant forcing, the mass density of the gas mixture has the same constant value everywhere.

During exhalation, a volumetric gas flow $\dot{V}_i$ leaves alveolar compartment i. A mass balance of the gas mixture in that compartment simplifies to

$$\frac{dV_i(t)}{dt} = -\dot{V}_i \quad (i = 1,2) \tag{22.3-1}$$

We consider the conducting airway tree as a single tube with cross-sectional area A(z) that changes with axial position but not with time. For an expired volumetric flow $\dot{V}$ of gas along the axis of such a rigid tube with no gas flow across the tube wall, a one-dimensional mass balance (Eq. 15.3-25) leads to

$$\frac{\partial \dot{V}}{\partial z} = 0 \tag{22.3-2}$$

where z is axial position measured from the mouth. Thus, $\dot{V}(t)$ is independent of z and equals the sum of the exiting alveolar flows, $\dot{V} = \dot{V}_1 + \dot{V}_2$. With a flow of zero at the start and end of exhalation, we approximate $\dot{V}$ as a sinusoidal function of time over a half cycle with an expiration period t_E and a maximum value $\dot{V}_{max}$:

$$\dot{V} = \dot{V}_{max} \sin\left(\pi \frac{t}{t_E}\right) \quad (t_E \geq t \geq 0) \tag{22.3-3}$$

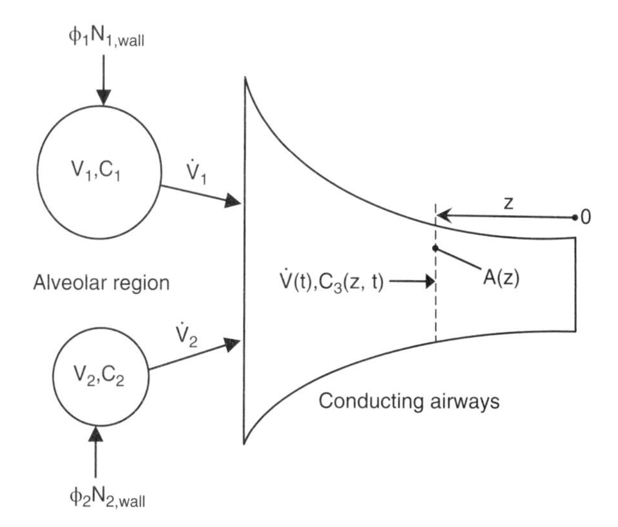

Figure 22.3-1 Lung model of ethanol transport from pulmonary blood to exhaled gas.

Thus, we rewrite Equation 22.3-1 as

$$\frac{dV_i(t)}{dt} = -\left(\frac{\dot{V}_i}{\dot{V}}\right)\dot{V}_{max}\sin\left(\pi\frac{t}{t_E}\right) \tag{22.3-4}$$

During a single exhalation, we assume that $\dot{V}_1$ and $\dot{V}_2$ are in phase with $\dot{V}$ such that the ratio $\dot{V}_i/\dot{V}$ is constant. We integrate Equation 22.3-4 with the initial condition that $\dot{V}_i(0) = \dot{V}_{io}$.

$$V_i = V_{io} - \left(\frac{\dot{V}_{max}t_E}{\pi}\right)\left(\frac{\dot{V}_i}{\dot{V}}\right)\left[1-\cos\left(\pi\frac{t}{t_E}\right)\right] \quad (i = 1, 2) \tag{22.3-5}$$

Since the airways are rigid, the total exhaled volume V_E is the difference between the total alveolar volume at the start and end of exhalation:

$$V_E = \sum_{i=1}^{2}[V_{io} - V_i(t_E)] = \frac{2\dot{V}_{max}t_E}{\pi} \tag{22.3-6}$$

With this relationship, the decrease of each alveolar volume during exhalation can be expressed as

$$V_i = V_{io} - \left(\frac{V_E}{2}\right)\left(\frac{\dot{V}_i}{\dot{V}}\right)\left[1-\cos\left(\pi\frac{t}{t_E}\right)\right] \quad (i = 1,2) \tag{22.3-7}$$

Ethanol Concentration Dynamics

In a well-mixed alveolar compartment i, the change of ethanol concentration $C_i(t)$ during exhalation is derived from a molar species balance in combination with Equation 22.3-1:

$$\frac{d(V_iC_i)}{dt} = -\dot{V}_iC_i + S_iN_{i,wall} \;\Rightarrow\; \frac{dC_i}{dt} = \phi_iN_{i,wall} \quad (i = 1,2) \tag{22.3-8a, b}$$

where ϕ_i is the alveolar surface-to-volume ratio and $N_{wall,i}$ is the diffusion flux of ethanol from pulmonary blood into the alveolar gas:

$$N_{i,wall} = K_i\left(\lambda^{G,B}C_B - C_i\right) \quad (i = 1,2) \tag{22.3-9}$$

Here, C_B is a constant ethanol concentration in blood; $\lambda^{G,B}$ is the ethanol gas–blood partition coefficient; K_i $(i = 1,2)$ is an overall mass transfer coefficient across the alveolar–capillary membrane. Substituting this flux equation into Equation 22.3-8b, we get

$$\frac{dC_i}{dt} = K_i\phi_i\left(\lambda^{G,B}C_B - C_i\right) \quad (i = 1,2) \tag{22.3-10}$$

We model transport of ethanol through the airway tube with a one-dimensional species balance (Eq. 15.3-24). Noting that the airway cross section does not depend on time and ethanol is nonreactive and neglecting ethanol absorption into the airway wall, the dynamics of ethanol concentration $C_3(z,t)$ are governed by:

$$\frac{\partial C_3}{\partial t} + \frac{1}{A}\frac{\partial(N_3A)}{\partial z} = 0 \tag{22.3-11}$$

where the axial flux $N_3(z,t)$ can be expressed in terms of convection, diffusion $(\mathcal{D})$, and dispersion $(\mathcal{D}^*)$:

$$N_3 = uC_3 - (\mathcal{D} + \mathcal{D}^*)\frac{\partial C_3}{\partial z} \tag{22.3-12}$$

The axial velocity $u(z,t)$, defined as positive in the z direction, is equivalent to $-\dot{V}(t)/A(z)$ during expiration. We can combine Equations 22.3-11 and 22.3-12 to obtain

$$\frac{\partial C_3}{\partial t} - \frac{\dot{V}}{A}\frac{\partial C_3}{\partial z} = \frac{1}{A}\frac{\partial}{\partial z}\left[(\mathcal{D} + \mathcal{D}^*)A\frac{\partial C_3}{\partial z}\right] \tag{22.3-13}$$

For the airway tree, changes in total cross-sectional area with axial position are extremely large. To make spatial variations more gradual, we express them in terms of a cumulative gas volume, $\nu \equiv \int_0^z A(\varsigma)d\varsigma$, that varies from zero at the mouth to V_3 at the distal end of the airways. That is, $d\nu = Adz$ so that Equation 22.3-13 becomes

$$\frac{\partial C_3}{\partial t} - \dot{V}\frac{\partial C_3}{\partial \nu} = \frac{\partial}{\partial \nu}\left[(\mathcal{D} + \mathcal{D}^*)A^2\frac{\partial C_3}{\partial \nu}\right] \quad (V_3 > \nu > 0) \tag{22.3-14}$$

Initial and Boundary Conditions

Initially, the alveolar compartments contain residual ethanol. However, the airway tube has none because the air that filled the airways during the previous inhalation is ethanol-free:

$$t = 0: \quad C_1 = C_{1o}, \quad C_2 = C_{2o}, \quad C_3 = 0 \quad (V_3 > \nu > 0) \tag{22.3-15}$$

At the boundary between the airway tube and alveolar compartments, we assume that axial diffusion and dispersion are small compared to convection. Thus, convective transport of ethanol is continuous across the boundary:

$$\nu = V_3: \quad \dot{V}_1C_1 + \dot{V}_2C_2 = \dot{V}C_3 \tag{22.3-16}$$

At the mouth, we also assume that the ethanol concentration gradient is negligible:

$$\nu = 0: \quad \frac{\partial C_3}{\partial z} = 0 \Rightarrow \frac{\partial C_3}{\partial \nu} = 0 \tag{22.3-17}$$

Variable Coefficients

Due to geometric variations of alveoli and conducting airways, the coefficients in Equations 22.3-10 and 22.3-14 change substantially with position and time. During exhalation when alveolar volume decreases, alveolar tissues tend to come together. This decreases the effective surface area for transport between capillary blood and alveolar gas. Approximating alveoli in compartment i as spheres, their diameter is proportional to $V_i^{1/3}$, their surface is proportional to $V_i^{2/3}$, and their surface-to-volume ratio $\phi_i(t)$ is proportional to $V_i^{2/3}/V_i = V_i^{-1/3}$. Since the proportionality constant is the same at t = 0 as it is at any later time,

$$\frac{\phi_i(t)}{\phi_{io}} = \frac{V_i(t)^{-1/3}}{V_{io}^{-1/3}} \Rightarrow \phi_i(t) = \phi_{io}\left[\frac{V_{io}}{V_i(t)}\right]^{1/3} \quad (i = 1,2) \tag{22.3-18a,b}$$

where $\phi_{io} \equiv \phi_i(t = 0)$ and $V_{io} \equiv V_i(t = 0)$. In the airway tube, the summed cross-sectional area increases exponentially between the mouth where $A(\nu = 0) \equiv A_m$ and any more distal position $V_3 \ge \nu > 0$:

$$A(\nu) = A_m \exp(\gamma_A \nu) \tag{22.3-19}$$

In contrast, we expect the dispersion coefficient to be a multiple of the diffusion coefficient $\mathcal{D}$ that diminishes exponentially from the mouth toward zero at the distal end of the airways:

$$\mathcal{D}^*(\nu) = \mathcal{D}\{\exp[\gamma_{\mathcal{D}}(V_3 - \nu)] - 1\} \tag{22.3-20}$$

Here, we ignore the possible effect of a time-varying flow on $\mathcal{D}^*$.

Analysis

Dimensionless Forms

The model equations can be expressed in dimensionless form using the dimensionless variables:

$$t \equiv \frac{\text{t}}{\text{t}_E}, \quad \xi \equiv \frac{\nu}{V_3}, \quad C_i \equiv \frac{C_i}{\lambda^{G,B} C_B} \quad (i = 1,2,3), \quad V_i \equiv \frac{V_i}{V_E} \quad (i = 1,2) \tag{22.3-21a-d}$$

where C_i is ethanol concentration relative to the value that would be in equilibrium with pulmonary blood. The dimensionless volume and ethanol concentrations in the alveolar compartments change as

$$V_i(t) = h_i V_o - \frac{f_i}{2}[1 - \cos(\pi t)] \quad (1 \ge t \ge 0;\ i = 1,2) \tag{22.3-22}$$

$$\frac{dC_i}{dt} = K_i \phi_i(t)(1 - C_i) \quad (i = 1,2) \tag{22.3-23}$$

The dimensionless ethanol equation over the domain $1 > \xi > 0$ in the airways becomes

$$\frac{\partial C_3}{\partial t} - \frac{\dot{V}(t)}{V_3}\frac{\partial C_3}{\partial \xi} = \frac{1}{Pe V_3^2}\frac{\partial}{\partial \xi}\left[g(\xi)\frac{\partial C_3}{\partial \xi}\right] \tag{22.3-24}$$

The dimensionless variable coefficients in these governing equations are

$$\phi_i(t) \equiv \left(\frac{V_{io}}{V_E}\right)^{1/3} \frac{\phi_i(t)}{\phi_{io}} = \frac{1}{V_i(t)^{1/3}} \quad (i = 1,2)$$

$$\dot{V}(t) \equiv \frac{t_E}{V_E}\dot{V}(t) = \frac{\pi}{2}\sin(\pi t) \tag{22.3-25a-c}$$

$$g(\xi) \equiv \frac{\mathcal{D}A^2 + \mathcal{D}^* A^2}{\mathcal{D}A_m^2} = \exp[\gamma_{\mathcal{D}} + (2\gamma_A - \gamma_{\mathcal{D}})\xi]$$

For the dimensionless initial and boundary conditions, we obtain

$$t = 0: \quad C_1 = C_{1o}, \;\; C_2 = C_{2o}, \;\; C_3 = 0 \tag{22.3-26}$$

$$\xi = 0: \quad \frac{\partial C_3}{\partial \xi} = 0 \tag{22.3-27}$$

$$\xi = 1: \quad f_1 C_1 + f_2 C_2 = C_3 \tag{22.3-28}$$

The constant dimensionless groups in this model are

$$C_{io} = \frac{C_{io}}{\lambda^{G,B} C_B}, \quad K_i \equiv \phi_{io} K_i t_E \left(\frac{V_{io}}{V_E}\right)^{1/3}, \quad f_i \equiv \frac{\dot{V}_i}{\dot{V}}, \quad h_i \equiv \frac{V_{io}}{V_{1o} + V_{2o}} \quad (i = 1,2)$$

$$V_o \equiv \frac{V_{1o} + V_{2o}}{V_E}, \quad V_3 \equiv \frac{V_3}{V_E}, \quad Pe = \frac{(V_E/t_E A_m)(V_E/A_m)}{\mathcal{D}} \tag{22.3-29a-i}$$

$$\gamma_A \equiv \gamma_A V_3, \quad \gamma_{\mathcal{D}} \equiv \gamma_{\mathcal{D}} V_3$$

Here, f_i is a measure of flow inhomogeneity and h_i is a measure of volume inhomogeneity. Since $f_1 + f_2 = 1$ and $h_1 + h_2 = 1$, an ideal lung in which flow and volume among alveoli are homogeneously distributed is defined by values of $f_1 = f_2 = h_1 = h_2 = 0.5$. Whether or not a lung is ideal, Equation 22.3-22 indicates that f_i must be less than $h_i V_o$ in order to avoid negative values of alveolar volume at the end of exhalation.

Simulation

To simulate dynamic behavior, we numerically solve the model equations with equal initial alveolar concentrations $C_{1o} = C_{2o} \equiv C_o$ and equal alveolar–capillary mass transfer parameters $K_1 = K_2 \equiv K$. Other parameter values are fixed at values appropriate for typical adult human lung during a five second exhalation to residual volume: $Pe = 80{,}000$, $\gamma_A = 3.8$, $\gamma_{\mathcal{D}} = 5.6$, $V_3 = 0.05$, and $V_o = 1.6$.

Figure 22.3-2 shows the distribution of ethanol concentration $C_3(\xi,t)$ in the conducting airway tube of an ideal lung with an expiratory mass transfer parameter of $K = 0.4$ and with C_o equal to either 1 or 0.5. Since V_o has been set to 1.6 and $f_i = h_i = 0.5$, the $f_i < h_i V_o$ constraint is satisfied in these simulations.

A value of $C_o \equiv C_o/\lambda^{G,B}C_B = 1$ (left graph) occurs when mass transfer across the alveolar–capillary membrane is so rapid during inhalation that phase equilibrium is established between pulmonary blood and alveolar gas at the beginning of exhalation. In that case, ethanol in the distal airways maintains equilibrium with blood throughout exhalation such that $C_3(1,t) = 1$ at all t. Simultaneously, $C_3(0,t)$ at the mouth increases from zero toward one as equilibrated gas from the alveolar compartments displaces ethanol-free gas that initially filled the airways. Overall, the dimensionless concentration distribution in the airway tube reaches a spatially uniform equilibrium value of $C_3(\xi,t) = 1$ for $t \geq 0.3$.

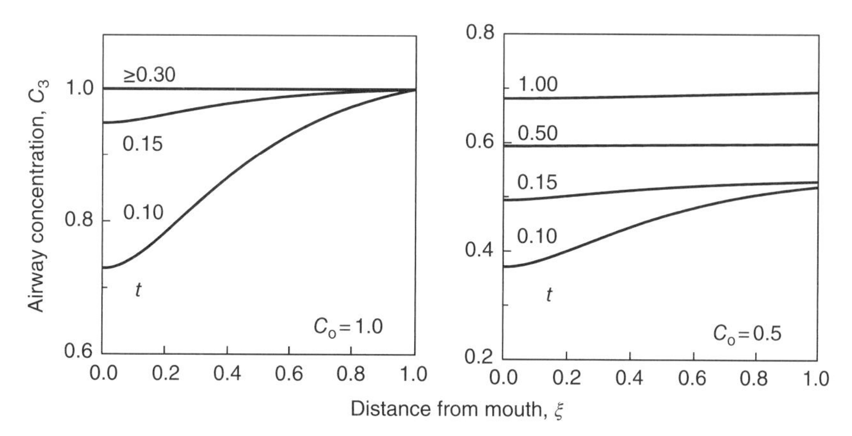

Figure 22.3-2 Ethanol distribution in conducting airways of an ideal lung.

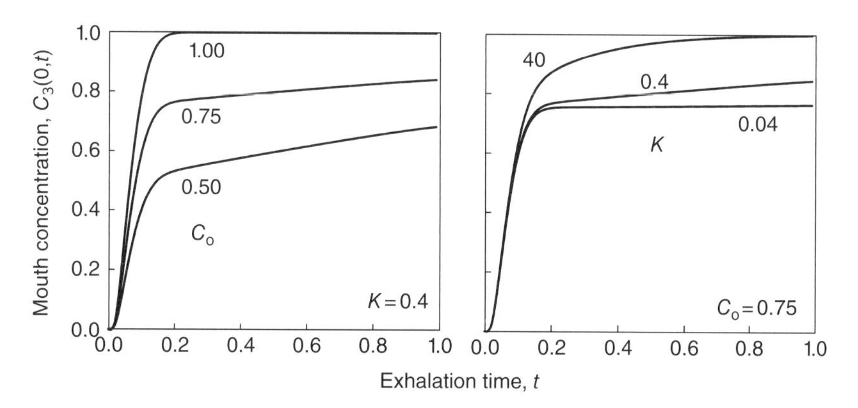

Figure 22.3-3 Expired ethanol concentration at the mouth of an ideal lung.

A value of $C_o = 0.5$ (right graph) implies that an alveolar–capillary diffusion restriction during inhalation only allows alveolar gas to reach half its equilibrium concentration with pulmonary blood. During the following exhalation, the ethanol distribution becomes uniform at $C_3 = 0.5$ when t reaches 0.15. Thereafter, C_3 remains uniform at a level that increases with t as equilibration between pulmonary blood and alveolar gas progresses. At the end of expiration, the concentration at the mouth is well below 1.0.

Figure 22.3-3 examines the simulated effect of C_o and K on dynamic changes in ethanol concentration at the mouth of an ideal lung. In inferring BAC with a breathalyzer, it is assumed that end-expired ethanol concentration measured at the mouth is equivalent to equilibrium blood concentration, that is, $C_3(0,1) = 1$. According to the simulations, however, $C_3(0,1) < 1$ except when $C_o = 1$ or $K \gg 1$. Thus, a breathalyzer measurement in an ideal lung underestimates BAC unless there is sufficiently rapid ethanol transport between pulmonary blood and alveolar gas during inhalation and/or exhalation. This error is exaggerated if a test breath is prematurely terminated at a $t < 1$.

Figure 22.3-4 shows the effect of a heterogeneous flow distribution, $f_1 \neq 0.5$ when $h_1 = 0.5$, or a heterogeneous alveolar volume distribution, $h_1 \neq 0.5$ when $f_1 = 0.5$. In both cases,

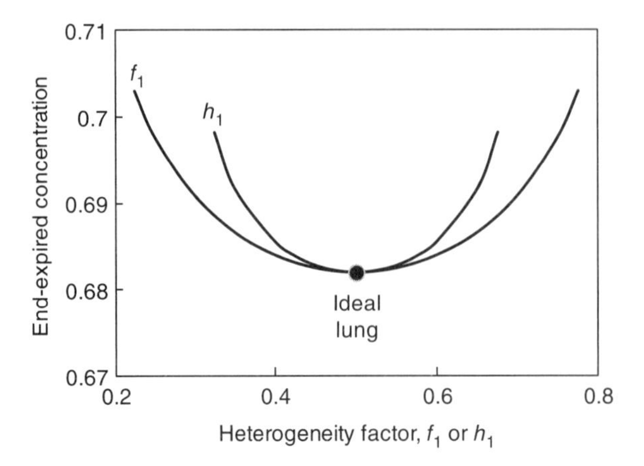

Figure 22.3-4 Ethanol distribution with inhomogeneity in flow ($f_1 \neq 0.5, h_1 = 0.5$) or volume ($h_1 \neq 0.5, f_1 = 0.5$).

$C_o = 0.5, K = 0.4$, and $V_o = 1.6$. Since any f_1 value in alveolar compartment 1 must result in the same expired concentration pattern as a complementary value of $f_2 = (1 - f_1)$ in alveolar compartment 2, the f_1 curve is symmetric about $f_1 = 0.5$. By the same reasoning, the h_1 curve is symmetric about $h_1 = 0.5$. In order to satisfy the $f_i < h_i V_o$ constraint with $V_o = 1.6$ and $h_i = 0.5$, the flow heterogeneity is confined to the range $0.8 > f_1 > 0.2$. Similarly, the volume heterogeneity must be in the range $0.7 > h_1 > 0.3$. With the parameter values chosen (and also because we assumed synchronous flow between the dead space and the two alveolar compartments), simulations indicate only a small effect of flow and volume heterogeneities on the end-expired ethanol concentration $C_3(0,1)$, that is, the estimated BAC.

22.4 Oxygen Uptake and Utilization in Exercising Muscle

A key measure of the potential to do work, even tasks of daily living, is the rate of O_2 uptake by muscle cells during exercise. This depends on overall cardiorespiratory function as well as cellular metabolism. In humans, evaluation of O_2 utilization rate by skeletal muscle during exercise is based on the noninvasive measurement of the volumetric O_2 uptake rate from respired air. To relate these rates and gain a quantitative understanding of O_2 transport at the cellular, tissue, and whole-body levels, we develop a multiscale mechanistic model. This model incorporates O_2 transport from the airway opening to working muscle and includes an oxidative pathway of ATP synthesis in cells.

Model Formulation

Structure and Assumptions

The model (Fig. 22.4-1) comprises three compartments: pulmonary (p), skeletal muscle (m), and other organs (o). The pulmonary compartment, consisting of alveolar and capillary subcompartments, has O_2 inputs via inspired gas and mixed venous blood. Oxygen outputs occur via expired gas and arterialized blood. Extravascular pulmonary tissue is ignored since it contains very little O_2 compared to capillary blood. The muscle and other organs compartments each consist of a capillary and an extravascular tissue subcompartment. In addition to an O_2

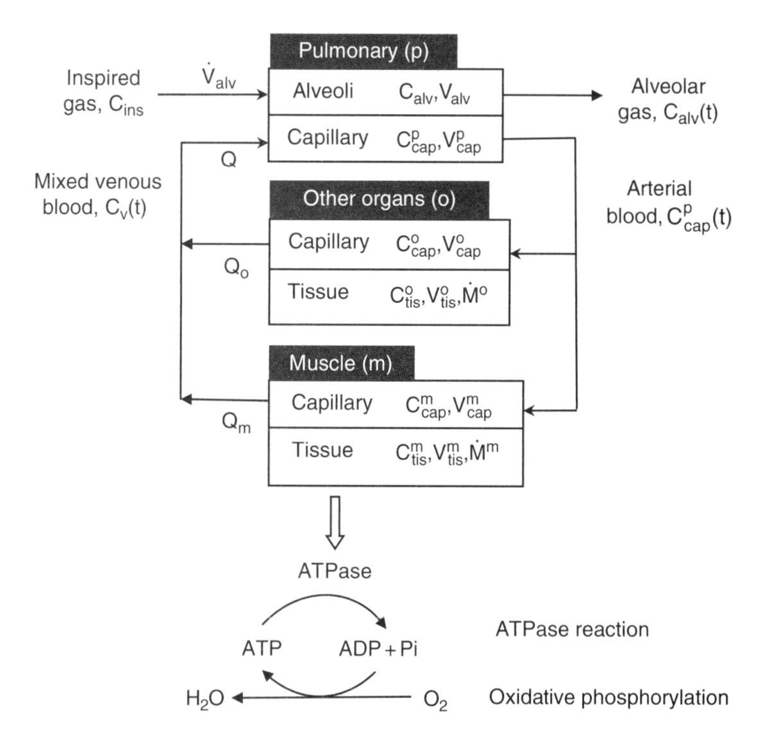

Figure 22.4-1 Compartment model of whole-body O_2 transport and muscle metabolism.

input via arterial blood and an output via venous blood, both of these compartments experience a metabolic oxygen loss in their tissue subcompartments.

We make several assumptions regarding the properties and behavior of this system. The volume of each subcompartment is constant. All subcompartments have internal fluid compositions that are uniform. The alveolar and all capillary subcompartments have outflow compositions that are equal to their internal fluid compositions. Mass density is constant and equal in blood and tissues throughout the system. Oxygen in capillary blood and in tissues is physically dissolved. Oxygen is also present in bound form, primarily as oxyhemoglobin in blood and as myoglobin in muscle tissue. Oxygen binding in the tissue of other organs is much less important than it is in the pulmonary and muscle tissue. Diffusion between subcompartments in the pulmonary and other organs compartments is rapid compared to convection through subcompartments. In such flow-limited transport, dissolved O_2 is in equilibrium between subcompartments. Transport in the muscle compartment is not necessarily flow limited. Oxygen is depleted by metabolic processes in the tissue subcompartments of muscle and other organs, but is not metabolized in the blood. The principal responses to changes in exercise intensity occur in alveolar ventilation, blood flow, and extravascular O_2 metabolism associated with the muscle compartment.

Mixed Venous Flows and O_2 Concentrations

The outflows from the other organs and muscle compartments combine to form the mixed venous blood that enters the pulmonary compartment. Since these three bloodstreams have equal densities, a solution mass balance around the mixing point yields

$$Q(t) = Q_o + Q_m(t) \tag{22.4-1}$$

where Q is the pulmonary blood flow (essentially the cardiac output), Q_m is the blood flow through the capillaries of the muscle, and Q_o is the blood flow through the other organs. Whereas Q and Q_m increase in response to exercise, Q_o remains approximately constant. A molar balance on total O_2 about the same mixing point yields

$$Q(C_{v,d} + C_{v,b}) = Q_o\left(C^{o}_{cap,d} + C^{o}_{cap,b}\right) + Q_m\left(C^{m}_{cap,d} + C^{m}_{cap,b}\right) \tag{22.4-2}$$

Here, $C_{v,j}$ refers to either the dissolved (j = d) or bound (j = b) O_2 in the mixed venous blood, and $C^{i}_{cap,j}$ refers to their molar concentrations in the capillaries of compartment i.

Pulmonary O_2 Concentrations

Separate molar balances on the dissolved and the bound O_2 in the pulmonary capillary subcompartment are:

$$V^{p}_{cap}\frac{dC^{p}_{cap,d}}{dt} = Q\left(C_{v,d} - C^{p}_{cap,d}\right) - \dot{R}^{p}_{cap,d} + \dot{N}^{p} \tag{22.4-3}$$

$$V^{p}_{cap}\frac{dC^{p}_{cap,b}}{dt} = Q\left(C_{v,b} - C^{p}_{cap,b}\right) + \dot{R}^{p}_{cap,b} \tag{22.4-4}$$

where V^{p}_{cap} is the capillary volume, $\dot{R}^{p}_{cap,j}$ (j = d,b) is the rate of formation of dissolved or bound O_2, and $\dot{N}^{p}$ is the molar rate of molecular O_2 diffusion from the gas-filled alveoli to capillary blood. Since the molar rate of formation of O_2 bound to a heme group is equal to the rate of loss of O_2 by binding, $\dot{R}^{p}_{cap,b} = -\dot{R}^{p}_{cap,b}$ and we add these equations to obtain

$$V^{p}_{cap}\frac{d\left(C^{p}_{cap,d} + C^{p}_{cap,b}\right)}{dt} = Q\left[(C_{v,d} + C_{v,b}) - \left(C^{p}_{cap,d} + C^{p}_{cap,b}\right)\right] + \dot{N}^{p} \tag{22.4-5}$$

Over several identical breaths, we write an average O_2 molar balance on alveolar gas as if the O_2 input and output rates are each continuous in time:

$$V_{alv}\frac{dC_{alv}}{dt} = \dot{V}_{alv}(C_{ins} - C_{alv}) - \dot{N}^{p} \tag{22.4-6}$$

Here, V_{alv} is the alveolar volume; $\dot{V}_{alv}$ is an effective continuous alveolar gas flow that depends on breathing frequency, tidal volume, and dead space volume (Eq. 19.1-36); C_{ins} is the O_2 concentration of inspired air; and $C_{alv}(t)$ is the average continuous O_2 concentration leaving the alveolar compartment during expiration.

Since the pulmonary compartment is flow limited, we relate the dissolved O_2 in capillary blood and in alveolar gas with an equilibrium partition coefficient $\lambda^{cap,alv} \equiv C^{p}_{cap,d}/C_{alv}$. Incorporating this equilibrium relation in Equation 22.4-6 and then adding the result to Equation 22.4-5, $\dot{N}^{p}$ is eliminated and we obtain an O_2 transport equation for the pulmonary compartment:

$$\frac{d}{dt}\left[\left(\frac{V_{alv}}{\lambda^{cap,alv}} + V^{p}_{cap}\right)C^{p}_{cap,d} + V^{p}_{cap}C^{p}_{cap,b}\right] = Q\left[(C_{v,d} + C_{v,b}) - \left(C^{p}_{cap,d} + C^{p}_{cap,b}\right)\right] + \dot{V}_{alv}\left(C_{ins} - \frac{C^{p}_{cap,d}}{\lambda^{cap,alv}}\right) \tag{22.4-7}$$

Oxygen Concentrations in Other Organs Compartment

Assuming an equilibrium partition coefficient of one between tissue and blood, the concentrations of dissolved O_2 in the other organs subcompartments are equal, $C^{o}_{tis,d} = C^{o}_{cap,d}$. In that case, we can derive a total O_2 molar balance for the other organs compartment by adding molar balances in tissue and blood subcompartments:

$$\frac{d}{dt}\left[\left(V^{o}_{tis} + V^{o}_{cap}\right)C^{o}_{cap,d} + V^{o}_{cap}C^{o}_{cap,b}\right] = Q_o\left[\left(C^{p}_{cap,d} + C^{p}_{cap,b}\right) - \left(C^{o}_{cap,d} + C^{o}_{cap,b}\right)\right] - \dot{M}^{o} \tag{22.4-8}$$

where $\dot{M}^{o}$ is the molar rate of O_2 utilization in the tissue. We assume that $\dot{M}^{o}$ and Q_o are constant since they change little between rest and exercise.

Oxygen Concentrations in Muscle Compartment

Allowing for a nonequilibrium O_2 transport rate $\dot{N}^{m}$ from capillary to muscle subcompartments, the molar balance of O_2 in muscle capillaries is

$$V^{m}_{cap}\frac{d\left(C^{m}_{cap,d} + C^{m}_{cap,b}\right)}{dt} = Q_m\left[\left(C^{p}_{cap,d} + C^{p}_{cap,b}\right) - \left(C^{m}_{cap,d} + C^{m}_{cap,b}\right)\right] - \dot{N}^{m} \tag{22.4-9}$$

Visualizing this transport rate as a membrane process driven by the difference in dissolved O_2 concentrations and assuming that the equilibrium partition coefficient between tissue and blood is one:

$$\dot{N}^{m} = PA_m\left(C^{m}_{cap,d} - C^{m}_{tis,d}\right) \tag{22.4-10}$$

where PA_m is the permeability of a capillary bed times its surface area. The O_2 molar balance in the muscle capillary subcompartment now becomes

$$V^{m}_{cap}\frac{d\left(C^{m}_{cap,d} + C^{m}_{cap,b}\right)}{dt} = Q_m\left[\left(C^{p}_{cap,d} + C^{p}_{cap,b}\right) - \left(C^{m}_{cap,d} + C^{m}_{cap,b}\right)\right] - PA_m\left(C^{m}_{cap,d} - C^{m}_{tis,d}\right) \tag{22.4-11}$$

In the extravascular muscle tissue, the total O_2 molar balance for dissolved plus bound O_2 is

$$V^{m}_{tis}\frac{d\left(C^{m}_{tis,d} + C^{m}_{tis,b}\right)}{dt} = \dot{N}^{m}_{d} - \dot{M}^{m}_{tis} = PA_m\left(C^{m}_{cap,d} - C^{m}_{tis,d}\right) - \dot{M}^{m} \tag{22.4-12}$$

where $\dot{M}^{m}$ is the molar rate of O_2 utilization in the muscle tissue.

Reaction Equilibrium for O_2 Binding

Since the reversible reaction rate of dissolved and bound O_2 is so fast compared to other rate processes, we assume local O_2 reaction equilibrium occurs in all subcompartments where binding occurs. Consequently, algebraic relationships exist between the concentrations of bound and dissolved O_2. In all three capillary subcompartments and in venous blood, we use the Hill equilibrium relationship between dissolved O_2 and O_2 bound to hemoglobin (Eq. 5.5-11):

$$C^i_{cap,b} = \frac{C_{Hb}\left(\kappa_{Hb}C^i_{cap,d}\right)^n}{1+\left(\kappa_{Hb}C^i_{cap,d}\right)^n} \quad (i = p, o, m)$$
$$C_{v,b} = \frac{C_{Hb}(\kappa_{Hb}C_{v,d})^n}{1+(\kappa_{Hb}C_{v,d})^n} \tag{22.4-13a,b}$$

where C_{Hb} is the molar O_2 concentration when all heme binding sites are occupied, κ_{Hb} is a binding equilibrium constant, and n is the number of O_2 molecules that bind simultaneously to hemoglobin. In the tissue subcompartment of striated muscle, we assume that O_2 binding to myoglobin is a monovalent process such that

$$C^m_{tis,b} = \frac{C_{Mb}\left(\kappa_{Mb}C^m_{tis,d}\right)}{1+\left(\kappa_{Mb}C^m_{tis,d}\right)} \tag{22.4-14}$$

where C_{Mb} is the molar O_2 concentration at complete saturation of myoglobin and κ_{Mb} is an equilibrium binding constant. We relate the time derivatives of bound and dissolved concentrations by using the chain rule:

$$\frac{dC^i_{cap,b}}{dt} = \left\{\frac{n\kappa_{Hb}C_{Hb}\left(\kappa_{Hb}C^i_{cap,d}\right)^{n-1}}{\left[1+\left(\kappa_{Hb}C^i_{cap,d}\right)^n\right]^2}\right\}\frac{dC^i_{cap,d}}{dt} \quad (i = p, o, m)$$
$$\frac{dC_{v,b}}{dt} = \left\{\frac{n\kappa_{Hb}C_{Hb}(\kappa_{Hb}C_{v,d})^{n-1}}{[1+(\kappa_{Hb}C_{v,d})^n]^2}\right\}\frac{dC_{v,d}}{dt} \tag{22.4-15a-c}$$
$$\frac{dC^m_{tis,b}}{dt} = \left[\frac{\kappa_{Mb}C_{Mb}}{\left(1+\kappa_{Mb}C^m_{tis,d}\right)^2}\right]\frac{dC^m_{tis,d}}{dt}$$

Muscle Metabolism

The change in the metabolic rate of O_2 with exercise is significant for the skeletal muscle where O_2 metabolism occurs primarily in its tissue subcompartment. There, the hydrolysis of ATP to ADP provides the energy necessary for forming the actin–myosin cross-bridges that lead to muscle contraction. This reaction is enzymatically coupled to a regeneration of ATP by the oxidative phosphorylation (OxPhos) of ADP. For simplicity, we omit other aspects of energy metabolism such as cycling between ADP and ATP by creatine kinase and the production of ATP by glycolysis.

The stoichiometry of the ATP – ADP cycling reactions is then simplified:

$$\bar{r}_{ATPase}: \quad ATP \xrightarrow{ATPase} ADP + Pi$$
$$\bar{r}_{OxPhos}: \quad O_2 + \beta ADP + \beta Pi \rightarrow \beta ATP \tag{22.4-16a,b}$$

where $\bar{r}_i$ represents the inherent rates of the individual reactions and $\beta \approx 3$ is the phosphorus–oxygen stoichiometric ratio. The ATP hydrolysis rate is

$$\bar{r}_{ATPase} = k_{ATPase}C_{ATP} \tag{22.4-17}$$

Here, the reaction rate coefficient k_{ATPase} depends on the concentration of the ATPase, which varies with exercise intensity. We express the rate of oxidative phosphorylation such that it reaches saturation at high dissolved O_2 concentration $C^m_{tis,d}$ and ADP concentration C_{ADP}:

$$\bar{r}_{OxPhos} = \nu_{OxPhos}\left(\frac{C^m_{tis,d}}{K_{O_2} + C^m_{tis,d}}\right)\left(\frac{C_{ADP}}{K_{ADP} + C_{ADP}}\right) \tag{22.4-18}$$

where ν_{OxPhos} is the maximum reaction rate per unit tissue volume and K_{O_2} and K_{ADP} are substrate–enzyme Michaelis constants. With this reaction rate per unit volume of muscle tissue, the metabolic O_2 consumption in the entire subcompartment is

$$\dot{M}^m = V^m_{tis}\nu_{OxPhos}\left(\frac{C^m_{tis,d}}{K_{O_2} + C^m_{tis,d}}\right)\left(\frac{C_{ADP}}{K_{ADP} + C_{ADP}}\right) \tag{22.4-19}$$

Within muscle tissue, a material balance on ADP leads to

$$\frac{dC_{ADP}}{dt} = \bar{r}_{ATPase} - \beta\bar{r}_{OxPhos} \tag{22.4-20}$$

Consolidated Balance Equations

Combining the ATP hydrolysis and phosphorylation reaction rate equations with the ATP concentration equation and noting that the total adenosine concentration ($C_T = C_{ADP} + C_{ATP}$) does not change, we get

$$\frac{dC_{ADP}}{dt} = k_{ATPase}(C_T - C_{ADP}) - \beta\nu_{OxPhos}\left(\frac{C^m_{tis,d}}{K_{O_2} + C^m_{tis,d}}\right)\left(\frac{C_{ADP}}{K_{ADP} + C_{ADP}}\right) \tag{22.4-21}$$

After substituting the oxyhemoglobin equilibrium relation into the O_2 balance, the model for the venous mixing point becomes

$$\begin{aligned} C_{v,d} + \frac{C_{Hb}(\kappa_{Hb}C_{v,d})^n}{1 + (\kappa_{Hb}C_{v,d})^n} &= \frac{Q_o}{Q}\left[C^o_{cap,d} + \frac{C_{Hb}\left(\kappa_{Hb}C^o_{cap,d}\right)^n}{1 + \left(\kappa_{Hb}C^o_{cap,d}\right)^n}\right] \\ &\quad + \left(1 - \frac{Q_o}{Q}\right)\left[C^m_{cap,d} + \frac{C_{Hb}\left(\kappa_{Hb}C^m_{cap,d}\right)^n}{1 + \left(\kappa_{Hb}C^m_{cap,d}\right)^n}\right] \end{aligned} \tag{22.4-22}$$

Similarly, we obtain the governing equations for the pulmonary compartment:

$$\begin{aligned} &V^p_{cap}\left\{1 + \frac{V_{alv}}{\lambda^{cap,alv}V^p_{cap}} + \frac{n\kappa_{Hb}C_{Hb}\left(\kappa_{Hb}C^p_{cap,d}\right)^{n-1}}{\left[1 + \left(\kappa_{Hb}C^p_{cap,d}\right)^n\right]^2}\right\}\frac{dC^p_{cap,d}}{dt} \\ &= Q\left[C_{v,d} - C^p_{cap,d} + \frac{C_{Hb}(\kappa_{Hb}C_{v,d})^n}{1 + (\kappa_{Hb}C_{v,d})^n} - \frac{C_{Hb}\left(\kappa_{Hb}C^p_{cap,d}\right)^n}{1 + \left(\kappa_{Hb}C^p_{cap,d}\right)^n}\right] + \dot{V}_{alv}\left(C_{ins} - \frac{C^p_{cap,d}}{\lambda^{cap,alv}}\right) \end{aligned} \tag{22.4-23}$$

for the other organs compartment:

$$V_{cap}^{o}\left\{1+\frac{V_{tis}^{o}}{V_{cap}^{o}}+\frac{n\kappa_{Hb}C_{Hb}\left(\kappa_{Hb}V_{cap}^{o}C_{cap,d}^{o}\right)^{n-1}}{\left[1+\left(\kappa_{Hb}C_{cap,d}^{o}\right)^{n}\right]^{2}}\right\}\frac{dC_{cap,d}^{o}}{dt}$$
$$=Q_{o}\left[C_{cap,d}^{p}-C_{cap,d}^{o}+\frac{C_{Hb}\left(\kappa_{Hb}C_{cap,d}^{p}\right)^{n}}{1+\left(\kappa_{Hb}C_{cap,d}^{p}\right)^{n}}-\frac{C_{Hb}\left(\kappa_{Hb}C_{cap,d}^{o}\right)^{n}}{1+\left(\kappa_{Hb}C_{cap,d}^{o}\right)^{n}}\right]-\dot{M}^{o} \tag{22.4-24}$$

and for the muscle capillary subcompartment:

$$V_{cap}^{m}\left\{1+\frac{n\kappa_{Hb}C_{Hb}\left(\kappa_{Hb}C_{cap,d}^{m}\right)^{n-1}}{\left[1+\left(\kappa_{Hb}C_{cap,d}^{m}\right)^{n}\right]^{2}}\right\}\frac{dC_{cap,d}^{m}}{dt}=PA_{m}\left(C_{tis,d}^{m}-C_{cap,d}^{m}\right)$$
$$+\left(Q-Q_{o}\right)\left[C_{cap,d}^{p}-C_{cap,d}^{m}+\frac{C_{Hb}\left(\kappa_{Hb}C_{cap,d}^{p}\right)^{n}}{1+\left(\kappa_{Hb}C_{cap,d}^{p}\right)^{n}}-\frac{C_{Hb}\left(\kappa_{Hb}C_{cap,d}^{m}\right)^{n}}{1+\left(\kappa_{Hb}C_{cap,d}^{m}\right)^{n}}\right] \tag{22.4-25}$$

Employing the metabolic rate equation in addition to the O_2 myoglobin binding relation, the governing equation for the tissue subcompartment in muscle becomes

$$V_{tis}^{m}\left[1+\frac{C_{Mb}\kappa_{Mb}}{\left(1+\kappa_{Mb}C_{tis,d}^{m}\right)^{2}}\right]\frac{dC_{tis,d}^{m}}{dt}=PA_{m}\left(C_{cap,d}^{m}-C_{tis,d}^{m}\right)$$
$$-V_{tis}^{m}\nu_{OxPhos}\left(\frac{C_{tis,d}^{m}}{K_{O_2}+C_{tis,d}^{m}}\right)\left(\frac{C_{ADP}}{K_{ADP}+C_{ADP}}\right) \tag{22.4-26}$$

The model equations, Equations 22.4-21–22.4-26, require an initial value for the total adenosine concentration and for each of the dissolved O_2 concentration variables:

$$t=0:\quad \begin{array}{l} C_{ADP}=C_{ADP}(0),\;\; C_{v,d}=C_{v,d}(0),\;\; C_{cap,d}^{p}=C_{cap,d}^{p}(0) \\ C_{cap,d}^{o}=C_{cap,d}^{o}(0),\;\; C_{cap,d}^{m}=C_{cap,d}^{m}(0),\;\; C_{tis,d}^{m}=C_{tis,d}^{m}(0) \end{array} \tag{22.4-27}$$

At the onset of exercise at $t = 0$, the system is at a resting steady state. Thus, these six initial concentrations must satisfy the steady-state limits of the six model equations.

Time-Dependent Functions in Response to Exercise

As a consequence of exercise onset at $t = 0$, there is an upregulation of ATPase that is approximately proportional to the increased energy demand:

$$k_{ATPase}(t)=k_{ATPase}(0)+\Delta k_{ATPase}\left(1-e^{-t/t_{ATP}}\right) \tag{22.4-28}$$

Here, $k_{ATPase}(0)$ is the rate coefficient at the initial resting steady state; Δk_{ATPase} is the ultimate increase at the steady state generated by exercise; and t_{ATP} is the exponential time constant during which this increase occurs.

To compensate for the decrease in ATP concentration resulting from the upregulation of ATPase, there must be an increase in the oxidative phosphorylation of ADP. This requires greater oxygen delivery to muscle, provided in part by an increase in blood flow from its initial steady-state value:

$$Q(t) = Q(0) + \Delta Q\left(1 - e^{-t/t_Q}\right) \tag{22.4-29}$$

To accommodate this increased blood flow, additional capillaries are recruited. This provides more surface for O_2 diffusion from capillary blood into myocytes. We reflect this in the PA_m product as:

$$PA_m = PA_m(0) + \Delta PA_m\left(1 - e^{-t/t_{PA}}\right) \tag{22.4-30}$$

Also contributing to more oxygen delivery to skeletal muscle is greater alveolar ventilation:

$$\dot{V}_{alv}(t) = \dot{V}_{alv}(0) + \Delta\dot{V}_{alv}\left(1 - e^{-t/t_{alv}}\right) \tag{22.4-31}$$

Analysis

Outcome Variables

The numerical solution of the model equations (Eqs. 22.4-21–22.4-26) with dynamic parameter variations (Eqs. 22.4-28–22.4-30) provides simulations for the dynamics of the compartmental concentrations. In these simulations, we examine the response to a change in energy demand from a resting state indicated by $k_{ATPase}(0)$ to a steady level of exercise indicated by $k_{ATPase}(0) + \Delta k_{ATPase}$. An important aspect of these dynamics is the relation between the dynamic response of O_2 utilization by skeletal muscle and the dynamic response of pulmonary O_2 uptake.

Once the model equations have been solved for $C^p_{cap}(t)$, the equivalent alveolar gas concentration, $C_{alv}(t) = C^p_{cap}(t)/\lambda^{cap,alv}$, can be used in Equation 22.4-6 to obtain volumetric O_2 uptake into the pulmonary capillaries from the environment:

$$\dot{V}_p \equiv \frac{\dot{N}^p}{c^o_G} = \dot{V}_{alv}\left(C_{ins} - \frac{C^p_{cap}}{\lambda^{cap,alv}}\right) - \frac{V_{alv}}{\lambda^{cap,alv}}\frac{dC^p_{cap}}{dt} \tag{22.4-32}$$

where c^o_G is the standard molar gas density. Often, $\dot{V}_p$ is estimated from breath-by-breath measurements of C_{alv} at the airway opening by assuming that transport in the pulmonary compartment is pseudo-steady so that the time derivative can be ignored:

$$\dot{V}_{p,est} \equiv \frac{\dot{N}^p}{c^o_G} = \dot{V}_{alv}\left(C_{ins} - \frac{C^p_{cap}}{\lambda^{cap,alv}}\right) \tag{22.4-33}$$

The volumetric O_2 uptake rate in the other organs compartment has a constant value:

$$\dot{V}_o = \frac{1}{c^o_G}\dot{M}^o \tag{22.4-34}$$

The volumetric uptake in muscle can be obtained from Equation 22.4-19:

$$\dot{V}_m(t) = \frac{\dot{M}^m}{c^o_G} = \frac{V^m_{tis}\nu_{OxPhos}}{c^o_G}\left[\frac{C^m_{tis,d}(t)}{K_{O_2} + C^m_{tis,d}(t)}\right]\left[\frac{C_{ADP}(t)}{K_{ADP} + C_{ADP}(t)}\right] \tag{22.4-35}$$

Total O_2 utilization in the body is the sum of the contributions from other organs and muscle metabolism $\dot{V}_{m+o} = \dot{V}_m + \dot{V}_o$. When O_2 transport is at steady state throughout the model, either initially or at long times, O_2 is not accumulated in any of the compartments so that $\dot{V}_{m+o} = \dot{V}_p$. Also, since $\dot{V}_o$ is constant in the model, dynamic changes in $\dot{V}_{m+o}$ are identical to dynamic changes in $\dot{V}_m$ alone.

Simulation

In numerical solutions of the model equations, we use the constant parameter values in Table 22.4-1, adapted in part from Lai *et al.* (2007). To obtain initial concentrations that correspond to steady-state resting conditions (Table 22.4-2), we solved Equations 22.4-21–22.4-26 with zero time derivatives and $\Delta k_{ATPase} = \Delta Q = \Delta \dot{V}_{alv} = \Delta PA_m = 0$. In all dynamic simulations, we used the same increase in reaction rate coefficient $\Delta k_{ATPase} = 1\ min^{-1}$ to reflect a large upregulation of ATP hydrolysis in the muscle during a heavy workload.

Figure 22.4-2 shows the dynamics of pulmonary O_2 uptake, $\dot{V}_p$ (Eq. 22.4-32), and total O_2 utilization by muscle and other organs, $\dot{V}_{m+o} \equiv \dot{V}_m + \dot{V}_o$ (Eqs. 22.4-34 and 22.4-35). Also included in these graphs is the pulmonary uptake estimate $\dot{V}_{P,est}$ obtained by neglecting the time derivative in Equation 22.4-32. The only difference among the three graphs in this figure is the exercise-induced increase in pulmonary blood flow, ΔQ, that was imposed in the simulations. The exercise-induced increases in alveolar ventilation and capillary permeability parameters were fixed at $\Delta \dot{V}_{alv} = 9 L/min$ and $\Delta PA_m = 5000\ L/min$, respectively.

In the left graph obtained at the lowest flow increase of $\Delta Q = 3\ L/min$, all three O_2 consumption curves have exponential shapes that originate at the same initial steady state of 0.22 L(STP)/min and merge again at $t = 2$ min when the system reaches its exercise steady

Table 22.4-1 Parameter Values for Model Simulation

Physiology and Anatomy		*Metabolic Reactions*		*Blood–Gas Equilibria*	
$Q(0)$	6.00 L/min	C_T	8.20 mM	$\lambda^{cap,alv}$	0.0252
Q_o	5.90 L/min	$k_{ATPase}(0)$	0.01 min^{-1}	n	2.8
$\dot{V}_{alv}(0)$	5.00 L(STP)/min	ν_{OxPhos}	14.80 mM/min	C_{Hb}	7.96 mM
V_{alv}	3.00 L	K_{O_2}	0.0007 mM	κ_{Hb}	31.0 mM^{-1}
V^p_{cap}	0.08 L	K_{ATP}	0.058 mM	C_{Mb}	0.653 mM
V^o_{cap}	1.85 L	β	3	κ_{Mb}	157 mM^{-1}
V^o_{tis}	27.00 L	**Exercise Dynamics**		c^o_G	44.4 mM
V^m_{cap}	0.65 L	t_{ATP}	0.1 min		
V^m_{tis}	9.10 L	t_Q	0.4 min		
C_{ins}	8.06 mM	t_{alv}	0.4 min		
$\dot{M}^o$	8.98 mmol/min	t_{PA}	0.4 min		
$PA_m(0)$	115 L/min				

Table 22.4-2 Initial Conditions from Solution of Steady-State Equations

0.157 mM	0.0429 mM	Cv,d(0)	0.0539 mM
0.0541 mM	0.0408 mM	CADP(0)	0.109 μM

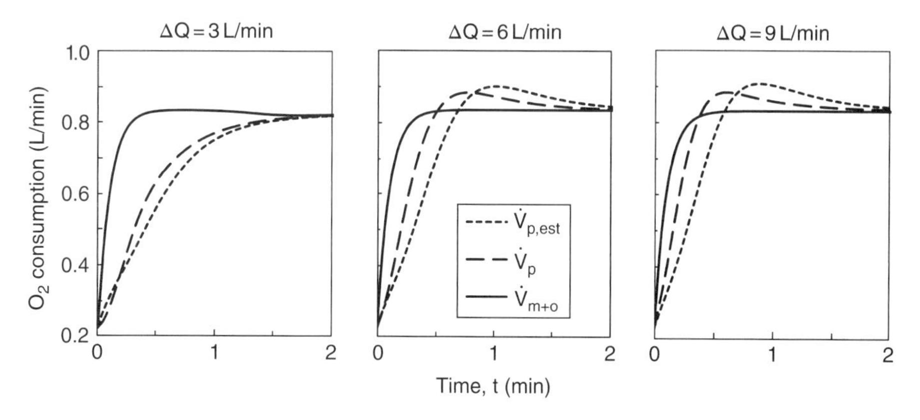

Figure 22.4-2 Dynamics of the model when $\Delta\dot{V}_{alv} = 9L/min$ and $\Delta PA_m = 5000\ L/min$.

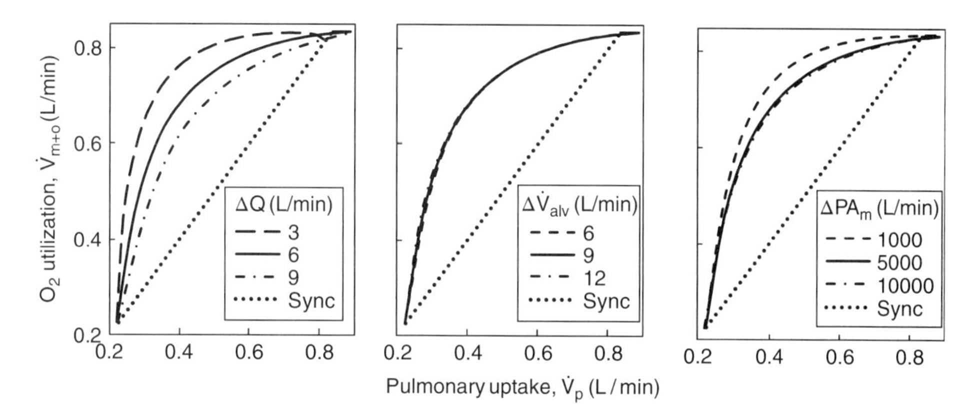

Figure 22.4-3 Phase plane diagrams of dynamic response to exercise.

state. The two pulmonary uptake curves have similar overdamped dynamics that are far more sluggish in response to exercise than is the total utilization curve. This occurs because the time constants for increases in Q, $\dot{V}_{alv}$, and PA_m are much longer that the time constant t_{ATP} for ATP upregulation.

Comparing the three graphs in this figure reveals that $\dot{V}_p$ and $\dot{V}_{p,est}$ dynamics become underdamped and less sluggish compared to $\dot{V}_{m+o}$ dynamics as ΔQ becomes larger so that arterial blood more rapidly delivers O_2 from the pulmonary to the muscle compartment. Also with increasing ΔQ, the $\dot{V}_p$ and $\dot{V}_{p,est}$ curves more closely approximate the $\dot{V}_{m+o}$ curve. No matter what the value ΔQ, however, $\dot{V}_{p,est}(t)$ always lags $\dot{V}_p(t)$. This is due to O_2 depletion in the pulmonary compartment that is not accounted for in the estimated value (i.e., a positive value of the $-dC^P_{cap}/dt$ term in Eqs. 22.4-32).

The dynamic differences between $\dot{V}_p(t)$ and $\dot{V}_{m+o}(t)$ for different changes in exercise parameters are shown by the cross-plots in Figure 22.4-3. The left graph shows the effect of different ΔQ at fixed values of $\Delta\dot{V}_{alv} = 9L/min$ and $\Delta PA_m = 5000\ L/min$; the middle graph shows the effect of different $\Delta\dot{V}_{alv}$ at fixed values of $\Delta Q = 6\ L/min$ and $\Delta PA_m = 5000\ L/min$; the right graph shows the effect of different ΔPA_m at fixed values of $\Delta Q = 6\ L/min$

and $\Delta\dot{V}_{alv} = 9L/\min$. The dotted "sync" lines correspond to a hypothetical situation in which dynamic changes in $\dot{V}_p(t)$ are able to keep up with those of $\dot{V}_{m+o}$ so the two quantities are equal at all times. In fact, the simulation curves only lie on the sync line at two points. These points correspond to the initial resting steady state and the final exercise steady state.

In general, model simulations lie to the left of the sync line because pulmonary uptake usually lags O_2 total utilization such that $\dot{V}_p(t) < \dot{V}_{m+o}(t)$. Whereas these asynchronies become more pronounced as either ΔQ or ΔPA_m decreases, changes in $\Delta\dot{V}_{alv}$ have virtually no effect. Some simulations in the upper right-hand corner of the graphs lie to the right of the sync line. This occurs when pulmonary uptake overshoots the exercise steady state such that $\dot{V}_p(t) > \dot{V}_{m+o}(t)$.

We conclude that the estimation of O_2 utilization by muscle from pulmonary measurements of O_2 retention is practical only when O_2 transport is close to steady state throughout the body. And even when relying on pulmonary measurements, there is a significant discrepancy between the true O_2 uptake into pulmonary capillaries and the uptake that is typically approximated by assuming pseudo-steady transport in the pulmonary compartment.

22.5 Tracer Analysis with Pet Imaging

Positron emission tomography (PET) is a medical imaging technique for noninvasive assessment of tissue dysfunction. A PET scanner detects gamma rays from a tissue in which a biologically active molecule of interest has been tagged with a positron-emitting radionuclide. A gamma ray is produced when an emitted positron collides with an electron in the tissue. By recording the radiation from a multitude of positions surrounding the body, a PET system can reconstruct three-dimensional images of the tracer within various tissue regions of interest. The radioactively tagged molecule chosen for PET imaging depends on the biological process under study or the disorder to be diagnosed. For example, oxygen-15 has been used as an indirect measure of blood flow in studying brain activity. Molecular analogs of glucose have been used as markers of metabolic activity associated with myocardial ischemia and with cancer metastasis. Tagged ligands that bind to neuroreceptors have been used to study schizophrenia, mood disorder, and other psychiatric conditions. The distinction between normal and abnormal physiological function is often determined by the density of cell receptors, which is inferred from the distribution of multiple nuclide species.

As an illustration of this, we consider the transport and reaction processes of endogenous glucose (g) and two ^{18}F-tagged analogs, 2-fluorodeoxyglucose (a2) and 6-fluorodeoxyglucose (a6), after the two radionuclides are injected intravenously. Upon reaching a capillary bed, these molecules diffuse into the extravascular space. These are transported across cell membranes by binding to the receptor of a GLUT carrier molecule. Once within a cell, glucose g and analog a2 are phosphorylated to form gp and a2p, respectively, which cannot leave the cell. Since analog a6 does not undergo phosphorylation, it can leave the cell.

We model the transport and reaction of glucose and its radionuclides in a small tissue region-of-interest (ROI). When a rapid injection of a6 is followed at a much later time by a rapid injection of a2, a6 is largely washed out before a2 is introduced. By simulating a dynamic sequence of PET scans, we can determine how GLUT receptor density affects the difference between gamma emissions from a6 at short times and a2 at later times and how a deficiency in GLUT receptors affects the nuclide responses.

Model Formulation

Model Structure

Our representation of the ROI (Fig. 22.5-1) consists of well-mixed capillary, interstitial, and intracellular compartments of volumes V_{cap}, V_{int}, and V_{cell} (Huang *et al.*, 2011). This system is perfused by capillary blood at a volumetric flow Q.

Separate solutions of the two nuclides are prepared for injection at $t = 0$. These a6 and a2 solutions are injected sequentially into a large blood vessel outside of the ROI at times $t = \tau_{a6}$ and $t = \tau_{a2} \gg \tau_{a6}$, respectively. This results in capillary input concentrations $C^{cap}_{a6,in}(t;\tau_i)$ and $C^{cap}_{a2,in}(t;\tau_i)$ within the ROI. Endogenous glucose ($i = g$) is also present in the capillary compartment. The tagged ligands and glucose diffuse from capillary blood to the interstitial fluid compartment at a molar rate $\dot{N}_i^{int}$. They move by facilitated membrane transport from the interstitial compartment into the intracellular compartment at a molar rate $\dot{N}_i^{cell}$. Within the cells, ligands g and a2 are converted at a molar rate $\dot{R}_{ip}^{cell}$ to phosphorylated species ($i = gp$, $a2p$) that cannot leave the cells.

Solute Concentration Dynamics

During the limited time of the experiment, endogenous glucose concentrations in capillary blood C_g^{cap}, interstitial fluid C_g^{int}, and intracellular spaces C_g^{cell} remain close to their steady-state levels. These quantities therefore appear as constant parameters in this model.

The molar concentrations of the radionuclides in the capillary compartment change according to

$$V_{cap}\frac{dC_i^{cap}}{dt} = QC^{cap}_{i,in} - QC_i^{cap} - N_i^{int} \quad (i = a2, a6) \tag{22.5-1}$$

Since the convective transport rates are much greater than both the transient term and the interphase transport rate, $C^{cap}_{i,in} \approx C_i^{cap}$. Thus, C_i^{cap} serves as the input variable to our model.

In the interstitial compartment, molar concentrations C_i^{int} of the ligands are described by

$$V_{int}\frac{dC_i^{int}}{dt} = \dot{N}_i^{int} - \dot{N}_i^{cell} \quad (i = g, a2, a6) \tag{22.5-2}$$

In the cellular compartment, their concentrations C_i^{cell} behave according to

$$V_{cell}\frac{dC_i^{cell}}{dt} = \dot{N}_i^{cell} - \dot{R}_{ip}^{cell} \quad (i = g, a2, a6) \tag{22.5-3}$$

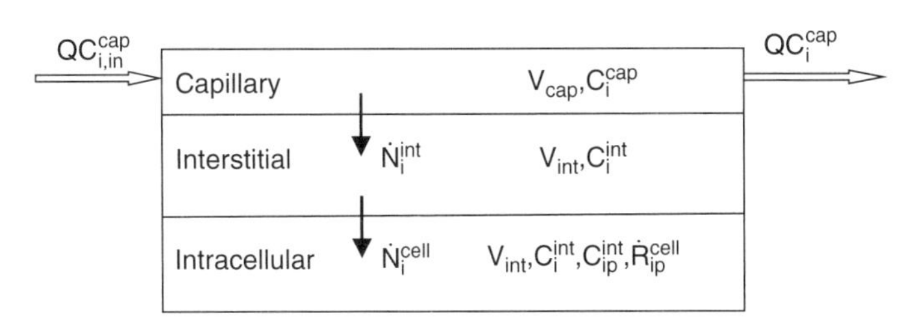

Figure 22.5-1 Tracer transport of glucose analogs for PET image analysis.

The intracellular concentrations of the phosphorylated species are

$$V_{cell}\frac{dC_{ip}^{cell}}{dt} = \dot{R}_{ip}^{cell} \quad (i = g, a2) \tag{22.5-4}$$

Transport and Reaction Rates

For endogenous glucose at steady state, the time derivatives in these three equations vanish so that

$$\dot{N}_g^{int} = \dot{N}_g^{cell} = \dot{R}_{gp}^{cell} \tag{22.5-5}$$

For glucose and its analogs, the transport rate between the capillary and interstitial compartments is determined by capillary membrane diffusion:

$$\dot{N}_i^{int} = P_{cap}S_{cap}\left(C_i^{cap} - \lambda C_i^{int}\right) \quad (i = g, a2, a6) \tag{22.5-6}$$

where P_{cap} is a solute permeability coefficient, S_{cap} is the surface area of the capillary walls, and λ is an equilibrium partition coefficient between capillary blood and interstitial fluid. Because of the similarity of glucose and its analogs, the values of P_{cap} and λ are the same for all of them.

The three nonphosphorylated ligands compete for the same binding site on the GLUT transporter during cotransport across cell membranes. With radionuclide concentrations much smaller than glucose concentration, interactions between a2 and a6 are minimal compared to their interactions with glucose. Therefore, the transport rates of a2 and a6 can be represented by Equation 11.3-6:

$$\dot{N}_i^{cell} = \frac{C_G P_G S_{cell}}{2}\left(\frac{C_i^{int}}{C_g^{int} + \kappa} - \frac{C_i^{cell}}{C_g^{cell} + \kappa}\right) \quad (i = a2, a6) \tag{22.5-7}$$

Because glucose concentration is relatively high, glucose interactions with the radioactive analogs do not affect its transport rate that can be expressed by Equation 11.3-7:

$$\dot{N}_g^{cell} = \frac{C_G P_G S_{cell}}{2}\left(\frac{C_g^{int}}{C_g^{int} + \kappa} - \frac{C_g^{cell}}{C_g^{cell} + \kappa}\right) \tag{22.5-8}$$

These two facilitated transport equations assume that (i) the GLUT transporter concentration C_G is fixed, (ii) the same translocation rate constant P_G governs the translocation of free transporter and its occupied forms in both directions across the membrane, and (iii) the transporter binding reactions have the same equilibrium binding constant κ for glucose and its analogs.

Within the cell, phosphorylation is described by Michaelis–Menten kinetics for both glucose and analog a2, which are assumed to be governed by the same kinetic parameters:

$$\dot{R}_{ip}^{cell} = \begin{cases} V_{cell}\left(\dfrac{V_m C_i^{cell}}{C_i^{cell} + K_m}\right) & \text{when } i = g, a2 \\ 0 & \text{when } i = a6 \end{cases} \tag{22.5-9}$$

Here, V_m is the maximum phosphorylation rate per unit volume of the intracellular compartment, and K_m is the concentration at which the phosphorylation rate reaches one-half its maximum.

Consolidated Equations

Endogenous glucose concentrations, which are constant, are related by the steady-state equation.

$$P_{cap}S_{cap}\left(C_g^{cap}-\lambda C_g^{int}\right)=\frac{C_G P_G S_{cell}}{2}\left(\frac{C_g^{int}}{C_g^{int}+\kappa}-\frac{C_g^{cell}}{C_g^{cell}+\kappa}\right)=\frac{V_{cell}V_m C_g^{cell}}{K_m+C_g^{cell}} \quad (22.5\text{-}10)$$

The interstitial concentration dynamics of the a2 and a6 analogs are

$$\frac{dC_i^{int}}{dt}=\frac{P_{cap}S_{cap}}{V_{int}}\left(C_i^{cap}-\lambda C_i^{int}\right)-\frac{C_G P_G S_{cell}}{2V_{int}}\left(\frac{C_i^{int}}{C_g^{int}+\kappa}-\frac{C_i^{cell}}{C_g^{cell}+\kappa}\right) \quad (i=a2,a6) \quad (22.5\text{-}11)$$

In the cell, the concentrations of the a2 and a6 analogs change according to

$$\frac{dC_i^{cell}}{dt}=\frac{C_G P_G S_{cell}}{2V_{cell}}\left(\frac{C_i^{int}}{C_g^{int}+\kappa}-\frac{C_i^{cell}}{C_g^{cell}+\kappa}\right)-\begin{cases}\dfrac{V_m C_i^{cell}}{K_m+C_i^{cell}} & \text{when } i=a2\\ 0 & \text{when } i=a6\end{cases} \quad (22.5\text{-}12)$$

Also in the cell, the concentration of analog a2p changes as:

$$\frac{dC_{a2p}^{cell}}{dt}=\frac{V_m C_{a2p}^{cell}}{C_{a2p}^{cell}+K_m} \quad (22.5\text{-}13)$$

Initially, the glucose analogs are not present in the interstitial and intracellular compartments:

$$t=0:\quad C_{a2p}^{cell}=0,\quad C_i^{int}=C_i^{cell}=0 \quad (i=a2,a6) \quad (22.5\text{-}14)$$

Radioactivity Output

While the ^{18}F incorporated into the glucose analogs is emitting positrons, it decays with half-life of $T_{1/2}=109.8$ min to ^{18}O, a stable isotope of oxygen. Because sequential PET measurements are taken over 60 min or more, we must account for radio-decay when converting total ligand concentration simulated with the transport model to the corresponding radioactivity.

A single PET image provides the total count of nuclear disintegrations from the three compartments in an ROI over a short time interval about a time t. Specific radioactivity $[y^*(t)]_i^j$ is the count rate at a time t due to nuclide i in compartment j divided by the total moles of species i in both nuclide and nonradioactive forms [Bq/mol i]. The absolute count rate is

$$\gamma_i^j=[y^*(t)]_i^j C_i^j V_j \quad (22.5\text{-}15)$$

where V_j is compartment volume, and C_i^j is the total molar concentration of ligand i. After accounting for the exponential decay of nuclide i over a time interval {0,t}, this equation can be written in terms of the specific radioactivity initially originating from compartment j, $[y^*(0)]_i^j$:

$$[y^*(t)]_i^j=\exp\left[-\frac{\ln(2)}{T^{1/2}}t\right][y^*(0)]_i^j \;\Rightarrow\; \gamma_i^j=\exp\left[-\frac{\ln(2)}{T^{1/2}}t\right][y^*(0)]_i^j C_i^j V_j \quad (22.5\text{-}16)$$

Here, $\mathcal{A}$ is Avogadro's number, and $T_{1/2}$ is the half-life of the nuclide.

Because radioactive and nonradioactive forms of species i behave in essentially the same manner, we expect that $[y^*(0)]_i^j$ will deviate little from its value in the injection solutions that are prepared at t = 0. The $[y^*(0)]_i^j$ will then be the same for all compartments and will be designated as $y_i^*(0)$. The total count rate produced by radionuclide i in all three compartments of an ROI can then be written as

$$\gamma_i(t) = y_i^*(0)\exp\left[-\frac{\ln(2)}{T_{1/2}}t\right]\left[V_{cap}C_i^{cap} + V_{int}C_i^{int} + V_{cell}\left(C_i^{cell} + C_{ip}^{cell}\right)\right] \tag{22.5-17}$$

When ligands a6 and a2 are simultaneously present, the total activity in an ROI is the sum of their individual radioactivities:

$$\begin{aligned}\gamma^{PET}(t) = \exp\left[-\frac{\ln(2)}{T_{1/2}}t\right]&\left[y_{a6}^*(0)\left(V_{cap}C_{a6}^{cap} + V_{int}C_{a6}^{int} + V_{cell}C_{a6}^{cell}\right)\right.\\ &\left.+ y_{a2}^*(0)\left(V_{cap}C_{a2}^{cap} + V_{int}C_{a2}^{int} + V_{cell}C_{a2}^{cell} + V_{cell}C_{a2p}^{cell}\right)\right]\end{aligned} \tag{22.5-18}$$

Analysis

To simplify the simulations, we assume that the starting solutions prepared at t = 0 have the same specific activities of a2 and a6, that is, $y_{a2}^*(0) = y_{a6}^*(0) = y^*(0)$. As a further simplification, we assume that these solutions also contain the same concentration of ligand. In that case, the dynamic input concentrations of a6 and a2 are essentially equal but are shifted in time, that is, $C_{a6}^{cap}(t + \tau_{a6}) = C_{a2}^{cap}(t + \tau_{a2}) \equiv C^{cap}(t)$. This allows us to write separate input concentrations of a ligand as

$$C_i^{cap}(t) = C^{cap}(t - \tau_i) \quad (i = a2, a6) \tag{22.5-19}$$

Dimensionless Forms

The model equations can be made dimensionless with the following variables:

$$\begin{aligned}&t \equiv \frac{P_{cap}S_{cap}}{V}t, \quad \gamma^{PET} = \frac{\gamma^{PET}}{\hat{A}VC_o^{cap}}, \quad C_{a2p}^{cell} \equiv \frac{C_{a2p}^{cell}}{C_o^{cap}}\\ &C_i^j \equiv \frac{C_i^j}{C_o^{cap}} \quad (i = a2, a6;\ j = int, cell)\\ &C^{cap}(t - \tau_i) \equiv \frac{C^{cap}(t - \tau_i)}{C_o^{cap}} \quad (i = a2, a6)\end{aligned} \tag{22.5-20a-e}$$

where V is the ROI volume and C_o^{cap} is a scaling concentration associated with the radionuclide input function $C^{cap}(t)$. In the interstitial space, the dimensionless concentrations for the glucose analogs change as

$$\frac{dC_i^{int}}{dt} = \frac{1}{\varepsilon_{int}}\left[C^{cap}(t - \tau_i) - \lambda C_i^{int}\right] - \frac{C_G}{\varepsilon_{int}}\left(\frac{C_i^{int}}{C_g^{int} + \kappa} - \frac{C_i^{cell}}{C_g^{cell} + \kappa}\right) \quad (i = a2, a6) \tag{22.5-21}$$

In the cell, the dimensionless concentrations of nonphosphorylated analogs change as

$$\frac{dC_i^{cell}}{dt} = \frac{C_G}{\varepsilon_{cell}}\left(\frac{C_i^{int}}{C_g^{int}+\kappa} - \frac{C_i^{cell}}{C_g^{cell}+\kappa}\right) - \begin{cases} \dfrac{V_m C_i^{cell}}{C_i^{cell}+K_m} & \text{when i = a2} \\ 0 & \text{when i = a6} \end{cases} \tag{22.5-22}$$

The concentration of the phosphorylated analog a2 changes as

$$\frac{dC_{a2p}^{cell}}{dt} = \frac{V_m C_{a2}^{cell}}{C_{a2}^{cell}+K_m} \tag{22.5-23}$$

The dimensionless initial conditions on these equations are

$$t = 0: \quad C_{a2p}^{cell} = 0, \quad C_i^{int} = C_i^{cell} = 0 \quad (\mathrm{i = a2, a6}) \tag{22.5-24}$$

The dimensionless total radioactivity is

$$\begin{aligned} \gamma^{PET} = \exp\left(-\frac{t}{T_{1/2}}\right)\Big\{&\left[\varepsilon_{cap}C^{cap}(t-\tau_{a6}) + \varepsilon_{int}C_{a6}^{int} + \varepsilon_{cell}C_{a6}^{cell}\right] \\ &+ \left[\varepsilon_{cap}C^{cap}(t-\tau_{a2}) + \varepsilon_{int}C_{a2}^{int} + \varepsilon_{cell}\left(C_{a2}^{cell} + C_{a2p}^{cell}\right)\right]\Big\} \end{aligned} \tag{22.5-25}$$

In these equations, the dimensionless parameters are

$$\begin{aligned} &C_g^j \equiv \frac{C_g^j}{C_o^{cap}}, \quad \varepsilon_j \equiv \frac{V_j}{V} \quad (\mathrm{j = cap, int, cell}) \\ &\tau_i \equiv \frac{P_{cap}S_{cap}}{V}\tau_i \quad (\mathrm{i = a2, a6}) \\ &V_m \equiv \frac{VV_m}{P_{cap}S_{cap}C_o^{cap}}, \quad K_m \equiv \frac{K_m}{C_o^{cap}}, \quad C_G \equiv \frac{1}{2}\frac{C_G P_G S_{cell}}{C_o^{cap}P_{cap}S_{cap}} \\ &\kappa \equiv \frac{\kappa}{C_o^{cap}}, \quad T_{1/2} \equiv \frac{P_{cap}S_{cap}T_{1/2}}{V\ln(2)} \end{aligned} \tag{22.5-26a-g}$$

where $\varepsilon_{cap} + \varepsilon_{int} + \varepsilon_{cell} = 1$. The dimensionless equation for endogenous glucose provides additional relations between the parameters:

$$\left(C_g^{cap} - \lambda C_g^{int}\right) = C_G\left(\frac{C_g^{int}}{C_g^{int}+\kappa} - \frac{C_g^{cell}}{C_g^{cell}+\kappa}\right) = \frac{\varepsilon_{cell}V_m C_g^{cell}}{C_g^{cell}+K_m} \tag{22.5-27}$$

Simulation

The input concentration equation $C^{cap}(t-\tau_i)$ depends on the details of ligand injection and the physiological characteristics of an individual patient. We model this by a concentration equation inferred from the activity-time behavior of a sequence of blood samples taken from one individual subsequent to nuclide injection (Fang and Muzic, 2008). By normalizing $C^{cap}(t-\tau_i)$ with the scaling parameter $C_o^{cap} = 0.04\,\mathrm{mM}$, we force the dimensionless input concentration $C^{cap}(t-\tau_i)$ to have a peak value of one:

$$\begin{aligned} C^{cap}(t-\tau_i) = H(t-\tau_i)\{&(770(t-\tau_i)-0.46)\exp[-400(t-\tau_i)] \\ &+ 0.31\exp[-20(t-\tau_i)] + 0.15\exp[-0.5(t-\tau_i)]\} \end{aligned} \tag{22.5-28}$$

where $H(t-\tau_i)$, the unit step function applied at $t = \tau_i$, insures that $C^{cap} = 0$ prior to injection. Using the C_o^{cap} value with published parameter values for rat skeletal muscle (Huang *et al.*,

2011), we determined the following baseline values of the dimensionless parameters: $\varepsilon_{cap} = 0$, $\varepsilon_{int} = 0.25$, $\varepsilon_{cell} = 0.75$, $\kappa = 90$, $K_m = 3.3$, $C_g^{cap} = 150$, $C_g^{int} = 130$, $C_g^{cell} = 1.2$, and $T_{1/2} = 6.3$; and the scaling factor for time is $P_{cap}S_{cap}/V = 2.4$ h. After choosing $\lambda = 1$, the remaining dimensionless parameters values of $C_G = 40$ and $V_m = 120$ are computed from the steady-state glucose equations (Eq. 22.5-27).

The upper graphs of Figure 22.5-2 show the results of model simulations for a two hour experiment for which $\tau_{a6} = 0$ and $\tau_{a2} = 2.4$. As seen in the figures (and expected from the time derivative in Eq. 22.5-21), increases in interstitial concentrations C_i^{int} of both nonphosphorylated ligands are most rapid soon after injection when C_i^{cap} is at its largest and C_i^{int} is very small. Both ligands reach a maximum C_i^{int} at about 0.3 dimensionless time units after initiating their injection. This corresponds to an actual time interval of 8 min. Thereafter, C_i^{int} tends to equalize with the rapidly declining capillary concentration C_i^{cap}. Whereas the a6 ligand continually accumulates within the cell, the rapid phosphorylation of the a2 ligand diminishes its intracellular concentration to a very low level. Intracellular phosphorylation of a2 is also evident from the continuous increase in intracellular a2p.

The lower graphs of Figure 22.5-2 show simulations of radionuclide concentration dynamics when there is an insufficiency in GLUT transporter density, as can occur with diabetes.

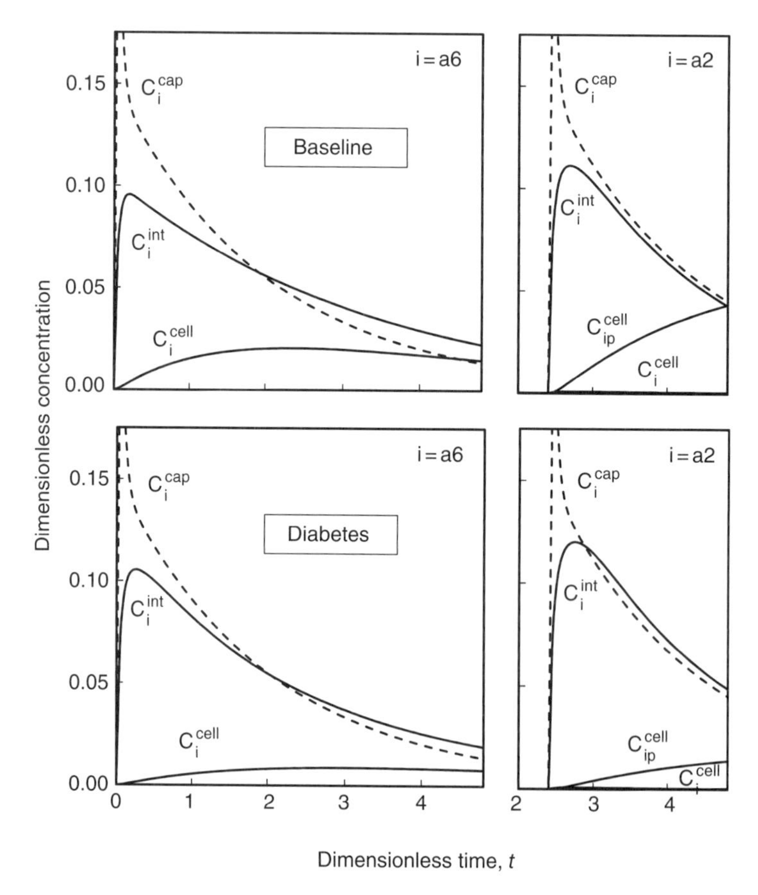

Figure 22.5-2 Concentration dynamics of radionuclides in ROI compartments.

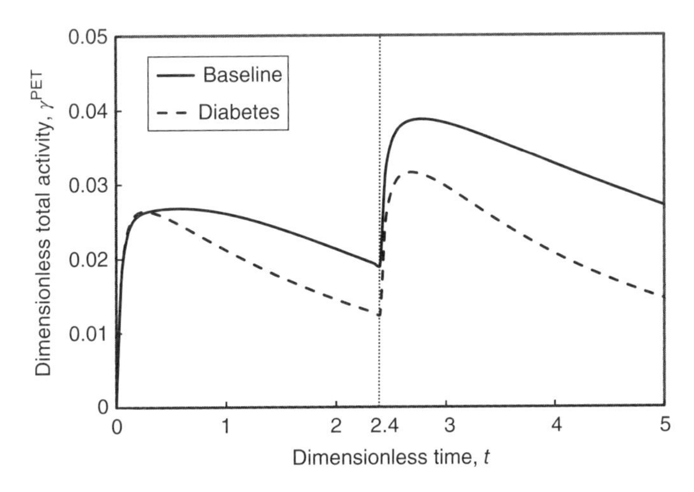

Figure 22.5-3 Sequential PET recordings of radioactive counts in an ROI.

This situation was portrayed by reducing the GLUT transporter concentration C_G to half of its baseline value and increasing the endogenous blood glucose concentration C_g^{cap} to twice its baseline level. To maintain steady-state glucose transport (Eq. 22.5-27), it was also necessary to increase interstitial glucose to $C_g^{int} = 290$ and decrease intracellular glucose to $C_g^{cell} = 0.74$. The elevated C_g^{int} results in more competition with glucose analogs for GLUT transporter in the interstitial space. This process, in addition to the lower C_G, causes a reduction in cellular uptake of a2 and a6 compared to the baseline graphs. Consequently, the intracellular formation of a2p is also reduced.

Figure 22.5-3 compares the simulated radioactivity γ^{PET} of the glucose analogs under baseline and diabetic conditions. Following a6 injection at $t = 0$ and a2 injection at $t = 2.4$, the γ^{PET} curves exhibit a rapid initial rise to a peak value, paralleling the early portion of the interstitial concentration curves. The decline in total radioactivity occurring after the γ^{PET} peaks is due to a loss of the radionuclides to capillary blood as well as to the radio-decay of retained radionuclides. This decline is more rapid when GLUT receptor density is reduced, probably because less radionuclide accumulates in the cellular space, which contributes to the largest volume fraction among the three model compartments to Equation 22.5-25. As a result, the γ^{PET} peak height during the second hour relative to that occurring during the first hour decreases from a value of about 1.4 for the baseline curves to a value of 1.1 when the GLUT receptor density is reduced.

22.6 Cancer Cell Migration with Cell–Cell Interaction

Migration of cancer cells is a key determinant of metastasis, which is correlated with a poor prognosis in patients. The motility of cancer cells is regulated by their interaction with stromal cells (e.g., fibroblasts). In a tumor, a mixture of cancer and stromal cells is surrounded by extracellular fluid (ECF) containing matrix molecules (ECM) and soluble factors. One of these factors, matrix metalloproteinase (MMP), which is produced by cancer cells, digests

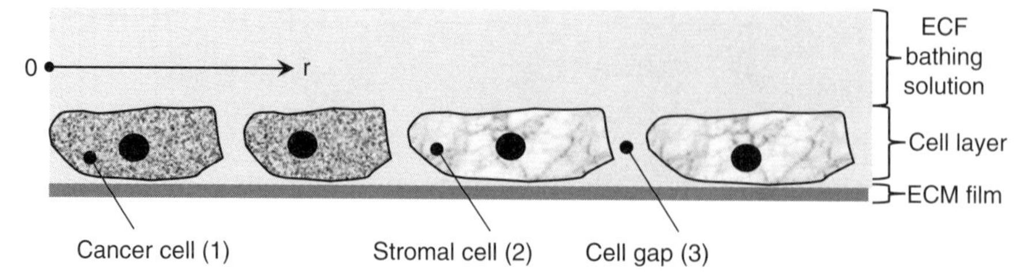

Figure 22.6-1 *In vitro* assay for cancer cell migration in tissue.

local matrix molecules. The resulting spatial gradient in matrix concentration enhances cancer cell migration into normal tissue by haptotaxis.

Incorporating this mechanism of directed migration, we model an *in vitro* assay of cancer cell transport in a thin layer. The layer initially consists of a circular region containing a uniform number density Ω_{1o} of cancer cells (i) surrounded by a uniform number density Ω_{2o} of stromal cells (ii). The ECF that fills the gaps between cells (iii) is also present as a bathing layer above the cells. A film of ECM molecules such as collagen or fibronectin anchors the cells to a planar support below the cell layer (Fig. 22.6-1).

Model Formulation

We expect transport in the cell layer to be axisymmetric with cancer cell migration affected by interactions with stromal cells as well as with cell gaps. To account for random motility in this ternary system, we will model it as a multicomponent molecular diffusion process.

The MMP excreted by cancer cells into ECF is transported by pseudo-binary diffusion, primarily in the radial direction. We assume that at any radial position, MMP concentration is equally distributed among the bathing solution, cell gaps, and the ECM film. Because ECM molecules are firmly attached to the underlying substrate, they do not undergo transport.

Number Density Relationships

The mixture of cells and cell gaps forms a planar layer whose discrete components have local volume fractions $\varepsilon_i(r,t)$. The spatial distribution of cells is given by their number densities, $\Omega_1(r,t)$ and $\Omega_2(r,t)$. If $\hat{\Omega}_i$ represents the maximum packing density of the cells, then their number densities can be written as $\Omega_1 = \varepsilon_1\hat{\Omega}_1$ and $\Omega_2 = \varepsilon_2\hat{\Omega}_2$. We assume that the cell gaps have a constant volume and their number distribution $\Omega_3(r,t)$ has a maximum packing density $\hat{\Omega}_3$. Since the volume fractions of the cells plus their gaps must sum to one at any r position,

$$\sum_{i=1}^{3}\varepsilon_i = 1 \;\Rightarrow\; \sum_{i=1}^{3}\Omega_i/\hat{\Omega}_i = 1 \qquad (22.6\text{-}1a,b)$$

For simplicity, we assume approximately equal packing densities such that $\hat{\Omega}_i \rightarrow \hat{\Omega}$ and $\Omega_i \rightarrow \varepsilon_i\hat{\Omega}$. Consequently,

$$\sum_{i=1}^{3}\left(\Omega_i/\hat{\Omega}\right) = 1 \;\Rightarrow\; \Omega_1 + \Omega_2 + \Omega_3 = \hat{\Omega} \qquad (22.6\text{-}2a,b)$$

With this equation, we can determine $\Omega_3(r,t)$ indirectly from known values of $\hat{\Omega}$, $\Omega_1(r,t)$, and $\Omega_2(r,t)$.

Component Dynamics

Transport of cells occurs by radial fluxes associated with random motility (Θ_i^{rand}) and by haptotaxis along the ECM gradient in the surface film (Θ_i^{taxis}). For a relative short experiment in which cell fate processes (e.g., proliferation or death) are absent, changes in cell densities can be determined from number balances involving only transport processes in the cell (Eq. 17.1-1). For transport in the r direction alone, a cell number balance in the cell layer leads to

$$\frac{\partial \Omega_i}{\partial t} = -\frac{1}{r}\frac{\partial\left(r\Theta_i^{rand}\right)}{\partial r} - \frac{1}{r}\frac{\partial\left(r\Theta_i^{taxis}\right)}{\partial r} \quad (i = 1,2) \tag{22.6-3}$$

Since ECM does not move, its concentration changes only by a degradation reaction that is proportional to the concentrations of ECM and MMP. A molar balance on ECM in the surface film yields

$$\frac{\partial C_{ECM}}{\partial t} = -k_{ECM} C_{ECM} C_{MMP} \tag{22.6-4}$$

The concentration of MMP changes because of diffusion, production by cancer cells, and loss of enzyme activity:

$$\frac{\partial C_{MMP}}{\partial t} = \mathcal{D}_{MMP}\frac{1}{r}\frac{\partial}{\partial r}\left(r\frac{\partial C_{MMP}}{\partial r}\right) - k_{MMP} C_{MMP} + k_1 \Omega_1 \tag{22.6-5}$$

where $\mathcal{D}_{MMP}$ is a diffusion coefficient, k_{MMP} is a loss coefficient, and k_1 is a production coefficient.

Cell Flux Relationships

In modeling the random motility of cancer cells, stromal cells, and cell gaps in the cell layer, we relate fluxes Θ_i^{rand} to number density driving forces $\partial\Omega_i/\partial r$ by analogy to ternary diffusion. A basic assumption is that the process of random motility is distinct (i.e., thermodynamically uncoupled) from chemotaxis.

For a solution containing molecular species i = 1,2,3 with a constant molar density $c = C_1 + C_2 + C_3$, the multicomponent diffusion equations (Eqs. 6.3-17a and 6.3-18) indicate that the molar fluxes of species 1 and 2 are related to the gradients of their molar concentration gradients as follows:

$$\begin{bmatrix} \mathbf{J}_1 \\ \mathbf{J}_2 \end{bmatrix} = -\begin{bmatrix} B_{11} & B_{12} \\ B_{21} & B_{22} \end{bmatrix}\begin{bmatrix} \nabla C_1 \\ \nabla C_2 \end{bmatrix} \tag{22.6-6}$$

We are concerned only with radial fluxes, which are governed by the r component of this vector-matrix equation:

$$\begin{bmatrix} J_{1,r} \\ J_{2,r} \end{bmatrix} = -\begin{bmatrix} B_{11} & B_{12} \\ B_{21} & B_{22} \end{bmatrix}\begin{bmatrix} \dfrac{\partial C_1}{\partial r} \\ \dfrac{\partial C_2}{\partial r} \end{bmatrix} \tag{22.6-7}$$

Here, the components of the [B] matrix are defined by Equations 6.3-14 and 6.3-15 as:

$$\begin{bmatrix} B_{11} & B_{12} \\ B_{21} & B_{22} \end{bmatrix} = \frac{1}{c} \begin{bmatrix} \dfrac{c - C_2}{\mathcal{D}_{13}} + \dfrac{C_2}{\mathcal{D}_{12}} & C_1\left(\dfrac{1}{\mathcal{D}_{13}} - \dfrac{1}{\mathcal{D}_{12}}\right) \\ C_2\left(\dfrac{1}{\mathcal{D}_{23}} - \dfrac{1}{\mathcal{D}_{12}}\right) & \dfrac{c - C_1}{\mathcal{D}_{23}} + \dfrac{C_1}{\mathcal{D}_{12}} \end{bmatrix} \tag{22.6-8}$$

In this equation, $\mathcal{D}_{ij} = \mathcal{D}_{ji}$ is the binary diffusion coefficient in a solution of i and j alone. Since the sum all three diffusion fluxes $J_{i,r}$ must sum to zero, the diffusion flux of species 3 can be obtained from the fluxes of the other two species:

$$J_{3,r} = -\left(J_{1,r} + J_{2,r}\right) \tag{22.6-9}$$

For the random migration in the analogous three-component cell layer, we identify $C_i \rightarrow \Omega_i$, $J_{i,r} \rightarrow \theta_i^{rand}$, $(\mathcal{D}_{ij} = \mathcal{D}_{ji}) \rightarrow \left(\mu_{ij}^{rand} = \mu_{ji}^{rand}\right)$, and $c \rightarrow \hat{\Omega}$. Consequently, Equations 22.6-7–22.6-9 become

$$\begin{bmatrix} \theta_1^{rand} \\ \theta_2^{rand} \end{bmatrix} = -\begin{bmatrix} B_{11} & B_{12} \\ B_{21} & B_{22} \end{bmatrix}^{-1} \begin{bmatrix} \dfrac{\partial \Omega_1}{\partial r} \\ \dfrac{\partial \Omega_2}{\partial r} \end{bmatrix} \tag{22.6-10}$$

where

$$\begin{bmatrix} B_{11} & B_{12} \\ B_{21} & B_{22} \end{bmatrix} = \frac{1}{\hat{\Omega}} \begin{bmatrix} \dfrac{\hat{\Omega} - \Omega_2}{\mu_{13}^{rand}} + \dfrac{\Omega_2}{\mu_{12}^{rand}} & \Omega_1\left(\dfrac{1}{\mu_{13}^{rand}} - \dfrac{1}{\mu_{12}^{rand}}\right) \\ \Omega_2\left(\dfrac{1}{\mu_{23}^{rand}} - \dfrac{1}{\mu_{12}^{rand}}\right) & \dfrac{\hat{\Omega} - \Omega_1}{\mu_{23}^{rand}} + \dfrac{\Omega_1}{\mu_{12}^{rand}} \end{bmatrix} \tag{22.6-11}$$

and

$$\Theta_3^{rand} = -\left(\Theta_1^{rand} + \Theta_2^{rand}\right) \tag{22.6-12}$$

Using the methods in Appendix C2, we find the inverse of the B matrix:

$$\begin{bmatrix} B_{11} & B_{12} \\ B_{21} & B_{22} \end{bmatrix}^{-1} = \frac{\begin{bmatrix} \dfrac{\hat{\Omega} - \Omega_1}{\mu_{23}^{rand}} + \dfrac{\Omega_1}{\mu_{12}^{rand}} & -\Omega_1\left(\dfrac{1}{\mu_{13}^{rand}} - \dfrac{1}{\mu_{12}^{rand}}\right) \\ -\Omega_2\left(\dfrac{1}{\mu_{23}^{rand}} - \dfrac{1}{\mu_{12}^{rand}}\right) & \dfrac{\hat{\Omega} - \Omega_2}{\mu_{13}^{rand}} + \dfrac{\Omega_2}{\mu_{12}^{rand}} \end{bmatrix}}{\dfrac{\hat{\Omega}}{\mu_{13}^{rand}\mu_{23}^{rand}} + \Omega_1\left(\dfrac{1}{\mu_{13}^{rand}\mu_{12}^{rand}} - \dfrac{1}{\mu_{13}^{rand}\mu_{23}^{rand}}\right) + \Omega_2\left(\dfrac{1}{\mu_{12}^{rand}\mu_{23}^{rand}} - \dfrac{1}{\mu_{13}^{rand}\mu_{23}^{rand}}\right)} \tag{22.6-13}$$

Random motility is a process that occurs between cells and cell gaps, but not between cells so that $\mu_{12}^{rand} \rightarrow 0$. In that case, the inverse $[B]^{-1}$ reduces to

$$\begin{bmatrix} B_{11} & B_{12} \\ B_{21} & B_{22} \end{bmatrix}^{-1} \equiv \begin{bmatrix} B_{11}^{-1} & B_{12}^{-1} \\ B_{21}^{-1} & B_{22}^{-1} \end{bmatrix} = \frac{1}{\Omega_1/\mu_{13}^{rand} + \Omega_2/\mu_{23}^{rand}} \begin{bmatrix} \Omega_1 & \Omega_1 \\ \Omega_2 & \Omega_2 \end{bmatrix} \tag{22.6-14}$$

Substituting this into Equation 22.6-10, we find the random motilities of the cellular components

$$\begin{aligned} \Theta_1^{rand} &= -\frac{\Omega_1}{\Omega_1/\mu_{13}^{rand} + \Omega_2/\mu_{23}^{rand}} \frac{\partial(\Omega_1 + \Omega_2)}{\partial r} \\ \Theta_2^{rand} &= -\frac{\Omega_2}{\Omega_1/\mu_{13}^{rand} + \Omega_2/\mu_{23}^{rand}} \frac{\partial(\Omega_1 + \Omega_2)}{\partial r} \end{aligned} \qquad (22.6\text{-}15a,b)$$

from which the motility Θ_3^{rand} of cell gaps can be obtained with Equation 22.6-12. The r component of primary haptotaxis imposed by the ECM gradient on the cancer cells (Eq. 17.2-2) is

$$\Theta_1^{taxis} = \left(\mu^{taxis} \frac{\partial C_{ECM}}{\partial r}\right) \Omega_1 \qquad (22.6\text{-}16)$$

where μ^{taxis} is a taxis rate coefficient. Note that there is no haptotaxis associated with stromal cells. Substituting these flux equations for random migration and cancer cell haptotaxis into the cell density equations (Eq. 22.6-3), we obtain

$$\begin{aligned} \frac{\partial \Omega_1}{\partial t} &= \frac{1}{r}\frac{\partial}{\partial r}\left[\frac{r\Omega_1}{\Omega_1/\mu_{13}^{rand} + \Omega_2/\mu_{23}^{rand}} \frac{\partial(\Omega_1 + \Omega_2)}{\partial r}\right] - \frac{\mu^{taxis}}{r}\frac{\partial}{\partial r}\left(r\Omega_1 \frac{\partial C_{EM}}{\partial r}\right) \\ \frac{\partial \Omega_2}{\partial t} &= \frac{1}{r}\frac{\partial}{\partial r}\left[\frac{r\Omega_2}{\Omega_1/\mu_{13}^{rand} + \Omega_2/\mu_{23}^{rand}} \frac{\partial(\Omega_1 + \Omega_2)}{\partial r}\right] \end{aligned} \qquad (22.6\text{-}17a,b)$$

Initial Conditions

In the cell layer, a circle of radius "a" initially contains a uniform distribution of cancer cells of number density Ω_{10}, but no stromal cells. Surrounding this cancer cell region is a uniform distribution Ω_{20} of stromal cells with an inner radius "a" and a much larger outer radius ($r \gg a$):

$$t = 0: \quad \begin{aligned} \Omega_1 &= \Omega_{1o}[H(r-0) - H(r-a)] \\ \Omega_2 &= \Omega_{2o} H(r-a) \end{aligned} \qquad (22.6\text{-}18)$$

where $H(r - k)$ is one for $r > k$ and zero otherwise. In addition, the ECM film initially contains a uniform ECM concentration C_{ECMo}. No MMP initially exists in the system:

$$t = 0: \quad C_{ECM} = C_{ECMo}, \quad C_{MMP} = 0 \qquad (22.6\text{-}19)$$

Boundary Conditions

Spatial symmetry dictates that gradients of cell number density and MMP concentration are zero at $r = 0$. Far from the origin, these gradients are zero because spatial changes are negligible:

$$\left.\begin{aligned} r &= 0 \\ r &\to \infty \end{aligned}\right\}: \quad \frac{\partial \Omega_1}{\partial r} = \frac{\partial \Omega_2}{\partial r} = \frac{\partial C_{MMP}}{\partial r} = 0 \qquad (22.6\text{-}20a,b)$$

Analysis

Dimensionless Forms

We can express this model using the following dimensionless variables:

$$t \equiv \frac{t\mu_{13}^{rand}}{a^2},\quad r \equiv \frac{r}{a},\quad C_{MMP} = \frac{\mu_{13}^{rand}}{a^2 k_1 \hat{\Omega}} C_{MMP},\quad C_{ECM} = \frac{C_{ECM}}{C_{ECMo}}$$
$$\Omega_i \equiv \frac{\Omega_i}{\hat{\Omega}} \quad (i = 1,2) \tag{22.6-21a-e}$$

The dimensionless governing equations for the cell number densities are

$$\frac{\partial \Omega_1}{\partial t} = \frac{1}{r}\frac{\partial}{\partial r}\left[\frac{r\Omega_1}{\Omega_1 + \mu^{rand}\Omega_2}\frac{\partial(\Omega_1 + \Omega_2)}{\partial r}\right] - \frac{\mu^{taxis}}{r}\frac{\partial}{\partial r}\left(r\Omega_1 \frac{\partial C_{ECM}}{\partial r}\right) \tag{22.6-22}$$

$$\frac{\partial \Omega_2}{\partial t} = \frac{1}{r}\frac{\partial}{\partial r}\left[\frac{r\Omega_2}{\Omega_1 + \mu^{rand}\Omega_2}\frac{\partial(\Omega_1 + \Omega_2)}{\partial r}\right] \tag{22.6-23}$$

For ECM and MMP, changes in dimensionless concentrations are

$$\frac{\partial C_{ECM}}{\partial t} = -k_{ECM} C_{ECM} C_{MMP} \tag{22.6-24}$$

$$\frac{\partial C_{MMP}}{\partial t} = D_{MMP}\frac{1}{r}\frac{\partial}{\partial r}\left(r\frac{\partial C_{MMP}}{\partial r}\right) + \Omega_1 - k_{MMP} C_{MMP} \tag{22.6-25}$$

The dimensionless initial conditions are

$$t = 0:\quad \begin{aligned} &C_{ECM} = 1,\ C_{MMP} = 0 \\ &\Omega_1 = H(r-0) - H(r-a) \\ &\Omega_2 = H(r-a) \end{aligned} \tag{22.6-26}$$

The dimensionless boundary conditions are

$$\left.\begin{aligned} r &= 0 \\ r &\to \infty \end{aligned}\right\}:\quad \frac{\partial \Omega_1}{\partial r} = \frac{\partial \Omega_2}{\partial r} = \frac{\partial C_{MMP}}{\partial r} = 0 \tag{22.6-27a,b}$$

The independent dimensionless parameter groups are

$$D_{MMP} = \frac{\mathcal{D}_{MMP}}{\mu_{13}^{rand}},\quad k_{ECM} = \frac{k_{ECM} k_1 a^4 \Omega}{\left(\mu_{13}^{rand}\right)^2},\quad k_{MMP} = \frac{k_{MMP} a^2}{\mu_{13}^{rand}},\quad \mu^{rand} = \frac{\mu_{13}^{rand}}{\mu_{23}^{rand}}$$
$$\mu^{taxis} = \frac{\mu^{taxis} C_{ECMo}}{\mu_{13}^{rand}},\quad \Omega_{io} = \Omega_{io}\hat{\Omega} \quad (i = 1,2) \tag{22.6-28a-f}$$

Simulation

The model equations are solved numerically to simulate the spatial distribution dynamics of cancer and stromal cells. To provide computational stability, the initial spatial step changes of Ω_1 and Ω_2 are approximated by continuous functions:

$$\Omega_1(0) = \Omega_{1o}\left(1 - \frac{1}{1 + e^{-2(r-1)}}\right),\quad \Omega_2(0) = \frac{\Omega_{2o}}{1 + e^{-2(r-1)}} \tag{22.6-29a,b}$$

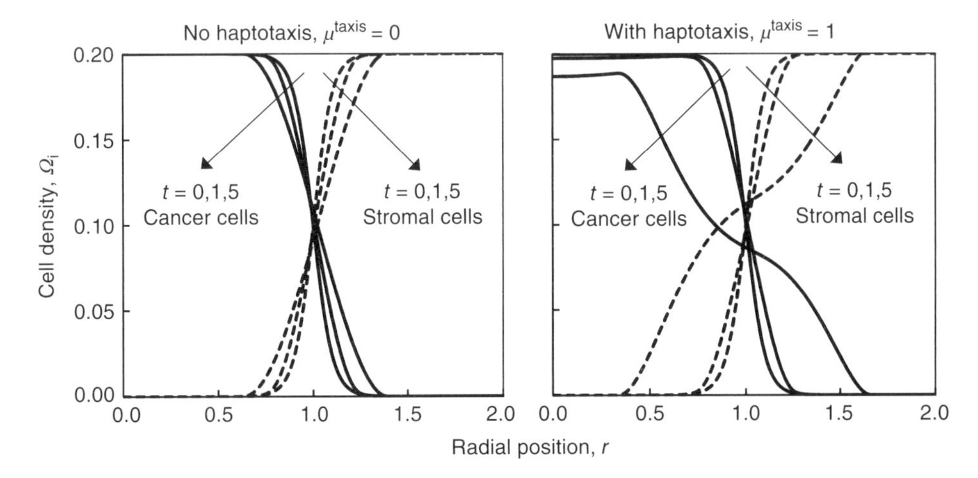

Figure 22.6-2 Dynamic distributions of cancer and stromal cells.

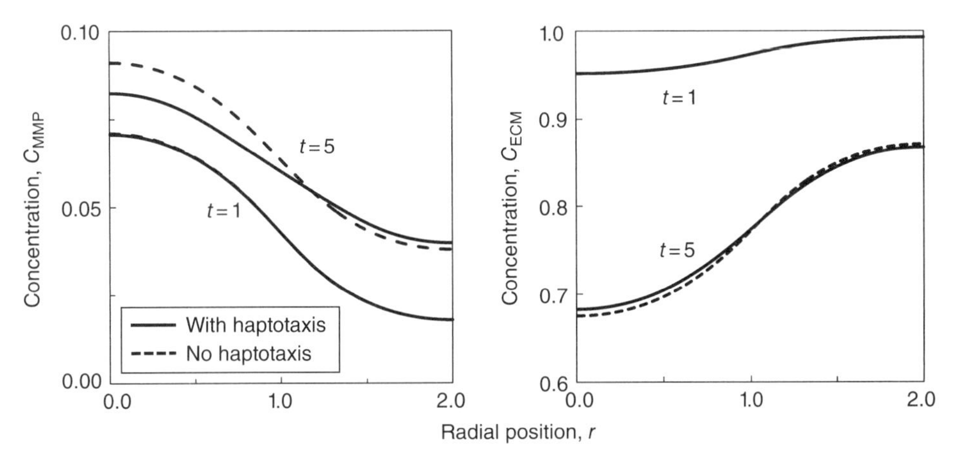

Figure 22.6-3 Dynamic distributions of MMP and ECM.

One of the key aspects of model behavior is the effect of cancer cell haptotaxis on the distribution of cancer and stromal cells in the same spatial domain. We evaluate this by comparing simulations with haptotaxis ($\mu^{taxis} > 0$) to those without haptotaxis ($\mu^{taxis} = 0$). Realistic values of dimensional model parameters are available from Sarkar *et al.* (2011). However, to allow processes other than haptotaxis to be of equal importance, we let $D_{MMP} = k_{ECM} = k_{MMP} = \mu^{rand} = 1$. We arbitrarily set the initial cell seeding fractions at $\Omega_{1o} = \Omega_{2o} = 0.2$.

Figure 22.6-2 shows the changes of number density distributions of cancer and stromal cells with radial position in the cell layer at dimensionless times $t = 0,1,5$. Simulations with random motility and no haptotaxis (left graph) indicate that there is similar invasion of cancer cells into the stromal region at $r > 1$ and stromal cells into the cancer region at $r < 1$. With haptotaxis of cancer cells (right graph), these effects are enhanced, particularly for the cancer cells.

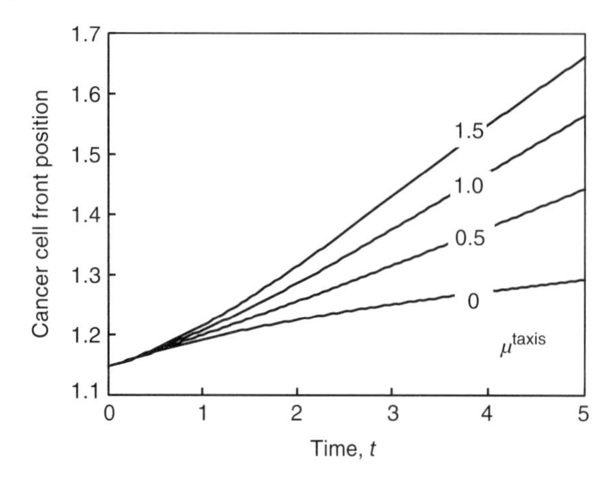

Figure 22.6-4 Effect of haptotaxis on cancer cell migration.

Figure 22.6-3 shows the spatial distributions of MMP concentration in ECF and ECM concentration in the surface film. The MMP concentration decreases with distance from the system center at $r = 0$ because of the radial decrease in the density of cancer cells that produce MMP. In contrast, C_{ECM} increases as r increases, because of the slower rates of ECM breakdown at the lower MMP concentrations. As time progresses, cancer cell haptotaxis causes a reduction in C_{MMP}, especially closer to the center where the number density of cancer cells is most reduced by the directed motility. Haptotaxis has a smaller effect on C_{ECM} than on C_{MMP}.

Figure 22.6-4 shows the time-varying position of a "cancer cell migration front," defined as the r location where Ω_1 first exceeds a threshold value of 0.01. With cancer cell haptotaxis at $\mu^{taxis} > 0$, the front position is an increasing function of time that is greater than with no haptotaxis at $\mu^{taxis} = 0$. The faster the front position moves, the larger is the haptotaxis rate constant.

References

Fang YD, Muzic RF. Spillover and partial-volume correction for image-derived input functions for small-animal ^{18}F-FDG PET studies. J Nucl Med. 2008; 49:606–614.

Huang H-M, Ismail-Beigi F, Muzic RM. A new Michaelis–Menten-based kinetic model for transport and phosphorylation of glucose and its analogs in skeletal muscle. Med Phys. 2011; 38:4587–4599.

Kim T-J, Firouztale E, Ultman JS. Simultaneous measurement of liquid velocity and oxygen tension with catheter-tip electrodes. IEEE Trans Biomed Eng. 1981; 28:342–348.

Lai N, Camesasca M, Saidel GM, Dash RK, Cabrera ME. Linking pulmonary oxygen uptake, muscle oxygen utilization and cellular metabolism during exercise. Ann Biomed Eng. 2007; 35:956–969.

Sarkar S, Bustard BL, Welter JF, Baskaran H. Combined experimental and mathematical approach for development of microfabrication-based cancer migration assay. Ann Biomed Eng. 2011; 39:2346–2359.

Appendix A

Units and Property Data

A.1 American National Standard for SI Units

Table A1-1 Base Units

Quantity	Name	Symbol
Length	Meter	m
Mass	Kilogram	kg
Time	Second	s
Electric current	Ampere	A
Absolute temperature	Kelvin	K
Amount of substance	Mole	mol
Luminous intensity	Candela	cd

From IEEE/ASTM (2002).

Table A1-2 Prefixes

Multiplication Factor	Name	Symbol
10^{12}	Tera	T
10^{9}	Giga	G
10^{6}	Mega	M
$10^{3} = 1000$	Kilo	k
$10^{2} = 100$	Hecto	h
$10^{1} = 10$	Deka	da
$10^{-1} = 0.1$	Deci	d
$10^{-2} = 0.01$	Centi	c
$10^{-3} = 0.001$	Milli	m
10^{-6}	Micro	μ

(continued overleaf)

Biomedical Mass Transport and Chemical Reaction: Physicochemical Principles and Mathematical Modeling,
First Edition. James S. Ultman, Harihara Baskaran, and Gerald M. Saidel.

Table A1-2 *(continued)*

Multiplication Factor	Name	Symbol
10^{-9}	Nano	n
10^{-12}	Pico	p
10^{-15}	Femto	f

From IEEE/ASTM (2002).

Table A1-3 Derived Units

Quantity	Name	Symbol	Derived Units	Base Units
Angle, plane	Radian	rad	—	$m{\cdot}m^{-1} = 1$
	Degree	°	$(\pi/180)$ rad	$(\pi/180)\ m{\cdot}m^{-1}$
	Revolution	rev	(2π) rad	$(2\pi)\ m{\cdot}m^{-1}$
Radioactivity (rate of nuclear decay)	Becquerel	Bq	—	s^{-1}
Concentration and molar density	Molar	M	mol/L	$kmol/m^3$
	Millimolar	mM	—	mol/m^3
Electric charge	Coulomb	C	—	s·A
Electric conductance	Siemens	S	A/V = 1/Ohm	$m^{-2}{\cdot}kg^{-1}{\cdot}s^3{\cdot}A^2$
Electric potential	Volt	V	W/A	$m^2{\cdot}kg{\cdot}s^{-3}{\cdot}A^{-1}$
Electric resistance	Ohm	Ω	V/A	$m^2{\cdot}kg{\cdot}s^{-3}{\cdot}A^{-2}$
Energy, work, heat	Joule	J	N·m	$m^2{\cdot}kg{\cdot}s^{-2}$
Force	Newton	N	—	$m{\cdot}kg{\cdot}s^{-2}$
Frequency	Hertz	Hz	—	s^{-1}
Molecular weight	Daltons	Da	g/mol	kg/kmol
Power	Watt	W	J/s	$m^2{\cdot}kg{\cdot}s^{-3}$
Pressure, stress	Pascal	Pa	N/m^2	$kg{\cdot}m^{-1}{\cdot}s^{-2}$
Volume	Liter	L	—	$10^{-3}\ m^3$

From IEEE/ASTM (2002).

Table A1-4 Conversion to SI Units from Other Commonly Used Units

Quantity	Name	Symbol	Value in SI Units
Distance	Angstrom	Å	Å = 0.1 nm = 10^{-10} m
Energy	erg	erg	erg = 10^{-7} J
Force	dyne	dyn	dyn = 10^{-5} N
Force	Kilogram-force	kgf	kgf = 9.80665 N
Heat	Calorie	cal	cal = 4.148 J
Pressure	Standard atmosphere	atm	atm = 101.325 kPa
Pressure	Millimeter of mercury	mmHg or torr	mmHg = 133.3 Pa
Viscosity	Poise	P	P = $dyn{\cdot}s/cm^2$ = 0.1 Pa·s

From IEEE/ASTM (2002).

A.2 Definitions of Concentration

Table A2-1 Concentration Measures in Different Phases

A. Bulk-phase properties

$\rho \equiv$ mass density $[kg/m^3]$

$c \equiv$ molar density, subscript G for gas and L for liquid $[mol/dm^3 \equiv mol/L \equiv M]$

$M \equiv \rho/c \equiv$ average molecular weight $[g/mol \equiv Da]$

B. Component i in a gas or vapor phase

$y_i \equiv$ mole fraction of i [mole i/mol mixture]

$p_i \equiv$ partial pressure of i [kPa]

$C_i \equiv$ molar concentration of i $[mol\ i/dm^3 mixture \equiv mol/L \equiv M]$

Useful ideal gas relationships:

$p_i = y_i P;\ \ C_i \equiv y_i c_G;\ \ C_i = p_i/\mathcal{R}T$

$$\sum_{i=1}^{I} p_i = P\sum_{i=1}^{I} y_i = P;\ \ \sum_{i=1}^{I} C_i = (1/\mathcal{R}T)\sum_{i=1}^{I} p_i = c_G$$

C. Component i in a liquid or solid phase

$x_i \equiv$ mole fraction of i [mol i/mol mixture]

$C_i \equiv$ molar concentration of i $[mol\ i/dm^3 mixture \equiv mol/L \equiv M]$

$\omega_i \equiv$ mass fraction of i [kg i/kg mixture]

$\rho_i \equiv$ mass concentration $[kg\ i/dm^3 mixture]$

$\hat{C}_i \equiv$ content of gas i $[m^3\ gas/m^3\ liquid]$

Useful relationships:

$C_i = x_i c_L;\ \ \rho_i = \omega_i \rho_L;\ \ \rho_i = MC_i;\ \ \hat{C}_i = C_i/c_G^o$

$$\sum_{i=1}^{I} C_i = c_L\sum_{i=1}^{I} x_i = C;\ \ \sum_{i=1}^{I} \rho_i = \rho_L$$

Table A2-2 Concentration Measures for Different Applications

A. Molarity [1 mol/L ≡ 1 M]

Used for molar balances. Based on a mole(mol) that is defined as an Avogadro's number of molecules

B. Normality [1 eq/L ≡ 1 N]

Used for electric charge balances. Based on the equivalent (eq) that is defined as one mole divided by the combining capacity (CC). For our purposes, CC is equal to the total charge number of the positive or the negative ions into which a strong electrolyte dissociates

$NaCl \rightarrow Na^{+} + Cl^{-}$ $\quad CC = (1)(1) = (1)(1) = 1$

$CaCl_2 \rightarrow Ca^{+2} + 2Cl^{-}$ $\quad CC = (1)(2) = (2)(1) = 2$

$Al_2(SO_4)_3 \rightarrow 2Al^{+3} + 3SO_4^{-2}$ $\quad CC = (2)(3) = (3)(2) = 6$

C. Osmolarity [1 osmol/L ≡ 1 OSM]

Used in water flux computations. Based on the osmole(osmol), the moles of separate entities formed by dissolved substances. For an undissociated substance, osmole ≡ mole. For a strong electrolyte, osmole = (#ions) (mole)

(continued overleaf)

Table A2-2 *(continued)*

$NaCl \rightarrow Na^{+} + Cl^{-}$	1 mole $NaCl = (1+1)$ osmles
$CaCl_2 \rightarrow Ca^{+2} + 2Cl^{-}$	1 mole $CaCl_2 = (1+2)$ osmles
$Al_2(SO_4)_3 \rightarrow 2Al^{+3} + 3SO_4^{-2}$	1 mole $Al_2(SO_4)_3 = (2+3)$ osmles

D. Comparison

NaCl 1 M = 1 N = 2 OSM

$CaCl_2$ 1 M = 2 N = 3 OSM

$Al_2(SO_4)_3$ 1 M = 6 N = 5 OSM

Glucose 1 M = 1 OSM

A.3 Thermodynamic Properties

Table A3-1 Partition Coefficients $\lambda_s^{A,B}$ of Liquid (l) and Solid (s) Solutes between Organic (A) and Aqueous (B) Solvents at 25°C

Phases Solutes	A = Olive oil B = Water	A = Octanol B = Water	A = Lipid B = Water
Erythritol (s)	3.0×10^{-5}	0.0012	0.026
Ethanediol (l)	4.9×10^{-4}	0.012	0.12
Ethanol (l)	3.6×10^{-2}	0.48	0.44
Glycerol (l)	7.0×10^{-5}	0.0028	0.050
Methanol (l)	9.5×10^{-3}	0.18	0.21
n-Hexanol (l)	7.6	110	—
n-Propanol (l)	1.4×10^{-1}	2.2	1.3
Sucrose (s)[a]	3.0×10^{-5}	—	—
Thiourea (s)	1.2×10^{-3}	0.072	—
Urea (s)	1.5×10^{-4}	0.0022	0.23
Water (l)	1.3×10^{-3}	0.041	—

From Leib and Stein (1986).
[a] Davson (1964).

Table A3-2 Henry's Law Constant, h_s [10^9 Pa] of Gases in Water

	T (°C)					
Gas Solute s	25	30	35	37	40	45
Acetylene (C_2H_2)	0.135	0.148	—	0.166[a]	—	—
Air	7.29	7.81	8.34	8.53[b]	8.81	9.23
Carbon dioxide (CO_2)	0.166	0.188	0.212	0.221[b]	0.236	0.260
Carbon monoxide (CO)	5.88	6.28	6.68	6.83[b]	7.05	7.39
Helium (He)	—	12.6	—	12.3[b]	12.2	—
Hydrogen (H_2)	7.16	7.38	7.52	7.55[b]	7.61	7.70

Table A3-2 *(continued)*

Gas Solute s	T (°C) 25	30	35	37	40	45
Nitric oxide (NO)	2.91	3.14	3.35	3.44[b]	3.57	3.77
Nitrogen (N_2)	8.76	9.36	9.98	10.21[b]	10.54	11.04
Nitrous oxide (N_2O)	0.228	0.262	0.306	0.325[a]	—	—
Oxygen (O_2)	4.44	4.81	5.14	5.25[b]	5.42	5.70
Ozone	0.463	0.606	0.829	0.949[b]	1.216	—

From Perry *et al.* (1963).
[a] Extrapolated values.
[b] Interpolated values.

Table A3-3 Bunsen Solubility Coefficients α_s of O_2 and CO_2

Gas Solute	Oxygen			Carbon Dioxide	
	ml gas(STP)/(dL liquid-kPa)			ml(STP)/(dL-kPa)	
T (°C)	Isotonic Saline[a]	Human Plasma	Human Blood[b]	Isotonic Saline[a]	Ox Plasma
20	0.0295	0.0273	0.0282	0.850	0.777
25	0.0268	0.0254	0.0261	0.732	0.672
30	0.0246	0.0235	0.0241	0.640	0.593
31	0.0243	0.0231	0.0238	0.625	0.579
32	0.0240	0.0227	0.0234	0.611	0.566
33	0.0236	0.0223	0.0231	0.596	0.553
34	0.0233	0.0220	0.0228	0.582	0.540
35	0.0230	0.0217	0.0225	0.568	0.528
36	0.0227	0.0214	0.0222	0.556	0.518
37	0.0224	0.0211	0.0219	0.543	0.508
38	0.0222	0.0209	0.0217	0.531	0.497
39	0.0219	0.0207	0.0214	0.516	0.486
40	0.0216	0.0205	0.0211	0.506	0.476

Modified from Altman and Dittmer (1971, pp. 20–21).
[a] Solution of 0.155 mol NaCl per liter.
[b] Based on a hemoglobin content of 15 g hemoglobin/dL (α_{O2} increases by about 0.3% per increase of 1 g/dL blood in hemoglobin content).

Table A3-4 Partition Coefficients of Gases at 37–38°C

Gas Solute	Phases A : B	$\lambda_i^{A,B}$	Gas Solute i	Phases A : B	$\lambda_i^{A,B}$
Acetone	Blood–gas	333.0	Helium	Water–gas	0.0097
				Blood–gas	0.0098
				Oil–water	1.7
Acetylene	Water–gas	0.850	Hydrogen	Water–gas	0.018
	Blood–gas	0.795		Oil–water	3.1
Argon	Water–gas	0.0295	Nitrogen	Water–gas	0.0144
	Oil–water	5.3		Blood–gas	0.0147
				Brain–blood	1.1
				Liver–blood	1.1
				Fat–blood	5.2
				Oil–water	5.2
Chloroform	Water–gas	4.6	Nitrous oxide	Water–gas	0.440
	Blood–gas	7.3		Blood–gas	0.466
	Brain–blood	1.1		Brain–blood	1.0
	Liver–blood	0.9		Heart–blood	1.0
	Oil–water	110.0		Oil–water	3.2
Ethylene	Water–gas	0.089	Sulfur hexafluoride	Water–gas	0.00437
	Blood–gas	0.140			
	Brain–blood	1.2			
	Heart–blood	1.0			
	Oil–water	14.4			
Ethyl ether	Water–gas	15.5	Xenon	Water–gas	0.097
	Blood–gas	14.9		Oil–water	20.0
	Brain–blood	1.14			
	Oil–water	3.2			

From Altman and Dittmer (1971, pp. 20–21).

Table A3-5 Standard Free Energy of Reaction at Body Conditions (T = 37°C, P = 101.3 kPa)

Reaction	ΔG_r^* (kJ/mole substrate)	pH
Glucose $\rightleftharpoons$ 2 Lactate$^-$ + 2H$^+$	−198	7.0
Glucose + ATP^{4-} $\rightleftharpoons$ Glucose 6P^{2-} + ADP^{3-} + H$^+$	−21	7.0
ATP^{4-} + H_2O $\rightleftharpoons$ ADP^{3-} + HPO_4^{2-} + H$^+$	−37	7.5
ADP^{3-} + H_2O $\rightleftharpoons$ AMP^{2-} + HPO_4^{2-} + H$^+$	−40	7.5
Enolpyruvate 2P^{3-} + ADP^{3-} + H$^+$ $\rightleftharpoons$ Pyruvate + ATP^{4-}	−25	7.0
Pyruvate$^-$ + DPNH + H$^+$ $\rightleftharpoons$ Lactate$^-$ + DPN$^+$	−23	7.0
Succinate^{2-} + ½ O_2 $\rightleftharpoons$ Fumarate^{2-} + H_2O	−151	7.0
DPNH + ½ O_2 + H$^+$ $\rightleftharpoons$ DPN$^+$ + H_2O	−219	7.0
Glucose + 6O_2 $\rightleftharpoons$ 6CO_2 + 6H_2O	−2872	—
Glycylglycine + H_2O $\rightleftharpoons$ 2 Glycine	−15	—

From Snell *et al.* (1965, p. 255).

Table A3-6 Dissociation Equilibria of Some Common Amino Acids[a]

Amino Acid	pK_1	pK_2	pK_3	pI
Glycine	2.34 (COOH)	9.60 (NH_3^+)	—	5.97
Aspartic acid	2.09 (COOH)	3.86 (COOH)	9.82 (NH_3^+)	2.97
Lysine	2.18 (COOH)	8.95 (NH_3^+)	10.53 (NH_3^+)	9.74
Tyrosine	2.20 (COOH)	9.11 (NH_3^+)	10.07 (OH)	5.65
Cysteine	1.71 (COOH)	8.33 (SH)	10.78 (NH_3^+)	5.02
Proline	1.99 (COOH)	10.6 (NH)	—	6.10

From Mahler and Cordes (1968).
[a] Parenthesis contains the protonated form of an acid side group in the amino acid.

A.4 Transport Properties

Table A4-1 Diffusion Coefficients for Binary Gases at 101.3 kPa

System	$\mathcal{D}_{12}$ (10^{-4} m^2/s)	T (°K)
Air–carbon dioxide	0.177	317.2
Air–ethanol	0.145	313
Air–helium	0.765	317.2
Air–water	0.288	313
Carbon dioxide–argon	0.133	276.2
Carbon dioxide–helium	0.612	298
Carbon dioxide–nitrogen	0.167	298
Carbon dioxide–nitrous oxide	0.128	312.8
Carbon dioxide–sulfur dioxide	0.064	263
Carbon dioxide–water	0.198	307.2
Helium–benzene	0.610	423
Helium–ethanol	0.821	423
Helium–water	0.902	307.1
Hydrogen–ammonia	0.783	298
Hydrogen–benzene	0.404	311.3
Hydrogen–cyclohexane	0.319	288.6
Hydrogen–methane	0.694	288
Hydrogen–sulfur dioxide	1.23	473
Hydrogen–water	1.121	328.5
Nitrogen–ammonia	0.2230	298
Nitrogen–benzene	0.102	311.3
Nitrogen–carbon monoxide	0.318	373
Nitrogen–helium	0.687	298
Nitrogen–hydrogen	0.784	298
Nitrogen–sulfur dioxide	0.104	263

(continued overleaf)

Table A4-1 *(continued)*

System	$\mathcal{D}_{12}$ (10^{-4} m^2/s)	T (°K)
Nitrogen–water	0.256	307.5
Oxygen–benzene	0.101	311.3
Oxygen–carbon dioxide	0.153	293.2
Oxygen–carbon tetrachloride	0.0749	296
Oxygen–helium	0.729	298
Oxygen–nitrogen	0.181	273.3
Oxygen–water	0.352	352.3

From Hines and Maddox (1985).

Table A4-2 Diffusion Coefficients at Infinite Dilution in Water

Solute s	$\mathcal{D}_s^\infty$ (10^{-9} m^2/s)	T (°C)
Acetone	1.16	25
Air	2.00	25
Ammonia	1.64	25
Argon	2.00	25
Benzene	1.02	25
Carbon dioxide	1.92	25
Carbon monoxide	2.03	25
Chlorine	1.25	25
Helium	6.28	25
Hydrogen	4.50	25
Hydrogen sulfide	1.41	25
Methane	1.49	25
Nitric oxide	2.60	25
Nitrogen	1.88	25
Oxygen	2.10	25
Ethanol	0.84	25
Methanol	0.84	25
n-Butanol	0.77	25
Propanol	0.87	25
Acetic acid	1.21	25
Benzoic acid	1.00	25
Formic acid	1.50	25
Nitric acid	2.60	25
Propionic acid	1.06	25
Sulfuric acid	1.73	25
Human serum albumin	0.061	20
Fibrinogen	0.020	25
Glucose	0.673	20
Glycine	1.06	25

Table A4-2 *(continued)*

Solute s	$\mathcal{D}_s^\infty$ (10^{-9} m^2/s)	T (°C)
Hemoglobin	0.069	25
Ovalbumin	0.078	25
Phenylalanine	0.705	25
Ribonuclease	0.102	20
Sucrose	$(0.5228–0.265C_s)^a$	25
Tryptophan	0.659	25
Urea	$(1.380 - 0.0782C_s + 0.00464C_s^2)^a$	25
Urease	0.035	25
Valine	0.83	25

From Cussler (1997) and Snell *et al.* (1965, p. 204).
[a] Diffusion coefficients as a function of solute concentration C_s[mol/L].

Table A4-3 Oxygen and Carbon Dioxide Diffusion in Fluids and Tissues

Medium	$\mathcal{D}_s$ (10^{-9} m^2/s)	α_s 10^{-5} ml(STP)/(ml-Pa)	T^a (°C)
s = Oxygen			
Water	2.30	0.031	20
Water	2.85	0.026	30
Water	3.30	0.024	37
Gelatin (15%)	$\alpha_s \mathcal{D}_s = 4.6 \times 10^{-16}$ m^2/(s-Pa)		20
Methemoglobin (8% sol'n.)	1.87	0.025	25
Methemoglobin (33% sol'n.)	0.70	0.033	25
Serum protein (8% sol'n.)	1.85	0.025	25
Serum protein (30% sol'n.)	0.77	—	25
Connective tissue (dog)	0.97	0.023	37
Connective tissue (frog)	0.62	0.031	20
Erythrocytes (human)	$\alpha_s \mathcal{D}_s = 2.1 \times 10^{-16}$ m^2/(s-Pa)		20
Lung tissue (rat)	2.30	0.018	37
Muscle (frog)	0.75	0.031	20
Muscle (frog)	1.17	0.023	37
Myocardial tissue (rat)	1.50	0.021	20
Serum (ox)	1.87	0.021	37
s = Carbon dioxide			
Water	1.85	0.684	25
Water	2.18	0.521	35
Water	2.55	0.560	37
Hemoglobin (33% sol'n.)	0.83	0.65	22
Hemoglobin (33% sol'n.)	1.17	0.43	37
Brain tissue	1.00	0.96	22

(continued overleaf)

Table A4-3 (continued)

Medium	$\mathcal{D}_s$ (10^{-9} m^2/s)	α_s 10^{-5} ml(STP)/(ml-Pa)	T^a (°C)
Connective tissue (frog)	0.88	0.76	20
Diaphragm (dog)	0.60	0.72	22
Erythrocytes (ox)	0.83	0.65	22
Muscle (dog)	1.00	0.77	22
Muscle (frog)	1.95	0.77	22
Nerve	0.12	0.77	22
Skin (frog)	0.70	0.72	22
Smooth muscle (cat)	1.07	0.77	22

From Altman and Dittmer (1971, pp. 21–22).
[a] $\mathcal{D}_s$ increases by about 2% per increase of 1°C in temperature.

References

Altman PL, Dittmer DS. Respiration and Circulation. Bethesda: Federation of American Societies for Experimental Biology; 1971.

Cussler EL. Diffusion: Mass Transfer in Fluid Systems. Cambridge: Cambridge University Press; 1997, p 112.

Davson H. A Textbook of General Physiology. Boston: Little, Brown and Company; 1964, p 286.

Hines AL, Maddox RN. Mass Transfer Fundamentals and Applications. Englewood Cliffs: Prentice Hall. 1985, p 24–25.

IEEE/ASTM. American National Standard for Use of the International System of Units (SI): The Modern Metric System. New York/West Conshohocken: Institute of Electrical and Electronics Engineers/ASTM International; 2002.

Leib WR, Stein WD. Non-Stokesian nature of transverse diffusion within human red cell membranes. J Membrane Biol. 1986; 92:114–119.

Mahler HR, Cordes EH. Basic Biological Chemistry. New York: Harper and Row; 1968, p 35–37.

Perry JH, Chilton CH, Kirkpatrick SD. Chemical Engineers' Handbook. 4th ed. New York: McGraw-Hill; 1963, p 3–7, Sec 14.

Snell FM, Shulman S, Spencer RP, Moos C. Biophysical Principles of Structure and Function. Reading: Addison-Wesley; 1965.

Appendix B

Representing Transport Processes in Complex Systems

B.1 Vector and Tensor Operations

In this section, we briefly describe the vector analysis that pertains to the mathematical developments in the main text. For a more detailed and complete presentation, the reader is referred to other sources such as Temam and Miranville (2001) and Wegner and Haddow (2009).

B.1.1 Algebraic Operations

A scalar such as concentration (C) has a magnitude but no direction. A vector such as velocity ($\mathbf{u}$), force ($\mathbf{f}$), or mass flux $\mathbf{n}_i$ has both magnitude and direction. A tensor such as stress ($\mathbf{T}$) has magnitude and is associated with two different directions.

Figure 9.1-1 illustrates some common orthogonal coordinate systems, each consisting of mutually perpendicular unit base vectors $\mathbf{i}_j (j = 1,2,3)$ aligned with the directions of the coordinates x_j. A vector $\mathbf{u}$ at a position (x_1,x_2,x_3) in orthogonal coordinates can be written as the sum of its components u_j in the three coordinate directions:

$$\mathbf{u}(x_1,x_2,x_3) = \sum_{i=1}^{3} u_i(x_1,x_2,x_3)\mathbf{i}_i = \sum_{i=1}^{3} u_i(\mathbf{x})\mathbf{i}_i \tag{B1-1}$$

The dot product between two vectors $\mathbf{u}$ and $\mathbf{v}$ is a scalar:

$$\mathbf{u}\cdot\mathbf{v} \equiv |\mathbf{u}||\mathbf{v}|\cos\theta \tag{B1-2}$$

where $|\mathbf{u}|$ and $|\mathbf{v}|$ are the magnitudes of the vectors and θ is the angle between them.

Since the base vectors of a rectangular coordinate system are mutually perpendicular and have a magnitude of one, their dot products are given by

$$\mathbf{i}_i\cdot\mathbf{i}_j = \begin{Bmatrix} 1 & \text{when } i = j \\ 0 & \text{when } i \neq j \end{Bmatrix} \equiv \delta_{ij} \tag{B1-3}$$

where δ_{ij} is the Kronecker delta. The dot product of two vectors can be written in terms of their orthogonal coordinate components:

Biomedical Mass Transport and Chemical Reaction: Physicochemical Principles and Mathematical Modeling, First Edition. James S. Ultman, Harihara Baskaran, and Gerald M. Saidel.

$$\mathbf{u}\cdot\mathbf{v} = \left(\sum_{i=1}^{3} u_i \mathbf{i}_i\right)\cdot\left(\sum_{j=1}^{3} v_j \mathbf{i}_j\right) = \sum_{i=1}^{3}\sum_{j=1}^{3} u_i v_j (\mathrm{i}_i \cdot \mathrm{i}_j) = \sum_{i=1}^{3}\sum_{j=1}^{3} u_i v_j \delta_{ij} = \sum_{i=1}^{3} u_i v_i \qquad \text{(B1-4)}$$

Note that the dot product of a vector **u** with itself is $(u_1^2 + u_2^2 + u_3^2)$ and its direction cosine is cos(0)=1. Then, according to Equation B1-2, the magnitude of a vector can be determined directly from its components as $|\mathbf{u}| = \sqrt{u_1^2 + u_2^2 + u_3^2}$.

We will represent the components of any vector or tensor quantity by using square brackets. The components of vector **u**, which are written as [**u**], can be arranged in row or a column array depending on the situation:

$$[\mathbf{u}] = [u_1 \ \ u_2 \ \ u_3], \quad [\mathbf{u}] = \begin{bmatrix} u_1 \\ u_2 \\ u_3 \end{bmatrix} \qquad \text{(B1-5a, b)}$$

The dot product between **u** and **v** can be represented in two ways:

$$[\mathbf{u}\cdot\mathbf{v}] = [u_1 \ \ u_2 \ \ u_3]\begin{bmatrix} v_1 \\ v_2 \\ v_3 \end{bmatrix}, \quad [\mathbf{u}\cdot\mathbf{v}] = [\mathbf{v}\cdot\mathbf{u}] = [v_1 \ \ v_2 \ \ v_3]\begin{bmatrix} u_1 \\ u_2 \\ u_3 \end{bmatrix} \qquad \text{(B1-6a, b)}$$

A second-order stress tensor **T** at a position **x** can be expressed as the sum of nine components T_{ij} associated with a vector product of the base vectors $\mathbf{i}_i \otimes \mathbf{i}_j$:

$$\mathbf{T}(\mathbf{x}) = \sum_{i=1}^{3}\sum_{j=1}^{3} T_{ij}(\mathbf{x})\mathbf{i}_i \otimes \mathbf{i}_j \qquad \text{(B1-7)}$$

The nine components of **T** can be represented by a square matrix:

$$[\mathbf{T}] = \begin{bmatrix} T_{11} & T_{12} & T_{13} \\ T_{21} & T_{22} & T_{23} \\ T_{31} & T_{32} & T_{33} \end{bmatrix} \qquad \text{(B1-8)}$$

A tensor can also be formed from a vector product:

$$\mathbf{u}\otimes\mathbf{v} = \sum_{i=1}^{3}\sum_{j=1}^{3} (u_i v_j)\mathbf{i}_i \otimes \mathbf{i}_j \qquad \text{(B1-9)}$$

For convenience, we may also define $\mathbf{u}\otimes\mathbf{v} = \mathbf{uv}$. This tensor has the components

$$[\mathbf{uv}] = \begin{bmatrix} u_1 v_1 & u_1 v_2 & u_1 v_3 \\ u_2 v_1 & u_2 v_2 & u_2 v_3 \\ u_3 v_1 & u_3 v_2 & u_3 v_3 \end{bmatrix} \qquad \text{(B1-10)}$$

A second-order tensor transforms a vector **u** into another vector **v**:

$$\mathbf{v} = \mathbf{T}\cdot\mathbf{u} = \sum_{k=1}^{3} T_{ik} u_k \mathbf{i}_i \Leftrightarrow \begin{bmatrix} v_1 \\ v_2 \\ v_3 \end{bmatrix} = \begin{bmatrix} T_{11} & T_{12} & T_{13} \\ T_{21} & T_{22} & T_{23} \\ T_{31} & T_{32} & T_{33} \end{bmatrix}\begin{bmatrix} u_1 \\ u_2 \\ u_3 \end{bmatrix} \qquad \text{(B1-11a, b)}$$

This differs from the scalar product of a vector **u** with a tensor **T**:

$$\mathbf{v} = \mathbf{u}\cdot\mathbf{T} = \sum_{k=1}^{3} u_k T_{ki}\mathbf{i}_i \Leftrightarrow [v_1 \ v_2 \ v_3] = [u_1 \ u_2 \ u_3]\begin{bmatrix} T_{11} & T_{12} & T_{13} \\ T_{21} & T_{22} & T_{23} \\ T_{31} & T_{32} & T_{33} \end{bmatrix} \tag{B1-12a,b}$$

Note that **u**·**T** and **T**·**u** are only equal if **T** is symmetric, that is, $T_{ij} = T_{ji}$.

For representing the components of vector and tensor operations, it is often convenient to use a convention that eliminates the summation sign. In this index summation notation, a duplicate subscript (or dummy index) is understood to mean summation, while an unpaired subscript (or free index) indicates coordinate direction. The operations shown above would then be written in terms of a dummy index k and a free index i as

$$\begin{aligned} \mathbf{u}\cdot\mathbf{v} &= \sum_{k=1}^{3} u_k v_k \quad &&\Leftrightarrow \quad [\mathbf{u}\cdot\mathbf{v}] = u_k v_k \\ \mathbf{T}\cdot\mathbf{u} &= \sum_{k=1}^{3} T_{ik}u_k\mathbf{i}_i \quad &&\Leftrightarrow \quad [\mathbf{T}\cdot\mathbf{u}] = T_{ik}u_k \\ \mathbf{u}\cdot\mathbf{T} &= \sum_{k=1}^{3} u_k T_{ki}\mathbf{i}_i \quad &&\Leftrightarrow \quad [\mathbf{u}\cdot\mathbf{T}] = u_k T_{ki} \end{aligned} \tag{B1-13a-f}$$

B.1.2 Derivative Operations

The del vector operator ∇ is a directional derivative with components ∇_i in rectangular coordinates given by

$$\nabla_i = \frac{\partial}{\partial x_i} \Rightarrow \nabla = \sum_{i=1}^{3} \frac{\partial}{\partial x_i}\mathbf{i}_i \tag{B1-14a,b}$$

The scalar product of two del vectors is the Laplacian operator, which in rectangular coordinates is expressed as

$$\nabla^2 \equiv \sum_{i=1}^{3}\sum_{j=1}^{3}\left(\frac{\partial}{\partial x_i}\mathbf{i}_i\right)\cdot\left(\frac{\partial}{\partial x_j}\mathbf{i}_j\right) = \sum_{i=1}^{3}\sum_{j=1}^{3}\left(\frac{\partial}{\partial x_i}\right)\left(\frac{\partial}{\partial x_j}\right)\delta_{ij} = \sum_{i=1}^{3}\frac{\partial^2}{\partial x_i^{2}} \tag{B1-15}$$

Differential vector operations in commonly used coordinate systems are shown in Tables B1-1, B1-2, and B1-3.

Table B1-1 Gradient Operation on a Scalar Function C

Coordinates	Components of ∇C
Rectangular	x: $\frac{\partial C}{\partial x}$, y: $\frac{\partial C}{\partial y}$, z: $\frac{\partial C}{\partial z}$
Cylindrical	r: $\frac{\partial C}{\partial r}$, θ: $\frac{1}{r}\frac{\partial C}{\partial \theta}$, z: $\frac{\partial C}{\partial z}$
Spherical	r: $\frac{\partial C}{\partial r}$, θ: $\frac{1}{r}\frac{\partial C}{\partial \theta}$, ϕ: $\frac{1}{r\sin\theta}\frac{\partial C}{\partial \phi}$

Table B1-2 Divergence Operation on a Vector Field **u**

Coordinates	$\nabla \cdot \mathbf{u}$
Rectangular	$\frac{\partial u_x}{\partial x}+\frac{\partial u_y}{\partial y}+\frac{\partial u_z}{\partial z}$
Cylindrical	$\frac{1}{r}\frac{\partial}{\partial r}(ru_r)+\frac{1}{r}\frac{\partial u_\theta}{\partial \theta}+\frac{\partial u_z}{\partial z}$
Spherical	$\frac{1}{r^2}\frac{\partial}{\partial r}(r^2u_r)+\frac{1}{r\sin\theta}\frac{\partial}{\partial \theta}(u_\theta \sin\theta)+\frac{1}{r\sin\theta}\frac{\partial u_\phi}{\partial \phi}$

Table B1-3 Laplacian Operation on a Scalar Function C

Coordinates	$\nabla^2 C$
Rectangular	$\frac{\partial^2 C}{\partial x^2}+\frac{\partial^2 C}{\partial y^2}+\frac{\partial^2 C}{\partial z^2}$
Cylindrical	$\frac{1}{r}\frac{\partial}{\partial r}\left(r\frac{\partial C}{\partial r}\right)+\frac{1}{r^2}\frac{\partial^2 C}{\partial \theta^2}+\frac{\partial^2 C}{\partial z^2}$
Spherical	$\frac{1}{r^2}\frac{\partial}{\partial r}\left(r^2\frac{\partial C}{\partial r}\right)+\frac{1}{r^2\sin\theta}\frac{\partial}{\partial \theta}\left(\sin\theta\frac{\partial C}{\partial \theta}\right)+\frac{1}{r^2\sin^2\theta}\frac{\partial^2 C}{\partial \phi^2}$

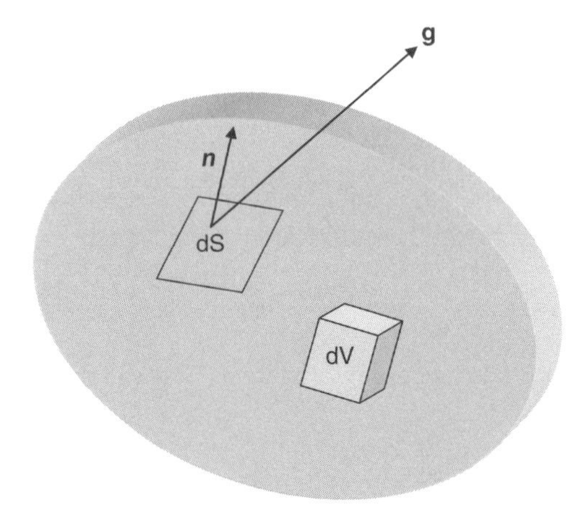

Figure B1-1 Transport domain showing internal volume element dV and surface element dS with unit outward normal ***n*** and acted on by a vector **g**.

B.1.3 Key Theorems

Two important theorems are applied in the development of mass and momentum balance equations: the divergence theorem and the Reynolds transport theorem. Consider a transport domain with a surface S enclosing a volume V in which S is a collection of differential surface elements dS and V is composed of volume elements dV (Fig. B1-1). The divergence (or Gauss) theorem relates a surface integral of a vector or tensor quantity **g** acting on S to the volume integral of the divergence of **g**. If **g** is a vector such as velocity **u** or mass flux $\dot{\mathbf{n}}$, then this theorem is written in vector and in summation notation as

$$\int_S \mathbf{g}\cdot \boldsymbol{n}\,dS = \int_V \nabla\cdot \mathbf{g}\,dV \Leftrightarrow \int_S g_i n_i\,dS = \int_V \nabla_i g_i\,dV \tag{B1-16a,b}$$

where $\boldsymbol{n}$ is a unit normal pointing outward from the surface. If $\mathbf{g}$ is a tensor such as the stress tensor $\mathbf{T}$ or the velocity product $\mathbf{uv}$, then

$$\int_S \mathbf{g}\cdot \boldsymbol{n}\,dS = \int_V \nabla\cdot \mathbf{g}^T\,dV \Leftrightarrow \int_S g_{ik} n_k\,dS = \int_V \nabla_k g_{ki}\,dV \tag{B1-17a,b}$$

where $\mathbf{g}^T$ represents the transpose of $\mathbf{g}$.

The Reynolds transport theorem (or generalized Leibnitz rule) is applied to a time derivative of a time-varying volume integral. For any scalar or vector quantity $\mathbf{f}$, the transport theorem can be expressed as

$$\frac{d}{dt}\int_{V(t)} \mathbf{f}\,dV = \int_{V(t)} \left[\frac{\partial \mathbf{f}}{\partial t} + \nabla\cdot(\mathbf{fu})\right] dV \tag{B1-18}$$

If $\mathbf{f} = \rho$, the local mass density, the transport theorem reduces to the scalar equation:

$$\frac{d}{dt}\int_{V(t)} \rho\,dV = \int_{V(t)} \left[\frac{\partial \rho}{\partial t} + \nabla\cdot(\rho\mathbf{u})\right] dV \tag{B1-19}$$

In summation notation, this equation is

$$\frac{d}{dt}\int_{V(t)} \rho\,dV = \int_{V(t)} \left[\frac{\partial \rho}{\partial t} + \nabla_k(\rho u_k)\right] dV \tag{B1-20}$$

If $\mathbf{f} = \rho\mathbf{u}$, the local momentum vector, then the transport theorem leads to a vector equation:

$$\frac{d}{dt}\int_{V(t)} (\rho\mathbf{u})\,dV = \int_{V(t)} \left[\frac{\partial(\rho\mathbf{u})}{\partial t} + \nabla\cdot(\rho\mathbf{uu})\right] dV \tag{B1-21}$$

In summation notation,

$$\frac{d}{dt}\int_{V(t)} (\rho u_i)\,dV = \int_{V(t)} \left[\frac{\partial(\rho u_i)}{\partial t} + \nabla_k(\rho u_k u_i)\right] dV \tag{B1-22}$$

B.1.4 Vector–Tensor Calculus

Several vector–tensor operations are of particular importance in the development of mass and momentum balance equations. A term appearing in the mass concentration equation for a species i is the dot product $\nabla\cdot(\rho_i\mathbf{u})$. In summation notation, this scalar term can be written in rectangular coordinates as

$$\nabla_k(\rho_i u_k) = \frac{\partial(\rho_i u_k)}{\partial x_k} = \rho_i \frac{\partial u_k}{\partial x_k} + u_k \frac{\partial \rho_i}{\partial x_k} \tag{B1-23}$$

In vector notation, this relation is equivalent to

$$\nabla \cdot (\rho_i \mathbf{u}) = \rho_i \nabla \cdot \mathbf{u} + \mathbf{u} \cdot \nabla \rho_i \tag{B1-24}$$

A term appearing in the momentum balance equation is the dot product $\nabla \cdot (\rho \mathbf{uu})$. The rectangular component of this vector operation is

$$\nabla_k(\rho u_k u_i) = \frac{\partial(\rho u_k)u_i}{\partial x_k} = u_i \frac{\partial(\rho u_k)}{\partial x_k} + \rho u_k \frac{\partial u_i}{\partial x_k} \tag{B1-25}$$

Expressed in vector notation, this equation is equivalent to

$$\nabla \cdot (\rho \mathbf{uu}) = \mathbf{u} \nabla \cdot (\rho \mathbf{u}) + \rho \mathbf{u} \cdot \nabla \mathbf{u} \tag{B1-26}$$

The frictional force term of a Newtonian fluid is proportional to $\nabla \cdot \left[\nabla \mathbf{u} + (\nabla \mathbf{u})^T\right]$ where superscript T indicates the transpose of the tensor quantity $\nabla \mathbf{u}$. The rectangular components of this term can be expressed as

$$\nabla_k(\nabla_k u_i + \nabla_i u_k) = \frac{\partial}{\partial x_k}\left(\frac{\partial u_i}{\partial x_k}\right) + \frac{\partial}{\partial x_k}\left(\frac{\partial u_k}{\partial x_i}\right) \tag{B1-27}$$

Since the derivatives are commutative, this result can also be written as

$$\nabla_k(\nabla_k u_i + \nabla_i u_k) = \frac{\partial^2 u_i}{\partial x_k^2} + \frac{\partial}{\partial x_i}\left(\frac{\partial u_k}{\partial x_k}\right) \tag{B1-28}$$

In vector notation, this is equivalent to

$$\nabla \cdot \left[\nabla \mathbf{u} + (\nabla \mathbf{u})^T\right] = \nabla^2 \mathbf{u} + \nabla(\nabla \cdot \mathbf{u}) \tag{B1-29}$$

B.2 Nonequilibrium Thermodynamics

B.2.1 Entropy Generation Rate

In the development below, the entropy production rate of a dynamic (nonequilibrium) system is derived using a simplified approach patterned after Jou and Llebot (1990). A more rigorous development can be found in classic texts such as Fitts (1962).

Diffusion

Consider an isolated system that is subdivided into subsystems A and B by a partition (Fig. B2-1). This partition is permeable to all I substances in the system and conducts heat between the two subsystems. The partition is also flexible so that pressure–volume work can be performed by one subsystem on the other. We will analyze a process in which there is a disequilibrium between the subsystems that allows a differential amount of material dm to be transported across the partition from subsystem A to subsystem B. We will assume that dm is sufficiently small that the temperatures and pressures of each subsystem remain constant.

Whereas the system as a whole is not in equilibrium during this process, each subsystem can be treated as if it were in a local equilibrium state. According to Gibbs phase rule for a nonreacting system, equilibrium can be specified in each of the subsystems by (I + 1) intensive

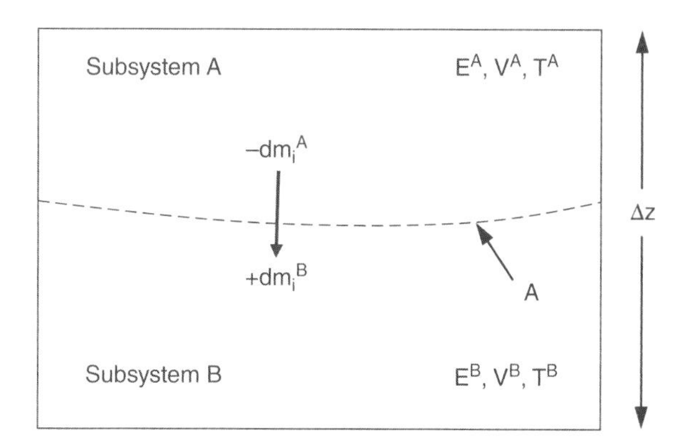

Figure B2-1 Isolated system for the formulation of entropy generation rate.

variables or, equivalently, by (I + 2) extensive variables. We will select the internal energies (E^A, E^B), the volumes (V^A, V^B) and the moles of each component $\left(m_i^A, m_i^B : i = 1, 2, \ldots I\right)$ to satisfy the (2I+4) degrees of freedom for the overall system.

The transport of material across the partition causes an increase in entropy that can be evaluated from the sum of the entropy changes in subsystems A and B

$$dS = dS^A + dS^B \tag{B2-1}$$

Applying the Gibbs equation of classical thermodynamics (Eq. 3.5-11) to the separate phases, we find the entropy change in each of the subsystems:

$$dS^A = \frac{1}{T^A} dE^A + \frac{P^A}{T^A} dV^A - \frac{1}{T_A} \sum_{i=1}^{I} \mu_i^A dm_i^A \tag{B2-2}$$

and

$$dS^B = \frac{1}{T^B} dE^B + \frac{P^B}{T^B} dV^B - \frac{1}{T^B} \sum_{i=1}^{I} \mu_i^B dm_i^B \tag{B2-3}$$

where dm_i^A and dm_i^B represent the change in the number of moles of component i in subsystems A and B, respectively, and μ_i^A and μ_i^B are the electrochemical potentials of the two subsystems. Since the overall system is isolated, the total internal energy and total volume are each constant, and the number of moles of component i that are transferred out of subsystem A must be transferred into subsystem B. Therefore, we can write that

$$dE^A = -dE^B, \quad dV^A = -dV^B, \quad dm_i^A = -dm_i^B \quad (i = 1, 2, \ldots I) \tag{B2-4a-c}$$

Combining Equations B2-1–B2-4, we obtain

$$dS = \left(\frac{1}{T^A} - \frac{1}{T^B}\right) dE^A + \left(\frac{P^A}{T^A} - \frac{P^B}{T^B}\right) dV^A - \sum_{i=1}^{I} \left(\frac{\mu_i^A}{T^A} - \frac{\mu_i^B}{T^B}\right) dm_1^A \tag{B2-5}$$

Suppose we restrict our attention to two subsystems that are in thermal and mechanical equilibrium in which case

$$T \equiv T^A = T^B, \quad P^A = P^B \tag{B2-6a,b}$$

The differential entropy increase then becomes

$$dS = \frac{1}{T}\sum_{i=1}^{I} \Delta\mu_i dm_i^A \tag{B2-7}$$

where $\Delta\mu_i$ is the difference in chemical potential of component i between subsystem B and subsystem A. For a system of total volume V, the entropy generation rate is defined in Equation 6.2-1 as $\vartheta \equiv (T/V)dS/dt$. Multiplying Equation B2-7 by T/V and dividing by the time dt in which the differential change in state occurs, we obtain

$$\vartheta = \frac{1}{V}\sum_{i=1}^{I} \Delta\mu_i \frac{dm_i^A}{dt} \tag{B2-8}$$

We now recognize that if the system has a cross-sectional area A and a length Δz, then $V = A\Delta z$ and

$$\vartheta = \sum_{i=1}^{I} \left(\frac{1}{A}\frac{dm_i^A}{dt}\right)\left(\frac{\Delta\mu_i}{\Delta z}\right) \tag{B2-9}$$

Imagine that a linear coordinate z crossing the partition from subcompartment A to subcompartment B. The total flux of substance i in this direction is given by

$$N_{i,z} = -\frac{1}{A}\frac{dm_i^A}{dt} \tag{B2-10}$$

The negative sign is necessary in this equation because a flux in the positive z direction requires a decrease in the moles of component i in compartment A. Substituting this equation into Equation B2-9, we have

$$\vartheta = -\sum_{i=1}^{I} N_{i,z}\left(\frac{\Delta\mu_i}{\Delta z}\right) \tag{B2-11}$$

To provide a continuum representation in which the system approaches a material point, we take the limit of both sides of Equation B2-11 as $\Delta z \rightarrow 0$. Recognizing that $(\Delta\mu_i/\Delta z)$ then becomes an ordinary derivative, we obtain

$$\vartheta = -\sum_{i=1}^{I} N_{i,z}\left(\frac{d\mu_i}{dz}\right) \tag{B2-12}$$

From this result, it is clear that the electrochemical gradient of a component i is the driving force conjugate to the molar flux of that component. Since ϑ must be a positive quantity, the negative sign associated with this equation indicates that a positive magnitude of the flux must occur in the direction of decreasing electrochemical potential. Molar flux and electrochemical gradient are generally three-dimensional vector quantities that can be represented as $\mathbf{N}_i$ and $\nabla\mu_i$, respectively. Equation B2-12 can thus be expressed in a more general form as

$$\vartheta = -\sum_{i=1}^{I} \mathbf{N}_i \cdot \nabla\mu_i \tag{B2-13}$$

where the dot product between $\mathbf{N}_i$ and $\nabla\mu_i$ is necessary because ϑ must be a scalar quantity.

Chemical Reaction

Chemical reaction can be accounted for in the entropy generation rate by a straightforward extension of the previous derivation. Suppose that a chemical reaction occurs in subsystem B of Figure B2-1. This reaction produces dm_i^r moles of a species i, which is a positive quantity for reaction products and a negative quantity for reactants. Instead of equating dm_i^B to $-dm_i^A$, as we did previously, an increase in the number of moles of any component i in subsystem B must now be equal to the decrease in the number of moles in subsystem A plus the number of moles produced in subsystem B by the chemical reaction:

$$dm_i^B = -dm_i^A + dm_i^r \quad (i = 1,2,\ldots I) \tag{B2-14}$$

With this modification, the same steps that led from Equation B2-1 to Equation B2-7 yield

$$dS = \frac{1}{T}\left(\sum_{i=1}^{I} \Delta\mu_i dm_i^A - \sum_{i=1}^{I} \mu_i^B dm_i^r\right) \tag{B2-15}$$

Applying the electroneutrality principle to the chemical reaction indicates that $\Sigma\mu_i^B dm_i^r = \Sigma\breve{\mu}_i^B dm_i^r$ where μ_i^B is electrochemical potential, whereas $\breve{\mu}_i^B$ is the chemical potential alone. Also using the relation $dm_i^r = \nu_i d\xi_r$ (Eq. 5.1-4b) leads to

$$dS = \frac{1}{T}\left(\sum_{i=1}^{I} \Delta\mu_i dm_i^A - d\xi_r \sum_{i=1}^{I} \nu_i \breve{\mu}_i^B\right) \tag{B2-16}$$

where ν_i are the modified stoichiometric coefficients and ξ is the extent of the reaction. With reaction affinity defined as $A_r^B \equiv -\Sigma\nu_i\breve{\mu}_i^B$ (Eq. 5.1-10), the entropy generation function defined as $\vartheta \equiv (T/V)dS/dt$ and the system volume rewritten as $V = A\Delta z$, this equation becomes

$$\vartheta = \sum_{i=1}^{I}\left(\frac{1}{A}\frac{dm_i^A}{dt}\right)\left(\frac{\Delta\mu_i}{\Delta z}\right) + \frac{A_r^B}{V}\frac{d\xi}{dt} \tag{B2-17}$$

Utilizing Equation B2-10 to define flux of component i between the compartments and taking the limit as $\Delta z \rightarrow 0$, we further obtain

$$\vartheta = -\sum_{i=1}^{I} N_{i,z}\left(\frac{d\mu_i}{dz}\right) + \frac{A_r}{V}\frac{d\xi}{dt} \tag{B2-18}$$

Notice that it is no longer necessary to associate the reaction affinity with subsystem B since we have taken a limit that essentially shrinks the two subsystems into a single volume element. Noting also that $(1/V)d\xi/dt$ is equivalent to the inherent reaction rate $\bar{r}$ (Eq. 8.1-4), the entropy generation function becomes

$$\vartheta = -\sum_{i=1}^{I} N_{i,z}\left(\frac{d\mu_i}{dz}\right) + \bar{r}A_r \tag{B2-19}$$

As before, we recognize that $N_{i,z}$ and $(d\mu_i/dz)$ can be written in a more general form as the vector quantities $\mathbf{N}_i$ and $\nabla\mu_i$. The final entropy generation rate can thus be written

$$\vartheta = -\sum_{i=1}^{I} \mathbf{N}_i \cdot \nabla\mu_i + \bar{r}A_r \tag{B2-20}$$

Curie's Theorem stipulates that forces and fluxes whose orders differ by an odd integer cannot interact in isotropic systems. Therefore, the transport rates $\mathbf{N}_i$ (i = 1,...I), which are vector quantities of order one, should not depend on reaction affinity A_r, which is a scalar of order zero. During active transport across a cell membrane, however, there is a coupling between the diffusion of a substance such as sodium ion and the affinity of an exergonic reaction such as ATP hydrolysis. This exception to Curie's theorem occurs because of the alignment of phospholipids and transmembrane proteins. This structural anisotropy restricts the direction of diffusion so that $\mathbf{N}_i$ is, in effect, a scalar quantity.

B.2.2 Gibbs–Duhem Equation

When state changes during diffusion are locally close to equilibrium, the differential changes in chemical potential $d\mu_i$ for different species i can be related by the Gibbs–Duhem equation of classical thermodynamics (Eq. 3.5-19). At constant T and P,

$$\sum_{i=1}^{I} m_i d\mu_i = 0 \tag{2.2-21}$$

If the total moles m of material in the system are also constant, then

$$\sum_{i=1}^{I} \left(\frac{m_i}{m}\right) d\mu_i = 0 \Rightarrow \sum_{i=1}^{I} x_i d\mu_i = 0 \tag{B2-22a,b}$$

Suppose that $\mathbf{dr}$ represents the displacement vector between two closely located points in the spatial domain of an electrochemical diffusion process. Recognizing that $\nabla\mu_i$ is a vector representing the directional derivative of μ_i, the differential change in electrochemical potential along the $\mathbf{dr}$ vector is given by the dot-product relationship

$$d\mu_i = \nabla\mu_i \cdot \mathbf{dr} \tag{B2-23}$$

With this relation, we rewrite Equation B2-22b as

$$\sum_{i=1}^{I} x_i \nabla\mu_i \cdot \mathbf{dr} = 0 \tag{B2-24}$$

Since the selection of $\mathbf{dr}$ is arbitrary, it is not necessarily equal to zero, and Equation B2-24 leads to the Gibbs–Duhem equation for a system with a spatially distributed composition that is in local equilibrium:

$$\sum_{i=1}^{I} x_i \nabla\mu_i = 0 \tag{B2-25}$$

B.3 Spatially Averaged Balances for Heterogeneous Tissue

In Section 7.4, we developed a volume-averaged approach for obtaining macroscopic transport equations in heterogeneous media. That development was limited to two-phase materials composed of an impermeable solid matrix containing fluid-filled pores. We now consider a tissue model consisting of three phases with immovable boundaries: capillaries,

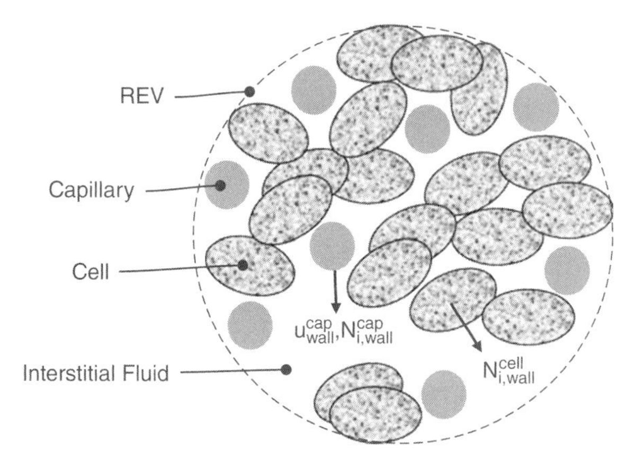

Figure B3-1 REV in a tissue composed of cell, interstitial, and capillary phases.

extravascular cells, and interstitial fluid surrounding the cells (Fig. B3-1). Solute as well as water can be transported between interstitial fluid and capillaries. Water is also free to flow within the capillaries and the interstitial space. Because they have closed fixed boundaries, there is no net water flow into the cells. We assume that cells transfer solute with interstitial fluid but not by direct contact with capillary walls or with each other.

B.3.1 Interstitial and Macroscopic Volume Averages

Because they have fixed boundaries, the three phases j in a representative elemental volume (REV) of our model each occupy a constant volume fraction ε_j and have a constant ratio of surface to tissue volume, ϕ_j. For a REV of volume ΔV^{REV}, the total volume of phase j is $\varepsilon_j \Delta V^{REV}$ and its total surface is $\phi_j \Delta V^{REV}$.

Suppose ψ is a spatially distributed, microscopic property such as concentration or velocity. We define the intrinsic average of ψ in the volume domain of phase j as

$$\overline{\psi}^j = \frac{1}{\varepsilon_j \Delta V^{REV}} \int_{\varepsilon_j \Delta V^{REV}} \psi dV \tag{B3-1}$$

We can also define a macroscopic average of a property ψ over the entire domain of the REV:

$$\overline{\psi} \equiv \frac{1}{\Delta V^{REV}} \int_{\Delta V^{REV}} \psi dV \tag{B3-2}$$

Decomposing the integral in this equation, we can relate the intrinsic averages to the macroscopic average:

$$\overline{\psi} \equiv \sum_{j=1}^{3} \left(\frac{1}{\Delta V^{REV}} \int_{\varepsilon_j \Delta V^{REV}} \psi dV \right) = \sum_{j=1}^{3} \varepsilon_i \left(\frac{1}{\varepsilon_j \Delta V^{REV}} \int_{\varepsilon_j \Delta V^{REV}} \psi dV \right) = \sum_{j=1}^{3} \varepsilon_j \overline{\psi}^j \tag{B3-3}$$

Identifying a local boundary element by its unit outward normal $\boldsymbol{n}$, we can also define the average of a vector property $\boldsymbol{\psi}$ over the phase j surface $\phi_j \Delta V^{REV}$ as

$$\psi^j_{wall} \equiv \frac{1}{\phi_j \Delta V^{REV}} \int_{\phi_j \Delta V^{REV}} (\boldsymbol{\psi} \cdot \boldsymbol{n}) dS \tag{B3-4}$$

In deriving macroscopic balance equations, we will make use of two averaging rules for averaged quantities (Bear and Bachmat, 1990). For a phase j with stationary boundaries,

$$\overline{\frac{\partial \psi^j}{\partial t}} = \frac{\partial \overline{\psi}^j}{\partial t} \quad \text{or} \quad \overline{\frac{\partial \psi}{\partial t}} = \frac{\partial \overline{\psi}}{\partial t} \quad \text{Time derivative} \tag{B3-5a, b}$$

$$\overline{\nabla \cdot \boldsymbol{\psi}}^j = \nabla \cdot \overline{\psi}^j + \frac{\phi_j}{\varepsilon_j} \psi^j_{wall} \quad \text{Divergence} \tag{B3-6}$$

Combining the divergence rule with Equation B3-3 indicates that

$$\overline{\nabla \cdot \boldsymbol{\psi}} = \sum_{j=1}^{3} \varepsilon_j \overline{\nabla \cdot \boldsymbol{\psi}}^j = \nabla \cdot \overline{\boldsymbol{\psi}} + \sum_{j=1}^{3} \phi_j \psi^j_{wall} \tag{B3-7}$$

B.3.2 Solution Balances

Intrinsic Balances

The microscopic solution balance on a material point in each of the three incompressible phases of our tissue model is given by

$$\nabla \cdot \mathbf{u}^j = 0 \tag{B3-8}$$

Taking the intrinsic volume average of this equation, we get

$$\overline{\nabla \cdot \mathbf{u}}^j = 0 \tag{B3-9}$$

The divergence rule indicates that

$$\overline{\nabla \cdot \mathbf{u}}^j = \nabla \cdot \overline{\mathbf{u}}^j + \frac{\phi_j}{\varepsilon_j} u^j_{wall} \tag{B3-10}$$

Combining this with Equation B3-9, we arrive at the intrinsic solution balance on a phase j:

$$\nabla \cdot \overline{\mathbf{u}}^j = -\frac{\phi_j}{\varepsilon_j} u^j_{wall} \tag{B3-11}$$

According to Equation B3-4,

$$u^j_{wall} \equiv \frac{1}{\phi_j \Delta V^{REV}} \int_{\phi_j \Delta V^{REV}} (\mathbf{u} \cdot \mathbf{n}) dS \tag{B3-12}$$

The product $\phi_j u^j_{wall}$ represents the volumetric solution outflow per unit tissue volume through the boundary of phase j. With no net water flow into cells, $\phi_{cell} u^{cell}_{wall} = 0$, and the water flow out of the incompressible capillary phase will be balanced by the water flow into the interstitial phase:

$$\phi_{cap} u_{wall}^{cap} = -\phi_{int} u_{wall}^{int} \tag{B3-13}$$

Equation B3-11 can now be written for the three phases as

$$\nabla \cdot \overline{\mathbf{u}}^{cell} = 0, \quad \nabla \cdot \overline{\mathbf{u}}^{int} = \frac{\phi_{cap} u_{wall}^{cap}}{\varepsilon_{int}}, \quad \nabla \cdot \overline{\mathbf{u}}^{cap} = -\frac{\phi_{cap} u_{wall}^{cap}}{\varepsilon_{cap}} \tag{B3-14a-c}$$

Macroscopic Balance

To obtain a macroscopic solution balance, we take the macroscopic volume average of Equation B3-8:

$$\overline{\nabla \cdot \mathbf{u}} = 0 \tag{B3-15}$$

By identifying ψ with $\mathbf{u}$ in Equation B3-7, we further obtain

$$\nabla \cdot \overline{\mathbf{u}} + \left(\phi_{cell} u_{wall}^{cell} + \phi_{int} u_{wall}^{int} + \phi_{cap} u_{wall}^{cap}\right) = 0 \tag{B3-16}$$

Since $u_{wall}^{cell} = 0$ and $\phi_{int} u_{wall}^{int} + \phi_{cap} u_{wall}^{cap} = 0$, the macroscopic solution balance is actually identical in form to a microscopic solution balance:

$$\nabla \cdot \overline{\mathbf{u}} = 0 \tag{B3-17}$$

B.3.3 Solute Balances

Intrinsic Balances

We can follow a similar procedure to obtain intrinsic and macroscopic solute balances starting with the microscopic balance for a solute species i:

$$\frac{\partial C_i}{\partial t} + \nabla \cdot \mathbf{N}_i = R_i \tag{15.1-7}$$

Taking the intrinsic average of this equation for either the cell, interstitial or capillary phase, we get

$$\overline{\frac{\partial C_i}{\partial t}}^j + \overline{\nabla \cdot \mathbf{N}_i}^j = \overline{R}_i^j \tag{B3-18}$$

Applying the derivative and divergence rules results in

$$\frac{\partial \overline{C}_i^j}{\partial t} + \nabla \cdot \overline{\mathbf{N}}_i^j = \overline{R}_i^j - \frac{\phi_j}{\varepsilon_j} N_{i,wall}^j \tag{B3-19}$$

According to Equation B3-4,

$$N_{i,wall}^j \equiv \frac{1}{\phi_j \Delta V_{REV}} \int_{\phi_j \Delta V_{REV}} (\mathbf{N}_i \cdot \mathbf{n}) dS \tag{B3-20}$$

The $\phi_j N^j_{i,wall}$ product represents the molar efflux of solute i from phase j per unit tissue volume. With no accumulation of solute i at the phase boundaries, the moles of solute transferred out of the cell and capillary surfaces must equal to the moles of solute transferred into the interstitial fluid:

$$\phi_{cell} N^{cell}_{i,wall} + \phi_{cap} N^{cap}_{i,wall} = -\phi_{int} N^{int}_{i,wall} \tag{B3-21}$$

This allows to write Equation B3-19 for the three phases as

$$\frac{\partial \overline{C}^{int}_i}{\partial t} + \nabla \cdot \overline{\mathbf{N}}^{int}_i = \overline{R}^{int}_i + \frac{1}{\varepsilon_{int}}\left(\phi_{cell} N^{cell}_{i,wall} + \phi_{cap} N^{cap}_{i,wall}\right) \tag{B3-22}$$

$$\frac{\partial \overline{C}^{cap}_i}{\partial t} + \nabla \cdot \overline{\mathbf{N}}^{cap}_i = \overline{R}^{cap}_i - \frac{N^{cap}_{i,wall}}{\varepsilon_{cap}} \tag{B3-23}$$

$$\frac{\partial \overline{C}^{cell}_i}{\partial t} + \nabla \cdot \overline{\mathbf{N}}^{cell}_i = \overline{R}^{cell}_i - \frac{N^{cell}_{i,wall}}{\varepsilon_{cap}} \tag{B3-24}$$

Macroscopic Balance

Taking the macroscopic average of the microscopic solute balance, we obtain

$$\frac{\partial \overline{C}_i}{\partial t} + \overline{\nabla \cdot \mathbf{N}_i} = \overline{R_i} \tag{B3-25}$$

This equation can be simplified by applying the derivative rule as well as Equation B3-7:

$$\frac{\partial \overline{C}_i}{\partial t} + \nabla \cdot \overline{\mathbf{N}}_i + \left(\phi_{cell} N^{cell}_{i,wall} + \phi_{cap} N^{cap}_{i,wall} + \phi_{int} N^{int}_{i,wall}\right) = \overline{R}_i \tag{B3-26}$$

Employing Equation B3-21, we obtain a macroscopic solute balance

$$\frac{\partial \overline{C}_i}{\partial t} + \nabla \cdot \overline{\mathbf{N}}_i = \overline{R}_i \tag{B3-27}$$

that has the same form as the microscopic solute balance in a homogeneous material.

B.3.4 Convection–Diffusion Equations

Intrinsic Form

The specification of a rate equation for molar flux is a critical step in the application of an interstitial or macroscopic solute balance equation. We previously developed an intrinsic flux equation for the simultaneous convection and diffusion of a solute through the pore of a heterogeneous medium. Although this analysis was performed for solutes that were confined to the pore, we will apply the results to each phase of our tissue model, as a first approximation. When stated for a particular phase j, the three-dimensional form of Equation 7.4-24 is

$$\overline{\mathbf{N}}^j_i = \beta^j_i \overline{\mathbf{u}}^j \overline{C}^j_i - \overline{\mathcal{D}}^j_i \nabla \overline{C}^j_i \tag{B3-28}$$

The parameter β_i^j is a convection correction, and $\overline{\mathcal{D}}_i^j$ is a diffusion coefficient in phase j. The result of combining the intrinsic flux equation (Eq. B3-28) with the intrinsic solute balance (Eq. B3-19) is

$$\frac{\partial \overline{C}_i^j}{\partial t} + \nabla \cdot \left(\beta_i^j \overline{\mathbf{u}}^j \overline{C}_i^j\right) = \nabla \cdot \overline{\mathcal{D}}_i^j \nabla \overline{C}_i^j + \overline{R}_i^j - \frac{\phi_j}{\varepsilon_j} N_{i,\text{wall}}^j \tag{B3-29}$$

Employing Equation B3-21, this equation can be written for the separate phases as

$$\frac{\partial \overline{C}_i^{\text{int}}}{\partial t} + \beta_i^{\text{int}} \nabla \cdot \left(\overline{\mathbf{u}}^{\text{int}} \overline{C}_i^{\text{int}}\right) = \overline{\mathcal{D}}_i^{\text{int}} \nabla^2 \overline{C}_i^{\text{int}} + \overline{R}_i^{\text{int}} + \frac{1}{\varepsilon_{\text{int}}} \left(\phi_{\text{cell}} N_{i,\text{wall}}^{\text{cell}} + \phi_{\text{cap}} N_{i,\text{wall}}^{\text{cap}}\right) \tag{B3-30}$$

$$\frac{\partial \overline{C}_i^{\text{cap}}}{\partial t} + \beta_i^{\text{cap}} \nabla \cdot \left(\overline{\mathbf{u}}^{\text{cap}} \overline{C}_i^j\right) - \nabla \overline{\mathcal{D}}_i^{\text{cap}} \overline{C}_i^{\text{cap}} = \overline{R}_i^{\text{cap}} - \frac{\phi_{\text{cap}}}{\varepsilon_{\text{cap}}} N_{i,\text{wall}}^{\text{cap}} \tag{B3-31}$$

$$\frac{\partial \overline{C}_i^{\text{cell}}}{\partial t} = \overline{\mathcal{D}}_i^{\text{cell}} \nabla^2 \overline{C}_i^{\text{cell}} + \overline{R}_i^{\text{cell}} - \frac{\phi_{\text{cell}}}{\varepsilon_{\text{cell}}} N_{i,\text{wall}}^{\text{cell}} \tag{B3-32}$$

where we have assumed that β_i^j and $\overline{\mathcal{D}}_i^j$ are constants.

Macroscopic Form

Finding the macroscopic average solute flux by operating on each term of Equation B3-28 in accordance with Equation B3-3, we get

$$\overline{\mathbf{N}}_i = \sum_{j=1}^{3} \varepsilon_j \overline{\mathbf{N}}_i^j = \sum_{j=1}^{3} \varepsilon_j \beta_i^j \overline{\mathbf{u}}^j \overline{C}_i^j - \sum_{j=1}^{3} \varepsilon_j \overline{\mathcal{D}}_i^j \nabla \overline{C}_i^j \tag{B3-33}$$

Given its complexity, it is impractical to substitute Equation B3-33 into the macroscopic material balance equation (Eq. B3-27) in order to derive a macroscopic convection–diffusion equation. Therefore, we resort to several assumptions that simplify the flux equation.

First, we assume that the solute size is very small compared to the characteristic dimension of each of the three phases j. In that case, solute convection behaves as if it is not affected by wall interactions so that $\beta_i^j \to 1$. With the added assumption that the intrinsic diffusion coefficients $\overline{\mathcal{D}}_i^j$ of solute are similar in the three phases, $\overline{\mathcal{D}}_i^j \to \overline{\mathcal{D}}_i$, and the flux equation becomes

$$\overline{\mathbf{N}}_i = \sum_{j=1}^{3} \varepsilon_j \overline{\mathbf{u}}^j \overline{C}_i^j - \overline{\mathcal{D}}_i \nabla \sum_{j=1}^{3} \varepsilon_j \overline{C}_i^j = \sum_{j=1}^{3} \varepsilon_j \overline{\mathbf{u}}^j \overline{C}_i^j - \overline{\mathcal{D}}_i \nabla \overline{C}_i \tag{B3-34}$$

Our final assumption is that the sum $\sum_j \varepsilon_j \overline{\mathbf{u}}^j \overline{C}_i^j$ can be approximated by $\overline{\mathbf{u}}\overline{C}_j$. For this special case, substitution of Equations B3-17 and B3-34 into Equation B3-27 yields

$$\frac{\partial \overline{C}_i}{\partial t} + \overline{\mathbf{u}} \nabla \overline{C}_i = \nabla \cdot \overline{\mathcal{D}}_i \nabla \overline{C}_i + \overline{R}_i \tag{B3-35}$$

This macroscopic convection–diffusion equation has the same form as a microscopic convection–diffusion equation in a homogeneous material.

B.4 Tables for Fluid Motion in Common Coordinate Systems

Table B4-1 Continuity Equation

Rectangular	$\frac{\partial\rho}{\partial t}+\frac{\partial(\rho u_x)}{\partial x}+\frac{\partial(\rho u_y)}{\partial y}+\frac{\partial(\rho u_z)}{\partial z}=0$
Cylindrical	$\frac{\partial\rho}{\partial t}+\frac{1}{r}\frac{\partial}{\partial r}(\rho r u_r)+\frac{1}{r}\frac{\partial(\rho u_\theta)}{\partial\theta}+\frac{\partial(\rho u_z)}{\partial z}=0$
Spherical	$\frac{\partial\rho}{\partial t}+\frac{1}{r^2}\frac{\partial}{\partial r}(\rho r^2 u_r)+\frac{1}{r\sin\theta}\frac{\partial}{\partial\theta}(\rho u_\theta\sin\theta)+\frac{1}{r\sin\theta}\frac{\partial(\rho u_\phi)}{\partial\phi}=0$

Table B4-2 Equation of Motion in Rectangular Coordinates

x	$\rho\left(\frac{\partial u_x}{\partial t}+u_x\frac{\partial u_x}{\partial x}+u_y\frac{\partial u_x}{\partial y}+u_z\frac{\partial u_x}{\partial z}\right)=-\frac{\partial\mathcal{P}}{\partial x}+\frac{\partial\tau_{xx}}{\partial x}+\frac{\partial\tau_{xy}}{\partial y}+\frac{\partial\tau_{xz}}{\partial z}$
y	$\rho\left(\frac{\partial u_y}{\partial t}+u_x\frac{\partial u_y}{\partial x}+u_y\frac{\partial u_y}{\partial y}+u_z\frac{\partial u_y}{\partial z}\right)=-\frac{\partial\mathcal{P}}{\partial y}+\frac{\partial\tau_{yx}}{\partial x}+\frac{\partial\tau_{yy}}{\partial y}+\frac{\partial\tau_{yz}}{\partial z}$
z	$\rho\left(\frac{\partial u_z}{\partial t}+u_x\frac{\partial u_z}{\partial x}+u_y\frac{\partial u_z}{\partial y}+u_z\frac{\partial u_z}{\partial z}\right)=-\frac{\partial\mathcal{P}}{\partial z}+\frac{\partial\tau_{zx}}{\partial x}+\frac{\partial\tau_{zy}}{\partial y}+\frac{\partial\tau_{zz}}{\partial z}$

Table B4-3 Navier–Stokes Equation in Rectangular Coordinates

x	$\rho\left(\frac{\partial u_x}{\partial t}+u_x\frac{\partial u_x}{\partial x}+u_y\frac{\partial u_x}{\partial y}+u_z\frac{\partial u_x}{\partial z}\right)=-\frac{\partial\mathcal{P}}{\partial x}+\mu\left(\frac{\partial^2 u_x}{\partial x^2}+\frac{\partial^2 u_x}{\partial y^2}+\frac{\partial^2 u_x}{\partial z^2}\right)$
y	$\rho\left(\frac{\partial u_y}{\partial t}+u_x\frac{\partial u_y}{\partial x}+u_y\frac{\partial u_y}{\partial y}+u_z\frac{\partial u_y}{\partial z}\right)=-\frac{\partial\mathcal{P}}{\partial y}+\mu\left(\frac{\partial^2 u_y}{\partial x^2}+\frac{\partial^2 u_y}{\partial y^2}+\frac{\partial^2 u_y}{\partial z^2}\right)$
z	$\rho\left(\frac{\partial u_z}{\partial t}+u_x\frac{\partial u_z}{\partial x}+u_y\frac{\partial u_z}{\partial y}+u_z\frac{\partial u_z}{\partial z}\right)=-\frac{\partial\mathcal{P}}{\partial z}+\mu\left(\frac{\partial^2 u_z}{\partial x^2}+\frac{\partial^2 u_z}{\partial y^2}+\frac{\partial^2 u_z}{\partial z^2}\right)$

Table B4-4 Equation of Motion in Cylindrical Coordinates

r	$\rho\left(\frac{\partial u_r}{\partial t}+u_r\frac{\partial u_r}{\partial r}+\frac{u_\theta}{r}\frac{\partial u_r}{\partial\theta}+u_z\frac{\partial u_r}{\partial z}\right)=-\frac{\partial\mathcal{P}}{\partial r}+\frac{1}{r}\frac{\partial(r\tau_{rr})}{\partial r}+\frac{1}{r}\frac{\partial\tau_{r\theta}}{\partial\theta}+\frac{\partial\tau_{rz}}{\partial z}$
θ	$\rho\left(\frac{\partial u_\theta}{\partial t}+u_r\frac{\partial u_\theta}{\partial r}+\frac{u_\theta}{r}\frac{\partial u_\theta}{\partial\theta}+\frac{u_r u_\theta}{r}+u_z\frac{\partial u_\theta}{\partial z}\right)=-\frac{1}{r}\frac{\partial\mathcal{P}}{\partial\theta}+\frac{1}{r^2}\frac{\partial(r^2\tau_{\theta r})}{\partial r}+\frac{1}{r}\frac{\partial\tau_{\theta\theta}}{\partial\theta}+\frac{\partial\tau_{\theta z}}{\partial z}$
z	$\rho\left(\frac{\partial u_z}{\partial t}+u_r\frac{\partial u_z}{\partial r}+\frac{u_\theta}{r}\frac{\partial u_z}{\partial\theta}+u_z\frac{\partial u_z}{\partial z}\right)=-\frac{\partial\mathcal{P}}{\partial z}+\frac{1}{r}\frac{\partial(r\tau_{zr})}{\partial r}+\frac{1}{r}\frac{\partial\tau_{z\theta}}{\partial\theta}+\frac{\partial\tau_{zz}}{\partial z}$

Table B4-5 Navier–Stokes Equation in Cylindrical Coordinates

r

$$\rho\left(\frac{\partial u_r}{\partial t}+u_r\frac{\partial u_r}{\partial r}+\frac{u_\theta}{r}\frac{\partial u_r}{\partial \theta}+u_z\frac{\partial u_r}{\partial z}\right)=$$

$$-\frac{\partial \mathcal{P}}{\partial r}+\mu\left[\frac{\partial}{\partial r}\left(\frac{1}{r}\frac{\partial (ru_r)}{\partial r}\right)+\frac{1}{r^2}\frac{\partial^2 u_r}{\partial \theta^2}-\frac{2}{r^2}\frac{\partial u_\theta}{\partial \theta}+\frac{\partial^2 u_r}{\partial z^2}\right]$$

θ

$$\rho\left(\frac{\partial u_\theta}{\partial t}+u_r\frac{\partial u_\theta}{\partial r}+\frac{u_\theta}{r}\frac{\partial u_\theta}{\partial \theta}+\frac{u_r u_\theta}{r}+u_z\frac{\partial u_\theta}{\partial z}\right)=$$

$$-\frac{1}{r}\frac{\partial \mathcal{P}}{\partial \theta}+\mu\left[\frac{\partial}{\partial r}\left(\frac{1}{r}\frac{\partial (ru_\theta)}{\partial r}\right)+\frac{1}{r^2}\frac{\partial^2 u_\theta}{\partial \theta^2}+\frac{2}{r^2}\frac{\partial u_r}{\partial \theta}+\frac{\partial^2 u_\theta}{\partial z^2}\right]$$

z

$$\rho\left(\frac{\partial u_z}{\partial t}+u_r\frac{\partial u_z}{\partial r}+\frac{u_\theta}{r}\frac{\partial u_z}{\partial \theta}+u_z\frac{\partial u_z}{\partial z}\right)=-\frac{\partial P}{\partial z}+\mu\left[\frac{1}{r}\frac{\partial}{\partial r}\left(r\frac{\partial u_z}{\partial r}\right)+\frac{1}{r^2}\frac{\partial^2 u_z}{\partial \theta^2}+\frac{\partial^2 u_z}{\partial z^2}\right]$$

Table B4-6 Equation of Motion in Spherical Coordinates

r

$$\rho\left(\frac{\partial u_r}{\partial t}+u_r\frac{\partial u_r}{\partial r}+\frac{u_\theta}{r}\frac{\partial u_r}{\partial \theta}+\frac{u_\phi}{r\sin\theta}\frac{\partial u_r}{\partial \phi}-\frac{u_\theta^2+u_\phi^2}{r}\right)=$$

$$-\frac{\partial \mathcal{P}}{\partial r}+\frac{1}{r^2}\frac{\partial (r^2\tau_{rr})}{\partial r}+\frac{1}{r\sin\theta}\frac{\partial (\sin\theta\,\tau_{r\theta})}{\partial \theta}+\frac{1}{r\sin\theta}\frac{\partial \tau_{r\phi}}{\partial \phi}-\frac{\tau_{\theta\theta}+\tau_{\phi\phi}}{r}$$

θ

$$\rho\left(\frac{\partial u_\theta}{\partial t}+u_r\frac{\partial u_\theta}{\partial r}+\frac{u_\theta}{r}\frac{\partial u_\theta}{\partial \theta}+\frac{u_\phi}{r\sin\theta}\frac{\partial u_\theta}{\partial \phi}+\frac{u_r u_\theta}{r}-\frac{u_\phi^2\cot\theta}{r}\right)=$$

$$-\frac{1}{r}\frac{\partial \mathcal{P}}{\partial \theta}+\frac{1}{r^3}\frac{\partial (r^3\tau_{\theta r})}{\partial r}+\frac{1}{r\sin\theta}\frac{\partial (\sin\theta\,\tau_{\theta\theta})}{\partial \theta}+\frac{1}{r\sin\theta}\frac{\partial \tau_{\theta\phi}}{\partial \phi}-\frac{\cot\theta}{r}\tau_{\phi\phi}$$

φ

$$\rho\left(\frac{\partial u_\phi}{\partial t}+u_r\frac{\partial u_\phi}{\partial r}+\frac{u_\theta}{r}\frac{\partial u_\phi}{\partial \theta}+\frac{u_\phi}{r\sin\theta}\frac{\partial u_\phi}{\partial \phi}+\frac{u_r u_\phi}{r}+\frac{u_\theta u_\phi\cot\theta}{r}\right)=$$

$$-\frac{1}{r\sin\theta}\frac{\partial \mathcal{P}}{\partial \phi}+\frac{1}{r^3}\frac{\partial (r^3\tau_{\phi r})}{\partial r}+\frac{1}{r\sin^2\theta}\frac{\partial (\sin^2\theta\,\tau_{\phi\theta})}{\partial \theta}+\frac{1}{r\sin\theta}\frac{\partial \tau_{\phi\phi}}{\partial \phi}-\frac{\cot\theta}{r}\tau_{\phi\phi}$$

Table B4-7 Navier–Stokes Equation in Spherical Coordinates[a]

r

$$\rho\left(\frac{\partial u_r}{\partial t}+u_r\frac{\partial u_r}{\partial r}+\frac{u_\theta}{r}\frac{\partial u_r}{\partial \theta}+\frac{u_\phi}{r\sin\theta}\frac{\partial u_r}{\partial \phi}-\frac{u_\theta^2+u_\phi^2}{r}\right)=$$

$$-\frac{\partial \mathcal{P}}{\partial r}+\mu\left(\Psi u_r-\frac{2}{r^2}u_r-\frac{2}{r^2}\frac{\partial u_\theta}{\partial \theta}-\frac{2}{r^2}u_\theta\cot\theta-\frac{2}{r^2\sin\theta}\frac{\partial u_\phi}{\partial \phi}\right)$$

θ

$$\rho\left(\frac{\partial u_\theta}{\partial t}+u_r\frac{\partial u_\theta}{\partial r}+\frac{u_\theta}{r}\frac{\partial u_\theta}{\partial \theta}+\frac{u_\phi}{r\sin\theta}\frac{\partial u_\theta}{\partial \phi}+\frac{u_r u_\theta}{r}-\frac{u_\phi^2\cot\theta}{r}\right)=$$

$$-\frac{1}{r}\frac{\partial \mathcal{P}}{\partial \theta}+\mu\left(\Psi u_\theta+\frac{2}{r^2}\frac{\partial u_r}{\partial \theta}-\frac{u_\theta}{r^2\sin^2\theta}-\frac{2\cos\theta}{r^2\sin^2\theta}\frac{\partial u_\phi}{\partial \phi}\right)$$

φ

$$\rho\left(\frac{\partial u_\phi}{\partial t}+u_r\frac{\partial u_\phi}{\partial r}+\frac{u_\theta}{r}\frac{\partial u_\phi}{\partial \theta}+\frac{u_\phi}{r\sin\theta}\frac{\partial u_\phi}{\partial \phi}+\frac{u_r u_\phi}{r}+\frac{u_\phi u_\phi\cot\theta}{r}\right)=$$

$$-\frac{1}{r\sin\theta}\frac{\partial \mathcal{P}}{\partial \phi}+\mu\left(\Psi u_\phi-\frac{u_\phi}{r^2\sin^2\theta}+\frac{2}{r^2\sin\theta}\frac{\partial u_r}{\partial \phi}+\frac{2\cos\theta}{r^2\sin^2\theta}\frac{\partial u_\theta}{\partial \phi}\right)$$

[a] $\Psi\equiv\frac{1}{r^2}\frac{\partial}{\partial r}\left(r^2\frac{\partial}{\partial r}\right)+\frac{1}{r^2\sin\theta}\frac{\partial}{\partial \theta}\left(\sin\theta\frac{\partial}{\partial \theta}\right)+\frac{1}{r\sin^2\theta}\frac{\partial^2}{\partial \phi^2}$

Table B4-8 The Deformation Rate Tensor, γ

	Rectangular 1 = x, 2 = y, 3 = z	Cylindrical 1 = r, 2 = θ, 3 = z	Spherical 1 = r, 2 = θ, 3 = φ
γ_{11}	$\frac{\partial u_x}{\partial x}$	$\frac{\partial u_r}{\partial r}$	$\frac{\partial u_r}{\partial r}$
γ_{22}	$\frac{\partial u_y}{\partial y}$	$\frac{1}{r}\frac{\partial u_\theta}{\partial \theta}+\frac{u_r}{r}$	$\frac{1}{r}\frac{\partial u_\theta}{\partial \theta}+\frac{u_r}{r}$
γ_{33}	$\frac{\partial u_z}{\partial z}$	$\frac{\partial u_z}{\partial z}$	$\frac{1}{r\sin\theta}\frac{\partial u_\phi}{\partial \phi}+\frac{u_r}{r}+\frac{u_\theta \cot\theta}{r}$
$\gamma_{12}=\gamma_{21}$	$\frac{1}{2}\left(\frac{\partial u_x}{\partial y}+\frac{\partial u_y}{\partial x}\right)$	$\frac{1}{2}\left[r\frac{\partial}{\partial r}\left(\frac{u_\theta}{r}\right)+\frac{1}{r}\frac{\partial u_r}{\partial \theta}\right]$	$\frac{1}{2}\left[r\frac{\partial}{\partial r}\left(\frac{u_\theta}{r}\right)+\frac{1}{r}\frac{\partial u_r}{\partial \theta}\right]$
$\gamma_{13}=\gamma_{31}$	$\frac{1}{2}\left(\frac{\partial u_x}{\partial z}+\frac{\partial u_z}{\partial x}\right)$	$\frac{1}{2}\left(\frac{\partial u_r}{\partial z}+\frac{\partial u_z}{\partial r}\right)$	$\frac{1}{2}\left[\frac{1}{r\sin\theta}\frac{\partial u_r}{\partial \phi}+r\frac{\partial}{\partial r}\left(\frac{u_\phi}{r}\right)\right]$
$\gamma_{23}=\gamma_{32}$	$\frac{1}{2}\left(\frac{\partial u_y}{\partial z}+\frac{\partial u_z}{\partial y}\right)$	$\frac{1}{2}\left(\frac{\partial u_\theta}{\partial z}+\frac{1}{r}\frac{\partial u_z}{\partial \theta}\right)$	$\frac{1}{2}\left[\frac{\sin\theta}{r}\frac{\partial}{\partial \theta}\left(\frac{u_\phi}{\sin\theta}\right)+\frac{1}{r\sin\theta}\frac{\partial u_\theta}{\partial \phi}\right]$

Table B4-9 Squared Characteristic Deformation Rate, $\gamma_{app}^2 = 2tr(\boldsymbol{\gamma}\cdot\boldsymbol{\gamma})$

	γ_{app}^2
Rectangular	$2\left[\left(\frac{\partial u_x}{\partial x}\right)^2+\left(\frac{\partial u_y}{\partial y}\right)^2+\left(\frac{\partial u_z}{\partial z}\right)^2\right]+\left(\frac{\partial u_y}{\partial x}+\frac{\partial u_x}{\partial y}\right)^2+\left(\frac{\partial u_z}{\partial y}+\frac{\partial u_y}{\partial z}\right)^2+\left(\frac{\partial u_x}{\partial z}+\frac{\partial u_z}{\partial x}\right)^2$
Cylindrical	$2\left[\left(\frac{\partial u_r}{\partial r}\right)^2+\left(\frac{1}{r}\frac{\partial u_\theta}{\partial \theta}+\frac{u_r}{r}\right)^2+\left(\frac{\partial u_z}{\partial z}\right)^2\right]+\left[r\frac{\partial}{\partial r}\left(\frac{u_\theta}{r}\right)+\frac{1}{r}\frac{\partial u_r}{\partial \theta}\right]^2+\left(\frac{1}{r}\frac{\partial u_z}{\partial \theta}+\frac{\partial u_\theta}{\partial z}\right)^2+\left(\frac{\partial u_r}{\partial z}+\frac{\partial u_z}{\partial r}\right)^2$
Spherical	$2\left[\left(\frac{\partial u_r}{\partial r}\right)^2+\left(\frac{1}{r}\frac{\partial u_\theta}{\partial \theta}+\frac{u_r}{r}\right)^2+\left(\frac{1}{r\sin\theta}\frac{\partial u_\phi}{\partial \phi}+\frac{u_r}{r}+\frac{u_\theta\cot\theta}{r}\right)^2\right]+\left[r\frac{\partial}{\partial r}\left(\frac{u_\theta}{r}\right)+\frac{1}{r}\frac{\partial u_r}{\partial \theta}\right]^2+\left[\frac{\sin\theta}{r}\frac{\partial}{\partial \theta}\left(\frac{u_\phi}{\sin\theta}\right)+\frac{1}{r\sin\theta}\frac{\partial u_\theta}{\partial \phi}\right]^2+\left[\frac{1}{r\sin\theta}\frac{\partial u_r}{\partial \phi}+r\frac{\partial}{\partial r}\left(\frac{u_\phi}{r}\right)\right]^2$

References

Bear J, Bachmat Y. Introduction to Modeling of Transport Phenomena in Porous Media. Boston: Kluwer Academic Publishers; 1990, Eqs. 2.3.10 and 2.3.29.

Fitts DD. Nonequilibrium Thermodynamics. New York: McGraw-Hill; 1962.

Jou D, Llebot JE. Introduction to the Thermodynamics of Biological Processes. Englewood Cliffs: Prentice Hall; 1990, (English Translation).

Temam R, Miranville A. Mathematical Modeling in Continuum Mechanics. Cambridge: Cambridge University Press; 2001, ch 2,5,7.

Wegner JL, Haddow JB. Elements of Continuum Mechanics and Thermodynamics. Cambridge: Cambridge University Press; 2009, ch 1,2.

Appendix C

Mathematical Methods

C.1 Dimensionless Forms and Scaling

A model in dimensionless form has the following advantages:

- The number of parameters is minimized.
- Variables and parameters are independent of units.
- Dimensionless parameter groups provide a natural quantitative criteria for neglecting terms in an equation.
- Equations are scaled for convenient comparison to experimental data.
- The number of experiments for establishing relationships or validating a model is reduced.

Model equations are made dimensionless by the following procedure:

1) Define dimensionless variables by dividing each dimensional variable with an arbitrary scaling parameter.
2) Substitute dimensionless variables in place of dimensional variables in the model equations as well as the boundary and initial conditions of the problem.
3) Group all parameters into dimensionless combinations such that every term in the model equations and conditions is dimensionless.
4) Define the arbitrary scale parameters in terms of actual model parameters such that the maximum values of the dimensionless parameter groups become unity.

C.1.1 Dimensionless Representation of a Spatially Lumped Model

Consider the example of a single-compartment model equation with its initial condition:

$$\begin{aligned} \frac{dC}{dt} &= \frac{Q}{V}(C_{in} - C) \quad (t > 0) \\ t = 0 &: C = C_0 \end{aligned} \qquad \text{(C1-1a,b)}$$

Biomedical Mass Transport and Chemical Reaction: Physicochemical Principles and Mathematical Modeling,
First Edition. James S. Ultman, Harihara Baskaran, and Gerald M. Saidel.

Following the procedures outlined above, we define dimensionless variables t = t/a and C = C/b, where a and b are arbitrary scale parameters. Substituting into Equation C1-1, we obtain

$$\frac{d(bC)}{d(at)} = \frac{Q}{V}(C_{in} - bC) \quad (at > 0)$$
$$at = 0 \ : \ bC = C_0 \qquad \text{(C1-2a,b)}$$

Since a and b are constants,

$$\frac{dC}{dt} = \frac{Qa}{V}\left(\frac{C_{in}}{b} - C\right) \quad (t > 0)$$
$$t = 0 \ : \ C = \frac{C_0}{b} \qquad \text{(C1-3a,b)}$$

Here, the dimensionless parameter groups are C_{in}/b, Qa/V, and C_0/b. Because there are only two arbitrary scaling parameters in this problem, we can only set two of the dimensionless groups equal to unity. We choose to let Qa/V = 1 and $C_{in}/b = 1$. Consequently, we obtain the time scale parameter a = V/Q and the concentration scale parameter $b = C_{in}$. For convenience, we define the remaining dimensionless group as $C_0/b = C_0/C_{in} = C_0$. Now, Equation C1-2 with every term in dimensionless form becomes

$$\frac{dC}{dt} = (1 - C) \quad (t > 0)$$
$$t = 0 \ : \ C = C_0 \qquad \text{(C1-4a,b)}$$

Alternatively, if we choose $b = C_0$ and define $C_{in}/b = C_{in}/C_0 = C_{in}$, then the dimensionless form of Equation C1-1 is

$$\frac{dC}{dt} = (C_{in} - C) \quad (t > 0)$$
$$t = 0 \ : \ C = 1 \qquad \text{(C1-5a,b)}$$

These dimensionless forms of the problem involve one dimensionless parameter, either C_0 or C_{in}. Dimensionless forms can be even simpler by making either the initial condition homogeneous or the differential equation homogeneous. By redefining $\bar{C} = (C - C_0)/b$ where $b = C_{in} - C_0$, the boundary condition in Equation C1-1b becomes homogeneous and the dimensionless form of the problem is

$$\frac{d\bar{C}}{dt} = (1 - \bar{C}) \quad (t > 0)$$
$$t = 0 \ : \ \bar{C} = 0 \qquad \text{(C1-6a,b)}$$

Alternatively, if we let $\bar{C} = (C - C_{in})/b$ where $b = C_0 - C_{in}$, the differential equation takes on another homogeneous form:

$$\frac{d\bar{C}}{dt} = -\bar{C} \quad (t > 0)$$
$$t = 0 \ : \ \bar{C} = 1 \qquad \text{(C1-7a,b)}$$

Dimensionless parameters do not appear in either of these two dimensionless representations of the problem.

C.1.2 Dimensionless Representation of a Spatially Distributed Model

Let us start with a one-dimensional convective–diffusion model describing a dynamic concentration distribution C(z,t) having an impulse source term at the z = 0 boundary:

$$\frac{\partial C}{\partial t}+u\frac{\partial C}{\partial z}=\mathcal{D}\frac{\partial^2 C}{\partial z^2}+\frac{m}{A}\delta(z)\delta(t)\quad (L\geq z\geq 0,\ t\geq 0) \tag{C1-8}$$

where u, $\mathcal{D}$, m/A, and L are constants and the δ's are unit impulse functions. Typically, there is no solute in the system initially and no solute at the upstream boundary thereafter:

$$\begin{aligned} t=0 &: C=0 \\ z=0 &: C=0 \end{aligned} \tag{C1-9a,b}$$

A typical condition on the downstream boundary is a zero concentration gradient:

$$z=L:\ \frac{\partial C}{\partial z}=0 \tag{C1-10}$$

Following the standard procedure to make the problem dimensionless, we define dimensionless variables:

$$\mathcal{C}=\frac{C}{C_0},\quad \mathcal{t}=\frac{t}{t_0},\quad \mathcal{z}=\frac{z}{z_0},\quad \delta(\mathcal{z})=z_0\delta(z),\quad \delta(\mathcal{t})=t_0\delta(t) \tag{C1-11a-e}$$

where C_0, t_0, and z_0 are arbitrary scale parameters. This leads to the dimensionless derivatives

$$\begin{aligned} \frac{\partial C}{\partial t}&=\frac{\partial(C_0\mathcal{C})}{\partial(t_0\mathcal{t})}=\frac{C_0}{t_0}\frac{\partial \mathcal{C}}{\partial \mathcal{t}} \\ \frac{\partial C}{\partial z}&=\frac{\partial(C_0\mathcal{C})}{\partial(z_0\mathcal{z})}=\frac{C_0}{z_0}\frac{\partial \mathcal{C}}{\partial \mathcal{z}} \\ \frac{\partial^2 C}{\partial z^2}&=\frac{\partial^2(C_0\mathcal{C})}{\partial(z_0\mathcal{z})^2}=\frac{C_0}{z_0^2}\frac{\partial^2 \mathcal{C}}{\partial \mathcal{z}^2} \end{aligned} \tag{C1-12a-c}$$

Substitution of the dimensionless variables and derivatives into the partial differential equation and its boundary condition yields

$$\begin{aligned} &\frac{\partial \mathcal{C}}{\partial \mathcal{t}}+\left(\frac{ut_0}{z_0}\right)\frac{\partial \mathcal{C}}{\partial \mathcal{z}}=\left(\frac{\mathcal{D}t_0}{z_0^2}\right)\frac{\partial^2 \mathcal{C}}{\partial \mathcal{z}^2}+\left(\frac{m}{C_0Az_0}\right)\delta(\mathcal{z})\delta(\mathcal{t})\quad (L/z_0>\mathcal{z}\geq 0,\ \mathcal{t}\geq 0) \\ &\mathcal{t}=0:\ \mathcal{C}=0 \\ &\mathcal{z}=0:\ \mathcal{C}=0 \\ &\mathcal{z}=\frac{L}{z_0}:\ \frac{\partial \mathcal{C}}{\partial \mathcal{z}}=0 \end{aligned} \tag{C1-13a-d}$$

In this form, the problem has four dimensionless groups:

$$\left(\frac{ut_0}{z_0}\right),\ \left(\frac{\mathcal{D}t_0}{z_0^2}\right),\ \left(\frac{m}{C_0Az_0}\right),\ \left(\frac{L}{z_0}\right)$$

These are not all independent because we can arbitrarily define the three scale factors (C_0, t_0, z_0) by setting three of the dimensionless groups equal to one:

$$\frac{ut_0}{z_0}=1,\quad \frac{m}{C_0Az_0}=1,\quad \frac{L}{z_0}=1 \tag{C1-14a-c}$$

so that $z_0 = L$, $t_0 = L/u$, and $C_0 = m/LA$. Consequently, only one independent dimensionless parameter group remains:

$$\frac{\mathcal{D}t_0}{z_0^2} = \frac{\mathcal{D}}{uz_0} = \frac{\mathcal{D}}{uL} \equiv \alpha \tag{C1-15}$$

Substitution of these relations in the dimensionless model yields

$$\frac{\partial C}{\partial t} + \frac{\partial C}{\partial z} = \alpha \frac{\partial^2 C}{\partial z^2} + \delta(z)\delta(t) \quad (1 > z \geq 0, t \geq 0) \tag{C1-16}$$

with the conditions

$$\begin{aligned} t = 0 &: C = 0 \\ z = 0 &: C = 0 \\ z = 1 &: \frac{\partial C}{\partial z} = 0 \end{aligned} \tag{C1-17a-c}$$

C.2 Inversion of Square Matrices

Consider a system of linear algebraic equations, $[B][x] = [y]$, in which an unknown $n \times 1$ matrix $[x]$ must be determined from a known $n \times n$ matrix $[B]$ and a known $n \times 1$ matrix $[y]$. This equation may be solved if $[B]$ can be inverted (i.e., $\det[B] \neq 0$). The formal solution is obtained by premultiplication by the inverse matrix $[B]^{-1}$ so that $[B]^{-1}[B][x] = [x] = [B]^{-1}[y]$. If the dimension of $[B]$ is low, then the adjoint method described below can be used for an analytical evaluation of $[B]^{-1}$. For matrices of high dimension, numerical methods are much more efficient.

Suppose that $[B]$ has elements B_{ij} ($i = 1,2,\ldots N$; $j = 1,2,\ldots N$). The cofactors of $[B]$ are defined as

$$C_{ij} = (-1)^{i+j} M_{ij} \quad (i = 1,2\ldots N; j = 1,2\ldots N) \tag{C2-1}$$

where M_{ij} is a minor that is obtained by eliminating the ith row and jth column from $[B]$ and then taking the determinant of the reduced matrix. The determinant of a matrix $[A]$ with elements A_{ij} can be expressed in terms of its cofactors:

$$\det[A] = \sum_{k=1}^{N} A_{kj} C_{kj} \quad \text{or} \quad \det[A] = \sum_{k=1}^{N} A_{ik} C_{ik} \tag{C2-2a, b}$$

The adjoint of matrix $[B]$ is defined as the transpose of the matrix of its cofactors such that

$$\text{adj}[B] = [C]^T \tag{C2-3}$$

Finally, the inverse of $[B]$ is obtained from the ratio of its adjoint matrix to its determinant:

$$[B]^{-1} = \frac{\text{adj}[B]}{\det[B]} \tag{C2-4}$$

To demonstrate this process of matrix inversion, consider the simplest case of a 2×2 matrix:

$$[B] = \begin{bmatrix} B_{11} & B_{12} \\ B_{22} & B_{22} \end{bmatrix} \tag{C2-5}$$

The four minors and four cofactors of [B] are given by

$$\begin{aligned} M_{11} &= \det[B_{22}] = B_{22}, \quad C_{11} = (-1)^2 B_{22} = +B_{22} \\ M_{12} &= \det[B_{21}] = B_{21}, \quad C_{12} = (-1)^3 B_{21} = -B_{21} \\ M_{21} &= \det[B_{21}] = B_{12}, \quad C_{21} = (-1)^3 B_{12} = -B_{12} \\ M_{22} &= \det[B_{11}] = B_{11}, \quad C_{22} = (-1)^4 B_{11} = +B_{11} \end{aligned} \tag{C2-6a-h}$$

The adjoint of [B] can now be found using the cofactors:

$$\text{adj}[B] = \begin{bmatrix} C_{11} & C_{12} \\ C_{21} & C_{22} \end{bmatrix}^T = \begin{bmatrix} C_{11} & C_{21} \\ C_{12} & C_{22} \end{bmatrix} = \begin{bmatrix} +B_{22} & -B_{12} \\ -B_{21} & +B_{11} \end{bmatrix} \tag{C2-7}$$

By selecting $j = 1$, we find the determinant of [B] from Equation C2-2a:

$$\det[B] = \sum_{k=1}^{2} B_{k1} C_{k1} = B_{11} C_{11} + B_{21} C_{21} = B_{11} B_{22} - B_{21} B_{12} \tag{C2-8}$$

Alternatively, we can compute det[B] by selecting $i = 1$ in Equation C2-2b:

$$\det[B] = \sum_{k=1}^{2} B_{1k} C_{1k} = B_{11} C_{11} + B_{12} C_{12} = B_{11} B_{22} - B_{12} B_{21} \tag{C2-9}$$

Finally, the inverse of [B] can be determined as

$$[B]^{-1} = \frac{\text{adj}[B]}{\det[B]} = \frac{\begin{bmatrix} +B_{22} & -B_{12} \\ -B_{21} & +B_{11} \end{bmatrix}}{B_{11} B_{22} - B_{12} B_{21}} \tag{C2-10}$$

or

$$\begin{bmatrix} B_{11} & B_{12} \\ B_{21} & B_{22} \end{bmatrix}^{-1} = \begin{bmatrix} \dfrac{+B_{22}}{B_{11} B_{22} - B_{12} B_{21}} & \dfrac{-B_{12}}{B_{11} B_{22} - B_{12} B_{21}} \\ \dfrac{-B_{21}}{B_{11} B_{22} - B_{12} B_{21}} & \dfrac{+B_{11}}{B_{11} B_{22} - B_{12} B_{21}} \end{bmatrix} \tag{C2-11}$$

C.3 Initial-value Problems

C.3.1 Classification

A system of ordinary differential equations with all conditions specified at one point in the time domain is an initial-value problem. Suppose that the state of a system is defined by a set of dependent (or state) variables $\{y_j(t): j = 1,2,\ldots J\}$. For a spatially lumped system, the dynamic changes in $y_j(t)$ in response to a set of inputs (or forcing functions) $\{x_j(t)\}$ are described by a system of J first-order differential equations accompanied by J initial conditions:

$$\begin{aligned} &\frac{dy_j}{dt} + F_i\left(y_1, y_2, \ldots . y_J\right) = x_j(t) \\ &t = t_0 \; : \; y_j = y_j^0 \end{aligned} \qquad (j = 1,2,\ldots J) \tag{C3-1a,b}$$

where F_j can be a nonlinear function. An example of such an initial-value problem is

$$\frac{dy_1}{dt} + a_1(y_1)^2 - a_2 y_2 = \sin(a_3 t)$$
$$\frac{dy_2}{dt} + b_1 t y_2 - b_2 y_1 = b_3 \qquad \text{(C3-2a-c)}$$
$$t = 0: \quad y_1 = a_4, \quad y_2 = b_4$$

where a_j and b_j (j = 1,2,3,4) are constants. Since the order of the system is the sum of the highest order derivative of each equation in the system, this is a second-order system. The term $(y_1)^2$ makes the first equation nonlinear, whereas the second equation is linear. When the input to a state equation is zero and all terms involve a state variable, the equation is homogeneous. The input $\sin(a_3t)$ makes the first equation inhomogeneous; the second equation is homogeneous when $b_3 = 0$. A system is time varying (or not time invariant) when there is at least one term that is a multiplicative function of both independent and dependent variables. The term b_1ty_2 makes the second equation and therefore the system itself time varying.

An initial-value problem can also involve higher-order equations. Consider the initial-value problem

$$\frac{d^3y}{dt^3} + 4\frac{d^2y}{dt^2} + 5\frac{dy}{dt} + 2y = \exp(-t) \qquad \text{(C3-3)}$$

$$t = 0: \quad \frac{d^2y}{dt^2} = \frac{dy}{dt} = 0, \quad y = 1 \qquad \text{(C3-4)}$$

This third-order system can be expressed in terms of three first-order equations by defining a set of three new dependent variables:

$$y_1 = y, \quad y_2 = \frac{dy}{dt}, \quad y_3 = \frac{d^2y}{dt^2} \qquad \text{(C3-5a-c)}$$

Equations C3-3 and C3-5a-c can now be written as

$$\frac{dy_1}{dt} = y_2, \quad \frac{dy_2}{dt} = y_3, \quad \frac{dy_3}{dt} + 4y_3 = -2y_1 - 5y_2 + \exp(-t) \qquad \text{(C3-6a-c)}$$

with initial conditions

$$t = 0 : \; y_3 = y_2 = 0, \quad y_1 = 1 \qquad \text{(C3-7)}$$

This system is linear because the state variables (y_1,y_2,y_3) in each term do not involve nonlinear expressions. This system is time invariant because none of the coefficients of the dependent variable terms is a function of time. Furthermore, the system is inhomogeneous because the term exp(−t), which appears without a dependent variable, is considered an input to the system.

C.3.2 Reduction of Order

A dynamic system consisting of two or more dependent variables that constitute an initial-value problem can sometimes be simplified by either of two closely related approximations—the "pseudo-steady-state assumption" or the "quasi-equilibrium hypothesis." The pseudo-steady-state assumption is applied when the characteristic times for changes of one or more independent variables differ by at least one-order of magnitude from the characteristic times of the other variables. As a consequence, the variables with the short characteristic times change very quickly and are much closer to their steady state than the other variables. The differential equations associated with these variables can then be approximated by an algebraic equation, thereby reducing the order of the initial-value problem. The quasi-equilibrium hypothesis is appropriate when some rate coefficients are very large relative to others, while the rates of change of the associated variables remain finite. This behavior is fairly common in multistep transport or chemical kinetic processes in which opposing processes affect changes of the independent variables.

To illustrate these alternative simplifications, we start with a second-order system:

$$\frac{dx}{dt} = -k_1 f(x,y) \qquad \frac{dy}{dt} = k_2 g(x,y) - k_3 h(x,y) \tag{C3-8a, b}$$

with initial conditions

$$t = 0: \quad x(0) = x_o > 0, \ y(0) = y_o > 0 \tag{C3-9}$$

If the characteristic time of y(t) is much shorter than that of x(t), then the pseudo-steady-state assumption is

$$\frac{dy}{dt} \approx 0 \Rightarrow k_2 g(x,y) - k_3 h(x,y) \simeq 0 \tag{C3-10a, b}$$

Consequently, y(t) is related to x(t) by an algebraic equation instead of a differential equation. In general, the initial condition for y(t) is not satisfied by this approximation.

Alternatively, if the rate coefficients k_2 and k_3 are arbitrarily large while the rate of change of y(t) is finite, then according to the quasi-equilibrium hypothesis:

$$\lim_{t\to\infty}\left[\frac{1}{k_2}\frac{dy}{dt}\right] \to 0 \Rightarrow g(x,y) - K_{2,3} h(x,y) = 0 \tag{C3-11a, b}$$

where $K_{2,3} = k_3/k_2$ is an equilibrium coefficient. Here too, the initial condition for y(t) is not satisfied.

In both cases, a second-order, initial-value problem has been reduced to a first-order problem by approximating one of the differential equations as an algebraic equation. The two approximations have the same form, but the parameters are represented differently.

C.3.3 Solution of a Linear, First-Order, Initial-Value Problem

A basic building block for understanding linear systems is the first-order, initial-value problem. For an arbitrary input $x(t) \geq 0$ to such a system, the output variable y(t) satisfies

$$\frac{dy}{dt} + a(t)y = x(t) \quad (t > t_0) \qquad t = t_0 : \ y = y_0 \tag{C.3-12a, b}$$

where $a(t) \geq 0$. For an impulse input $\delta(t - t_0)$ applied at $t = t_0$, the output response $g(t|t_0)$ satisfies

$$\frac{dg}{dt} + a(t)g = \delta(t - t_0) \quad (t \geq t_0) \qquad t < t_0 : \ g = 0 \tag{C.3-13a, b}$$

As will be shown in Section C5, the equivalent problem can be expressed with zero input and nonzero initial condition:

$$\frac{dg}{dt} + a(t)g = 0 \quad (t > t_0)$$
$$t = t_0 \;:\; g = 1 \tag{C.3-14a,b}$$

By separation of variables and integration, the solution is

$$g(t|t_0) = \exp\left[-\int_{t_0}^{t} a(w)dw\right] \quad (t \geq t_0) \tag{C.3-15}$$

where $0 \geq g(t|t_0) \geq 1$. In many textbooks, this impulse response function is called an "integrating factor." For a time-invariant system, a(t) = k (a constant) and the impulse response function becomes

$$g(t|t_0) = \exp[-k(t - t_0)] \equiv g(t - t_0) \tag{C.3-16}$$

In this context, time invariance means that the response function depends on the time difference $(t - t_0)$ from the initial time t_0, but not on the running time t.

The solution of the inhomogeneous problem given by Equations C.3-12a,b, which is derived in standard textbooks, can be expressed as

$$y(t) = y_0 g(t|t_0) + \int_{t_0}^{t} g(t|w)x(w)dw \tag{C.3-17}$$

This solution can be proved by substitution into the Equation C.3-12a. It consists of the superposition of two separate solutions: the first term is a homogeneous solution with $x(t) = 0$ and $y(t_0) = y_0$; the second term is an inhomogeneous solution with $x(t) > 0$ and $y_0 = 0$.

If the system is time invariant, then we can express Equation C.3-17 with a translated time $\tau = t - t_0$:

$$y(\tau) = y_0 g(\tau) + \int_{0}^{\tau} g(\tau - w)x(w)dw \tag{C.3-18}$$

If $y_0 = 0$, then this input–output relation becomes the convolution integral:

$$y(\tau) = \int_{0}^{\tau} g(\tau - w)x(w)dw \equiv x(\tau)^{*}g(\tau) \tag{C.3-19}$$

A change of variable from w to $v = \tau - w$ indicates that the convolution integral can also be written as

$$x(\tau)^{*}g(\tau) = \int_{0}^{\tau} g(v)x(\tau - v)dv \tag{C.3-20}$$

C.4 Laplace Transforms

Integral transform methods can be used to solve linear initial-value problems, especially differential equations with inputs that involve impulse functions and discontinuities. For time-variant systems, we use the Laplace transform, which is a linear integral operator defined over the domain $\{0, \infty\}$:

$$\boldsymbol{L}[f(t)] = \int_0^\infty f(t)e^{-st}dt = \tilde{f}(s) \tag{C4-1}$$

where the lower bound includes the zero value. An important property that stems from the linearity of the Laplace operator is

$$\boldsymbol{L}[af(t) + bg(t)] = a\tilde{f}(s) + b\tilde{g}(s) \tag{C4-2}$$

The Laplace transforms of some useful operations and functions are listed in Tables C4-1 and C4-2, respectively. By applying an operation from Table C4-1 to a function in Table C4-2, one can obtain the Laplace transform of another function of interest. For example, we find the Laplace transform of cos(ωt) by applying the first derivative operation to sin(ωt).

$$\cos(\omega t) = \frac{1}{\omega}\frac{d}{dt}(\sin\omega t) \Rightarrow \boldsymbol{L}[\cos(\omega t)] = \frac{\{s\boldsymbol{L}[\sin(\omega t)] - \sin(0)\}}{\omega} = \frac{s}{s^2 + \omega^2} \tag{C4-3a, b}$$

We can also combine functions according to the linearity property to obtain a new transform. For example, the Laplace transform of $\cosh(at) \equiv (e^{at} + e^{-at})/2$ is

$$\boldsymbol{L}[\cosh(at)] = \boldsymbol{L}\left[\frac{e^{at} + e^{-at}}{2}\right] = \frac{\boldsymbol{L}[e^{at}]}{2} + \frac{\boldsymbol{L}[e^{-at}]}{2} = \frac{1}{2(s-a)} + \frac{1}{2(s+a)} = \frac{s}{s^2 - a^2} \tag{C4-4}$$

A common form of the impulse and step responses in compartment systems is $Q_m(s)/P_n(s)$ where $Q_m(s)$ and $P_n(s)$ are polynomials of order m and n ($m < n$). For the special case that

Table C4-1 Laplace Transforms of Common Operations

Operation	Time Domain	Laplace Domain
First derivative	$f'(t)$	$s\tilde{f}(s) - f(0)$
n^{th} Derivative	$f^{(n)}(t)$	$s^n\tilde{f}(s) - s^{n-1}f(0) - s^{n-2}f'(0) - \ldots - sf^{n-2}(0) - f^{(n-1)}(0)$
Integration	$\int_0^t f(\tau)d\tau$	$\frac{\tilde{f}(s)}{s}$
Initial-value theorem	$\lim_{t\to 0} f(t)$	$\lim_{s\to\infty} s\tilde{f}(s)$
Final-value theorem	$\lim_{t\to\infty} f(t)$	$\lim_{s\to 0} s\tilde{f}(s)$
Shift of t	$f(t - t_o)$	$e^{-st_o}\tilde{f}(s)$
Shift of s	$e^{at}f(t)$	$\tilde{f}(s-a)$
Convolution integral	$\int_0^t f(t-\tau)g(\tau)d\tau$	$\tilde{f}(s) * \tilde{g}(s)$

Table C4-2 Laplace Transforms of Common Functions*

Time Domain, f(t)	Laplace Domain, $\tilde{f}(s)$
a	$\frac{a}{s}$
t, t^n (n = 1,2,...)	$\frac{1}{s}, \frac{n!}{s^{n+1}}$
$e^{at}, \ t^n e^{at}(n = 1,2,\ldots)$	$\frac{1}{s-a}, \frac{n!}{(s-a)^{n+1}}$
$\frac{e^{at}-e^{bt}}{a-b}, \ \frac{ae^{at}-be^{bt}}{b-a}$	$\frac{1}{(s-a)(s-b)}, \frac{s}{(s-a)(s-b)}$
sinh(at), cosh(at)	$\frac{a}{s^2-a^2}, \frac{s}{s^2-a^2}$
$\sin(\omega t + \phi), \cos(\omega t + \phi)$	$\frac{s\sin(\phi)+\omega\cos(\phi)}{s^2+\omega^2}, \frac{s\cos(\phi)-\omega\sin(\phi)}{s^2+\omega^2}$
$\text{erfc}\left(\frac{a}{2\sqrt{t}}\right)$	$\frac{e^{-a\sqrt{s}}}{s}$
Unit impulse: $\delta(t - t_o)$	e^{-st_o}
Unit step: $H(t - t_o)$	$\frac{e^{-st_o}}{s}$

*a and b are real distinct constants.

$P_n(s)$ has distinct roots $\lambda_i(i = 1,2,\ldots n)$, we can expand $Q_m(s)/P_n(s)$ in partial fractions to show that the inverse transform is the sum of exponentials:

$$\boldsymbol{L}^{-1}\left[\frac{Q_m(s)}{P_n(s)}\right] = \boldsymbol{L}^{-1}\left[\frac{Q_m(s)}{\prod_{i=1}^{n}(s-\lambda_j)}\right] = \sum_{i=1}^{n}\beta_i e^{\lambda_i t}; \ \beta_i = \left[\frac{Q_m(s)}{dP_n/ds}\right]_{s=\lambda_i} \quad \text{(C4-5a, b)}$$

For example, consider the Laplace transform of the step response of two flow-through compartments in series (Eq. 18.2-6), which has the form $1/s(s + a)(s + b)$. By inspection, $Q_m(s) = 1$, $P_n(s) = s(s + a)(s + b)$, $n = 3$, $\lambda_1 = 0$, $\lambda_2 = -a$, and $\lambda_3 = -b$. Also noting that $dP_n/ds = (s + a)(s + b) + s(s + a) + s(s + b)$, we can evaluate the step response in the time domain according to Equation C4-5:

$$\boldsymbol{L}^{-1}\left[\frac{1}{s(s+a)(s+b)}\right] = \sum_{i=1}^{3}\frac{e^{\lambda_i t}}{[(s+a)(s+b)+s(2s+a+b)]_{s=\lambda_i}} = \frac{1}{ab} + \frac{e^{-at}}{a(a-b)} - \frac{e^{-bt}}{b(a-b)} \quad \text{(C4-6)}$$

C.5 Alternative Representation of a Point Source

For an initial-value problem, which in general is nonlinear and not time invariant, a relationship exists between an impulse input and the initial condition. For example, the problem statement

$$\begin{aligned} &\alpha\frac{dC}{dt} + f(C,t) = \beta\delta(t-t_0) \quad (t \ge t_0) \\ &t = t_0 \ : \ C = C_0 \end{aligned} \quad \text{(C5-1a, b)}$$

is equivalent to

$$\alpha\frac{dC}{dt} + f(C,t) = 0 \quad (t > t_0)$$
$$t = t_0 \ : \ C = C_0 + \frac{\beta}{\alpha} \qquad \text{(C5-2a, b)}$$

where α and β are constants. In place of the impulse source, the initial condition of the equivalent problem statement has an added β/α term. Note that the domain of Equation C5-1a includes the point t_0, but the domain of C5-2a does not.

If these two problems have the same solution, then they are equivalent. Integration of Equation C5-2a including the point t_o yields

$$C(t) = C_0 + \frac{\beta}{\alpha} - \frac{1}{\alpha}\int_{t_0}^{t} f(C,u)du \qquad \text{(C5-3)}$$

Integration of Equation C5-1a starting at $t_o - \varepsilon$ $(t_o > \varepsilon > 0)$ such that the point t_0 is not included yields

$$C(t) = [C]_{t=t_0-\varepsilon} + \frac{\beta}{\alpha}\int_{t_0-\varepsilon}^{t} \delta(u - t_0)du - \frac{1}{\alpha}\int_{t_0-\varepsilon}^{t} f(C,u)du \qquad \text{(C5-4)}$$

Since the integral of the impulse function around the point t_0 is unity, the solution to Equation C5-4 in the limit as $\varepsilon \rightarrow 0$ is the same as Equation C5-3. This type of analysis applies also to any system of ordinary or partial differential equations that change as a consequence of an impulse input.

For the special case of a linear, time-invariant, initial-value problem with constant coefficients $(\alpha,\beta,\kappa,\sigma)$, the statement

$$\alpha\frac{dC}{dt} + \kappa C = \sigma + \beta\delta(t) \quad (t \geq 0)$$
$$t = 0 \ : \ C = C_0 \qquad \text{(C5-5a, b)}$$

is equivalent to

$$\alpha\frac{dC}{dt} + \kappa C = \sigma \quad (t > 0)$$
$$t = 0 \ : \ C = C_0 + \frac{\beta}{\alpha} \qquad \text{(C5-6a, b)}$$

since both problems have the same solution in the Laplace domain:

$$\tilde{C}(s) = \boldsymbol{L}\{C(t)\} = \frac{1}{s + \kappa/\alpha}\left(C_0 + \frac{\beta}{\alpha} + \frac{\sigma}{\alpha s}\right) \qquad \text{(C5-7)}$$

C.6 Similarity Transform of a Partial Differential Equation

Consider a model that describes a continuous function of two independent variables for which there are no characteristic scales. It is possible to simplify such a model by introducing a new independent variable that combines the original variables. This leads to a similarity function in the domain of the combined variable. To determine how the independent variables should be combined, we express the model in dimensionless form using arbitrary scaling parameters of the variables.

For example, consider a parabolic differential equation of the form

$$y^m \frac{\partial C}{\partial w} = k \frac{\partial^2 C}{\partial y^2} \tag{C6-1}$$

with boundary conditions given by

$$\begin{aligned} y = 0: &\quad C = C_o \\ y \to \infty: &\quad C = C_\infty \\ w = 0: &\quad C = C_\infty \end{aligned} \tag{C6-2a-c}$$

where k and m are positive constants, y is a spatial coordinate, and w represents either another spatial coordinate or time. We define dimensionless variables:

$$\mathcal{C} = \frac{C - C_\infty}{C_o - C_\infty}, \quad \mathcal{y} = \frac{y}{y_c}, \quad \mathcal{w} = \frac{w}{w_c} \tag{C6-3a-c}$$

where w_c and y_c are unknown scaling factors for w and y. Equation C6-1 then becomes

$$\beta \mathcal{y}^m \frac{\partial \mathcal{C}}{\partial \mathcal{w}} = \frac{\partial^2 \mathcal{C}}{\partial \mathcal{y}^2} \tag{C6-4}$$

where $\beta = y_c^{m+2}/kw_c$ is an arbitrary dimensionless parameter. The dimensionless boundary conditions are given by

$$\begin{aligned} \mathcal{y} = 0: &\quad \mathcal{C} = 1 \\ \mathcal{y} \to \infty: &\quad \mathcal{C} = 0 \\ \mathcal{w} = 0: &\quad \mathcal{C} = 0 \end{aligned} \tag{C6-5a-c}$$

In this problem, there is one dimensionless parameter group, but two arbitrary scaling factors, which can be related as

$$\beta = y_c^{m+2}/kw_c \Rightarrow \frac{y_c}{w_c^{1/(m+2)}} = (k\beta)^{1/(m+2)} \tag{C6-6a,b}$$

This relationship allows us to combine $\mathcal{w}$ and $\mathcal{y}$ to form a single independent variable that does not separately involve w_c and y_c:

$$\eta \equiv \frac{\mathcal{y}}{\mathcal{w}^{1/(m+2)}} = \frac{(y/y_c)}{(w/w_c)^{1/(m+2)}} = \frac{1}{(k\beta)^{1/(m+2)}} \frac{y}{w^{1/(m+2)}} \tag{C6-7}$$

We now transform Equation C6-4 from (w,y) variables to the similarity variable η by using the chain rule:

$$\frac{\partial C}{\partial w}=\frac{dC}{d\eta}\frac{\partial \eta}{\partial w}=\frac{dC}{d\eta}\left[-\frac{\eta}{(m+2)w}\right]$$
$$\frac{\partial C}{\partial y}=\frac{dC}{d\eta}\frac{\partial \eta}{\partial y}=\frac{dC}{d\eta}\left(\frac{\eta}{y}\right) \qquad \text{(C6-8a-c)}$$
$$\frac{\partial^2 C}{\partial y^2}=\frac{\partial}{\partial y}\left(\frac{dC}{d\eta}\frac{\partial \eta}{\partial y}\right)=\frac{d^2C}{d\eta^2}\left(\frac{\partial \eta}{\partial y}\right)^2+\frac{dC}{d\eta}\frac{\partial^2 \eta}{\partial y^2}=\frac{d^2C}{d\eta^2}\left(\frac{\eta}{y}\right)^2$$

Substituting Equations C6-8a and C6-8c into Equation C6-4, we obtain

$$\frac{d^2C}{d\eta^2}+\frac{\beta\eta^{m+1}}{(m+2)}\frac{dC}{d\eta}=0 \qquad \text{(C6-9)}$$

This equation must be solved with transformed boundary conditions obtained by substituting Equation C6-7 into Equation C6-5a-c. This collapses the three original boundary conditions into two boundary conditions:

$$\eta=0\ :\ C=1,\quad \eta\rightarrow\infty\ :\ C=0 \qquad \text{(C6-10a,b)}$$

This transformation changed the form of the problem from a partial differential equation with two independent variables into a second-order, ordinary differential equation with one independent variable. To solve this boundary-value problem, we define a new dependent variable:

$$F\equiv\frac{dC}{d\eta} \qquad \text{(C6-11)}$$

such that Equation C6-9 becomes

$$\frac{dF}{d\eta}+\left(\frac{\beta\eta^{m+1}}{m+2}\right)F=0 \qquad \text{(C6-12)}$$

After separation of variables, this equation is integrated to obtain

$$F=\frac{dC}{d\eta}=a_1\exp\left[-\frac{\beta\eta^{m+2}}{(m+2)^2}\right] \qquad \text{(C6-13)}$$

where a_1 is a constant. Integration of Equation C6-13 yields

$$C=a_1\int_0^{\eta}\exp\left[-\frac{\beta\xi^{m+2}}{(m+2)^2}\right]d\xi+a_2 \qquad \text{(C6-14)}$$

where ξ is a dummy variable and a_2 is a constant. To evaluate the constants of integration a_1 and a_2, we apply the two boundary conditions, which leads to

$$C=1-\frac{\int_0^{\eta}\exp\left[-\frac{\beta\xi^{m+2}}{(m+2)^2}\right]d\xi}{\int_0^{\infty}\exp\left[-\frac{\beta\xi^{m+2}}{(m+2)^2}\right]d\xi} \qquad \text{(C6-15)}$$

Since β is an arbitrary parameter, we let $\beta = (m + 2)^2$ so that

$$C = 1 - \frac{\int_0^{\eta} \exp\left(-\xi^{m+2}\right) d\xi}{\int_0^{\infty} \exp\left(-\xi^{m+2}\right) d\xi} \tag{C6-16}$$

In dimensional form, the solution is

$$C = C_o - (C_o - C_\infty) \frac{\int_0^{\eta} \exp\left(-\xi^{m+2}\right) d\xi}{\int_0^{\infty} \exp\left(-\xi^{m+2}\right) d\xi} \tag{C6-17}$$

where the upper limit η of integration is the combination of y and w given in Equation C6-7:

$$\eta = \left[k(m + 2)^2\right]^{-1/(m+2)} \frac{y}{w^{1/(m+2)}} \tag{C6-18}$$

Nomenclature

Conventions

Several naming conventions for symbols are used, with some exceptions, throughout this book.

Italic font: A dimensionless parameter such as Reynolds number *Re* or a dimensionless variable such as dimensionless molar concentration *C*.

Euclid Math One font: Either a universal constant such as the ideal gas constant $\mathcal{R}$, a diffusion coefficient such as $\mathcal{D}_i$ or modified pressure $\mathcal{P}$.

Bold font: A vector **a** with a component a_i in the ith coordinate direction or a second-order tensor **B** with a component B_{ij} associated with ith and jth coordinate directions. For example, the T_{ij} component of the stress tensor **T** represents the stress component in the ith direction acting on a surface with an outward normal in the jth direction.

Square brackets: A column or row matrix [b] or a square matrix [B]. Also, components of a vector **b** or of a tensor **B** arranged in matrix arrays.

Overdot: Refers to a rate variable such as volumetric flow rate $\dot{V}$, molar transport rate $\dot{N}$, or molar reaction rate $\dot{R}$.

Overbar: Used with a variable such as concentration $\bar{C}$ or velocity $\bar{u}$ that is spatially averaged over one or more dimensions.

Tilde: Indicates the Laplace transform of a time-dependent function, $\boldsymbol{L}\{f(t)\} = \tilde{f}(s)$.

Subscripts s and w: When used with transport quantities (e.g., concentration, flux), these refer to a solute and to the solvent, respectively, in a liquid solution.

Subscript i: When used with transport quantities, this indicates any chemical species in a gas mixture or in a liquid solution, whether it is a solute or solvent.

Superscript: This is usually used to designate the phase or transport domain in a system with multiple phases or domains.

Biomedical Mass Transport and Chemical Reaction: Physicochemical Principles and Mathematical Modeling, First Edition. James S. Ultman, Harihara Baskaran, and Gerald M. Saidel.

Symbols and Abbreviations

Universal parameters:

$\mathcal{A}$ Avogadro's number [molecules/mol]
$\mathcal{F}$ Faraday's constant [A-s/eq]
$\mathcal{G}, \mathcal{G}$ Gravitational acceleration vector and its magnitude [m/s^2]
$\mathcal{H}$ Planck constant [kg-m^2/s]
$\mathcal{K}$ Boltzmann's constant [J/(molecule-°K)]
$\mathcal{R}$ Ideal gas constant [J/mol-°K]

Common dimensionless groups:

Bi Biot number
Da Damköhler number
Gz Graetz number
j_D Chilton–Colburn j factor
Pe Péclet number
Re Reynolds number
Sc Schmidt number
Sh Sherwood number
St Strouhal number
Wo Womersley number

Roman symbols:

a Radius [m]
A Usually cross-sectional area, but sometimes surface area [m^2]
A_r Reaction affinity [J/mol]
c Molar density [mol/m^3]
c^o Molar density at standard temperature and pressure conditions [mol/m^3]
C_i Molar concentration of species i [mol/m^3]
$\hat{C}_i$ Volumetric content of gas species i in a liquid [m^3(STP)gas/m^3 liquid]
$\widehat{C}^b_{O_2,max}$ Volumetric binding capacity of blood [m^3(STP)/m^3]
d Diameter [d]
$\mathcal{D}^p_i, \bar{\mathcal{D}}_i$ Interstitial diffusion coefficient through pores, or hindered diffusion coefficient through an entire porous medium [m^2/s]
$\mathcal{D}_i, \mathcal{D}^*_i$ Pseudo-binary diffusion coefficient, or dispersion coefficient of species i [m^2/s]
$\mathcal{D}_{ij}$ Binary diffusion coefficient of species i through species j [m^2/s]
E Internal energy [J]
g Time-dependent impulse response function [s^{-1}]
G Gibbs free energy [J]
h Diffusion path length or thickness [m]
h_m Membrane thickness [m]
h_i Henry's law constant [Pa]
$\tilde{g}(s)$ Transfer function
H Enthalpy [J]
$H(t - t_o)$ Unit step (Helmholtz) function introduced at $t = t_o$

Hb, HbO_2	Heme and oxygenated heme groups, respectively, on a hemoglobin molecule
i	Electrical current [A]
[I], **I**	Identity matrix and identity tensor, both with components $I_{ij} = 1(i = j)$ and $I_{ij} = 0(i \neq j)$
$\mathbf{j}_i$, $\mathbf{J}_i$	Mass flux [kg/(s-m^2)] and molar flux [mol/(s-m^2)] of species i by diffusion
j	Imaginary number unit $\sqrt{-1}$
K_m	Substrate concentration at half the maximum rate V_m of an enzyme reaction [mol/m^3]
k_i, K_i	Individual and overall mass transfer coefficients for a species i [m/s]
k_r	Rate constant for a reaction of overall order ν [(m^3/mol)$^{\nu-1}$s^{-1}]
L	Length or thickness [m]
$\boldsymbol{L}$	Laplacian operator [s]
L_p	Hydraulic permeability of a membrane [m/(s-Pa)]
M_i, M	Molecular weight of species i and average molecular weight of a solution [Da]
m_i	Moles of species i [mol]
$\boldsymbol{n}$	Unit vector perpendicular to and pointing outward from a surface element
$\mathbf{n}_i$, $\mathbf{N}_i$	Overall mass flux [kg/(s-m^2)] and molar flux [mol/(s-m^2)] of species i
$\dot{\mathbf{n}}_i, \dot{\mathbf{N}}_i$	Overall mass transport rate [kg/s] and molar transport rate [mol/s] of species i
p_i	Partial pressure of species i [Pa]
p_i^*	Pure component vapor pressure of species i [Pa]
P	Absolute pressure [Pa]
$\mathcal{P}$	Modified pressure containing the effect of a uniform gravitational field [Pa]
P^o	Standard pressure [101.3 kPa]
P_i	Membrane permeability to species i or rate constant for solute-transporter translocation step i [m/s]
$\hat{P}_i$	Specific permeability of a membrane to species i [m^2/s]
P_i^G	Permeability of a membrane to gas species i [m/(s-Pa)]
PBPK	Physiologically based pharmacokinetic (model)
q	Heat [J]
Q	Volumetric flow rate of a liquid [m^3/s]
Q_s^C	Clearance of solute s from an artificial or natural kidney [m^3/s]
Q_F	Volumetric filtration rate through a membrane [m^3/s]
r, r	Radial position vector and radial coordinate [m]
$\bar{r}, \bar{r}_j$	Inherent reaction rate of a simple reaction and of reaction step j in a complex reaction [mol/(s-m^3)]
R_i	Intensive reaction rate [mol/(s-m^3)] or membrane retention coefficient of species i
$\dot{r}_i, \dot{R}_i$	Mass and molar rate of production species i by chemical reaction [kg/s, mol/s]
RBC	Red blood cell
REV	Representative elemental volume
RH	Percent relative humidity
s	Independent variable in Laplace domain [s^{-1}]
S	Entropy [J/°K] or surface area [m^2]
$\dot{S}, \dot{S}_m$	Molar input rate of a tracer source [mol/s] or mass input rate of a solution source [kg/s]
$\bar{S}$	Spatially averaged surface area [m^2]

S_{O_2}	Oxygen saturation fraction of hemoglobin
T	Temperature [°K] or total concentration of binding sites [mol/m^3]
T	Total stress tensor [N/m^2]
T^o	Standard temperature [273°K(STP)]
T_R	Total concentration of soluble [M] or surface [mol/m^2] binding sites
t	Time [s]
$\bar{t}_y$	Mean time of a concentration function $C_y(t)$ [s]
u, **u***	Mass and molar average solution velocities [m/s]
$\mathbf{u}_i$	Molecular average velocity of species i [m/s]
V	Volume [m^3]
V_m	Maximum (saturation) rate of an enzyme reaction [mol/m^3]
$\dot{V}_i, \dot{V}$	Volumetric flow rates of a gas species i and gas mixture [m^3/s]
$\hat{V}_i$	Partial molar volume of species i [m^3/mol]
w	Work [J]
x_i, y_i	Mole fraction of species i in a liquid solution and a gas mixture
z	Longitudinal position [m]
z_i	Electrical charge number on a species i

Greek symbols:

α_i	Bunsen solubility of gas component i in a liquid [m^3(STP)/(m^3-Pa)]
β_i	Convection correction factor for a solute i in a liquid
$\boldsymbol{\gamma}$, γ	Deformation rate tensor and a shear rate component [1/s]
γ_i	Thermodynamic activity coefficient of a species i
ΔG_r	Gibbs free energy of reaction [J/mol]
δ, δ^c	Momentum and concentration boundary layer thickness, respectively [m]
$\delta(t - t_o)$	Ideal unit impulse (Dirac delta function) introduced at time t_o[s^{-1}]
ε	Void fraction or volume fraction of a continuous phase in a suspension
ε_H	Volume fraction of red cells in blood (hematocrit)
ε_o	Electrical permittivity [A-s/(m^2-V)]
ζ_s	Efficiency of removal of solute s by an artificial or natural kidney
$\eta_s^{A,B}$	Nernst partition coefficient of solute s, phase A relative to phase B
η_i	Shape factor of a spheroidal molecule of species i
Θ_i	Transport flux of a particle of cell type i [s^{-1}m^{-2}]
θ	Angular position in cylindrical or spherical coordinates [rad], or transit time [s]
ϑ	Entropy generation function [J/(m^3-s)]
κ_c	Reaction equilibrium coefficient based on molar concentration [M$^{\Sigma\nu_i}$]
κ_p	Reaction equilibrium coefficient based on partial pressure [Pa$^{\Sigma\nu_i}$]
Λ_i	Electrical conductance of charged species i [V/A]
$\lambda_i^{A,B}$	Equilibrium partition coefficient of species i in phase A relative to phase B
μ	Shear viscosity [Pa-s]
μ_i, $\breve{\mu}_i$	Electrochemical and chemical potentials of species i [J/mol]
μ_i^o, μ_i^*	Standard state values of $\breve{\mu}_i$ in a gas mixture and liquid solution [J/mol]
$\mu_i^{rand}, \mu_i^{direct}$	Random and directed motility of cell type i [m^2/s]
$\mu_{i,y}$	The ith mathematical moment with respect to time of a concentration function $C_y(t)$ [s^i]

ν_i	Stoichiometric coefficient of species i in a chemical reaction; $\nu_i < 0$ for reactants and $\nu_i > 0$ for products
Ξ_i	Formation rate of a particle of cell type i [$s^{-1}m^{-3}$]
ξ_r	Extent of reaction [mol]
π	Osmotic pressure [Pa]
ρ_i, ρ	Mass concentration of species i and mass density of a solution [kg/m^3]
Σ_0	Surface charge density [$A\text{-}s/m^2$]
Σ_s	Membrane sieving coefficient of solute s
σ	Local charge density [$A\text{-}s/m^3$]
σ_i	Reflection coefficient of species i through a membrane
σ_y^2	Variance about the mean of a concentration function $C_y(t)$ [t^2]
$\boldsymbol{\tau}$, τ	Viscous stress tensor (total stress less the pressure contribution) and a shear stress component [Pa]
τ_p	Tortuosity of a pore in a heterogeneous material
υ	Kinematic viscosity [m/s]
$\hat{\upsilon}$	Atomic volume [m^3/kg atom]
Φ_w	Osmotic pressure coefficient
ϕ	Surface-to-volume ratio [1/m]
ψ	Electrical potential [V]
Ω	Rotational frequency [rad/s]
Ω_i	Number density of a particle or cell type i in suspension [$1/m^3$] or on a surface [$1/m^2$]
ω	Oscillatory frequency [rad/s]
ω_i	Mass fraction of species i
ϖ_i	Distribution coefficient of species i between a porous and a homogeneous medium

Index

Note: Page numbers in *italics* refer to Figures; those in **bold** to Tables.

Biomedical Mass Transport and Chemical Reaction: Physicochemical Principles and Mathematical Modeling, First Edition. James S. Ultman, Harihara Baskaran, and Gerald M. Saidel.